REACTOR DOSIMET

T0238730

REACTOR DOSIMETRY

Dosimetry Methods for Fuels, Cladding and Structural Materials

Proceedings of the Fifth ASTM-Euratom Symposium
on Reactor Dosimetry, GKSS Research Centre,
Geesthacht, F.R.G., September 24-28, 1984

edited by

J. P. GENTHON

Commissariat à l'Energie Atomique,
Centre d'Etudes Nucleaires, Cadarache, France

and

H. RÖTTGER

Commission of the European Communities,
Joint Research Centre, Petten Establishment, Petten, The Netherlands

SPRINGER-SCIENCE+BUSINESS MEDIA, B.V.

Library of Congress Cataloging in Publication Data

ASTM-Euratom Symposium on Reactor Dosimetry (5th: 1984: GKSS Research
  Centre Geesthacht)
  Reactor Dosimetry

  English and French.
  Includes bibliographies and index.
  1. Nuclear reactors–Materials–Effect of radiation on–Congresses. 2. Radiation dosimetry–Congresses. I. Genthon, Jean Pierre. II. Röttger, Heinz,
1927–   . III. American Society for Testing and Materials. IV. Euratom.
V. GKSS-Forschungszentrum Geesthacht. VI. Title.
TK9185.A1A24   1984   621.48'33   85–2342

ISBN 978-94-010-8873-2   ISBN 978-94-009-5378-9 (eBook)
DOI 10.1007/978-94-009-5378-9

Publication arrangements by
Commission of the European Communities
Directorate-General Information Market and Innovation, Luxembourg

Lay-out: Reproduction service J.R.C. PETTEN

EUR 9869

© 1985 Springer Science+Business Media Dordrecht
Originally published by D. Reidel Publishing Company in 1985
Softcover reprint of the hardcover 1st edition 1985

CONFERENCE ORGANIZATION

- **Co-sponsors**

  o ASTM Committee E10 on Nuclear Technology and Applications
  o Commission of the European Communities, Joint Research Centre
  o U.S. Department of Energy (DOE)
  o U.S. Nuclear Regulatory Commission (NRC)
  o U.S. Electric Power Research Institute (EPRI)

  The symposium was held in cooperation with the International Atomic Energy Agency (IAEA).

- **Organizing Committees**

  o ASTM-EURATOM Symposium Committee

J.P. Genthon	EURATOM Co-Chairman
W.N. McElroy	ASTM Co-Chairman

  o ASTM Program Committee

  F.B.K. Kam (Chairman), H. FARRAR IV (Vice-Chairman),
  E.B. Norris (Secretary)

  Members: S.L. Anderson, C.D. Bingham, J.R. Hawthorne,
  G.P. Lamaze, B.A. Magurno, E.D. McGarry,
  O. Ozer, K.C. Pearson (ASTM Liaison),
  A. Sekiguchi, C.Z. Serpan, Jr.,
  F.W. Stallmann

  o EURATOM Programme Committee

  J.P. Genthon (Chairman), W. Schneider (Vice-Chairman),
  H. Röttger (Scientific Secretary)

  Members: R. Dierckx, P. D'hondt, A. Fabry, A.J. Fudge,
  F. Hegedüs, P. Mas, M. Petilli, H. Tourwé,
  P. Wille, S.B. Wright, W.L. Zijp

- **Host Committee**

Mrs. U. Bockelmann	GKSS Geesthacht
H.F. Christiansen	GKSS Geesthacht
P. Wille	GKSS Geesthacht

PREFACE

Ouverts à l'ensemble de la communauté internationale, les symposia ASTM-EURATOM sur la dosimétrie des rayonnements de réacteur traitent de tous les sujets de dosimétrie dans tous les systèmes à neutron: dosimétrie des expériences en réacteur, codes d'ajustement, précision, étalons et intercomparaison, donneés nucléaires, techniques de mesure, corrélation de dommages radio-induits, échauffement nucléaire, etc.... appliqués principalement aujourd'hui aux problèmes des réacteurs à eau légère, des réacteurs à neutrons rapides et aux systèmes à fusion.

Les travaux en dosimétrie, tout comme l'ensemble du domaine de l'énergie atomique, sont moins caractérisés aujourd'hui par des idées scientifiques franchement nouvelles que par la nécessité d'échange, de coopération, de collaboration, appliqués à la satisfaction de besoins de type industriel ou quasi-industriel.

L'organisation de nos symposia a suivi cette évolution.Nombre de ceux qui y ont participé ont un souvenir ému du 1er symposium à Petten en 1975, qui fut une réussite complète.L'organisation et le dévouement du CCR de Petten y avait beaucoup contribué. Et puis, aussi, c'était un commencement, c'était le premier de nos symposia .....

Les symposia suivants, alternativement aux USA et en EUROPE, à rythme sensiblement bi-annuel, ont du faire face progressivement à un besoin plus grand d'échange et de discussion par petits groupes, ainsi qu'à afflux croissant de propositions de communications de valeur.

L'organisation a du s'adapter en conséquence. Ce 5ème symposium ASTM-EURATOM représente, de part sa belle réussite, une étape importante de cette évolution et un garant de la maturation correspondante.

Par la mise en place d'une organisation rigoureuse, en particulier l'appel à des sessions poster bien conçues ainsi que le développement d'ateliers, on a pu faire de cette session du 24 au 28 Septembre 1984 à GEESTHACHT un lieu et un moment privilégié d'échanges multiples, souples, denses, et d'une grande richesse.

Plus de cent personnes ont participe á ces échanges, en provenance des Communautés Européennes, des USA, du Japon, du Moyen-Orient, d'Europe Centrale, etc .....

Grâce à l'accueil compétent et non mesuré du Centre de GEESTHACHT, le parrainage toujours fidèle du CCR PETTEN, ainsi qu'à tous les organisateurs EURATOM et USA, et les participants, ce 5ème symposium constitue, en 1984 après le succès des commencements à PETTEN en 1975 et l'évolution qui a suivi, l'avènement réussi de la maturité.

J.P.G.

PREFACE

Open to the entire international community, the ASTM-Euratom symposia on reactor radiation dosimetry deal with all aspects of dosimetry in all neutron systems: dosimetry in in-pile experiments, adjustment codes, precision, standards and intercomparison, nuclear data, measuring techniques, correlation of radiation-induced damage, nuclear heating, etc., as mainly applied today to problems associated with light-water reactors, fast reactors and fusion systems.

Dosimetry work, like everything else in the field of atomic energy, is nowadays characterized less by genuinely new scientific ideas than by the need for the exchange of information and for cooperation and joint efforts aimed at satisfying the requirements of industrial-scale or quasi-industrial-scale operations.

This development is reflected in the way that our symposia are organized. Many of those who have participated in them cherish fond memories of the first symposium, held at Petten in 1975, which was a complete success. The organizing ability and the devotion of the JRC-Petten contributed greatly to it. And that was only a beginning, the first of our symposia!

The subsequent symposia, held alternately in the USA and Europe at intervals of approximately two years, had to cope progressively with a greater need for information exchange and for discussion by small groups and with the growing number of proposals for interesting communications.

The symposium organization had to be adapted accordingly. This fifth ASTM-Euratom symposium represents, by reason of its great success, an important stage in this development and is a sure sign of corresponding maturity.

Punctilious organization and, in particular, the use of carefully arranged poster sessions and workshops made the meeting at Geesthacht on 24–28 September 1984 an outstanding occasion for many wide-ranging exchanges of views and for intense and immensely fruitful discussion.

Over 100 persons from the European Community, the USA, Japan, the Middle East, Central Europe and elsewhere participated in these exchanges.

Thanks to the efficiency and boundless hospitality of the Geesthacht Centre, the loyal sponsorship of the JRC-Petten and the efforts of all the Euratom and American organizers, not to mention the enthusiasm of the participants, this fifth symposium, following the successful beginnings at Petten in 1975 and the developments that ensured, marks the advent of full maturity.

J.P.G.

Editorial Remark

Some contributions exceed the allowed maximum number of 8 pages per contribution considerably, without permission of the programme committees. To avoid a delay in the publication of the conference proceedings in-hand the editors have renounced to ask the authors concerned for a reduction of their contributions.

LIST OF CONTENTS - Volume 1

CONFERENCE ORGANIZATION v

PREFACE vii

PART I: LWR-PV SURVEILLANCE

ix

PART II: FAST REACTORS

PART III: FUSION AND SPALLATION

PART IV: TECHNIQUES

xvi

PART VII: BENCHMARKS, REFERENCE AND STANDARD SPECTRA

PART VIII: NUCLEAR DATA

PART IX: GENERAL INTEREST

PART VII: BENCHMARKS, REFERENCE AND STANDARD SPECTRA

PART VIII: NUCLEAR DATA

PART IX: GENERAL INTEREST

PART IV: TECHNIQUES

PART I
LWR-PV SURVEILLANCE

AMELIORATION DE LA SURVEILLANCE DE LA CUVE D'UN REACTEUR

A EAU PRESSURISEE DE LA S.E.N.A..

A.BEVILACQUA[1], M.CAMPANI[2], C.DUPONT[3], R.LLORET[1], J-C.NIMAL[3],
M.POITOU[4], R.RIEHL[5].

1-CEA Service des Piles de Grenoble. 85 X .38041 Grenoble Cedex.
2-EDF Service de Contrôle des Matériaux Irradiés.BP23 37420 Chinon.
3-CEA-SERMA-LEPF. CEN-SACLAY - 91191 Gif-sur-Yvette.
4-FRAMATOME. Tour Fiat. 1 Place de la Coupole - 92084 La Défense.
5-S.E.N.A. BP 60 - 08600 Givet.

ABSTRACT

This paper describes a new dosimetry, installed inside and outside the Pressure Vessel of CHOOZ Nuclear Power Plant of the Société d'Energie Nucléaire Franco-Belge des Ardennes (S.E.N.A.), during its 1982-83 operation cycle.

The inner dosimetry deals with a simulated capsule located like those of the previous program, under the reactor plate, and includes copper, nickel, iron, niobium, copper-cobalt, neptunium and uranium dosimeters. Its aim is to qualify the information given by the existing copper dosimetry which is up till now the only way to caracterise the fluence determinations of the specimens. The spectrum used with these measurements is obtained by the 1 D ANISN Code and BIP-N 2 library.

The outer dosimetry is the fluence determination along the outer wall of the vessel. Two tubes, equiped by neutron dosimeters, seven meters long, were fixed along the vessel. On the median plane, the results are compared to a 2 D DOT transport calculation.

Preliminary results are given which improve the vessel and specimens neutronic caracterisation.

INTRODUCTION

Afin d'apprécier les marges de sécurité vis-à-vis du risque de rupture de la cuve de la Centrale Nucléaire des Ardennes (SENA), située à CHOOZ, les autorités de sureté ont demandé de mieux évaluer les principaux paramètres qui s'y rapportent, et en particulier, la fluence de neutrons rapides intégrée par la cuve.

J. P. Genthon and H. Röttger (eds.), Reactor Dosimetry, 3–9.

Ce réacteur à eau sous pression, après quelques mois de fonctionnement, a subi en 1968-1970 des modifications de structures internes. La suppression de l'écran thermique a entraîné le retrait des éprouvettes de surveillance du métal de la cuve situées en périphérie du coeur. En remplacement, un autre programme de surveillance a été mis sur pied comportant de nouvelles capsules situées, celles-ci, sous le coeur, et équipées seulement de dosimètres en cuivre. L'exploitation de ce second programme après 12 ans d'irradiation, utilisant les spectres neutroniques disponibles, a mis en évidence que les flux neutroniques déterminés par la mesure, sont inférieurs de 30% aux flux calculés.

En vue d'affiner ces résultats, la démarche suivante est entreprise :
- reprise de l'étude du flux et du spectre de neutrons au droit des éprouvettes d'irradiation et de la paroi de la cuve, à l'aide de calculs actuels mieux adaptés.
- amélioration de la connaissance des flux et spectres de neutrons par la mise en place de deux mesures nouvelles, l'une dans une capsule et l'autre à l'extérieur de la cuve, basées sur un jeu de dosimètres couvrant plus largement le spectre.
- réévaluation de la fluence du métal de cuve.
Ce sont les étapes principales de cette démarche qui sont rapportées ici.

PROGRAMME DE TRAVAIL ADOPTE.

L'exploitation des capsules irradiées pendant 2, 5, 10 et 12 ans a fourni des résultats d'une grande régularité, mais d'un caractère peu représentatif. On a donc cherché à rendre significative la dosimétrie du programme de surveillance existant par une procédure de qualification des déterminations de flux et fluences neutroniques aussi bien au niveau des éprouvettes (sous le coeur) qu'à l'extérieur de la cuve. Ce travail a mis en jeu la coopération d'équipes de divers organismes, de l'exploitant de la centrale au fournisseur de la cuve, assistées de divers services de l'EDF et du CEA. La partie expérimentale eut lieu au cours du douzième cycle du réacteur, de septembre 1982 à avril 1983.

1-Evaluation du flux à l'extérieur de la cuve.

A - Calcul des flux et spectres.

Le calcul des flux de neutrons du coeur à l'extérieur de la cuve dans le plan médian horizontal a été effectué à l'aide du code DOT de transport multigroupe à deux dimensions et aux ordonnées discrètes. Il a été utilisé en géométrie R, θ, les contours du coeur et du baffle étant approximés par des portions de cyclindre. Le découpage en R a comporté 100 pas de R = 85 cm à R = 247 cm. Seuls 50 cm de la protection thermique en eau située autour de la cuve ont été pris en compte. Ils sont suffisants pour assurer la validité du spectre de neutrons autour de la cuve. Le découpage en θ a comporté 38 pas de θ= 0° à θ= 45°. La discrétisation en direction utilise un découpage en S6. L'anisotropie des sections efficaces est traitée en P1. Le découpage en énergie comporte 21 groupes de 10 MeV à l'énergie thermique dont 9 audessus de 1 MeV. Le programme DOT, en géométrie R,θ,suppose une distribution de source constante dans la direction z. La valeur utilisée est la valeur moyenne (source totale divisée par la hauteur du coeur). Un pic axial égal à 1,2 (intermédiaire entre début de vie et fin de vie-sans barres) a été utilisé pour obtenir les fluences maximales.

B - Mesures.

Deux dispositifs ont été fixés contre la face externe du calorifuge de la cuve suivant deux génératrices à 90° (fig 1). Ils se trouvent dans des positions azimutales équivalentes à flux maximum (distance combustible-cuve minimum) car des difficultés de positionnement n'ont pas permis d'atteindre un emplacement à flux minimum. Sur chacune des génératrices sont disposés, tous les 50 cm sur une hauteur de 7 m, un ensemble de dosimètres comprenant un fil d'alliage Al-(0,483%)Co alternativement nu et sous cadmium, un fil de nickel et un fil de fer. Les activités sont mesurées à mieux que 3% dans la zone centrale, là où le calcul du spectre est valide. Aux extrêmités, l'incertitude se dégrade jusqu'à 6% pour les plus faibles activités (par exemple en ^{54}Mn, 0,06 Bq.mg^{-1} de fer). Le rapport cadmium est obtenu par interpolation avec une précision de l'ordre de 10%.

C - Résultats.

La figure 2 donne un aperçu des résultats obtenus, à partir de sections efficaces moyennes calculées d'après IRDF 82 /1/.

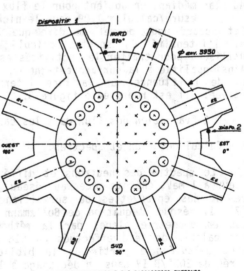

Fig 1 - IMPLANTATION RADIALE DES DISPOSITIFS EXTERNES

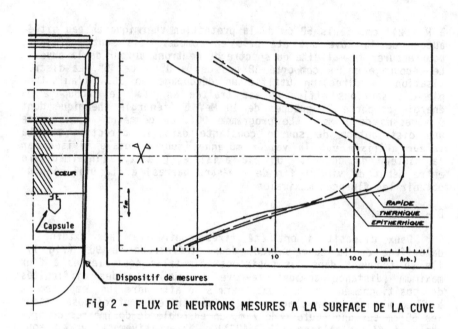

Dispositif de mesures

Fig 2 - FLUX DE NEUTRONS MESURES A LA SURFACE DE LA CUVE

Au plan médian, on obtient pour le flux supérieur à 1 MeV :
     mesure/calcul = 1,09 pour le nickel, 1,02 pour le fer.
Cet accord nous permet de dire que la cohérence entre la forme
et l'intensité du spectre calculé et les sections efficaces
utilisées pour traiter les activités est tout à fait satisfaisante.
Ainsi qualifié à la surface externe de la cuve, le calcul du spectre
et de la fluence sur la face interne voit sa représentativité
étayée. Il est représenté figure 3.

     2-Evaluation des flux dans les capsules de surveillance.

A - Calcul des flux et spectres.

     Le spectre des neutrons et différents paramètres (flux supé-
rieur à 1 MeV, énergie cédée au réseau), ont été calculés à l'emplacement des éprouvettes de surveillance à l'aide du code ANISN
(2). Il résoud l'équation de Boltzmann à une dimension (dans notre
cas en géométrie plane) par la méthode des ordonnées discrètes.
Les calculs ont été effectués en S16 P3. Les sections efficaces
sont calculées à partir de la bibliothèque binaires BIP-N2 (3)
tirée de ENDF/B IV dans un découpage à 100 groupes.

     La géométrie du calcul ANISN est constituée de 6 milieux
homogènes : coeur, zone des bouchons de crayons, embout d'assem-

blage, zone des ailettes d'entrée d'eau, plaque inférieure, eau
située sous la plaque. Notons que les éprouvettes ont été négli-
gées ce qui induit un léger durcissement du spectre en partie
basse. Un calcul de Monte-Carlo (4) aurait permis d'en tenir compte.
Par contre, le domaine d'énergie étendu jusqu'à 15 MeV couvre
mieux la zone de sensibilité du cuivre. Un calcul d'évolution
par APOLLO (5) a fourni les répartitions de fission sur les différents isotopes fissiles au cours de la vie du réacteur. Ceci
permet d'obtenir, en moyenne sur la vie du réacteur, un nombre
de neutrons émis par fission et leur spectre. Le spectre ANISN
est représenté fig.3.

B - Définitions des mesures.

Le dispositif est une capsule simulée, en acier, équipée
de jeux de dosimètres à trois niveaux. Les dosimètres à activation
(nickel, cuivre, fer, niobium, alliage Cu-0,1% Co) sont encapsulés
dans des boîtiers en aluminium. Les dosimètres fissiles (pastilles
de 20 mg de $^{238}U_3O_8$ et de $^{237}NpO_2$) sont encapsulés sous nickel
et 2 mm d'oxyde de cadmium. Un tel équipement de dosimètres répond
aux objectifs suivants :
    - qualifier les fluences mesurées par le cuivre seul.
    - assurer la cohérence de la dosimétrie du programme de sur-
veillance de la SENA avec ceux des réacteurs de l'EDF.
    - utiliser un moyen expérimental imposé par les circons-
tances pour obtenir un maximum d'information.

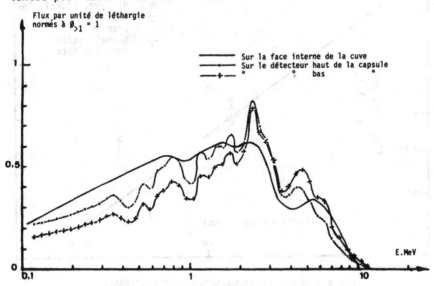

Fig 3 - SPECTRES DES NEUTRONS D'ENERGIE SUPERIEURE A 0,1 Mev

7

La capsule a été introduite lors de l'arrêt de tranche en août 1982, à la place occupée par celle retirée après le 1er cycle. Les sections moyennes ont été calculées sur les spectres de la figure 3 au moyen de IRDF 82.

C - Principaux résultats.

Certains résultats provisoires sont représentés figure 4. La mesure définitive des activités est prévue pour fin 1984. Elle sera suivie d'une analyse détaillée qui fournira l'évaluation des fluences reçues par les éprouvettes. Pour l'heure, les flux reportés sont déduits de mesures masquées par l'activité des boî-tiers en nickel. La tendance observée est la suivante :
- nickel/niobium = 1,05± 0,03
- fer/niobium   = 0,94± 0,02.
Cette cohérence est mise à profit pour adopter la moyenne (Nb-Ni-Fe) comme terme de comparaison. Comme on peut le voir sur la figure 4, le cuivre s'écarte de plus de 15% de cette moyenne dans le sens qui pourrait confirmer la tendance du calcul à trop durcir le spectre en partie basse. L'information des dosimètres fissiles

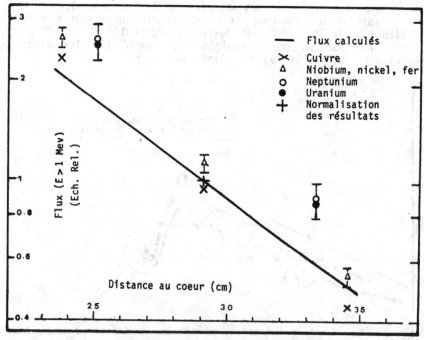

Fig 4 - RESULTATS PROVISOIRES DANS LA CAPSULE.

sera analysée sur les résultats définitifs. Néanmoins, on peut dire d'ores et déjà qu'il est justifié de procéder à une réévaluation des fluences reçues par toutes les éprouvettes sorties.

CONCLUSION.

La démarche qui a été rapportée ici avec quelques résultats provisoires obtenus avait pour objectif de mieux évaluer les paramètres d'irradiation de la cuve et éprouvettes de surveillance. Pour la cuve, les mesures réalisées à la surface extérieure qualifient le calcul à cet endroit et permettent de penser qu'il est représentatif sur la paroi interne. Pour les éprouvettes, l'évaluation est en cours à partir des informations recueillies. Elles permettront une réévaluation plus satisfaisante des fluences reçues par toutes les éprouvettes du programme.

Ce travail, réalisé par les actions concourantes d'équipes de plusieurs organismes (SENA, CEA, FRAMATOME, EDF), montre que l'on peut remédier avec succès à certains "défauts de jeunesse" de réacteur à eau pressurisée, même s'il faut utiliser des moyens de calcul et de mesure que leur finesse réserve en général aux études de maquette en laboratoire ou en réacteur de recherche.

REFERENCES

1 - D.E.Cullen, N. Kocherov, P.M.Mc Lauglin.   International Reactor Dosimetrie File. IAEA-NDS-42.R.

2 - C.Devillers.   Description et mode d'utilisation du programme aux ordonnées discrètes ANISN. Note CEA.N.1358 (1970).

3 - Bibliothèque binaire multigroupe neutron BIP-N2. Note SERMA/LEP/81.195 (1981).

4 - A.Baur, L. Bourdet, G. Dejonghe, J. Gonnord, A. Monnier, J-C.Nimal, T. VERGNAUD.   TRIPOLI 2 - Note CEA à paraître.

5 - A.Hoffman, F. Jeanpierre, A.Kavenoky, M. Livolant, H. Lorain. APOLLO. Code multigroupe de résolution de l'équation du transport Note CEA.N.1610.

PRESSURE VESSEL DOSIMETRY AT U.S. PWR PLANTS

C. O. Cogburn and J. G. Williams,   and N. Tsoulfanidis

University of Arkansas   University of Missouri

Fayetteville, AR 72701   Rolla, MO 65401

ABSTRACT

A series of Neutron Dosimetry experiments has been performed at three Pressurized Water Reactor Plants, representing each of the three U.S. PWR vendors. The study has included analysis of several sets of radiometric monitors irradiated in the pressure vessel cavity of each reactor, as well as monitors from in-vessel positions. In each case, measurements have been compared with transport calculations performed by the University of Missouri (Rolla) research team. Other labs have performed both independent measurements and calculations, in verification, and the N.B.S. has provided fluence calibrations. The aim of this research project has been to provide benchmark data and methodology for pressure vessel surveillance. The study has been underway for 7 years and has covered several fuel cycles, various operational conditions and different radial and azimuthal positions relative to the pressure vessels. Specimen results from each reactor are included and compared with calculations.

Research sponsored by Arkansas Power & Light Co., Duke Power Co. and the Electric Power Research Institute

INTRODUCTION

The pressurized water reactor plants now in service in the U.S.A. have been supplied by three vendors, namely; the Babcock & Wilcox Co. (B & W), the Combustion Engineering Co. (C-E) and the Westinghouse Co. (W). This paper summarizes a pressure vessel (PV) dosimetry study of a reactor from each of these vendors.

Each pressurized water reactor (PWR) vendor provided a set of in-vessel surveillance capsules as the principal means of monitoring radiation effects on the pressure vessel. These sealed capsules are mounted near the inner wall of the vessel at, typically, four azimuthal positions and spanning most of the vertical height of the core. The surveillance capsules contain metallurgical samples from the vessel, of base metal, weld material and heat affected zone (HAZ) materials, plus temperature monitors (alloys of different melting temperature) and radiation monitors (fission and activation materials). These capsules are removed for analysis at a refueling interval, after varying periods of total irradiation, depending on the calculated fluence and predicted shift in NDTT (nil ductility transition temperature). The design and practice generally follows, or exceeds, the recommendations of ASTM E-185.

In some of the U.S. surveillance capsule installations, failure of the mounting fixture occurred, due to flow-induced vibration. This required early removal of capsules from several reactors. Sample materials from these reactors were then loaded into a re-designed capsule and placed in a surrogate reactor of the same vendor model.

Several surveillance capsules have been removed from U.S. nuclear plants and analyzed for metallurgical properties (Charpy, tensile tests, etc.) neutron irradiation (fluence) and peak temperature reached. This program dates back several years (1,2). More recently, the in-vessel dosimetry has been supplemented by ex-vessel dosimetry in a BWR (3,4). The U.S. PWR ex-vessel dosimetry effort began with the placement of radiation monitors (foils) in the P.V. cavity of the Arkansas Nuclear One (ANO) Unit I (a B & W reactor) in 1977 and in ANO Unit II (a C-E reactor) in 1979 (5). These reactors have since been continuously monitored by P.V. cavity dosimetry, and it is now the owner's intention to continue monitoring for the life of each plant. Also, other PWR plant owners have recently commenced P.V. cavity dosimetry programs at their plants.

COMPARISON OF IN- AND EX-VESSEL DOSIMETRY

When one compares in- and ex-vessel dosimetry, it is clear that both methods have advantages and disadvantages relative to the other. Wagschal et al (6) concluded that, "...the combination of dosimetry and metallurgical specimens at an accelerated location along with dosimetry in the reactor cavity would ideally satisfy the requirements for surveillance." This conclusion was based on the need for the measurements to be done at locations which would be sensitive to all the possible error sources in calculation of the fluence parameters within the vessel steel. Other considerations are involved, and these are summarized as follows:

"Surveillance" Position Dosimetry (In-Vessel)

. Monitors neutrons incident on vessel
. Includes metallurgical specimens at an accelerated location
. Monitors neutron fluence from start-up
. Survivability is a problem (e.g., loose parts), and capsules are not easily replaced later
. A few predetermined positions are used
. Core changes, such as introduction of low leakage cores, not easy to assess.

Cavity Dosimetry (Ex-Vessel)

. Monitors neutrons passing through vessel
. Tests calculation up to and through vessel
. No metallurgical specimens
. Survivability is not usually a problem, and foils can be replaced
. Core changes can be monitored
. Flexible choice of position.

Combined In- and Ex-Vessel Dosimetry

The advantages of combined in- and ex-vessel dosimetry are therefore apparent in an improved ability to test the transport calculations, to follow core changes at a wide variety of axial and azimuthal locations and to provide a measure of redundancy or replaceability in the event of loss of some in-vessel dosimetry. An example of the benefits of including cavity dosimetry arises in the recent trend to introduce low leakage core configurations in some PWRs. In some instances, special fuel has been installed in outer assemblies which are positioned in portions opposite critical vessel welds. This results in a flux reduction at the weld.

13

Calculation of such effects is clearly a difficult task and needs to be supported by specially installed dosimetry. This has been recently done in two Westinghouse reactors, but the results are not yet available.

An in-vessel/ex-vessel dosimetry experiment/calculation, plus metallurgical analysis, has just been completed on ANO-II. This was a joint effort between the University of Arkansas, University of Missouri, Rolla (UMR), the Battelle Columbus Laboratory and the National Bureau of Standards. A similar arrangement is planned for an in-vessel/ex-vessel study of the McGuire-I reactor (W), the analysis of which is now underway (7,8).

CAVITY EXPERIMENTS

In each reactor studied, the P.V. cavity dosimetry experiments have been suspended vertically, parallel to the pressure vessel, at two or more azimuthal positions. A defined, fixed-geometry location is desired, and this has dictated radial distances from the vessel, varying from about 15 cm to about 75 cm in the different plants. Most of the irradiations have involved attaching foil packages at the desired elevations on pre-measured stainless steel chain, and lowering the assembly into ex-vessel detector wells (vertical pipes from top to bottom of the cavity). The stainless steel beaded chain is also used as a "flux-wire", with $^{54}Fe(n,p)$, $^{58}Ni(n,p)$ and three (n,γ) reactions. Each foil holder of thin-wall aluminum alloy (A9-6061) is sealed and contains three foil packages. These are: a cadmium box containing ^{235}U, ^{238}U and ^{237}Np; a second Cd box containing cobalt, iron and scandium; and an aluminum package containing copper, nickel, silver and titanium.

Irradiations are placed and removed at any convenient time during a plant outage, provided, on removal, that the experiment has had at least 30 EFPD (effective full power days) of irradiation. Some irradiations have been for this minimum time, others for various parts of the fuel cycle, and some irradiations have been for a full fuel cycle.

Typical dose rate at the surface of the experiment on removal is 100-200 m.Rem/hr. The entire experiment, including anchoring fixture, stainless steel chain and sealed foil holders, is immersed in the plant sonic-cleaner for decontamination. Most experiments are then transported by auto in a sealed box, to the University of Arkansas. The foil holders are opened and beads are snipped from the stainless steel chain at 30 cm (1 ft.) intervals and stored in a shield to await counting. Counting is done in a separate room by high-resolution HPGE detector/computer systems.

Data reduction takes into account reactor power variations (or outages) during the irradiation.

CALCULATIONS

There are four major sources of uncertainty in calculating neutron energy spectra for PWR plants. These are:

(a) Geometric modeling
(b) Power distribution and source calculation
(c) Flux leakage correction along the Z-direction
(d) Cross sections and group energy structure.

Geometric Modeling

The basic transport calculation is performed in 2-D geometry. The choices one has are R-θ or X-Y geometry. Since the major objective is to calculate as accurately as possible the flux at the location of the PV, the R-θ geometry is preferred since the PV is cylindrical. However, this choice of geometry is not the proper one for the core itself which consist of fuel assemblies with square cross sections on the horizontal plane, and for the core shroud which defines the edge of the core and thus follows the contour of the fuel assemblies. An additional argument for the use of R-θ geometry is the better representation of the azimuthal variation of the flux in the pressure vessel. We performed our calculations in R-θ geometry but used an adequate number of radial and angular intervals so that the error in modeling the fuel assemblies and the core shroud be minimized. For ANO-I, 52 angular intervals were used, providing an excellent representation of the shroud and resulting in acceptable details of the angular distribution of neutrons. In the core, the radial mesh intervals were chosen in such a way that every mesh belongs, for all practical purposes, to a single assembly (i.e. little crossing of assembly boundaries). We believe this choice of R-θ geometry minimizes the error, although it is very difficult to quantify the error.

Power Distribution and Source Calculation

Transport calculations were performed using the "source option" of the transport code DOT. Since the flux is directly proportional to the source, it is of paramount importance to use as accurate a source as possible. The source should not be the result of another calculation, but the result of measurement at the plant itself or, at least, of a combination of measurements and calculations. The procedure followed involved calculating a neutron source for each R-θ mesh interval in the core, based on the measured power distribution. In the case of ANO-I, the power was measured by Self Powered Neutron Detectors (SPND), placed in instrumented fuel assemblies, and the neutron source was normalized by the equation

$$p = \frac{\sum_g \Sigma_{fg} \phi_g}{2.989 \times 10^{16}} \quad MW/cm^3$$

For ANO-II and McGuire, a power distribution was provided by the utility, again based on direct measurement. For all three reactors, the Z-distribution of power was included in the experimental information, thus leading to a Z-distribution of the neutron source for each fuel assembly. The determination of the neutron source by this method would be correct if the power within the R-θ or X-Y space occupied by each assembly were constant. But, this is not the case. To take into account the radial variation of the power (and neutron source) across each assembly, we performed the DOT R-θ transport calculation twice. Using the neutron sources calculated as discussed above, the first DOT calculation was made. The output of this DOT run provides a neutron source for every R-θ mesh interval, a source which reflects the power distribution within each assembly. This source was reintroduced into DOT and the calculation was repeated keeping all other input information the same. It is the flux of this second DOT run which was used in the subsequent calculations.

Flux Leakage Correction Along the Z-Dimension

Since the R-θ calculation is based on two dimensions and the transport code does not recognize leakage along the third dimension, a correction had to be made for that effect. the correction was accomplished by executing not only the already mentioned R-θ computation but also performing a R-Z DOT and a 1-D ANISN calculation in cylindrical geometry. Axial "leakage correction factors" were obtained by dividing the DOT R-Z flux by the corresponding 1-D flux. Thus, for every group and location of interest, a factor LF (r) was obtained from

$$LF_g(r,z) = \Phi_g(r,z)_{DOT} / \Phi_g(r)_{1-D}$$

and a 3-D flux was reconstructed by writing

$$\Phi_g(r,\theta,z) = \Phi_g(r,\theta) \times LF_g(r,z)$$

Comparison of calculated and measured reaction rates indicates that whereas this correction may be adequate for the reactor midplane, it is not so accurate for points away from that location.

Cross Sections and Group Structure

The basic data for these calculations came from ENDF/B IV in the form of the VITAMIN C library. VITAMIN-C consists of 171 neutron groups and 36 gamma energy groups. Because such a large number of neutron groups would result in an inordinate amount of computer time, without increasing the accuracy of the results considerably, most investigators in this field choose a smaller number of groups. For this work 26 groups were chosen, paying more attention to the high rather than the low energy part of the spectrum.

The neutron group library used was obtained by running the code XSDRNPM to collapse the 171 VITAMIN-C library into the 26 group working library. The XSDRNPM 1-D transport code was run in cylindrical geometry using 1-D information from the reactor under study. Thus, the collapsing process was accomplised with a 1-D representation of the neutron energy spectrum of the reactor itself.

RESULTS

In ANO-I (B & W), eleven cavity dosimetry experiments have been irradiated at various cavity locations during the second to the sixth reactor fuel cycles. Transport calculations have been performed for core conditions representing the middle of cycle 4, the end of cycle 4 and the middle of cycle 5. Both measurements and calculations show that at each location the neutron spectrum has remained almost constant in shape from one fuel cycle to another and also during each cycle. The fluence rate, however, has changed, the measurements showing a drop of 26% from cycle 2 to cycle 3 and a further drop of 24% from cycle 3 to cycle 4. Subsequent changes through cycles 4 to 6 have been comparatively small. Measurements made at various heights in the cavity detector wells show that axial spectrum changes are small within the limits of the core height. This is also predicted by calculation, though the axial variations in the reaction rates away from the horizontal midplane are not well predicted. Table 1 shows the results from one of the experiments at the horizontal midplane compared with calculation. Various non-threshold reactions were also measured, but these are not shown here. The agreement between measurement and calculation for the non-threshold reactions is poorer than for the fast reactions, but this may be

Table 1

Reaction Rates in the Cavity of ANO-I, Middle of Fuel Cycle 4

Reaction	Experiment	Experiment Uncertainty (1 sd) %	Calculation	C/E
Cu-63(n,a)	1.70-19	6.6	2.01-19	1.18
Ti-46(n,p)	2.50-18	7.9	2.65-18	1.06
S-32(n,p)	1.02-17	6.0	1.03-17	1.01
Ni-58(n,p)	1.76-17	7.7	1.80-17	1.02
Fe-54(n,p)	1.27-17	7.8	1.29-17	1.02
Np-237(n,f)	1.23-15	8.0	8.88-16	0.72

attributed to the choice of cross section weighting and group structure which concentrated on the fast neutron energies. Among the threshold reactions, the fission rate for neptunium shows poorer agreement between measurement and calculation than the others. Photofission may be responsible for part of this discrepancy.

In ANO-II (C-E), seven cavity experiments have been completed in four detector wells during the first three reactor fuel cycles. At the end of the second fuel cycle, one of the in-vessel surveillance capsules was removed and taken for analysis to Battelle Columbus Laboratories (7). The dosimetry materials from these were analyzed by Battelle and also by the University of Arkansas. Table 2 shows the results from one of the surveillance capsule data sets using the Arkansas measurements and the UMR calculations, and Table 3 shows the results from the adjacent cavity detector well. These cavity results are for a position at the height of the top of the active fuel because this well was obstructed at lower elevations. Similar spectral trends were found at the horizontal midplane elevations in other detector wells, with somewhat better agreement between calculation and experiment. Comparison of Tables 2 and 3 shows that in both cases, the calculations overpredict the reaction rates and the discrepancy is slightly greater at the cavity position. The neptunium result, as in the case of ANO-I, gives a lower C/E ratio than the other reactions.

In McGuire-I (W), one cavity measurement has been analyzed so far. Additional foil sets at many locations and a surveillance capsule have been removed this year and analysis is in progress. Preliminary results from the first experiment are shown in Table 4.

Table 2

Reaction Rates at the Surveillance Capsule
(Top Compartment) of ANO-II

Reaction	Experiment	Calculation	C/E
Cu-63(n,a)	9.07-17	1.14-16	1.25
Ti-46(n,p)	1.66-15	1.96-15	1.18
Ni-58(n,p)	1.14-14	1.36-14	1.19
Fe-54(n,p)	8.15-15	1.06-14	1.29
U-238(n,f)	data was rejected, because of cadmium contamination		

Table 3

Reaction Rates at the Top of Active Fuel Elevation
in the Cavity of ANO-II

Reaction	Experiment	Calculation	C/E
Cu-63(n,a)	4.59-19	5.80-19	1.26
Ti-46(n,p)	7.05-18	8.74-18	1.24
Ni-58(n,p)	5.14-17	6.87-17	1.33
Np-237(n,f)	3.47-15	3.62-15	1.04

Table 4

Reaction Rates in the Cavity of McGuire-I,
Near the Horizontal Mid-Plane

Reaction	Experiment	Calculation	C/E
Cu-63(n,a)	5.13-19	5.49-19	1.07
Ti-46(n,p)	7.32-18	7.77-18	1.06
Ni-58(n,p)	4.93-17	5.64-17	1.14
Fe-54(n,p)	3.38-17	4.01-17	1.19

CONCLUSIONS

Experience of cavity dosimetry measurements and corresponding calculations is now available for PWRs built by each of the three U.S. vendors. In one of these, ANO-II, simultaneous in- and ex-vessel dosimetry has been carried out. From the results, it seems clear that cavity dosimetry is a valuable alternative or supplement to the conventional surveillance position dosimetry, and when combined with transport calculations can be expected to provide similar estimates of fast neutron fluence to positions within the reactor vessels.

REFERENCES

1. F. J. Rahn, "Power Reactor Pressure Vessel Benchmarks, an Overview", Neutron Cross Sections for Reactor Dosimetry, Vol. 1, IAEA-208 (1978).

2. S. Rothstein, "Survey of Utility Experience with Reactor Vessel Surveillance", ASTM STP 819, L. E. Steele, Ed. (1983).

3. G. C. Martin, "Browns Ferry Unit 3 Cavity Neutron Spectral Analysis", EPRI NP-1991 (1981).

4. G. C. Martin and C. O. Cogburn, "Special Considerations for LWR Neutron Dosimetry Experiments", this 5th ASTM-Euratom Symposium

5. W. E. Brandon, et al, "Neutron Dosimetry in the Pressure Vessel Cavity of Two Pressurized Water Reactors", 4th ASTM-Euratom Symposium, pp. 533-544 (1982).

6. J. J. Wagschal, R. E. Maerker and B. L. Broadhead, "Surveillance Dosimetry: Achievements and Disappointments", Proc. of the Fourth ASTM-Euratom Symposium on Reactor Dosimetry, NUREG/CP-0029 Vol. 1, CONF-82032/VI, pp. 79-92 (1982).

7. M. P. Manahan, et al, "Battelle's Columbus Laboratories Reactor Surveillance Service Activities", this 5th ASTM-Euratom Symposium

8. N. Tsoulfanidis, et al, "Neutron Energy Spectrum Calculations in Three PWRs", this 5th ASTM-Euratom Symposium

IMPROVEMENT OF LWR PRESSURE VESSEL STEEL EMBRITTLEMENT SURVEILLANCE : 1982-1983 PROGRESS REPORT ON BELGIAN ACTIVITIES IN COOPERATION WITH THE USNRC AND OTHER R & D PROGRAMS

A. Fabry, J. Debrue, L. Leenders, F. Motte,
G. Minsart, P. Gubel, R. Menil, P. D'hondt
G. and S. De Leeuw-Gierts, H. Tourwé
J. Van de Velde and Ph. Van Asbroeck

SCK/CEN, Mol, Belgium

SUMMARY

The objectives, direction and current achievements of LWR pressure vessel surveillance research and development activities in Belgium are re-visited. This is done in a perspective which accounts for the recent embodiment in Regulation of the concern about overcooling accidents susceptible to cause a pressurized thermal shock to the reactor vessel.

Supportive considerations are provided by a review of some selected aspects of the vessel safety analysis program at the Belgian BR3 plant. New dosimetry measurements have been performed on an array of specimens sampled out of the thermal shield at midplane and out of the vessel wall at various levels around the nozzle centerline; they are described and combined to transport theory as well as to surveillance measurements in order to develop a three-dimensional neutron exposure map and its uncertainties. The determination of actual material chemistries from the vessel scraps, most importantly of the copper and nickel contents (which strongly influence the radiation embrittlement), is discussed and the related uncertainties are assessed on basis of interlaboratory comparisons; also outlined is a novel neutron activation technique for copper analysis, successfully applied to BR3. Using

1) these dosimetry and chemistry data
2) an evaluation of the vessel fabrication procedure, and
3) the results of dedicated test reactor irradiations,

a structural beltline integrity analysis is performed and illus-

J. P. Genthon and H. Röttger (eds.), Reactor Dosimetry, 21–38.
© 1985 ECSC, EEC, EAEC, Brussels and Luxembourg

trated for hypothetized cases of severe steam-line break accidents. Particularly addressed is the impact of relevant sources of uncertainty, such as assumptions on flaw shape and distributions, material chemistry, neutron exposure, irradiation temperature, heat transfer coefficients and fracture toughness (including on the upper shelf). The discussion encompasses an attempt to separate the influence of general features from plant-specific ones, among which are a low irradiation temperature of 500 °F, some ECCS and steam-line isolation details, the vessel design, the significant effect of metal-stored energy on crack initiation and of vertical toughness gradients on crack-arrest - both stemming from core, thermal shield and vessel geometries.

The goals and the accuracy requirements for reactor physics and dosimetry activities are tentatively re-appraised at the light of the above analysis.

INTRODUCTION

This paper is a follow-on to a similar progress report presented at the 4th ASTM-EURATOM Symposium on Reactor Dosimetry /1/. To many respects, the introduction and overview section 1 of the former reference /1/ does still apply to-date, i.e. could be repeated here with little modification, and is consequently referenced. So the programme directions and objectives have not changed, while implementation has on the other hand continued actively, on the international and interlaboratory basis /2/ previously emphasized. Various papers at this Symposium do reflect well this reality /3/ to /9/.

The reactor physics-dosimetry aspects of the programme and some developments in the metallurgy-fracture mechanics area have, since 1982, received the support of and are now directly sponsored by the Belgian Utilities; this is true for ex. of the VENUS PWR Engineering Mock-up experiment /3/, as well as of such activities as work on broken Charpy-V specimen reconstitution and acquisition of a well-needed Belgian capability for modern J-integral upper-shelf and brittle-ductile transition K_{JC} testing on irradiated materials. Other parts of the programme remain supported solely by the SCK/CEN Laboratories, more specifically the R & D efforts related to the fracture-safety evaluation of the BR3 reactor pressure vessel (RPV). Because of its more general and interdisciplinary nature, this last topic has been singled-out herein as focus, with the intent of

contributing to place RPV dosimetry in the very context for
which it is needed /2/ : this extends well beyond current surveillance Regulatory requirements /10/.

REACTOR PRESSURE VESSEL FRACTURE SAFETY ANALYSIS :
BR3 PLANT ILLUSTRATIONS

This small*, one-loop PWR, the first WESTINGHOUSE plant to
operate in Europe, is mostly used for engineering experimental
purposes (Plutonium recycling, international testing of advanced
fuel concepts, etc. ...). Nevertheless, from a safety Regulatory viewpoint, it is governed by the same requirements /11/ as
the ones applicable to larger, more modern units such as the ones
at, for instance, the DOEL and TIHANGE sites. Summary features
of the pressure vessel have been reviewed previously /1/; a description and evaluation of fabrication and pre-service inspection
procedures has been performed /12/ as thoroughly as possible on
basis of documentation dating back to over more than a quarter
of century.

The metallurgical surveillance program is however inadequate, the vessel beltline is non-inspectable for shallow flaws,
even considering the latest advances in this field /13/, and
above all, the neutron exposure is large.

Since 1981, the current efforts to assess the vessel integrity have been significantly expanded into a coordinated program involving both in-home activities and consulting support
from various organizations, to the degree found necessary. Upon
request of the Belgian Authorities, the U.S. Nuclear Regulatory
Commission licensing branch has accepted to act as technical
advisor for this program; this help is sustainedly invaluable.

During Spring 1984, the vessel has been wet-annealed at
650 °F for 168 hr with core removal and within design margins.
The operation, an engineering success /14/ unprecedented in the
nuclear commercial community, was promoted by the will to enhance safety margins, even though it was considered that these
would remain sufficient till 1987. This last consideration has
not been completely demonstrated to-date, mostly because one
possible hypothetized scenario of non-pressurized thermal shock

---

*Gross electric power of 11.5 MW.

to the vessel (SBLOCA with break size in the range 1.5" to 3")
entails some doubt about the accuracy of its thermohydraulic
analysis (now submitted to counter–expertise). All illustrations
in this paper refer to the un-annealed vessel under the upper-
bound exposure conditions by 1987 : 5.10^{19} cm^{-2} >l MeV at the vessel
inner surface. Evaluation of the vessel anneal is overviewed in a
separate, initial document to be published soon /15/.

Normal Plant Operation Conditions

Safety-wise, the vessel embrittlement risk has been found to
be dominated by the vertical submerged arc-weld seam, a
nickel-bearing LINDE 80 flux, Oxweld 40 wire Babcock & Wilcox 1957
legacy /12/. Although the plate material (nickel-modified A302B
LUKENS STEEL) is also a crucial part of the investigations, the
weld only is addressed here. Figure 1 illustrates our current
(Summer 1984) evaluation of some upper bound properties for a
vessel inner surface fluence of 5.10^{19} cm^{-2}. All symbols are the
familiar ones. The upper part of the figure corresponds to the
usual Charpy-V impact data forming the present backbone of
surveillance programs, while the lower part displays fracture
toughness, i.e. the data that do actually matter for structural
integrity analysis. Established indexation procedures /16//17/
have been adopted to link the two parts of the figure. Note the
plant service temperature of 500 °F, a specific feature of
governing importance : detrimential for irradiation-embrittlement,
albeit favourable in terms of wet-anneal recovery. Below the
temperature marked DBL, the fracture mechanisms are brittle
(cleavage); at 500 °F however, all our calculations indicate that
K_{IC} becomes unattainable at BR3 (maximum stresses < cleavage
stress)/18/ so that potential cracking appears to be possible only
by ductile mechanisms. Consequently, under normal, test or up-set
operation conditions, BR3 remains in "upper-shelf" regime and
under J-controlled crack growth /17/.

The figures 2 and 3 are intended at illustrating further the
plant specificity from a fracture-mechanics standpoint. The
conservative vessel design does indeed entail low pressure-induced
stresses (favourable diameter-to-thickness ratio) /1/ and low
thermal loads (favourable thickness) /19/, see fig. 2. Another
helpful specificity is that the plant does not need to be brought
critical below the service temperature. Upper shelf safety
margins derived from two independent approaches /20//21/ agree
well; essentially the same, conservative materials properties were
used in both cases, and pressure related safety factors are well
above the regulatory requirement of two, even if no stable
crack-growth is allowed. Leak-before-break is insured. The lower
part of fig. 3 furthermore emphasizes that the use of more recent
J versus C_v-USE correlations does considerably enhance the
previously derived safety margins.

Hypothetized Pressurized Thermal-Shock (PTS) Accidents

Final Safety Analysis Reports (FSAR$_s$) have in the past primarily focused on potential thermohydraulic transients that could affect the fuel clad "safety barrier". In recent years, more attention has been paid at the possible PTS risk /22//23/ /24//25/ to the pressure vessel.

Four classes of potential event initiators are generally considered :
1. Primary loop failures, such as piping breaks, pressurizer surge line break, stuck-open relief valves reclosing when vessel is cooling, ...
2. Secondary loop failures, i.e. all accidents involving steam-line breaks with or without turbine trips etc. ...
3. Excessive feedwater supply to the steam generator(s) for ex. the March 1978 Rancho Seco event /22/
4. Steam-generator tube rupture, eventually combined with other deficiencies.

At BR3, class 3 entails low occurence probabilities ($< 10^{-4}$/ year) and mild thermohydraulic transients. Also, steam-generator integrity threats have been shown to be extremely remote /26/. Attention has consequently concentrated on the first two classes, using event-tree methods when appropriate, in order to identify the risk-bounding hypothetized thermohydraulic transient. The current conclusion is that the worse event would be a main steam-line break (MSLB) of $\sim$ 0.4 ft^2 size with the reactor coolant pumps tripped $\sim$ 1 min. after the safety injection signal (steamline isolated by upstream valve), and at full reactor power, not at hot shutdown. The importance of assessing accidental transients on a plant-specific basis could not be better illustrated.

Among BR3 plant system and operation specificities miti-gating the PTS risk, the following ones deserve mention :
 · Repressurization is limited by the low safety-injection shut-off head pressure of $\sim$ 1000 psi
 · Steam-generator feedwater isolation control valve is closed automatically or by procedure at no-load conditions (operator success probability as quantified in /27/, should the automatic actuation train fail).
 · Significant fraction ($\sim$ 40 %) of the total safety injection flow of 120 m^3/hr is diverted over the core by a spray basket designed to protect the fuel
 · Metal-stored energy has an important effect in reducing vessel chilling (dimensioning of internals).

Unfavourable are the pressurizer, surge-line and vessel geometrical features in case of SBLOCA. For break sizes in excess of 1.5", rapid and important voidage (up to 65 %) occurs, resulting in loop flow stagnation and potentially poor mixing conditions. However, the break size above which there is no more re-pressurization is small, ~ 0.5", and the associated voidage fraction less than 1 % : natural circulation insures good mixing.

For this case which still involves maximum re-pressurization, even a very conservative extrapolation of the thermohydraulics calculations /28/ does not lead to unacceptable vessel failure risk (Fig. 4), for two reasons :
· Warmprestressing (WPS) /29/ is effective in prohibiting crack initiation
· Even if WPS was not credited for, and although crack arrest is not effective here, an utterly unlikely flaw (> 1/4 T) would be required to result in crack initiation; this point will be addressed further below.

The base fracture mechanics analysis, such as the one on fig. 4, is carried out by means of the OCA-1 code /30/, adapted for the BR3 conditions. The code considers only surface flaws infinitely long, and of depth $\underline{a}$ variable from zero to the vessel thickness, $\underline{w}$; the corresponding LEFM stress-intensity factor is labelled here K_{I_∞}. The bottom part of fig. 4 is the time-dependent locus of fractional wall penetration values $(a/w)_c$ for which $K_{I_\infty} = K_{I_C}$ (initiation) or $= K_{I_a}$ (arrest). It is extremely rewarding to examine what happens at the most critical time in the transient, i.e. at $t = t_c$ for which $(a/w)_c$ is minimum. For the sake of this paper, we will illustrate such evaluation in the case of one of the worse MSLB transients. This is displayed on fig. 5, with emphasis on a specific "interface" problem between the thermohydraulics and fracture mechanics aspects for PTS. The fluid-metal heat transfer coefficient H in the downcomer is seen to play a major role : the dotted lines correspond to a value used for some generic calculations /22/ while the full lines are representative of our current assumption for this type of transient. In the first case, there is no crack initiation, while there would be vessel failure in the second. Actually, H depends on the flow regime, thus on time as well as on the downcomer geometry; therefore, it is also different when assessing the impact of metal-stored energy on the fluid temperature. We have handled this "interface" question by insisting on safety-conservatism in both "disciplines". The figure also shows the "penalty" associated to the use of dpa rather than $\emptyset_{> 1\ MeV}$ as exposure unit : there is negligible influence for crack initiation, but a sizeable impact on K_{I_a} - irrelevant here, but very significant for twice thicker (ASME III design) vessels that would approach the PTS "screening criterion".

The transient used as example here has also been analysed by means of the OCA-P code /31/. The deterministic part of the calculation is in agreement with the runs at MOL, as should be expected; the probabilistic fracture-mechanics part has given the following results :

Crack initiation probability : $2.10^{-3}/Y$
Vessel failure probability   : $\underline{9.10^{-4}/Y}$

These are conditional probabilities (i.e. assuming that the transient happens) based on flaw probabilities taken from the MARSHALL report /32/. The interesting points are :

1. <u>Generic</u> conditional failure probability : $\underline{5.10^{-2}/Y}$
2. About 80 % of flaws that initiate do arrest.

It appears thus that the deterministic results on fig. 5 may be unduly conservative - at least if credit is to be taken for crack arrest. Why not ? Should an overcooling accident develop, safety would not be jeopardized; the penalty would be economical.

Fig. 6 illustrates how much crack arrest considerations for BR3 are non-generic. For the considered MSLB transient at the critical time t_c, it is quite clear that the postulated flaw, if initiated, would not grow vertically very far. Prior to initiation, a flaw aspect ratio 2c/a of 6/1 is not likely to be exceeded, while before reaching vessel half-thickness 1/2 T, it is hard to conceive that it would attain 10/1 in this case. The following sketch does specify our notations.

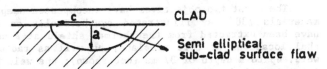

CLAD
Semi elliptical sub-clad surface flaw

The physical reason under the argument is the small fuel core-height of 1 meter, resulting in vertical exposure and fracture toughness gradients worth accounting for. This in turn entails crack tip driving forces well below K_{I_∞} /33/, as shown on the right-hand side of fig. 6. Even considering very pessimistically weld residual-stresses uniformly distributed and as large as 7.5 Ksi /34/, radial crack arrest will occur before 1/2 T and below upper shelf toughness levels. The remaining ligament is sufficient to prevent failure. Comparison with fig. 5 shows that this conclusion is not affected by the penalty inherent in using DPA as exposure unit, rather than $\varnothing_{> 1 \text{ MeV}}$ as done here.

Another beneficial consequence of fig. 6 (vertical K_{IC} curve) is that the BR3 <u>effective weld length</u> of concern is <u>less than 2 meter</u> (volume < 8000 cm³); therefore, the probability of fabrication defects is much less than for a generic plant beltline. This is our rationale for discarding the possibility of flaws exceeding 1/4 T (28 mm) : some literature estimates are gathered in the following table.

Table 1. BR3 : PROBABILITY FOR WELD FABRICATION
DEFECT EXCEEDING 1/4 T

Reference	Probability per meter
MARSHALL et al. /32/	$\sim 4.10^{-4}$
DUFRESNE et al. /35/	$< 10^{-6}$
NILSSON /36/	$\sim 10^{-4}$
HARRIS /37/	$<< 10^{-6}$

The probability for <u>undetected</u> fabrication defects of this
size is much less. It is believed that flaws deeper than 6 mm are
unlikely in this weld /12/.

SUPPORTIVE EXPERIMENTAL WORK

Three major areas have been emphasized :

1. Provide a pedigreed neutron and gamma physics-dosimetry
   characterization of the plant.

2. Establish the plant chemistry accurately for the plate,
   the weld and the heat-affected zone.

3. Develop plant-applicable metallurgical trend curves.

The first two objectives have encompassed sampling the vessel
materials /38/, as illustrated synthetically on fig. 7. Chips
have been extracted from the thermal shield at a number of azimuthal locations at core midplane, as well as radially toward the
weld, up to a depth of 37 mm (not shown); the weld itself has been
investigated in the same manner, at eight vertical elevations
above the thermal shield top. Typical physics-dosimetry and
chemical analysis results are displayed.

. ^{54}Fe(n,p) and ^{59}Co(n,γ) radiometric measurements agree well
  with transport theory insofar as the radial fluence distribution <u>shape</u> is concerned; as the methodology and material
  cross-section data have been validated at PCA /2/ and VENUS
  /3/, high confidence can be placed upon projections into the
  vessel wall

· There is however an absolute fast flux scale discrepancy,
  with an average C/E of 1.10 to 1.15 /39/, as well as
  departure between the measured and predicted azimuthal
  distributions, the former one being much flatter than
  expected, for unknown reasons; it must be cautioned however
  that the fuel configuration is complex, involving many test
  assemblies of different designs, and that the source
  distribution, which was independently provided to SCK/CEN,
  has been established by diffusion theory, an approach not
  yet benchmarked in VENUS /3/.

- The vertical fast flux distribution agrees well with theory, giving strong support to the vertical fracture toughness gradients accepted for PTS analysis

- Coupled neutron-gamma calculations indicate small energy-deposition in the vessel wall and negligible influence on the metal temperature

- The chemical definition, which addresses all metals as well as gazes, has been benchmarked carefully on similar materials by means of interlaboratory comparisons using different approaches, including wet chemistry. Worth to emphasize is a novel activation technique for copper assay, especially developed to monitor the vessel sampling in a short time and with good accuracy, as typically demonstrated by Table II.

The third objective above, metallurgical trend curves, is by far the most challenging, especially in the vessel post-anneal conditions. It has been addressed through a dedicated irradiation and testing program conducted by MATERIALS ENGINEERING Ass. (MEA) at the State University of New-York, at Buffalo, under the sponsorship of SCK/CEN. An overview will be given as part of a forthcoming publication /15/ and extensive documentation will be provided in future /40/. Only a few highlights of the approach can be considered herein. The test matrix employs five LINDE 80 archive weld metals covering a range of compositions bracketing BR3, as well as an archive plate selected to match closely the BR3 plate chemistry, at least insofar as all metallic constituents are concerned. In all cases, care was taken to insure that the fabrication procedures and heat treatments be also representative (the only exception is the plate which does not simulate unirradiated upper shelf conditions adequately). Most samples are of the conventional Charpy-V type, with a few tensile ones, except for one weld for which 0.5 T-CT compact tension specimens have been emphasized also. The first step has been to establish the irradiation temperature effect by means of controlled exposures for three of the welds at 550 °F and at 500 °F; this allows to take advantage of the broader data base available for these materials at 550 °F /41//42/. The relative irradiation temperature effect φ $(T)/\varphi$ (550 °F) = $(\Delta RT_{NDT})_T/(\Delta RT_{NDT})_{550}$ °F is well represented by either one of the two expressions

$$\varphi(T) = \frac{\Delta_o}{2} \left[1 - \tanh \left(\frac{T - \bar{T}}{c} \right) \right] \tag{1}$$

with $\bar{T}$ = 550 °F and c = 150 °F

29

Table II

NEUTRON ACTIVATION ANALYSIS[a] OF COPPER CONTENT IN
RPV STEEL SPECIMENS AND COMPARISON TO CURRENT METHODS

	Activation Analysis		Microprobe[d]
NBS Standard[b]	0.373	(± 0.0075)[b]	
Reference Weld WI-34	0.312	(± 0.0066)[c]	
BR3 Specimens			
Identification			
I-1	0.122	(± 0.0070)	0.16[e]
I-2	0.189	(± 0.0056)	0.15
II-1	0.158	(± 0.0079)	0.18
II-2	0.178	(± 0.0060)	–
III-1	0.180	(± 0.0033)	0.17
III-2	0.176	(± 0.0078)	0.16

[a] With reference to pure copper foils

[b] Low alloy steel A242 Mod. I.D.-C1285; value quoted by NBS is 0.37%

[c] Value recommended as a result of round robin test through subcontractors to B&W and MEA is 0.31% (including wet chemistry data).

[d] Examination of a few chips of < 10 mg each. The activation measurements were done on ~ 300 mg batches while each specimen extracted by Rolls Royce Associates weighs 1.5 to 2 g.

[e] One chip gave a result of 0.08%.

$$\text{or} \quad \phi(T) = \sqrt{e^{-0.75 \exp \frac{Q}{k}\left[\frac{1}{\overline{T}} - \frac{1}{T}\right]}} \tag{2}$$

with Q = defect production activation energy (0.5 eV)

k = Boltzmann constant ($8.615,10^{-5}$ eV/°K)

$\overline{T}$ = 530 °K = 495 °F.

Equation (2) is based on earlier modelling work by G.R. ODETTE. The two curves are undiscernable from each other below 630 °F and are compared on Fig. 8 to the present data and to literature.

The next step concentrates on annealing recovery measurements, which are analysed at the light of a correlation of all published data for both plates and welds; /43/ this incorporates an empirical correction for flux level acceleration effects in test reactors, as compared to power plants.

Finally, the re-embrittlement path is examined; IAR (irradiated-annealed-reirradiated) data generated by the program are linked to the pre-anneal trend curves through engineering-type empirism. The overall effort is not complete yet, but some of the data available at this time are synthetized on fig. 9 .

A simultaneous physically-based evaluation is in progress through consultation with Dr. Ing. D. PACHUR.

GENERAL CONCLUSIONS

Structural integrity analysis of a reactor pressure vessel calls for interdisciplinary communication and for realistic appraisal of the directions of efforts; in particular, accuracy goals must be set in context, whatever facet of the problem is considered. Clearly, the physics-dosimetry aspects constitute an important area. The concern over potential pressurized thermal shock (PTS) accidental risks has more than ever reinforced this recognition, while stressing the need for plant-specific versus generic considerations. In a Regulatory perspective, the PTS screening criterions/22/ may be too severe for certain plants, if considered as an operational limitation; such is not their meaning, but they certainly draw a "line" above which the need for caution

necessarily entails economic penalty, and in this sense, they
are of a generic nature. While long-range R & D programs are
expected to provide the potentials for more realistic assessment of all safety-sensitive parameters, only a handful among
them are accessible to immediately fruitful improvement in the
perspective of a plant operator : vessel exposure reduction by
core management, better dosimetry, better materials characterization, better thermohydraulics analysis are examples.
Narrowing even further the attention upon just four parameters :
neutron exposure, copper and nickel contents of steel, irradiation temperature, it is attempted to show on fig.10 that
vessel exposure accuracy goals of $\pm$ 20 % (1 σ) are both sufficient and consistent with current uncertainties on the other
three, restricted parameters. It is trusted that this level of
dosimetry accuracy requirement is near to be met and that, although a limited additional R & D effort in this field remains
most warranted for the next two years*, attention of the dosimetrist should turn more than before toward disseminating and
maintaining the know-how acquired in the past five years - a
process well advancing through the establishment of concensus dosimetry standards and the availability of permanent referencing facilities and procedures.

---

* Improved understanding of gamma-heating effects, of exposure
units and their through-vessel variation, of plutonium contribution as source in establishing core leakage, ...

REFERENCES

/1/ A. FABRY et al. - "Improvement of LWR Pressure Vessel Steel Embrittlement Surveillance ..." Proceedings Fourth ASTM-EURATOM Symposium on Reactor Dosimetry, NUREG/CP-0029, VI, 45 (1982).

/2/ W.N. McELROY et al. - "Surveillance Dosimetry of Operating Power Plants" Ibid.

/3/ A. FABRY et al. - "VENUS PWR Engineering Mock-up : Core Qualification, Neutron and Gamma Field Characterization" This Symposium.

/4/ M.L. WILLIAMS et al. - "Calculation of the Neutron Source Distributions in the VENUS PWR Mockup Experiment" This Symposium.

/5/ A. FERO - "Neutron and Gamma Ray Flux Calculations for the VENUS PWR Engineering Mockup" This Symposium.

/6/ W. MANNHART - "Spectrum-Averaged Neutron Cross Sections measured in the U-235 Fission-Neutron Field at MOL" This Symposium.

/7/ D.M. GILLIAM et al. - "Cross Section Measurements in the U-235 Fission Spectrum Neutron Field" This Symposium.

/8/ R. GOLD et al. - "Non-Destructive Determination of Reactor Pressure Vessel Exposure by Continuous Gamma-Ray Spectrometry and Non-Destructive Measurement of Neutron Exposure of BR3" This Symposium.

/9/ J.A. MASON - "Characterization of the Imperial College Gamma-Ray Reference Field" This Symposium.

/10/ US Code of Federal Regulations, Title 10, Energy, Part 50, Appendix H, Reactor Vessel Material Surveillance Program Requirements, revised (1983).

/11/ US Code of Federal Regulations, Title 10, Energy, Part 50, Appendix G, Fracture Toughness Requirements, revised (1983).

/12/ A.L. LOWE Jr, L.B. GROSS - "Evaluation of Fabrication Procedures for BR3 Reactor Vessel" BAW-1807, Sept. 1983.

/13/ Commission of the European Communities "Defect Detection and Sizing" Proceedings of a Specialist Meeting sponsored by OECD and IAEA, 3-6 May 1983, EUR 9066 EN, CSNI Report n° 75 (1983).

/14/ F. MOTTE et al. - Work to be published.

/15/ A. FABRY, F. MOTTE et al. - "Annealing of the Belgian BR3 Pressure Vessel" Paper for presentation at the 12th USNRC Safety Research Information Meeting, National Bureau of Standards, October 22-26, 1984.

/16/ ASME Boiler and Pressure Vessel Code Section XI "Rules for In-service Inspection of Nuclear Power Plant Components" (July 1, 1974).

/17/ R. JOHNSON et al. - "Resolution of the Reactor Vessel Materials Toughness Safety Issue" NUREG-0744 (1981).

/18/ IL MILNE, D.A. CURRY - "Ductile Crack Growth Analysis Within the Ductile-Brittle Transition Regime : Predicting the Permissible Extent of Ductile Crack Growth" Elastic-Plastic Fracture, Second Symposium, Vol. II, 278-290, ASTM-STP 803 (1981).

/19/ U.S. Nuclear Regulatory Commission Standard Review Plan, Report NUREG-75/087 and Branch Technical Position MTEB 5.2.

/20/ A. FABRY - "BR3 Pressure Vessel Safety Margins against Upper Shelf Failure by Plastic Instability or Cleavage Fracture" CEN/SCK 380/81-50 (Dec. 4, 1981).

/21/ J.M. BLOOM - "Safety Factor Calculations for the BR3 and Reference Vessels using the Failure Assessment Diagram Approach" Report BAW-1811 (Nov. 1983).

/22/ W.J. DIRCKS - Policy Issue (Notation Vote) SECY-82-465 to the USNRC Commissioners and Attachment "NRC Staff Evaluation of Pressurized Thermal Shock", November 23, 1982.

/23/ D.L. PHUNG, W.B. COTTRELL - "Analysing Precursors to severe Thermal Shock".

/24/ T.A. MEYER - "Summary Report on Reactor Vessel Integrity for w Operating Plants" WCAP-10019, Dec. 1981.

/25/ "B & W Owners Group Probabilistic Evaluation of PTS" Phase 1 Report, BAW-1791, June 1983.

/26/ F. MOTTE - "BR3 Steam Generator : Elements of Judgment on the Integrity of this Equipment" CEN/SCK Internal Report (1983).

/27/ J. BOUCAU, M. DUPREZ - "BR3 - Pressurized Thermal Shock Analysis, Steamline Break Transients" Report WENX/83/30 (1983).

/28/ G.J. FREEMAN - "LOCA Analysis of the BR3 Plant at Mol, Belgium" Report RRA 6784 (1984).

/29/ B.W. PICKLES, A. COWAN - "A Review of Warm Prestressing Studies" Int. J. Pres. Ves. & Piping 14, 95-131 (1983).

/30/ S.K. ISLANDER, R.D. CHEVERTON and D.G. BALL - "OCA-I, A Code for Calculating the Behavior of Flaws on the Inner Surface of a Pressure Vessel Subjected to Temperature and Pressure Transients" Report NUREG/CR 2113, ORNL/NUREG-84 (1981).

/31/ R.D. CHEVERTON, D.G. BALL - Private Communication, July 11, 1983.

/32/ "An Assessment of the Integrity of PWR Pressure Vessels" Second Report of UKAEA Study Group under the Chairmanship of Dr. W. MARSHALL (1983).

/33/ I.S. RAJU, J.C. NEWMAN Jr. - "Stress-Intensity Factor Influence Coefficients for Internal and External Surface Cracks in Cylindrical Vessels" J. Press. Vessel Techn. Trans. ASME V104 n° 4, 37-48 (1982).

/34/ D.P.G. LIDBURY - "The Significance of Residual Stresses in Relation to the Integrity of LWR Pressure Vessels" Report to Principal Working Group on Primary Circuit Integrity, Commission on the Safety of Nuclear Installations, CSNI, 1984.

/35/ J. DUFRESNE, A.C. LUCIA et al. - To be published (1984).

/36/ F. NILSSON - "A Model for Fracture Mechanical Estimation of the Failure Probability of Reactor Pressure Vessels" Third Int. Conf. on Pressure Vessel Technology, Tokyo (1977).

/37/ D.O. HARRIS - "A Means of Assessing the Effect of NDE on the Reliability of Cyclically Loaded Structures" Materials Evaluation p. 57-65, July 1977. See also Progr. in Nucl. Eng. (10), 1 (1982).

/38/ J. GOODMAN - "Reactor Material Sampling, BR3, Mol, Belgium" Report RRA/6605 (1983).

/39/ J. DEBRUE, G. MINSART - To be published (1984).

/40/ J.R. HAWTHORNE et al. - To be published (1984).

/41/ P.N. RANDALL - "NRC Perspective of Safety and Licensing Issues Regarding Reactor Vessel Steel Embrittlement - Criteria for Trend Curve Development" ANS Annual Meeting, Detroit, June 14 (1983).

/42/ A.S. HELLER, A.L. LOWE Jr. - Correlations for Predicting the Effects of Neutron Radiation on Linde 80 Submerged-Arc Welds" Report BAW-1803 (1983).

/43/ B. McDONALD - "A Note on Commercial Annealing Experiments" To be published (1984).

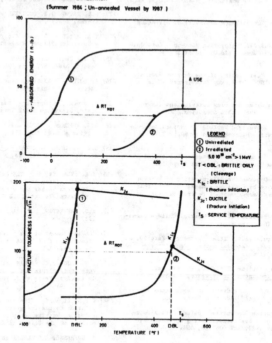

Fig.1
SCHEMATIC ILLUSTRATION OF UPPER BOUND IRRADIATION EFFECTS ON THE BR3
VERTICAL WELD SEAM : CHARPY-V IMPACT AND FRACTURE TOUGHNESS PROPERTIES
FROM CURRENT EVALUATION

(Summer 1984 ; Un-annealed Vessel by 1987)

LEGEND
① Unirradiated
② Irradiated
   5.0 10¹⁸ cm⁻² > 1 MeV
T < DBL : BRITTLE ONLY
         (Cleavage)
K_{Ic} : BRITTLE
       (Fracture Initiation)
K_{Jc} : DUCTILE
       (Fracture Initiation)
T_S SERVICE TEMPERATURE

Fig. 2 HEAT-UP AND COOLDOWN PRESSURE-TEMPERATURE LIMITATIONS : SCHEMATIC ANALYSIS OF FUNDAMENTAL PLANT-SPECIFICITY

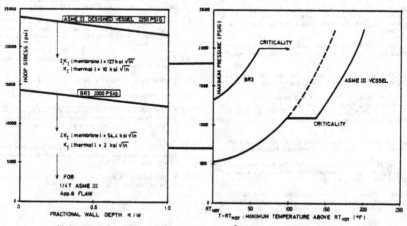

$$2K_I \text{ (membrane)} + K_I \text{ (thermal)} \leq K_{IR} = K_{Ic} = 26.8 + 1.223 \exp\left[0.0145(T - RT_{NDT} + 160)\right] \text{ (ksi}\sqrt{\text{in}}\text{)}.$$

ADDITIONAL 40°F MARGIN REQUIRED AT REACTOR CORE CRITICALITY.

34

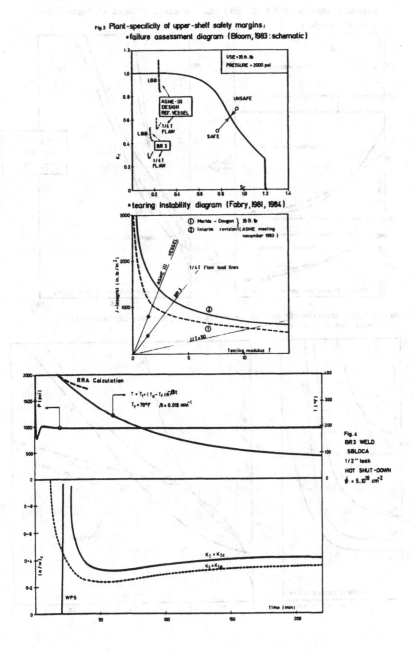

Fig 3 Plant-specificity of upper-shelf safety margins:
• failure assessment diagram (Bloom, 1983: schematic)

USE = 35 ft.lb
PRESSURE = 2000 psi

LBB

ASME-III DESIGN REF. VESSEL
1/4 T FLAW

LBB

BR 3
1/4 T FLAW

UNSAFE

SAFE

K_r

S_r

• tearing instability diagram (Fabry, 1981, 1984)

① Merkle – Dougan } 35 ft.lb
② Interim revision (ASME meeting november 1983)

ASME III VESSEL

1/4 T Flaw load lines

J-integral (in. lb/in²)

②

①

J/T = 50

Tearing modulus T

RRA Calculation

$T = T_f + (T_0 - T_f)e^{-\beta t}$
$T_f = 70°F$ $\beta = 0.015$ min^{-1}

P (psi)

T (°F)

Fig. 4
BR 3 WELD
SBLOCA
1/2" leak
HOT SHUT-DOWN
$\phi = 5.10^{18}$ cm^{-2}

$K_I = K_{Ic}$

$K_I = K_{Ia}$

(a / w)$_c$

WPS

Time (min)

35

Fig. 5
Influence of fluid-metal heat-transfer coefficient H and neutron exposure unit on a typical BR3 fracture safety analysis of hypothetized MSLB accident.

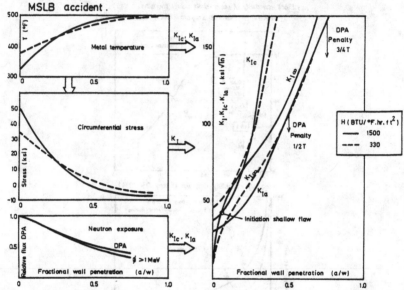

Fig. 6 REALISTIC CRACK-ARREST ANALYSIS FOR TYPICAL HYPOTHETIZED MSLB ACCIDENT AT BR3.
(Same Transient as on fig. 5)

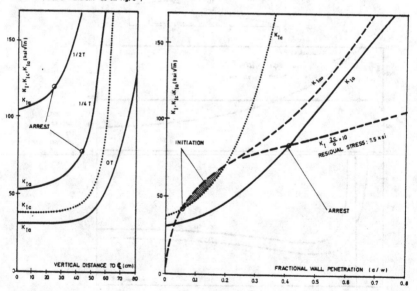

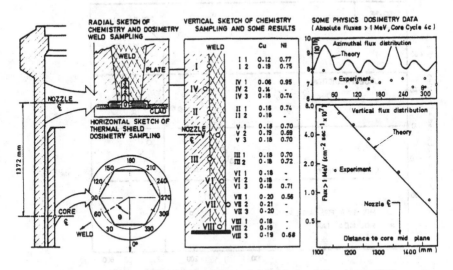

Fig 7

SCHEMATIC ILLUSTRATION OF BR3 REACTOR MATERIALS SAMPLING AND SELECTED REPRESENTATIVE RESULTS.

Fig. 8 Sensitivity of the ΔRT_{NDT} embrittlement index to irradiation temperature

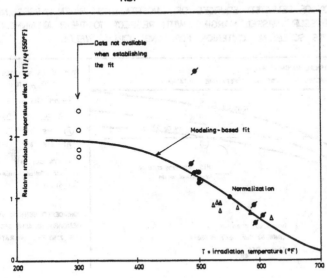

Fig. 9 COMPARISON BETWEEN OBSERVED NIL - DUCTILITY TRANSITION TEMPERATURE SHIFTS
AND PREDICTIONS BASED ON THE CORRELATION DEVELOPED FOR LINDE 80 FLUX WELDS

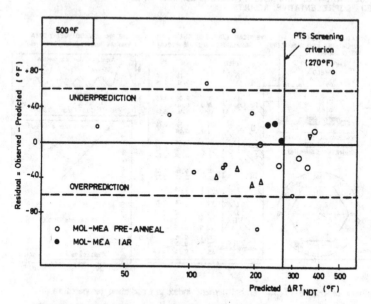

Fig 10 COMPARISON OF SELECTED SOURCES OF UNCERTAINTIES WHEN ASSESSING
OLDIER PRESSURE VESSEL MARGINS WITH RESPECT TO THEIR ATTAINMENT
OF THE PTS SCREENING CRITERION FOR LONGITUDINAL WELDS :

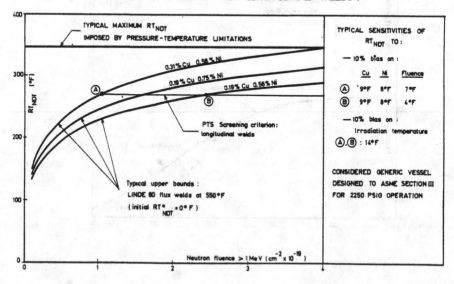

UTILITY PERSPECTIVES RELATED TO PRESSURE VESSEL

AND SUPPORT STRUCTURE SURVEILLANCE

S. P. Grant

Carolina Power & Light Co.

Raleigh, North Carolina USA

ABSTRACT

There are several efforts ongoing in the United States
and England to produce improved radiation damage trend curves.
The accuracy of neutron fluence determinations is an important
factor to the degree of improvement attainable in assessment
of damage as measured with Charpy V and other surveillance
specimens.

Dosimetry in capsules within the reactor vessel is limited
in vessel coverage and accessibility. Handling costs are
large. These disadvantages can be overcome through the use of
cavity dosimetry.

New types of dosimeters now supplement radiometric foils.
These include solid state track recorders, damage monitors and
helium accumulation monitors. An experiment which includes
all of these dosimeters has been completed at the H. B. Robinson
Unit No. 2 plant. In addition, the cavity at the Robinson
plant is being mocked up in a confirmatory experiment at the
NESDIP Winfrith facility.

Utilities making use of these new, improved methods will
have excellent characterization of neutron fluence profiles of
the vessel and its supports.

J. P. Genthon and H. Röttger (eds.), Reactor Dosimetry, 39–45.

INTRODUCTION

An accurate accounting for radiation damage to the reactor
vessel has become of primary importance since pressurized
thermal shock (PTS) was recognized as a potential threat to
some nuclear reactor vessels exposed to high fast neutron
fluence. Support structures receive much smaller neutron
doses, but a good estimate of damage may be necessary for
extension of life cases when the support is located in the
core region.

Improved radiation damage trend curves and fluence
determinations can be very helpful in obtaining maximum
service life for reactor vessels and supports.

RADIATION DAMAGE TREND CURVES AND SURVEILLANCE REQUIREMENTS

Utilities in the United States are attentive to developments
in trend curves which may soon be applicable to their reactor
vessels. Guthrie is statistically analyzing all Charpy V
surveillance data (1). Odette has developed a physically
based model which includes copper precipitation (2). Williams,
et al, has a large effort under way to produce well characterized data for model development (3). Grant and Earp found
manganese to be a factor in the increase in yield strength
during irradiation (4). The Carolina Power & Light Company
(CP&L) and the Tennessee Valley Authority are investigating
the role of nickel in sustaining radiation damage by copper
(4) and secondary contributions by manganese and other elements.

Surveillance capsules were adopted for the purposes of
monitoring radiation damage in reactor vessels in the early
1960s without a well defined method for accomplishing the
task. The samples which represented the vessel and which were
chosen for destructive examination usually were not the most
limiting; therefore, they cannot be used to establish nil
ductility limits for the vessels. In such cases special
radiation effects studies must be used to supplement surveillance results as required by the 1983 revision to Appendix G
of U. S. Rules and Regulation 10CFR50.

The ASTM recommended practice for surveillance tests
(E-185) did not cite the importance of residual amounts of

copper and phosphorus until 1973. The possibility that nickel, manganese, and other residual/alloying elements may contribute to damage is noted in the current 1982 edition. Meanwhile, Ayres (5) and Randall (6) have cited the large effect of nickel. Hawthorne (7) has determined the ranges of chemistry and fluence in which phosphorus is important.

REACTOR VESSEL SURVEILLANCE DOSIMETRY

Surveillance dosimetry has been located in the above mentioned surveillance capsules fixed to the vessel or thermal shield. Unless well fixed, the capsules came loose and the dosimetry was lost. When well fixed, insertion or removal can be difficult. Handling of these capsules for shipment has proven to be vexing and costly under conditions inherent to reactor sites.

An attractive alternative, or supplement, is now available in reactor cavity dosimetry. It offers a capability for locating the dosimetry at as many axial and azimuthal locations as desired, which permits complete mapping of the core region of the vessel. Moreover, lower cost and ease of handling are very attractive. Figures 1 and 2 describe dosimetry and holders being used in cavities.

It has been accepted for most power reactor vessel surveillance programs that radiation damage can be correlated sufficiently well with neutrons >1Mev and that the $54Fe(n,p)$ reaction adequately determines the specific activity. There have been proposals that neutrons of lesser energy may contribute significantly to radiation damage through the vessel wall and in the cavity where support structures may be found. To provide a measure of radiation damage for these several possibilities, NRC Research has supported the development of direct reading solid state track recorders (SSTRs) and damage monitors (8). Helium accumulation fission monitors (HAFMs) provide an accounting for helium generation by low energy neutron reaction with the boron (9).

41

CAVITY DOSIMETRY AT CP&L NUCLEAR REACTORS

SSTRs and HAFMs are being employed in cavity flux mapping at the H. B. Robinson 2 PWR and the Brunswick 2 BWR. Full axial coverage is obtained with SSTRs and gradient wires. Azimuthal coverage is assured by use of dosimetry strings at 228°, 270°, 282°, and 300°, and making the reliable assumption.

Preliminary results project greater an order of magnitude reduction in fast neutron flux from the internal surface of the reactor to the cavity. Unless low energy neutrons contribute significantly to damage of support steel at the outer vessel diameter, supports should not be a concern.

To establish the validity of the reactor cavity dosimetry, benchmarking with dosimetry inside the vessel and transport calculations is essential. For this purpose, a special internal capsule filled with advanced dosimetry supplied by the Hanford Engineering Development Laboratory was installed concurrently with the reactor cavity dosimetry. In addition, the Robinson 2 reactor cavity is being mocked up at the Winfrith, England, research facility in an agreement reached by the UKAEA and NRC Research and CP&L. Westinghouse will complete the benchmarking to the transport calculations and through damage reduction estimates.

CONCLUSIONS

Advanced dosimetry methods are now available for better reactor vessel coverage in the core region. They will also provide correlation with those new trend curve developments which are spectrum dependent.

REFERENCES

1. G. Guthrie, "Pressure Vessel Steel Irradiation Formulas Derived from PWR Surveillance Data", Trans American Nuclear Society, 44, 1983, pp. 222-223.

2. G. R. Odette," On The Dominant Mechanism of Irradiation Embrittlement of Reactor Pressure Vessel Steels", Scripta Metallurgica, Volume 17, 1983, pp. 1183-1188.

3. T. J. Williams, A. F. Thomas, R. A. Berrisfor, M. Austin, R. L. Squires, and H. J. Venables, "The Influence of Neutron Exposure, Chemical Composition and Metallurgical Condition on the Irradiation Shift of Reactor Pressure Vessel Steels", Effects of Irradiation on Materials: Eleventh Conference, ASTM 782, 1982.

4. S. P. Grant and S. L. Earp, "Methods for Extending Life of a PWR Reactor Vessel After Long Term Exposure to Fast Neutron Radiation", 12th Symposium on Effects of Radiation on Materials, ASTM STP 870, 1984.

5. D. J. Ayres, "Combustion Engineering Owners' Group Pressurized Thermal Shock Program", Presented to Nuclear Regulatory Commission, October 7, 1981.

6. P. N. Randall, "NRC Perspectives of Safety and Licensing Issues Regarding Reactor Vessel Steel Embrittlement, American Nuclear Society Annual Meeting", Detroit, Michigan, June 14, 1983.

7. J. R. Hawthorne, "Evaluation of Embrittlement Rate Following Annealing and Related Investigations on RPV Steels", 11th Water Reactor Safety Research Meeting, NUREG/CP-0048, Volume 4, October, 1983.

8. R. Gold, F. Ruddy and J. Roberts, "Standard Method for Application and Analysis of Solid State Track Recorder Monitors for Reactor Surveillance", ASTM E854-81.

9. H. Farrar and B. Oliver, "Standard Method for Application and Analysis of Helium Accumulation Fluence Monitors for Reactor Vessel Surveillance", ASTM E910-82.

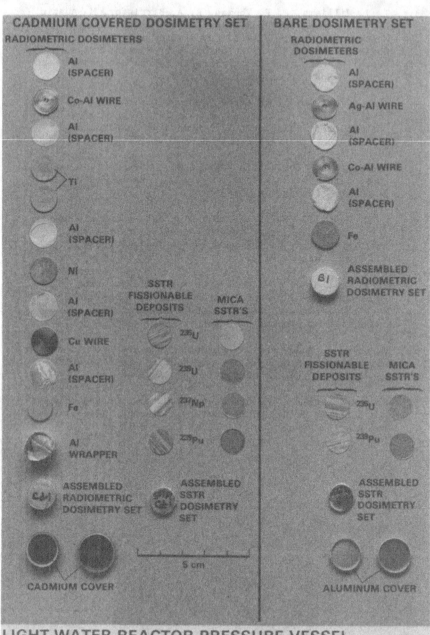

CADMIUM COVERED DOSIMETRY SET

RADIOMETRIC DOSIMETERS

Al (SPACER)
Co-Al WIRE
Al (SPACER)
Ti
Al (SPACER)
Ni
Al (SPACER)
Cu WIRE
Al (SPACER)
Fe
Al WRAPPER
ASSEMBLED RADIOMETRIC DOSIMETRY SET

SSTR FISSIONABLE DEPOSITS
^{235}U
^{238}U
^{237}Np
^{239}Pu

MICA SSTR'S

ASSEMBLED SSTR DOSIMETRY SET

5 cm

CADMIUM COVER

BARE DOSIMETRY SET

RADIOMETRIC DOSIMETERS

Al (SPACER)
Ag-Al WIRE
Al (SPACER)
Co-Al WIRE
Al (SPACER)
Fe
ASSEMBLED RADIOMETRIC DOSIMETRY SET

SSTR FISSIONABLE DEPOSITS
^{235}U
^{238}Pu

MICA SSTR'S

ASSEMBLED SSTR DOSIMETRY SET

ALUMINUM COVER

LIGHT WATER REACTOR PRESSURE VESSEL SURVEILLANCE CAVITY DOSIMETRY SET

FIG. 1
HEDL 8408-207

44

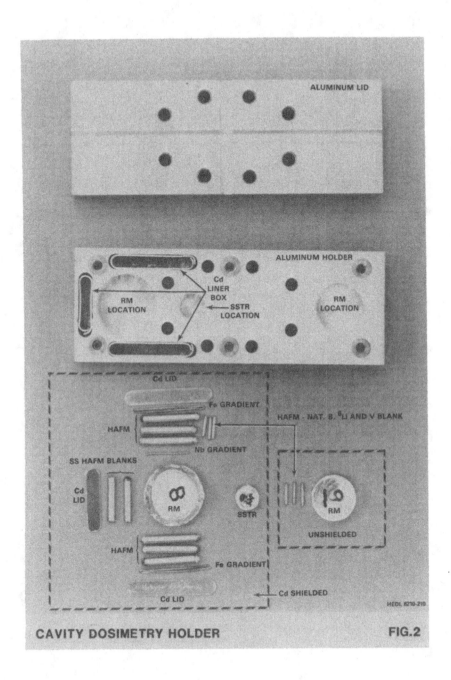

CAVITY DOSIMETRY HOLDER　　　　　　　　**FIG.2**

45

UNCERTAINTY CONSIDERATIONS IN DEVELOPMENT AND
APPLICATION OF CHARPY TREND CURVE FORMULAS

G. L. Guthrie

Hanford Engineering Development Laboratory

Richland, Washington, USA

ABSTRACT

A covariance treatment is described which takes
into account the estimated errors in the specific
independent variables in the application of Charpy
trend curve formulas developed at HEDL. Sources and
magnitudes of errors are discussed and covariance
matrices are supplied. Separate formulas are given
for plate and weld specimens. The covariance treatment assumes that the errors in the fluence are lognormal while errors in the chemical concentrations
and in the Charpy measurements are normal. The
uncertainties in the data base and in the application
are assumed to be independent. The present treatment
avoids the shortcomings of the usual procedure where
the trend formula standard deviation is used as a
complete error indicator. The more common treatment
contains the tacit assumption that the expected error
in the application is typical of the error in the
data base used to develop the trend curve formula,
but is not dependent on the particular values of the
independent variables associated with the application.

J. P. Genthon and H. Röttger (eds.), Reactor Dosimetry, 47–59.
© *1985 ECSC, EEC, EAEC, Brussels and Luxembourg*

INTRODUCTION

Covariance matrices have been derived for use with Charpy trend curve formulas developed at HEDL. The covariance matrices have been used to estimate uncertainties in calculated values for the irradiation induced increase in the Charpy 30 ft. lb. transition temperature for a number of surveillance capsule data points, using standard error propagation techniques and using estimates for the errors in the independent variables. A normal error distribution is assumed in the values of the copper and nickel concentrations as well as in the measured temperature shift while a log-normal distribution is assumed for the fluence errors.

A less rigorous but commonly used approach is to assume that the error or uncertainty in the application of a Charpy trend curve formula is nearly the same as the standard deviation found in the least squares fit to the original data. However, the independent input variables in the application may be known to an accuracy noticeably different than the average accuracy available in the data used to derive the trend curve formula. The method advocated here provides a procedure for giving credit or assessing a penalty for the difference in estimated accuracy.

The trend curve formulas and the resulting covariance matrices were derived (1) using Charpy input data originally supplied by Dr. P. N. Randall of the NRC. The fluence values have been modified using information supplied by R. L. Simons (2). Separate formulas were derived for plate and weld specimens. The independent variables chosen are the copper and nickel concentrations and the fluence (E > 1.0 MeV). Dpa is an optional exposure parameter preferred for deep wall positions, and is mathematically substituted in the formulas by using the relation 10^{19} n/cm^2 = 0.0160 dpa, which is an average relationship approximately valid for the input data used to derive the trend curves.

Derivation of the Trend Curve Formulas and Covariance Matrices

The object of the present work is to supply a method for estimating the standard deviation for the calculated Charpy shift on a case by case basis instead of using the standard deviation for the original fitting procedure as a universal error indicator. That is, we desire to estimate the square root of the expectation value of the square of the error in the calculated values of the irradiation-induced increase in the 30 ft lb Charpy transition temperature for each application.

This can be done for trend curve formulas recently developed at HEDL provided that several simplifying assumptions are made: 1) all errors and uncertainties are small enough that linear approximations are valid, 2) all distribution functions are "normal" in form, and 3) particular pairs of items are independent (uncorrelated). An example is the pair consisting of the expected error in the $\log_e$ of the fluence and the expected error in the copper concentration. When the indicated assumptions are made, the theory of linear least squares error estimation is available for application to the problem. To satisfy the assumption about normal distributions of errors, and to maintain consistency with previous work (3,4), $\log_e (\phi t)$ is used as the independent exposure variable, rather than ϕt itself. This is because the trend curve formulas have been derived under the assumption of a log-normal distribution in fluence errors (i.e., there is a normal distribution of errors in $\log_e$ of the fluence).

To briefly review the existing theory on error estimation, first assume:

$$\Delta T = f(P_i, V_j) \qquad (1)$$

where:

ΔT = Charpy shift
f = Some functional expression chosen by the analyst
P_i = Parameters (i = 1, 2, 3, etc.)
V_j = Variables (j = 1, 2, 3)

The parameters are the adjustable constants that have taken on "best" values found in a least squares fit to a data base, and the variables are items associated with the specimen and its irradiation. Possible independent variables are the chemical concentrations and $\log_e (\phi t)$ in a given application.

Taking differentials in Eq. (1) at any set of values for the P_i and V_j, the error in ΔT is:

$$\delta(\Delta T) = \sum \frac{\partial f}{\partial P_i} \delta(P_i) + \sum_j \frac{\partial f}{\partial V_j} \delta(V_j) \qquad (2)$$

The expectation value of $\delta^2(\Delta T)$ is then the sum of the expectation values of all the terms resulting from squaring the right-hand side of Eq. (2).

We find

$$\delta^2(\Delta T) = \sum \sum \frac{\partial f}{\partial P_k} \frac{\partial f}{\partial P_l} \overline{\delta P_k \cdot \delta P_l}$$

$$+ \sum_j \left(\frac{\partial f}{\partial V_j}\right)^2 \cdot \overline{\delta^2(V_j)} \qquad (3)$$

where the bar denotes an expectation value.

In the first set of terms, the factors $\partial f/\partial P_k$ and $\partial f/\partial P_l$ can be directly calculated from the formula f in any given application. The factor $\delta P_K \cdot \delta P_l$ is by definition the covariance matrix of the parameters in the formula. In the second set of terms, $(\partial f/\partial V_j)^2$ can be directly calculated from the trend curve formula, and $\delta^2(V_j)$ is an estimate of an expected value of the square of the errors in the variable V_j. This must be obtained from a knowledge of the uncertainties in chemistry, fluence, and any other relevant variables appearing in the function f. Cross terms of the type

$$\text{Cross Term (1)} = \frac{\partial f}{\partial V_m} \frac{\partial f}{\partial V_n} \overline{\delta(V_m) \cdot \delta(V_n)} \tag{4}$$

have been omitted on the assumption that V_m and V_n are independent, and the errors are uncorrelated so that the expectation value of the product is zero.

Cross terms of the type

$$\text{Cross Term (2)} = \frac{\partial f}{\partial P_k} \frac{\partial f}{\partial V_i} \overline{\delta(P_k) \cdot \delta(V_i)} \tag{5}$$

have been omitted since the parameters P_k are derived from a reference data base, while the V_i are the variables in a particular application. Thus, P_k and V_i are independent, so that each product term has an expectation value of zero.

The covariance matrix of the parameters, in a linear approximation, is available from the theory of linear least squares (5). For a linear problem, the covariance matrix is given by

$$C_{ij} = \frac{R_k W_{kl} R_l}{N-P} \left[\phi_{im}^T W_{mn} \phi_{nj} \right]^{-1} \tag{6}$$

where a repeated index indicates a summation and

W_{kl} = Weight matrix for the observations
R_k = Column matrix of residuals
ϕ = So-called "design" matrix, which is the matrix of the original set of overdetermined equations
ϕ^T = Matrix transpose

In Eq. (6), (N-P) is the number of equations minus the number of parameters (or, the degrees of freedom).

The quantity $\sigma^2 = R_k W_{kl} R_l/(N-P)$ \hfill (7)

is simply the estimate of the square of the standard deviation for the fit, and is a single scalar number.

It is worth noting that any arbitrary scale factor in W cancels in Eq. (6) since it appears in the numerator in the early part of Eq. (6) and appears as an inverse in the latter part of the same equation.

The ϕ_{im} matrix of the linear problem is replaced by the Jacobian matrix in a situation involving a linear approximation. A numerical approximation of the Jacobian is available in the output of commonly used nonlinear least squares computer packages.

The acquisition of the remaining required items in Eq. (3) is straight-forward, but perhaps tedious. The methods indicated above have been applied to a combined plate-weld formula developed at HEDL. The complete set of equations was developed using 126 data points for plate material and 51 data points for weld materials. The derivation is more completely described elsewhere (4,6) The trend curve formulas found by a nonlinear least squares technique were

$$\Delta T(welds) = [x(1) \cdot Cu + x(2) \cdot \sqrt{CuNi} + x(3) \cdot Ni] \cdot \frac{\phi t}{10^{19}}^{N} \quad (8a)$$

$$N(welds) = x(4) + x(5) \cdot \log_e \frac{\phi t}{10^{19}} \quad (8b)$$

$$\Delta T(plate) = x(6) + x(7) \cdot Cu + x(8) \, Cu \cdot \tanh [x(9) \cdot Ni/Cu]$$
$$\cdot \frac{\phi t}{10^{19}}^{N} \quad (9a)$$

$$N(plate) = x(10) + x(11) \log_e \frac{\phi t}{10^{19}} \quad (9b)$$

To put either the weld or plate formulas in a form consistent with the theory outlined above, the fluence factor must be rewritten so that the independent exposure parameter becomes $\log_e(\phi t/10^{19})$ rather than (ϕt). This substitution is necessary because of the assumption that there is a normal distribution of errors in $\log_e (\phi t)$, or a log-normal distribution of errors in (ϕt), and the error propagation theory being used assumes a normal distribution of errors in the independent variables. Therefore, we make the substitution

$$K = \log_e (\phi t/10^{19}) \quad (10)$$

or, using the weld fluence factor as an example,

$$(\phi t/10^{19})^{x(4) + x(5) \log_e (\phi t/10^{19})} = (e^K)^{x(4) + (x)5 \cdot K} \quad (11)$$

or

weld fluence factor $= \exp (x(4) \cdot K + x(5)K^2)$.

For the combined weld and plate equations, the entire set of adjustable parameters (i.e., x(1) through x(11) was determined in a single least-squares procedure, which adjusted all the parameters simultaneously. As with recent previous studies, the "true" value of the fluence was also adjustable. This is necessary to avoid biased (false low) values for the fluence exponent (4). The results of the fitting procedure were as follows:

x(1) =	624.0	x(6) =	-38.39
x(2) =	-333.1	x(7) =	555.6
x(3) =	251.2	x(8) =	480.1
x(4) =	0.28185	x(9) =	0.353
x(5) =	-0.0409	x(10 =	0.2661
		x(11) =	-0.0449

The weld covariance matrix [x(1) through x(5)] is given by

0.915+004	-0.140+005	0.509+004	-0.225+000	-0.459-001
-0.140+005	0.242+005	-0.941+004	0.229+000	-0.493-001
0.509+004	-0.941+004	0.388+004	-0.475-001	-0.315-001
-0.225+000	0.229+000	-0.475-001	0.517-003	-0.165-003
-0.459-001	-0.493-001	-0.315-001	-0.165-003	0.257-003

and the plate covariance matrix [x(6) through x(11)] is given by

0.387+002	-0.285+003	-0.168+003	0.311+000	-0.757-002	0.134-001
-0.285+003	0.494+004	-0.473+003	-0.659+001	-0.125+000	0.312-001
-0.168+003	-0.473+003	0.284+004	0.144-001	0.273+000	-0.224+000
0.311+000	-0.659+001	0.144-001	0.116-001	0.137-004	-0.243-003
-0.757-002	-0.125+000	0.273+000	0.137-004	0.394-003	-0.139-003
0.134-001	0.312-001	-0.224+000	-0.243-003	-0.139-003	0.185-003

In the trend curve formulas of Eqs. (9a), (9b), the parameters are consistent with Cu wt%, Ni wt%, fluence (n/cm², E > 1.0 MeV), and degrees Fahrenheit.

The standard deviations were found to be 28.2°F for the weld equation and 17.2°F for the plate equation. This is slightly larger than the similar values previously reported (4) for a smaller data base (106 plate data points and 41 weld data points).

The formal derivatives required in Eq. (3) for the weld and plate formulas can be found in a straightforward manner and will not be given here since they are available in Reference (4).

Application of Theory

The error estimation based on Eq. (3) can now be applied to a weld. Eq. (3) has two summations that contribute to $\delta^2(\Delta T)$. The first summation (double) came from the uncertainties in the parameters in the weld formulas, and the second came from the uncertainties in the variables associated with the given specimen and irradiation in any application.

The first summation (double) is a matrix multiplication. The analyst must take the data associated with the particular problem [Cu concentration, Ni concentration, $\log_e$ (fluence), etc.], and obtain a numerical matrix containing all the values of

$$D_{ke} = \frac{\partial f}{\partial P_k} \frac{\partial f}{\partial P_e} \, .$$

Then the D matrix is multiplied by the weld covariance matrix. To obtain the second term of Eq. (3), estimates are obtained for the squares of the uncertainties in the experimental variables, $(\delta v)^2$, and the formulas (from Reference 4) for the various $\partial f/\partial V_j$ are used for the computation.

When the formalism above is used to calculate an uncertainty in a Charpy shift value, it should be noted that the calculated uncertainty is that due to the uncertainties in the parameters and the stated independent variables. If the analyst wants the uncertainty in the value of a measured shift, then the uncertainty of the measurement must be added in quadrature. This remark also applies if there is an "inherent variability" in the material such that two identical materials with identical sets of input variables produce two noticeably different shifts.

The indicated procedure has been applied to 10 of the PWR surveillance capsule weld data points chosen from the data available for the original fit. The purpose of the exercise is to see what magnitudes of assumed input errors give a reasonable value of the total error, which variables contribute most heavily to the total, and what role is played by the parameter uncertainty. For the weld calculation, the assumed input uncertainties were 0.05 wt% for Cu and Ni, and a 25% error (log normal) for the fluence. This set of values was chosen after consideration of analysis accuracy and the difficulty of assigning a precise average value to the chemistry of an inhomogeneous weld. A similar exercise is shown in Table 2, using 10 PWR surveillance capsule plate data

TABLE 1

UNCERTAINTY IN PREDICTED CHARPY SHIFTS FOR VARIOUS Cu Ni CONCENTRATIONS IN WELD MATERIAL

Point Number	Fluence x 10^{19}	Cu wt%	Ni wt%	Calculated Uncertainty in Shift $\delta(\Delta T)$, °F	$\delta(\Delta T)$ (°F) Contribution From			
					δCu	δNi	δFluence	δParameters
Welds 1	0.61	0.36	0.78	25.2	14.8	11.2	15.5	7.0
2	0.62	0.35	0.60	24.8	16.0	11.3	13.6	6.9
3	7.85	0.36	0.78	37.8	25.5	19.3	7.4	18.7
4	0.11	0.34	0.71	17.1	7.3	5.5	10.5	9.9
5	5.8	0.19	0.08	40.3	34.2	18.7	3.5	10.0
6	0.74	0.34	0.65	25.6	16.4	11.9	13.9	7.0
7	0.73	0.25	0.59	23.0	15.2	11.9	11.3	5.2
8	0.31	0.19	0.56	16.8	10.2	8.7	9.0	5.7
9	0.65	0.20	0.77	21.5	11.6	11.4	12.2	4.6
10	0.28	0.27	0.74	18.9	10.1	8.4	12.0	6.4
11	0.68	0.08	0.07	22.6	19.2	11.6	2.7	1.8

TABLE 2

UNCERTAINTY IN PREDICTED CHARPY SHIFTS FOR VARIOUS (Cu Ni) CONCENTRATIONS IN PLATE MATERIAL

Point Number	Fluence x 10^{19}	Cu wt%	Ni wt%	Calculated Uncertainty in Shift $\delta(\Delta T)$, °F	$\delta(\Delta T)$ (°F) Contribution From δCu	δNi	δFluence	δParameters
Plate 1	0.61	0.15	0.59	15.5	13.6	2.1	6.6	2.9
2	1.79	0.15	0.59	19.7	18.0	2.8	6.1	4.1
3	1.79	0.18	0.71	20.7	18.1	2.8	7.7	5.5
4	0.62	0.18	0.71	16.6	13.6	2.1	8.4	4.0
5	6.1	0.25	0.53	21.4	17.1	7.6	5.7	8.6
6	7.85	0.15	0.59	23.8	22.6	3.5	2.9	6.6
7	0.60	0.14	0.68	18.9	14.9	1.3	6.3	3.0
8	0.11	0.17	0.64	14.3	6.7	1.2	5.9	4.8
9	0.15	0.13	0.50	8.8	7.8	1.3	4.4	3.1
10	5.8	0.18	0.20	18.5	15.5	9.9	3.1	4.4

sets, using the plate formula and plate covariance matrix given above, and using assumed input errors of 0.02% wt% for Cu, 0.05 wt% for Ni and a 25% err in the fluence. The consequences of the assumed error magnitude and the role of parameter uncertainty is discussed below.

Discussion of Results

The estimated error in the application of the trend curve formula comes from three sources: (1) errors due to incorrect parameters values derived in the least squares procedures, (2) errors due to incorrect values of the independent variables in the application and (3) the inherent variability of the material. A fourth source might be the error from a particular choice of functional form of the equation. If this consideration is dominant, it may be necessary to revert to the use of the σ of the fit as an error indicator. Differences in accuracies of independent variables can be still be accounted for by subtracting or adding to sigma squared a quantity intended to represent the change in the squared values of the contributions from the particular independent variable concerned. This approach also protects against "hidden variables." Presence of the latter is manifested by data scatter in the reference base which is too large to be explained by the anticipated sources of error. This is not apparent in the present case.

Source "one" (parameter uncertainties) is aggravated by random errors in the data base used to derive the parameters. However, as the quantity of data in the base increases, the error in the parameters decreases proportional to the square root of the number of data points. Although the sigma of the fit might remain constant, the elements of the covariance matrix became smaller as the quantity of data increases, and the error due to parameter uncertainty decreases accordingly.

From Table 2 we see that this source of error (parameter uncertainties) in the plate application is small. It is larger (but not dominant) for the welds. Source "one" is unaffected by changes in accuracy for the independent variables in the application data and only depends on the quantity and quality of the data in the original data base.

The magnitude of the second source of error (input variables in the application) is affected by the accuracy of the independent variables in the specific application and is not dependent on the quantity and quality of data in the data base used to derive the trend curve formulas. From Table 2 we see that for a plate application, a 0.02% error in the Cu concentration allows the copper contribution to source "two" to dominate the total error.

For the weld equation, with the assumed accuracies, the Cu and Ni contributions to the total error are more nearly equal and only slightly larger than the fluence contribution.

Source "three" (capricious material) is a recognition of the following. We have only derived the uncertainty in the shift between (a) a best Charpy transition curve for unirradiated data and (b) a best curve for irradiated data. A small piece of material the size of a Charpy bar still has its own capricious tendency to give a result that deviates from the smooth best curve drawn through a set of Charpy test data derived from supposedly identical specimens, and the analyst should be aware of this variability.

Source four (poor choice of analytic function) gives a bias. This effect is difficult to quantify and does not decrease as the quantity of data in the original base increases. It can be guarded against in the original derivation by plotting residuals as a function of the various independent variables and looking for any recognizable pattern in the plot. No such patterns were found in the present work (1).

For the plate examples in Table II the total errors are close enough to the observed sigma for the original data base to lend credibility to the over-all choice of error magnitudes. Further the assumed errors are in credible ratios. Then, from the table we see that parameter errors are not in general contributing heavily so that improvements in the accuracies in the data base would not be especially helpful. The fluence error is not a heavy contributor to the total uncertainty in ΔT either directly in the application nor through the parameter errors which might be caused by fluence errors in the data base. The most important item to control seems to be the copper error in the application.

For the welds, the situation is not so clear cut. The total uncertainty seems to lie in a range that is credible when compared to the observed sigma for the data base, and the various assumed errors for the input variables in the application are in a reasonable ratio. In general the errors due to parameter uncertainties are small, so that increased accuracy in the data base would give a limited improvement. However, for both plates and welds, the addition of information in regions of (Cu, Ni) space suffering from sparse data would give more confidence in the functional form. For welds as for plates, the main improvement would seem to come from more accurate chemistry information in the application.

It should be noted that the basic approach promoted here applies when the trend curve is given in tabular form as is planned for the current revision of Reg. Guide 1.99. That is, if the Charpy shift is given as a set of tables connecting ΔT and

the independent variables, the table can also include an estimated
error in each tabulated value. This estimated error will vary
inversely as the square root of the number of data points in each
region (and adjacent regions) of independent variable space, and
will vary directly as the sigma of the data in that region. This
uncertainty takes the part of the first error source (parameters)
in the work above. The second source (input variables error)
requires the use of derivatives which can be estimated numerically
from the tabular trend curve and the correspondence with the
analytical procedure is obvious. The comments about the third
source of error then apply unchanged. The fourth source of error
could conceivably dissappear if the table contains a full set of
important independent variables.

CONCLUSIONS

The method described has the advantage that the estimated
error depends on the values and estimated uncertainties of the
independent input variables in each individual application. It is
not assumed that the uncertainty in the variables in an application is the same as the uncertainties in the variables in the data
base used to derive the trend curve formulas. The major sources
of uncertainty appear to be due to uncertainties in the application chemistry (mainly copper), coupled with point-to-point
material variability, particularly in welds.

ACKNOWLEDGMENTS

The author wishes to acknowledge useful discussions with
G. R. Odette, P. N. Randall, and T. U. Marston. Thanks are due to
P. N. Randall for supplying the original data, and to R. L. Simons
for supplying early information regarding irradiation exposure
values.

REFERENCES

1. G. L. Guthrie, "Charpy Trend Curves Based on 177 PWR Data
   Points," LWR-PV-SDIP Quarterly Progress Report, April 1983 -
   June 1983, NUREG/CR-3391, Vol. 2, HEDL-TME 83-22, Hanford
   Engineering Laboratory, Richland, WA, HEDL-3 - HEDL-15,
   April 1984.

2. R. L. Simons, L. S. Kellogg, E. P. Lippincott and
   W. N. McElroy, "Updated Re-Evaluation of the Dosimetry for
   Reactor Pressure Vessel Surveillance Capsules," Fifth
   ASTM-EURATOM Symposium on Reactor Dosimetry, Geesthacht,
   Federal Republic of Germany, September 1984.

3. G. L. Guthrie, "Charpy Trend-Curve Development Based on PWR Surveillance Data," Proc. of the 11th WRSR Information Meeting, Gaithersburg, MD, October 24-28, 1983, NUREG/CP-0048, NRC, Washington, DC.

4. G. L. Guthrie, "Charpy Trend Curve Formulas Derived from an Expanded Surveillance Data Base," LWR-PV-SDIP Quarterly Progress Report, October - December 1982, NUREG/CR-2805, Vol. 4, HEDL-TME 82-21, Hanford Engineering Development Laboratory, Richland, WA, HEDL-3 - HEDL-13 (July 1983).

5. B. R. Martin, Statistics for Physicists, Sec. 8.1.2, Academic Press, New York, NY (1971).

6. G. L. Guthrie, "Error Estimations in Applicatons of Charpy Trend Curve Formulas," LWR-PV-SDIP Quarterly Progress Report, January - March 1983, NUREG/CR-3391, Vol. 1, HEDL-TME 83-21, Hanford Engineering Development Laboratory, Richland, WA, HEDL-3 - HEDL-13 (November 1983).

BABCOCK & WILCOX REACTOR VESSEL SURVEILLANCE SERVICE ACTIVITIES

L. A. Hassler and A. L. Lowe, Jr.

Nuclear Power Division
Babcock & Wilcox, A McDermott Company
Lynchburg, Virginia U.S.A.

ABSTRACT

To ensure the continued licensability of Babcock & Wilcox operating reactors, B&W established testing and analytical procedures to support the required reactor vessel surveillance program. These procedures included the design of dosimeters, gamma spectroscopy of dosimeters, design of dosimetry installations, and evaluations and analyses to verify the reactor vessel fluences. Machining capabilities for irradiated materials were developed and a program was established to test both unirradiated and irradiated materials to determine mechanical properties. Advanced techniques were developed in linear elastic fracture mechanics high temperature single specimen J-R fracture testing, and related neutron fluence analysis. The experience gained from testing sixteen surveillance capsules from B&W reactors has been extended to a service program that is available for other vendors' capsules.

INTRODUCTION

To ensure the continued licensability of Babcock & Wilcox (B&W) operating reactors, B&W designed and implemented complete reactor vessel surveillance programs in accordance with NRC regulations. B&W then extended the experience and expertise gained from the programs to offer a service program for other reactors. Services include the mechanical testing of irradiated material, gamma spectroscopy of dosimeters, design of dosimeters, design of dosimetry installations and the required evaluations and analyses. With the support of the B&W 177 FA Owners Group, advanced methods were developed in technical areas such as linear elastic fracture mechanics, high temperature single specimen J-R

J. P. Genthon and H. Röttger (eds.), Reactor Dosimetry, 61–67.
© 1985 ECSC, EEC, EAEC, Brussels and Luxembourg

fracture testing, and neutron fluence analysis. To date, sixteen capsules from B&W reactors and two capsules from other reactors have been analyzed.

Fluence exposure of material specimens and reactor vessels is calculated with a multi-energy, multi-dimensional particle transport code based on a geometric model of the surveillance capsule and the related reactor vessel structures. Analytical fluence values are normalized to measured reaction rates from capsule dosimetry. Plant-specific calculations indicate that fluence rate is dependent on fuel loading distribution and capsule location. A program to evaluate vessel fluence from passive dosimetry mounted in the cavity region outside the vessel is currently under development.

Data from Charpy testing of materials are correlated with fluence to produce a method to predict reference temperature shift as a function of chemical composition. Fracture mechanics evaluations using compact fracture toughness test data have shown that vessels can be operated at rated power until the end of design life. The use of high temperature J-integral testing techniques to determine fracture toughness data for elastic-plastic fracture mechanics analysis is particularly significant. Advanced analytical techniques have been developed to evaluate postulated accident conditions and to determine whether flaws found during inspection of pressure boundary components are acceptable.

DOSIMETER AND FLUENCE ANALYSIS

As used in the reactor vessel surveillance program (RVSP), fluence analysis has three objectives: (1) to determine maximum fluence at the pressure vessel as a function of reactor opera-tion, (2) to predict future pressure vessel fluence, and (3) to determine test specimen fluence within the surveillance capsule. Vessel fluence data are used to evaluate changes in the reference transition temperature and upper shelf energy levels, to estab-lish pressure-temperature operation curves, and, in conjunction with data from the capsule test specimens, to develop a corre-lation between changes in material properties and fluence.

Fluence data are obtained from flux distributions calculated with a computer model of the reactor. The accuracy of the calculated fast flux ($E > 1$ MeV) is enhanced by normalizing it to the measured activities of the capsule dosimeters. The fast fluence is measured from the Fe-54(n,p), Ni-58(n,p), U-238(n,f), and Np-239(n,f) reaction products (2.5, 2.3, 1.1, and 0.5 MeV thresholds respectively). Due to the long half-life (~ 30 years) and effective threshold energies, the measured activity of Cs-137 production from fissions in Np-237 and U-238 is more directly applicable to analytically determining the fast fluence for multiple fuel cycles than other dosimeter reactions. These other reactions are useful primarily as corroborating data for shorter time intervals and/or higher energy fluxes.

Two dimensional discrete ordinates transport calculations (R-θ) are used to obtain the energy-dependent flux distribution throughout the reactor system. Reactor conditions are selected that are representative of an average over the irradiation time period. Geometric detail is selected to explicitly represent the surveillance capsule and the reactor vessel. To date, sixteen capsules from B&W reactors and two from other reactors have been analyzed. The analyses have provided the capsule and reactor flux, and flux gradients in the capsule specimens, radially through the vessel, and azimuthally at the vessel surface as a function of reactor and fuel cycle. A brief description of the analytic procedure used in the fluence analysis is given below. The accuracy of this procedure has been verified by an independent re-evaluation of the vessel fluence based on B&W capsule dosimeters (1). The fluence values calculated by B&W agreed to within 15% of fluence values obtained in the re-evaluation.

Energy-dependent neutron fluxes throughout the reactor are determined by a discrete ordinates solution of the Boltzmann transport equation using the two-dimensional code DOTIV (2). The reactor is modeled from the core to the primary concrete shield in R-θ geometry based on a plane view along the axial core midplane and eighth core symmetry in the azimuthal direction. All material regions from the core to the concrete shield are explicitly modeled. Input data to DOTIV includes a PDQ calculated pin-by-pin, time averaged power distribution, CASK23E 22-group P3 cross sections, and S8 angular quadrature. Reactor conditions, i.e. power distribution, temperature and pressure, are averaged over the irradiation period.

Two basic reactor models are used. One includes a specific model of the surveillance capsule and the other deletes the capsule to assess the shadowing effect of the capsule on the vessel flux. Axial power distribution effects are integrated into the calculation by applying an axial shape factor. This factor (local to average axial RPD ratio) is obtained from predicted fuel burnup distributions in the peripheral fuel assemblies nearest the vessel.

The detailed model of the surveillance capsule serves two purposes. In the vessel surveillance program, it provides detailed fast fluence values to correlate test specimen material changes with fluence. In respect to the calculational procedure, it allows the calculated fast flux to be normalized to measured values. These calculated fluxes at the dosimeter locations are used to obtain calculated dosimeter activities from the product of the energy dependent flux, group average reaction cross sections, and a saturation factor dependent upon the reactor power history. A normalization factor is obtained from the ratio of the measured activity to the calculated activity (using an average of the Np-137 and U-238 activities). Although this factor is strictly correct only at the capsule location, it is assumed to apply to the vessel flux. This assumption is based on the unlikelihood of any significant change occurring in the ~ 15 cm at water separating the capsule and the vessel.

Extrapolation of fluence to future operation periods is based on available calculated fluence data, available PDQ data for future cycles that have been designed for reload cores, and future fuel cycle designs for which no PDQ analyses exist. Data for the latter two periods are related to that of period 1 by the premise that excore flux is proportional to the fast flux that escapes the core boundary. Thus, the vessel flux for fuel cycle C is obtained as:

$$\phi_{v,c} = \frac{\phi_{e,c}}{\phi_{e,R}} \times \phi_{v,R}$$

where the subscripts are defined as v = vessel, c = future cycle, e = core escape, and R = reference cycle. The core escape flux is available from PDQ output. For those future cycles for which no PDQ data is available, the last fuel cycle for which a design has been made is assumed to be the "equilibrium" cycle that represents all future cycles. This method enables relatively accurate extrapolations based on all available cycle data, updated with each new analysis. While this method has proved suitable for fluence extrapolation, a new method is being developed based on the proportionality of excore flux to a product of geometrical weighting factors and peripheral fuel assembly power. This method has the potential to significantly improve the accuracy of fluence extrapolation.

Because of the complexity of the fluence calculations, no comprehensive uncertainty limits exist for the results obtained by the above procedure. However, estimates of the fluence values were obtained based on comparison to available benchmark experiments, estimated and measured variations in input data, and engineering judgment. Typical uncertainties associated with reported fluence values are $\sim \pm 18\%$ for capsule fluence, $\sim \pm 24\%$ for maximum vessel fluence, and $\sim \pm 26\%$ for end-of-life maximum vessel fluence values.

The B&W Owners' Group has developed plans for an ex-vessel cavity dosimetry program. This program calls for placing B&W, HEDL, and University of Arkansas dosimeters at various azimuthal

64

and axial locations in the cavity of a 177 FA reactor. The measured data from this cavity irradiation and in-vessel surveillance dosimetry will be used to quantify the flux gradient in the vessel. It would also benchmark an analytic procedure being developed to related cavity dosimetry results to vessel fluence.

MATERIAL PROPERTIES ANALYSIS

The material properties analysis part of the reactor vessel surveillance program has three primary objectives: (1) to determine the material sensitivity to neutron fluence damage, (2) to verify material behavior in accordance with accepted theory and (3) to develop data to refine techniques for predicting irradiation behavior. Regulations require that all surveillance capsules contain tension and Charpy V-notch specimens. In cases where fracture toughness data may be needed for future analysis of reactor vessel integrity, it is recommended that compact toughness specimens for materials with a predicted high sensitivity to neutron radiation damage also be included. Evaluation of surveillance capsules are performed in accordance with ASTM Practice for conducting Surveillance Tests for Light Water Cooled Nuclear Power Reactor Vessels (E185).

Tension specimens are fabricated from the reactor vessel shell course plate and weld metal. Two sizes of specimens are included in the B&W RVSPs. A large size with a reduced section 4.445 cm (1.750 in.) long by 0.907 cm (0.357 in.) in diameter. A small-Charpy size with a reduced section of 2.400 cm (0.945 in.) long by 0.533 cm (0.210 in.) in diameter. Test conditions are in accordance with the applicable engineering requirements. For each material type and/or condition, specimens are tested from room temperature to 315C (600F). All test data for the tension specimens are transferred to permanent storage and future re-analysis.

Charpy V-notch impact tests are conducted on an impact tester certified to meet Watertown standards. Tests specimens are of the Charpy V-notch type. Before testing, specimens are temperature-conditioned in a chamber designed to cover the temperature range from -65 to 388C° (-85 to +550F). Specimens are transferred from the conditioning chamber to the test frame anvil and precisely pretest-positioned with a fully automated, remotely controlled apparatus. Transfer times are less than 3 seconds and are repeatable within 0.1 second. The impact test is fully instrumented and all recorded data is transferred to discs for permanent storage and future reanalysis. Computer techniques are available for statistical evaluation of the data and for curve fitting if required. Instrumented precracked Charpy testing is performed and has been benchmarked through an EPRI sponsored program.

Fracture toughness testing is performed on both unirradiated and irradiated material using either linear-elastic or elastic plastic test procedures. These tests are performed on compact toughness specimen, or Charpy size specimen using the three point bend procedure. The testing and evaluation of fracture data for elastic-plastic fracture analysis is especially significant for reactor surveillance programs. A single-specimen J-integral test procedure was specifically developed to test surveillance program compact fracture toughness specimens. The single-specimen technique is important for surveillance work because it makes maximum use of minimal material, which is important because of the limited number of specimens in a typical surveillance capsule. The surveillance program uses 1.000 cm (0.394 in.), 1.270 cm (0.500 in.) and 2.377 cm (0.936 in.) thick compact fracture toughness specimens for both unirradiated and irradiated conditions. The specimens are precracked and are side-grooved a total of 20% to promotes straight crack front during testing. These are in accordance with recommended practice, verified by B&W testing.

For a more extensive description, the reader is referred to other B&W documents prepared expressly for this purpose (3-5).

The J-R test is a computer-guided, load-line displacement-controlled, single-specimen J-integral procedure. Crack extension (Δa) is calculated using the load-unload compliance technique, where the load line displacement is related to compliance and crack length using currently accepted methods (6). The output from a test is a series J-Δa data points plotted in a semi-continuous R-curve. Each data point represents a place at which the test specimen was partially unloaded. At the unload, Δa is calculated based on measurement of the unload compliance, while J is calculated from an equation based on area under the load-displacement diagram, specimen dimensions, a tensile correction factor, and a moving crack correction. The J calculation is in accordance with ASTM Test Method for J_{Ic}, a Measure of Fracture Toughness (E813).

Once the J-Δa data points are established, valid data are used to calculate J_{Ic} using the method detailed in ASTM Standard E813, a leastsquares fit of the valid R-curve data at the point where it intercepts the material blunting line. The tearing modules is calculated using the slope of the R curve. Test parameters and data are permanently stored on magnetic disks so that test can be reanalyzed at a later date using different input parameters.

The test techniques and analytical procedures developed for J-R testing are capable for use in testing and evaluation of WOL specimens and have been used for testing of Charpy size specimens using the three-point bend technique.

The data from Charpy testing of the materials are correlated with the fluence to predict the reference temperature shift as a function of material chemical composition. These valves are used for indexing the analytical calculations to determine the pressure-temperature operation limitations of the reactor vessel during normal service conditions. The tension and fracture toughness data as correlated with fluence is used for the evaluation of postulated accident conditions and the acceptability of flaws found during inspections of pressure boundary components.

REFERENCES

1. R.L. Simons, et al., "Re-evaluation of the Dosimetry for Reactor Vessel Surveillance Capsules," Proceedings of the Fourth ASTM-EURATOM Symposium on Reactor Dosimetry, Gaithersburg, Maryland, March 1982, Vol. 2, pp. 903.

2. DOTIV, Two Dimensional Disrete Ordinates Radiation Transport Code System," W.A. Rhodes, et al., Oak Ridge National Laboratory.

3. J.D. Aadland and A.L. Lowe, Jr., "Fracture Toughness Test Procedure for Elastic-Plastic Fracture Evaluation Techniques," BAW-1633, Babcock & Wilcox, April 1981.

4. Van Der Sluys, W.A. and Futato, R.J., "Computer-Controlled Single-Specimen J-Test," Elastic-Plastic Fracture: Second Symposium, Volume II, ASTM STP 803, 1983, pp. II-464-II-482.

5. J.D. Aadland and A.L. Lowe, Jr., "The B&W J-R Test Procedure: Development of the R-Curve, Effect of Input Variables, and Variability of the J_{Ic} Value," BAW-1696, Babcock & Wilcox, March 1982.

6. G.A. Clark, et al., " Procedure for the Determination of Ductile Fracture Toughness Values Using J-Integral Techniques," J. Test Eval., Vol. 7 No. 1 (1979), pp. 49.

NOTCH DUCTILITY AND TENSILE STRENGTH DETERMINATIONS

FOR REFERENCE STEELS IN PSF SIMULATED SURVEILLANCE

AND THROUGH-WALL IRRADIATION CAPSULES

J. R. Hawthorne and B. H. Menke

Materials Engineering Associates, Inc.

Lanham, Maryland, USA

ABSTRACT

The Light Water Reactor Pressure Vessel Surveillance Dosimetry Improvement Program of the Nuclear
Regulatory Commission (NRC) has irradiated mechanical
property test specimens of several steels in a pressure
vessel wall/thermal shield mock-up facility. The
investigation is part of a broad NRC effort to develop
neutron physics-dosimetry-metallurgy correlations for
making highly accurate projections of radiation-induced
embrittlement to reactor vessels. This report presents
notch ductility and tensile properties information
developed with specimen irradiations at 288°C in
simulated surveillance and in-wall locations. The
steels studied represent a wide range of radiation
embrittlement sensitivity and include plates, forgings
and submerged-arc weld deposits.

One objective of the Light Water Reactor Pressure Vessel
(LWR-PV) Surveillance Dosimetry Improvement Program established by
the Nuclear Regulatory Commission (NRC), is the development of key
information for the accurate projection of radiation-induced
mechanical properties changes in reactor vessel walls [1]. The
total effort represents a multilaboratory program with international participation. Materials Engineering Associates (MEA) was
given responsibility for the development and analysis of
mechanical properties data required for the study.

J. P. Genthon and H. Röttger (eds.), Reactor Dosimetry, 69–77.
© 1985 ECSC, EEC, EAEC, Brussels and Luxembourg

This report is a sequel to Reference 2 which presented the initial set of postirradiation mechanical properties data from the Program. Initial determinations were for two reference plates irradiated in the form of Charpy-V (C_v), compact tension (CT) and tension (T) test specimens. For the remaining materials reported upon here, only C_v and T specimens were irradiated. The present report provides a complete summary of C_v and T data for all of the materials investigated.

Additional background on the objectives and approach of the Program is presented in References 2 and 3. In brief, the investigations were designed to study through-wall toughness gradients produced by 288°C irradiation, and to determine the relative irradiation effect at surveillance capsule vs. in-wall locations. Six materials (steel plates, forgings and submerged-arc weld deposits) were provided by USA and overseas laboratories and represent a broad range of radiation embrittlement sensitivities such as that found among reactor vessels produced before 1972.

A pressure vessel mock-up facility was specially constructed for the Program for performing through-wall neutron dosimetry investigations and neutron exposures of metallurgical specimens. The facility, known as the Pool-Side Facility (PSF), simulates a relatively large segment of a reactor thermal shield and vessel wall as illustrated in Fig. 1. The steel thermal shield is 5.9-cm thick, the pressure vessel wall simulator is 22.5-cm thick. Five capsules were irradiated at 288°C [3]. Two capsules (designated SSC-1 and SSC-2) respectively represent surveillance capsules taken from a pressurized water reactor plant after about 15 years and 30 years of operation. The remaining three capsules (Wall 1, 2 and 3) represent vessel surface (OT), quarter-wall thickness (1/4T) and half-wall thickness (1/2T) positions. The lead factor, that is, ratio of neutron flux levels, between the surveillance capsule location and the wall surface location was about eight for the particular PSF configuration used.

Further details of this study including more complete descriptions of the PSF facility and the bases for its particular design configuration are given in References 3 and 4.

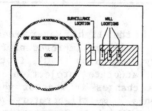

Fig. 1 Schematic illustration of the PSF Facility showing the locations of the specimen capsules.

MATERIALS AND SPECIMEN DESIGNS

The materials investigated are identified by type, code number, supplier and chemical composition in Table 1. Initial yield strength levels ranged from 407 to 489 MPa. Heat treatment conditions are documented in Reference 2. The submerged arc weld, Code R, was water (spray) quenched and tempered after welding; the weld, Code EC, was stress relief annealed only. The A 302-B plate, Code F23, has seen extensive use in reactor vessel surveillance applications and in test reactor studies [4,5]. The A 533-B plate (Codes 3PT and 3PU) has been applied as a reference material also [5]. The respective plates are considered representative of early vessel manufacture and more recent vessel fabrication.

Table 1 Materials

Material	A 533-B Plate[a]	A 302-B Plate[a]	A 533-B S/A Weld[b]	A 533-B S/A Weld[c]	22NiMoCr37 Forging[d]	A 508-3 Forging[e]
Code	3PT, 3PU	F23	R	EC	K	MO
Thickness (mm)	305	152	160	235	295	238
C	0.20	0.24	0.05	0.11	0.18	0.20
Si	0.25	0.23	0.45	0.52	0.16	0.28
Mn	1.26	1.34	1.54	1.57	0.72	1.43
P	0.011	0.011	0.009	0.007	0.009	0.008
S	0.018	0.023	0.008	0.011	0.004	0.008
Cr	0.10	0.11	0.12	0.02	0.45	-----
Mo	0.45	0.51	0.34	0.48	0.63	0.53
Ni	0.56	0.18	1.58	0.64	0.96	0.75
Cu	0.12	0.20	0.23	0.24	0.12	0.05
Al	-----	0.04	0.01	0.008	0.031	0.031

Suppliers: [a] NRL (USA) [b] Rolls Royce & Assoc. (UK) [c] EPRI (USA) [d] KFA (FRG) [e] CEN/SCK (Belgium)

In Table 1, the materials are seen to differ considerably in their contents of copper (an impurity) and nickel (an alloying element). A high copper content in pressure vessel steels is known to be detrimental to radiation embrittlement resistance at 288°C. Nickel alloying in amounts of 0.4% Ni or more has been found to reinforce or magnify the deleterious effect of a high copper content.

Standard C_v (ASTM Type A) and 4.52-mm diameter tension specimens were used for making the notch ductility and tensile strength determinations. The locations of specimen blanks in parent materials are documented in Reference 4. Specimens of the A 533-B and 22NiMoCr37 forging were oriented in the transverse (TL, weak) direction; those of A 302-B plate and A 508-3 forging were oriented in the longitudinal (LT, strong) orientation.

MATERIAL IRRADIATION AND SPECIMEN TESTING

Capsule construction, irradiation and disassembly operations were conducted by the Oak Ridge National Laboratory (ORNL) for the NRC. Primary responsibility for neutron dosimetry and fluence determinations is shared by ORNL and the Hanford Engineering Development Laboratory (HEDL).

The irradiation histories and target fluence conditions of the five capsules are summarized in Table 2. The exposure time of capsule SSC-1 was adjusted to provide a fluence matching that of the Wall-2 capsule located in the quarter-wall thickness position. The exposure time of capsule SSC-2 was similarly adjusted to match its fluence against that of the Wall-1 capsule located at the wall inner surface. The Wall-1, Wall-2 and Wall-3 capsules were irradiated simultaneously and were exposed for the same time period. In turn, the spread in fluences between these capsules should reflect normal flux attenuation conditions through a vessel wall thickness. Except for capsule SSC-1, fluence determinations given in this report are preliminary values based, in part, on SSC-1 dosimetry findings. Adjustments to listed fluences are not expected to be greater than 10% Accordingly, the conclusions

Table 2 Capsule Irradiation Conditions

Capsule No.	PSF Location	Irradiation Time (Hours at Power)	MW Hours Exposure	Target Neutron Fluence $(n/cm^2, E > 1$ MeV$)$
SSC-1	Thermal Shield	1,291	32,000	3×10^{19}
SSC-2	Thermal Shield	2,845	64,700	6×10^{19}
Wall-1	Simulator (Surface, OT)	18,748	430,000*	$\sim 6 \times 10^{19}$
Wall-2	Simulator (Quarter T)	18,748	430,000	$\sim 3 \times 10^{19}$
Wall-3	Simulator (Half T)	18,748	430,000	$\sim 1.5 \times 10^{19}$

* Approximate

drawn from capsule intercomparisons will remain unchanged. For a given material, the same specimen locations were reserved in all five capsules.

Energy absorption and lateral expansion were determined in each C_v test; in addition, applied load vs. time-of-fracture records were made using a Dynatup system for future NRC studies. With the tensile specimens (button head design), elongation of the gage section was monitored from test machine actuator displacement. Specimen load vs. actuator defection was recorded simultaneously on two X-Y plotters, one of which provided an expanded load vs. deflection record. This in turn was analyzed on-line by a computer-controlled data acquisition system.

RESULTS

Experimental results for the materials are summarized in Figs. 2, 3 and 4. Space limitations preclude a detailed review of individual material findings in this document; this is provided in Reference 4. Relative radiation embrittlement was determined primarily from the C_v 41 J transition temperature elevation, based on the mean curve drawn through the data.

Referring first to the SSC-1 results (Fig. 2), very large material-to-material differences in radiation embrittlement sensitivity are observed. For individual materials, the fluence variation among specimens typically was less than 10%; the fluence difference between two materials, however, was much higher in many cases and was taken into account in making the material comparison below. As expected from its higher copper content and its somewhat higher fluence, the A 302-B plate showed a greater embrittlement than the A 533-B plate. The low sensitivity to irradiation

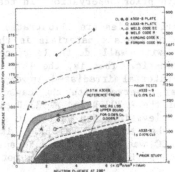

Fig. 2   Data from capsules SSC-1 and SSC-2 compared against trends of C_v 41 J transition temperature change with irradiation observed with in-core test reactor experiments.

73

of the forging code MO was a manifestation of its low (0.05%) copper content, its low (0.008%) phosphorus content, and in the opinion of the authors, its prior forging (metal working) history. The very high radiation embrittlement sensitivity of the weld code R may be attributed to its high (0.23%) copper content and its nickel alloying (1.58%Ni). However as discussed below, it is believed that more is involved in the high apparent sensitivity of this weld. The absolute sensitivity level of weld code R was somewhat unexpected but comparably high – radiation sensitivity has been observed previously with other weld compositions (NiCrMo). On balance, the postirradiation data for SSC-1 provided no surprises in regard to the ranking of materials by sensitivity level. The same ranking was given by the SSC-2 findings.

A primary observation from the SSC-1 vs. SSC-2 data is that a doubling of the fluence exposure of the materials provided only a small (and in many cases, negligible) additional 41 J temperature elevation. The 41 J temperatures of the A 302-B plate and the weld code EC were further increased by only 12°C; those of the A 533-B plate and the forging code MO were further increased by 19 to 20°C. Weld code R again was the exception. Here, the SSC-2 specimens indicated a 67°C greater transition elevation by irradiation than specimens from SSC-1. The percentage increase (30%) however is about the same as that observed for the A 533-B (33%); thus, the 67°C elevation is not disproportionately large.

In Fig. 2, the SSC-1 and SSC-2 results are compared against data trends observed with in-core test reactor experiments at 288°C. The small difference in irradiation effect between SSC-1 and SSC-2 is predicted well by the data trends. Also, the data for the A 302-B are found to agree well with the in-core trend which was developed with numerous experiments in several test reactors. Findings for the A 533-B plate and weld code EC also show general agreement with prior observations in test reactor studies.

Data for the wall capsules are illustrated in Fig. 3; primary observation for five of the six materials is that the increase in fracture resistance with wall depth (surface to mid-wall) is neither rapid nor dramatic. That is, the 41 J transition temperature elevation for material irradiated in capsule Wall-3 is not much lower than that observed with the capsule Wall-1 irradiation. In the case of the A 302-B and A 533-B plates, for example, the 41 J temperature elevations for Wall-3, respectively, are only 31°C and 22°C lower than those of Wall-1. Likewise, the difference for weld code EC was only 25°C. Because of inexplicably large data scatter, a similar analysis was not attempted for the forging code K.

Of additional interest, the A 302-B and A 533-B plates experienced about the same transition temperature elevations,

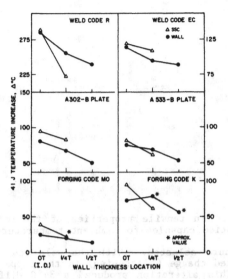

Fig. 3 Comparison of C_v data from simulated surveillance capsules vs. in-wall capsules (triangle vs. circle points).

i.e., about equal irradiation sensitivities, when exposed in the wall capsule locations. This is a clear contrast to the significant difference in embrittlement sensitivities observed with the SSC-1 and SSC-2 surveillance capsule irradiations. Whether or not the change in relative behavior arose from the much longer exposure time of the wall capsules is a key question.

Referring next to surveillance vs. wall capsule embrittlement indications (Fig. 3), the data are in good agreement relative to demonstrating the difference in transition temperature elevation between the lowest fluence and highest fluence condition. That is, the fluence attenuation with wall depth between surface and quarter-thickness positions did not translate to a dramatic gradient in fracture toughness in the general sense. The nonconformer is the weld code R but this weld composition has not been used in commercial reactor vessels. With few exceptions, the surveillance capsule data either predict well or overpredict the transition temperature elevation for the wall location. For the A 302-B and A 533-B plates and weld code EC, the amount of overprediction is 20°C or less. Upper shelf changes also were predicted well.

Exclusive of weld code R, the data sets indicated close agreement of 41 J, 68 J and 0.89-mm transition temperature elevations, within 14°C. Accordingly, the ranking of the irradiation effect for the materials is quite independent of the C_v

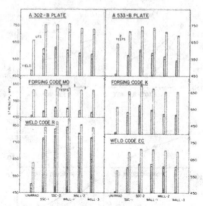

Fig. 4 Variation in tensile properties of the materials between irradiation capsules for ambient temperature tests.

indexing procedure selected. With the weld code R, irradiation produced a marked change (flattening) in the shape of the transition curve. This alteration produced a 34°C difference between transition elevations measured at 41 J vs. 69 J.

Tensile strength elevations (Fig. 4) in general followed closely, observations made in the notch ductility tests.

DISCUSSION

The very high radiation embrittlement sensitivity observed for the weld code R may be due to more than just the synergism of nickel content with copper content. It is conceivable that, when the nickel content is high (> 1% Ni), an independent contribution of nickel to radiation sensitivity development occurs. The code R data also could be reflecting some non-nuclear time-at-temperature effect. This suspicion derives from the lower embrittlement observed for the capsule SSC-1 specimens compared to the capsule Wall-2 specimens coupled with their large difference in exposure times, i.e., 1291 hrs vs. 18478 hrs. Capsule SSC-2 which was exposed for 2845 hrs on the other hand, did produce comparable embrittlement to its mating capsule, Wall-1. Thermal control data unfortunately are not available to help resolve questions of one vs. two operating mechanisms.

CONCLUSIONS

Primary conclusions and observations drawn from the results for the materials, except for code R, were:

1. The survellance capsule data indicate reasonably well the irradiation effect to the wall surface and quarter-wall

thickness locations. With one exception the C_v surveillance results proved conservative where significant ($> 10°C$) differences were observed; predictions of the 41 J transition temperature were within 20°C.

2. The in-wall toughness gradient produced by irradiation, indexed to the transition temperature, was small. The difference in 41 J temperature between wall surface and midthickness locations ranged between 31°C and 11°C.

3. In parallel with 2, the doubling of the fluence to the materials produced only a small additional 41 J temperature elevation in the surveillance capsule irradiations (SSC-2 vs. SSC-1).

4. Irradiation sensitivity levels of the materials are in accord with initial predictions based on material copper and nickel contents.

5. Tensile test findings support the notch ductility trend determinations.

High radiation embrittlement sensitivity was observed for the weld code R which contained 0.23% Cu and 1.58% Ni. The unusually high level of sensitivity suggests contributions by two or more embrittlement mechanisms, in addition to the normal irradiation effect.

REFERENCES

1. C. Z. Serpan, "USNRC Surveillance Dosimetry Improvement Program," Proceedings of the 6th MPA Seminary -- Safety of the Pressure Boundary of Light Water Reactors, Staatliche Materialprunfungsantalt, Universitat Stuttgart, 9/10, Oct. 1980.

2. J. R. Hawthorne, B. H. Menke and A. L. Hiser, "Notch Ductility and Fracture Toughness Degradation of A 302-B and A 533-B Reference Plates from PSF Simulated Surveillance and Through-Wall Irradiation Capsules," USNRC Report NUREG/CR-3295, Vol. 1, April 1984.

3. W. N. McElroy, et. al., "LWR Pressure Vessel Surveillance Dosimetry Improvement Program - 1982 Annual Report," USNRC Report NUREG/CR-2805, Vol. 3; HEDL-TME-82-20, Hanford Engineering Development Laboratory, Jan. 1983.

4. J. R. Hawthorne and B. H. Menke, "Postirradiation Notch Ductility and Tensile Strength Determinations for PSF Simulated Surveillance and Through-Wall Specimen Capsules," NUREG/CR-3295, Vol. 2, MEA-2017, Materials Engineering Associates, April 1984.

5. Metal Properties Council, "Prediction of the Shift in the Brittle/Ductile Transition Temperature of LWR Pressure Vessel Materials," ASTM Journal of Testing and Evaluation, July 1983, pp. 237-260.

TRANSPORT CALCULATION OF NEUTRON FLUX AND SPECTRUM
IN SURVEILLANCE CAPSULES AND PRESSURE VESSEL OF A PWR

A. KODELI, M. NAJŽER, I. REMEC

"J. STEFAN" Institute

Ljubljana, YUGOSLAVIA

ABSTRACT

Neutron flux, neutron exposure rate and iron displacement
cross-section averaged over the neutron spectrum in a two-loop,
632 MWe PWR were calculated. The transport code DOT 3.5 was used.
Calculations were performed separately for the reactor and the
surveillance capsule. Presented are spatial distributions as well
as absolute values at most important locations within the capsule
and the pressure vessel for BOL and EOL of the first fuel cycle.

INTRODUCTION

Reactor pressure vessel of a nuclear power plant is exposed
to neutron irradiation and therefore subject to encreasing radiation embrittlement during plant life. A surveillance programme,
aiming to experimentally determine the embrittlement, is carried
out at operating power plants. Specimens made of pressure vessel
material are irradiated, together with neutron flux and fluence
monitors, in capsules located behind the thermal shield of PWR
reactors. At prescribed time intervals capsules are removed from
the reactor. Specimens are mechanically tested and activities of
neutron monitors measured to determine the shift in Nil Ductile
Temperature and neutron exposure respectively.

Data obtained in this way refer to the capsule position while
the embrittlement of the pressure vessel is of interest. As neutron
environments are far from identical, the extrapolation from capsules to the pressure vessel can only be done on the basis of detailed neutron transport calculation. In the following methods, models
and results of such calculation for a 2 loop, 632 MWe PWR are presented.

CALCULATION

The neutron transport calculation was performed using 2-dimensional, discrete-ordinates computer code DOT 3.5 in P3, S_4 and S_6 approximation. Cross-sections were generated from DLC-2D library and collapsed to 27 energy groups, covering the energy range up to 15 MeV.

Reactor was described in R-θ and R-Z geometry alternatively. Geometrical model consisted of 10 radial zones with constant material composition: 3 zones in the core with different enrichment and fission products densities, core buffle, core barrel, thermal shield, reactor vessel and layers of borated water between the zones. In R-Z calculation, reflector was added on the top and bottom.

The core was described as fixed fission neutron source in non-multiplying medium. The fission density in each mesh point was derived from 2-dimensional, 2 energy groups codes FASVER 2 and CEBIS, used in core design and burn-up calculations. 61x33 mesh points were used in R-θ and 33x36 in R-Z geometrical model. Calculations were performed for the beginning (BOL) and end (EOL) of the first fuel cycle.

Capsules, which are 28x28 mm in cross section and 1600 mm long, were not included in reactor calculations due to too coarse

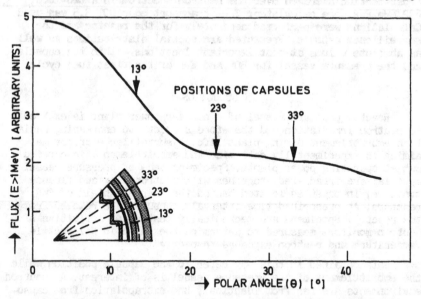

Fig. 1.: Variation of the neutron flux above 1 MeV at radius corresponding to locations of the surveillance capsules in the reactor (r=156.7 cm) with the polar angle (θ)

mesh. Therefore the flux distribution in the horizontal section
through the capsule was separately calculated in X-Y geometry using
a finer mesh. Capsule was considered as perturbation to flux within
the   capsule and in the surroundings. Boundary of computation
area was several neutron free paths from the capsule. The source
of neutrons was determined with angular neutron fluxes at the boundary, taken from R-θ reactor calculation. Calculations were performed with the capsule and without the capsule (capsule was replaced with borated water), to determine perturbation of the flux.

RESULTS

Neutron group fluxes in the capsules and in the pressure vessel were obtained. Fission density was normalized to nominal thermal power of the reactor to get absolute values. From neutron group fluxes, exposure rates, expressed in terms of displacement rate
per atom (dpa/s) and neutron flux above 1 MeV, were calculated.

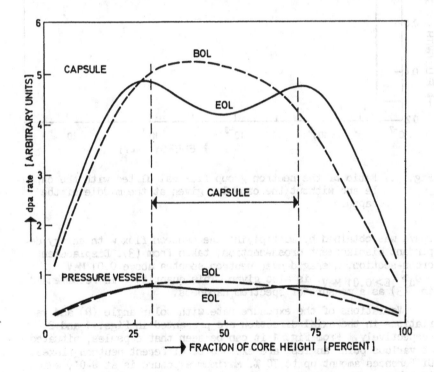

Fig. 2.: Axial dependence of displacement rate per atom for BOL
         and EOL at capsule position and at one quarter thickness
         of the pressure vessel

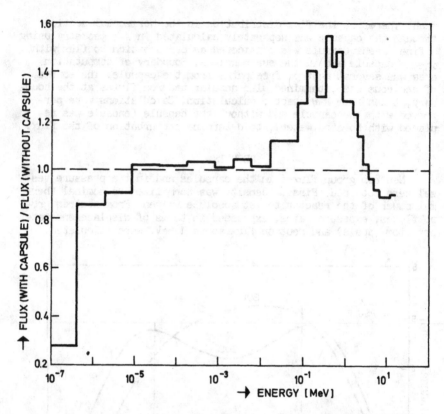

Fig. 3.: Ratio of the neutron group flux calculated with the capsule and without the capsule, given at the middle of the capsule

Dpa/s was obtained by multiplying the neutron flux with energy dependant displacement cross-section, taken from (3). Displacement cross-section, averaged over neutron spectra above 0.01 MeV $\langle \sigma_d \rangle_{E > 0.01 \text{ MeV}}$, is also given. This quantity is recommended in (3) as a measure of spectrum hardness.

Variations of the exposure rate with polar angle (θ) and variations in the axial direction are presented in Figs. 1 and 2 respectively. From Fig. 1 it can be seen that capsules, situated at various polar angles, are exposed to different neutron fluxes. Differences amount up to 70 %. Maximum exposure is at θ-0°, which is taken as the reference angle, where pressure vessel exposure is given.

In axial direction, displacement rate varies for $\pm$ 10 % at

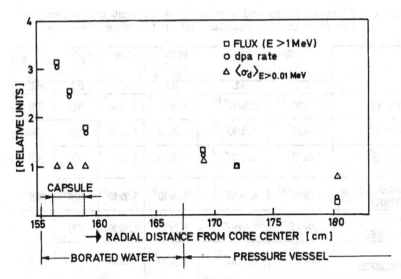

Fig.4: Neutron displacement rates per atom, integrated neutron fluxes above 1 MeV and average displacement cross-sections at various positions in the capsule and in the pressure vessel.
Values refer to the capsule at polar angle $\theta=33^{\circ}$ and to maximum reactor vessel exposure at $\theta=0^{\circ}$. All values are normalized to 1 at one quarter thickness of the pressure vessel.
They are given at reactor midplane.

BOL and $\pm$ 7 % at EOL. Differences are due to axial variation of
the flux magnitude, spectrum of neutrons remaining practically
constant in axial direction as confirmed by only slight variations
of $<\sigma_d>$ $_{E> 0.01\ MeV}$.

Axial dependence of dpa/s significantly changes with burn-up.
Maximum exposure is at EOL shifted from midplane and maximum difference in exposure rate, averaged over the first fuel cycle is reduced to less than $\pm$ 5 %.

Comparison of neutron group fluxes, calculated with the capsule and without the capsule shows significant differences (Fig.3).
They are due to different iron and water neutron cross-sections.
As a consequence the exposure rate in the centre of capsules is
roughly 15 % higher and $<\sigma_d>$ $_{E> 0.01\ MeV}$ is 5 % lower.

High exposure rate variations within the capsule have been
found. In the radial direction, dpa/s and flux above 1 MeV change
for up to 80 % (Fig. 4) and in the transverse direction for 5 %.
During evaluation of data, obtained from mechanical tests on specimens and from measured activation rates of neutron monitors, it
is therefore necessary to take into account their actual positions

Table I.: Exposure rates and average displacement cross-section for iron in the capsules and in the pressure-vessel

		dpa/s (s^{-1})		FLUX (E > 1 MeV) ($cm^{-2} s^{-1}$)		$<\sigma_d>$ E > 0.01 MeV (barn)	
		BOL	EOL	BOL	EOL	BOL	EOL
CAPSULES *	$\theta = 13°$	3.4×10^{-10}	2.9×10^{-10}	1.9×10^{11}	1.7×10^{11}	373	374
	$\theta = 23°$	2.1×10^{-10}	1.8×10^{-10}	1.2×10^{11}	1.1×10^{11}	395	396
	$\theta = 33°$	2.0×10^{-10}	1.7×10^{-10}	1.2×10^{11}	1.0×10^{11}	389	389
PRESSURE VESSEL **	INNER SURFACE	1.0×10^{-10}	9.3×10^{-11}	6.3×10^{10}	5.7×10^{10}	424	425
	1/4 THICKNESS	8.3×10^{-11}	7.5×10^{-11}	4.7×10^{10}	4.3×10^{10}	379	379
	3/4 THICKNESS	3.0×10^{-11}	2.7×10^{-11}	1.3×10^{10}	1.2×10^{10}	307	308

* average over the length of the capsule

** maximal value corresponding to polar angle of zero and peak in axial direction

within the capsule.

Values of dpa/s, flux above 1 MeV and average displacement cross-section in the capsules at various polar angles and in the pressure vessel are presented in Tab. I.

Good agreement in $<\sigma_d>$ E > 0.01 MeV between BOL and EOL values can be noticed, indicating that neutron spectra above 0.01 MeV does not significantly change with burn-up. Exposure rates in capsules at various polar angles are for 2.4, 2.5 and 4.0 times higher than at one quarter thickness of the pressure vessel. Only minor relative differences exist between neutron exposure expressed in dpa/s and flux above 1 MeV.

CONCLUSIONS

Large differences approaching a factor of two were found in the neutron exposure rate within the capsule. As a consequence a considerable spread of data obtained in mechanical testing of specimens is expected.

Neutron exposure rates differ among the capsules themselves and among the capsules and the pressure vessel at one quarter thickness for a factor of 1.7 and 2.4 - 4.0 respectively. Therefore an accurate and reliable neutron transport calculation is mandatory to extrapolate experimental data, obtained from flux monitors and steel specimens, irradiated in the capsules, into the pressure vessel.

Differences of up to 5 % were found in exposure rates within the capsule and the pressure vessel when expressed in dpa/s or flux above 1 MeV. This fact as well as difference in the $<\sigma_d>$ E > 0.01 MeV indicate that neutron spectra in respective positions are significantly different.

Exposure rate differences in axial direction are small when compared to those discussed above. They are further smoothed out (to less than $\pm$ 5 %) during the first fuel cycle. Contrary to usual assumptions the maximum exposure is shifted from the core midplane.

REFERENCES

1. F.B. Mynatt, et. al.: The DOT III Two Dimensional Discrete Ordinates Transport Code, ORNL-TM-4280, September 1973

2. DLC-2D, '100 Group neutron cross-section data based on ENDF/B', ORNL, RSIC Data Library Collection

3. Characterizing exposures in ferritic steels in terms of displacement per atom, ASTM, Standard Practice E 693-79

4. M.Ravnik, S.Slavič, FASVER 2 - Description of a FORTRAN program for 2D - two group diffusion calculation of power density in PWR core, IJS report - IJS-DP-1698, May 1979.

CALCULATED SPECTRAL FLUENCES AND DOSIMETER ACTIVITIES FOR THE
METALLURGICAL BLIND TEST IRRADIATIONS AT THE ORR-PSF*

R. E. Maerker and B. A. Worley

Oak Ridge National Laboratory

Oak Ridge, TN   37831

ABSTRACT

Fluence rate, fluence, and activity calculations
were performed for each of the three exposures (two surveillance capsules and a pressure vessel capsule) performed during the two-year metallurgical blind test
experiment at the ORR-Poolside Facility in Oak Ridge.
Motivation for these calculations was prompted by differences of up to 25% between dosimetry measurements
performed in the earlier startup scoping experiment and
the two-year experiment.

Comparisons of the dosimeter end-of-irradiation
activities with HEDL measurements indicate agreement
generally within 15% for the first surveillance capsule, 5% for the second, and 10% for three locations in
the pressure vessel capsule, which are as good as (if
not somewhat better than) comparisons in the startup
experiment. The calculations thus validate the trend
of the measurements in both the startup and the two-
year experiments, and confirm the presence of a significant cycle-to-cycle variation in the core leakage.
The tape containing the unadjusted spectral fluences
for each of the three exposures that can be used in the
metallurgical analysis is thus considered to be
accurate to within about 10%.

*Work performed for the NRC sponsored Surveillance Dosimetry
Improvement Program under DOE Interagency Agreements 40-551-75
and 40-552-75, NRC FIN No. B0415, under Contract DE-AC05-840R21400
with Martin Marietta Energy Systems, Inc.

J. P. Genthon and H. Röttger (eds.), Reactor Dosimetry, 87–94.
© *1985 ECSC, EEC, EAEC, Brussels and Luxembourg*

INTRODUCTION

The purpose of the two-year irradiation experiment performed at the Oak Ridge Reactor Poolside Facility (ORR-PSF) during the period April 1980 - June 1982 is to serve as an international metallurgical benchmark for evaluating the accuracy of existing damage predictions in PWR pressure vessels. This experiment is a programmatic follow-on to both the low-intensity Pool Critical Assembly (PCA) and high-intensity PSF startup experiments. The PCA was intended primarily as a neutron transport benchmark to test the accuracy of existing calculational methods in estimating measured dosimeter activities throughout simulated PWR geometries. The PSF startup experiment was intended as a short duration (1 cycle) scoping experiment, again involving only dosimetry, to establish a suitable geometry and the irradiation times necessary to yield the desired fluence levels in the long-term metallurgical benchmark. Since the geometries of the startup and metallurgical experiments are essentially identical, saturated activity measurements (1,2) and calculations (3) performed for the short-term experiment were expected to serve equally well for the long-term one. Subsequent HEDL dosimetry measurements (4) performed during the long-term experiment indicated, however, the fluence rate levels to average between 5 and 25 percent below those of the scoping experiment. It was immediately suspected that time variations of the core fission source might account for these differences. Consequently, new calculations were initiated which took into account any significant cycle-to-cycle variations that might exist over the course of the complete exposure which involved 52 cycles in all. These calculations should thus supplant the earlier ones of the startup experiment[†] in the analysis of the metallurgical specimens. The present analysis is limited to a calculation of the spectral fluences in each of 38 groups down to 0.1 MeV accumulated by the specimens and comparisons with some of the measured dosimetry activities at the end of irradiation. It is anticipated that several of the countries participating in this experiment will apply the results of these fluence calculations to the analysis of their own metallurgical specimens.

[†]Ref. 3 describes the almost identical geometry and details an iterative procedure for solving the coupled XY and YZ Boltzmann equations introduced by the presence of the finite surveillance capsule. Subsequently it was found that even for detector locations up to 30 cm removed from the horizontal midplane the coupling is very weak and may be disregarded, thus greatly simplifying the calculations and reducing the procedure to the one outlined in Ref. 8.

THE CALCULATIONS

Many of the details of the calculations, omitted here for brevity, appear in a recently written report (5). Summarizing, the first surveillance capsule (SSC-1) was irradiated for 45 days and 4 cycles, the second surveillance capsule (SSC-2) for 92 days and 8 cycles, and the pressure vessel capsule (SPVC) for 603 days and 52 cycles. All but six of the 52 cycles contributing to the exposures were individually analyzed with the 3-D diffusion code VENTURE (6) to obtain middle-of-cycle fission source distributions. The distribution from each cycle was then integrated over the core height, the core width, and both the core height and core width to yield 2-D and 1-D source distributions to be used in subsequent 2-D and 1-D transport calculations (7), completely analogous to a procedure described previously (8). Instead of performing three transport calculations for each cycle, however, the source distributions from several of the cycles were often weighted and combined in a way that preserved the sensitivities of all the dosimeter activities at the ends of irradiation to the presence of relatively large down time intervals between cycles. Not more than a few cycles were combined at a time in order to preserve effects of cycle-to-cycle variations, and little or no combining was done near the end of an irradiation period (i.e., time dependence was retained in detail for cycles active immediately prior to the removal of the SSC-1, SSC-2, or the SPVC). Thus, for cycle group G, the combined source $\bar{S}_G$ was related to the individual cycle sources S_i, iϵG, in the following way:

$$\bar{S}_G = [\sum_{i=1}^{N(G)} (\frac{S_i}{k_{eff_i}})(\frac{\bar{P}_i}{30})\Delta t_{up_i}]/[t_N^G(retracted)-t_1^G(inserted)] , \qquad (1)$$

where the VENTURE calculations were always normalized to 30Mwt, $\bar{P}_i$ is the average power during cycle i, k_{eff_i} is the calculated criticality constant for cycle i, Δt_{up_i} is the length of time cycle i was up and operational, and the denominator in Eq. (1) represents the total clock time (including down time between the N cycles) that the reduced source $\bar{S}_G$ is to be considered operational. Seventeen cycle groups were chosen to represent the original 52 cycles, and 48 transport calculations were performed using all but one of these groups.

Three-dimensional fluence rates were synthesized from the results of each triad of discrete ordinates transport calculations using a procedure previously described in Ref. 8:

$$\phi_g^G(x,y,z) = \phi_g^G(x,y)\phi_g^G(y,z)/\phi_g^G(y) , \qquad (2)$$

with corresponding spectral fluences Φ of

$$\phi_g^G(x,y,z) = \phi_g^G(x,y,z)[t_N^G(\text{retracted})-t_1^G(\text{inserted})] \ . \tag{3}$$

In these expressions, x refers to the crosswise dimension, z to the axial, and y to the radial dimension. Saturated activities RR_G^d for each cycle group G were also synthesized using available edits from the transport runs:

$$RR_G^d(x,y,z) = [RR_G^d(x,y)RR_G^d(y,z)/RR_G^d(y)] \ \cdot$$

$$\cdot \ \{[t_N^G(\text{retracted})-t_1^G(\text{inserted})]/ \sum_{i=1}^{N(G)} \Delta t_{up_i}\} \ , \tag{4}$$

and the activities at the end of irradiation A_G^d from cycle group G were derived from the saturated activities calculated from Eq. (4):

$$A_G^d(x,y,z) = RR_G^d(x,y,z)\{ \sum_{i=1}^{N(G)} \Delta t_{up_i}/[t_N^G(\text{retracted})-t_1^G(\text{inserted})]\}\cdot Y \cdot$$

$$\cdot \ [e^{-\frac{0.69315}{\tau_{1/2}^d}(t_{EOI}-t_N^G)} -e^{-\frac{0.69315}{\tau_{1/2}^d}(t_{EOI}-t_1^G)}] \ , \tag{5}$$

where Y is the yield of the reaction product. The reciprocal of the "up" time fraction appearing in the brackets in Eq. (4) is necessary to renormalize the reduced fluence rate levels calculated using the source derived from Eq. (1) to their proper levels during actual irradiation, while this factor must be removed using Eq. (5) in order to reestablish the reduced fluence rate levels assumed constant over the extended time interval $(t_N^G-t_1^G)$ so that decay rates at the end of irradiation can be calculated.

RESULTS OF THE CALCULATIONS

Table 1 illustrates the cycle group-to-cycle group variation of three saturated activities at a half-depth location in the simulated pressure vessel capsule, although the _relative variations_ are about the same for any location from the sur-veillance capsules outward. The spectrum is seen to be essen-tially independent of cycle group (see last column), but up to a 40% variation in the fluence rate level is to be observed, almost all of which arises as a result of different fuel loadings in the outermost two rows of the ORR core. The corresponding activities for the startup experiment indicate that average fluence rate levels in the startup exposure exceed those of the two-year exposures by anywhere from 3 to 17 percent, depending on the capsule.

Table 1. Cycle Group-to-Cycle Group Variation of Some Saturated Activities at the T/2 Location in the SPVC

Cycles	^{54}Fe(n,p)	^{63}Cu(n,α)	^{237}Np(n,f)	Np/Cu
153B+153C	7.59-15*	5.87-17*	6.17-13*	1.05+4
153D	7.58-15	5.87-17	6.16-13	1.05+4
153F	7.38-15	5.71-17	5.99-13	1.05+4
153G-154C	7.83-15	6.05-17	6.35-13	1.05+4
154D-154J	7.47-15	5.79-17	6.06-13	1.05+4
155B-155F	9.15-15	7.06-17	7.42-13	1.05+4
156C-157B	8.65-15	6.68-17	6.99-13	1.05+4
157C-157E	8.82-15	6.80-17	7.14-13	1.05+4
158C+158D	9.65-15	7.45-17	7.83-13	1.05+4
158E-158G	8.24-15	6.36-17	6.64-13	1.04+4
158H-158K	8.14-15	6.33-17	6.50-13	1.03+4
159A-159C	8.42-15	6.54-17	6.73-13	1.03+4
159D-160C	7.83-15	6.10-17	6.24-13	1.02+4
160D+160E	7.27-15	5.69-17	5.76-13	1.01+4
161B	7.14-15	5.62-17	5.65-13	1.01+4
161C	6.86-15	5.40-17	5.41-13	1.00+4
Startup	8.84-15	7.05-17	7.04-13	1.00+4

*Units are reactions per atom per second normalized to 30 Mw.

Tables 2 and 3 present comparisons of some measured end-of-irradiation activities with calculated values obtained by summing Eq. (5) over all pertinent cycle groups for the SSC-1 and SSC-2 exposures respectively. The agreement lies within the range of 15 to 20% for the SSC-1 and 5 to 10% for SSC-2, except for the ^{238}U(n,f)X activities in the SSC-2. Although the ^{238}U(n,f)X measurements in Tables 2 and 3 have been corrected for ^{235}U contamination, they do not contain any corrections for photofission or ^{239}Pu burn-in. Subsequent calculations by both HEDL (9) and ORNL (10) suggest that the ^{239}Pu burn-in corrections are significant for the longer-term SSC-2 exposure.

Table 2. Comparison of Some Measured and Calculated Activities at the End of Irradiation for Several Locations in the SSC-1

Reaction	Measured	Calculated	C/E	Reaction	Measured	Calculated	C/E
^{54}Fe(n,p)	3.92-14†	3.42-14	0.87	^{238}U(n,f)^{103}Ru	7.70-14	6.22-14	0.81
^{46}Ti(n,p)	1.67-14	1.31-14	0.78	^{238}U(n,f)^{140}Ba	1.06-13	9.05-14	0.85
^{58}Ni(n,p)	2.00-13	1.68-13	0.84	^{237}Np(n,f)^{137}Cs	3.29-15	2.61-15	0.79
^{63}Cu(n,α)	4.47-17	3.63-17	0.81	^{237}Np(n,f)^{95}Zr	3.75-13	3.07-13	0.82
^{238}U(n,f)^{137}Cs	4.04-16	3.34-16	0.83	^{237}Np(n,f)^{103}Ru	5.18-13	4.15-13	0.80
^{238}U(n,f)^{95}Zr	4.35-14	3.67-14	0.84	^{237}Np(n,f)^{140}Ba	8.04-13	6.25-13	0.78

†Read 3.92x10^{-14} disintegrations per second per atom, etc.

Table 3. Comparison of Some Measured and Calculated Activities at the End of Irradiation for Several Locations in the SSC-2

Reaction	Measured	Calculated	C/E	Reaction	Measured	Calculated	C/E
^{54}Fe(n,p)	7.71-14†	7.36-14	0.95	^{237}Np(n,f)^{95}Zr	5.91-13	5.42-13	0.92
^{46}Ti(n,p)	2.77-14	2.43-14	0.88	^{237}Np(n,f)^{103}Ru	7.24-13	6.52-13	0.90
^{58}Ni(n,p)	3.15-13	2.99-13	0.95	^{238}U(n,f)^{137}Cs	1.11-15	7.75-16	0.70
^{63}Cu(n,α)	9.04-17	8.40-17	0.93	^{238}U(n,f)^{95}Zr	9.05-14	6.37-14	0.70
^{237}Np(n,f)^{137}Cs	6.81-15	6.16-15	0.90	^{238}U(n,f)^{103}Ru	1.48-13	9.57-14	0.65

†Read 7.71x10^{-14} disintegrations per second per atom, etc.

Finally, in Table 4 similar comparisons between calculation and measurement are presented for three different locations in the SPVC - near the inside surface ("OT"), at a quarter-depth into the pressure vessel (T/4), and at a half-depth into the pressure vessel (T/2). For these calculations, data from all 52 cycles were utilized.

An inspection of Table 4 shows that agreement averages within about 5% at both OT and T/4 and 10% at T/2 with the exception of the ^{238}U(n,f)X activities, where again subsidiary calculations (10) have indicated significant contributions to the measurements from ^{239}Pu burn-in. These contributions are important for a location of higher fluence such as OT but are less important for the remaining locations. The increasing disagreement in most of the remaining reactions with depth into the pressure vessel is consistent with earlier comparisons made in the PCA and startup experiment that indicated inaccuracies in the iron inelastic cross sections used in

Table 4. Comparison of Some Measured and Calculated Activities at the End of Irradiation for Three Locations in the SPVC

Reaction	OT Measured	Calculated	C/E	T/4 Measured	Calculated	C/E	T/2 Measured	Calculated	C/E
^{54}Fe(n,p)	3.40-14†	3.46-14	1.02	1.51-14	1.45-14	0.96	5.87-15	5.29-15	0.90
^{46}Ti(n,p)	5.90-15	5.53-15	0.94	2.57-15	2.26-15	0.88	9.45-16	8.10-16	0.86
^{58}Ni(n,p)	5.44-14	5.39-14	0.99	2.45-14	2.31-14	0.94	9.61-15	8.64-15	0.90
^{63}Cu(n,α)	8.24-17	7.70-17	0.93	3.64-17	3.70-17	1.02	1.39-17	1.19-17	0.86
^{237}Np(n,f)^{137}Cs	5.10-15	4.47-15	0.88	3.06-15	2.76-15	0.90	1.69-15	1.51-15	0.89
^{237}Np(n,f)^{95}Zr	8.62-14	8.20-14	0.95	5.13-14	5.03-14	0.98	2.84-14	2.76-14	0.97
^{237}Np(n,f)^{103}Ru	8.54-14	7.96-14	0.93	5.09-14	4.90-14	0.96	2.84-14	2.67-14	0.94
^{238}U(n,f)^{137}Cs	9.17-16	5.96-16	0.65	3.56-16	2.87-16	0.81	1.43-16	1.20-16	0.84
^{238}U(n,f)^{95}Zr	1.67-14	1.03-14	0.62	5.74-15	4.94-15	0.86	2.32-15	2.05-15	0.88
^{238}U(n,f)^{103}Ru	2.31-14	1.24-14	0.54	7.71-15	6.00-15	0.78	2.96-15	2.48-15	0.84

†Read 3.40x10^{-14} disintegrations per second per atom, etc.

the calculations. Water leakage into the void box at an early
stage of the exposure prevented any meaningful comparisons from
being made in the simulated void box capsule (SVBC).

CONCLUSIONS

An important conclusion derived from this study is that
cycle-to-cycle variations in the ORR core leakages are as much as
40% and can account for most of the difference in fluence rate
levels observed between the startup and two-year experiments.
Secondly, the C/E comparisons indicate better general agreement in
the present experiment than either the PCA or the startup experiments, and the resulting calculated spectral fluences available
for metallurgical analysis and/or adjustment should be accurate to
within about 10%. Finally, the seeming importance of ^{239}Pu burn-in
as yet another necessary correction to ^{238}U(n,f)X dosimeter measurements in regions of relatively high fluence has been demonstrated, perhaps further compromising the usefulness of this
dosimeter for some applications.

Typical axial profiles for the calculated fluences above 1
MeV are shown in Table 5. They are excerpted from the spectral
fluence tape available to anyone who wishes to use it (11).

Table 5. Typical Axial Profiles for Fluence Above 1 MeV at Several
Locations for the Two-Year Metallurgical Blind Test Experiment

z(cm)[*]	SSC-1	SSC-2	OT	T/4	T/2
27.94	7.02+18[†]	1.62+19	2.08+19	1.14+19	5.43+18
23.02	9.44+18	2.19+19	2.65+19	1.44+19	6.84+18
20.165	1.37+19	3.21+19	3.01+19	1.63+19	7.82+18
18.735	1.68+19	3.94+19	3.17+19	1.75+19	8.40+18
17.465	1.70+19	3.98+19	3.39+19	1.94+19	9.25+18
16.195	1.68+19	3.94+19	3.33+19	1.83+19	8.71+18
13.97	1.84+19	4.32+19	3.44+19	1.87+19	8.81+18
11.43	1.96+19	4.64+19	3.55+19	1.92+19	8.97+18
9.69	2.02+19	4.79+19	3.61+19	1.94+19	9.08+18
8.42	2.07+19	4.93+19	3.67+19	1.97+19	9.17+18
6.20	2.13+19	5.09+19	3.77+19	2.02+19	9.32+18
3.66	2.17+19	5.20+19	3.83+19	2.04+19	9.42+18
1.27	2.19+19	5.26+19	3.86+19	2.06+19	9.47+18
-0.85	2.18+19	5.25+19	3.86+19	2.05+19	9.44+18
-3.665	2.17+19	5.23+19	3.85+19	2.04+19	9.38+18
-6.625	2.10+19	5.08+19	3.73+19	1.98+19	9.13+18
-8.56	2.04+19	4.93+19	3.63+19	1.93+19	8.91+18
-10.0	1.97+19	4.76+19	3.54+19	1.88+19	8.71+18
-11.6	1.90+19	4.59+19	3.44+19	1.83+19	8.51+18
-13.97	1.76+19	4.26+19	3.26+19	1.75+19	8.19+18
-16.195	1.58+19	3.83+19	3.03+19	1.67+19	7.91+18

[*]Measured from a plane located 5.08 cm below the horizontal midplane.
[†]Read 7.02x10^{18} neutrons/cm^2, etc.

REFERENCES

1. D. J. Ketema, H. J. Nolthenius and W. L. Zijp, Neutron Metrology in the ORR: ECN Activity and Fluence Measurements for the LW Pressure Vessel Surveillance Dosimetry Program, ECN-80-164, Petten, the Netherlands (1980).

2. D. J. Ketema, H. J. Nolthenius and W. L. Zijp, Neutron Metrology in the ORR: Second Contribution of ECN Activity and Fluence Measurements for the LWR Pressure Vessel Surveillance Dosimetry Program, ECN-81-097, Petten, the Netherlands (1981).

3. M. L. Williams and R. E. Maerker, "Calculations of the Startup Experiments at the Poolside Facility," Proc. Fourth ASTM-EURATOM Symposium on Reactor Dosimetry, Vol. 1, NUREG/CP-0029, CONF-82032/V1, NBS (1982).

4. Informal Communication, E. P. Lippincott and L. S. Kellogg, HEDL, to Distribution.

5. R. E. Maerker and B. A. Worley, Activity and Fluence Calculations for the Startup and Two-Year Irradiation Experiments Performed at the Poolside Facility, NUREG/CR-3886, ORNL/TM-9265 (1984).

6. D. R. Vondy, T. B. Fowler and G. W. Cunningham, III, The Bold Venture Computation System for Nuclear Reactor Core Analysis Version III, ORNL-5711, Oak Ridge National Laboratory (1981).

7. W. A. Rhoades and R. L. Childs, An Updated Version of the DOT4 One-and Two-Dimensional Neutron/Photon Transport Code, ORNL-5851 (1982).

8. R. E. Maerker and M. L. Williams, "Calculations of the Westinghouse Perturbation Experiment at the Poolside Facility," Proc. Fourth ASTM-EURATOM Symposium on Reactor Dosimetry, Vol. 1, NUREG/CP-0029, CONF-820321/V1, NBS (1982).

9. Informal Communication, E. P. Lippincott and R. L. Simons, HEDL, to Distribution.

10. F. W. Stallmann, Determination of Damage Exposure Values in the PSF Metallurgical Irradiation Experiment, NUREG/CR-3814, ORNL/TM-9166 (1984).

11. Contact F. B. K. Kam, Oak Ridge National Laboratory, P.O. Box X, Oak Ridge, Tennessee 37831.

BATTELLE'S COLUMBUS LABORATORIES REACTOR

VESSEL SURVEILLANCE SERVICE ACTIVITIES

M. P. Manahan, A. R. Rosenfield, C. W. Marschall, and M. P. Landow

Battelle's Columbus Laboratories

Columbus, Ohio 43201 U.S.A.

ABSTRACT

This paper describes the current methodology in use at Battelle's Columbus Laboratories for obtaining and processing pressure vessel surveillance data, extrapolating mechanical behavior, and calculating reactor coolant pressure-temperature operating curves. Recent results from the Arkansas Nuclear One--Unit 2 (ANO-2) benchmark cavity dosimetry experiment are reported. The results indicate that it is possible to calculate the flux in ex-vessel locations with accuracies on the order of 10-15 percent. Also, end of life metallurgical predictions for the Poolside Facility (PSF) Blind Test materials are compared with experimental data.

Another Battelle-Columbus research activity related to pressure vessel surveillance is fracture initiation and arrest toughness determination using miniature specimens. Small-sized laboratory specimens provide data which can be extremely non-conservative compared with large in-service structures and which exhibit considerable scatter. Battelle-Columbus has examined these effects and has developed a simple method for minimizing them.

The radiation-induced temperature shift of crack arrest toughness has been measured and found to be significantly less for high-copper materials than estimated from the shift in the Charpy 40.7 Joule temperature or from U.S. Nuclear Regulatory Commission (NRC) Regulatory Guide

J. P. Genthon and H. Röttger (eds.), Reactor Dosimetry, 95–111.
© *1985 ECSC, EEC, EAEC, Brussels and Luxembourg*

1.99 (Rev. 1) calculations. A method has been developed for estimating crack arrest toughness as a function of fluence which is still conservative for high-copper materials but significantly less restrictive.

1.0 INTRODUCTION

Battelle-Columbus maintains a strong commitment to service activities as well as to research and development to ensure continued safe operation of the world's nuclear power plants. In the early days of reactor operation, comparatively little was known about the detrimental effects of neutron exposure of ferritic steels. Fortunately, surveillance programs were implemented early and the necessary data to characterize the embrittlement mechanisms were obtained. Unique surveillance program testing challenges continue to emerge today as older reactors approach the end of design life, as fracture mechanics continues to develop, and as new test methods are perfected. This paper presents a summary of existing Battelle-Columbus surveillance activities as well as several results from new research and development thrusts begun over the past few years.

2.0 CURRENT LABORATORY METHODOLOGY

This section describes the general procedures and methodology used in pressure vessel surveillance capsule testing and analysis. In particular, we discuss the general methodology used to estimate the neutron exposure parameter, a method for establishing conservative estimates of the adjusted reference nil-ductility temperatures, and the procedures for generating the set of reactor operating pressure-temperature curves.

2.1 Neutron Dosimetry Analysis

ASTM procedures (1-9) are followed in the measurement of neutron dosimeter activities. Once the activities have been determined, the flux of neutrons with energies above a given level, E_C (corrected for decay time between exposure and counting), can be calculated using standard equations (10). In order to determine the effective cross-section to be used in the above calculations, the cross-section and the neutron flux as a function of energy must be known. The neutron flux and spectrum are calculated using the DOT computer code (11). This code solves the two-dimensional Boltzmann transport equation using the method of discrete ordinates. The reactor geometrical design configuration is modeled to simulate the core structure, intervening structures, and pressure vessel.

Calculations are performed using third order scattering (P_3) and S_8 angular quadrature with 47 neutron group cross-sections from the DLC-75 library (12).

2.2 Mechanical Behavior Measurements

Standard tensile and impact measurements are carried out in accordance with the appropriate ASTM standards (13-16). The impact machine is calibrated periodically as specified in ASTM E23-82 and verified using a set of standard Charpy specimens obtained from the U.S. Army Materials and Mechanics Research Center (AMMRC). Instrumented impact tests are conducted using a tup (hammer) to which strain gage instrumentation has been added.

Currently, Charpy data are analyzed by a computer program which is based on the familiar hyperbolic-tangent (tanh) algorithm. Although we feel that computer analysis removes the human bias in determining the transition curves, some interpretation is still required by the analyst. For example, the tanh analysis occasionally leads to curves with negative values for the lower shelf. In these cases, the data are reanalyzed using a modification of the tanh analysis where the lower shelf energy is set to a fixed value.

Alternatively, a new algorithm developed at Battelle-Columbus provides a method of removing the ambiguities of the tanh procedures. (17) We use a Weibull formulation in which both the central tendency and variation of the data are allowed to be temperature-dependent via a series representation. An iterative procedure is then used to minimize errors in the coefficients of the series. This procedure appears quite promising and we intend to continue research in the area.

2.3 Extrapolation of Mechanical Behavior

The allowable internal vessel pressure for a specific coolant temperature is a function of several key variables including the adjusted nil-ductility reference temperature (ART_{NDT}). The NRC Regulatory Guide 1.99 (Rev. 1) (18) suggests two basic alternative approaches for predicting the RT_{NDT} shift depending on whether or not credible surveillance data are available. Reference (18) has been sharply criticized over the past few years as more analyses have been performed and more data have accumulated. Since Reference (18) is now under revision, Battelle-Columbus uses recently developed trend curves for ΔRT_{NDT} predictions. The results of a recent analysis of available trend curves for use in the PSF Blind Test were reported in Reference (19). In order to evaluate the current trend curve correlations, several criteria were identified. The criteria are as follows:

(1) The correlation should be based on a data base that has been scrubbed (i.e., where chemistry and fluence are known and the fluence values reported have been verified).

(2) The data base should be divided into weld metal and plate.

(3) The standard deviation should be low and approach the experimental uncertainty limit for Charpy testing ($\sim$8.3 - 11.1 C).

(4) The correlation should be consistent with experimental observations and current hypotheses of physical mechanisms of radiation damage in steel.

(5) Reasonable extrapolations of ΔRT_{NDT} ($\lesssim$1 x 10^{20} n/cm^2) beyond the data should give physically meaningful results.

(6) Uncertainties in fluence and temperature shift should be considered in the regression analysis.

Application of these criteria eliminated all of the correlations examined except for those of Guthrie (20) and Odette (21). The Odette physically-based methodology was judged to be very promising and represents an important contribution toward understanding the fundamental mechanisms of steel embrittlement. However, we feel that the model is not yet complete since it does not contain all of the microstructural mechanisms possible and, therefore, the Guthrie correlation which is a function of copper, nickel, and fast fluence was chosen. Further results from the PSF Blind Test are presented in Section 3.0.

Subsequent to completion of our PSF analysis, we learned (22) that the draft of Revision 2 of NRC Regulatory Guide 1.99 is a combination of the Reference (20) and Reference (21) correlations. We are currently using the draft Revision 2 version of NRC Regulatory Guide 1.99 in our ΔRT_{NDT} predictions.

2.4 Pressure-Temperature Curve Development

To ensure safe operation of a nuclear power plant during heatup, cooldown, and hydrotest conditions, it is necessary to conservatively calculate allowable stress loadings for the ferritic pressure vessel materials. These allowable loadings can be conveniently presented as a plot of measured coolant pressure versus measured coolant temperature (P-T curves). Appendix G (23) of the ASME Boiler and Pressure Vessel Code presents a procedure for obtaining the allowable loadings for ferritic pressure-retaining materials in Class 1 components. This procedure is based on the principles of linear elastic fracture mechanics. The Battelle-Columbus calculative methodology is based on this procedure and the guidance presented in References (24) and (25). The hoop stress can be calculated using either the thick-wall analytical equation or a thin-wall approximation. The thin-wall approximation is recommended in Reference (24). However, Table I shows that the thin-wall approximation is non-conservative for the inner regions of the reactor pressure vessel

TABLE I. COMPARISON OF THIN-WALL AND THICK-WALL HOOP STRESS CALCU-
LATIONS FOR A 20.3-CM THICK PRESSURE VESSEL WALL

Vessel Inner Radius (cm)	Percent Error Resulting From Use of Thin-Wall Approximation		
	Maximum Stress Location	1/4 Thickness	3/4 Thickness
152.4	-7.07	-3.27	+3.33
203.2	-5.24	-2.46	+2.52
254.0	-4.15	-1.98	+2.01
304.8	-3.44	-1.65	+1.68
355.6	-2.94	-1.42	+1.44

wall. Battelle-Columbus currently uses the thick-wall equation in
P-T calculations because it provides a more accurate representation
of the wall membrane stresses for relatively small vessel radii.
The vessel material temperature used in the calculation is the temperature measured in the cold leg of the steam generator (PWR). This
temperature is conservative since the actual reactor beltline temperature will be higher due to gamma and coolant friction heating.
The instrument error also is included in a conservative manner. The
pressure is typically measured in the pressurizer (PWR) or top dome
(BWR) and must be corrected to the pressure at the bottom of the
vessel beltline.

Use of a single value for the stress intensity for membrane
stressing which corresponds to the highest loading over the thermal
ramp results in overly conservative allowable pressure loadings for
a given temperature. Therefore, we use the analytical representation
for the membrane stress intensity reported in Reference (26). Using
this relation, the following equation is derived for heatup and cooldown calculations:

$$P = \sqrt{\frac{1.243\left\{[K_{IR} - M_T \Delta T](B^2 - A^2)\sigma_y\right\}^2}{(A^2 + B^2)^2\left\{4.39(B - A)\sigma_y^2 + 0.212[K_{IR} - M_T \Delta T]^2\right\}}} - P_D - P_H \quad [1]$$

where

σ_y = yield stress (MPa)
P = measured coolant pressure (MPa)
B = vessel outer radius (m)
A = vessel inner radius (m)
P_D = core pressure drop (MPa), and
P_H = static head (MPa).

Similarly, the equation for hydrotest conditions is:

$$P = \sqrt{\frac{1.243[K_{IR}(B^2 - A^2)\sigma_y]^2}{(A^2 + B^2)^2[2.47\sigma_y^2(B - A) + 0.212 K_{IR}^2]}} - P_D - P_H \qquad [2]$$

3.0 APPLICATIONS

In Section 2.0 we described the general methodology used by Battelle-Columbus in surveillance capsule analysis. In this section we present the key results from two programs recently performed by Battelle-Columbus that have expanded our range of services for fluence calculations and ΔRT_{NDT} predictions.

3.1 ANO-2 Cavity Dosimetry Results

Under contract to the University of Missouri-Rolla, Battelle-Columbus has participated in a research program funded by the Electric Power Research Institute (EPRI) to assist in the benchmarking of methods of analysis used in pressure vessel surveillance programs for commercial nuclear reactors. The program has been performed for dosimeters irradiated in both in-vessel and ex-vessel locations at the ANO-2 nuclear plant. The specific objectives of this work were to provide benchmark information that would allow: (1) synthesized three-dimensional neutron transport analyses of the flux through the vessel wall to be compared with experimental measurements, (2) comparison of in-vessel and ex-vessel surveillance measurements, and (3) the use of counting standards to be tested as a correction factor to predictions of the fluence in the vessel.

To calibrate the results, a number of NBS standard foils which had been irradiated in a fission spectrum also were counted. These results were used to establish "experimental" threshold reaction total cross-sections (σ_e) for these reactions. The total cross-sections (σ_c) for these reactions in a fission spectrum were also calculated using the DETAN code (27). The ratio of these cross-sections (σ_e/σ_c) was used as a calibration factor to correct foil activities calculated using DOT code fluxes and DETAN calculated cross-sections.

The DOT 4.3 (11) transport analysis code was used to do both an R-θ calculation and an R-Z calculation. The modeling of the reactor, especially the outer portion of the core, was done in great detail. In the R-θ model of the core, there were 2250 meshes and six types of fuel. To facilitate preparing the DOT input, a computer code was developed to determine what material is in each mesh and to

TABLE II. C/E RATIOS CALCULATED FOR THRESHOLD REACTION ACTIVITIES
IN ANO-2 AFTER APPLYING THE CALIBRATION FACTOR

| Monitor | Calibration Factor (σ_e/σ_c) | In-Vessel Surveillance Capsule | | | Ex-Vessel Dosimeters |
		Top Comp.	Middle Comp.	Bottom Comp.	
Fe	1.0015	0.9869	1.0076	1.0433	---
Ni	1.0454	0.9224	1.0553	0.9983	0.9275
Cu	---	1.0695	1.1224	1.0748	1.0466
Ti	---	0.8555	0.9679	0.9618	0.8765
U-238	0.9496	2.9085	4.4308	4.5367	0.6984
Np-237	---	---	---	---	0.7677

assign the material numbers by θ-row for input to DOT. In the 1/8
core model for the R-θ calculation, there are over 5000 fuel pins
in 22 1/8-fuel assemblies. Thus, a computer code was developed to
assign pins to meshes, to calculate the total power in the meshes,
and to assign powers by θ-row. For the R-Z DOT run, a similar computer code was developed. The results of the two DOT runs were
synthesized to give an R-θ-Z representation of the flux using a pro-
prietary Battelle-Columbus code that was developed for this purpose.
The following relation was derived to synthesize the two-dimensional
fluxes:

$$\Phi(r,z,\theta) = C(r)\ \Phi(R,\theta)\ \Phi(r,z) \quad , \quad [3]$$

where

$$C(r) = \frac{2\pi}{\int_0^{2\pi}\Phi(r,\theta)d\theta} \quad , \text{ and}$$

$$\Phi = \text{neutron flux (n/cm}^2\text{ sec).}$$

The calculated activities of the threshold reaction foils were
compared with the measured activities. The ratios of calculated-to-
measured activities (C/E) at the in-vessel and ex-vessel positions
for the various foils are given in Table II.

The results in Table II suggest that it is feasible to calculate
the flux at positions outside the pressure vessel wall with accuracies of 10 to 15 percent or better. The rather large discrepancies
in the cases of the neptunium and uranium foils could be the result
of one of the following:

(1) Use of incorrect constants in the calculation of fissions/second from the measured counting rate.

(2) Leakage through the cadmium cover on the foils which may have been too thin or nonuniform, permitting some thermal neutrons to penetrate.

(3) Contamination of the U-238 and Np-237 foils with transuranic isotopes.

(4) Photofission adding to the fission rate.

Investigations at the University of Arkansas are under way to find which mechanism(s) is responsible for the discrepancy. Further details concerning this program are provided in papers published in these proceedings. (28,29)

3.2 _ PSF Blind Test Results

The Nuclear Regulatory Commission (NRC) in cooperation with EPRI is conducting a blind test to assess the accuracy of end of life (EOL) predictions for several types of pressure vessel steels. These steels have been irradiated in the Oak Ridge Reactor (ORR) PSF to doses that represent approximately 15 years ($\sim$1/2 EOL) and 30 years ($\sim$EOL) of operation for a PWR implementing a low neutron leakage fuel management scenario. The test assembly consists of two Simulated Surveillance Capsules (SSC-1 and SCC-2) and a Simulated Pressure Vessel Capsule (SPVC) which is representative of a 22.6-cm thick pressure vessel steel wall. The metallurgical conditions present after irradiation in the SCC-2 and SPVC 0-thickness (0-T), 1/4-T, and 1/2-T positions for six steels are to be predicted. Part I of the test has been designed to represent the analysis procedures normally found in surveillance reports. Battelle-Columbus results are presented in detail in Reference (19). Only selected results will be presented here.

The exposure parameter widely used in surveillance reports prepared by Battelle-Columbus is fluence (E>1.0 MeV). We also offer sponsors the option of having displacements per atom (DPA) reported as well. We believe that DPA is a superior damage parameter since it accounts for displacement damage caused by neutrons scattered below 1.0 MeV. Other relevant arguments for the use of DPA have been presented in Reference (30). Our predictions using DPA are presented in Part III of the blind test report (19). However, in order to make comparisons with other participants, we are presenting here results based on fast fluence calculations.

An example comparison of Battelle-Columbus predictions with both measured values and with predictions of the other blind test participants is shown in Figure 1. The scatter bars are representative of all six blind test materials. The Battelle-Columbus results

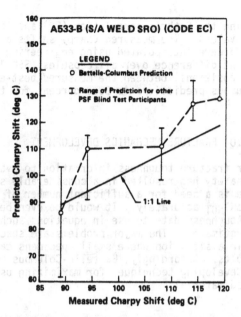

FIGURE 1. COMPARISON OF PREDICTED AND MEASURED CHARPY SHIFT (40.7 J)
FOR A533-B WELD MATERIAL IRRADIATED IN PSF BLIND TEST

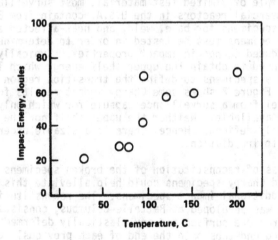

FIGURE 2. SOME SURVEILLANCE CAPSULE RESULTS SHOWING THE DIFFICULTY
THAT CAN OCCUR IN DRAWING A TRANSITION CURVE WITH A
LIMITED NUMBER OF SPECIMENS

were obtained using the Guthrie correlation (20) and the HEDL best-estimate fluence values. The measured Charpy shifts are reported in Reference (31) and were interpolated using engineering judgment. The average percent difference over all capsules (SSC-1, SSC-2, O-T, 1/4-T, and 1/2-T positions) between the measured best-estimate and the Battelle-Columbus predictions is 20.3 percent for the 40.7 Joule shift.

4.0 FRACTURE MECHANICS DEVELOPMENT

The need for fracture toughness information to evaluate reactor pressure vessel safety has resulted in a considerable surveillance challenge. There is a need for a sufficient number of impact specimens to measure ΔRT_{NDT} accurately. It would be even better to have actual fracture toughness data to use in equations such as [1] and [2] rather than estimates. The major problems are space and material limitation, in a situation where small specimens can yield non-conservative results. Accordingly, Battelle-Columbus research has concentrated on developing techniques for maximizing usable information given limited material.

4.1 Ductile/Brittle Transition Temperature

As an example of limited test material, most surveillance capsules from commercial reactors in the U.S.A. contain from 8 to 12 Charpy impact specimens for base, weld, and heat-affected zone materials. These specimens must be tested in order to determine the irradiation-induced change in impact properties. Typically, three specimens are used to obtain the upper shelf energy which leaves only from 5 to 9 specimens to define the transition region of the Charpy curve. Figure 2 shows some Charpy impact results for an irradiated steel from a surveillance capsule for which only eight specimens were available. Neither the upper shelf nor the transition region is clearly defined. Hence, there is a sizable uncertainty in the shift due to irradiation.

The process of reconstitution of the broken specimens to produce full-sized Charpy specimens could help alleviate this problem by providing additional impact specimens. The reconstitution process, which was developed at Battelle-Columbus, consists of first removing the fracture surface and any plastically deformed material, arc-stud welding end-tabs onto the end of each previously broken specimen half, and machining and notching the reconstituted specimen to the size of the standard Charpy specimen. Using this process, two additional Charpy specimens can be reconstituted from a single original. Test results to date show excellent agreement between the impact results from the original Charpy specimens and the reconstituted specimens. (32,33)

4.2   Fracture Toughness

In addition to the normal impact test, the reconstituted Charpy specimens could be tested to obtain fracture toughness data directly. Standard K_{IC} and J_{IC} tests can be performed on pre-cracked reconstituted specimens to obtain fracture toughness. Dynamic fracture toughness has also been obtained at Battelle-Columbus by testing the pre-cracked specimen in a Charpy machine that has been equipped with an instrumented tup. However, there is a significant problem in interpreting the test records. Researchers in a number of laboratories have shown that small-sized laboratory specimens can provide data which are extremely non-conservative compared to large model cylinders undergoing thermal shock. (34,35) In addition, the small-specimen toughness values display extreme scatter. The result is that a large number of surveillance specimens are required to represent the fracture resistance of a reactor pressure vessel realistically.

Research at Battelle-Columbus has examined the physical origins of the scatter and developed a simple method of eliminating much of it. (35) The method is based on the observation that compact specimens will often display some stable crack growth prior to unstable cleavage fracture, which is the failure mechanism being guarded against. The argument is that the energy expended in stable growth is not available to drive the unstable crack and its influence must be eliminated from the experimental record in order for the data to be representative of a vessel, where the stable growth stage is unlikely to occur.

The difference between small specimens and large structures is in part a reflection of the effect of size on the stress concentrating power of a crack, as discussed by Merkle (34) in developing an earlier correction factor. In effect, the high local stresses associated with thick sections favor cleavage over stable growth. Figure 3 provides an illustration of this effect. Here the crosses represent conventional analysis of compact-specimen data, while the open symbols represent corrected data using both the Battelle-Columbus and Merkle methods. The solid point is the result from a large cylinder subjected to thermal shock. Note that Figure 3 shows that the small-specimen data are very unrepresentative of the cylinder unless there is no stable growth and/or a correction is made. Unfortunately, irradiated specimen data have not been treated by either of these correction techniques, so that currently it is difficult to judge how much of existing crack-initiation data bases are representative of reactor pressure vessels.

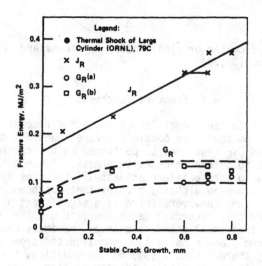

FIGURE 3. CRACK-GROWTH RESISTANCE CURVES. THE ELASTIC COMPONENT
OF J WAS ESTIMATED BY (A) THE MERKLE METHOD AND (B) THE
BATTELLE-COLUMBUS METHOD.

4.3 Crack-Arrest Toughness

Battelle-Columbus crack-arrest research has provided a statistical analysis for unirradiated base plate (36) which can provide a first approximation of the trend for irradiated materials. The basic results are that crack-arrest toughness averages 5/3 of the ASME K_{IR} curve and that the standard deviation is approximately 20 percent of the mean for an individual plate. (36,37)

Further research has provided a basis for using Charpy data to realistically estimate the detrimental effects of radiation on crack-arrest toughness. Radiation effects on K_{Ia} cannot be obtained directly from Charpy surveillance specimens. Instead, Section XI of the ASME Boiler and Pressure Vessel Code requires that the K_{Ia}-versus-temperature curve for unirradiated material, assumed to be the ASME K_{IR} curve in the absence of K_{Ia} data, be shifted upward in temperature by the amount of the radiation-induced shift in RT_{NDT}. The latter shift is commonly estimated as the shift in the 40.7 Joule energy obtained from Charpy surveillance specimens. Marschall, et al (38), have shown that for plates and welds that contain from 0.20 to 0.29 percent copper, the shift in 40.7 Joule energy can overestimate the shift in the K_{Ia}-versus-temperature curve by as much as 55 C. They showed also that a more realistic estimate of radiation effects on K_{Ia} for high-copper materials can be obtained from the observed shift in the Charpy fracture-appearance-transition-temperature (FATT), which is conservative by only 5 to 10 C.

In view of the foregoing correlation between the Charpy-FATT shift and the K_{Ia} curve shift, it would appear important that the FATT be determined from Charpy surveillance specimens. When the quantity of specimens is insufficient to accomplish this, reconstitution of previously tested specimens could be carried out.

5.0 CONCLUSIONS

The main conclusions, based on our pressure vessel surveillance service research activities, are as follows:

• We believe DPA is a superior damage parameter since it accounts for displacement damage caused by neutrons scattered below 1.0 MeV in a mechanistic fashion.

• We believe that computer analysis of Charpy data is superior to interpolation using engineering judgment since it minimizes the human bias and enhances reproducibility. However, interpretation is still required by the analyst for some materials.

• Use of the thin-wall approximation for the vessel hoop stress calculation results in non-conservative results for the inner regions of the vessel wall.

• The C/E ratios for the nickel, copper, and titanium foils were of comparable magnitude for the ANO-2 in-vessel and ex-vessel locations. It is feasible to calculate the flux at positions outside the pressure vessel wall with accuracies of 10 to 15 percent or better.

• The cavity dosimetry C/E results for the fission foils differed substantially from 1.0. Additional research is necessary to determine the mechanism(s) responsible for the discrepancy. Investigations of alternative dosimeters(e.g., niobium) to possibly replace U-238 and Np-237 is also warranted.

• Reconstitution of broken Charpy specimens can significantly increase the amount of data available from a given volume of material. The method has been benchmarked and is being used in surveillance capsule analysis.

• A technique which corrects for the energy expended in stable crack growth in miniature specimens has been developed. The corrected data shows excellent agreement with experimental data from large cylinders subjected to thermal shock.

A more realistic estimate of radiation effects on K_{Ia} for high-copper materials can be obtained from the observed shift in the Charpy FATT, which is conservative by only 5 to 10 C.

6.0 ACKNOWLEDGMENT

This paper has reported the results of several research programs. We are grateful to our sponsors for their support and permission to publish these findings. In particular, we wish to express gratitude to the Heavy-Section Steel Technology Program (MMES Subcontract 85X-17624C), to Arkansas Power and Light, the Electric Power Research Institute, the University of Arkansas, and the University of Missouri-Rolla.

7.0 REFERENCES

1. "Standard Method for Measuring Neutron Flux, Fluence, and Spectra by Radioactivation Techniques", ASTM Designation E261-77, Annual Book of ASTM Standards, Part 45, pp 930-941 (1982).

2. "Standard Method for Determining Fast-Neutron Flux Density by Radioactivation of Iron", ASTM Designation E263-82, Annual Book of ASTM Standards, Part 45, pp 951-956 (1982).

3. "Standard Guide for Application of Neutron Transport Methods for Reactor Vessel Surveillance", ASTM Designation E482-82, Annual Book of ASTM Standards, Part 45, pp 1088-1092 (1982).

4. "Standard Method for Calibration of Germanium Detectors for Measurement of Gamma-Ray Emission of Radionuclides", ASTM Designation E522-78, Annual Book of ASTM Standards, Part 45, pp 1139-1144 (1982).

5. "Standard Method for Determining Fast-Neutron Flux Density by Radioactivation of Copper", ASTM Designation E523-82, Annual Book of ASTM Standards, Part 45, pp 1145-1152 (1982).

6. "Measuring Thermal Neutron Flux by Radioactivation Techniques", ASTM Designation E262-77, Annual Book of ASTM Standards, Part 45, pp 942-950 (1982).

7. "Determining Fast-Neutron Flux Density by Radioactivation of Nickel", ASTM Designation E264-82, Annual Book of ASTM Standards, Part 45, pp 957-961 (1982).

8. "Measuring Fast-Neutron Flux by Radioactivation of Titanium", ASTM Designation E526-82, Annual Book of ASTM Standards, Part 45, pp 1153-1156 (1982).

9. "Measuring Fast-Neutron Flux Density by Radioactivation of Uranium-238", ASTM Designation E704-79, Annual Book of ASTM Standards, Part 45, pp 1244-1248 (1982).

10. W. S. Brasher and D. H. Timmons, "Comparative Analysis of the Pressure Vessel Surveillance Requirements for Commonwealth Edison Power Reactors", Final Report to Commonwealth Edison Co. (November 1, 1974).

11. "DOT IV Version 4.3, One- and Two-Dimensional Transport Code System", CCC-429, Radiation Shielding Information Center, Oak Ridge National Laboratory, Oak Ridge, Tennessee.

12. "Bugle 80 Coupled 47 Neutron, 20 Gamma Ray, P_3, Cross Section Library for LWR Shielding Calculations", DLC-75, Radiation Shielding Information Center, Oak Ridge National Laboratory, Oak Ridge, Tennessee.

13. "Standard Methods and Definitions for Mechanical Testing of Steel Products", ASTM Designation A370-77, Annual Book of ASTM Standards, Part 10, pp 28-83 (1982).

14. "Standard Methods for Notched Bar Impact Testing of Metallic Materials", ASTM Designation E23-82, Annual Book of ASTM Standards, Part 10, pp 277-300 (1982).

15. "Standard Methods of Tension Testing of Metallic Materials", ASTM Designation E8-81, Annual Book of ASTM Standards, Part 10, pp 197-217 (1982).

16. "Standard Recommended Practice for Elevated Temperature Tension Tests of Metallic Materials", ASTM Designation E21-79, Annual Book of ASTM Standards, Part 10, pp 267-276 (1982).

17. T. A. Bishop, A. J. Markworth, and A. R. Rosenfield, "Analyzing Statistical Variability of Fracture Properties", Metall. Trans. A., Vol. 14A, pp 687-693 (1983).

18. Regulatory Guide 1.99, Revision 1, "Effects of Residual Elements on Predicted Radiation Damage to Reactor Vessel Materials", U.S. Nuclear Regulatory Commission, Washington, D.C. (April 1977).

19. M. P. Manahan, et al, "Final Report on PSF Blind Test", October 7, 1984; to be published in NUREG-CR-3320, Vol. 1, HEDL-TME 84-3 (scheduled for release February 1985).

20. E. P. Lippincott and W. N. McElroy, "LWR Pressure Vessel Sur-
    veillance Dosimetry Improvement Program: Quarterly Progress
    Report October 1982-December 1982", NUREG-CR-2803, Vol. 4,
    HEDL-TME 18-21, Hanford Engineering Development Laboratory,
    Richland, Washington (1983).

21. G. R. Odette, Private Communication to M. P. Manahan, Battelle's
    Columbus Laboratories, dated September 21, 1983.

22. N. Randall, Private Communication to M. P. Manahan, Battelle's
    Columbus Laboratories, dated February 1984.

23. ASME Boiler and Pressure Vessel Code, Section III, Appendix G
    for Nuclear Power Plant Components, Division 1, "Protection
    Against Nonductile Failure", pp 543-552, 1980 Edition.

24. U. S. Nuclear Regulatory Commission Standard Review Plan, Sec-
    tion 5.3.2, "Pressure-Temperature Limits", Revision 1, Office
    of Nuclear Reactor Regulation (July 1981).

25. Code of Federal Regulation, Title 10, Part 50, Appendix G,
    "Fracture Toughness Requirements", pp 424-427 (1982).

26. PVRC ad hoc Task Group on Toughness Requirements, "PVRC Recom-
    mendations on Toughness Requirements for Ferritic Materials",
    Welding Research Council Bulletin 175 (August 1972).

27. "DETAN", a Fortran program used to calculate broad group thresh-
    old reaction cross-sections; supplied by Charles Eisenhauer of
    National Bureau of Standards (NBS).

28. N. Tsoulfanidis, et al, "Neutron Energy Spectrum Calculations
    in Three PWR's", University of Missouri-Rolla; presented at
    5th ASTM-EURATOM Symposium on Reactor Dosimetry, September 24-
    28, 1984, Geesthacht, FRG.

29. C. O. Cogburn and J. G. Williams, "Pressure Vessel Dosimetry at
    U.S. PWR Nuclear Plants", University of Arkansas; presented at
    5th ASTM-EURATOM Symposium on Reactor Dosimetry, September 24-28,
    1984, Geesthacht, FRG.

30. W. N. McElroy and F. B. K. Kam, HEDL letter on "PSF Blind Test
    Instructions and Data Packages", to Distribution, dated April 13,
    1983.

31. J. R. Hawthorne and B. H. Menke, "Light Water Reactor Pressure
    Vessel Surveillance Dosimetry Improvement Program: Post-Irradia-
    tion Notch Ductility and Tensile Strength Determinations for
    PSF Simulated Surveillance and Through-Wall Specimen Capsules",
    NUREG/CR-3457, MEA-2026 (August 1983).

32. J. S. Perrin, et al, "Preparation of Reconstituted Charpy V-Notch Impact Specimens for Generating Pressure Vessel Steel Fracture Toughness Data", Effects of Radiation on Materials: Eleventh Conference, ASTM STP 782, H. R. Brager and J. S. Perrin, Eds., American Society for Testing and Materials (1982).

33. J. S. Perrin, et al, "Reconstituted Charpy Impact Specimens", NP-2759 Final Report, Fracture Control Corporation (December 1982).

34. J. G. Merkel, "An Examination of the Size Effects and Data Scatter Observed in Small-Scale Cleavage Fracture Toughness Testing", NUREG-CR-3672, ORNL/TM-9088, Oak Ridge National Laboratory, Oak Ridge, Tennessee (1984).

35. A. R. Rosenfield and D. K. Shetty, "Cleavage Fracture of Steel in the Ductile-Brittle Transition Region", presented at the ASTM Symposium on User's Experience with Elastic-Plastic Test Methods, Louisville, Kentucky (1983).

36. A. R. Rosenfield, et al, "Battelle-Columbus HSST Support Program", Heavy Section Steel Technology Program, Semi-Annual Progress Report for October 1983-March 1984, NUREG/CR-3744, Vol. 1, ORNL/TM-9154/V1 (1984).

37. A. R. Rosenfield, et al, "Recent Advances in Crack-Arrest Technology", in Fracture Mechanics: Fifteenth Conference, R. J. Stanford, Ed., ASTM-STP 833 (1984).

38. C. W. Marschall, et al, "Crack-Arrest Behavior of Pressure Vessel Plates and Weldments as Influenced by Radiation and Copper Content", presented at the Twelfth International Symposium on Effects of Radiation on Materials, June 1984, Williamsburg, Virginia.

29. Parris, et al., "Correlation of Recrystallized Primary
    Through Wall Specimens for Degrading Pressure Vessel Steel
    Fracture Toughness Data," Fracture Radiation in an Materials,
    ASTM Symposium, ASTM STP 782, H. L. Sprague and D. S. Harkins,
    Eds., American Society for Testing and Materials, (1982).

30. J. C. Silverman, et al., Fracture Toughness, Impact, Department
    of Energy final Report, Westinghouse R and Corporation, (the nrc
    ..... 180.

31. J. G. Merkle, "An Examination of The Size Effects and On-Set
    of stable Crack Growth in Small Scale Charpy Fracture Toughness
    Specimen," ORNL/TM-6399, Oak Ridge National Lab-
    oratory, Oak Ridge, Tennessee (1981).

32. W. A. Logsdon and W. G. Shogan, "Dynamic Fracture of Steel
    in the Ductile-Brittle Transition Region", presented at the
    ASTM Symposium on Fracture Mechanics: Elastic-Plastic Fract-
    ure Mechanics, (1982)

33. R. A. Rosenfield et al., "Parallel Evaluations of Subcritical
    crack Heavy Section Steel Technology Program, Semi-Annual
    Program Report for October 1982-March 1983, NUREG/CR-0345,
    ORNL/TM-5627, (1982).

34. J. A. Ruschau, "Identification of Fracture Parameters for Elastic-
    Plastic in Fracture Mechanics," Eleventh Symposium, R. M.
    Ford, Ed., ASTM-STP 833, (1984).

35. C. A. Hippsley et al., "Grain Boundary Behavior of Type stain-
    less steel and Dynamic as Influenced by Radiation and copper,
    present, presented by The Mechanical Properties Symposium
    Effects of Radiation on Materials, June-July, Williamsburg,
    Virginia.

TREND CURVE EXPOSURE PARAMETER DATA

DEVELOPMENT AND TESTING

W. N. McElroy, G. L. Guthrie, R. L. Simons,
E. P. Lippincott and R. Gold

Westinghouse Hanford Company
Hanford Engineering Development Laboratory
Richland, WA 99352

S. L. Anderson

Westinghouse Nuclear Technology Division
Pittsburgh, PA 15230

ABSTRACT

An important aspect of the Light Water Reactor
Pressure Vessel Surveillance Dosimetry Improvement
Program (LWR-PV-SDIP) is the effort to develop and
test trend curve exposure parameter data. Progress
in these trend curve-data correlation analysis activities at HEDL is described. The exposure parameters
of primary interest are those associated with the
production of displaced atoms and helium in different
PV steels, particularly, A302B, A533B, and A508. In
order to determine the effect(s) of helium, the production of helium from boron (n,α), iron (n,α), nickel
(n,α) and low concentrations of impurity elements are
being investigated. Preliminary results of these HEDL
investigations and related studies suggest that both
displaced iron atoms and total helium content need to
be considered in the future development, testing, and
application of revisions of selected ASTM standards
and Reg. Guide 1.99 and for the generation of PWR and
BWR plant specific trend curves.

J. P. Genthon and H. Röttger (eds.), Reactor Dosimetry, 113–135.
© *1985 ECSC, EEC, EAEC, Brussels and Luxembourg*

INTRODUCTION

As discussed in this paper and elsewhere (1-4), a new development in the study and selection of trend curve exposure parameters has been the establishment of embrittlement curves that contain terms to account for the separate and combined effects of fast, intermediate, and thermal neutrons. If the magnitude of the thermal neutron contribution to elevated temperature (~288°C) PV steel embrittlement suggested by the present and previous work is shown to be real, the impact of this work will be quite important for future revisions of the new set of 21 LWR ASTM standards, Reg. Guide 1.99 (5), and licensing and regulatory issues and actions related to the new Nuclear Regulatory Commission's (NRC)· screening criteria requirements associated with pressurized thermal shock and other PV integrity issues (6-8).

The neutron exposure parameters of primary interest are those associated with the production of displaced atoms and helium in different PV steels, particularly A302B, A533B, and A508. In order to determine the effect(s) of helium, the production of helium from boron (n,α), iron (n,α), nickel (n,α) and low concentrations of impurity elements are being investigated. Preliminary results of these HEDL investigations and related studies suggest that both displaced iron atoms and total helium content need to be considered in the future development, testing, and application of revisions of selected ASTM standards and Reg. Guide 1.99 and for the generation of Pressurized Water Reactor (PWR) and Boiling Water Reactor (BWR) plant specific trend curves.

ASTM STANDARDS

ASTM Standard Practice E853, "Analysis and Interpretation of Light-Water Reactor Surveillance Results," covers the methodology (summarized in Annex Al of that standard) to be used in the analysis and interpretation of neutron exposure data obtained from LWR-PV surveillance programs, and, based on the results of that analysis, establishes a formalism to be used to evaluate the present and future condition of the PV and its support structures. This practice relies on, and ties together, the application of several supporting ASTM standard practices, guides, and methods that are in various stages of completion (1,9).

ASTM Standard Practice E185 (9) describes the criteria that should be considered in planning and implementing surveillance test programs and points out precautions that should be taken to ensure that: 1) surveillance capsule exposures can be related to beltline exposures, 2) materials selected for the surveillance program are samples of those materials most likely to limit the operation of the reactor vessel, and 3) the tests yield results useful for the evaluation of radiation effects on the reactor vessel.

The ASTM Standard Practices E853, E482 "Application of Neutron Transport Methods," the new E560 "Extrapolating Reactor Vessel Surveillance Results," and the Standard Guide E900 "Predicting Neutron Radiation Damage to Reactor Vessel Materials" address the neutron transport methodology and extrapolation of physics-dosimetry-metallurgy results in both the design of the surveillance program and the analysis of individual surveillance capsules (9). The neutron field information obtained from transport computations and dosimetry measurements should be used to project damage gradients within the PV wall. Currently, all such projections are based on neutron fluence (E > 1 MeV). However, it is recommended in these standards that supplementary projections based on dpa maps throughout the PV beltline region/surveillance capsule geometry be included in the surveillance report.

With these introductory comments in mind, subsequent sections of this paper address the current status of physics-dosimetry related trend curve exposure parameter data development, testing and applications while other papers being presented at this symposium address the metallurgical aspects of the problem; i.e., the development, testing, and application of power and test reactor trend curves based on fast (E > 1 MeV) and/or dpa exposure parameter terms for both generic and plant specific PWR and BWR power reactor conditions.

TREND CURVE EQUATIONS

Reference 2 presents the most recent results of the joint NRC (Randall) and HEDL (Guthrie) efforts to establish improved Charpy trend curve equations to support the 1984 revision of Reg. Guide 1.99. Randall anticipates that the 1984 revision will use Cu and Ni as independent chemical variables and that the Charpy shift will be a product of a chemistry factor and a fluence factor (7,8). Earlier studies by Guthrie (2) showed that satisfactory C_v weld trend curve results were obtained for equations of the type

$$\Delta T_{NDT}(°F) = \left[x(1) \cdot Cu + x(2) \cdot \sqrt{CuNi} + x(3) \cdot Ni\right] x \left(Dose\right)^N \qquad (1a)$$

$$N = x(4) + x(5) \cdot \log_e \left(Dose\right) \qquad (1b)$$

where $x(1)$ through $x(5)$ are a set of adjustable constants, Cu and Ni are the weight % Cu and % Ni in the steel, and Dose is the neutron exposure term given as the fast (E > 1 MeV) fluence in units of 10^{19} n/cm^2 = 1. A single least-squares procedure was used to adjust all the $x(1)$ through $x(5)$ parameters simultaneously.

Subsequent studies by Guthrie et al. (3) included the investigation of a dose term that included an additional constant $x(6)$ and dose terms of the form

Dose = Fluence (E > 1 MeV) + $x(6) \cdot$ Thermal Fluence   (2a)

and

Dose = Dpa + $x(6) \cdot$ Thermal Fluence.   (2b)

The results of the fitting procedure based on a 31-point PWR weld data base are given in Ref. 3. The results of a parametric study of the application of the derived equations for predicting the Charpy shift for a selected number of surveillance capsules with-drawn from PWR and BWR power plants are given in Ref. 4. The best fits occurred for the cases in which the exposure parameter was a mixture of fast fluence (E > 1 MeV) (or dpa) and thermal fluence. The F tests showed a significant improvement over the case in which only fast fluence (E > 1 MeV) or dpa were used. The values were 5.5 for the addition of a thermal term to the fast fluence and 6.6 for the addition to dpa. An improvement of this amount (or better) occurs at a frequency of ~4% by chance.

TREND CURVE EXPOSURE PARAMETERS

The present study has included the investigation of a larger number of separate and combined exposure parameter terms, as shown in Table 1. The results of the fitting procedure using a 30-point PWR weld data base are given in Tables 1 and 2. Clearly, there is a significant improvement (reduction) in the standard deviation of the fit (Table 1) for Eqs. 7 and 8, which include the effect of low energy thermal neutrons. Similarly, there is an improvement when the dose term, Eq. 9, includes the combined effect of dis-placed iron atoms and total helium production from boron (B_0) and high energy (n,α) reactions in the PV steel.

116

It is noted that in the derivation of the $x(1)$ through $x(6)$ constants for Eq. 9, that the B_0 value was set equal to 1 wt ppm of natural boron for all 30 weld data points since the actual values were not known. This value of 1 wt ppm was selected on the basis of Rockwell International (RI) results for the boron content of a number of the PSF space-compatible compression steel specimens (1). As reported by Oliver and Farrar (10), measured boron contents of 0.65, 0.68, 0.54, 0.43, 0.52, 0.54, and 1.27 wt ppm were found for seven of these steel specimens.

It is of further interest to observe (Table 1) that the existing 30-point PWR weld data base can be correlated equally well using a single thermal (E < 0.4 eV), fast (E > 1 MeV), or fast (E > 6 MeV) exposure term. The use of just the dpa, fast (E > 0.1 MeV), or intermediate (0.4 eV < E < 1 MeV) exposure parameter, however, does not show as good a fit.

The possible effect of helium production from boron and high energy (n,α) reactions in the PV steel and the result of varying the boron concentration (B_0) in Eq. 9 are illustrated by the results presented in Table 3. That is by comparison of the Charpy shift calculated-to-experimental ratio (C/E) versus the wt ppm boron content of the Charpy weld material for a selected number of PWR and BWR surveillance capsules. The predicted percent boron-10 converted to helium, percent contribution of helium to the total dose term (the underlined value), and the wt ppm of natural boron (identified by the boxed values) required to give a C/E ratio within ~5% of unity are tabulated. Generally, boron contents in the range of ~0.1 up to 5 wt ppm will provide good predicted-to-measured values of the Charpy shift.

Not shown in Table 3 are the separate contributions of thermal (T), intermediate (I), and fast (F6) neutrons to the production of helium. With reference to Figure 1, if the surveillance capsule is partially (or fully) embedded in the back face of the thermal shield, with a resulting low T/F ratio, the high energy fast F6 may be equal to or larger than the intermediate energy I helium contribution, depending on the boron content of the PV steel and the surveillance capsule design. For surveillance capsules placed next to or near the thermal shield (or PV wall) or placed at intermediate water positions, the T/F ratio will be higher and the I may be equal to or larger than the F6 helium contribution, again depending on the boron content of the steel and the surveillance capsule design. In most cases, the the combined I and F6 helium contributions to the total dose are small (in the range of 3 up to 15%) compared to that from thermal neutrons and dpa. Within the PV, however, the high energy F6 can be as high or higher than the T+I helium contributions, depending on the boron content of the PV steel.

It is of value to compare the results of this study, Eq. 7, Tables 1 and 2, with results obtained by Alberman (11-13) for A537 steel irradiated at 60°C. Alberman found an experimental relationship between the thermal and fast (equivalent iron fission) fluences that indicated an 0.45% thermal-relative-to-fast neutron contribution to the Charpy shift for a thermal-to-fast ratio of unity, or 4.5% for a ratio of 10. For the A537 steel used in his experiment, the derived Charpy shift equation was

$$\Delta T(°C) = 145 \left[(\phi t)_{Fe} + 0.0045 \cdot (\phi t)_{T} \right]^{1/2} \qquad (3a)$$

where $(\phi t)_{Fe}$ is the equivalent iron fission and $(\phi t)_{T}$ is the thermal neutron fluence in units of 10^{19} n/cm² = 1. The fast, $(\phi t)_{Fe}$, and thermal, $(\phi t)_{T}$, fluence ranges for the irradiated Charpy specimens were from $\sim 3 \times 10^{17}$ to 2×10^{18} and $\sim 9 \times 10^{17}$ to 12×10^{20} n/cm², respectively. For comparing Eq. 3a with Eq. 7, $(\phi t)_{Fe} \approx 4.1$ Fl, so Eq. 3a may be rewritten as

$$\Delta T(°C) \approx 293 [Fl + 0.0011 \cdot T]^{1/2}. \qquad (3b)$$

In this case, a value of the constant $x(6) = 0.0011$ is found for a $\sim 60°C$ irradiation temperature as compared with the Eq. 7 value of 0.41172 for a $\sim 288°C$ irradiation temperature.

In Ref. 12, and based on calculations of dpa in iron, Alberman has estimated that thermal neutrons contribute $\sim 1.5\%$ to the total dpa dose for equal $(\phi t)_{Fe}$ and $(\phi t)_{T}$ fluences. Neglecting thermal neutron temperature differences and using the Fl and T fluence units of this paper, the 1.5% contribution changes to $\sim 0.4\%$. This theoretically based value of 0.4% is a factor of ~ 4 times higher than the experimentally determined value of 0.11% (Eq. 3b) found by Alberman for A537 steel. It follows from these observations that past theoretical estimates of the low-energy thermal neutron contribution to the energy dependent dpa cross section being used and recommended in the ASTM E693 DPA Standard (9) may be significantly too high for the correlation of PV steel power and test reactor irradiation effects data.

If this conclusion is correct, it could help in explaining why a better Std. Dev. of fit (27.58 vs. 28.09) is obtained with a fast (E > 1 MeV) [Fl] rather than a dpa dose term, Table 1; i.e., even for a very high T/F ratio, of say 5, thermal neutron-induced displaced iron atoms would contribute only about 0.6% to the total integrated dpa dose.

The above leads to the conclusion that the current ASTM E693 dpa cross section should not be used to correlate highly thermalized light or heavy water moderate power or test reactor irradiation effects data because it significantly over estimates the low energy thermal neutron dpa contribution.

From the foregoing, it is further concluded that:

1) Any signficant thermal neutron contribution to PV steel embrittlement is, most probably, a result of (n,α) reactions in boron-10 rather than neutron-induced Fe (n,γ) recoil reactions.

2) A knowledge of the actual boron content of PV steels and use of a trend curve model, Eq. 9, that makes use of a dose term, which includes the total production of dpa in iron and helium, could make a significant improvement in reducing the Std. Dev. of the fit for the existing PWR surveillance capsule metallurgical weld data base.

In support of the above conclusions is information a) reported by Barnes (14) on low fast fluence (3×10^{17} to 8×10^{18} n/cm²), low temperature ($\sim 40°C$), test reactor irradiations of stainless steel and nickel-based alloys and b) more recent studies by Alberman et al. (11) and Serpan et al. (15), in which it has been reported that the A302B steel results of Serpan et al., for irradiation temperatures <116°C (240°F), supported a thermal neutron contribution to damage for research reactor test locations with high-thermal-to-fast (E > 1 MeV) neutron ratios (> about 10); but the nonboron-containing A533B steel results of Alberman et al. did not for an irradiation temperature of $\sim 100°C$ (212°F). Alberman et al. did observe, however, a substantial thermal neutron effect at $\sim 100°C$ for steel specimens with boron concentrations up to 5 ppm, irradiated in high-thermal-to-fast-neutron flux ratios. Above the 5-ppm level, the increased boron content appeared to have little further influence on any increases in measured mechanical property.

DAMAGE GRADIENTS

The ORR-PSF experiment (1,16-18) was designed to simulate (at $\sim 288°C$) the surveillance capsule-pressure vessel configuration in power reactors and to test the validity of procedures and data being recommended in selected ASTM standards (1,9). In this section, preliminary results of this very specialized PSF experiment will be used together with the Table 1 set of C_v trend curve model equations to test the applicability of individual and combined exposure parameter terms for projecting damage gradients in PV steel walls.

With reference to Figures 2 and 3, the FERRET-SAND-II experimentally derived PSF simulated PV-wall (288°C) T/F ratio was 4.3 at the notch for the first layer of Charpy specimens at the OT (0.17") surface position and was 1.3 for the second layer at the OT (1.01") position; corresponding values for the 1/4T (1.94" and 2.78") and 1/2T (3.94" and 4.81") positions were 0.53 and 0.24, and 0.16 and 0.14, respectively; the front and back values for the SSC1 and SSC2 simulated surveillance capsules were both 0.65 and 1.03, respectively. These values are consistent with similarily derived values for PWR and BWR surveillance capsules as given in Table 3 and shown in Figure 1.

For the Brown's Ferry BWR, the GE experimentally derived T/F ratios between the barrel and PV wall are in the range of ∿ 4 up to 6 for a specially designed set of in-situ physics-dosimetry measurements reported by Martin (19); measurements were not performed any closer than within ∿4 cm of the shroud and ∿7 cm of the PV wall. For a Westinghouse designed PWR with pads (a segmented thermal shield) and the region between the pads (i.e., no thermal shield), the T/F ratio is calculated to vary from a low value of ∿2 just behind the barrel and just in front of the PV to a peak value around 9, midway between the barrel and the PV.

FERRET-SAND-II exposure parameter values (T, I, FP1, F1, F6 and Dpa) based on ORNL transport calculations and HEDL dosimetry measurements for the PSF experiment were derived for the Charpy notch positions shown in Figure 2 for the six PV steels identified in Figure 3. These values were then used as input to Eqs. 1-6 and 9 to predict the relative Charpy shift, renormalized to unity at the 0.17" OT notch position, that should be experimentally obtained if the energy-dependent neutron reactions associated with an individual or combined [dpa (ASTM E693) + helium] exposure term were totally responsible for the observed damage gradient.

The Hawthorne and Menke reported (17) PSF Experiment measured relative PV-in-wall ΔRT_{NDT} shift gradients for two A508 forgings, two A302B and A533B plates, and two A533B weld materials are shown in Figure 3, again renormalized to unity at the OT 0.17" notch position for each of the six PV steels.

If fast neutrons (E >1 MeV) were predominately responsible for the embrittlement of each of the six PV steels, then all of the Figure 3 damage gradients (slopes) should be the same as that shown for the fast (F1) exposure parameter dose term of Figure 2. A similar statement would apply for each of the other Figure 2 dose terms.

Studies by Stallmann (20) and Guthrie (21) are in progress to derive independent sets of Charpy shift values and the associated uncertainties that need to be assigned to the individual measured points of Figure 3. This work is intended to complement and supplement the results of the study already completed by Hawthorne and Menke (17) and used in Figure 3. Until the results of these new studies are available, little can be said with confidence about which equation (or equations) and which dose term (or terms) should be used to best represent the Figure 3 measured in-wall damage gradients for the six PV steels.

What can be concluded at this time, from Figures 2 and 3, is that:

1) For Reg. Guide 1.99, Rev. 2, the tentative selection by Randall of the equation $[\Delta RT_{NDT}(x) = (\Delta RT_{NDT} \cdot \text{surface}) \cdot \exp(-0.067x)]$ to represent the PV-in-wall gradient is most closely approximated by the Eq. 4, Table 1, derived gradient; this equation makes use of a fluence (E >1 MeV) [F1] exposure parameter term, see Figure 2.

2) None of the Table 1 equations, except perhaps Eq. 3 (based on fluence (E > 0.1 MeV), appear to adequately bound all of the six observed PSF damage gradient curves shown in Figure 3. [This conclusion can not be made more definitive until after error bars are assigned to the individual measured data points of Figure 3 and the through-wall metallurgical results for the PSF space compatible specimens are available (1,22).]

3) The slopes of at least three of the observed damage gradient curves of Figure 3 are steep enough that they would justify the use of the Eq. 9 embrittlement trend curve, Figure 2, that makes use of a combined [dpa (ASTM E693) + helium] exposure parameter dose term. This would apply for some plant specific A302B, A533B, and A508 PV steels with known boron (B_0) contents in the range of ~0.1 up to 5 wt ppm.

Figure 4 provides information on the application of the Eqs. 1, 4 and 7 trend curve models, Table 1, that make use of thermal (T), fast (E >1 MeV) (F1), and combined (F1 + 0.41172·T) exposure parameter dose terms, respectively. The predicated values of the Charpy shift versus exposure parameter dose are shown for these three equations for the Table 3 Maine Yankee (MY) weld material containing 0.36% Cu and 0.78% Ni. The lower dose W263 wall capsule data point (with a T/F ratio of 4.71) and the two higher dose accelerated capsule data points (with T/F ratios of 1.7 and 1.6) correlate well with either the Eq. 4 fast fluence or Eq. 7 fast plus thermal fluence models. The log-log straight line (dashed curve) fit provides the best correlation for the three MY data points, but does not allow or reflect saturation of the trend curve at high fluences. These three MY data points were used in helping to set the upper bound truncated shape for the ΔRT_{NDT} α constant $\cdot (Fl)^{1/2}$ trend curve for Reg. Guide 1.99, Rev. 1 (5).

For these discussions, it is helpful to show, Figure 4, the measured data points for Gundremmingen (23), which when fit with a smooth curve, also do not show a saturation effect. From about 3% up to 68% of the boron-10 present in the PV steels are expected to be converted to helium for these three MY and four Gundremmingen surveillance capsule data points (see Table 3). Assuming 1 wt ppm of natural boron in the steels, this indicates that the helium content of these steels will reach values up to about 0.1 wt ppm. It would appear unlikely that such quantities of helium would not have some effect on the embrittlement of these steels.

Neither the MY or Gundremmingen weld data results, Figure 4, or any of the higher fluence weld data sets (PB2, SO1, REG1, ZION1, and QC1, Table 3), except for PB1, support a saturation effect at high fluences above $\sim 1 \times 10^{19}$ n/cm^2. This may explain why the Eq. 9 model C/E ratio values for Gundremmingen, Table 3, are so low (0.60 to 0.65) for all values of boron up to 20 wt ppm; i.e., the Eq. 9 model requires saturation at fluences above $\sim 1 \times 10^{19}$ n/cm^2. This might also be an explanation for the very low C/E ratio results for SO1 capsule F and QC1 capsule B, rather than assuming higher steel boron contents. Consequently, the existing Reg. Guide 1.99.1 upper bound (truncated) trend curve model shape (or plant specific curves, such as those shown in Figure 4 for MY and Gundremmingen) may have to be used for high fluence embrittlement predictions for PV steel welds, and perhaps, forgings and plates; i.e., for fast (E > 1 MeV) fluence values above $\sim 1 \times 10^{19}$ n/cm^2.

In Ref. 14, Barnes of the UK shows that atomic concentrations as low as 3×10^{-9} of helium in the nickel-based (PE16) steel containing 0.00027% wt boron are sufficient to cause a detectable embrittlement for fast fluence exposures as low as 3×10^{17} n/cm^2, an irradiation temperature of $\sim 40°C$, and for tensile testing at 750°C. In these studies, it was found (for fast neutron irradiations) that hardening and a consequent decrease in ductility were detectable up to temperatures of $\sim 500°C$; whereas, at higher temperatures the effect annealed and the mechanical properties became indistinguishable from those of equivalent unirradiated samples.

Samples of the same steel were irradiated in reactor positions with various fast-to-thermal neutron flux ratios, some with and some without a cadmium shield. These UK experiments showed conclusively that, whereas, the low temperature embrittlement could be correlated with the fast neutron flux, as expected, the high temperature embrittlement correlated with the thermal neutron flux, confirming that atoms displaced by fast neutrons were not producing the effect and concentrating attention on the possible neutron transmutation effects, which would be expected to persist at these high temperatures.

Barnes further concluded that precipitation effects were not the prime cause of the thermal neutron-induced effects. This led him to conclude that boron-10 was the most likely isotope to cause the observed effects because of its high thermal neutron cross-section and in turn led to the conclusion that the effect was associated with helium bubbles. He also states that it was assumed that the gas is in the form of bubbles at grain boundaries, just big enough to grow under the ultimate applied stress; the necessary concentrations are so low, however, that even with less optimum conditions and some gas in the grains, embrittlement would ensue. He also did not completely rule out some effect from the boron (n,α) reaction recoil lithium atoms but pointed out, that much smaller amounts of helium are equally effective in producing damage.

This review of selected PWR and BWR surveillance capsule and the PSF experiment calculated and experimental information on the embrittlement of PV steels and through-wall damage gradients indicates that: 1) the proper assignment of uncertainties to the Charpy shift ΔRT_{NDT} data points is an essential next step, as well as the completion of the analysis of the PSF through-wall space compatable specimen metallurgical results; 2) hardness test results for the broken C_V and space compatable specimens should be obtained and analyzed to supplement and complement the Charpy and tensile results; 3) more complete knowledge of the the boron content and its effect on individual PSF and surveillance capsule Charpy specimens is needed; and 4) improved knowledge of the effects of other variables will be required to more confidently determine which trend curve model or models, such as Eqs. 1 through 9, can be used to best correlate the existing experimental results for the six PSF steels, Figure 3; and any other selected set of PWR and BWR plant specific surveillance capsule results, Table 3.

Finally, PV steel metallurgical and dosimetry specimens are being obtained by NRC from the Gundremmingen vessel wall for the direct study of the PV through-wall embrittlement and toughness. For ~ 10 years of neutron exposure, the estimated Charpy shifts and fast fluence dose for the ~ 5" thick PV wall are shown in Figure 4. Because of the anticipated small in-wall neutron exposures and resultant small ΔRT_{NDT} shifts, it will be important to correlate the Gundremmingen surveillance capsule and through-wall physics dosimetry-metallurgy results with those obtained from the PSF experiment. This should improve the overall value of the Gundremmingen plant specific and PSF data bases for predicting PV through-wall damage gradients. Further, these Gundremmingen results will supplement the data obtained from the PSF experiment in providing in-wall results in the fluence (E > 1 MeV) range of $\sim 2 \times 10^{18}$ to 1×10^{19} n/cm^2 and surveillance capsule results in the range of $\sim 5 \times 10^{18}$ to $\sim 2 \times 10^{20}$ n/cm^2. The PSF data are limited to the range of ~ 1 up to 4×10^{19} n/cm^2.

CONCLUSIONS

The main conclusion of the present investigation is that if the magnitude of the thermal neutron contribution to PV steel embrittlement at 288°C suggested by this and previous studies is shown to be real, the impact of this work will be quite important for future revisions of the new set of 21 LWR ASTM standards, Reg. Guide 1.99, and licensing and regulatory issues and actions related to the new NRC screening criteria requirements associated with pressurized thermal shock.

Based on the Eq. 9 trend curve model (Table 1), the PSF results (Figure 3), PWR and BWR surveillance results (Table 3) and existing surveillance capsule designs (Figure 1), additional conclusions are:

1) There is a significant improvement (reduction) in the Std. Dev. of the fit for weld Charpy shift trend curves that include the effect of low energy thermal neutrons. For the 30 point weld data set, improvements of the amounts observed could occur at a frequency of ∿4% by chance (See Table 1).

2) A knowledge of the actual boron content of PV steels and the use of the trend curve, Eq. 9, that makes use of an exposure parameter dose term that includes the total production of dpa in iron and helium could make significant improvements in lowering the Std. Dev. of the fit for the existing PWR surveillance capsule metallurgical weld data base (See Table 3).

3) Based on the Eq. 9 trend curve model, for both PWR and BWR power plants, up to about 80% of the SS clad/PV steel wall interface and surveillance capsule specimen dose term values can be attributed to helium production in PV steels, depending on the particular surveillance capsule design, Charpy specimen placement, steel boron content, and power plant operating conditions (See Table 3).

4) Existing PWR and BWR surveillance capsule derived embrittlement trend curves [based on the use of just fast fluence (E > 1 MeV) or dpa for the exposure term] cannot be expected to give reliable predictions of the combined fast and thermal neutron contributions to the Charpy shift at the SS clad/PV steel wall interface, 1/4T, 1/2T, 3/4T, or 1T locations (See Figures 2 and 3). [It is noted that the PSF experiment provides physics-dosimetry-metallurgy data for predicting the Charpy shift in PV steels at deep in-wall locations, such as the 1/4T, 1/2T, and 3/4T positions where the T/F ratios are in the very low range of ∿0.14 to ∿0.53. However, even for these very low ratios, helium from both boron and steel high energy (n, α) reactions may still contribute 5 to 30% to the exposure parameter dose term value.]

5) None of the Charpy shift trend curve equations, Table 1, except perhaps Eq. 3 (based on fluence E > 0.1 MeV), appear to properly bound all of the six PV steel observed PSF damage gradient curves (See Figure 3). [Based on the French simulated PV-wall Dompac Experiment (1,13), Alberman concluded that for low temperature (<100°C) irradiations, fast fluence (E > 1 MeV) is too "optimistic" and is not, therefore, a conservative neutron exposure parameter. That, at low temperature, 95% of the measured damage (based on tungsten and graphite damage monitor results) comes from neutrons with energy E > 0.1 MeV. This led him to conclude that the exposure parameter, fluence (E > 0.1 MeV), is perhaps "pessimistic", but has the advantage of being the lower threshold of all (displacement) damage models and thus it takes into account all neutrons which create (displacement) damage.]

6) The plant specific weld data sets used in this study, except for one, do not support a saturation effect at high fluences above $\sim 1 \times 10^{19}$ n/cm^2 (E > 1 MeV). Consequently, the existing Reg. Guide 1.99.1 upper bound (truncated) trend curve model shape (or plant specific curves) may have to be used for high fluence embrittlement predictions for PV steel welds, and perhaps forgings and plates.

7) Any significant thermal neutron contribution to PV steel embrittlement is, most probably, a result of (n,α) reactions in boron-10 rather than by neutron-induced $Fe(n,\gamma)$ recoil reactions.

8) It appears that the current ASTM E693 dpa cross section should not be used to correlate highly thermalized light or heavy water moderated power or test reactor irradiation effects data because it significantly overestimates the low energy thermal neutron dpa contribution.

9) The PV-wall SS clad/PV steel interface surface T/F ratio for PWR and BWR power plants is expected to be in the range of 2 to 6 on the basis of surveillance capsule measurements, Westinghouse transport calculations, GE measurements, and the PSF experiment physics-dosimetry results (See Figure 1).

10) Individual Charpy specimens (with natural boron content of ~ 0.4 up to perhaps 5 wt ppm) in PWR and BWR surveillance capsules will be subject to ₵ neutron exposures with T/F ratios in the range of ~ 0.5 to 5, depending on the surveillance capsule design, its placement and the reactor operating conditions (See Figure 1). The thermal-to-fast ratio (T/F) variation for individual Charpy specimens, therefore, could be an important parameter for the correlation of a set of Charpy specimen results and derived ΔRT_{NDT} values.

11) From this study, that of Grant and Earp (25), and others discussed in Ref. 1, a final conclusion is: the PSF experiment and PWR and BWR surveillance program results clearly show that comparison of the effects of radiation damage on yield strength, hardness, RT_{NDT} and USE will be needed to aid in improving and refining our knowledge of trend curves and PV wall damage gradients. Implicit in this are the current observations that the establishment of separate trend curves for welds, forgings, and plates will give increased understanding and accuracy in projections of the present and future metallurgical condition of PV steels.

RERERENCES

1. W. N. McElroy et al., LWR-PV-SDIP 1983 Annual Report, NUREG/CR-3391, Vol. 3, HEDL-TME 83-23, NRC, Washington, DC (January 1984).

2. G. L. Guthrie, "Charpy Trend Curves Based on 177 PWR Data Points," LWR-PV-SDIP Quarterly Progress Report - April-June 1983, NUREG/CR-3391, Vol. 2, HEDL-TME 83-22, NRC, Washington, DC, p. HEDL-3 (April 1984).

3. G. L. Guthrie et al., "Effects of Thermal Neutrons in Irradiation Embtittlment of PWR Pressure Vessel Plates and Welds," LWR-PV-SDIP Quarterly Progress Report - April-June 1983, NUREG/CR-3391, Vol. 2, HEDL-TME 83-22, NRC, Washington, DC p. HEDL-16 (April 1984).

4. W. N. McElroy et al., "Thermal-Relative-to-Fast-Neutron Contribution to Charpy Shift for PWR and BWR Surveillance Capsule Weld Materials," LWR-PV-SDIP Quarterly Progress Report - April-June 1983, NUREG/CR-3391, Vol. 2, HEDL-TME 83-22, NRC, Washington, DC p. HEDL-22 (April 1984).

5. Regulatory Guide 1.99, Effects of Residual Elements on Predicted Radiation Damage to Reactor Vessel Materials, Rev. 1, NRC, Washington, DC (April 1977).

6. P. N. Randall, "Status of Regulatory Demands in the U.S. on the Application of Pressure Vessel Dosimetry," Proc. of the 4th ASTM-EURATOM Symposium on Reactor Dosimetry, Gaithersburg, MD, March 22-26, 1982, NUREG/CP-0029, NRC, Washington, DC, Vol. 2, pp. 1011-1022 (July 1982).

7. P. N. Randall, "NRC Perspective of Safety and Licensing Issues Regarding Reactor Vessel Steel Embrittlement," from the ANS Special Session on Correlations and Implications of Neutron Irradiation Embrittlement of Pressure Vessel Steels, Detroit, MI, June 12-16, 1983, Trans. Am. Nucl. Soc. 44, p. 220 (1983).

8. W. J. Dircks, Pressurized Thermal Shock (PTS), and Enclosure A, "NRC Staff Evaluation of PTS," SECY-82-465, NRC, Washington, DC (November 1982).

9. 1984 Annual Book of ASTM Standards, Section 12, Volume 12.02, "Nuclear (II), Solar, and Geothermal Energy," American Society for Testing and Materials, Philadelphia, PA (1984).

10. B. M. Oliver and H. Farrar, "Application of Helium Accumulation Fluence Monitors (HAFM) to LWR Surveillance, LWR-PV-SDIP Quarterly Progress Report - April-June 1983, NUREG/ CR-3391, Vol. 2, HEDL-TME 83-22, NRC, Washington, DC, pp. RI-3 - RI-5 (April 1984).

11. A. A. Alberman et al., "Damage Function for the Mechanical Properties of Steels, "Nucl. Technol. 36, p. 336 (1977).

12. A. A. Alberman et al., "Influence des Neutrons Thermiques sur la Fragilisation de l'Acier de Peau d'Etancheite des Reacteurs a Haute Temperature (H.T.R.)," Proc. of the 4th ASTM-EURATOM Symposium on Reactor Dosimetry, Gaithersburg, MD, March 22-26, 1982, NUREG/CP-0029, NRC, Washington, DC, Vol. 2, p. 839 (July 1982).

13. A. A. Alberman et al., DOMPAC Dosimetry Experiment Neutron Simulation of the Pressure Vessel of a Pressurized-Water Reactor, Characterization of Irradiation Damage, CEA-R-5217, Centre d'Etudes Nucleaires de Saclay, France (May 1983).

14. R. S. Barnes, "Embrittlement of Stainless Steels and Nickel-Based Alloys at High Temperature Induced by Neutron Radiation," Nature, 206, p. 1307 (June 1965).

15. C. Z. Serpan, Jr, "Engineering Damage Cross Sections for Neutron Embrittlement of A302B Pressure Vessel Steel," Nucl. Eng. Design 33, pp. 19-29 (1975).

16. F. B. K. Kam et al, "LWR-PV-PDIP: PSF Metallurgical Blind Test Results", Proc. of the NRC 12th Water Reactor Safety Information Meeting, NBS, Washington DC (October 1984).

17. J. R. Hawthorne and B. H. Menke, "Postirradiation Notch Ductility and Tensile Strength Determinations for PSF Simulated Surveillance and Through-Wall Specimen Capsules," NUREG/CR-3295, Vol. 2, MEA-2017, NRC, Washington, DC (April 1984).

18. J. R. Hawthorne, B. H. Menke, and A. L. Hiser, "Notch Ductility and Fracture Toughness Degradation of A302B and A533B Reference Plates from PSF Simulated Surveillance and Through-Wall Irradiation Capsules," NUREG/CR-3295, Vol. 1, MEA-2017, NRC, Washington, DC (April 1984).

19. G. C. Martin, "Browns Ferry Unit 3 In-Vessel Neutron Spectral Analysis", NEDO-24793 (August 1980).

20. F. W. Stallmann, "ORNL Evaluation of the ORR-PSF Metallurgical Experiment and "Blind Test," Proc. of the 5th ASTM-Euratom Symposium on Reactor Dosimetry, GKSS Geesthacht, Federal Republic of Germany, September 24-28, 1984.

21. G. L. Guthrie et al, "HEDL Analysis of the PSF Experiment," Proc. of the 5th ASTM-EURATOM Symposium on Reactor Dosimetry, GKSS Geesthacht, Federal Republic of Germany, September 24-28, 1984.

22. G. L. Guthrie, K. Carlson, and G. R. Odette, "Embrittlement of Compression Specimens Irradiated in the SSC-1 and SSC-2 Capsules of the PSF Experiment," LWR-PV-SDIP Semiannual Progress Report - October 1983-March 1984, NUREG/CR-3746, Vol. 1, HEDL-TME 84-20, NRC, Washington, DC, p. HEDL-20 (September 1984).

23. Von N. Eickelpasch and R. Seepolt, "Experimentelle der Neutronendosis des KRB-Drukgefasses und deren Betriebliche Bedeutung", Atomkernenergie (ATKE) Bd. 29, Lfg. 2, p. 149 (1977).

24. S. E. Yanichko et al., Analysis of the Third Capsule from the Commonwealth Edison Company, Quad-Cities Unit 1 Nuclear Plant Reactor Vessel Radiation Surveillance Program, WCAP-9920, Westinghouse Electric Corp., Pittsburgh, PA (August 1981).

25. S. P. Grant and S. L. Earp, "Methods For Extending Life of a PWR Reactor Vessel After Long-Term Exposure to Fast Neutron Radiation", Proc. of the 12th ASTM Symposium on Effects of Radiation on Materials, ASTM STP 870, Williamsburg, VA (June 1984).

TABLE 1

RESULTS OF EQUATIONS 1a AND 1b REGRESSION ANALYSIS USING DIFFERENT DOSE TERMS AND A 30-WELD POINT PWR DATA SET

Eq. No.	Dose Term*	Std Dev of Fit (°F)
1	Thermal Neutrons (E < 0.4 eV) [T]	27.50
2	Intermediate Neutrons (0.4 eV < E < 1.0 MeV) [I]	29.95
3	Fast Neutrons (E > 0.1 MeV) [FP1]	29.86
4	Fast Neutrons (E > 1.0 MeV) [F1]	27.58
5	Fast Neutrons (E > 6.0 MeV) [F6]	27.58
6	Dpa/0.016	28.09
7	F1 + 0.41172 · T	25.41
8	Dpa/0.016 + 0.46553 · T	25.45
9	Dpa/0.016 + 21.375 [B_0·(1-exp-(0.02457·T + 0.000256·I) + 0.1321·F6]	25.44

*The "Dose Term" input fluence values (T, I, FP1, F1 and F6) are in units of 10^{19} n/cm² = 1, Dpa is in units of displaced iron atoms, and the boron content (B_0) is given as weight ppm of natural boron. In the derivation of the constants for Equation 9, the B_0 values for all 30 weld data points were set equal to unity since measured values of B_0 were not available.

TABLE 2

DERIVED VALUES OF CONSTANTS X1(J) THROUGH X6(J) FOR EQUATIONS 1a AND 1b

Eq. No. J	Values of Constants					
	X1(J)	X2(J)	X3(J)	X4(J)	X5(J)	X6(J)
1	788.25	-685.04	369.82	0.27584	-0.028339	0.00000
2	335.33	-140.35	136.51	0.25636	-0.0062751	0.00000
3	379.18	-164.96	161.17	0.27737	-0.016126	0.00000
4	569.39	-313.17	262.02	0.27351	-0.045065	0.00000
5	877.59	-651.34	431.76	-0.18681	-0.074362	0.00000
6	533.78	-286.81	247.36	0.26248	-0.03504	0.00000
7	574.84	-401.09	272.12	0.32026	-0.043383	0.41172
8	566.11	-391.60	264.61	0.31419	-0.038275	0.46553
9	547.91	-381.40	256.58	0.32870	-0.039688	21.375

TABLE 3

CHARPY SHIFT PREDICTIONS BASED ON COMBINED DPA AND HELIUM PRODUCTION EXPOSURE TERMS

Type	Power Plant*	Surveillance Capsule	CS** (°F)	Cu (%)	Ni (%)	dpa	E>1.0 Mev	Ratio T/F***	% Boron-10 Converted to Helium	% Contribution of Helium to Total Dose	0.1	0.3	0.5	1.0	1.5	2.0	3.0	5.0	10	20
PWR	HBR2	V	200	0.34	0.65	11.5	0.724	1.24	2.27	44, 53(b)				0.94	0.95					
PWR	TP3	T	162	0.31	0.57	10.9	0.701	0.73	1.33	33				0.97						
PWR	TP4	T	225	0.30	0.60	13.0	0.754	1.11	2.17	40, 25, 85				0.76				0.93	1.09	
PWR	FC	W225	238	0.35	0.60	8.79	0.583	5.31	7.39	75, 82				0.91	0.98					
PWR	NY	1	270	0.36	0.78	28.5	1.79	1.68	7.33	49			0.96	1.03						
PWR	NY	2	345	0.36	0.78	121.0	7.73	1.55	26.24	46				1.05						
PWR	NY	W263	222	0.36	0.78	8.4	0.567	4.71	6.41	50, 73				1.08						
PWR	OC1	E	80	0.32	0.58	2.08	0.150	1.75	0.65	34, 42, 56		1.06	1.12	1.26						
PWR	OC2	C	40	0.30	0.48	1.48	0.101	1.53	0.39	22, 52	1.47			1.87(c)						
PWR	TMI1	E	117	0.34	0.71	1.51	0.109	1.75	0.47	56, 64, 70				0.85	0.93					
PWR	PB1	S	165	0.21	0.57	14.6	0.845	1.42	3.07	45, 60	1.06		1.11	0.91						
PWR	PB1	R	165	0.21	0.57	43.2	2.29	1.24	7.20	11, 27, 40				1.16						
PWR	PB2	V	165	0.25	0.59	12.1	0.728	1.50	2.77	46, 55	1.02			0.95	1.01					
PWR	PB2	T	145	0.25	0.59	15.7	0.940	1.57	3.71	14, 47				1.18						
PWR	PB2	R	230	0.25	0.59	46.0	2.52	1.87	11.39	48, 57				0.96	1.00					
PWR	CY (HAMA)	A	95	0.22	0.05	4.82	0.316	0.80	0.66	36, 44, 51, 60, 71, 83, 90				0.75	0.80	0.83	0.90			
PWR	SO1	A	80	0.19	0.08	48.6	2.87	0.72	5.34	8, 30, 26, 39, 60, 74, 85	1.36			1.44(c)		0.92				
PWR	SO1	F	145	0.19	0.08	95.5	5.73	0.52	7.89	45, 78, 31, 44				0.89						
PWR	SUR1	T	165	0.25	0.68	10.2	0.286	1.25	0.92			0.96		0.73						
PWR	SUR2	X	95	0.19	0.56	4.73	0.303	1.20	0.93				0.96	1.04						
PWR	P11	V	25	0.13	0.17	10.2	0.603	1.53	2.35	44, 79, 88, 94				2.46(c)						
PWR	P12	V	60	0.08	0.07	11.7	0.675	1.44	2.49	44, 55				0.62(c)						
PWR	REG1	R	175	0.23	0.56	21.5	1.17	1.58	4.64	44, 55		0.98		0.98						
PWR	REG1	V	144	0.23	0.56	10.2	0.593	2.31	3.42	50				1.03						
PWR	KEWA	V	175	0.20	0.77	11.4	0.641	1.92	3.09	15, 43	1.00			1.00	1.01					
PWR	COOK1	T	85	0.27	0.74	4.45	0.271	1.20	0.84	20, 39	1.30			1.52(c)						
PWR	IP3	T	143	0.15	1.02	5.20	0.323	0.97	0.81	13, 40				1.09						
PWR	ZION1	T	105	0.35	0.57	4.88	0.304	1.04	0.62	25, 36	1.08	0.59		1.26						
PWR	ZION1	U	188	0.35	0.57	16.6	1.01	0.88	2.32	61				1.04						
PWR	ZION2	U	128	0.28	0.55	4.46	0.280	1.36	0.97	46, 55, 61			0.96	0.89						
PWR	PAL	A240	350	0.24	0.95	97.2	6.06	1.20	17.05	41, 56				0.96						
BWR	GUND	A	72	0.18	0.21	8.80(f)	0.55	2.00	2.77	29, 39, 54	1.05(d)	1.05		1.15				0.66(e)		
BWR	GUND	B	86	0.18	0.21	17.6(f)	1.10	2.00	5.46	29, 38, 54	1.06(d)	1.09		1.17						
BWR	GUND	C	115	0.18	0.21	38(f)	2.37	2.00	14.19	57, 37, 53	1.09(d)	1.03		1.09		0.66(e)				
BWR	GUND	D	234	0.18	0.21	360(f)	22.5	2.00	68.25	77, 93				0.65(e)						
BWR	QC1	A	265	0.31	0.65	37.9(f)	2.37	5.00	25.59	56, 71, 70, 82, 92, 96	1.20(d) 1.18	1.26		1.05		0.87	0.73			0.60(e)
BWR	QC1	B	175	0.17	0.28	57.0(f)	3.56	5.00	35.85	28, 48, 59, 73	1.20(d) 1.20	1.28		0.83						0.92
BWR	QC1	C	195	0.31	0.65	360(f)	0.720	5.00	8.59	81, 57, 58, 73				1.42						
BWR	QC1	D	80	0.17	0.28	14.2(f)	0.890	5.00	10.51	28, 47, 58, 73				1.43						

TABLE 3 FOOTNOTES

*Power plant identification:

CY	Connecticut Yankee (HANA)	PI1	Praire Island 1
COOK1	D. C. Cook 1	PI2	Praire Island 2
GUND	Gundremmingen	QC1	Quad Cities 1
FC	Fort Calhoun	REG1	R. E. Ginna 1
HBR2	H. B. Robinson 2	SO1	San Onofre 1
IP3	Indian Point 3	SUR1	Surry 1
KEWA	Kewaunee	SUR2	Surry 2
MY	Maine Yankee	TMI1	Three Mile Island 1
OC1	Oconee 1	TP3	Turkey Point 3
OC2	Oconee 2	TP4	Turkey Point 4
PAL	Palisades	ZION1	Zion 1
PB1	Point Beach 1	ZION2	Zion 2
PB2	Point Beach 2		

**CS = Measured Charpy shift; from Ref. 2 for PWRs and Refs. 23 and 24 for BWRs.

***Thermal ($E < 0.4$ eV) fluence per fast ($E > 1$ MeV) fluence (T/F) ratio.

(a) Fast fluence ($E > 1.0$ MeV) is given in units of 10^{19} n/cm² = 1 and dpa in units of millidpa.

(b) The first value (e.g., 44%) corresponds to the C/E ratio of 0.94 and the second value (e.g., 53%) corresponds to the value of 0.99. The underlined value corresponds to a selected boxed C/E ratio that is within ∿5% of unity.

(c) The measured Charpy shift value is probably in error because the measured shift is not very large; changing the helium production (boron content) is, therefore, an illustrative rather than a realistic step.

(d) It is likely that the true T/F ratios are different than 2.0 and 5.0 for Capsules A, B, C, and D for Gundremmingen and QC1, respectively; these results, therefore, can only be considered as illustrative for BWRs; measured thermal fluence data are not available.

(e) Since the slope of the Charpy shift Equation (J = 9), Table 1 is ∿0 at a dpa of 0.36 [or fast ($E > 1.0$ MeV) fluence of 22.5×10^{19} n/cm²], further changes in the total dpa or helium production can have little or no effect on the shift, see Figure 4.

(f) Value is based on conversion factor of 0.016 dpa/unit fluence $E > 1.0$ MeV.

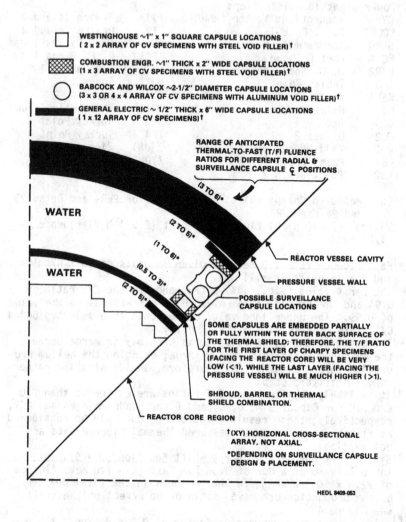

FIGURE 1. PWR and BWR Anticipated ₵ Surveillance Capsule Thermal/Fast Neutron Ratios.

132

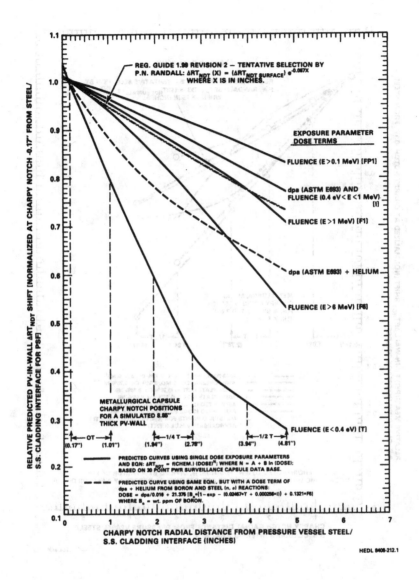

FIGURE 2. PSF Experiment Predicted PV-in-Wall ΔRT_{NDT} Shift Gradients Using Different Exposure Parameters and Derived Trend Curves Based on a 30-Point PWR Surveillance Capsule Weld Data Base.

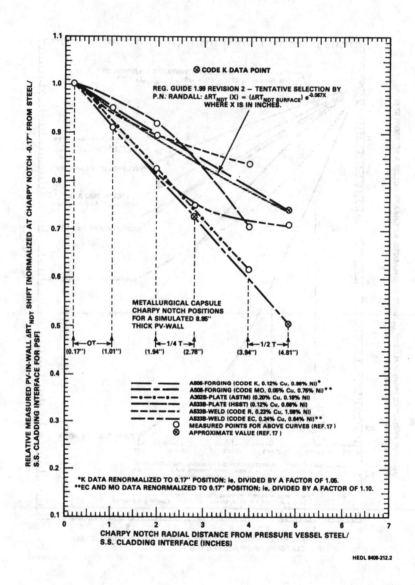

FIGURE 3. PSF Experiment Measured PV-in-Wall ΔRT_{NDT} Shift Gradients for Two Forgings, Two Plates, and Two Weld Materials.

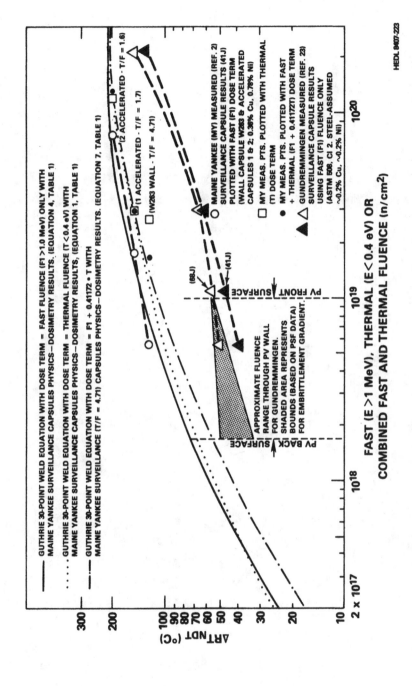

FIGURE 4. Predicted and Measured ΔRT_{NDT} Shifts Based on Fast, Thermal, or Combined Fast and Thermal Fluence Exposure Terms for Maine Yankee and Gundremmingen.

EFFECT OF THERMAL POWER AVERAGING METHOD ON THE

DETERMINATION OF NEUTRON FLUENCE FOR LWR-PV SURVEILLANCE

E. B. Norris

Southwest Research Institute

San Antonio, Texas

ABSTRACT

Irradiation in a light water reactor involves varying neutron intensity levels. The calculation of neutron flux density is based on the assumption that the neutron intensity is proportional to the reactor thermal power. Ideally, the total irradiation time should be broken down into periods of uniform thermal power levels; however, analysts use approximations such as average power during operating periods between shutdowns, or monthly average power.

This study compares the values of A/A_S obtained with several commonly-used averaging methods, such as those mentioned above, with that obtained using a daily thermal power generation history. All calculations will be made using actual light water reactor operating histories for several of the common dosimeter materials.

INTRODUCTION

The allowable loadings on a light water reactor (LWR) pressure vessel operating in the United States are determined by applying the rules in Appendix G, "Fracture Toughness Requirements", of 10 CFR Part 50 (1). The pressure temperature (P-T) limits for hydrotest, normal and upset operation depend on the reference stress intensity (K_{IR}) curve which is indexed to the reference nil ductility temperature (RT_{NDT}). The K_{IR} curve is presented in Appendix G, "Protection Against Non-ductile Failure", of Section III of the ASME Code (2). Further, the materials in the beltline region of the LWR pressure vessel must be monitored for radiation-induced changes in RT_{NDT} per the requirements of Appendix H, "Reactor Vessel Material Surveillance Program Requirements", of 10 CFR Part 50.

Appendix H requires that vessel material surveillance capsules, containing tensile and Charpy V-notch specimens representative of the LWR pressure vessel beltline materials, be placed in the vessel for periodic removal, testing and evaluation. These capsules also contain various neutron dosimeters for the determination of the capsule exposure conditions. The reactor vessel surveillance results are extrapolated to locations of interest along and within the pressure vessel walls to provide the bases for determining the allowable P-T limits for selected periods of operation. Accurate and validated measurement and analysis procedures are required to minimize the uncertainties in the predicted properties of the pressure vessel steel.

The US Nuclear Regulatory Commission (NRC) Light Water Reactor Pressure Vessel Surveillance Dosimetry Improvement Program (LWR-PV-SDIP), operating for nearly a decade in cooperation with other research programs worldwide, provides the major thrust for improving the prediction of the metallurgical condition of LWR pressure vessel surveillance programs by conducting a series of analytical and experimental studies to establish and certify the precision and accuracy of the recommended methods.

One step in the analysis of neutron dosimeters, which is not specifically addressed by LWR-PV-SDIP, is the conversion of the isotopic activity at the end of the irradiation, A, to the saturated activity, A_s. A_s corresponds to the steady-state condition when the rate of production is equal to the rate of loss by decay and transmutation. For irradiations at a constant flux density, A and A_s are related as follows:

$$A/A_s = (1-\exp-\lambda't_i) (\exp-\lambda't_d) \tag{1}$$

where λ = effective decay constant,
    t_i = irradiation time, and
    t_d = decay time following irradiation.

LWR pressure vessel surveillance capsules are exposed during reactor operations at various power levels, including zero-power periods during forced and planned outages. ASTM E261, "Standard Method for Determining Neutron Flux, Fluence, and Spectra by Radioactivation Techniques" (3), directs that the analysis of surveillance capsule dosimetry data be accomplished by dividing the irradiation into continuous series of periods during which the energy-dependent neutron flux density is essentially constant. However, the exact method of dividing the irradiation into periods is left to the analyst. This paper compares the values of A/A_s produced by several schemes for describing the irradiation history.

REACTOR POWER HISTORY DESCRIPTIONS

One method commonly used in the analysis of power reactor surveillance dosimetry data is to divide the operating history into outages of at least one day's duration and operating periods extending between each of these outages (4). Another method divides the reactor power history into monthly operating periods (5,6). In both of these methods, the fraction of full power during each period is taken as the quotient of the average and full power thermal power generation rates.

A method which approaches the ideal power history description (e.g. constant power level during period) is to utilize daily operating periods. On the other hand, using an average power level computed over a complete fuel cycle should be completely inadequate.

A commercial nuclear reactor thermal power generation history covering nearly a ten-year period has been described by each of the above four methods. The between-outages, the monthly and the daily average thermal power generation periods for fuel cycle 1 produce strikingly different power histories, see Figure 1.

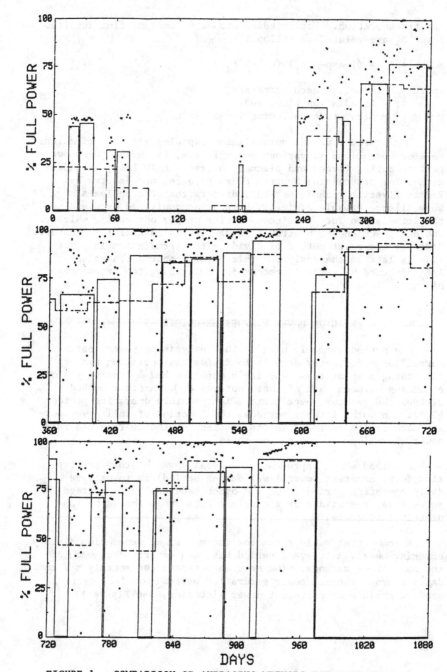

FIGURE 1. COMPARISON OF AVERAGING METHODS FOR FUEL CYCLE 1

CALCULATION OF A/A$_s$

Equation (1) applies to a single-period irradiation. For multiple-period irradiations, the full-power saturated activity can be approximated by:

$$A/A_s = \Sigma P_m \ (1-\exp-\lambda'T_m) \ (\exp-\lambda't_m) \tag{2}$$

where P_m = fraction of full power during period m,
$\quad T_m$ = duration of irradiation period m, and
$\quad t_m$ = decay time following irradiation period m,
if it is assumed that the neutron flux density at the detector location is proportional to the reactor thermal power and that the neutron spectrum is constant at all power levels.

Values of A/A$_s$ were calculated for Fe-59, Co-58, Sc-46, Mn-54, Co-60, and Cs-137 using each of the four power history description schemes applied to the first fuel cycle, the fifth fuel cycle and all five fuel cycles. The results are given in Table I.

The between-outages and the monthly average-power period descriptions produce nearly identical values of A/A$_s$ for all cases examined. Both of these data sets agree with the daily-period values for the longer-lived products Co-60 and Cs-137; the maximum disagreement (½%) occurs for Fe-59, the shortest half-life isotope examined.

As expected, using a complete fuel cycle as an operating period produces unsatisfactory results for all cases examined. However, the error is greater for the products having longer half lives.

CONCLUSIONS

The two power-averaging methods commonly encountered in power reactor surveillance reports (4-6) produce values of A/A$_s$ which are in very good agreement with those calculated by the daily operating period method when applied to a commercial reactor operating history. In general, the monthly average-power period method was slightly superior to the between-outages method, but only for isotopic half lives of about one year or less. For the power histories studied, and excluding the completely inadequate fuel cycle average-power method, the maximum discrepancy between the daily operating period method and the two power-averaging methods was 1/2% for Fe-59.

TABLE I

COMPARISON OF A/A$_{SAT}$ VALUES

Reaction Product	Half Life (days)	Calc. Method	A/A$_{SAT}$		
			1st Fuel Cycle	5 Fuel Cycles	5th Fuel Cycle
Fe-59	44.6	Day	0.827	0.801	0.801
		Period	0.823	0.798	0.798
		Month	0.824	0.798	0.798
		Cycle	0.540	0.800	0.800
Co-58	70.85	Day	0.791	0.817	0.816
		Period	0.789	0.815	0.814
		Month	0.790	0.816	0.815
		Cycle	0.537	0.784	0.784
Sc-46	83.85	Day	0.778	0.814	0.812
		Period	0.776	0.813	0.810
		Month	0.777	0.813	0.811
		Cycle	0.533	0.771	0.769
Mn-54	312.5	Day	0.576	0.690	0.538
		Period	0.575	0.689	0.537
		Month	0.575	0.690	0.538
		Cycle	0.369	0.589	0.463
Co-60	1925.	Day	0.162	0.418	0.129
		Period	0.162	0.418	0.129
		Month	0.162	0.418	0.129
		Cycle	0.092	0.315	0.104
Cs-137	11,021.	Day	0.031	0.109	0.024
		Period	0.031	0.109	0.024
		Month	0.031	0.109	0.024
		Cycle	0.017	0.079	0.019

REFERENCES

1. Title 10, Code of Federal Regulations, Part 50, "Licensing of Production and Utilization Facilities," 1983 Revision.

2. ASME Boiler and Pressure Vessel Code, Section III, "Nuclear Power Plant Components," 1980 Edition through the Summer 1982 Addenda.

3. ASTM E 261, "Standard Method for Determining Neutron Flux, Fluence, and Spectra by Radioactivation Techniques," ASTM Annual Book of Standards, Vol. 12.02, 1983.

4. E. B. Norris, "Reactor Vessel Material Surveillance Program for Zion Unit No. 2; Analysis of Capsule T," SwRI Report 06-6901, July 6, 1983.

5. S. E. Yanichko, et al, "Analysis of Capsule U from the Commonwealth Edison Company Zion Nuclear Plant Unit 1 Reactor Vessel Radiation Surveillance Program (WCAP-9890)," EPRI Research Project 1021-3 Topical Report, March 1981.

6. J. S. Perrin, et al, "Zion Nuclear Plant Reactor Pressure Vessel Program: Unit No. 1 Capsule T, and Unit No. 2 Capsule U," BCL-585-4, March 25, 1978.

ACKNOWLEDGEMENT

The author wishes to express his appreciation to Janet P. Buckingham for her indispensible help in performing the computations of A/A_s.

DOSIMETRY MEASUREMENTS AND CALCULATION OF FAST NEUTRON FLUX

IN THE REACTOR CAVITY OF A 3-LOOP PRESSURIZED WATER REACTOR

D. Rombouts and M.L. Perez-Griffo

Westinghouse Nuclear International

73 rue de Stalle, 1180 Brussels, Belgium

ABSTRACT

Westinghouse has developed a dosimetry measurement
program in the reactor cavity of an European commercial
3-loop Pressurized Water Reactor (PWR).
The dosimetry material which includes Fe, bare and
Cd-shielded Al-Co wires, was installed during a refueling
shutdown. The dosimeter arrangement was selected to obtain
information on both the axial and azimuthal flux gradients
within the reactor cavity.
Radial and axial assemblywise power distributions in the
core were obtained from monthly flux mappings performed
with the in-core fission detectors while out-of-core water
temperature, was obtained from the cold leg Resistance
Temperature Detector (RTD) measurement records.
The dosimeters were removed at the next refueling shutdown
and fast dosimetry results were compared with
two-dimensional discrete ordinates transport calculations
employing P_3 neutron cross sections generated from the
ENDF/B-IV data. Axial peaks of Fe-54 (n,p) Mn-54 reaction
rates were predicted within 15 percent of the measured
values.

INTRODUCTION

Assessment of Pressure Vessel (PV) integrity of Pressurized
Water Reactors (PWR) requires on one hand the knowledge of the
neutron induced mechanical property changes versus the neutron
exposure and on the other hand the prediction of the projected

neutron exposure at the critical locations through the PV. As part of the current PV surveillance programs, the correlation of the material fracture toughness behavior versus neutron exposure is based upon the mechanical testing of material samples submitted to accelerated irradiation and on the associated dosimetry measurement analyses. The determination of the neutron exposure through the various courses and welds of the PV relies on neutron transport physics calculations.

The evaluation of the uncertainty associated with neutron transport calculations is very complex but nevertheless very important to the design and safety of PWR's. There are many sources of uncertainty which can be classified into four categories : methodology, physics, manufacturing and operating.

Some uncertainties are related to the method approximation such as the transport approximation, the group structure, the angular quadrature and the reactor modeling. Uncertainties lie also in the knowledge of physics parameters such as the neutron cross sections, the fission spectra, the energy released per fission, and the number of neutrons emitted per fission. Manufacturing uncertainties are due to the mechanical tolerances of the reactor internals and the PV. Last, operating uncertainties are those which pertain to the specific operating conditions of the reactor. This category involves the uncertainties in the core power distributions, the power history, the density of in-core and downcomer water and, the geometrical uncertainty due to the thermal expansion of the fuel assemblies, the reactor internals and the PV.

The dosimeters located in the surveillance capsules provide information against which the analytical tools predicting the neutron fluxes may be evaluated. However, the value of the capsule dosimetry results for this purpose is limited due to a number of considerations. First the capsules are usually located in regions with large neutron flux azimuthal gradients which are difficult to predict. The dosimetry is also liable to significant flux perturbations caused by the capsule itself. More importantly, the capsules are located closer to the core than the PV so that a validation of the analytical calculations at the capsule locations would not prevent the neutron flux predictions from significantly deviating at deeper penetration through the PV wall.

In order to understand the uncertainties associated with the neutron transport calculations and contribute to their validation, various benchmark neutron field experiments have been set up during the last years in clean mock up facilities as well as in commercial operating reactors (1). This report describes a dosimetry program developed by Westinghouse in the reactor cavity of an European commercial 3-loop PWR and compares the fast flux dosimetry results at the core elevation with two-dimensional neutron transport calculations.

DOSIMETRY MEASUREMENT PROGRAM

The objective of this measurement program was to obtain axial and azimuthal gradients of fast, intermediate and thermal fluxes within the reactor cavity of a 3-loop PWR. This program is intended to ultimately assess the capability of the analytical tools in predicting the radial neutron penetration through the reactor internals and the PV as well as the subsequent axial neutron streaming within the reactor cavity. Such a program will contribute to improve the PV irradiation prediction by fast neutron dosimetry in the vicinity of the vessel beltline region. However, other disciplines like nuclear heat generation rate calculations, shielding and health physics are also believed to get some benefits through such a program provided appropriate efforts are devoted to the interpretation of the measurements results.

The dosimetry material consisted in Fe wires and, bare and Cd-shielded Al-Co one percent wires of a diameter of 0.5 mm. This material was installed in the reactor cavity during a refueling shutdown and removed at the next refueling after 300 Equivalent Full Power Days (EFPD) of reactor operation.

Figure 1 shows a cross-sectional view of the core, the PV and the reactor cavity. The wires were hooked at the two azimuthal locations (0 and 45 degrees) where the maximum and minimum in the azimuthal flux variation are expected to occur.

At both azimuthal locations, three sections of dosimeters were installed covering the axial range level from the reactor vessel flange down to the sump area below the reactor vessel until a level of approximately 2.5 meters above the reactor cavity bottom as depicted in Figure 2.

The first section of dosimeters was attached to a support of the reactor cavity seal ring hanging down to the level of the reactor vessel support ring. The second section of dosimeters was hanging from a support fitting the ring girder and was weighted on the reactor cavity bottom to prohibit significant displacement forced by the reactor cavity ventilation flow. A third section of dosimeters was placed in the circular holes which are used for loading the ex-core detectors from the bottom of the refueling pit to their operating position. The purpose of this third section is to provide a better understanding of the neutron streaming in the upper part of the reactor cavity and particularly determining the impact of the large circular vent area that is designed in the biological shield around the reactor vessel nozzles.

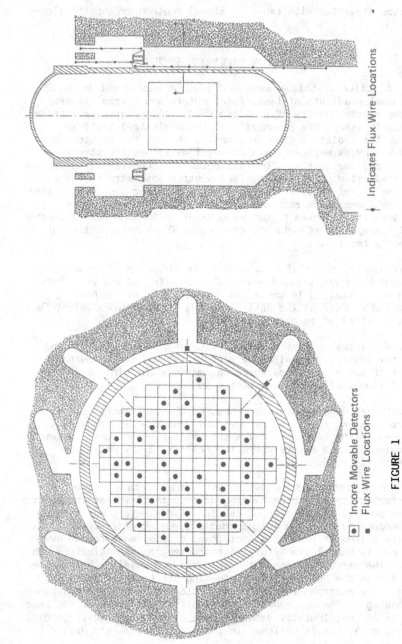

☑ Incore Movable Detectors
■ Flux Wire Locations

FIGURE 1

Cross Sectional View of the Core,
Pressure Vessel and Reactor Cavity

⊕ Indicates Flux Wire Locations

FIGURE 2

Axial View of the Reactor Cavity

PLANT OPERATING PARAMETERS

The plant operating parameters which most affect the results of these experiments are the reactor power history, the core power distributions and the temperature of the water flowing between the core and the PV.

The radial and axial assemblywise power distributions were obtained from monthly flux mappings performed with the in-core instrumentation system. This system consists of five miniature fission chambers which are moved by separate drive units into any of the retractable thimbles which penetrate the bottom of the PV and extend up to the top of the core into the central guide tube of the fifty fuel assemblies shown in Figure 1.

The continuous voltage output of the detectors is digitized and recorded for processing by the program INCORE which converts them into three-dimensional full-core assemblywise power distributions. The power in non-instrumented assemblies is derived from the measured fission reaction rates of the detectors in instrumented neighboring assemblies using the ratio of predicted power in the considered assembly to predicted reaction rates in the surrounding detectors. The power and reaction rate predictions needed by INCORE were calculated by a modified version of the two-group diffusion and depletion code TURTLE (2).

The temperature of the water flowing between the core and the PV is dependent upon the actual primary coolant flow. The coolant temperatures at the vessel inlets and outlets are measured during plant operation by Resistance Temperature Detectors (RTD) located in the cold and hot legs of the primary system. The recorded values were averaged over the plant cycle to get the in-core and out-of-core water temperatures to be used in the calculations.

The daily core energy output is determined from integration of calorimetric balances across the secondary side of the steam generators using measurements of the feedwater temperature, feedwater flow and steam pressure.

CALCULATIONAL METHODOLOGY

In the reactor cavity the neutron field is essentially dictated by axial streaming resulting from multiple scattering on the PV outer surface and the primary concrete shield inner surface. In the vicinity of the core centerplane elevation however, considering the height of the active core (3.66 m), the neutron axial leakages are negligible and the neutron fluxes in the reactor cavity are mainly controlled by the radial leakages from the core and subsequent radial transport through the reactor internals and the PV.

Therefore for the purpose of predicting axial peak reaction rate in the reactor cavity, radial two-dimensional transport calculations without any axial leakage term are expected to be adequate.

For this analysis, the Westinghouse version of the two-dimensional multigroup discrete ordinates transport code DOT-3 (3) was used. The calculations were carried out in an (R, ϑ) one octant model with one-eight core symmetry boundary conditions and used a P_3 cross section matrix expansion, an S_8 symmetric angular quadrature and a 13 energy group structure in the fast range above 0.821 MeV as shown in Table 1.

The program ETOG (4) was used to process the ENDF/B-IV library and generate P_3 microscopic cross section matrices for all the elements within the reactor geometry. Fuel cell weighted cross sections for the core region were obtained by the one-dimensional transport program ANISN. The fuel isotopic inventories were calculated by the point model cell- homogenization, neutron spectrum isotopic depletion program ARK which has evolved from the code LEOPARD (5).

A single equivalent rodwise power distribution for use as a fixed source in DOT was devised with the objective of meaningful comparison between calculated and measured Fe-54 reaction rates. To this end, the monthly assembly power distributions derived from the in-core measurements were time-weighted to account for the reactor power history and the radioactive decay constant of Mn-54. On these assembly powers in the peripheral assemblies, were surimposed rodwise power gradients as calculated by the two-group diffusion program TURTLE (2).

RESULTS AND CONCLUSIONS

The comparison between calculation and measurements is performed in terms of Fe-54 (n,p) Mn-54 reaction rates at nominal reactor power. The Fe-54 (n,p) Mn-54 multigroup dosimetry cross sections used for the calculations of reaction rate are given in Table 1. They were generated by weighting the ENDF/B-IV dosimetry cross section in the ENDF/B-IV U-235 fission spectrum.

The calculated radial distributions of Fe-54 (n,p) Mn-54 reaction rate at nominal power, both at 0 and 45 degrees through the PV-concrete gap at the elevation of the axial peak are plotted in Figure 3. For comparison, the measured axial peak reaction rates are depicted by boxes which sides represent the measurement and the radial location uncertainties ($\pm 2 \sigma$).

TABLE 1

Fast Energy Group Structure and Fe-54(n,p) Mn-54
Dosimetry cross sections

Group Nr.	Lower Energy (Mev)	σ(E)
1	14.19	2.716 × 10⁻¹
2	11.62	4.627 × 10⁻¹
3	10.0	5.430 × 10⁻¹
4	7.79	5.852 × 10⁻¹
5	6.07	5.602 × 10⁻¹
6	4.72	4.421 × 10⁻¹
7	3.68	2.860 × 10⁻¹
8	2.87	1.729 × 10⁻¹
9	2.23	7.135 × 10⁻²
10	1.74	1.369 × 10⁻²
11	1.35	1.209 × 10⁻³
12	1.05	9.683 × 10⁻⁶
13	0.821	0.0

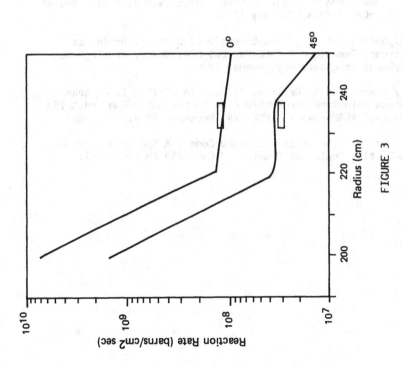

FIGURE 3

Comparison between Calculated and Measured
Fe-54(n,p) Mn-54 - Reaction Rates at the
Peak Location within the Reactor Cavity

151

The calculated axial peaks of reaction rates are obtained by multiplying the DOT results by the peak-to-average ratio of the axial power distribution measured by the in-core detectors in the peripheral fuel assembly which is the closest to the dosimeters. The measured core axial power distribution and the measured axial reaction rate distribution both peak at identical axial elevation.

Calculated versus measured Fe-54 (n,p) Mn-54 reaction rate ratio is within 0.9 at 0 degree and 1.15 at 45 degrees.

An evaluation of the uncertainties associated with the results of this calculation is beyond the scope of this paper. These results show however that the calculational methodologies employed here have successfully stood the proof of predicting the axial peak exposures within 15 percent outside the PV of a 3-loop reactor.

REFERENCES

1. W.N. McElroy et al., "Surveillance Dosimetry of Operating Power Plants", Proceedings of the 4th ASTM-EURATOM Symposium on Reactor Dosimetry, NUREG/CP-0029, Vol.1, March 22-26, 1982

2. S. Altomare and R.F. Barry, "The TURTLE 24.0 Diffusion Depletion Core", WCAP-7758-A, January 1975.

3. R.G. Soltesz et al., "Two-Dimensional Discrete Ordinates Transport Technique," WANL-PR-(LL) 034, Vol.5, Westinghouse Astronuclear Laboratory, August 1970.

4. D.E. Kusner and S. Kellman, "ETOG-1, A FORTRAN IV Program to Process Data from the ENDF/B File to the MUFT, GAM and ANISN Formats," WCAP-3845-1, ENDF 114, December 1969.

5. R.F. Barry, "The Revised LEOPARD Code - A Spectrum Dependent Non-Spatial Depletion Program, "WCAP-2759 (March 1965).

STANDARDIZED PHYSICS-DOSIMETRY FOR

US PRESSURE VESSEL CAVITY SURVEILLANCE PROGRAMS

F. H. Ruddy, W. N. McElroy, E. P. Lippincott, L. S. Kellogg
R. Gold, J. H. Roberts and C. C. Preston

Westinghouse Hanford Company
Hanford Engineering Development Laboratory
Richland, WA USA

J. A. Grundl and E. D. McGarry

National Bureau of Standards
Washington, DC USA

H. Farrar IV and B. M. Oliver

Rockwell International Corporation
Canoga Park, CA USA

S. L. Anderson

Monroeville Nuclear Center-478
Westinghouse - Nuclear Energy Systems
Pittsburgh, PA USA

ABSTRACT

Standardized Physics-Dosimetry procedures and data
are being developed and tested for monitoring the neutron doses accumulated by reactor pressure vessels (PV)
and their support structures. These procedures and
data are governed by a set of 21 ASTM standard practices, guides, and methods for the prediction of

J. P. Genthon and H. Röttger (eds.), Reactor Dosimetry, 153–164.
© 1985 ECSC, EEC, EAEC, Brussels and Luxembourg

neutron-induced changes in light water reactor (LWR) PVs and support structure steels throughout the service life of the PV. This paper summarizes the applications of these standards to define the selection and deployment of recommended dosimetry sets, the selection of dosimetry capsules and thermal neutron shields, the placement of dosimetry, the methods of measurement of dosimetry sensor reaction products, data analysis procedures, and uncertainty evaluation procedures. It also describes the validation of these standards both by in-reactor testing of advanced PV cavity surveillance physics-dosimetry and by data development. The use of these standards to guide selection and deployment of advanced dosimetry sets for commercial reactors is also summarized.

INTRODUCTION

Federal codes (1,2) require that reactor coolant pressure boundaries have sufficient margin to ensure that the boundary behaves in a non-brittle manner when stressed under operating, maintenance, testing, and postulated accident conditions and that the probability of rapidly propagating fracture is minimized. These requirements necessitate the prediction of the amount of radiation damage to the reactor vessel throughout its service life, which in turn requires that the neutron exposure to the PV be monitored.

In 1977, the Nuclear Regulatory Commission (NRC) established the Light Water Reactor Pressure Vessel Surveillance Dosimetry Improvement Program (LWR-PV-SDIP) to improve, standardize, and maintain dosimetry, damage correlation, and the associated reactor analysis procedures used for predicting the integrated effects of neutron exposure to PVs. This program is in the process of establishing a series of 21 ASTM Standard Practices, Guides, and Methods(3). The application of standardized physics-dosimetry for PV cavity surveillance programs will rely on these standards for: 1) Selection and deployment of recommended dosimetry sets [made up of an appropriate balance of advanced Radiometric Monitors (RMs), Solid-State Track Recorders (SSTRs), Helium Accumulation Fluence Monitors (HAFMs), and Damage Monitors (DMs)]; 2) Selection of dosimetry capsules and thermal neutron shields; 3) Placement of dosimetry; 4) Methods of measurement of dosimetry sensor reaction products; 5) Data analysis procedures; and 6) Uncertainty evaluation procedures. ASTM Standard E706, "Master Matrix for LWR-PV Surveillance Standards," (4) describes these standard practices, guides and methods, while Reference 3 discusses the current status of their preparation and reference documentation. The reader is referred to ASTM E853 [E706-(IA)], "Standard Practice for Analysis and Interpretation of Light-Water Reactor Surveillance Results" for a discussion of current recommendations for reactor cavity dosimetry for verification of reactor physics computations for surveillance programs(5).

Data development and testing for advanced standardized PV physics-dosimetry for cavity surveillance is underway. The status and results of this work, which are discussed and summarized in Reference 3, are discussed further in this paper.

The use of individual ASTM standards to guide in selection and deployment of advanced cavity dosimetry sets for operating LWRs is also reviewed below.

RECOMMENDED DOSIMETRY SETS

ASTM Standard Guide E844 [E706-(IIC)], "Sensor Set Design and Irradiation for Reactor Surveillance," (6) deals with the overall selection, design, irradiation, and post-irradiation handling of RMs, SSTRs, HAFMs, DMs and temperature monitor (TM) sensors and sensor sets. Specific information on TM sensors is not, as yet, provided in E844.

Radiometric Monitors. ASTM Standard Method E1005 [E706-(IIIA)], "Analysis of Radiometric Monitors for Reactor Vessel Surveillance," (7) describes the use of RMs for neutron dosimetry in LWR applications. The useful flux-fluence range and limitations for the application of RMs for time-integrated LWR reaction rate measurements are specified. This standard contains a list of proposed RMs for reactor surveillance along with the neutron response and decay characteristics of each RM. This method should be used in conjunction with Standard Guide E844 for selection of appropriate fission and non-fission RMs for reactor surveillance. In the case of cavity surveillance, primary concerns are lower neutron fluence and the high degree of thermalization of the neutron spectrum.

Solid State Track Recorders. ASTM Standard Method E854 [E706-(IIIB)], "Analysis of Solid State Track Recorder Monitors for Reactor Vessel Surveillance," (8) describes the use of SSTRs in LWR applications. The useful flux-fluence range and limitations for the application of SSTRs for time-integrated LWR reaction and reaction rate measurements are specified. The uses of both fission and non-fission [(n,α),(n,p)] reactions are described. The production of ultra low-mass fissionable deposits for surveillance applications is summarized. This standard is also used in conjunction with ASTM Standard Guide E844 for selection of appropriate monitor sets.

Helium Accumulation Fluence Monitors. ASTM Standard Method E910 [E706-(IIC)], "Analysis of Helium Accumulation Fluence Monitors for Reactor Vessel Surveillance," (9) describes the use of HAFMs in LWR applications. The useful flux-fluence range and limitations for the application of HAFMs for time-integrated LWR reaction and reaction rate measurements are specified. Proposed HAFM sensors for LWR applications are given. This method uses high-sensitivity helium gas mass spectrometry coupled with isotope dilution methods to accurately measure helium produced by selected (n,α) reactions. Selection of appropriate (n,α) reaction dosimeters is made using this method in conjunction with ASTM Standard Guide E844.

Damage Monitors. ASTM Standard Method E706-(IIID), "Analysis of Damage Monitors for Reactor Vessel Surveillance," (10) will describe the use of DMs for LWR applications. Emphasis is placed on the use of sapphire, which has great potential as a direct monitor of neutron damage because of its increase in optical absorbance with increased neutron bombardment. The uses of quartz, graphite, iron, and tungston sensors, as well as the development of new DMs, will be referenced and/or described (see Reference 3).

DOSIMETRY CAPSULES AND THERMAL NEUTRON SHIELDS

ASTM Standard Guide E844 also provides information on design of dosimetry capsules and thermal neutron shields. Figure 1 shows a typical dosimetry capsule fabricated of stainless steel with a gadolinium liner for thermal neutron absorption for inside containment LWR applications. The space inside the shield is used for the placement of an appropriate mix of RM, SSTR, HAFM, and DM. The mix of sensors is chosen to give spectral sensitivity over the desired range of neutron energies as well as time history (exposure time and time-integration) information. When thermal neutron-induced reactions are to be monitored, the gadolinium shield is omitted.

PLACEMENT OF DOSIMETERS

ASTM Standard Practice E185 [E706-(IB)], "Surveillance Tests for Nuclear Reactor Vessels" (11), E706-(IG), "Determining Radiation Exposure for Nuclear Reactor Support Structures" (4), and E853, "Analysis and Interpretation of LWR Surveillance Results" (5) should be used in conjunction with ASTM Standard guide E706-(IIC) to guide the selection, number, and placement of the

156

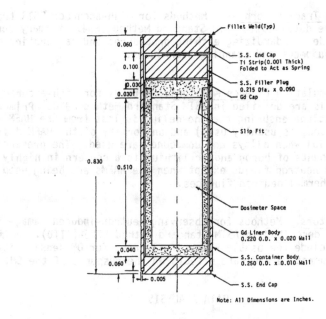

Fillet Weld(Typ)

S.S. End Cap
Ti Strip(0.001 Thick)
Folded to Act as Spring
S.S. Filler Plug
0.215 Dia. x 0.090
Gd Cap

Slip Fit

Dosimeter Space

Gd Liner Body
0.220 O.D. x 0.020 Wall
S.S. Container Body
0.250 O.D. x 0.010 Wall

S.S. End Cap

Note: All Dimensions are Inches.

FIGURE 1. Typical Dosimetry Capsule for Light Water Reactor
Pressure Vessel Surveillance Applications.

dosimetry sets. The primary concerns are that the dosimetry capsule exposures can be related to the exposures of the beltline
welds and that the dosimetry data provide results useful for evaluation of the radiation effects on the reactor vessel and its support structure.

MEASUREMENT OF DOSIMETRY SENSOR REACTION PRODUCTS

Radiometric Monitors. Methods for absolute high resolution gamma-
ray spectrometric analysis of RM reaction products for fission and
non-fission sensors are described and referenced in ASTM Standard
Method E1005. General procedures for measuring the amounts of
radioactive nuclides produced in RMs by nuclear reactions induced
during surveillance exposures are outlined with emphasis on the
use of high resolution gamma-ray spectrometry. Primary concerns
include the presence of interfering reactions or radionuclide
activities, sample size effects, and absolute counter efficiency
calibrations.

157

Solid State Track Recorders. Methods for high-accuracy SSTR track counting are detailed in ASTM Standard Method E854. Primary concerns include standardizing etching procedures and calibrating of scanning equipment and scanning techniques.

Helium Accumulation Fluence Monitors. Methods for high-accuracy HAFM analysis are detailed in ASTM Standard Method E910. Primary concerns include ensuring that no helium is lost from the HAFM (encapsulation, is usually used) and uniformity of the HAFM dosimetry material when alloys and compounds are used. The presence of small amounts of boron and/or lithium is a concern in highly thermalized neutron fields except when the HAFMs are being used to determine thermal neutron fluences.

Damage Monitors. Methods for observing neutron-induced damage in DMs will be described in ASTM Standard Method E706-(IIID). Primary concerns include the usable range of fluences for DM sensitivity and the temperature and rate effects on the response of the DM.

DATA ANALYSIS

Neutron Transport. ASTM Standard Guide E482 [E706-(IID)], "Application of Neutron Transport Methods for Reactor Vessel Surveillance," (12) describes the methodology for performing radiation transport calculations to determine the neutron and gamma flux spectra at surveillance positions and within the PV. These procedures are applicable for the evaluation and extrapolation of both research and surveillance program measurement results obtained in operating reactors. The reader is referred to the ASTM Standard Practice [E706-(II)], "Analysis and Interpretation of Physics Dosimetry Results for Test Reactors" (4) for additional and supplementary information related to research reactors.

Spectral Adjustment. ASTM Standard Guide E944 [E706-(IIA)], "Application of Neutron Spectrum Adjustment Methods," (13) describes the procedures and codes recommended for use in determining neutron flux-fluence spectra from multiple sensor dosimetry measurements. ASTM Standard Guide E706- (IIB), "Application of ENDF/A Cross-Section and Uncertainty Files," (14) is directly related to and should be used in conjunction with guide E706-(IIA). This guide covers the establishment and use of an ASTM ENDF/A cross-section and uncertainty tape file for analysis of multiple sensor measurements in LWR neutron fields. Also included are calculations of spectral-averaged damage cross sections for iron, quartz, and other sensors that might be used as DMs.

Documentation. Documentation in the form of as-built reports should be provided with each dosimetry set. The as-built report should document the type and number of dosimeters, dosimeter materials (supplier, batch number, isotopic composition, chemical form, weight, dimensions, impurity content, etc.), dosimeter capsules (materials, dimensions, characteristics of thermal neutron shields, etc), capsule support structure (materials, dimensions, etc), location of dosimetry in the reactor, and any other pertinent information. Following the exposure, the reactor power and sensor set flux time history should be documented along with analysis results and interpretation (including spectral adjustment calculations and damage calculations).

UNCERTAINTY EVALUATION

Benchmark Referencing. ASTM Standard Guide E706-(IIE), "Benchmark Testing of Reactor Vessel Dosimetry," (15) will describe methodology for using the known characteristics of selected neutron fields to calibrate neutron sensors or to validate procedures used to derive neutron field information from measurements of their response. Of particular interest to cavity surveillance measurements are the calibration measurements performed at the Surveillance Measurements Dosimetry Facility (SDMF) at Oak Ridge (16,17) in the ANO-1 (Arkansas) reactor cavity (18,19) and other related benchmark fields.(3)

The SDMF serves as a realistic mockup of operating power reactors. This facility has been used for a series of dosimetry tests, some of which were designed to replicate some details of cavity dosimetry. ANO-1, ANO-2, Browns Ferry 3, H. B. Robinson, Maine Yankee, and BR-3 (Belgium), on the other hand, are operating commercial power reactors that have been the site of numerous cavity dosimetry measurements and related neutronic calculations.(3,18,19)

Experimental Uncertainties. The maximum uncertainties desirable for LWR-PV steel exposure definition are in the range of ±10% to 15%. In order to achieve these accuracy goals, the dosimeter reaction and reaction rates must, in general, be measurable to within ±5% (1σ). Experimental design and analysis should provide a margin such that overall errors (response uncertainty, extrapolation, interpolation, etc.) are within these bounds.

COMMERCIAL LWR CAVITY SURVEILLANCE DOSIMETRY SETS

Advanced surveillance dosimetry sets have been chosen, assembled, and deployed by LWR-PV Surveillance Dosimetry Improvement Program participants and US utilities in the cavities of operating commercial reactors (see Table 2.27 of Reference 3). These sets contain a mix of RMs, SSTRs and HAFMs. The layout of a typical set is shown in Figure 2. RMs consisting of silver (in the form of Ag-Al alloy wires), cobalt (in the form of Co-Al alloy wires), and iron are deployed as thermal neutron monitors. Included in this so-called bare (aluminum wrapped) set are ultra low-mass fissionable deposits of ^{235}U and ^{239}Pu with mica SSTRs to measure the integral neutron-induced fission rates. After radiometric analysis of the Ag and Co dosimeters for their (n,γ) reaction products, they may be subjected to analysis as HAFMs for their helium content [Al(n,α) integral reaction rates]. The thermal neutron shielded (cadmium-covered) dosimeter set contains Co, Ti, Ni, Cu, and Fe RMs. The Co, Ti, Fe, Ni and Cu dosimeters can also serve as HAFMs. ^{235}U, ^{239}Pu, ^{237}Np and ^{238}U ultra low-mass fissionable deposits and mica SSTRs are also included. The former two isotopes provide epithermal neutron response, whereas the latter provide threshold neutron response. Additional HAFM sensors are currently being developed to supplement existing RMs and SSTRs and are discussed in ASTM Standard Method E910 (9). HAFMs have a particular advantage for those LWR applications where long durations are involved.

Figure 3 shows a typical dosimetry holder for emplacement of the assembled dosimetry capsules in the reactor cavity. Capsules are typically located at three locations (above midplane, midplane, and below midplane) at several azimuthal positions around the PV. Dosimetry sets of the type shown in Figure 2 that have been deployed (or have been planned for deployment) in commercial power reactors are as shown in Table 1. In the case of SSTRs, replicate sets of SSTR fissionable deposits are made to facilitate replacement of the dosimetry at the end of an operating cycle. The pairs of SSTR dosimetry sets are exchanged with the ex-reactor set being subjected to mass calibration overchecks before re-insertion with new mica SSTRs in the following cycle.

CONCLUSIONS

A matrix of ASTM Standard Methods, Guides and Practices is being established to document the procedures to be followed for LWR surveillance programs. These standards are in turn being validated by data development and testing of the dosimetry in reference neutron fields, such as SDMF, ANO-1, H. B. Robinson, and Maine Yankee. The standards are currently being put into practice for dosimetry

160

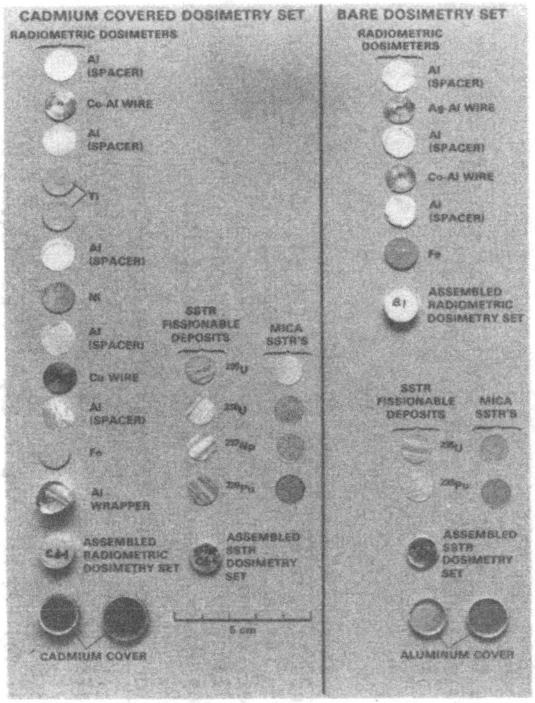

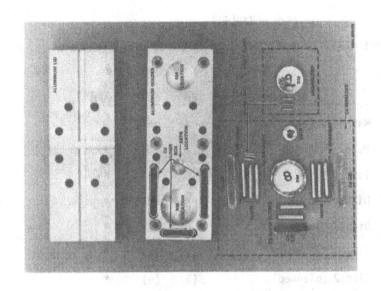

FIGURE 3. Cavity Dosimetry Holder for Light Water Reactor Pressure Vessel Surveillance.

FIGURE 2. Advanced Cavity Dosimetry Set for Light Water Reactor Pressure Vessel Surveillance.

161

TABLE 1

ADVANCED CAVITY SURVEILLANCE DOSIMETRY IN OPERATING LWRs

Location	Number of Sets		
	RM	SSTR	HAFM
Maine Yankee	6	5	3
H. B. Robinson	6	6	*
Turkey Point 3	6	6	*
Diablo Canyon 1	6	6	*
Diablo Canyon 2	6	6	*
Zion 1 (planned)	(6)	(6)	*
Zion 2 (planned)	(6)	(6)	*
Brunswick (BWR) (planned)	(2)	(2)	*

*No separate HAFM sets are included. However, after radiometric analysis, selected RMs can be analyzed for helium (see text).

measurements in commercial reactor cavities for accurate and cost effective determination of PV neutron exposure.

REFERENCES

1. Code of Federal Regulations, "Domestic Licensing of Production and Utilization Facilities," 10 CFR 50; "General Design Criteria for Nuclear Power Plants," Appendix A; "Fracture Toughness Requirements," Appendix G; "Reactor Vessel Material Surveillance Program Requirements," Appendix H; US Government Printing Office, Washington, DC, current edition.

2. Regulatory Guide 1.99, "Effects of Residual Elements on Predicted Radiation Damage to Reactor Vessel Materials," Rev. 1, US Nuclear Regulatory Commission, Washington, DC (April 1977).

3. W. N. McElroy et al., LWR-PV-SDIP 1983 Annual Report, NUREG/CR-3391, Vol. 3, HEDL-TME 83-23, NRC, Washington, DC (June 1984).

4. Annual Book of ASTM Standards, Vol. 12.02, (Current Edition), American Society for Testing and Materials, Philadelphia, PA, 1254 (1981), ASTM Standard E706, "Master Matrix for LWR Pressure Vessel Surveillance Standards."

5. Ibid, ASTM Standard E853, "Standard Practice for Analysis and Interpretation of Light-Water Reactor Surveillance Results."

6. Ibid, ASTM Standard E844, "Sensor Set Design and Irradiation for Reactor Surveillance."

7. Ibid, ASTM Standard E1005, "Analysis of Radiometric Monitors for Reactor Vessel Surveillance."

8. Ibid, ASTM Standard E854, "Analysis of Solid State Track Recorder Monitors for Reactor Vessel Surveillance."

9. Ibid, ASTM Standard E910, "Analysis of Helium Accumulation Fluence Monitors for Reactor Vessel Surveillance."

10. ASTM Standard E706-IIID, "Analysis of Damage Monitors for Reactor Vessel Surveillance," (in preparation), see References 3 & 4.

11. Ibid, ASTM Standard E185, "Surveillance Tests for Nuclear Reactor Vessels."

12. Ibid, ASTM Standard E482, "Application of Neutron Transport Methods for Reactor Vessel Surveillance."

13. Ibid, ASTM Standard E944, "Application of Neutron Spectrum Adjustment Methods."

14. ASTM Standard E706-IIB, "Application of ENDF/A Cross Section and Uncertainty Files," (in preparation), see References 3 & 4.

15. ASTM Standard E706-IIE, "Benchmark Testing of Reactor Vessel Dosimetry," (in preparation), see References 3 & 4.

16. H. Tourwé and G. Minsart, "Surveillance Capsule Perturbation Studies in the PSF 4/12 Configuration," Radiation Metrology Techniques, Data Bases, and Standardization: Proc. of the 4th ASTM-EURATOM Symposium on Reactor Dosimetry, NUREG/CP-0029, CONF-820321, Vol. 1 (1982).

163

17. R. E. Maerker and M. L. Williams, "Calculations of the Westinghouse Perturbation Experiment at the Poolside Facility," Radiation Metrology Techniques, Data Bases, and Standardization: Proc. of the 4th ASTM-EURATOM Symposium on Reactor Dosimetry, NUREG/CP-0029, CONF-820321, Vol. 1, 131 (1982).

18. W. E. Brandon, C. O. Cogburn, R. R. Culp, J. L. Meason, W. W. Sallee and J. C. Williams, "Neutron Dosimetry in the Pressure Vessel Cavity of Two Pressurized Water Reactors," Radiation Metrology Techniques, Data Bases, and Standardization: Proc. of the 4th ASTM-EURATOM Symposium on Reactor Dosimetry, NUREG/CP-0029, CONF-820321, Vol. 1 (1982).

19. M. Petilli, "A New Analysis of the Experiment for Measurement of ϕ > 1 MeV in the Pressure Vessel Cavity of US Light Water Reactor Arkansas," Radiation Metrology Techniques, Data Bases, and Standardization: Proc. of the 4th ASTM-EURATOM Symposium on Reactor Dosimetry, NUREG/CP-0029, CONF-820321, Vol. 1 (1982).

20. W. N. McElroy et al., "Surveillance Dosimetry of Operating Power Plants," Radiation Metrology Techniques, Data Bases, and Standardization: Proc. of the 4th ASTM-EURATOM Symposium on Reactor Dosimetry, NUREG/CP-0029, CONF-820321, Vol. 1 (1982).

RE-EVALUATION OF THE PHYSICS-DOSIMETRY FROM PWR AND BWR

REACTOR PRESSURE VESSEL SURVEILLANCE PROGRAMS

R. L. Simons, L. S. Kellogg, E. P. Lippincott and W. N. McElroy

Westinghouse Hanford Company
Hanford Engineering Development Laboratory
Richland, WA USA

ABSTRACT

The goals of this Hanford Engineering Development Laboratory (HEDL) Reactor Dosimetry Center work are (1) to develop, test and apply new physics-dosimetry analysis procedures and data being recommended in ASTM standards for light water power plant surveillance programs, and (2) to provide improved neutron exposure parameter values for the metallurgical data bases that support the analysis of the integrity of reactor pressure vessels during their service life. Neutron dosimetry data from surveillance capsules from several pressurized water reactors (PWR) and boiling water reactors (BWR) were re-evaluated, resulting in revised values of the exposure parameters of fluence and displacements per atom (dpa) and their uncertainties. Improvements not previously reported include: (1) the use of improved a priori neutron spectra, (2) corrections relating to spectrum weighting of the thermal neutron cross sections, (3) the inclusion of thermal neutron fluences and their uncertainties, and (4) calculated helium content for a generic pressure vessel steel chemical composition. Helium production is dominated by the $^{10}B(n,\alpha)$ reaction and ranged from 2 to 100 atomic parts per billion for most capsules. However, the actual helium concentrations will depend on the actual boron concentration in the steel.

J. P. Genthon and H. Röttger (eds.), Reactor Dosimetry, 165–173.
© 1985 ECSC, EEC, EAEC, Brussels and Luxembourg

INTRODUCTION

The Light Water Reactor Pressure Vessel Surveillance Dosimetry Improvement Program was established by the NRC in order to improve and standardize the neutron dosimetry analysis, damage correlation, and reactor analysis procedure and data used to predict current and end of life condition of light water reactor (LWR) pressure vessels.(1,2) As part of this program a number of new and updated ASTM standards (1,3) are being established to facilitate the analysis of LWR power plant surveillance data. The need for re-evaluation of the neutron damage exposures of surveillance capsules stems from the fact that predictions based on the use of Regulatory Guide 1.99 (4) had indicated (5-7) that the safe operating life of several older reactor pressure vessels could be comparable to the planned service life of the pressure vessels. Consequently, improved accuracy of the prediction of the metallurgical condition of pressure vessels is necessary in order to assure that the planned service life of both PWR and BWR nuclear power plants can be achieved.

A previously reported analysis of the dosimetry from 42 surveillance capsules described several improvements in the analysis.(8) The improvements were due to using a consistent set of auxiliary parameters such as cross sections, decay constants, neutron spectra, and correction for burn-in and burn-out of important reaction products. The improvements not previously reported include: (1) the use of improved a priori neutron spectra, (2) corrections related to the spectrum weighting of the thermal neutron cross sections, (3) the inclusion of thermal neutron fluences and their uncertainties, and (4) estimates of the helium content and sources of helium in a typical pressure vessel steel. The thermal neutron exposure (and corresponding helium production from boron) may be significant in the analysis of power and test reactor radiation effects data as reported elsewhere.(1,9-11)

ANALYSIS PROCEDURE

The basic analysis method used in this work is the same as reported in earlier work.(8) In summary the analysis involves determining the correct reaction rates at a central location in the capsule after accounting for the location of the specific neutron sensors, reactor power time history, and correction for initial or irradiation induced impurities. The reaction rates along with the a priori calculated spectrum and reaction cross sections are input to the generalized least square fitting code FERRET.(12)

RESULTS

The results of the present analysis of 47 surveillance capsules are tabulated in Table 1. The table includes an identification of the reactor and capsules, the service laboratory that originally analyzed the capsule, the old, new, and ratio of new/old fluence >1 MeV, the thermal neutron fluence <0.414 eV, dpa, dpa/ϕt > 1 MeV, displacement rate, helium content in units of atomic parts per billion (appb), and the effective full power seconds of irradiation. The results of the physics-dosimetry analysis for five additional surveillance capsules are included in Table 1. The five new results include capsules from Arkansas nuclear one, (13) Beznau 2, (14) Dresden 3, (15) and Quad Cities 1 and 2.(16,17)

The ^{238}U neutron sensor in the Beznau 2 capsule R required a 45% correction for burn-in of ^{239}Pu fission products. Although this was the largest correction required in Table 1, the corrected value was consistent with the reaction rates of the remaining sensors in the surveillance capsule. The ^{238}U sensor responds to an energy range (>1.5 MeV) and that makes its use important in reducing the uncertainty in derived damage exposure parameter values.

The surveillance capsules removed from General Electric (GE) plants contain only Cu, Ni and Fe neutron sensors. Consequently, only neutrons above ~2 MeV are monitored. As a result, the derived damage exposure parameter values are highly sensitive to the a priori spectrum uncertainties. This accounts for the large uncertainty assigned to the thermal neutron fluences. The capsules from the GE plants are above the reactor midplane in a region that shows a flux gradient of more than a factor of four variation over the length of the capsule. Consequently, the fluences for the GE capsule are reported for each basket within a capsule.

The helium concentrations given in Table 1 are based on an A302B PV steel chemistry and an assumed 0.55 appm boron concentration.(18) Table 2 provides a summary of the estimated helium production for the 47 capsules. The table includes the type of capsule, helium/dpa ratio, the range of helium concentration and the percent helium due to the 0.55 appm concentration of boron. There were five basic capsule types. Embedded capsules are from Westinghouse plants where the capsules are embedded into the backside of the thermal shield. Accelerated capsules refer to standard capsules from Westinghouse or Combustion Engineering plants in which the capsule is attached to the back side of the thermal shield. Babcock and Wilcox capsules are in water at an intermediate position some distance from the vessel wall. The wall capsules are from Combustion Engineering plants with capsules attached to

TABLE 1

RE-EVALUATED EXPOSURE VALUES AND THEIR UNCERTAINTIES FOR LWR PRESSURE VESSEL SURVEILLANCE CAPSULES

Plant	Unit	Capsule	Service Lab*	Fluence ($\phi t > 1$ MeV) (n/cm²) Old	New [% (1s)] ***	New/Old	Fluence (E < 0.414 eV) (n/cm²)	dpa [% (1σ)]	New dpa/ϕt	dpa/s	hpa (appb)†	Exposure** Time (s)
Westinghouse												
Conn. Yankee	3	A	BMI	2.08 E+18	3.16 E+18 (12)	1.53	2.54 E+18 (18)	0.00482 (12)	1.52 E-21	9.06 E-11	6	5.233 E+07
Conn. Yankee		F	BMI	4.04 E+19	6.06 E+19 (24)	1.50	5.43 E+18 (32)	0.00949 (27)	1.56 E-21	1.24 E-10	13	7.651 E+07
Conn. Yankee		H	W	1.79 E+19	2.00 E+19 (24)	1.12	2.33 E+19 (19)	0.0324 (27)	1.62 E-21	1.36 E-10	52	2.390 E+08
San Onofre		A	SwRI	1.20 E+19	2.86 E+19 (22)	2.38	2.05 E+19 (23)	0.0486 (27)	1.70 E-21	8.35 E-10	43	5.824 E+07
San Onofre		D	SwRI	2.36 E+19	5.62 E+19 (26)	2.38	3.76 E+19 (23)	0.0944 (29)	1.68 E-21	1.06 E-09	80	8.881 E+07
San Onofre		F	W	5.14 E+19	5.73 E+19 (14)	1.11	2.99 E+19 (28)	0.0955 (20)	1.67 E-21	3.92 E-10	73	2.438 E+08
Turkey Point	3	S	SwRI	1.41 E+19	1.62 E+19 (24)	1.15	1.34 E+19 (24)	0.0255 (27)	1.57 E-21	2.33 E-10	33	1.095 E+08
Turkey Point	4	T	W	5.68 E+18	7.01 E+18 (10)	1.23	5.12 E+18 (58)	0.0109 (12)	1.55 E-21	4.73 E-10	14	2.302 E+07
Turkey Point	4	S	SwRI	1.25 E+19	1.31 E+19 (25)	1.05	1.31 E+19 (25)	0.0213 (27)	1.63 E-21	1.97 E-10	37	1.079 E+08
Turkey Point	4	T	SwRI	6.05 E+18	7.54 E+18 (13)	1.25	8.40 E+18 (21)	0.0130 (13)	1.72 E-21	3.48 E-10	20	3.728 E+07
H. B. Robinson	2	S	W	3.02 E+18	3.91 E+18 (24)	1.29	8.81 E+18 (18)	0.00615 (13)	1.57 E-21	1.06 E-10	19	4.209 E+07
H. B. Robinson	2	V	SwRI	4.51 E+18	7.24 E+18 (22)	1.61	8.96 E+18 (20)	0.0115 (25)	1.59 E-21	1.09 E-10	21	1.050 E+08
Surry	1	T	BMI	2.50 E+18	2.86 E+18 (9)	1.14	3.57 E+18 (20)	0.00449 (12)	1.57 E-21	1.33 E-10	8	3.378 E+07
Surry	1	X	BMI	3.02 E+18	3.03 E+18 (11)	1.00	3.64 E+18 (20)	0.00473 (13)	1.56 E-21	1.28 E-10	9	3.687 E+07
North Anna	1	V	B&W	2.49 E+18	2.72 E+18 (9)	1.09	5.80 E+18 (14)	0.00411 (11)	1.51 E-21	1.15 E-10	11	3.570 E+07
Beznau	2	R	EIFR	1.70 E+19	1.34 E+19 (9)	1.27	2.27 E+19 (26)	0.0198 (14)	1.48 E-21	1.16 E-10	49	1.714 E+08
Pr. Island	1	V	W	5.21 E+18	6.03 E+18 (11)	1.16	9.21 E+18 (21)	0.0102 (16)	1.69 E-21	2.41 E-10	20	4.248 E+07
Pr. Island	2	V	W	5.49 E+18	6.74 E+18 (10)	1.23	9.75 E+18 (26)	0.0117 (13)	1.74 E-21	2.67 E-10	21	4.394 E+07
R. E. Ginna	1	R	W	7.60 E+18	1.17 E+19 (14)	1.54	1.84 E+19 (14)	0.0215 (14)	1.83 E-21	2.59 E-10	38	8.328 E+07
R. E. Ginna		V	W	4.90 E+18	5.93 E+18 (14)	1.21	1.37 E+19 (59)	0.0102 (22)	1.78 E-21	2.20 E-10	29	4.612 E+07
Kewaunee	1	S	W	5.59 E+18	6.41 E+18 (10)	1.15	1.23 E+19 (23)	0.0114 (13)	1.78 E-21	2.82 E-10	26	4.057 E+07
Point Beach	1	S	W	7.05 E+18	8.45 E+18 (10)	1.20	1.20 E+19 (19)	0.0146 (13)	1.73 E-21	1.25 E-10	27	1.163 E+08
Point Beach	2	V	BMI	2.22 E+19	2.29 E+19 (11)	1037	2.85 E+19 (22)	0.0408 (13)	1.78 E-21	2.50 E-10	61	1.632 E+08
Point Beach	2	R	W	4.74 E+18	7.28 E+18 (11)	1.54	1.09 E+19 (18)	0.0121 (13)	1.66 E-21	2.52 E-10	23	4.805 E+07
Point Beach	2	T	W	9.45 E+18	9.40 E+18 (10)	0.99	1.48 E+19 (21)	0.0157 (12)	1.67 E-21	1.44 E-10	32	1.087 E+08
Point Beach		R	W	2.01 E+19	2.52 E+19 (10)	1.25	4.71 E+19 (26)	0.0460 (14)	1.83 E-21	2.81 E-10	93	1.640 E+08
D. C. Cook	1	T	SwRI	1.80 E+18	2.71 E+18 (22)	1.51	3.26 E+19 (19)	0.00445 (25)	1.64 E-21	1.12 E-10	77	3.991 E+07
Indian Point	2	T	SwRI	2.02 E+18	3.28 E+18 (22)	1.62	4.01 E+18 (44)	0.00537 (27)	1.64 E-21	1.20 E-10	91	4.473 E+07
Indian Point	3	T	BMI	2.92 E+18	3.23 E+18 (22)	1.11	3.13 E+18 (21)	0.00520 (25)	1.61 E-21	1.23 E-10	74	4.211 E+07
Zion	1	T	BMI	1.80 E+18	3.04 E+18 (10)	1.69	3.17 E+18 (21)	0.00488 (12)	1.61 E-21	1.29 E-10	82	3.789 E+07
Zion	1	U	W	8.92 E+18	1.01 E+19 (10)	1.13	8.87 E+18 (24)	0.0166 (13)	1.64 E-21	1.47 E-10	21	1.123 E+08
Zion	2	U	BMI	2.00 E+18	2.80 E+18 (15)	1.40	3.80 E+18 (15)	0.00446 (13)	1.59 E-21	1.11 E-10	10	4.007 E+07
Salem	1	T	W	2.56 E+18	2.84 E+18 (22)	1.11	3.26 E+18 (19)	0.00460 (25)	1.62 E-21	1.34 E-10	7	3.426 E+07

*BMI = Battelle Memorial Institute; W = Westinghouse; SwRI = Southwest Research Institute; CE = Combustion Engineering; ET = Effects Technology; B&W = Babcock and Wilcox, EIFR Eidg. Institute für Reaktorforschung.

**Equivalent constant power level exposure time.

***3.16 E+18 (12) means 3.16 x 10¹⁸ with a 12% (1σ) uncertainty.

†Calculated for A302B steel with a nominal concentration of 0.55 appm boron present.

TABLE 1 (Cont'd)

Plant	Unit	Capsule	Service Lab*	Fluence (φt > 1 MeV) (n/cm²)			Fluence (E < 0.414 eV) (n/cm²)	dpa [% (1σ)]	New dpa/φt	dpa/s	hpa (appb)†	Exposure** Time (s)
				Old	New [% (1σ)]	New/Old						
Combustion Engineering												
Palisades		A240	BMI	4.40 E+19	6.06 E+19 (23)	1.38	7.26 E+19 (61)	0.0972 (28)	1.60 E-21	1.36 E-09	170	7.130 E+07
Fort Calhoun		M225	CE	5.10 E+18	5.83 E+18 (14)	1.14	3.09 E+19 (60)	0.00879 (18)	1.51 E-21	1.07 E-10	63	8.191 E+07
Maine Yankee		1	ET	1.30 E+19	1.76 E+19 (19)	1.35	3.00 E+19 (29)	0.0285 (23)	1.62 E-21	1.03 E-09	62	2.777 E+07
Maine Yankee		2	W	8.84 E+18	7.73 E+19 (13)	0.87	1.20 E+20 (23)	0.121 (18)	1.57 E-21	8.38 E-10	230	1.446 E+08
Maine Yankee		M263	BMI	7.10 E+18	5.67 E+18 (2)	0.82	2.67 E+19 (21)	0.00843 (14)	1.49 E-21	5.83 E-11	55	1.446 E+08
Babcock & Wilcox												
Oconee	1	F	B&W	8.70 E+17	6.98 E+17 (21)	0.80	1.00 E+18 (13)	0.000959 (19)	1.37 E-21	3.65 E-11	3	2.629 E+07
Oconee	1	F	B&W	1.50 E+18	1.50 E+18 (10)	1.00	2.61 E+18 (15)	0.00208 (10)	1.39 E-21	4.01 E-11	7	5.186 E+07
Oconee	2	C	B&W	9.43 E+17	1.01 E+18 (10)	1.07	1.55 E+18 (15)	0.00148 (11)	1.47 E-21	3.88 E-11	4	3.802 E+07
Oconee	3	A	B&W	7.39 E+17	8.05 E+17 (10)	1.09	1.34 E+18 (11)	0.00113 (11)	1.40 E-21	3.79 E-11	3	2.983 E+07
Three Mile Is.	1	E	B&W	1.07 E+18	1.09 E+18 (9)	1.02	1.90 E+18 (11)	0.00151 (9)	1.39 E-21	3.75 E-11	5	4.036 E+07
Arkansas Nuclear	1	E	B&W	7.27 E+17	8.18 E+17 (8)	1.13	6.32 E+17 (9)	0.00117 (8)	1.43 E-21	3.92 E-11	2	2.981 E+07
General Electric												
Dresden	3	4G14	W	2.06 E+19	1.86 E+19 (17)		1.51 E+20 (62)	0.0285 (17)	1.53 E-21	3.35 E-10	290	8.483 E+07
Dresden	3	4G15		1.50 E+18	1.35 E+19 (17)	0.89	1.19 E+20 (62)	0.0209 (17)	1.55 E-21	2.46 E-10	240	8.483 E+07
Dresden	3	4G16		1.20 E+19	1.08 E+19 (17)		9.70 E+19 (62)	0.0168 (17)	1.55 E-21	1.98 E-10	200	8.483 E-07
Dresden	3	4G17		5.16 E+18	4.51 E+18 (17)		5.20 E+19 (62)	0.00733 (18)	1.63 E-21	8.64 E-11	120	8.438 E+07
Quad Cities	2	3G14	W	4.14 E+19	4.28 E+19 (16)		2.45 E+20 (62)	0.0611 (17)	1.43 E-21	4.29 E-10	400	1.422 E+08
Quad Cities	2	3G15		3.48 E+19	3.60 E+19 (16)	1.03	2.13 E+20 (62)	0.0516 (17)	1.43 E-21	3.63 E-10	370	1.422 E+08
Quad Cities	2	3G16		2.43 E+19	2.52 E+19 (16)		1.55 E+20 (62)	0.0362 (17)	1.44 E-21	2.54 E-10	290	1.422 E+08
Quad Cities	2	3G17		2.32 E+19	2.37 E+19 (17)		1.49 E+20 (62)	0.0342 (17)	1.44 E-21	2.41 E-10	290	1.422 E+08
Quad Cities	1	3G5	W	4.04 E+19	4.23 E+19 (17)		2.41 E+20 (62)	0.0604 (17)	1.43 E-21	4.85 E-10	400	1.243 E+08
Quad Cities	1	3G6		3.08 E+19	3.12 E+19 (17)	1.01	1.92 E+20 (62)	0.0450 (17)	1.44 E-21	3.62 E-10	340	1.243 E+08
Quad Cities	1	3G7		2.37 E+19	2.47 E+19 (17)		1.54 E+20 (62)	0.0356 (17)	1.44 E-21	2.86 E-10	290	1.253 E+08
Quad Cities	1	3G8		1.24 E+19	1.17 E+19 (17)		1.03 E+20 (62)	0.0180 (17)	1.54 E-21	1.45 E-10	210	1.243 E+08
						Avg 1.25						

*BMI = Battelle Memorial Institute; W = Westinghouse; SwRI = Southwest Research Institute; CE = Combustion Engineering; ET = Effects Technology; B&W = Babcock and Wilcox, EIFR Eidg. Institute für Reaktorforschung.

**Equivalent constant power level exposure time.

***3.16 E+18 (12) means 3.16 x 10^18 with a 12% (1σ) uncertainty.

†Calculated for A302B steel with a nominal concentration of 0.55 appm boron present.

TABLE 2

SUMMARY OF HELIUM PRODUCTION IN LIGHT WATER REACTOR SURVEILLANCE CAPSULES

Capsule Type	helium/dpa	Helium (appb)	% Helium From Boron
Embedded	1.1 ± 0.3	6 to 86	73 to 85
Accelerated	1.9 ± 0.4	8 to 91	77 to 92
Babcock and Wilcox	2.8 ± 0.4	2 to 7	77 to 81
Vessel Wall	6.8 ± 0.4	55 to 63	92 to 94
General Electric	9. ± 4.	120 to 400	91 to 93

the vessel wall. The GE capsules are located above the reactor midplane and are placed near, but in front of the shroud.

The helium to dpa ratios are similar for each of the capsule types because of the relative amounts of water around the capsule. The embedded capsules are surrounded by steel on three sides and therefore have the lowest ratio of thermal to fast neutron flux and ratio of helium concentration to dpa. The standard accelerated capsule has water on three sides but the neutron source is shielded by iron and therefore it shows the next highest ratio of helium to dpa. As the amount of surrounding water and depth of penetration through water increases, the helium/dpa ratio correspondingly increases. The sensitivity of the helium/dpa ratio to surrounding water is evident by the fact that 80 to 90% of the helium, even at the 0.55 appm boron level, is from the boron. The total concentration of helium ranges from 2 to 91 appm for PWRs while the BWRs have 100-400 appb. However, the uncertainty in the helium concentration for the GE capsules is at least 60 to 70% due to the lack of cadmium covered and bare thermal neutron sensors and the resulting poor characterization of the thermal neutron fluence.

The primary source of helium is boron. Thus, the helium content should scale almost proportional to the boron content. Heat to heat variation of the boron content in steel can also result in variation in the helium production. The remaining source of helium is due to high energy (E > 6 MeV) threshold reactions. The distribution of helium production between various elements from threshold reactions is nearly independent of the neutron spectrum. About 87% of the threshold reaction source is from iron. The balance is 5% each due to carbon and silicon, ~2% due to nickel, and ≤ 1% due to the remaining elements. Consequently, it is not surprising that the helium production can be characterized as being from predominantly two sources, boron and iron.

CONCLUSIONS

Some additional improvements in the results and procedures used for the derivation of damage exposure parameters for light water reactor surveillance have been made. Among these are the use of improved a priori neutron spectra. Values of the damage parameters fluence (E > 1 MeV), dpa, fluence <0.414 eV, and helium content are reported. Results from five new capsules are included in the present report. The maximum correction to account for burn-in of ^{239}Pu in the ^{238}U neutron sensor was 45%. Helium concentration estimates ranged from 2 to 100 appb in PWR and 100 to 400 for BWR capsules. The principle sources ($\sim$98%) of helium are from (n,α) reactions in the boron and iron.

REFERENCES

1. W. N. McElroy et al., "LWR Pressure Vessel Surveillance Dosimetry Improvement Program, 1983 Annual Report," NUREG/CR-3391, HEDL-TME 83-23, NRC, Washington, DC January 1984.

2. W. N. McElroy et al., "Surveillance Dosimetry of Operating Power Plants," Proceedings of the 4th ASTM-Euratom International Symposium on Reactor Dosimetry, March 22-26, 1982, Natural Bureau of Standards, Gaithersburg, MD, HEDL-SA-2546, (February 1982).

3. ASTM E706, "Standard Master Matrix for LWR-PV Surveillance Standards," ASTM Annual Book of Standards, Section 12, Volume 12.02, "Nuclear (II), Solar, and Geothermal Energy," Philadelphia, PA 1984.

4. Regulatory Guide 1.99, Effects of Residual Elements on Predicted Radiation Damage to Reactor Vessel Materials, Rev. 1, NRC, Washington, DC (April 1977).

5. P. N. Randall, "Status of Regulatory Demands in the U.S. on the Application of Pressure Vessel Dosimetry," Proc. of the 4th ASTM-EURATOM Symposium on Reactor Dosimetry, Gaithersburg, MD, March 22-26, 1982, NUREG/CP-0029, NRC, Washington, DC, Vol. 2, pp. 1011-1022 (July 1982).

6. P. N. Randall, "NRC Perspective of Safety and Licensing Issues Regarding Reactor Vessel Steel Embrittlement," from the ANS Special Session on Correlations and Implications of Neutron Irradiation Embrittlement of Pressure Vessels Steels, Detroit, MI, June 12-16, 1983, Trans. Am. Nucl. Soc. 44, p. 220 (1983).

7. W. J. Dircks, Pressurized Thermal Shock (PTS), and Enclosure A, "NRC Staff Evaluation of PTS," SECY-82-465, NRC, Washington, DC (November 1982).

8. R. L. Simons et al., "Re-evaluation of the Dosimetry for Reactor Pressure Vessel Surveillance Capsules," 4th ASTM-Euratom Symposium on Reactor Dosimetry, March 22-26, 1982, Washington, DC.

9. G. L. Guthrie et al., "Effects of Thermal Neutrons in Irradiation Embrittlement of PWR Pressure Vessel Plates and Welds," NUREG/CR-3391, Vol. 2, HEDL-TME 83-22, p. HEDL-16 (April 1984).

10. W. N. McElroy et al., "Thermal-Relative-to-Fast-Neutron Contribution to Charpy Shift for PWR and BWR Surveillance Capsule Weld Materials," NUREG/CR-3391, Vol. 2, HEDL-TME 83-22, p. HEDL-22 (April 1984).

11. W. N. McElroy et al., "Trend Curve Exposure Parameter Data Development and Testing," Proc. of the 5th ASTM-EURATOM Symposium on Reactor Dosimetry, Gesthalt, Federal Republic of Germany, September 24-28, 1984.

12. F. A. Schmittroth, "FERRET Data Analysis Code," HEDL-TME 79-40, Hanford Engineering Development Laboratory, Richland, WA, (September 1979).

13. A. L. Lowe Jr, et al., "Analysis of Capsule ANI-E From Arkansas Power and Light Company Arkansas Nuclear One-Unit 1," BAW-1440, Babcock and Wilcox, Lynchburg, VA, (April 1977).

14. G. Ulrich, B. Bürgissen, F. Hegedues, "Nachbestrahlungsunlersuchungen An Nokreaktordrunckgefässmaterial Der Kernkraftwerke Beznau II/2, Kapsel R," PB-ME-8015, Erdg, Institut für Reaktorforschung, Baden, Switzerland, (May 1980).

15. S. E. Yanichko et al., "Analysis of the Fourth Capsule From the Commonwealth Edison Company Dresden Unit 3 Nuclear Plant Reactor Vessel Radiation Surveillance Program," WCAP-10030, Westinghouse Electric Corporation, Pittsburgh, PA, (January 1982).

16. S. E. Yanichko et al., "Analysis of the Third Capsule From the Commonwealth Edison Company Quad Cities Unit 1 Nuclear Plant Reactor Vessel Radiation Surveillance Program," WCAP-9920, Westinghouse Electric Corporation, Pittsburgh, PA, (August 1981).

17. S. E. Yanichko et al., "Analysis of the Third Capsule From the Commonwealth Edison Company Quad Cities Unit 2 Nuclear Plant Reactor Vessel Radiation Surveillance Program," WCAP-10064, Westinghouse Electric Corporation, Pittsburgh, PA, (April 1981).

18. B. M. Oliver and H. Farrar, "Application of Helium Accumulation Fluence Monitors (HAFM) to LWR Surveillance, LWR-PV-SDIP Quarterly Progress Report, April 1983 - June 1983, NUREG/CR-3391, Vol. 2, HEDL-TME 83-22, Hanford Engineering Development Laboratory, Richland, WA, pp. RI-3 - RI-5 (April 1984).

ORNL EVALUATION OF THE ORR–PSF METALLURGICAL

EXPERIMENT AND "BLIND TEST"

F.W. Stallmann

Oak Ridge National Laboratory

Oak Ridge, Tennessee, USA

ABSTRACT

A methodology is described to evaluate the dosimetry
and metallurgical data from the two-year ORR-PSF metallurgical irradiation experiment. The first step is to
obtain a three-dimensional map of damage exposure parameter values based on neutron transport calculations and
dosimetry measurements which are obtained by means of the
LSL-M2 adjustment procedure. Metallurgical test data are
then combined with damage parameter, temperature, and
chemistry information to determine the correlation between
radiation and steel embrittlement in reactor pressure
vessels including estimates for the uncertainties.
Statistical procedures for the evaluation of Charpy data,
developed earlier, are used for this investigation. The
data obtained in this investigation provide a benchmark
against which the predictions of the "PSF Blind Test" can
be compared. The results of this investigation and the
Blind Test comparison will be discussed.

INTRODUCTION

The two-year Pressure Vessel Simulator (PVS) metallurgical
irradiation experiment at the Oak Ridge Research Reactor (ORR)
Poolside Facility (PSF) was performed in order to simulate, as
closely as possible, the irradiation conditions in commercial
reactor pressure vessels and surveillance capsules (Fig. 1). Of
primary interest was the question whether results obtained from

J. P. Genthon and H. Röttger (eds.), Reactor Dosimetry, 175–183.
© 1985 ECSC, EEC, EAEC, Brussels and Luxembourg

surveillance capsule evaluations in commercial power reactors can be extrapolated safely. Since there are considerable differences in fluence rate and fluence spectrum between the pressure vessel wall and the surveillance capsules, possible effects of these factors on the irradiation damage need to be investigated. The magnitude of these effects, if any, may also be different for different types of materials, e.g., plate material vs. welds or between materials of different chemical compositions.

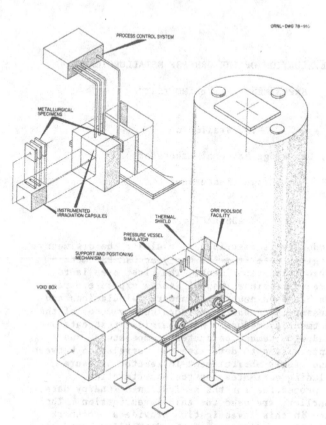

ORNL-DWG 78-910

PROCESS CONTROL SYSTEM

METALLURGICAL SPECIMENS

INSTRUMENTED IRRADIATION CAPSULES

THERMAL SHIELD

ORR POOLSIDE FACILITY

PRESSURE VESSEL SIMULATOR

SUPPORT AND POSITIONING MECHANISM

VOID BOX

Fig. 1. View of the PSF-PVS Facility.

In order to answer these questions, five capsules were irradiated in the PSF-PVS experiment, each containing metallurgical specimens of the same mix of plate and weld materials (see Fig. 2). Two of the capsules received high-fluence rates, characteristic of surveillance capsules (SSC capsules), and the other three were irradiated to about the same total fluence but at lower fluence rates and over a longer time period (two years vs. one to two months).

In order to serve its intended purpose, namely as a benchmark for testing damage prediction methodologies, both damage fluences

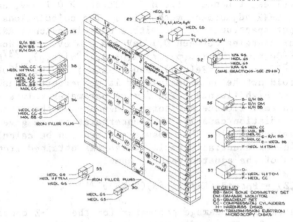

Fig. 2. Illustration of dosimeter and metallurgical specimen location in the irradiation capsules.

and materials damage must be determined with very high accuracy. Reliable estimates for the uncertainties of all data must be provided to ascertain whether differences between predictions and benchmark results are significant or attributable to measuring errors. High accuracies and reliable uncertainties are particularly important in determining possible effects of fluence rate and spectra, since these effects are likely to be small. A careful statistical evaluation of the PSF–PVS experiment is therefore necessary, details of which are reported in the next section.

MATERIALS AND PROCEDURES

The following input data were used for this analysis:

Neutron physics calculation. A detailed description of the calculation is given in Refs. 1,2.

Dosimetry. All dosimetry data used in this analysis were radiometric measurements provided by HEDL (2). Additional dosimetry provided by other U.S. and European Laboratories were not considered at this time.

Metallurgical tests. The metallurgical test results from this experiment are published in Refs. 3,4.

The first step in the analysis was the determination of the damage parameter values, fluence > 1.0 MeV, fluence > 0.1 MeV, and dpa using the LSL-M2 adjustment method (5). The calculations (1) were used as input spectrum together with a one-dimensional ANISN calculation for energies between 0.1 MeV and 0.4 eV. The dosimeters used were $^{63}Cu(n,\alpha)$, $^{46}Ti(n,p)$, $^{54}Fe(n,p)$, $^{58}Ni(n,p)$, $^{237}Np(n,f)$, $^{235}U(n,f)$, $^{59}Co(n,\gamma)$, $^{58}Fe(n,\gamma)$, and $^{45}Sc(n,\gamma)$. The $^{238}U(n,f)$ reaction could not be used because of large perturbations due to Pu burn-in at high fluence. The damage parameter values – including thermal fluence and total fluence – at the centers of the five irradiation capsules with uncertainties are given in Table 1. Values outside the capsule centers can be calculated from a cosine-exponential formula, which was obtained from least squares fits of the ajusted values, namely

$$P(X,Y,Z) = P_0 \cos B_X(X-X_0) \cos B_Z(Z-Z_0) e^{-\lambda(Y-Y_0)} \qquad (1)$$

P(X,Y,Z) is the given damage parameter for the X-Y-Z-coordinate system in Fig. 3. The coefficients of formula (1) are given in Table 2. More details can be found in Ref. 6.

Six different plate and weld materials were irradiated in each of the five metallurgical capsules. Type and chemical composition for each material is listed in Table 3. The raw Charpy data from Ref. 4 were processed with the CV81 Charpy curve fitting procedure (7) to obtain the values for ΔNDT and upper shelf drop with uncertainties for each irradiation capsule and material. The results are listed in Table 4. They agree within uncertainties with the values in Ref.4 (see Table 5). Details are given in Ref. 7.

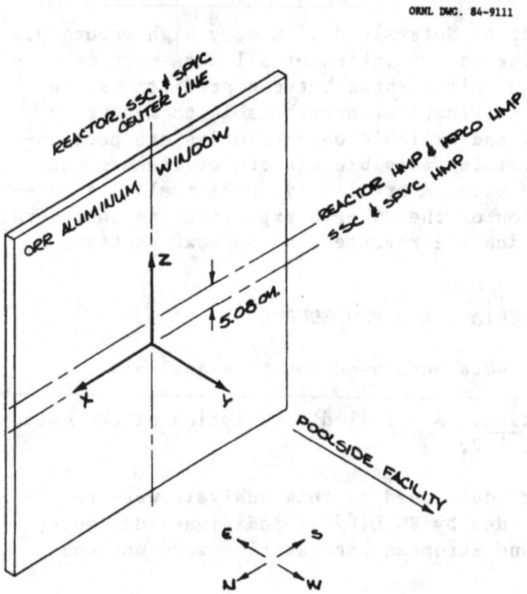

ORNL DWG. 84-9111

Fig. 3. Coordinate system for the ORR-PSF metallurgical experiment.

Table 1. Fluences and dpa at capsule centers

	$\phi>1.0$ MeV	Std. dev. (%)	$\phi>0.1$ MeV	Std. dev. (%)	$\phi<0.4$ eV	Std. dev. (%)	ϕ_{total}	Std. dev. (%)	dpa (10^{-2})	Std. dev. (%)
SSC1 H4	2.56*	5.1	7.74	5.8	1.26	7.4	14.20	5.8	4.07	4.9
SSC2 H9	5.50	5.1	16.84	5.8	2.79	7.4	30.55	5.5	8.80	4.9
0-T H14	4.10	5.1	12.26	5.8	6.29	7.6	27.66	5.8	6.56	4.9
1/4T H19	2.21	5.2	8.98	6.0	0.84	7.9	14.75	5.5	4.13	5.2
1/2T H24	1.05	5.4	5.83	6.0	0.27	8.3	9.17	5.6	2.39	5.4

*Read values for $\phi>1.0$ MeV, $\phi>0.1$ MeV, $\phi<0.4$ eV, and ϕ_{total} as 2.56 x 10^{19} neutrons/cm^2, etc.

Table 2. Fitting parameters for formula (1)

	P_0*	B_X (cm^{-1})	X_0 (cm)	B_Z (cm^{-1})	Z_0 (cm)	λ (cm^{-1})	Y_0 (cm)
SSC1							
$\phi t>1.0$ MeV	2.500E+19	0.0499	0.41	0.0436	0.97	0.176	13.29
$\phi t>0.1$ MeV	7.607E+19	0.0507	0.37	0.0464	0.80	0.134	13.29
dpa	3.995E-02	0.0502	0.38	0.0449	0.90	0.156	13.29
SSC2							
$\phi t>1.0$ MeV	5.341E+19	0.0528	-0.95	0.0457	0.03	0.176	13.29
$\phi t>0.1$ MeV	1.648E+20	0.0539	-0.88	0.0484	-0.02	0.134	13.29
dpa	8.580E-02	0.0533	-0.91	0.0470	0.02	0.156	13.29
0-T							
$\phi t>1.0$ MeV	3.924E+19	0.0517	-0.69	0.0395	0.72	0.107	24.05
$\phi t>0.1$ MeV	1.214E+20	0.0522	-0.64	0.0432	0.71	0.042	24.05
dpa	6.452E-02	0.0516	-0.67	0.0414	0.71	0.079	24.05
1/4T							
$\phi t>1.0$ MeV	2.143E+19	0.0478	-0.96	0.0378	1.30	0.134	28.56
$\phi t>0.1$ MeV	8.823E+19	0.0486	-0.86	0.0425	1.14	0.070	28.56
dpa	4.037E-02	0.0481	-0.91	0.0407	1.21	0.097	28.56
1/2T							
$\phi t>1.0$ MeV	1.016E+19	0.0441	-0.94	0.0349	1.94	0.146	33.70
$\phi t>0.1$ MeV	5.727E+19	0.0452	-0.79	0.0413	1.48	0.089	33.70
dpa	2.333E-02	0.0450	-0.83	0.0395	1.59	0.107	33.70

*Values for $\phi t > 1.0$ MeV and $\phi t > 0.1$ MeV are in neutrons/cm^2.

Table 3. List of materials and chemical compositions (wt-%)

Material	Heat code	Supplier	P	Ni	Cu
A302-B (ASTM reference plate)	F23	NRL	0.011	0.18	0.20
A533-B (HSST plate 03)	3PS, 3PT, 3PU	NRL	0.011	0.56	0.12
22NiMoCr37 forging	K	KFA	0.009	0.96	0.12
A508-3 forging	MO	MOL	0.008	0.75	0.05
Submerged arc weld (single vee type, A533-B base plate)	EC	EPRI	0.007	0.64	0.24
Submerged arc weld (single vee type A533-B base plate)	R	Rolls-Royce & Assoc. Ltd.	0.009	1.58	0.23

Table 4. 41J Charpy shift vs. fluence > 1.0 MeV for different materials in the PSF experiment

	SSC1	SSC2	0-T	1/4T	1/2T
A302-B Plate					
$\phi t > 1.0$ MeV (10^{19} n/cm^2)	2.59	5.41	3.73	2.15	1.05
ΔNDT- 41J (°C)	78	94	77	65	52
A533-B Plate					
$\phi t > 1.0$ MeV (10^{19} n/cm^2)	2.35	4.97	3.47	1.99	0.98
ΔNDT- 41J (°C)	71	84	71	69	52
22NiMoCr37 forging					
$\phi t > 1.0$ MeV (10^{19} n/cm^2)	1.64	3.44	2.80	1.51	0.73
ΔNDT- 41J (°C)	52	109	81	66	66
A508-3 forging					
$\phi t > 1.0$ MeV (10^{19} n/cm^2)	1.79	3.72	2.97	1.61	0.77
ΔNDT- 41J (°C)	15	39	27	23	22
Submerged arc weld (EC)					
$\phi t > 1.0$ MeV (10^{19} n/cm^2)	1.73	3.57	2.90	1.59	0.76
ΔNDT- 41J (°C)	112	123	125	96	94
Submerged arc weld (R)					
$\phi t > 1.0$ MeV (10^{19} n/cm^2)	2.47	5.06	3.59	2.08	1.03
ΔNDT- 41J (°C)	230	309	294	270	242

METALLURGICAL BLIND TEST

The metallurgical "Blind Test" (8) was initiated in order to test current damage prediction methodologies for pressure vessel surveillance against the established test data of the PSF-PVS benchmark experiment. The Blind Test participants were supplied with dosimetry and metallurgical test data from the SSC1 capsule, a detailed description of the experimental configuration, and a complete irradiation-temperature-time history. The participants were asked to determine the damage fluence at locations of the metallurgical specimen and to predict from it the amount of radiation damage in the specimen. A first summary of the Blind Test results is published in Ref. 9. Results were submitted by eight participants. The predicted values of fluence > 1.0 MeV were compared with the ORNL evaluation. All participants came to within +30% of the ORNL data and more than 60% of the predictions were within +10%. Neither over nor under prediction of fluences appears to be correlated to over or under prediction of damage by the same laboratory. Table 5 presents a summary of the Blind Test prediction of Charpy shifts compared to results of this evaluation. The predictions are fairly symmetrically scattered around the experimentally based evaluation with bounds which are roughly twice the variance of the statistical evaluation. Substantial and consistent under prediction of damage can be found only in the code R weld material, which appears to be more radiation-sensitive than other materials of similar chemical composition. Thus, the Blind Test results show neither consistent biases or other glaring defects of current damage prediction methodologies. Adequate safety margins must still be provided to avoid non-conservative predictions. The following suggestions for improvement are offered:

- Procedures for determining and reporting uncertainties should be included in the damage prediction methodology. Many Blind Test participants failed to report uncertainties altogether or did so only indirectly (e.g., by referring to contributing uncertainties). Uncertainties are important for meaningful benchmark comparisons and for the establishment of safety margins.

- Predictions based on fluence and chemical composition alone are not always reliable. Any new material should be tested in a variety of fluences to establish material-dependent trend curves with uncertainties which can be used for the prediction methodology.

CONCLUSION

Comparison of current damage prediction methodology with results from the PSF-PVS experiment shows no substantial biases or deficiencies, although improvements are desirable, particularly in

Table 5. Comparison between experimentally determined
Charpy shift and Blind Test predictions

	Determined from Charpy curves			Smallest and largest values predicted by Blind Test participants		Difference Blind Test – CV81	
	CV81* (°C)	Std. (°C)	MEA** (°C)	Min. (°C)	Max. (°C)	Min. (°C)	Max. (°C)
A302-B							
SSC1	78	+12	82	71	98	-7	+20
SSC2	94	+11	94	75	112	-19	+18
0-T	77	+10	81	71	96	-6	+19
1/4T	65	+18	67	65	81	0	+16
1/2T	52	+10	50	45	66	-7	+14
A533-B							
SSC1	71	+11	61	45	69	-24	-2
SSC2	84	+10	81	62	99	-22	+15
0-T	71	+13	75	60	87	-11	+16
1/4T	69	+ 9	69	54	63	-15	-6
1/2T	52	+10	53	26	52	-26	0
22NiMoCr37							
SSC1	52	+16	61	57	77	-5	+25
SSC2	109	+14	94	65	110	-44	+1
0-T	81	+16	72	63	97	-18	+16
1/4T	66	+18	78	52	76	-14	+10
1/2T	66	+13	56	45	64	-21	-2
A508-3							
SSC1	15	+ 7	20	6	43	-9	+18
SSC2	39	+ 7	39	11	53	-28	+14
0-T	27	+ 7	25	10	49	-17	+22
1/4T	23	+ 6	20	8	42	-15	+19
1/2T	22	+ 7	14	6	35	-16	+13
Submerged arc weld (EC)							
SSC1	112	+33	108	99	118	-13	+6
SSC2	123	+60	119	130	153	-7	+30
0-T	125	+50	124	121	135	-4	+10
1/4T	96	+18	94	91	115	-5	+19
1/2T	94	+20	89	63	103	-31	+9
Submerged arc weld (R)							
SSC1	230	+12	222	218	227	-12	-3
SSC2	309	+38	289	246	319	-63	+10
0-T	294	+15	286	239	288	-55	-6
1/4T	270	+25	256	180	218	-90	-52
1/2T	242	+44	239	143	189	-99	-53

*ORNL evaluation.
**Evaluation in Ref. 4.

the field of uncertainty analysis. Space does not permit a more thorough discussion of the experimental results, such as effects of fluence rate and spectrum. A detailed discussion is given in Ref. 7.

REFERENCES

1. R.E. Maerker and B.A Worley. "Calculated Spectral Fluences and Dosimeter Activities for the Metallurgical Blind Test Irradiations at the ORR-PSF," to be presented at the 5th ASTM-EURATOM Symposium on Reactor Dosimetry, Geesthacht, FRG, September 24-28, 1984.

2. W.N. McElroy and F.B.K. Kam, eds., PSF Blind Test SSC, SPVC Physics, Dosimetry, and Metallurgy Data packages, distributed to Blind Test participants, February 17, 1984.

3. J.R. Hawthorne, B.H. Menke, and A.L. Hiser, Light Water Reactor Pressure Vessel Surveillance Dosimetry Improvement Program: Notch Ductility and Fracture Toughness Degradation of A 302-B and A 533-B Reference Plates from PSF Simulated Surveillance and Through-Wall Irradiation Capsules, NUREG/CR-3295, MEA-2017, Vol. 1, U.S. Nuclear Regulatory Commission, Washington, DC, 1983.

4. J.R. Hawthorne and B.H. Menke, Light Water Reactor Pressure Vessel Surveillance Dosimetry Improvement Program: Notch Ductility and Tensile Strength Determinations for PSF Simulated Surveillance and Through-Wall Irradiation Capsules, NUREG/CR-3295, MEA-2017, Vol. 2, U.S. Nuclear Regulatory Commission, Washington, DC, 1983.

5. F.W. Stallmann, "LSL-M1 and LSL-M2: Two Extensions of the LSL Adjustment Procedure for Including Multiple Spectrum Locations," to be presented at the 5th ASTM-EURATOM Symposium on Reactor Dosimetry, Geesthacht, FRG, September 24-28, 1984.

6. F.W. Stallmann, Determination of Damage Exposure Parameter Values in the PSF Metallurgical Irradiation Experiment, NUREG/CR-3814, ORNL/TM-9166, U.S. Nuclear Regulatory Commission, Washington, DC, 1984.

7. F.W. Stallmann, Statistical Evaluation of the Metallurgical Test Data in the ORR-PSF-PVS Irradiation Experiment, NUREG/CR-3815, ORNL/TM-9207, U.S. Nuclear Regulatory Commission, Washington, DC, 1984.

8. W.N. McElroy and F.B.K. Kam, eds., PSF Blind Test Instructions and Data Packages, distributed to Blind Test participants, March 22, 1983.

9. G.L. Guthrie, E.P. Lippincott, and E.D. McGarry, Light Water Reactor Pressure Vessel Surveillance Dosimetry Improvement Program: PSF Blind Test Workshop Minutes, HEDL, April, 1984.

STATUS OF REGULATORY ISSUES AND RESEARCH IN

LIGHT WATER REACTOR SURVEILLANCE DOSIMETRY IN THE U.S.

A. Taboada, P. N. Randall, and C. Z. Serpan, Jr.

U. S. Nuclear Regulatory Commission

Washington, D. C.   20555 U.S.A.

ABSTRACT

Operating experience with light water reactors has generated regulatory needs for more precise information on neutron radiation embrittlement of vessel beltline material. Accurate prediction of fluence exposure continues to be required for setting pressure-temperature limits for start-up and cooldown operations of reactors and for improved surveillance programs. But, the emergence of the pressurized thermal shock issue has established a new impetus for more precise and accurate flux and fluence determinations and predictions at specific locations of the vessel wall. Such issues have led NRC to establish a cooperative research program to validate and improve methodologies used to predict radiation damage in pressure vessel steels. The program includes the validation of experimental and analytical methods of fluence determination in surveillance programs, data bases, and estimates of uncertainties in predictive methods using well-defined, reproducible benchmark facilities traceable to standard fields. Finally, the program culminates in a set of ASTM standards for analysis of LWR vessel irradiation surveillance.

INTRODUCTION

A major safety concern of the U.S. Nuclear Regulatory Commission (NRC) in the regulation of nuclear power plants is

J. P. Genthon and H. Röttger (eds.), Reactor Dosimetry, 185–193.
© *1985 ECSC, EEC, EAEC, Brussels and Luxembourg*

the assurance of continuing integrity of the reactor's primary system. In this regard the NRC has issued regulations[1] on fracture prevention requiring that the reactor coolant pressure boundary, which includes the reactor pressure vessel, be designed in a manner that during operation and under accident conditions; the boundary behaves in a nonbrittle manner, and the probability of rapidly propagating fracture is minimized.

Because of the embrittling effect of neutron irradiation on reactor pressure vessel steel, the loss of fracture toughness and ductility during operation must be carefully evaluated and controlled to assure that, despite such loss, there is always enough reserve to preclude crack initiation and uncontrolled propagation from accidental loading.

To evaluate these neutron embrittlement effects, the NRC requires a pressure vessel surveillance program[2] that monitors changes in fracture toughness of the vessel material and permits prediction of the end-of-life embrittlement for the vessel beltline region with maximum neutron exposure. In this program, metallurgical test specimens made of reactor vessel material and dosimetry sensors are placed in several capsules at or near the vessel inner wall. The capsules are removed after different periods of time so that the final capsule fluence corresponds to the vessel end-of-life fluence. Results are used to verify or readjust predictions of embrittlement. Two measures of radiation damage are obtained from Charpy V-notch impact test specimens in the capsule; the shift in the brittle-to-ductile transition temperature measured at 30 foot-pounds, and the decrease in the Charpy upper-shelf energy level.

REGULATORY ISSUES

NRC Guidance for Predicting Radiation Damage

If the surveillance capsules provide sufficient data to make a reliable trend line pattern of behavior, future embrittlement of the vessel can be reasonably predicted by extrapolation. But if adequate data are not available, the NRC provides a more conservative approach in Regulatory Guide 1.99, "Radiation Damage to Reactor Vessel Materials." This guide provides procedures and empirically determined trend curves based on the overall results of power reactor surveillance capsules reported to date. The trend curves account for major chemical elements found to have a significant effect on embrittlement.

186

Presently, Regulatory Guide 1.99 is being updated[3] to reflect recent studies of the physical basis for neutron radiation damage and more recent efforts to correlate damage to chemical composition and fluence. Draft revision 2, which forms the basis for present NRC licensing reviews on the subject, contains several significant changes. Welds and base metal are treated separately. The calculative procedure gives the adjustment of reference temperature as the product of a chemistry factor and a fluence factor as before, but nickel content is added as a variable in addition to copper, and phosphorus is removed. The fluence factor has been modified to provide a flatter trend curve slope, and one that is steeper at low fluences than at high fluences. Also guidance is given for calculating attenuation of damage throughout the vessel wall using displacements per atom as the damage function.

Pressurized Thermal Shock

Initially, the main function of the pressure vessel surveillance programs, was to provide the basis for setting upper limits for pressure as a function of temperature during heatup and cooldown operations of a reactor. Operating conditions of these reactors normally contained sufficient conservatism to permit relatively large uncertainties in damage predictions and vessel fluence calculations for this purpose.

More recently, however, the NRC has identified the issue of pressurized thermal shock (PTS) as a generic problem which poses a much more significant challenge to long term operation for certain Pressurized Water Reactors (PWR). The issue of PTS arises because transients and accidents can occur that result in severe overcooling, and coupled with internal pressure could develop high stresses on the vessel inner wall. As long as the fracture resistance of the material is relatively high, such events are not expected to cause failure. However, if the fracture toughness of the vessel material has been reduced sufficiently by neutron irradiation, severe PTS could cause propagation of small flaws near the inner surface into a crack that extends completely through the vessel wall and is large enough to threaten vessel integrity and, therefore, core cooling capacity.

After considerable study on the subject the NRC has developed a tentative resolution to the problem and has published a proposed amendment to its regulations[4] which is intended, if adopted, to produce an improvement in the safety of PWR vessels. It identifies those corrective actions that may be required to prevent or mitigate potential PTS events. The proposed amendment would: (1) establish screening criteria related to fracture resistance of PWR vessels during PTS events; (2) require analyses and implementation of flux reduction programs to avoid exceeding the screening criteria; and (3) require detailed safety evaluations to be performed before plant operation beyond the screening criteria will be considered.

The screening criteria established are an RT_{NDT} value of 270°F for plates, forgings and axial welds of the vessel and 300°F for circumferential welds. On the basis of available information it appears that most, if not all, plants can avoid reaching the screening criteria by timely implementation of a flux reduction program.

The flux reduction proposals evaluated by the NRC involve fuel management techniques such as rearranging core fuel elements or replacing those elements that contribute most to damage accumulation with spent fuel or nonfuel bearing elements. One of the original studies on the subject was presented by Guthrie et al[5] at the 4th ASTM-EURATOM Symposium on Dosimetry in Gaithersburg. Such an approach requires a new series of neutron flux calculations for the vessel wall and the surveillance capsule locations and may identify new flux peak locations. This is important because it may change the plate or weld that is limiting in the radiation damage analysis.

Dosimetry Improvement Research

As early as 1977 the NRC recognized the need for accurate and validated fluence measurements and data analysis procedures for predicting radiation damage to vessel material, at which time it established the LWR Pressure Vessel Surveillance Dosimetry Improvement Programs (LWR-PV-SDIP)[6]. The overall objective of this program is to establish the data, analysis methods and procedures required to predict the integrated effect of neutron exposure on reactor vessels and related structures with reasonable certainty. The goal of this program is for accuracies in the 5% to 15% range.

A vigorous research effort attacking these problems also exists worldwide and a strong cooperative link exists between the NRC supported research and research at CEN/SCK (Mol, Belgium), KFA (Julich, Germany), EPRI (Palo Alto, USA) and several UK laboratories, as well as numerous industrial organizations[7].

These combined efforts are culminating in a set of 21 standards[8] being prepared by the ASTM E-10 Committee on Nuclear Technology and Application, which will provide a complete basis for:

o   Calculating neutron flux-spectra and exposure parameters
o   Performing and analyzing neutron dosimetry
o   Evaluating and correlating radiation damage in surveillance specimens
o   Applying results to current and end-of-life predictions of the condition of reactor vessel materials

In support of the validation and calibration studies needed to accomplish the program objective, simulated reactor vessel environments have been established throughout the world at research reactors as well as power reactors. Table 1 provides summary information on the research reactor fields. Experiments associated with these high- and low-flux versions of PWR vessel mock-ups are in progress in the US, Belgium, France and the United Kingdom. The US low-flux version is known as the ORNL Poolside Critical Assembly[7] (PCA), and the high-flux version is known as the Oak Ridge Reactor Poolside Facility (ORR-PSF), both located at Oak Ridge, Tennessee. These facilities provide well-characterized neutron environments for dosimetry studies, neutron field calculations and controlled metallurgical tests on radiation damage. Experiments have been completed at both facilities.

Poolside Critical Assembly Experiment

The PCA experiment was initiated to establish a low power experimental/calculational benchmark for PWR pressure vessel configurations to verify radial neutron exposure gradients and lead factors. Four clean, low neutron flux benchmark arrangements[9] of a simulated LWR thermal shield and pressure vessel have been studied, using various size water gaps between the core edge, the thermal shield and the reactor vessel, to evaluate transport theory methods. A blind test was conducted by eleven laboratories on calculations of neutron flux spectra and reaction rates of dosimetry sensors at specific locations in the PCA. The following variables were investigated: Plant dimensions; core power distribution; reactor physics computations; selection neutron exposure units; and dosimetry measurements.

TABLE 1

LWR-PV BENCHMARK FIELD FACILITIES

Benchmark Field Facility	Location	Anticipated Operation Schedule	Main Purpose
$^{252}Cf/^{235}U$	NBS, US	1975-Open	Standard fields for cross-section testing and validation; emphasis is on equivalent fission flux calibrations and RM fluence counting standard.
PCA-PV	ORNL, US	1978-84	Data base for the "PCA Physics-Dosimetry Blind Test": Low-power experimental/calculational benchmark for different LWR-PV configurations; emphasis is on verification of radial neutron exposure gradients and lead factors; i.e., confirmation of radial through-wall fracture toughness and embrittlement predictions.
PSF-PV	ORNL, US	1980-84	Data base for the "PSF Physics-Dosimetry-Metallurgy Blind Test": High-power LWR-PV physics-dosimetry-metallurgical test; emphasis is on high-temperature and high-fluence simulation of PWR environmental conditions and verification of neutron damage gradients; i.e., confirmation of radial through-wall fracture toughness and embrittlement predictions.
PSF-SDMF	ORNL, US	1979-Open	High-power LWR-PV benchmark: Emphasis is on verification of surveillance capsule perturbations; specific RM, SSTR, HAFM, and DM verification tests, and quality assurance evaluations of commercial dosimetry materials and services; i.e., confirmation of the physics-dosimetry methods, procedures, and data recommended for in-situ in- and ex-vessel surveillance programs.
VENUS	CEN/SCK, Mol, Belgium	1982-Open	Low-power LWR-PV core source boundary benchmark: Emphasis is on verification of effects of new and old fuel management schemes and accuracy of azimuthal lead factors; i.e., confirmation of azimuthal PV-wall fracture toughness and embrittlement predictions.
NESDIP	AEEW, Winfrith, UK	1982-Open	Low-power LWR-PV cavity benchmark: Emphasis is on different PWR configurations and verification, via cavity measurements, of neutron exposure gradients and lead factors; i.e., confirmation of radial through-wall fracture toughness and embrittlement predictions.
DOMPAC	CEA, Fontenay, France	1980-1983	Low-fluence experimental/calculational benchmark for a specific PWR configuration: Emphasis is on verification of surveillance capsule perturbations and PV-wall neutron exposure and damage gradients; i.e., confirmation of radial PV-wall fracture toughness and embrittlement predictions.

Results of PCA studies and the blind test provide the benchmark data necessary for assessing the accuracy of procedures for analyzing surveillance results[9]. These studies indicate that routine LWR power plant calculations of flux, fluence and spectrum using current transport methods can be as accurate as ± 15% (1σ) if properly modeled and benchmarked. Otherwise errors in predictions may be as great as a factor of 2 or more.

Oak Ridge Reactor Poolside Facility Experiments

The ORR-PSF experiment[9] was designed to simulate the surveillance capsule-pressure vessel configurations in power reactors and to test the validity of procedures which determine the radiation damage in the vessel from test results of surveillance specimens. This 2-year irradiation experiment consisted of dosimetry and metallurgical specimens of several alloys irradiated to a maximum of 6×10^{19} n/cm^2, E >1.0 MeV. A unique feature was the arrangemement of the specimens to simulate radial locations through the vessel wall to permit confirmation of through-wall fracture toughness. Emphasis was on damage correlation to test current embrittlement prediction methodologies. For this purpose a PSF metallurgical blind test was initiated in which individual participants were given only the information normally contained in a surveillance report. The goal was to predict, from this limited information, the metallurgical test results of the pressure vessel wall specimens.

The experimental metallurgical data from the PSF blind test indicate that, in most cases, the surveillance capsule results gave accurate prediction of the wall position material properties and that the in-pressure-vessel toughness gradient is small[10]. Preliminary blind test results indicate that the prediction formulas are adequate as rough approximations but do not handle well the correlation between embrittlement and such variables as fluence, fluence rate, neutron spectrum, chemistry and others. It appears that realistic estimates of uncertainties need to be established in conjunction with prediction formulas and that prediction formulas that were derived from a data base may not be valid for materials with composition outside that range.

CONCLUSION

Operating experience with light water reactors and concern over the pressurized thermal shock question have generated a need for more precise information on neutron radiation embrittlement of reactor vessel beltline material. More precise and accurate flux and fluence determinations are required as well as improved fluence and damage predictions at specific locations in the beltline region and throughout the vessel wall. Such issues have led the NRC to establish a cooperative research program to validate and improve, through benchmark experiments, methodologies used to predict radiation damage in pressure vessel steels. Blind test experiments have been completed in the PCA and PSF facilities which indicate that routine LWR Power plant calculations of flux, fluence and spectrum using current transport methods can be as accurate as ± 15% (1 σ) if properly modeled and benchmarked. Preliminary PSF results indicate that surveillance capsule results provide an adequate basis for prediction of wall position material properties and that prediction formulas for embrittlement of pressure vessel steel are adequate as rough approximations.

ACKNOWLEDGEMENTS

This paper reviews work done by NRC staff and by NRC contractors whose dedication to completing this work was outstanding. A special acknowledgement should go to the members of the representatives and participants of those organizations who have cooperated in accomplishing these efforts. Individuals are too numerous to mention but have been identified as contributors to the series of LWR PV-SDIP NUREG reports published by the NRC.

REFERENCES

1. Code of Federal Regulations, Title 10, Part 50, Appendix A, "General Design Criteria for Nuclear Power Plants" and Appendix G, "Fracture Toughness Requirements"

2. Code of Federal Regulations, Title 10, Part 50, Appendix H, "Reactor Vessel Material Surveillance Program Requirements"

3. P. N. Randall, "Basis for Revision 2 of U.S. NRC Regulatory Guide 1.99," to be presented at the IAEA Specialist Meeting on Radiation Embrittlement, Vienna, October 10, 1984 and to be published by ASTM.

4. "Analysis of Potential Pressurized Thermal Shock Events," U.S. Federal Register, Volume 49, No. 26, Page 4498, February 7, 1984.

5. G. L. Guthrie, et al, "A Preliminary Study of the Use of Fuel Management Techniques for Slowing Pressure Vessel Embrittlement" Proceedings of the 4th ASTM-Euratom International Symposium on Reactor Dosimetry, March 22-26, 1982, National Bureau of Standards, Gaithersburg, Maryland. NUREG/CP-0029, Volume 1.

6. W. N. McElroy, et al, "LWR Pressure Vessel Surveillance Dosimetry Improvement Program, 1980 Annual Report," NUREG/CR 1747, HEDL-TME 80-73 (April 1979)

7. W. N. McElroy, et al, "Surveillance Dosimetry of Operating Power Plants" Proceedings of the 4th ASTM-Euratom International Symposium on Reactor Dosimetry, March 22-26, 1982, National Bureau of Standards, Gaithersburg, Maryland. NUREG/CP-0029, Volume 1.

8. ASTM E706-81a, "Standard Master Matrix for LWR PV Surveillance Standards, " ASTM Annual Book of Standards, Volume 12.02, 1983

9. W. N. McElroy, et al, "LWR Pressure Vessel Surveillance Dosimetry Improvement Program: PCA Experiments, Blind Test and Physics-Dosimetry Support for the PCA Experiments," NUREG/CR 3318, HEDL-TME-84-1 (September 1984)

10. W. N. McElroy, et al, "LWR Pressure Vessel Surveillance Dosimetry Improvement Program, 1983 Annual Report," NUREG/CR 3391, Volume 3 HEDL-TME 83-23 (June 1984)

PART II
FAST REACTORS

SUPER PHENIX

SURVEILLANCE DES STRUCTURES SOUMISES A IRRADIATION

PROGRAMME DE MESURES ET CALCULS NEUTRONIQUES

J.C. CABRILLAT(*) - G. ARNAUD(*) - D. CALAMAND(*)
D. MAIRE(**) - G. MANENT(***) - A.A. TAVASSOLI(****)

*   CEA/IRDI/DRNR - CEN CADARACHE
      13115 - SAINT PAUL LEZ DURANCE (FRANCE)
**   NOVATOME - La Boursidière 92-LE PLESSIS ROBINSON
***   CEA/IRDI/DRE - CEN CADARACHE
**** CEA/IRDI/D.Tech - CEN Saclay
      91 - GIF SUR YVETTE (FRANCE)

RESUME

Pour le réacteur à neutrons rapides SUPER-PHENIX, l'évolution en fonction de l'irradiation des propriétés mécaniques de l'acier constituant le sommier -structure fixe assurant l'alimentation et le supportage du coeur- fait l'objet d'études et d'un suivi particulier.

La mise en oeuvre s'appuie sur des irradiations d'éprouvettes en acier, effectuées actuellement dans le réacteur PHENIX, et projetées dans le réacteur SUPER PHENIX dès les premiers cycles de fonctionnement. Pour des raisons liées à l'interprétation des résultats, elles sont accompagnées d'un important programme de dosimétrie couplant calcul et mesure.

Les motivations, la définition de la structure, de la conduite et des matériels mis en oeuvre dans ce programme de dosimétrie ainsi que les premiers résultats d'une comparaison calcul-mesure fait l'objet de cette présentation.

INTRODUCTION

Pour les conditions de fonctionnement (température, contraintes mécaniques, flux neutronique) rencontrées dans les réacteurs à neutrons rapides, l'évolution des propriétés mécaniques des matériaux constituant certaines structures fixes internes à la cuve est entâchée d'incertitudes importantes ; elles résultent du concours d'incertitudes sur l'évaluation des doses reçues et d'incertitudes sur le comportement des matériaux soumis à irradiation.

J. P. Genthon and H. Röttger (eds.), Reactor Dosimetry, 197–208.
© 1985 ECSC, EEC, EAEC, Brussels and Luxembourg

Cette remarque prend toute son importance, vis-à-vis de la durée de fonctionnement et de la sûrete de la centrale SUPER PHENIX, pour le sommier (Fig. 1) assurant le supportage et l'alimentation du coeur. Elle est à la base d'un important programme d'étude et de suivi visant :

1/ la levée par anticipation sur la réalité, des incertitudes sur l'évolution des propriétés mécaniques de l'acier du sommier.

2/ la levée des incertitudes sur les valeurs des flux et des dommages vus par le sommier.

La réalisation pratique s'appuie :

a/ d'une part sur des irradiations d'éprouvettes d'acier, prélevées dans le matériau du sommier, pour des conditions proches de celles du fonctionnement de la structure sous surveillance soit :
- une campagne préliminaire dans le réacteur PHENIX se terminant en 1984
- une campagne "in situ" dans le réacteur SUPER PHENIX, prévue dès les premiers cycles de fonctionnement (1986)

b/ d'autre part des mesures du niveau et spectre de flux vus par les éprouvettes et par le sommier, plus particulièrement par sa plaque supérieure (Fig. 1).

En se situant du point de vue neutronique, cette présentation définit et justifie les options de ce programme concernant :

- le paramètre quantifiant les dommages dûs au bombardement neutronique et assurant la corrélation des diverses observations et leurs transcriptions au sommier
- les emplacements, durées et déroulement des irradiations
- la dosimétrie.

Une comparaison préliminaire calcul-expérience est ensuite discutée pour les premiers résultats de mesures obtenus dans PHENIX.

PARAMETRE DE CORRELATION

Le nombre d'atomes déplacés par atome cible (D.P.A.) a été préféré à des grandeurs intégrales, telle, en particulier, la fluence supérieure à 0,1 Mev.

Le modèle en usage au C.E.A. détaillé en (1) est inspiré des travaux de Lindhard (2). Pour le sommier de SUPER-PHENIX ce choix trouve ses motivations dans les faits suivants :

- la population neutronique est estimée en majorité (94 %) en dessous de 0,1 Mev (tableau 1)
- la contribution des neutrons en dessous de 0,1 MeV aux déplacements d'atomes est importante (50 %) (tableau 1)
- ce modèle donne une réponse cohérente, dans son principe avec les phénomènes supposés responsables (3) de la fragilisation de l'acier à la température de fonctionnement du sommier (400° C) : les créations de paires lacunes-intersticiels.

PLAGE D'ENERGIE	CENTRE DU COEUR		SOMMIER	
	Flux %	D.P.A. (%)	Flux %	D.P.A. (%)
E > 1 Mev	7,1	43,5	0,1	2,1
1 Mev>E>0,1 MeV	40,6	46	6,1	48,1
0,1 Mev>E>0,01 MeV	25,4	8,6	19,7	43,0
0,01 Mev>E>1 Kev	18,8	1,9	18,0	6,3
1 Kev > E > 100 eV	2,9	-	22,4	0,3
100 ev > E	5,2	-	33,7	0,2
TOTAL	100,0	100,0	100,0	100,0

TABLEAU 1 : REPARTITION DE LA POPULATION NEUTRONIQUE ET DES DOMMAGES EN FONCTION DE L'ENERGIE

Toutefois ce choix appelle quelques remarques conditionnant la conduite des irradiations :

- le D.P.A. se réfère à un modèle calculé : sa détermination "expérimentale" implique une bonne connaissance du spectre neutronique.
- les incertitudes liées à la transcription des résultats au sommier (incertitudes dues au modèle) seront minimales si les spectres neutroniques vus par les éprouvettes sont proches de celui vu par la structure étudiée.

Dans ces conditions la dose maximale attendue (toutes incertitudes comprises) est de 1.4 D.P.A. pour la plaque supérieure du sommier en fin de vie de la centrale.

DESCRIPTION DES IRRADIATIONS

Emplacements d'irradiation

Leur détermination s'appuie sur des valeurs de flux et D.P.A. issues de calculs de propagation neutronique à une et deux dimensions cylindriques [formulaire PROPANE DO (4)] dans la cuve des réacteurs PHENIX et SUPER PHENIX ; ils sont complétés par des calculs en géométrie "hexagonale-plane" du coeur afin de préciser les effets locaux de géométrie dans les zones de couvertures radiales (formulaire CARNAVAL IV (5)).

Trois zones vérifient les conditions requises : bonne accessibilité, niveau de flux permettant une anticipation des phénomènes vus par le sommier, spectre dégradé. Leurs caractéristiques d'irradiation, comparées à celles du sommier sont données tableau (2). Il s'agit de :

1. PHENIX   zone de stockage interne (Fig. 2) (irradiation DINOSAURE)
2. SUPER-PHENIX   1ère rangée d'assemblages de protection (Fig. 3) (irradiations DIPLODOCUS)
3. SUPER-PHENIX   Dernière rangée d'assemblages de couverture (Fig. 3) (irradiations IGUANODON) avec toutefois un spectre plus dur, mais des temps de séjour courts.

REACTEUR	SUPER PHENIX	PHENIX	SUPER PHENIX	SUPER PHENIX
Emplacement	Sommier	Stockage interne	Protect. latérale	Couvert. radiales
Irradiation		DINOSAURE	DIPLODOCUS	IGUANODON
Section DPA 1 groupe (barns) (indicateur de spectre)	44	100	170	> 200
T^*(JEPP)	~ 8800	700-800	4300	640-1500

TABLEAU 2 : CARACTERISTIQUES DES EMPLACEMENTS D'IRRADIATION

* T : temps moyen permettant d'atteindre la dose maximale de 1,4 DPA attendue sur la plaque supérieure du sommier en 30 ans (~ 8800 JEPP)

JEPP : journée équivalente à pleine puissance du réacteur.

Durées des irradiations

Les temps longs d'irradiation interdisent pratiquement la répétition d'une mesure en cas d'objectifs non atteints ou dépassés (dose) ce qui conduit, entre autres causes, à structurer les mesures en fonction des incertitudes importantes :

a) l'incertitude qui touche les évaluations prévision-nelles de dose (± 40 à 50 %) implique
- une mesure directe de flux et donc de débit de dose à hauteur du sommier de SUPER PHENIX (tableau 3 - Dosimètrie du sommier)
- la mise en place d'irradiations parallèles, exploitées en un temps compatible avec une modification du déroulement de l'irradiation principale. Cette stratégie est adoptée pour DINOSAURE (PHENIX) et DIPLODOCUS (SUPER PHENIX). Ces irradiations secondaires sont mises à profit pour compléter la prise d'information (tableau 3).

b)   l'incertitude sur le paramètre de corrélation motive
     en partie l'irradiation IGUANODON afin de cerner
     l'influence des spectres neutroniques, par rapport
     aux résultats attendus de l'irradiation principale
     DIPLODOCUS (tableau 3).

DOSIMETRIE

Phénix : Dinosaure

La dosimètrie de l'irradiation principale (détecteurs,
fer, nickel, cobalt, Niobium) dont la maitrise incombe à
DEBENE (PETTEN) est donc complétée sur l'initiative du CEA par
une irradiation d'isotopes lourds (DINOSAURE 2) dans une position similaire à l'irradiation principale (Fig. 2). Compte
tenu des informations attendues (débit de dose, puissance en
zone de stockage), les isotopes choisis couvrent la totalité
du spectre énergétique:U235, Pu239, U238, Pu242, Np237.

La détermination des taux de fission (spectrométrie sur
les produits de fission) est complétée par une gammamétrie sur
les conteneurs en acier des isotopes lourds.

Super-Phénix

Les principales contraintes du choix, de la nature et du
conditionnement des dosimètres sont, des temps d'irradiation
et de récupération des détecteurs longs (plusieurs années, 1
an minimum respectivement), une mise en oeuvre en sodium à
température élevée (400 à 550° C), une dynamique sur les flux
importante, la volonté d'une information fiable.

La nature et la présentation des dosimètres sont donnés
dans le tableau 4.

Ces détecteurs sont conditionnés sous atmosphère neutre
dans des boîtiers en acier inoxydable avec une double enceinte
pour le Niobium et le Neptunium.

Pour les irradiations DIPLODOCUS et IGUANODON, ils sont
regroupés par lot de cinq détecteurs de natures différentes ;
ces lots sont répartis tous les 25 cm environ dans les 7
capsules concernées sur une hauteur de 2 mètres environ.
L'interprétation de ces mesures est ainsi susceptible d'attri-
buer à chaque éprouvette un nombre d'atomes déplacés.

TABLEAU 3
PANORAMA DES IRRADIATIONS

IRRADIATIONS (type) . Dénomination . Réacteur . Emplacement	NOMBRE D'IRRADIATIONS ELEMENTAIRES ET DENOMINATIONS		TEMPS DE SEJOUR	INSTRUMENTATION
<u>DINOSAURE</u>	2.			Eprouvettes
- PHENIX	DINOSAURE 1		800 JEPP	Dosimètres
- STOCKAGE INTERNE	DINOSAURE 2		(1 à 1,4 DPA)	Détecteurs de température
<u>IGUANODON</u>	3.			
- SUPER PHENIX	IGUANODON 1		640 JEPP	Eprouvettes
			(1 DPA)	Dosimètres
- COUVERTURES	IGUANODON 2		640 JEPP	Détecteurs de
RADIALES			(2 DPA)	température
	IGNANODON 3		1440 JEPP	
			(2 DPA)	
<u>DOSIMETRIE DU SOMMIER</u>	3.			
		1	∼ 480 JEPP	
- SUPER PHENIX		2	480 JEPP	Dosimètres
- COEUR INTERNE		3	640 JEPP	
<u>DIPLOCODUS</u>	4.			
- SUPER PHENIX	DIPLODOCUS	1	1400 JEPP (0,5 DPA)	Eprouvettes
- PROTECTION	DIPLODOCUS	2	3000 JEPP (0,9 DPA)	Dosimètres
LATERALE	DIPLODOCUS	3	4320 JEPP (1.4 DPA)	Détecteurs de température
	DIPLODOCUS	4	(réserve)	

NATURE	REACTION	PRESENTATION
Fer	^{54}Fe (n,p)	Feuille
Ni	^{58}Ni (n,p)	Feuille
Co	^{59}Co (n,γ)	Feuille
Nb	^{93}Nb (n,n')	Fil
Np	^{237}Np (n,f)	Fil (alliage neptunium vanadium)

TABLEAU 4

La détermination de la dose au niveau de la plaque supérieure du sommier s'effectue à partir d'une interpolation des valeurs mesurées par des détecteurs, placés dans les évidements pratiqués dans les pièces massives de la partie basse de l'assemblage (tableau 3). La connaissance de la pente de variation du flux et donc des dommages, très importante dans cette zone (30 %/cm), est appréhendée par un empilement de détecteurs dans le pied et une répartition des dosimètres en plusieurs étages au niveau de la portée sphérique (Fig. 4).

De plus un recoupement avec les valeurs mieux connues de flux au niveau de la colonne combustible est obtenu par une répartition, dans des capsules instrumentant les assemblages affectés à cette mesure, du lot des cinq dosimètres précédemment défini.

PHENIX : PREMIERS RESULTATS

L'interprétation première des résultats de l'essai DINOSAURE (isotopes lourds) a été de conforter le temps de séjour de l'irradiation principale, obtenu en son temps à partir de calculs simples (une dimension cylindrique). Elle constitue de plus la première étape de travaux poursuivis actuellement dans le sens :

* d'une détermination précise des débits de dose et de puissance en zone de stockage, pouvant s'appuyer sur des techniques de type "unfolding" et incluant les résultats de gammamétrie.
* d'une qualification des méthodes de calcul susceptibles de rendre compte des niveaux de flux et de puissance des assemblages en zone de stockage.

Cette interprétation préliminaire s'appuie sur une comparaison calcul-mesure des taux de fission des isotopes lourds.

Les calculs de base, images des chargements du coeur et de la zone de stockage interne pendant l'irradiation, possèdent les caractéristiques suivantes :
- 25 groupes d'énergie
- géométrie hexagonale plane
- théorie de la diffusion
- formulaire CARNAVAL IV
- sections efficaces de fission :
  U235, Pu239 : origine : formulaire CARNAVAL IV (5)
  Pu242 : origine : ENDFB V
  Np237 : origine (6).

La comparaison donnée tableau 5 montre une sous-estimation des valeurs calculées, plus accentuée en moyenne aux hautes énergies.

ISOTOPES	TAUX DE FISSION MESURÉS ÉVENEMENTS/SEC. (INCERT. en %)	TAUX DE FISSION CALCULÉS ÉVENEMENTS/SEC. (INCERT. $\pm 50\%^a$)	$\dfrac{E-C}{C}$ (%)
U 235	$1,1.10^{-09}$ (4,3 %)	$1,0.10^{-09}$	9 %
Pu 239	$1,42.10^{-09}$ (3,6 %)	$1,1.10^{-09}$	29 %
Pu 242	$1,11.10^{-11}$ (19 %)	$9,4.10^{-12}$	18 %
Np 237	$1,5.10^{-11}$ (18 %)	$1,1.10^{-11}$	36 %

TABLEAU 5 : COMPARAISON DES TAUX DE FISSION MESURES ET CALCULES

a) L'incertitude tient compte des approximations de calcul (théorie de la diffusion, ...).

Finalement, la valeur "expérimentale" du taux de déplacement par atome est estimée à l'aide d'un ajustement du flux calculé, condensé à 6 groupes d'énergie, par une méthode de moindre x^2. La comparaison de la valeur centrée ainsi obtenue ($1,37.10^{-7}$ D.P.A./JEPP) à la valeur estimée de façon préliminaire ($1,7.10^{-3}$ DPA/JEPP) ne suscite pas de modification fondamentale de la durée de l'irradiation principale.

CONCLUSION

Un important programme expérimental, en cours d'exécution dans le réacteur PHENIX, et projeté dans le réacteur SUPER PHENIX, est mis en oeuvre afin de préciser le comportement de l'acier du sommier de SUPER PHENIX sous irradiation, jusqu'à la dose limite vue par cette structure.

Du point de vue neutronique, les incertitudes touchent :

- la corrélation des diverses expériences, malgré le choix d'un paramètre incluant un maximum d'informations sur la nature du spectre neutronique et sur les dommages créés au matériau (déplacement par atomes : DPA)
- les calculs et estimations des dommages vus par le sommier et les diverses éprouvettes irradiées.

La réponse proposée est un programme volumineux de mesure de flux permettant une prise d'informations la plus complète possible :

a) par l'utilisation de détecteurs couvrant le mieux possible le spectre neutronique :
   . isotopes lourds pour les irradiations dans PHENIX
   . ^{54}Fe, ^{58}Ni, ^{59}Co, ^{93}Nb, ^{297}Np pour les irradiations dans SUPER PHENIX.

b) par une répartition géographique permettant de déterminer la dose effectivement vue par chaque éprouvette.

c) par une mesure directe du flux à hauteur du sommier.

De plus, la structure et le déroulement des irradiations permet des prises d'informations intermédiaires autorisant d'éventuels réajustements des temps de séjour des irradiations principales.

REFERENCES

/1/   M. LOTT et al.
"Sections efficaces de création de dommage"
Nuclear data in Science and Technology - Vienne 1973
IAEA SM 170.65.

/2/   J. LINDHARD and al
"Integral equation gouverning radiation effects"
MAT. PHYS. MEDD DAN VID SELSK 33 N° 10 - 1903.

/3/   E. E. BLOOM
"An Investigation of Fast Neutron Radiation
Damage in an Austenetic Stainless Steel"
DRNL-4580.

/4/   J.C. ESTIOT, M. SALVATORES, J.P. TRAPP
"Basic Nuclear Data and the fast reactor shielding
design - Formulaire PROPANE DO"
Conf. Nucl. Cross Sections and Technology
Knoxville 22-26 October 1979.

/5/   J.P. CHAUDAT and al
"Data adjustment for Fast Reactor Design"
N. Trans. Am. Nucl. Soc 27.877 (1977).

/6/   H. DERRIEN, J.P. DOAT, E. FORT, D. LAFOND
"Evaluation of 237Np Neutron Cross sections in the
energy range from 10^{-5} eV to 14 MeV"
Rapport INDCI (FR) 42/L).

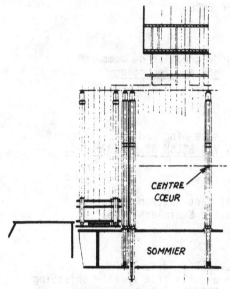

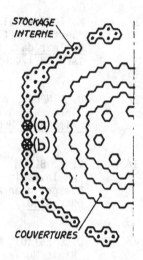

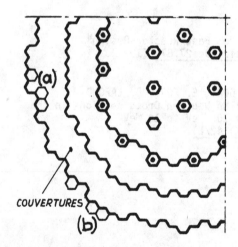

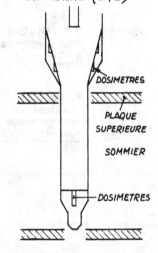

FIG.1 SUPER-PHENIX
LOCALISATION DU SOMMIER

FIG.2. PHENIX EMPLACEMENTS
DES IRRADIATIONS
DINOSAURE (a,b)

FIG.3. SUPER-PHENIX - EMPLACEMENTS
CARACTERISTIQUES DES IRRADIATIONS
DIPLODOCUS (a) ET IGUANODON (b)

FIG.4. DOSIMETRIE DU SOMMIER
PIED SPECIAL D'ASSEMBLAGE
CONTENANT DES DOSIMETRES

FLUENCE, DOSIMETRY, AND STEEL-DPA RATES IN EBR-II*

D. Meneghetti and D. A. Kucera

Argonne National Laboratory

Argonne, Illinois   60439, U.S.A.

ABSTRACT

Sensitivities of steel displacements-per-atom (dpa) rates to fluence-rate spectra in regions of the Experimental Breeder Reactor II (EBR-II) are presented. Low sensitivities in EBR-II of ratios of dpa-to-fission rates assuming ^{240}Pu as a dosimeter suggests its possible use for adjusting calculated dpa-rates for effects of errors in calculated fluence-spectra. Extension of the method to outer regions, having more degraded spectra, by use of ^{10}B-shielded ^{240}Pu dosimeters is also suggested.

INTRODUCTION

Calculated values of displacements-per-atom (dpa) rates for stainless steel in a reactor environment are useful both for determining correlations between observed volumetric expansions and exposures and for predicting volumetric expansions with exposures. If the values for the displacement cross section are correct, the accuracy of a calculated dpa rate is dependent upon the accuracy of the calculated neutron-flux spectrum at the location.

Neutron spectra at locations in a representative current configuration of the Experimental Breeder Reactor-II (EBR-II) [1, 2]

---

*Work supported by the U. S. Department of Energy under contract W-31-109-Eng-38.

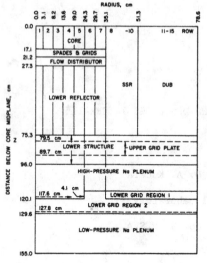

Fig. 1. Half-height Calculational Diagram.

were calculated using the half-height geometry description shown in Figure 1. Figure 2 shows the calculated flux spectra, in units of neutron fluence per unit lethargy per megawatt-day-thermal of reactor operations, at center of core and at a midline axial location near the middle of the subassembly-supporting grid-assembly region. These figures also show the multigroup values of the displacement cross sections of stainless steel used to obtain the shaded regions representing the group contributions to the total dpa-rates. The displacement cross sections are based on reference [3].

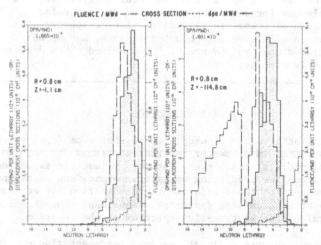

Fig. 2. Lethargy Dependence of Fluence and of Steel dpa Rates at Two Locations.

SPECTRAL-SHAPE MODIFICATIONS

The sensitivities of dpa rates to neutron spectra uncertainties were studied by modifying the calculated flux spectra in four different ways. The modifications simulate "real" spectra for comparisons with calculated "errored" spectra. (It should not be construed that any of these "real" spectra is in fact the actual spectra and that the calculated spectra are incorrect. These terminologies are solely for these sensitivity analyses.) The four spectral modifications are: group soften, group harden, skew soften, and skew harden. The group soften and group harden modifications approximate shifts in calculated multigroup spectra by one-quarter lethargy width toward smaller energies and larger energies. The skew soften modification approximates a 6% decrease of the calculated multigroup spectra per unit decrease of lethargy from a maximum of 16.5 (0.68 eV). The skew harden modification approximates a 6% decrease of the calculated multigroup spectra per unit increase of lethargy from zero (10 MeV). For the modified spectra the total fluence rates were normalized to equal the total fluence rates of the corresponding calculated spectra.

At the central location the calculated dpa-rate differs by about +16%, -16%, +40%, and -6% from the respective modified-spectra dpa-rates. At the grid-assembly location the calculated dpa-rate differs by about +14%, -18%, +91%, and -35% from the respective modified-spectra dpa-rates.

Relative spectral shapes of the skew type modifications together with the calculated spectra, here all normalized to unity for comparisons, are shown in Figure 3 for the central and for the grid-assembly locations.

RATIOS OF DPA RATES TO FISSION RATES

Calculations were made to determine the possibility of reducing errors in calculated dpa-rates in EBR-II by multiplying simulated-measured fission rates by ratios of calculated dpa rates to calculated fission rates. The isotopes whose fission reactions were scoped are ^{232}Th, ^{234}U, ^{236}U, ^{238}U, ^{237}Np, ^{240}Pu, and ^{242}Pu. Ratios based on the ^{240}Pu fission reaction were the least sensitive to simulated-variations of spectral shapes in core, radial reflector and blanket, and in axial reflector regions within about 30 cm of the core axial-interfaces. Comparison of the relative shapes of the steel displacement cross sections and of the fission cross sections of ^{240}Pu are shown in Figure 4. The displacement-to-fission ratios, D/F, at various axial and radial locations are shown in the top set of curves of Figure 5 for the calculated spectral-shapes and for the four modified spectral-shapes.

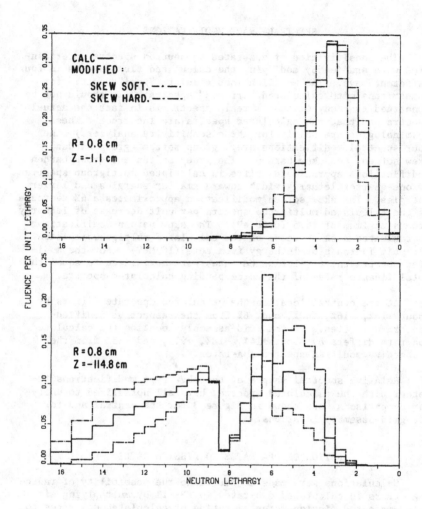

Fig. 3. Calculated and Modified Spectral Shapes at Two Locations.

The ratios from the unmodified spectra to those from the modified spectra in core, radial-reflector, radial-blanket, and at axial-reflector regions within about 30 cm of core bottom are such that the calculated (unmodified) ratios together with simulated measured fission rates correct the calculated (unmodified) dpa rates for stainless steel to well within 10% of these simulated "real" values in core locations and within about 15% in these out-of-core locations. (Because the dpa-to-fission ratios are dependent only on spectral shapes, these same percentages are retained irrespective of the magnitude factors by which the unmodified spectra differ.) With increasing distance into the lower regions of steel and sodium the errors become prohibitively large for some spectral modifications because of the contributions of the sub-threshold fissions.

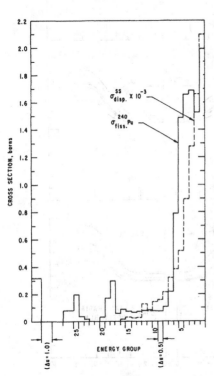

To ascertain whether consideration of removal of sub-threshold fission components from measurements would be helpful, fission-truncated ratios, D/F^{trun}, were calculated assuming no sub-threshold fissions below 20.3 keV (corresponding to lethargy 8.5) which is the demarcation between energy multigroups numbers 18 and 19. (Because of the very small displacement cross section values for the lower energy groups, 18→29, non-truncation or truncation of the displacement cross sections result in essentially identical numerators in these ratios.) These ratios, shown in the middle set of curves in Figure 5, indicate that corrected calculated dpa rates within 15% should be obtained also in the deeper out-of-core axial regions, if a method can be devised for estimating the sub-threshold fissions of a fission measurement.

Fig. 4. ^{240}Pu Fission Cross Sections (ENDF/B-V) Compared with Steel Displacement Cross Sections (based on [3]).

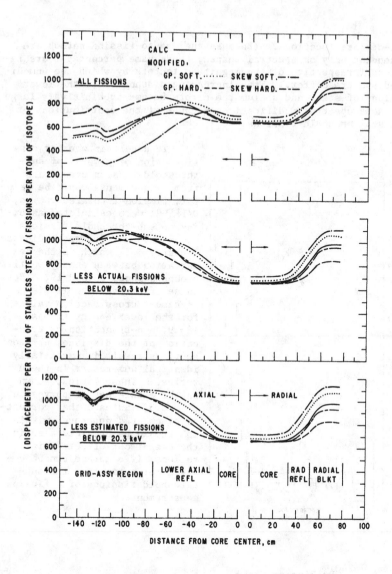

Fig. 5. Spectral Sensitivities of Steel dpa to ^{240}Pu Fission
with and without Sub-threshold Fission.

ESTIMATING SUB-THRESHOLD FISSIONS

A possible method for estimating the sub-threshold rate, F_{est}^{sub}, of a measured fission rate, M_o, is here very-briefly described. The method assumes that the unknown sub-threshold portion of a neutron spectrum can be estimated from the calculated sub-threshold portion of a neutron spectrum, in such regions, by lethargy dependence of the form

$$\phi_{est_j} = \phi_{calc_j} A \exp[a(\bar{u}_j - \bar{u}_{18})]$$

for groups $j \geq 18$. The constants, "A" and "a", for that location, are found by solution of a set of equations using as input values the measured fission rate of the isotope and the measured fission rates of the isotope when shielded by thin-absorbers of ^{10}B of two different thicknesses. The ^{10}B absorbers, however, must be thick enough and sufficiently different in thicknesses so that the constants can be determined; but, the absorbers must be thin enough so that the attenuation of the above-threshold fissions are minimal. Minimal means that the reduction of above-threshold fission rate due to a shield is sufficiently small that this above-threshold reduction can be assumed to be that of a spectrum having the above-threshold spectral-shape of the calculated spectrum for purposes of the accounting of the above-threshold reductions in the fission-rate measurements due to the shields. The above-threshold fluence-rate spectrum impinging on the shields (and on the unshielded detector) is for purposes of loss of fissions due to ^{10}B capture assumed to be

$$\phi_{est_j} = \phi_{calc_j} [(M_o - F_{est}^{sub})/F_{calc}^{trun}], \text{ for } j = 1 \rightarrow 17,$$

where

$$F_{calc}^{trun} = \sum_{j=1}^{17} \sigma_{F_j} \phi_{calc_j} \text{ and } F_{est}^{sub} = \sum_{j=18}^{29} \sigma_{F_j} \phi_{est_j}.$$

To test this method of estimating the sub-threshold fission rate of a measured fission rate and to simulate its application, simulated-measured values of bare and shielded fission rates based on the various modified spectral types were used as measured values and solutions were obtained using this method. The ^{10}B shields were assumed to be of thicknesses 0.05 cm and 0.10 cm. A comparison of the estimated sub-threshold fissions at various axial locations, with the calculated sub-threshold fissions and with the "true" sub-threshold fissions of the skew-softening spectral modification, is shown in Figure 6. (The ordinate units in these figures are given as these quantities divided by the calculated above-threshold fission rate i.e., the calculated truncated-fission rate.) The values of the estimated sub-threshold fissions

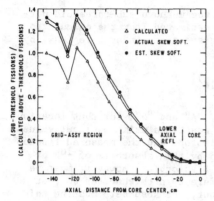

are seen to be considerably closer to those of the corresponding values of the modified-spectra sub-threshold fissions than to the values of the calculated sub-threshold fissions in those locations having large differences from the calculated values.

Fig. 6.   Relative Values of Sub-threshold ^{240}Pu Fissions.

The unshielded simulated-measured fission rates based on the various spectral-modifications were then reduced by the corresponding estimated values of the sub-threshold fissions (obtained using the above method) to obtain the estimated simulated-measured fission values without sub-threshold fissions. The ratios of the dpa-rates divided by the corresponding values of estimated above-threshold fission rates are shown in the lower set of curves in Figure 5. The close similarity between this set of curves with the middle set, representing idealized fission truncation, indicates the possible feasibility of this method for approximation of the truncation and thus enabling the truncation method to be utilized for correction of calculated dpa rates in such regions of EBR-II if necessary.

REFERENCES

1. L. J. Koch et al., Proc. 2nd UN Int. Conf. PUAE 9 (1958) 323.

2. F. S. Kirn, "EBR-II as a Fast Reactor Irradiation Facility," Nuclear News (March 1970).

3. D. G. Doran and N. J. Graves, "Displacement Cross Sections and PKA Spectra: Tables and Applications," HEDL-TME 76-70, Hanford Engineering Development Laboratory (1976).

TRITIUM PRODUCTION FROM THE ^{10}B(n,t)2α REACTION IN FFTF

B. M. Oliver and Harry Farrar IV
Rockwell International Corporation
Canoga Park, California 91304, USA

and

J. A. Rawlins and D. W. Wootan
Westinghouse Hanford Company
Richland, Washington 99352, USA

ABSTRACT

Tritium production from the ^{10}B(n,t)2α reaction has been measured using platinum-encapsulated natural boron crystalline samples irradiated in a core location in the Fast Flux Test Facility (FFTF). Tritium getters and barriers were used to prevent tritium migration into or out of the samples. The tritium was determined indirectly by the measurement of its decay product, ^{3}He, by high-sensitivity gas mass spectrometric analysis of vaporized samples. The tritium data are compared with a calculated value based on an adjusted neutron spectrum derived from a FERRET analysis of 12 other reaction rate measurements. For the calculations and adjustments, ENDF/B-V cross section and covariance data were used. The measured tritium production was 18% higher than calculated, although the uncertainty in the calculation was ~22%.

INTRODUCTION

Eleven capsules were included by Rockwell International in the FFTF-Reactor Characterization Program (RCP) 8-day high-power test (1). Four of these capsules contained threshold helium accumulation fluence monitors (HAFMs) and other materials for neutron dosimetry applications (2). The remaining seven capsules, the subject of this paper, were included as part of a joint Rockwell-HEDL experiment to measure the tritium produced by

J. P. Genthon and H. Röttger (eds.), Reactor Dosimetry, 217–224.
© 1985 ECSC, EEC, EAEC, Brussels and Luxembourg

the $^{10}B(n,t)2\alpha$ reaction at several locations in FFTF. A major source of tritium in FFTF is that produced in the B_4C control rods.

In the present experiment, the tritium produced by this reaction was determined indirectly by the measurement of ^3He, the decay product of tritium. Although two alpha particles also result from this reaction, the resulting ^4He was not used to determine the reaction rate because this helium could not be distinguished from that produced by the dominant $^{10}B(n,\alpha)^7Li$ reaction.

IRRADIATION CAPSULES AND HAFMs

Details of the seven stainless steel irradiation capsules are given in Table 1. Capsules CT1-CT3 contained boron crystals sealed in miniature vanadium capsules, Capsules CT4-CT6 contained vanadium-encapsulated LiF crystals, and Capsule CT7 contained vanadium-, platinum-, and stainless steel-encapsulated boron and LiF crystals. Capsules CT1, CT4, and CT7 were located close to a B_4C control assembly in a neutron environment similar to that in a region of primary tritium production.

The reason for inclusion of LiF samples in the irradiation was that the $^6Li(n,\alpha)t$ reaction provides a convenient mechanism to determine the effectiveness of the various tritium getters and barriers included in the experiment. This reaction generates a one-to-one production of ^4He and tritium, and thus a measurement of the ^3He (from the decay of tritium) and the ^4He at any given time after irradiation provides a measurement of any tritium gain or loss.

TABLE 1
ROCKWELL CAPSULE LOCATIONS IN FFTF

			FFTF Location			
Rockwell Capsule Number	HAFM Capsule Material	HAFM Sensor Material	Characterizer Assembly Row	Pin No.	z (cm)	Mean Neutron Energy $\bar{E}$ (MeV)
CT1	V	B	4	01B	-7.5	0.50
CT2	V	B	1	FP1	-16.6	0.52
CT3	V	B	7	FP2	-10.0	0.28
CT4	V	LiF	4	01B	-4.0	0.50
CT5	V	LiF	1	FP1	+12.6	0.52
CT6	V	LiF	7	FP2	-13.0	0.28
CT7	V, Pt, SS	B, LiF	4	01B	-11.0	0.50

The special HAFM capsules for the tritium experiment were manufactured at Rockwell International. Each capsule was 1.27-mm OD, 0.79-mm ID, and 6.35 mm in length. Capsule materials were vanadium, platinum, and Type 316 stainless steel. The boron crystals were from Rockwell Lot 2A, which is 20.02 ± 0.06% ^{10}B, with a purity >99.8 ± 0.2% boron. The LiF crystals were from Rockwell Lot 1, which is 99.10 ± 0.10% ^6Li with a purity >99.9% LiF.

The boron, LiF, and empty "control" HAFM capsules also included small amounts of Ti and Zr as getters to prevent or reduce tritium migration out of the HAFMs. Additional Ti and Zr foils were used to wrap the outside of the HAFMs and the inside of the stainless steel irradiation capsules respectively, to further block tritium movement either from the HAFMs or from the reactor environment. In addition, the seven stainless steel irradiation capsules were coated on their inner surfaces with a special ceramic tritium barrier originally developed at Rockwell for SNAP-8 reactor fuel cladding. The ceramic consisted of a mixture of ~46% SiO_2, ~45% BaO, ~8% Al_2O_3, and ~1% Cr_2O_3.

HELIUM CONCENTRATION RESULTS

Following irradiation, helium analyses were performed by vaporizing the complete HAFM capsules and then measuring the helium released using high-sensitivity isotope-dilution mass spectrometric techniques described previously (3,4). As most of the specially designed HAFMs included in the experiment used vanadium as the encapsulating material, these were used for the initial scoping analyses. Analyses were first conducted on ^6LiF HAFMs from Capsules CT4, 5, and 6 to determine the tritium retention capabilities of the vanadium. The results, however, showed that even with Ti and Zr getters inside and outside, the vanadium exhibited very poor retention of the internally generated tritium, with tritium retention values ranging from only ~2 to ~90%. Even empty vanadium HAFMs from Capsules CT4, 5, and 6 contained significant amounts of tritium. This was not the case with identical empty HAFMs from Capsules CT1, 2, and 3, indicating that the measured tritium had emanated from the ^6LiF HAFMs rather than from the general reactor environment. Some boron HAFMs were also analyzed from Capsules CT1, 2, and 3 but, as expected from the ^6LiF HAFM data, the results showed significant scatter.

As a result, additional analyses were then conducted only on the stainless steel- and platinum-encapsulated ^6LiF HAFMs included in Capsule CT7. The results showed only slight improvement in the tritium retention of the stainless steel, but the two platinum-encapsulated ^6LiF HAFMs showed excellent retention.

TABLE 2

HELIUM GENERATION IN PLATINUM-ENCAPSULATED HAFMs IN CAPSULE CT7

HAFM Name	Sensor Mass (mg)	Measured Helium		Tritium Content* (10^{13} atoms)	Ratio of Tritium to ^{4}He[†]
		^{3}He (10^{12} atoms)	^{4}He (10^{13} atoms)		
A5-6LiF	0.1169	745.8	1175.	1181.	1.005
B1-6LiF	0.0957	634.0	960.9	933.5	0.972
D1-MT	empty	0.0849	0.414	0.1245	-
A1-NB	0.0886	1.357	1052.	1.990	0.001774
A2-NB	0.2190	3.270	2567.	4.638	0.001759

*Tritium content at end of irradiation determined from the measured ^{3}He using a tritium half-life of 12.33 years.

[†]Tritium/^{4}He ratio calculated from the Column 4 and 5 data, and from the empty "blank."

Because of the latter results, additional analyses were conducted on the few remaining Pt-encapsulated boron and empty HAFMs from Capsule CT7.

The results of all the Pt-encapsulated HAFM helium measurements are given in Table 2. For the ^{6}LiF HAFMs, the ^{4}He results have been corrected by ~0.2 and ~1.0% respectively, to account for ^{4}He generation from ^{19}F(n,He) and ^{6}Li(n,n'd)α reactions. The ^{19}F correction was based on Rockwell measurements from numerous threshold HAFMs irradiated at nearby locations in FFTF during the same high-power test (2). Uncertainty in the two corrections is estimated to contribute ±0.1 and ±0.5% uncertainty, respectively, to the final results.

Also given in Table 2 (Column 5) is the tritium content in each HAFM at the end of the irradiation, based on the measured ^{3}He content (from Column 3) and the elapsed time between the irradiation and the helium analysis. The mean irradiation date was November 15, 1981, and the analyses were conducted between January 13 and May 11, 1983.

The final column in Table 2 gives the ratio of the tritium-to-^{4}He values measured for the ^{6}LiF and natural boron HAFMs. These ratios were calculated from the measured ^{4}He and tritium data in Columns 4 and 5 after small background correction had been made for the measured amounts of ^{4}He and tritium in the empty HAFM. In the case of the ^{6}LiF HAFMs, this ratio is equivalent to the fraction of tritium retained by the capsule.

DISCUSSION OF HELIUM RESULTS

As discussed above, analysis of the ^{6}LiF HAFMs provides
information on the ability of the miniature Pt HAFM capsules to
retain their tritium. Examination of the Column 6 data in
Table 2 indicates that the two platinum-encapsulated ^{6}LiF HAFMs
exhibited good tritium retention. The mean tritium retention was
0.98 ± 0.02%. Further, the tritium level measured in the blank
platinum HAFM is several orders of magnitude lower than that
measured in an empty vanadium HAFM irradiated at the same reactor
location, but in capsule CT4. Thus, the combination of the Ti
and Zr getters and the platinum encapsulating material provided a
good tritium barrier.

The question arises, however, as to whether this tritium
retention can also be expected for the Pt-encapsulated natural
boron HAFMs, which had significantly lower internal tritium partial pressures. Tritium diffusion through metals is generally
dependent on the square root of the tritium partial pressure difference across the metal (5). Therefore, for the lower partial
pressures of tritium in the boron HAFMs, the proportional tritium loss might be increased. Whether tritium loss in highly
irradiated materials can be expected to depend solely on diffusion, however, is not clear. Also, tritium gettering by the internal Ti and Zr foil getters would be expected to be more efficient for lower tritium pressures, and this could well reverse
any possible proportional increase in tritium loss by diffusion.

For the two Pt-encapsulated boron HAFMs irradiated in Capsule CT7, the measured tritium/^{4}He ratios are in excellent
agreement, yielding a mean value of 0.001767 with a standard
deviation between the two values of ±0.000011 (0.6%). This
agreement, although somewhat better than might be expected based
on measurement and other uncertainties, is additional evidence
for the good retention of tritium in the platinum HAFM capsules.
For these two samples, the background tritium measured in the
single empty Pt-encapsulated HAFM was small, representing only 3
to 6% of the total tritium generation.

ESTIMATED UNCERTAINTIES IN HELIUM RESULTS

Uncertainty in the measured boron and ^{6}LiF HAFM analyses,
based on the cumulative uncertainties in spike size, sample mass,
and isotope ratio measurement, is estimated to be ~1% of the
helium values for this series of runs. Uncertainty in the empty
"blank" HAFM analysis is slightly higher, ~2-3%, because this
analysis was run without any added ^{3}He spike.

In assessing uncertainty in the boron tritium generation measurements, two other factors must be considered in addition to the helium measurement uncertainty discussed above. These factors are: (1) possible variability in the background tritium data from the empty Pt HAFM, and (2) possible tritium loss from the boron HAFMs. Only one empty platinum HAFM (with both Ti and Zr getters) was irradiated in Capsule CT7. Obviously with only a single value for this background, it is not possible to assess the variability in this background in any rigorous sense. However, based on the good consistency in the helium data from the four Pt-encapsulated ^{6}LiF and boron HAFMs, it is reasonable to assume that variability in the background helium data from the blank platinum HAFM is probably small. Arbitrarily assigning an uncertainty of ±50% to this background would result in an additional uncertainty in the mean measured tritium/^{4}He ratio of ~2%.

Possible tritium loss mechanisms from the Pt-encapsulated boron HAFMs have been discussed earlier. Although a clear assessment of the uncertainty from possible tritium loss is not possible, there does exist strong evidence to indicate that tritium retention in the boron HAFMs is at least as good as that measured in the two ^{6}LiF HAFMs. For those HAFMs, the tritium retention was >98%.

COMPARISON WITH CALCULATED ^{10}B(n,t) REACTION RATES

To compare the measured data with calculations, the tritium contents given in Table 2 were converted to reaction rates. These calculations involved the measured tritium concentration at the end of irradiation and the power-time-history. The data were also adjusted for axial gradients in order to compare with other dosimeters located at core midplane in Pin 01B. Based on these calculations, the final measured tritium production rate in Pin 01B at core midplane at the average full power level of the irradiation is 2.26 ± 0.07 x 10^{-11} reactions/sec/nucleus.

Although the test was designed to determine the tritium production rate for boron in FFTF, the results can also be used to assess the adequacy of calculated tritium production rates in FFTF, as well as other facilities. Since the measurement capsule was not located near a calculational mesh point, the tritium results were compared with reaction rates measured at core midplane in Pin 01B using the FERRET code (6). A 53 group neutron spectrum with relatively large uncertainties, calculated with ENDF/B-V data and three dimensional diffusion theory (7,8), was used as input to FERRET. The dosimetry reaction rates and ENDF/B-V cross section and covariance data were then used to adjust the calculated spectrum for subsequent comparisons. The dosimetry set included the threshold reactions ^{237}Np(n,f),

$^{240}Pu(n,f)$, $^{238}U(n,f)$, $^{54}Fe(n,p)$, $^{46}Ti(n,p)$, and $^{55}Mn(n,2n)$, which span a region from ~0.3 MeV to 13 MeV. By comparison, 94% of the tritium production is calculated to occur in the range 1.4 to 7.4 MeV. Thus, the adjusted spectrum was well quantified in the experimental region of interest.

The calculated tritium production rate is $1.92 \pm 0.42 \times 10^{-11}$ reactions/sec/nucleus. The estimated uncertainty associated with the $^{10}B(n,t)$ cross section is ±20%. Combining this in quadrature with an adjusted neutron spectrum uncertainty of ±10%, gives a total quadrature uncertainty of ±22%. Comparison of calculated and measured values therefore gives a C/E ratio of 0.85 ± 0.19.

CONCLUSIONS AND RECOMMENDATIONS

The calculated tritium production rate in FFTF Row 4 fuel agrees with the measured value within estimated uncertainties, assuming 100% tritium retention by the platinum capsules. Analysis indicates that an upward adjustment of ~18% in the tritium production cross section would enable tritium production to be calculated in FFTF with an accuracy of ~10%. However, it should be noted that although the platinum-encapsulated HAFMs were adjacent to a B_4C control assembly, no assessment has been made of whether the results can be successfully applied to the inner regions of a control assembly where the tritium production actually takes place. In addition, because of the limited number of platinum capsules available for analysis, the 3% measurement uncertainty is difficult to fully assess. Additional experimental data are needed to (1) verify negligible tritium loss from platinum HAFM capsules under conditions of lower internal tritium pressure, and (2) verify low tritium background in the HAFMs from the reactor environment.

These questions could be resolved by conducting another irradiation, in FFTF or in a similar neutron spectrum, in and near B_4C assemblies using platinum as the encapsulating material, and including samples depleted in ^6Li to simulate lower tritium generation pressures. This irradiation would also provide valuable additional data on the $^{10}B(n,t)$ reaction. Simultaneous analysis of $^{10}B(n,t)$ data from a number of different reactor spectra would point to the neutron energy regions where adjustments to the ENDF/B-V ^{10}B cross section might be appropriate.

ACKNOWLEDGEMENTS

The authors wish to acknowledge technical contributions to this work by J. F. Johnson and J. T. Roberts of Rockwell International, and K. D. Dobbin, F. A. Schmittroth and R. E. Schenter of Westinghouse Hanford Company. We also wish to express our thanks to P. B. Hemmig and J. W. Lewellen of the U.S. Department of Energy for their continued interest and support of this work. This work was supported under DOE Contracts DE-AT03-81SF11561 at Rockwell International and DE-AC14-76FF02170 at Westinghouse Hanford Company.

REFERENCES

1. J. A. Rawlins, J. W. Daughtry, and R. A. Bennett, "FFTF Reactor Characterization Program Review," Proc. 4th ASTM-EURATOM Symp. on Reactor Dosimetry, NUREG/CP-0029, Vol. 1, p 245 (1982).

2. B. M. Oliver, Harry Farrar IV, J. A. Rawlins and D. W. Wootan, Threshold Helium Generation Reaction Rate Measurements in FFTF, Rockwell International Report ESG-DOE-13497 (September 1984).

3. H. Farrar IV, W. N. McElroy, and E. P. Lippincott, "Helium Production Cross-Section of Boron for Fast-Reactor Neutron Spectra," Nucl. Technol. 25, 305 (1975).

4. B. M. Oliver, James G. Bradley, and Harry Farrar IV, "Helium Concentrations in the Earth's Lower Atmosphere," Geochim. et. Cosmochim. Acta (in press, 1984).

5. Scientific Foundations of Vacuum Technology, S. Dushman, ed., John Wiley and Sons, New York, Chapter 8 (1962).

6. F. A. Schmittroth, FERRET Data Analysis Code, Hanford Engineering Development Laboratory Report HEDL-TME 79-40 (September 1979).

7. F. M. Mann, FTR Set 500, A Multigroup Cross Section Set for FTR Analysis, Hanford Engineering Development Laboratory Report HEDL-TME 81-30 (February 1982).

8. R. W. Hardie and W. W. Little, Jr., 3DB, A Three Dimensional Diffusion Theory Burnup Code, Battelle Northwest Laboratory Report BNWL-1264 (March 1970).

RECENT JAPANESE ACTIVITIES IN THE FAST BREEDER REACTOR PROGRAM
AND REACTOR DOSIMETRY WORKS

A. Sekiguchi[1] , Y. Matsuno[2] , I. Kimura[3] , T. Kodaira[4] ,
H. Susukida[5] and M. Nakazawa[1]

(1) University of Tokyo, Faculty of Engineering

(2) Power Reactor and Nuclear Fuel Development Corporation

(3) Kyoto University Research Reactor Institute

(4) Japan Atomic Energy Research Institute

(5) Mitsubishi Heavy Industries, Limited

ABSTRACT

From a viewpoint of reactor neutron irradiation and
dosimetry activities in JAPAN, two important recent
progresses are reviewed, one is the Japanese fast breeder
reactor development program and another is the reactor
dosimetry research activities in the special committee of
the Atomic Energy Society of JAPAN. The experimental fast
reactor JOYO has been operated successfully from December,
1981 and is now applied as the fuel and material irradiation
bed for the next prototype fast breeder reactor MONJU, which
is expected for its initial criticality on March, 1991. The
AESJ special committee on "Evaluation of Neutron Irradiation
Data" has a main role for the review of reactor dosimetry
research and developments in JAPAN and recently made a
participation to the PSF Blind Test of the LWR-PV-SDIP as
the Japanese task group.

J. P. Genthon and H. Röttger (eds.), Reactor Dosimetry, 225–231.
© 1985 ECSC, EEC, EAEC, Brussels and Luxembourg

INTRODUCTION

Here are summarized two Japanese activities relating to the reactor neutron irradiations and their dosimetry works, one is a brief summary of the national program of the Japanese fast breeder reactor developments which is primarily responsible to the Power Reactor and Nuclear Fuel Corporation (PNC) of Japan, where a new irradiation bed has been developed in the JOYO reactor.

Another is a comment on dosimetry relating activity in the special committee on "Evaluation of Neutron Irradiation Data" (chairman, Prof. A. Sekiguchi) in the Atomic Energy Society of JAPAN, where many review and discussions have been made on the topical issues in this field. Recently this committee made a participation to the PSF Blind Test of the current LWR-PV-SDIP as the Japanese task group.

DEVELOPMENTS OF FAST BREEDER REACTORS

According to the Japanese fast breeder development program, the fast experimental reactor "JOYO" has been completed as the first liquid sodium cooled reactor in Japan at April, 1977. The first role of this JOYO reactor is to get and accumulate many experiences in the fast reactor engineerings such as design, construction, fabrication, operation and maintenance, that have been successfully completed at the December, 1981. This first stage of the JOYO reactor had a breeding MK-I core for the fundamental LMFBR physics test and plant experience and had been operated at the rated output of 50 MW_t and 75 MW_t . (1)

The second stage of the JOYO reactor has been opened at March, 1983 as the fast neutron irradiation bed for the fuel and material testing, the core of which is called as MK-II core and is well operated at the reactor power of 100 MW_t according to schedule.

The JOYO MK-I operation experience is summarized in Table 1 and the JOYO MK-II operation schedule is shown in Table 2. Details of the irradiation and the dosimetry in JOYO reactor will be explained later.

The next goal of Japanese fast breeder reactor program is the construction and operation of the prototype fast breeder reactor named "MONJU". The safety evaluation work for MONJU by the Nuclear Safety Commission of Japan has been completed in May, 1983. Pre-construction works in the site are now being conducted and the contract of the plant between PNC and manufactures is expected to be made soon. Achievment of the initial criticality of the reactor is scheduled on March, 1991.

After the MONJU plant, the 1000 MWe class Demonstration FBR plant is thought necessary in order to demonstrate its performance, reliability and safety as a commercial-scale power reactor and in order to confirm especially the economic prospect for future commercialization. Its design studies have been made in progress and it is expected to begin construction in early 1990s. Because one of the current issues of the FBR development is pointed out in its high capital cost, present design studies are focused to get an inexpensive fast breeder reactor.

IRRADIATION AND DOSIMETRY IN JOYO

PNC/DOE Collaborative Dosimetry Test

In the JOYO MK-I core experiments, developments of the dosimetry method have been made in co-operation with the dosimetry group in the University of Tokyo. The main subjects were
(1) selection of dosimeters for the JOYO reactor,
(2) measurements of the reaction-rate and
(3) development of spectrum unfolding code NEUPAC.
These JOYO dosimetry technique has been well validated through the several inter-comparison works such as between PNC and HEDL dosimetry group as the PNC/DOE collaborative dosimetry test using both reactors of JOYO and EBR-II, that has been presented in this 4th Symposium. (2)

JOYO MK-II Irradiation bed

The present JOYO reactor is called as MK-II core-configuration and is utilized as fuels and materials irradiation bed for the next fast breeder development. Typical core-pattern is given in Fig. 1, where several kinds of irradiation rigs are introduced. Fuels are irradiated using special fuel assemblies called as UNIS (Uninstrumented Irradiation Subassembly), where three kinds of UNIS are used such as the name A, B and C depending on the fuel pin specimen numbers inside each UNIS. Figs. 2 and 3 show the whole view and cross-sectional view of the UNIS-B.
Materials are irradiated in the JOYO MK-II core using material irradiation rig such as AMIR (Absorber Materials Irradiation Rig), CMIR (fuel Cladding Material Irradiative Rig), SMIR (Structural Material Irradiation Rig). Some other irradiations can be made using INTA (INstrumented Test Assembly), UPR (UPper irradiation Rig plug) and some in-pile closed loop test sections.

JOYO MK-II Dosimetry

In each UNIS assemblies and material irradiation rigs, JOYO dosimeter set is loaded basing on the 13 kinds of reactions such as $^{93}Nb(n,n')^{93}mNb$, $^{58}Fe(n,\gamma)^{59}Fe$, $^{63}Cu(n,\alpha)^{60}Co$, $^{58}Ni(n,p)^{58}Co$, $^{59}Co(n,\gamma)^{60}Co$, $^{103}Rh(n,p)^{103}Ru$, $^{181}Ta(n,\gamma)^{182}Ta$, $Ti(n,p)Sc$,

^{237}Np(n,fission), ^{235}U(n,fission), Natural-U(n,fission), ^{232}Th(n,fission) and ^{238}U(n,fission). Using these dosimetry data, neutron fluences in the JOYO MK-II irradiation experiment are planned to be estimated through the NEUPAC adjustment calculations combined with the neutron transport calculations in the whole core configuration. (3)

As the initial characterization work of the JOYO fast neutron irradiation field, there have been carried out neutron flux distribution measurements around the core using the fission dosimeters such as ^{235}U, ^{238}U, ^{239}Pu, ^{232}Th and ^{237}Np and several activation foils of ^{197}Au and ^{58}Ni.

Some advanced dosimetry in the JOYO reactor is planned for the SSTR (Solid State Track Recorder), HAFM (Helium Accumulation Fluence Monitor) and some kinds of neutron damage monitors.

ACCURACY REQUIREMENTS FOR JOYO MK-II DOSIMETRY

One typical example of the JOYO MK-II irradiation experiment is a verification test of a radiation-resistant cladding material for the prototype FBR "MONJU", which will accumulate fast neutron fluences up to 2.3×10^{23}(n/cm^2), and due to this high fluence irradiation of fast neutrons, many swelling effect will be caused in this material.

This high fluence irradiation test needs about 20 cycle operations in JOYO MK-II irradiation bed and from a practical point of view for planning of these irradiation cycle schedules, accuracies of JOYO MK-II dosimetry are needed at least better than $1/20 = 5$ % , because each specimen should be taken out just after the planned fluences of irradiation.

At present, accurate neutron flux calculations can be expected within about a few percent in the reactor core-region, that is almost validated through the comparisons of physics experiments in the initial low-power operation. In the reflector region outside of the core, however, accuracies of present estimates are thought to be larger than 10 % , where additional experimental characterization works are required for actual dosimetry works.

This high fluence irradiation test is relating to lengthen the span of life of the fuel cladding materials, that is important for the cost reduction of the FBR power plant. This example shows that the more accurate dosimetry in the fast reactor is required for the current cost-reduction studies of the FBR power plant.

ACTIVITIES IN JAPANESE DOSIMETRY GROUPS

There are many dosimetry research groups in Japan. For the research and/or experimental fission reactor dosimetry works, there are main four groups such as JAERI (Japan Atomic Energy Research Institute), PNC, Kyoto-University Research Reactor Institute and University of Tokyo YAYOI group. They are making technical developments and dosimetry works in their own reactors of JMTR, JOYO, KUR and YAYOI, and dosimetry intercomparison studies between these four groups have been carried out, which has been also presented in the previous symposium. (4)

In these eight years, the special committee on "Reactor Neutron Dosimetry" (1977-1981), "Evaluation of Neutron Irradiation Data" (1981-1985) has a main role for the review and discussion on the reactor dosimetry relating researches and developments under the Atomic Energy Society of Japan. In this committee, there have been metallurgists and structural mechanics researchers with especial interests in the radiation damage estimations of reactor pressure-vessel steels, and they are belonging to the Universities, Research Institutes, Steel and Reactor Manufactures and Electric Power Companies.

Recently, this committee made an answer to the PSF Blind Test of the LWR-PV-SDIP, where corresponding task group has been formed to discuss and practice this interesting exercise of LWR-PV surveillance dosimetry. It is also useful for the discussions on what kind of data and procedures are and should be applied practically in the Japanese reactor pressure vessel surveillance test.

Some new activities on the fusion neutron dosimetry researches have been started for the 14 MeV neutron irradiation studies and for the fusion neutron physics experiments for the blanket/shield studies, which has been also included in the AESJ Reactor Dosimetry Committee. This committee has a periodical meeting in every two monthes, and the present members are 43 in all.

CONCLUSIVE REMARKS

Japanese two activities relating to the reactor neutron irradiation and dosimetry studies have been reviewed. The new irradiation bed in the JOYO MK-II reactor will be successfully applied to the high fluence fast neutron irradiation testing of the fuel and materials for the future economic fast breeder reactors.

Reactor Dosimetry Committee under the Atomic Energy Society of JAPAN has made a role as a corresponding group to the ASTM E10 committee in USA and to the EWGRD in Euratom, this activities are expected to be continued in co-operation with the USA and the Euratom groups.

REFERENCES

1. S. Nomoto, H. Yamamoto and Y. Sekiguchi, "Physical experiments at the Start-up of JOYO", IAEA-SM-244/8, (Sept. 24-28, 1979)

2. S. Suzuki, et al., "PNC/DOE Collaborative Dosimetry Test in JOYO", NUREG/CP-0029, Vol. 1, p 171 (1982)

3. A. Sekiguchi, et al., "Dosimetry Experiments in JOYO", ibid, Vol. 2, p 1111 (1982)

4. M. Nakazawa, et al., "The YAYOI Blind Intercomparison on Multiple-Foil Reaction Rate Measurements", ibid, p 1179, (1982)

Table 1 Summary of JOYO MK-I Operation Experience

• ACCUMULATED OPERATION TIME		12,968 HOUR
• ACCUMULATED HEAT GENERATION		673,333 MWH
• MAXIMUM FUEL BURN UP		
	MAX. FUEL PIN	40,500 MWD/T
	AVE. ASSEMBLY	40,100 MWD/T
• NUMBER OF STARTUPS (CRITICALITY TESTS INCLUDED)		260
• NUMBER OF FUEL ASSEMBLIES LOADED		
	CORE FUEL	107
	TEST FUEL	9
	BLANKET FUEL	220
• UNSCHEDULED PLANT OUTAGES		
	LOSS OF POWER	5
	COMPONENT FAILURE	12

Table 2 Experimental Fast Reactor "JOYO-MK-II" Operation Schedule

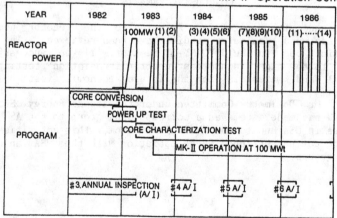

230

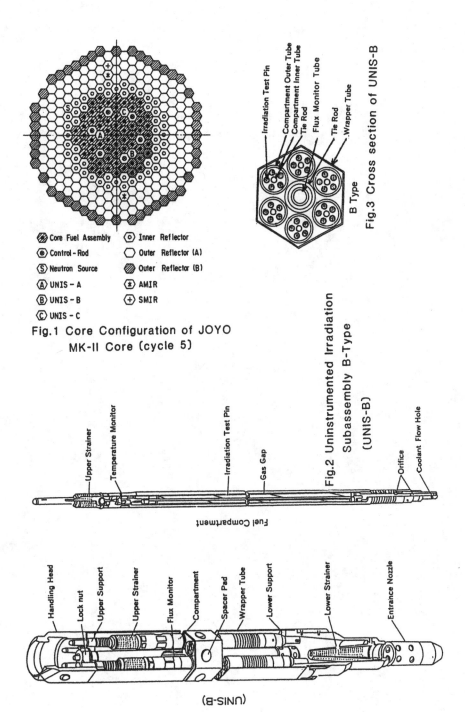

Fig.1 Core Configuration of JOYO MK-II Core (cycle 5)

Core Fuel Assembly Inner Reflector
Control-Rod Outer Reflector (A)
Neutron Source Outer Reflector (B)
UNIS – A AMIR
UNIS – B SMIR
UNIS – C

Fig.3 Cross section of UNIS-B

Irradiation Test Pin
Compartment Outer Tube
Compartment Inner Tube
Tie Rod
Flux Monitor Tube
Tie Rod
Wrapper Tube

B Type

Fig.2 Uninstrumented Irradiation Subassembly B-Type (UNIS-B)

Upper Strainer
Temperature Monitor
Irradiation Test Pin
Gas Gap
Orifice
Coolant Flow Hole

Fuel Compartment

(UNIS-B)

Handling Head
Lock nut
Upper Support
Upper Strainer
Flux Monitor
Compartment
Spacer Pad
Wrapper Tube
Lower Support
Lower Strainer
Entrance Nozzle

PART III
FUSION AND SPALLATION

NICKEL SAMPLE IRRADIATION IN THE BR2 REACTOR AT MOL IN THE FRAMEWORK OF THE OPTIMIZATION STUDY OF THE TIME-DEPENDENT HELIUM PRODUCTION/DAMAGE RATIO FOR IN-PILE FUSION MATERIALS TESTING

Ch. De Raedt, P. D'hondt

SCK/CEN, Mol, Belgium

ABSTRACT

Irradiations of fusion reactor structural materials
in materials testing fission reactors lead to time-
varying helium production over radiation damage
ratios (appm He/dpa) that are generally smaller than
the previsions for fusion reactors. Indeed, in
nickel-containing first wall materials such as AISI-
316 most of the helium formation during irradiation
in a materials testing reactor occurs in the thermal
neutron energy domain through the two-step reaction
$^{58}Ni(n,\gamma)^{59}Ni(n,\alpha)^{56}Fe$. The helium production is
hence non-linear with fluence and especially small
at the beginning of the irradiation. The dpa production on the other hand is proportional to the
fast fluence. In nickel-containing structural material samples higher helium productions and flatter
appm He/dpa ratios with respect to irradiation time
could be obtained by pre-irradiating them in a high
thermal flux during a well chosen time. To check
this assertion, pure nickel samples were irradiated
in BR2 according to various histories, in well characterized fluxes followed by the determination of
the Ni isotopic composition and of the He production.

INTRODUCTION

In fusion reactor first walls and other structural mate-

J. P. Genthon and H. Röttger (eds.), Reactor Dosimetry, 235–244.
© 1985 ECSC, EEC, EAEC, Brussels and Luxembourg

235

rials, the damages due to helium formation and to atom displacements occur simultaneously, with a time-independent appm He/dpa ratio. Values between 10 and 25 are generally cited for this ratio for the first wall.

In materials testing reactors, the dpa formation occurs mostly at high neutron energies and is a linear function of the neutron fluence. The helium formation, on the contrary, occurs, at least in nickel-containing materials, principally in the thermal energy domain because of the two-step reaction in nickel

$$^{58}Ni(n,\gamma)^{59}Ni(n,\alpha)^{56}Fe$$

which leads to a helium production according to a more or less quadratic function of the thermal fluence (at large fluences third and higher order terms become important also). Hence the ratio appm He/dpa is not at all time-independent : it is very small at the start of the irradiation and increases gradually as the irradiation proceeds (see curves " 0 days preirradiation" on fig. 1).

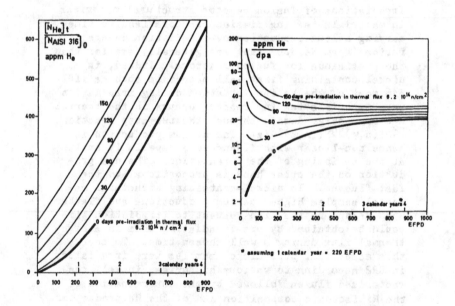

Fig. 1. He production and appm/dpa ratios in AISI-316, as a function of the irradiation time in BR2 ($\emptyset > 0.1$ MeV = $5.6 \ 10^{14} n/cm^2 s$; $\emptyset_{th} = 3.1 \ 10^{14} n/cm^2 s$), for various preirradiation durations.

A rather easy way to improve this situation could consist in the preirradiation of the test samples in a thermal flux, hence enhancing the helium formation practically without formation of dpa's, before the irradiation proper in the mixed spectrum [1]. Figure 1 indicates the calculated helium formation and corresponding appm He/dpa curves as a function of the irradiation time in the materials testing reactor BR2 for various nearly pure-thermal flux preirradiation periods. The appm He/dpa curves with preirradiation are flatter than without, due to the fact that

- some helium has already been formed in the preirradiation
- some ^{59}Ni has already been formed, which further produces helium by the one-step reaction ^{59}Ni(n,α)^{56}Fe.

To study these phenomena quantitatively, on an experimental basis, pure nickel samples have been irradiated in BR2 in 1982 and 1983 under various irradiation conditions, together with a series of dosimeters. The results obtained will be discussed in the present note.

The work was performed as a support to the irradiation programmes of fusion reactor structural materials carried out in BR2.

IRRADIATION SCHEME

Eight small capsules have been irradiated in BR2. Each contained :
- pure natural nickel discs
- cobalt dosimeters
- iron dosimeters
- niobium dosimeters.

The irradiation scheme is indicated in fig. 2. Hence thermal irradiations (in the central beryllium plug of BR2) during about 90 and 180 days, mixed irradiations (in a fast plus thermal flux viz in the axis of a BR2 fuel element) and thermal irradiations followed by mixed irradiations have been carried out, all both at high and at lower flux levels. The high flux levels correspond to irradiation positions in a channel axially near the maximum flux plane of the reactor, the lower flux levels correspond to positions shifted axially towards the lower end of the active core.

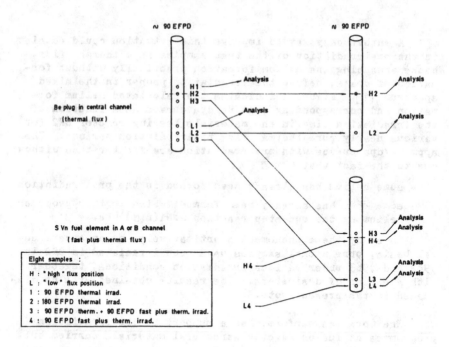

Fig. 2. Nickel irradiations in BR2 (1982 - 1983)

POST-IRRADIATION MEASUREMENTS

After the irradiations all eight capsules were dismantled :

- small parts of the nickel disks were sent to Rockwell
  International for precise helium content determination.
  Also the ratio $^{59}Ni/^{58}Ni$ was measured, by mass-spectrometry
  at SCK/CEN, in order to compare the real ^{59}Ni contents with
  the calculated values

- the cobalt dosimeters were counted. They yield the thermal and epithermal fluence values, account being taken of
  the cadmium ratios

- the iron and niobium dosimeters yield the fission fluence
  values.

MEASURED AND CALCULATED HELIUM PRODUCTION AND $^{59}Ni/^{58}Ni$ RATIOS

The helium contents in the eight samples, as measured by
R.I., are indicated in table I, 2nd column.

Table I. Measured and calculated helium production
in the eight samples

Sample N°	(appm He)$_{meas.}$	(appm He)$_{calc.}$ Mughabghab cross-sections (ref. 2)	(appm He)$_{calc.}$ "evaluated" cross-sections (ref. 3)
H1	257.2	241.3	246.0
H2	866.1	706.7	718.5
H3	743.4	653.8	667.3
H4	131.3	110.7	113.9
L1	87.38	85.62	87.45
L2	308.4	286.9	292.5
L3	171.3	(154.9)	(158.7)
L4	35.00	(28.50)	(29.39)

The helium formation in the nickel samples occurs accor-
ding to two* formation schemes : $[N_{He}]_t = [N_{He}]_{at} + [N_{He}]_{bt}$:

a. threshold reactions in nickel :

$$[N_{He}]_{at} = [N_{Ni}]_o \frac{\sigma_{n,\alpha\ Ni}}{\sigma_{Ni}} \left(1 - e^{-\sigma_{Ni}\emptyset t}\right) \quad (1)$$

b. the two-step reaction in ^{58}Ni :

$$[N_{He}]_{bt} = [N_{Ni58}]_o\ \sigma_{n,\alpha Ni59} \frac{\sigma_{Ni58}\left(1 - e^{-\sigma_{Ni59}\emptyset t}\right) - \sigma_{Ni59}\left(1 - e^{-\sigma_{Ni58}\emptyset t}\right)}{\sigma_{Ni59}\left(\sigma_{Ni58} - \sigma_{Ni59}\right)} \quad (2)$$

σ_{Ni}, σ_{Ni58} and σ_{Ni59} are absorption cross-sections. All cross-
sections are averaged over the whole neutron flux spectrum. $\emptyset$
is the total flux $\int_0^\infty \emptyset(E)\ dE$.

The helium formation in the nickel samples irradiated in
BR2 was calculated assuming, for all reaction types i :

$$\sigma_i\emptyset = \int_0^\infty \sigma_i(E)\ \emptyset(E)\ dE \approx \sigma_{i\ fiss} \times \emptyset_{fiss} \quad , \text{in (1)}$$

$$\sigma_i\emptyset = \int_0^\infty \sigma_i(E)\ \emptyset(E)\ dE \approx \sigma_{i\ 2200} \times nv_o + RI_i \times \emptyset_{epi} \quad , \text{in (2)}$$

---

*other reactions such as thermal n,α reactions in ^{58}Ni and ^{61}Ni
are negligible in view of the small thermal cross-sections.

With the thermal and epithermal fluences deduced from the $^{59}Co(n,\gamma)^{60}Co$ activities and the fission fluences deduced from the $^{54}Fe(n,p)^{54}Mn$ activities, the helium quantities indicated in the last two columns of table I were calculated. The penultimate column pertains to values obtained with the Mughabghab cross-sections of [2]. The values given in the last column were obtained with the "evaluated" cross-sections indicated in [3]. Both sets of cross-sections are indicated in table II, taken from [3]. The threshold (n,α) reaction cross-section used in the calculations was 4.36 mbarns (for the element Ni) [4].

Table II. Cross-sections used for the helium production and $^{59}Ni/^{58}Ni$ ratio calculations

Reaction	Mughabghab cross-sections (barns)		"evaluated" cross-sections [3] (barns)	
	σ_{2200}	RI*	σ_{2200}	RI
$^{58}Ni(n,\gamma)^{59}Ni$	4.6 ± 0.3	2.2 ± 0.2	4.81	2.27
$^{59}Ni(n,\gamma)^{60}Ni$	77.7 ± 4.1	(116.6)	81.04	124.2
$^{59}Ni(n,p)^{59}Co$	2.0 ± 0.5	(4.0)	1.96	4.22
$^{59}Ni(n,\alpha)^{56}Fe$	12.3 ± 0.6	(17.4)	11.99	18.51
$^{59}Ni_{abs}$	92 ± 4	138 ± 8	94.99	146.93

The measured and calculated appm He values are drawn in fig. 3, left part, as a function of the thermal fluence values issuing from the cobalt activity measurements. Samples H1, H2, L1 and L2, submitted to fluxes with practically the same (strongly thermalized) spectrum lay on a smooth curve, roughly proportional to $(nv_o \cdot t)^2$.

In fig. 3, right part, the measured and calculated $^{59}Ni/^{58}Ni$ ratios are indicated, as a function of the thermal fluence. The experimental values issue from mass-spectrometric measurements of the irradiated samples. The two values indicated for each sample correspond to the error margin resulting from a

---

*The values given between brackets, not indicated in [3], were deduced from $^{59}Ni_{abs}$, assuming the same relative contribution from n,γ , n,p and n,α as for the "evaluated" cross-sections.

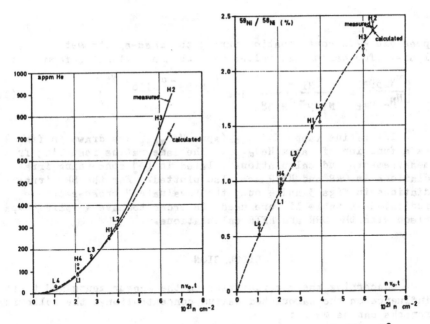

Fig. 3. Measured and calculated He productions and $^{59}Ni/^{58}Ni$ ratios in the samples as a function of the thermal fluence.

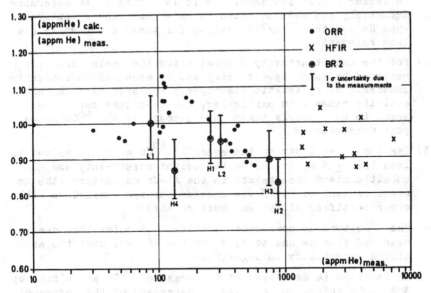

Fig. 4. Ratio calculated over measured He production in Ni samples irradiated in ORR, HFIR and BR2, as a function of the measured He production.

possible ^{59}Co contamination during the mass-spectrometric analysis. The calculated values were obtained with the formula :

$$\left[\frac{N_{Ni59}}{N_{Ni58}}\right]_t = \frac{\sigma_{Ni58}}{\sigma_{Ni59} - \sigma_{Ni58}} \left(1 - e^{-(\sigma_{Ni59} - \sigma_{Ni58})\,\emptyset t}\right) \qquad (3)$$

The ratios (appm He)$_{calc}$/(appm He)$_{meas}$ are drawn in fig. 4
as a function of (appm He)$_{meas}$. The same ratios resulting from
measurements and calculations related in [3] concerning irradiations in ORR and HFIR are also plotted. For the BR2 irradiations in figs 3 and 4 only the "evaluated" cross-sections
indicated in table II were used, in order to have a valid comparison with the ORR and HFIR calculations.

DISCUSSION

Concerning the uncertainties on the measurements and their
influence on the calculated helium concentrations, the following
remarks can be made :

1) The samples L3 (for the 2nd 90 EFPD period) and L4 were irradiated at an axial level in BR2 corresponding to the fuel -
   reflector transition zone. As it is difficult to determine
   accurately the cadmium ratio in this zone, the calculated
   appm He and the ^{59}Ni/^{58}Ni ratios for these two samples are
   less relevant

2) For the cobalt activity determination the dosimeters were
   measured by gamma spectrometry and by means of an ionization
   chamber. No systematic discrepancy appeared over the whole
   activity range : in particular, the dead time correction
   seems to be properly taken into account in the ^{60}Co gamma
   peak measurements

3) The 1 σ uncertainty on the specific ^{60}Co activity values is
   less than $\pm$ 3 %, including the weight uncertainty and the
   cobalt content uncertainty in the Al-Co dosimeters (the nominal content of 0.1 % was checked through comparison with
   other certified alloys and pure cobalt)

4) The ^{59}Co burn-up correction amounts to 4 % for the lowest
   measured fluence and to 12 % for the highest one; the associated uncertainty is negligible

5) An increase (a decrease) of the measured ^{60}Co activities by
   3 % would induce an increase (a decrease) of the corresponding calculated appm He values by about 6 %, because of the
   approximately quadratic nature of the appm He versus fluence
   curve

6) The uncertainty on the cadmium ratio values*, and therefore on the epithermal/thermal flux ratios, induces also an uncertainty on the calculated (appm He) values; as examples :

- for the L1 sample, a 10 % variation on R_{Cd} (11 % on $\emptyset_{epi}/\emptyset_{th}$), produces a 0.8 % variation on the calculated He content

- for H4, a 10 % variation on R_{Cd} (13 % on $\emptyset_{epi}/\emptyset_{th}$) produces a 2.3 % variation on the calculated He content.

It can also be mentioned that the helium production due to epithermal neutrons amounts to 7 % and 20 % respectively

7) The helium production due to threshold (n,α) reactions is only significant for the H3 and H4 samples; it amounts respectively to 1.5 % and 6.9 % with respect to the total He production (0.3 - 0.7 % for the other samples, excl. L3, L4).

8) The 1 σ uncertainty on the measured (appm He) values is estimated to $\pm$ 1 % .

Summarizing, the total uncertainty on the (appm He)$_{calc}$/ (appm He)$_{meas}$ ratios due to the measurements is

- 8 % for the L1, L2, H1, H2 samples

- 9 % for the H3 sample

- 10 % for the H4 sample.

Taking into account these error margins, it is observed that most of the calculation/experiment (C/E) ratios are lower than unity (average : 0.92, excluding L3 and L4) and that the underestimation in the calculation seems to increase with the fluence (or the helium production). Variations in the ^{58}Ni and ^{59}Ni(n,α) cross-sections lead to practically proportional variations in the calculated He production.

The BR2 results agree reasonably well with the ORR - HFIR results, using the same nickel cross-section set in the analysis (fig. 4).

CONCLUSION

Concerning the acceleration of the helium production through preirradiation mentioned above, one can compare the

---

*which were actually not measured during these irradiations but deduced from previous numerous measurements performed in similar conditions.

samples H3 and H4, although the present irradiation exercise was not conceived to match fusion reactor He/dpa exposure values.

The He content in H3 amounts to 743 appm; for a stainless steel sample containing 10 % nickel, preirradiated and irradiated in the same conditions, the appm He/dpa ratio would be 35. Referring to H4 (131 appm), this ratio would be 6 only for the same steel sample without preirradiation. Actually, the preirradiation should be shorter than the irradiation proper to reach some pseudo-equilibirium as shown on figure 1.

ACKNOWLEDGEMENTS

The nickel irradiations at BR2 were prepared by H. TOURWE and P. VANMECHELEN. The latter and J. LACROIX were responsible for the counting of the detectors. The $^{59}Ni/^{58}Ni$ ratios were measured by P. DE REGGE. H. FARRAR IV at Rockwell International performed the helium determinations. Last but not least are acknowledged the useful discussions with J. DEBRUE.

REFERENCES

[1] Ch. De Raedt, "Neutronic aspects of the utilization of nickel enriched in ^{59}Ni in samples for in-pile testing of fusion reactor first wall candidate materials," Salt Lake City Conference on fast, thermal and fusion reactor experiments, April 12-15, 1982.

[2] S.F. Mughabghab, M. Divadeenam, N.E. Holden, "Neutron cross-sections," Volume I - Neutron resonance parameters and thermal cross-sections, Part A, Z=1-60, Academic Press (1981).

[3] L.R. Greenwood, D.W. Kneff, R.P. Skowronski, F.M. Mann, "A comparison of measured and calculated helium production in nickel using newly evaluated neutron cross-sections for ^{59}Ni," Third Topical Meeting on Fusion Reactor Materials, Albuquerque, September 19-22, 1983.

[4] E.P. Lippincott, W.N. McElroy, H. Farrar IV, "Helium production in reactor materials," Washington Conference 1975.

REFERENCE SPECTRUM OF d(13)-Be AT THE INPE CYCLOTRON

I.Garlea, Cr.Miron-Garlea, F.Tancu, S.Dima, M.Macovei

Central Institute of Physics

Bucharest, ROMANIA

ABSTRACT

The neutron spectrum at U-120 cyclotron using d(13)-Be was determined by time of flight (TOF) and multiple foil methods. TOF measurements have been performed using a thick Be target and 13 MeV deuteron beam pulsed at 1.26 MHz. The neutron spectrum was measured in the range 0.3 - 17.25 MeV and angular distribution between 0-120°. The neutron field has been also characterized over range 10^{-10}-18MeV by the multiple foil method, using a large number of fission and activation rates, the unfolding SANDII code and Dosimetry file of ENDF/B V library.

TOF MEASUREMENTS

The U-120 cyclotron of Institute for Nuclear Physics and Engineering (INPE) has been used to produce a high intensity neutron source for various applications. The 13 MeV accelerated deuterons have bombarded a thick Be target (166 mg/cm²), at 20° relative to the incident beam. The Be target provided with an appropriate electron suppresion electrode and a Ta slit, has been indirectly water-cooled.

The neutron spectrum has been measured at first by TOF method (1) by using a base length of 4.5 m. The NE102 scintillator (used as neutron detector) has been calibrated in energy by using several standard γ sources. The data (2) concerning the light output

J. P. Genthon and H. Röttger (eds.), Reactor Dosimetry, 245–250.
© *1985 ECSC, EEC, EAEC, Brussels and Luxembourg*

of the plastic scintillator NE102 to electrons and protons were
used to find the equivalent neutron energies. The detection threshold has been established at 0.3 MeV. The efficiency calculation
has been performed by KURZ code (3), using the scintillator composition and nuclear data of reaction. The broad energy range com-
pelled us to use a pulse beam frequency as high as 1.26 MHz in
order to avoid the overlapping of two successive pulses on the
target.

The TOF spectrum was converted in energy units by usual methods. Some corrections (for dead-time, background and detection
efficiency) were applied to the data. The neutron spectrum measured by TOF at 0° is shown in Fig.1 together with the spectral
shape determined by unfolding (4).

FLUX-SPECTRUM OBTAINED BY UNFOLDING METHOD

The reaction rates have been absolutely measured at the distances of 5, 10 and 15 cm relative to target. The used activation
detectors and fission chambers have been placed on the geometrical axis of the neutron beams (0°). In the same plane with detectors have been mounted the three fission chambers at 120° (the
distance relative to axis being 6 cm), used as permanent monitors
of flux. The activation detector sandwiches and a fission chamber
containing ^{238}U have been simultaneously irradiated. In the Table
1 are given the diameters of activation foils and fissionable deposits in chambers, together their purities.

TABLE 1

Reaction	Detector	Diameter (mm)	Purity (%)	(mb)
$^{238}U(n,f)FP$	FC	15	100	590.7 ± 4.7%
$^{232}Th(n,f)FP$	FC	8	100	179.1 ± 6.5%
$^{237}Np(n,f)FP$	FC	15	100	1648.1 ± 5.4%
$^{235}U(n,f)FP$	FC	15	99.89	1392.2 ± 5.1%
$^{239}Pu(n,f)FP$	FC	15	99.88	2066.0 ± 5.8%
$^{115}In(n,n')$	In	19.05	99.99	288.0 ± 5.9%
$^{59}Co(n,\alpha)$	Co	19.05	99.99	3.8 ± 5.5%
$^{24}Mg(n,p)$	Mg	12.7	99.78	33.8 ± 6.1%
$^{27}Al(n,\alpha)$	Al	19.05	99.99	18.0 ± 5.2%
$^{56}Fe(n,p)$	Fe	12.7	99.82	20.4 ± 5.8%
$^{58}Ni(n,p)$	Ni	12.7	99.73	377.0 ± 5.2%

The absolute determination of the fission rates has been performed by means of the calibrated fission chambers (5), using
J.Grundl's procedure (6) and the associated electronic chain. The
activation rates have been determined by high resolution γ spectro-

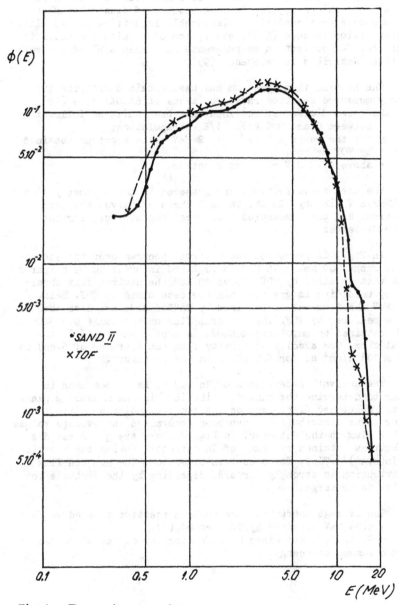

Fig. 1.- The neutron spectrum

metry (7) by means of Ge-Li crystal, calibrated in the absolute efficiency. The source-crystal distance is 5 cm. The γ peak
processing has been made by means of SAMPO code (8). The experi -
mental corrections applied for the results of fission chamber
measurements are: dead-time, fissionable impurities in deposits,
extrapolation to zero (ETZ), absorption of fission fragments in
deposits. For activation measurements have been applied the corrections described in reference (9).

The neutron flux-spectrum has been obtained from the absolutely measured reaction rates, by means of SANDII code (10),
for the three distances. The input has been built as follows:
 - between 0 and 300 keV - 1/E extrapolation;
 - for the range 300 keV - 17.2 MeV - the spectrum obtained
   by TOF method;
 - above 17.2 MeV - fusion spectrum.

The used cross sections are gathered in the Dosimetry file
of ENDF/B V library. In the Table 1 there are given the used
reactions and their measured cross-sections averaged for the
three distances.

In Fig.1 is presented the neutron spectrum over the energe-
tical range 300 keV - 18 MeV, normalized in order to be compared
with that obtained by TOF. Above 10 MeV the neutron flux obtained by unfolding is greater than one determined by TOF. Below
5 MeV the flux values obtained by SANDII code are less than
those measured by TOF. The integral flux on the range 0 - 300
keV, obtained by unfolding method, is approximately 1% in the
total flux. The absolute intensity flux (at distance of 5 cm) is
5.7×10^9 n/cm^2 s, for 8.6 μC /s on the Be target.

The activation detectors of In and Au have been used in
order to determine the flux distributions in the measuring plane,
at the mentioned distances, on the two perpendicular directions.
Using these distributions have been determined the average values
of the flux on the detector. In Fig. 2 there are given the distributions obtained by means of In detectors (using the reaction
^{115}In(n,n')^{115m}In), for a certain direction. The neutron flux
distribution is strongly forward, depending by the distance relative to Be target.

The average energy for the neutron spectrum created at INPE
is: - 5.24 MeV obtained by TOF method (1),
     - 5.352 MeV determined by unfolding (4) on the above men-
tioned ranges of energy.

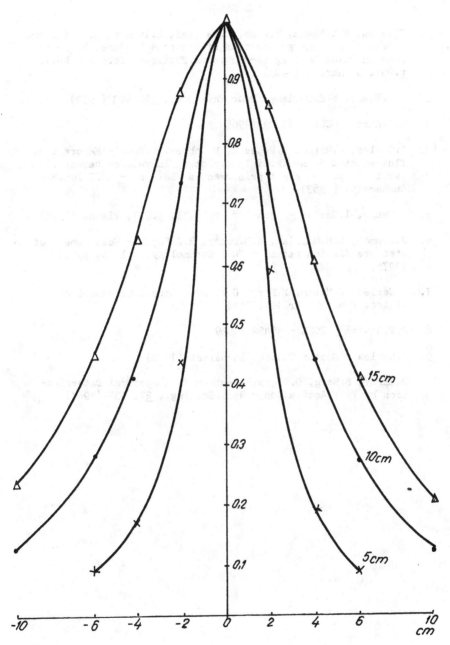

Fig. 2.-   The flux distributions

REFERENCES

1. F.Tancu, M.T.Magda, S.Dima, M.Macovei, E.Ivanov, R.Dumitrescu C.Stan-Sion - Energy Spectrum of Neutrons Produced by Deuterons on Thick Be Target - Rev.Roum.Physique, Tome 28, No.10 p.857, Bucharest (1983)

2. H.C.Evans, E.H.Bellamy, Proc.Phys.Soc., 74, 483 (1959)

3. R.J.Kurz - UCRL - 11339 (1965)

4. I.Garlea, C.Miron, S.Rapeanu, D.Dobrea, C.Roth - Masurari de flux-spectru de neutroni la ciclotron in vederea caracterizarii spectrului creat prin reactia $^9Be(d,n)$ - INPE Int.Rep., Bucharest, (1983)

5. A.Fabry, I.Garlea - IAEA-208, vol. II, p.291, Vienna (1978)

6. J.Grundl, D.M.Gilliam, N.D.Dudey, R.J.Popek - Measurement of absolute fission rates, - Nucl.Technology, vol.25, p.237 (1975)

7. I.Garlea, C.Miron, F.Popa, G.Fodor - Studii Cercetari de Fizica, Tom. 30, Nr.10 (1981)

8. J.T.Routti - UCRL - 19452 (1969)

9. I.Garlea - Doctor Thesis, Bucharest (1979)

10. McElroy, S.Berg, G.Gigas - Neutron Flux Spectral Determination by Foil Activation - Nucl.Sci.Eng., 27, 533 (1967)

RECENT DEVELOPMENTS IN NEUTRON DOSIMETRY AND DAMAGE

CALCULATIONS FOR FUSION MATERIALS IRRADIATIONS

L. R. Greenwood and R. K. Smither

Argonne National Laboratory

Argonne, Illinois USA

ABSTRACT

Fusion materials experiments require accurate
neutron dosimetry measurements and damage calculations.
Neutron cross section measurements are described for
application at fission reactors, accelerator neutron
sources, fusion reactors, and spallation neutron sources.
For fission reactors work is in progress to resolve
integral/differential discrepancies and to develop new
reactions for dosimetry. Reactions for fusion reactors
are being used for plasma diagnostics and waste generation.
Spallation reactions are being measured to extend our
techniques to 800 MeV for spallation neutron sources.
The SPECTER computer code has been developed to routinely
calculate displacement damage, gas production, recoil
spectra, and total dose for 38 elements for any user-
specified neutron spectrum. New helium measurements and
calculations are described. A new procedure is discussed
for the thermal production of helium by nickel including
previously neglected damage from the energetic ^{56}Fe
recoils.

INTRODUCTION

Fusion materials experiments span a diverse variety of
facilities, from fission reactors to high energy particle accelerators. In order to compare property changes in materials between

such different environments, it is first necessary to measure each radiation environment as accurately as possible and then to calculate damage parameters which can be used to correlate materials effects. Although each type of facility presents new challenges, all have the same basic needs, namely, the measurement of nuclear data, the development of new techniques, the improvement of computer codes and procedures, and the need for interlaboratory comparisons and standardization. The following sections will describe some of the recent developments in the U.S. fusion materials program. Neutron cross section measurements and integral testing is in progress for fission reactors, fusion reactors, 14 MeV and broad-spectrum accelerator sources, and spallation neutron sources. New measurements and calculations of helium production are reported, especially concerning a new procedure for calculating the thermal production of helium from nickel and the previously neglected displacement damage from this process. The SPECTER computer code has been developed to routinely calculate displacement damage and gas production for a user specified neutron spectrum.

CROSS SECTIONS FOR FISSION REACTORS AND ACCELERATOR SOURCES

Most fusion materials experiments are being conducted in either mixed-spectrum or fast fission reactors due to the availability of large volumes and high damage rates in these facilities. Superconductors and insulators have been intensively studied at 14 MeV sources and some limited measurements have been conducted at Be(d,n) broad-spectrum sources in anticipation of the Fusion Materials Irradiation Test Facility now under construction. We have been engaged in a program to measure and integrally test neutron activation dosimetry cross sections in a variety of neutron spectra. Both monoenergetic and integral measurements are being made with the eventual goals of providing new reactions for dosimetry and of correcting discrepancies between integral and differential measurements. Of particular interest for fission reactors is the $^{93}Nb(n,n')$ (14.6 y) reaction due to the low threshold and long-lived activity. Samples have been irradiated in a number of different spectra and we are currently engaged in developing counting facilities and procedures. Several reactions which have notable discrepancies between differential and integral measurements include $^{47}Ti(n,p)^{47}Sc$, $^{54}Fe(n,\alpha)^{51}Cr$, $^{60}Ni(n,p)^{60}Co$, $^{59}Co(n,p)^{59}Fe$, $^{58}Ni(n,2n)^{57}Ni$, $^{63}Cu(n,\alpha)^{60}Co$, and $^{65}Cu(n,p)^{65}Ni$. These reactions, as well as many others, have recently been measured in a well-known neutron field at Argonne in collaboration with Don Smith. These data, in conjunction with previous measurements at 14 MeV [1] and various Be(d,n) sources [2,3], as well as new monoenergetic measurements now being planned, will be used to adjust these cross sections. As an example, a preliminary test has been conducted for the $Ti(n,x)^{47}Sc$ reaction. The STAY'SL [4] computer code was used to readjust the

cross section as shown in Figure 1. The ENDF/B-V values were used
as the starting function and integral activities and spectra from
seven different reactors and accelerators were used in the readjust-
ment. The adjusted cross sections are in good agreement with new
differential measurements now in progress [5] demonstrating the
utility of this technique. Similar adjustments are now in progress
for many other reactions and this technique promises to significantly improve the accuracy and availability of dosimetry cross
sections over a wide range of energies. Coupled with calculations
and selected differential measurements, this technique could be used
to provide the numerous cross sections needed for dosimetry,
especially at higher neutron energies.

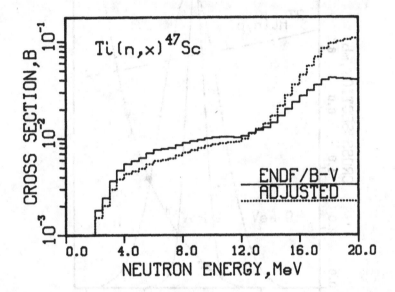

Figure 1.   Cross section adjustment for the Ti(n,x)^{47}Sc reaction
using the STAY'SL computer code. The results agree well
with new measurements [5]. The adjustment started with
ENDF/B-V cross sections and considered new integral
activities and spectra.

FUSION REACTOR CROSS SECTION MEASUREMENTS

Cross section measurements are being made for fusion reactor
diagnostics, in collaboration with Princeton University.   In
particular, the ^{27}Al(n,2n)^{26m}Al (6 s) [6] and ^{54}Fe(n,2n)^{53}Fe
(10 m) [7] cross sections have been measured near the 14 MeV

thresholds. The steep rise in yield with neutron energy and the short half-lives make these reactions quite sensitive to the plasma ion temperature as shown in Figure 2. These reactions could thus be used as a very sensitive plasma diagnostic for a single shot of current reactors such as the Tokamak Fusion Test Facility at Princeton.

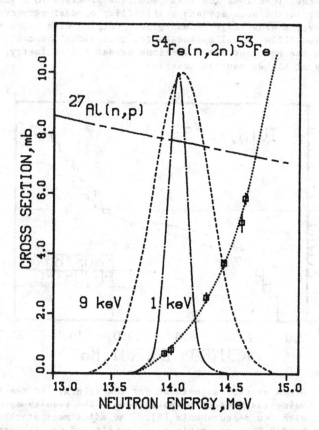

Figure 2. Cross section measurements are shown for the $^{54}Fe(n,2n)$-^{53}Fe reaction. Superimposed on the cross section data are two neutron energy distributions for d-t plasmas with ion temperatures of 1 keV and 9 keV (shown in arbitrary flux units). ^{53}Fe can be used for plasma diagnostics. Since ^{53}Fe decays to ^{53}Mn, the reaction is also of interest to fusion waste management.

Measurements are also being made of very long-lived activities which are of interest in fusion activation, dosimetry, and waste management. In particular, the ^{27}Al(n,2n)^{26}Al (7.3 E+5 y) has been studied by intense irradiations at the Rotating Target Neutron Source II at Lawrence Livermore National Laboratory followed both by direct gamma counting and by a relatively new technique, accelerator mass spectrometry [6]. The ^{54}Fe(n,2n) reaction mentioned above is also of interest since ^{53}Fe decays to the long-lived ^{53}Mn (3.7 E+6 y). Cross sections to both reactions have been found to be significantly lower (by about a factor of two) than predicted, thereby reducing their concern for fusion activation and waste generation. Other long-lived isotopes now being studied include ^{92}Nb (3.2 E+7 y), ^{59}Ni (7.5 E+4 y), ^{93}Zr (1.5 E+6 y), and ^{93}Mo (3000 y).

SPALLATION CROSS SECTION MEASUREMENTS

Low temperature (4 K) radiation damage experiments are being conducted at the Intense Pulsed Neutron Source (IPNS) at Argonne, a new facility is being prepared at the Los Alamos Meson Physics Facility (LAMPF) at Los Alamos National Laboratory, and even more powerful spallation sources are being proposed both in the U.S and Europe. These facilities produce high energy neutrons (500-800 MeV) requiring new nuclear data for dosimetry measurements and radiation damage calculations. Spallation cross sections themselves are ideal for dosimetry since they have very high thresholds and reasonably large magnitudes. We have thus measured spallation cross sections from Al to ^{7}Be, ^{22}Na, and ^{24}Na and from Cu to 20 different spallation products [8,9]. Foil packages were placed directly in the proton beam of the IPNS and activation measurements were made at eight energies between 30 and 450 MeV. Selected cross sections are shown in Figure 3 along with spallation data for iron [10]. These reactions are now being tested using the spallation sources at IPNS and LAMPF [11]. Figure 4 shows some adjusted spectra at these facilities. Radiation damage calculations are also being studied at these higher neutron energies [12]. The cross sections have also been compared to semiempirical models of Rudstam [13] and of Silberberg and Tsao [14]. These models only appear to be reliable within about a factor of two but are useful for predicting weak reactions and general systemmatic trends in the data. Further differential and integral measurements are now being planned to develop these cross sections for routine dosimetry. Other measurements have also been made to determine the proton flux directly in the beam envelope and the secondary proton flux at the neutron irradiation positions [11,15].

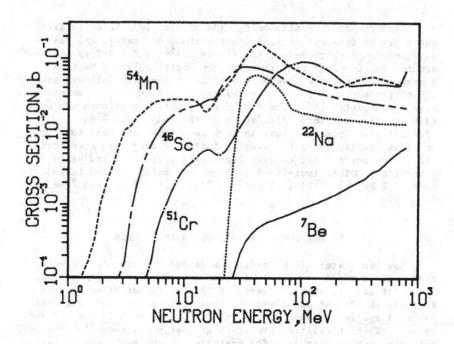

Figure 3.   Spallation cross sections used for neutron spectral
            adjustments at high energy neutron sources. Reactions
            plotted include Fe to ^{54}Mn and ^{51}Cr, Ti to ^{46}Sc, and Al
            to ^{22}Na and ^{7}Be.

HELIUM PRODUCTION CROSS SECTIONS

Gas production and other transmutation is of particular
interest in radiation damage studies since these products may
strongly influence the evolution of materials property changes,
especially swelling. Helium production is of particular interest
and is being routinely measured in fusion materials experiments by
Rockwell International. Samples have been jointly irradiated in all
types of facilities permitting the direct comparison of radiometric
and helium cross sections, especially at 14 MeV [16] and Be(d,n)
[17] neutron sources. Recent measurements also indicate serious
disagreements with ENDF/B-V [18] for several elements irradiated in
fission reactors. In particular, the ratio of calculated-to-measured
helium production in the Oak Ridge Research Reactor and in the High
Flux Isotopes Reactor has been determined to be about 2.3 for Ti and
about 0.60 for Cu [19]. Many other materials are now being studied
with the eventual goals of improving helium production cross sections
and of developing these data for dosimetry applications.

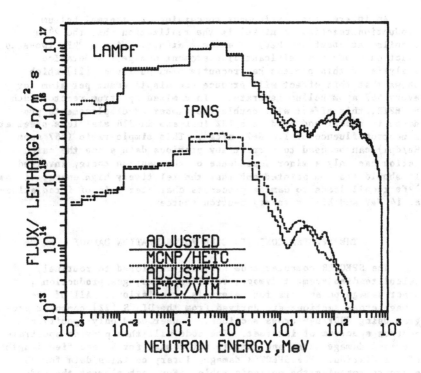

Figure 4. Calculated and adjusted neutron spectra are compared for IPNS and LAMPF. Note that the adjustment can now be done up to 800 MeV using spallation cross sections for dosimetry.

One of the most important helium producing reactions is for nickel since this thermal neutron effect allows us to obtain the high helium-to-displacement damage ratio characteristic of fusion reactors during irradiations in mixed spectrum fission reactors. A new evaluation of the capture cross sections on ^{59}Ni has recently been published [20] allowing us to directly calculate helium production in any neutron spectrum. Joint radiometric and helium measurements in a number of fusion materials experiments have been used to test these new cross sections. Forty-five samples of nickel have been irradiated in three different reactors to fluences between 1.0 E+21 and 1.0 E+23 n/cm^2. Excellent agreement is found between the calculated and measured helium content with an average ratio of 0.982 and a standard deviation of 6% [20].

An interesting development concerning the thermal helium production reaction in nickel is the realization that the ^{56}Fe recoils (at about 340 keV) from the ^{58}Ni$(n,\gamma)^{59}$Ni$(n,\alpha)^{56}$Fe two-step reaction produce significant displacement damage. A detailed analysis of this process has recently been published [21] which shows that this effect will produce one displacement per atom for every 567 appm helium generated. In a mixed spectrum reactor such as HFIR, this effect will double the number of displacements produced in nickel and produce a 13% increase in 316 stainless steel at a neutron fluence of 1.0 E+23 n/cm^2. This simple ratio (567 appm He/dpa) can be used to correct any previous data since the calculation has only a minor dependence on the neutron energy involved. It should also be pointed out that the relatively high energy of the ^{56}Fe recoil leads to damage processes characteristic of irradiations at 14 MeV and higher energy neutron sources.

THE SPECTER COMPUTER CODE FOR RADIATION DAMAGE

The SPECTER computer code has been developed to routinely calculate displacement damage, recoil spectra, gas production , and other damage parameters for materials irradiations. All of the damage cross sections are derived from the DISCS [22] computer program using ENDF/B-V cross sections. A user need only specify a neutron spectrum of interest. The code will then provide spectral-averaged damage rates and compute net damage for a specified length of irradiation. The SPECTER damage library contains data for 38 elements spanning the periodic table. For each element the code lists displacements for elastic, inelastic, (n,xn), (n,px), and (n,αx) reactions. Recently we have added damage calculations for (n,gamma) capture and any subsequent beta decay [23]. Gas production is included for each element; however, the new ENDF/B-V gas production file is now included separately [17]. Total dose values (Kerma) are also calculated for each irradiation using the MACKLIB program [24]. Recoil atom distributions for each element and each reaction are also presented in both differential and integral form. Provision is made for specifying a covariance uncertainty matrix for the input spectrum; this is routinely done for fusion experiments using the output of our spectral adjustment program STAY'SL [4]. SPECTER will then determine the uncertainty in each damage calculation due to the uncertainty in the flux spectrum. Of course, this does not consider the uncertainty in the damage cross sections themselves, a deficiency of all present damage calculations which must eventually be addressed.

At present, only six elements have been extended beyond the 20 MeV cutoff in ENDF for use at accelerator neutron sources. These files were extended using the available data and calculations in the literature, although some reactions had to be extrapolated due to the general lack of adequate data beyond 14 MeV. For FMIT-type sources as well as lower energy spallation sources such as IPNS this deficiency is not so serious since only a small fraction of the damage (5-10%) is caused by neutrons with energies above 20 MeV. However, recent measurements at LAMPF (800 MeV protons on a copper target) indicated that for the spectrum shown in Figure 4, about 26% of the dpa and nearly 95% of the helium in copper is generated by neutrons above 20 MeV. In this case the VNMTC code was used to compute the higher energy damage [12]. However, the SPECTER and VNMTC calculations do not match very well indicating a serious need to address the problem of damage calculations at intermediate energies in the range from 20 to 800 MeV.

CONCLUSIONS AND FUTURE DIRECTIONS

Recent measurements of neutron cross sections are being used to improve and to expand our capabilities for dosimetry measurements and damage calculations. Some outstanding discrepancies between differential and integral experiments will be resolved and new reactions will be developed for routine use. However, there still remains a paucity of reliable data above 14 MeV and further experiments and calculations are badly needed, especially for spallation sources. Several new techniques are very promising, especially the use of spallation reactions at energies above 40 MeV and accelerator mass spectrometry for the measurement of very long-lived or stable isotopes. This latter technique offers the promise of measuring transmutation to stable products directly, a feature which may be of particular importance in fusion reactors.

Dosimetry at fusion reactors will soon become practical as new facilities such as TFTR and JET approach breakeven conditions. Some of the reactions which we are developing could be very useful both for dosimetry and as a plasma diagnostic. However, testing of these and other reactions must await the high power (d,t) operation.

Nuclear data is of course needed for many elements over a wide range of energies to improve the quality of the damage calculations. A carefully planned mix of calculations and selected measurements is needed. The theory and computer codes must also be improved to consider alloys and insulators; at present only pure elements are included in an assumed metallic form. Finally, uncertainties should be assigned to the damage calculations. This is not easily done at present since uncertainty data is only included for a few reactions in ENDF. Assuming the inclusion of such data in future versions of ENDF and using sensitivity studies of model uncertainties, we plan to calculate uncertainties in damage rates and to interface these uncertainties with flux and spectral uncertainties to obtain more realistic estimates of damage parameters.

REFERENCES

[1] L. R. Greenwood, J. Nucl. Mater. 108, 21-27 (1982).

[2] L. R. Greenwood, R. R. Heinrich, R. J. Kennerley, and
R. Medrzychowski, Nucl. Technol. 41, 109-128 (1978).

[3] L. R. Greenwood, R. R. Heinrich, M. J. Saltmarsh, and
C. B. Fulmer, Nucl. Sci. Eng. 72, 175-190 (1979).

[4] F. G. Perey, ORNL/TM-6062 (1977); modified by L. R. Greenwood
(1979).

[5] D. L. Smith, Argonne National Laboratory, private
communication (1984).

[6] R. K. Smither and L. R. Greenwood, J. Nucl. Mater. 122,
1071-1077 (1984).

[7] R. K. Smither and L. R. Greenwood, Damage Analysis and
Fundamental Studies Quarterly Progress Report, DOE/ER-0046/17,
May 1984.

[8] L. R. Greenwood and R. K. Smither, Damage Analysis and
Fundamental Studies Quarterly Progress Report, DOE/ER-0046/14,
August 1983.

[9] L. R. Greenwood and R. K. Smither, Damage Analysis and
Fundamental Studies Quarterly Progress Report, DOE/ER-0046/18,
August 1984.

[10] R. Michel and R. Stuck, J. Geophysical Res. 89, Supl. B673-684
(1984).

[11] D. R. Davidson, L. R. Greenwood, R. C. Reedy, and W. F. Sommer,
Measured Radiation Environment at the LAMPF Irradiation
Facility, Proc. Twelfth International Symposium on the Effects
of Radiation on Materials, Williamsburg, VA, June 18-20, 1984.

[12] D. R. Davidson, L. R. Greenwood, W. F. Sommer, and
M. S. Wechsler, Calculation of Displacement and Helium
Production at the LAMPF Irradiation Facility, ibid.

[13] G. Rudstam, Z. Naturforsch. 21A, 1027 (1966).

[14] R. Silberberg and C. H. Tsao, Astrophy. J. Suppl. 25, 315-367
(1973); parameters updated in succeeding International Cosmic
Ray Conferences.

[15] L. R. Greenwood and R. J. Popek, Proc. of the Sixth Meeting on the Int. Coll. on Adv. Neutron Sources, ICANS VI, ANL-82-80, 605-618, 1982.

[16] D. W. Kneff, H. Farrar IV, and L. R. Greenwood, J. Nucl. Mater. 103,1451 (1981); a comprehensive report of 14-MeV helium cross sections is being prepared for publication.

[17] D. W. Kneff, H. Farrar IV, L. R. Greenwood, and M. W. Guinan, Symp. on Neutron Cross Sections from 10-50 MeV, BNL-NCS-51245, 113-132, (1980).

[18] Evaluated Neutron Data File Part B, Version V, National Neutron Cross Section Center, Brookhaven National Laboratory, 1979.

[19] B. M. Oliver, D. W. Kneff, and R. P. Skowronski, Damage Analysis and Fundamental Studies Quarterly Progress Report, DOE/ER-0046-18, August 1984.

[20] L. R. Greenwood, D. W. Kneff, R. P. Skowronski, and F. M. Mann, J. Nucl. Mater. 122, 1002-1010 (1984).

[21] L. R. Greenwood, J. Nucl. Mater. 115, 137-142 (1983).

[22] G. R. Odette and D. R. Dorion, Nucl. Technol. 29, 346 (1976).

[23] R. K. Smither and L. R. Greenwood, Proc. Fourth ASTM-EURATOM Symp. on Reactor Dosimetry, NUREG/CP-0029, 793-805 (1982).

[24] Y. Gohar and M. A. Abdou, ANL-FPP/TM-106 (1978).

MATERIALS IRRADIATION TESTING IN SUPPORT

OF THE EUROPEAN FUSION PROGRAMME

D.R. Harries and J.-M. Dupouy

The NET Team, Max-Planck-Institut für Plasmaphysik,

8046 Garching bei München, FRG

ABSTRACT

The objectives and planned schedule of the Next
European Torus (NET) programme are outlined and the
selection of materials for some of the principal components (first wall and tritium breeding blanket, first
wall protection, impurity control system and insulators)
is discussed in the light of the currently envisaged
parameters. The materials irradiation programmes which
are either being implemented or are proposed to provide essential data for the detailed design and construction of NET and for assessing component performance and endurance are summarized. Finally, the
longer-term programme which should be undertaken to
enhance the knowledge of the behaviour of materials in
a fusion environment and to guide the development of
materials for a DEMO fusion power reactor is indicated.

I. INTRODUCTION

The 1982-86 European Fusion Technology is currently being implemented within the CEC (EURATOM) Fusion Association Laboratories
and Joint Research Centre (JRC) and in two non-member countries,
Sweden and Switzerland. The development of the materials part of
this programme, in which the emphasis was initially placed on providing the required data base for INTOR, has been reviewed previously (1,2). The INTOR workshop was set up by the IAEA in 1978
with the aim of establishing the technical objectives and nature

of a large fusion research device of the Tokamak type which might be constructed internationally by the participants, namely, Europe, U.S.A., Japan and U.S.S.R.

However, the Community and Fusion Associations decided in 1982 to initiate the definition in Europe of the Next European Torus (NET) (3), envisaged as the sole step between the Joint European Torus (JET) and a fusion power demonstration reactor (DEMO). The NET Team was subsequently established at Garching to commence the technological developments required for its design and construction as well as these needed in the longer term for DEMO. The NET Team, in collaboration with the Brussels Fusion Office, Fusion Association Laboratories and JRC, have now formu- lated a 1985-89 NET and Technology Programme in which the tasks initiated during the 1982-86 phase have been extended and re- oriented towards generating essential data for NET. This 1985-89 programme, covering physics, magnets, blanket and first wall en- gineering, tritium technology, structural materials, tritium breeding materials, maintenance, and safety and environment, is awaiting approval by the European Council of Ministers.

The NET objectives, schedule and some of the principal para- meters are outlined in the second section of this paper. The third part summarizes the factors determining the choice of materials for the first wall, tritium breeding blanket and other components and the programmes devised to provide the materials data, particu- larly in terms of irradiation effects, necessary for the NET design and construction and for optimizing component performance and endurance. The longer-term programme required to guide the ma- terials development for DEMO is considered in the final section.

2. NET OBJECTIVES, SCHEDULE AND PARAMETERS

The primary objectives of the NET programme are the realisation of a D-T fusion device of the Tokamak type capable of:

(i)   demonstrating adequate plasma performance.
(ii)   proving the feasibility of the basic technology.
(iii) assessing the safe and reliable operation.
(iv)   providing, in association with the European Fusion
       Technology Programme, adequate information for the design
       of DEMO.

The current schedule envisages that the detailed design phases for the basic machine and blanket will begin in 1988/89 and 1989/90 respectively, with the construction of the basic machine commencing in 1992 and that of the blanket following two years later. Start of operation is scheduled for 1999.

The NET parameters have not yet been finalised and several alternative designs, for example, of the first wall and blanket, impurity control system, etc., are being evaluated and as many

options as possible are being retained during the present 1984-87
pre-design phase. However, the following parameters have been
assumed in respect of materials selection and property requirements:

Thermal power: 600 MW.
Burn pulse: 200 s.
Neutron wall loading: 1.3 MW.m.$^{-2}$
Integrated wall loading: 2.5 MWy.m.$^{-2}$
Availability: 20 - 25 % (max.).
Number of cycles: 2.5 x 10^5.
Tritium breeding ratio: 0.6 - 0.8.
First wall and blanket temperatures: 350° - 500° C.
Coolants: High pressure water or helium.

3. SELECTION OF MATERIALS AND IRRADIATION PROGRAMMES

3.1 First Wall and Breeder Structural Components

The materials selected for the first wall and breeder structure
must be readily fabricated and welded in the required geometries
and be compatible with the coolants and tritium breeding materials.
In addition, they must retain a high degree of dimensional stability and mechanical integrity under the onerous service conditions,
as frequent remote replacement of the components would adversely
affect the system availability.

Alternating thermal and mechanical stresses will be induced in
the first wall and blanket structure as a result of the high heat
loads and pulsed nature of reactor operation, coolant pressure, magnetic loads and temperature gradients. The materials selected for
these applications must initially possess the appropriate balance
of thermo-mechanical properties at the relevant temperatures to
withstand these stresses as well as adequate resistance to radiation damage which can detrimentally affect the properties.

The principal D-T plasma fusion reaction is:

$$D_1^2 + T_1^3 \rightarrow He_2^4 + n_0^1$$
$$(3.52 \text{ MeV}) \quad (14.08 \text{ MeV}).$$

Bulk radiation damage will be produced in the first wall and
blanket structural materials as a result of the formation of va-
cancy-interstitial pairs by the highly energetic neutron-atom
collisions and helium and hydrogen gases by (n,α) and (n,p) reactions respectively. These effects can lead to dimensional instability (phase changes, void swelling and irradiation creep and
growth) and hardening and embrittlement, manifested as changes in
tensile and creep-rupture strengths and ductilities, reduced
fatigue and creep-fatigue strengths and fracture toughness, and
enhanced fatigue crack growth rates. Furthermore, hydrogen embrittlement of the first wall structure could be induced as a

result of the hydrogen formed by (n,p) reactions and the hydrogen isotopes (D and T) injected from the plasma.

Two commercial steels have been selected as potential first wall and blanket structural materials for NET:

(i) Solution annealed or cold worked Type 316L austenitic steel (typical analysis: 0.03%C; 17%Cr; 12.5%Ni; 2.5%Mo; balance-essentially Fe).

The data generated in fast breeder reactor programmes show that the wrought 316L steel has comparable or superior resistance to irradiation-induced void swelling, in-reactor creep and changes in post-irradiation tensile and creep-rupture ductilities and similar unirradiated fatigue and creep-fatigue properties compared to other austenitic steels. However, additional information is required on the fatigue, fatigue crack growth and fracture toughness characteristics of welds after irradiation.

(ii) Quenched and tempered 1.4914 martensitic steel (typical analysis: 0.17%C; 10.5%Cr; 0.85%Ni; 0.55%Mo; 0.25%V; 0.20%Nb; low N; balance-essentially Fe).

This and other 10 - 12 % Cr ferritic/martensitic steels are being developed in Europe and elsewhere as alternatives to austenitic steels for fast reactor core component applications. These steels have higher strengths at temperatures of $\leqslant 500°$ C, better resistance to thermal stress development and superior resistance or immunity to irradiation-induced void swelling, high temperature (helium) embrittlement and, possibly, 'in-reactor' creep compared to the austenitic steels (4). The potential limitations of these steels are:

(a) In common with ferritic steels, they exhibit ductile-brittle transition temperatures (DBTT). It is necessary to demonstrate for the irradiation and temperature conditions appropriate to the first wall in NET, that the fracture toughnesses of the wrought 1.4914 steel and welds, as evidenced by increased DBTTs. and reduced upper shelf energies, are not excessively impaired, so that the possibility of failure under normal operating and fault conditions is minimized.

(b) They are ferromagnetic; however, initial analyses suggest that this will not be a serious problem in magnetically-confined tokamak fusion systems.

(c) Their compatibility with high pressure water may be inferior to that of austenitic steels and the water chemistry will have to be closely controlled and monitored to maintain the corrosion at acceptable levels.

The calculated displacement damage and gas productions in austenitic and martensitic steel first walls in NET are given in Table I.

Table I

Displacement Damage and Helium and Hydrogen Gas Productions in NET First Wall at 2.5 MWy.m^{-2}.

	Type 316 Steel	Martensitic Steel
Displacement damage (dpa)	28	33
Helium concentration (appm)	370	260
Hydrogen concentration (appm)	1420	1360
He: dpa	13	7.7

A high 14 MeV neutron flux irradiation test facility, in which the required displacement damage and gas productions could be achieved simultaneously in the 316L and 1.4914 steels, is not available. Consequently, data on the effects of irradiation on the properties and behaviour of the first wall and breeder structural materials have to be obtained in fission reactor experiments or using charged particle bombardments in accelerators, cyclotrons and high voltage microscopes.

The calculated helium and hydrogen concentrations versus displacement dose for Type 316 austenitic steel and FV 448 (similar composition to 1.4914) martensitic steel irradiated in a high thermal flux reactor (HFIR, Oak Ridge), in the core of a fast reactor (PFR, Dounreay) and in the first wall of a tokamak reactor are compared in Fig. 1.

It is evident that the He: dpa and H: dpa ratios appropriate to a fusion reactor first wall cannot be simulated in these thermal and fast reactor irradiation positions. Nevertheless, the required fusion reactor first wall He:dpa ratio for the high nickel 316 type steel can be approached by sequential irradiations in high thermal and fast neutron flux positions in mixed spectrum reactors such as BR-2, Mol and by irradiating in positions outside the core and in the reflector in fast reactors. However, prolonged irradiations (up to 6 years duration) would be required to achieve the NET end-of-life target displacement doses and helium concentrations. This is not the case for the low nickel martensitic steel as irradiations in thermal or fast reactors result in He:dpa ratios considerably lower than that for a fusion reactor first wall. Consequently, the effects of high helium concentrations on the properties and behaviour of the martensitic steel have to be elucidated either from fission reactor irradiations of samples previously implanted with helium or by sequential or simultaneous α- and charged-particle irradiations.

The principal irradiation studies on the austenitic and martensitic steels in the current 1982-86 and proposed 1985-89 programmes may be summarised as follows:

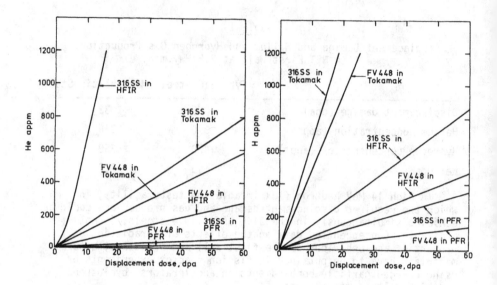

Fig. 1 Computed helium and hydrogen concentrations versus displacement dose.

(i) Tensile, fatigue, fatigue crack growth, creep-fatigue and fracture toughness tests on plates and welds following irradiation in thermal reactors (BR2, Mol; HFR, Petten and R2, Studsvik), initially to 10 dpa and subsequently to ∿ 30 dpa at 250° and/or 430° C, and in fast reactors.

(ii) In- (thermal) reactor fatigue, fatigue crack growth, creep and creep-fatigue tests to 10 dpa at 250° and 430° C. (316L steel only).

(iii) Post-bombardment and 'in-beam' charged particle irradiation tests at a variety of temperatures and doses to establish the individual and combined influences of displacement damage and helium on the structure, void swelling and mechanical properties (fatigue, fatigue crack growth, creep and rupture and creep-fatigue).

The work on the post-irradiation tensile, fatigue and fatigue crack growth and in-reactor fatigue and creep of the 316L austenitic steel is being implemented as part of an IEA co-operative project involving the European Community (including Sweden), Japan and the U.S.A. In addition to irradiations in the European reactors, experiments are being carried out in HFIR and ORR at Oak Ridge on the 316L steel, U.S. 316 steel and Japanese prime candidate alloy, one important objective being to compare data obtained in post-irradiation and 'in-reactor' tests.

These investigations will be augmented by studies of the
effects of hydrogen, initially introduced by cathodic charging, on
the structures and mechanical properties of the steels in the unirradiated and irradiated conditions. An additional requirement is
the development of surface barrier layers to inhibit the permeation of tritium injected from the plasma in the first wall and
from the breeding material through the coolant tubes, thereby minimizing its release into the coolant and inventory in the system.
Several potential barrier layers and production techniques are
currently being considered and it is necessary to demonstrate that
the coatings possess good adhesion and thermal shock resistance
before and after irradiation as well as the ability to reduce the
tritium permeation by at least two orders of magnitude to maintain
the inventory at tolerable levels.

3.2 Protection of the First Wall.

Surface radiation damage will also occur in the first wall due
to the ions and energetic neutral atoms from the plasma, possibly
resulting in blistering, sputtering and excessive surface erosion,
with the latter effect being exacerbated by plasma disruptions. The
possibility of protecting the plasma side of the first wall using
mechanically attached or bonded tiles of graphite, possibly impreg-
nated with SiC, or SiC in massive form or sprayed on to a metallic
substrate, is being considered. The use of such protective layers
would enable the thickness of a bare first wall to be reduced, as
the need for a sacrificial layer for erosion would be obviated,
thereby decreasing the thermal stress development.

The principal advantages of graphite are that its sputtered
atoms are less detrimental in cooling the plasma than those of the
heavier elements and it has high resistance to plasma disruptions
because it sublimates and does not melt. On the other hand, it exhibits high anisotropic growth at low irradiation fluences and unfavourable hydrogen isotope recycling characteristics.

However, additional investigations are required to establish
the erosion characteristics of the graphite and SiC using particles
of the appropriate energies and their physical and mechanical properties and dimensional stabilities at the relevant irradiation
temperatures before these options can be pursued further.

3.3 Tritium Breeding Materials

The tritium breeding materials considered for NET are the liquid $Li_{17}Pb_{83}$ eutectic and the solid ceramics, Li_2O, $LiAlO_2$,
Li_2SiO_3, Li_2TiO_3 and Li_2ZrO_3.

Pure lithium has not been included because it would require a
thicker blanket than is possible in NET to realize its superior

tritium breeding potential and it has other limitations with respect to safety and other considerations. There is little or no information on which to make a choice at this stage between the solid breeder materials; however, the aluminate and silicate have been selected as the prime candidate materials for NET with the oxide, titanate and zirconate being retained as longer term options.

In addition to fabrication and characterization studies, physical and thermodynamic property measurements and investigations of the compatibilities with the breeder structural materials and coolants, the principal aim of the programme on the breeding materials is to determine the tritium production and release as a function of irradiation temperature in SILOE and HFR.

With the exception of the LiPb eutectic which contains its own multiplier (lead), the solid breeder materials require neutron multipliers to achieve a local tritium breeding ratio in excess of unity. Beryllium, in the form of individual metallic rods, has been selected in preference to the other alternatives (lead and PbZr compound) on the basis of design, physical and mechanical property and neutron multiplication considerations. In addition to performing studies of its compatibility with the breeding, structural and coolant materials, it is planned to undertake an investigation of the effects of irradiation in fast reactors, where the relevant fusion environment He:dpa ratio can be obtained and to higher fluences than have been studied hitherto, on its high temperature mechanical properties.

3.4 Impurity Control System

A decision has not yet been made as to whether to use a divertor or limiter for plasma impurity control in NET and the designs of both options are being pursued. A preliminary investigation of the behaviour of a number of potential limiter materials during exposure in TEXTOR is underway at KfA Jülich and a programme on divertor materials is now being finalised.

The current divertor design envisages the use of a low sputtering coefficient tungsten alloy (W - 5%Re or W - 26%Re with superior fracture toughness characteristics compared to unalloyed tungsten) for the collector plates which are brazed to a cooled heat sink material. A high strength,high thermal conductivity Cu - 0.65%Cr - 0.08%Zr alloy has been chosen for the coolant tubes with OFHC copper as the heat sink material. The proposed programme involves:

(i) studies of the effects of thermal reactor irradiation, initially to 10 dpa and later to 30 dpa, at temperatures up to 600° C on the structural stabilities, mechanical properties and thermal conductivities of the Cu and Cu alloy and on the fracture toughnesses of the tungsten alloys.

(ii) charged particle irradiation experiments to establish the influence of high helium concentrations on the properties of the Cu and Cu alloy and to investigate the irradiation creep of the tungsten alloys.

(iii) the development of brazing techniques for the Cu-Cu alloy and Cu-W alloy joints followed by tests of the thermo-mechanical properties of the joints in the unirradiated and irradiated conditions.

3.5 Insulators

Ceramic insulators are required for RF heating and magnet systems. The former application is the most onerous as the insulators are located in an intense neutron and ionising radiation environment; the magnets are located outside the shield and the insulators are thus exposed to lower irradiation fluxes.

Several inorganic materials, including spinel $MgAl_2O_4$, single- and poly-crystalline Al_2O_3 and Si_3N_4, have been selected for initial evaluation, with AlON, AlN, BeO and fused SiO_2 also being retained in the programme. The current and proposed programmes include investigations of the effects of thermal reactor and charged particle irradiations on their structural, mechanical and electrical (during and after irradiation) properties. In addition, consideration is being given to mounting a programme on the effects of ionizing radiation on the structural stabilities of various organic insulators for magnet applications.

3.6 Other Components

It is envisaged that irradiation studies will also be required on the fracture toughness and crack growth characteristics of super-conducting and normal conducting magnet materials and, possibly, on the stability and mechanical properties of potential shielding materials. However, these programmes have not yet been formulated.

4. LONG TERM PROGRAMME

The principal aims in the programmes summarized in the previous section are to provide the data necessary to enable the materials selections to be optimised at the time when the NET detailed design phases are scheduled to commence and the likely component performance and endurance to be assessed. It is necessary to support these activities by a broader based and longer term programme involving:

(a) Increased emphasis on theoretical studies, including developing further the theory of microstructural evolution, modelling the effects of irradiation and hydrogen isotopes on mechanical properties, deformation and fracture, and stress and lifetime analyses.

(b) Experimental studies on (i) alternative commercially available materials to serve as an insurance against inadequate performance of those initially selected for NET, and (ii) advanced material concepts, including the identification and development of alloys with enhanced radioactive decay characteristics so as to permit material handling, storage and recycling scenarios to be established.

(c) Fundamental studies, with strong theoretical support, of the effects of exposure in a simulated fusion environment on the properties and behaviour of existing, alternative and advanced metallic and non-metallic materials.

Such a generic programme is essential to guide the development of materials to withstand the higher neutron and integrated wall loadings (typically, 3 MW.m^{-2} and 15 MWy.m^{-2} respectively) anticipated in DEMO.

REFERENCES

1. J. Nihoul, J.Nucl.Mat., 103/104, 57 (1981).

2. J. Darvas and R. Verbeek, ASTM Special Technical Publication No. 782, pp. 1140-1158 (1982).

3. R. Toschi, "The Next European Torus", 13th Symp. on Fusion Technology, Varese, September 1984, Proc. to be published.

4   D.R. Harries, Topical Conf. on "Ferritic Alloys for Use in Nuclear Energy Technologies", Snowbird, June 1983, Proc. to be published.

INVESTIGATION OF THE NEUTRON ENERGY TRANSFER BY NUCLEAR CASCADE

REACTIONS FOR IMPROVING NEUTRON DAMAGE DATA UP TO 40 MeV

G. Hehn, M. Mattes, G. Prillinger, M.K. Abu Assy;
Institut für Kernenergetik und Energiesysteme
University of Stuttgart, Germany
W. Matthes; JRC, EURATOM, Ispra, Italy

ABSTRACT

For simulating radiation damage of fusion reactors with high energetic neutrons the knowledge of
neutron cross-sections has to be improved especially
above 20 MeV. In this energy region the energy transfer is dominated by nuclear cascade reactions. Around
the fusion peak at 15 MeV about thirty percent of the
damage energy in iron is transferred by cascade reactions with a large (n,2n)-contribution. Above 20 MeV
the damage energy by cascade reactions dominates.

For iron and the main additional elements of
stainless steel like chromium and nickel the energy
transfer has been determined for each type of cascade
reaction separately. Using the evaporation model the
energy spectra of the heavy recoil atoms as well as
that of the smaller emitted particles were calculated
with a Monte Carlo procedure.

The cascade code SPALLS has been implemented in
our modular system RSYST and tested with neutron emission spectra available up to 20 MeV. Essential improvements have been made to determine the energy transfer by cascade reactions.

J. P. Genthon and H. Röttger (eds.), Reactor Dosimetry, 273–279.
© *1985 ECSC, EEC, EAEC, Brussels and Luxembourg*

NEED FOR IMPROVED DAMAGE DATA

For measuring neutron damage in materials needed for the design of fusion reactors the application of high energetic neutrons from generators and spallation sources are planned.The transfer of the experimental results to a different neutron spectrum in the real situation of a fusion reactor needs a detailed understanding at least of the primary energy transfer processes. First of all the knowledge of neutron cross-sections has to be improved especially above 20 MeV. Since measurements of partial cross-sections are rare, the evaluations are based primarily on predictions with nuclear model codes like GNASH from Los Alamos or STAPRE from Vienna /1, 2/. The results of both programs were used to split up the nonelastic cross-sections of iron and of the main additional elements of stainless steel into the partial cross-sections needed for studying the neutron energy transfer. The codes used for processing the evaluated partial cross-sections to produce neutron damage data like DON and the HEATR modul of NJOY are capable for treating the energy transfer of elastic scattering, inelastic scattering and absorption processes /3, 4/. But there are problems in calculating the energy transfer of nuclear cascade reactions. Around the fusion peak at 15 MeV about thirty percent of the damage energy in iron is due to cascade reactions with a large (n,2n)-contribution. The importance of all cascade reactions increases with rising neutron energy to become the dominating part above 20 MeV.

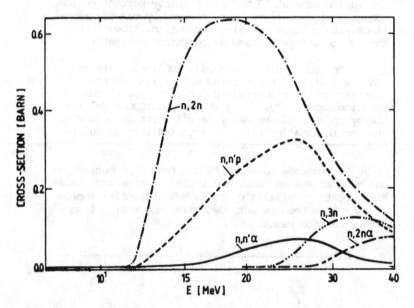

Figure 1: Cascade reactions of iron-56 up to 40 MeV

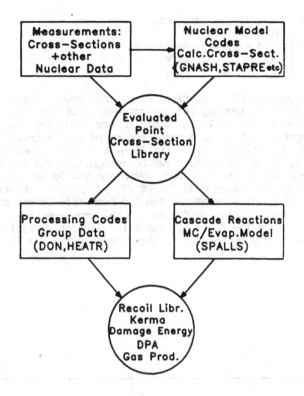

Figure 2: Neutron damage data up to 40 MeV

CASCADE PROCESSING CODE SPALLS

To study the neutron energy transfer by nuclear cascade reactions the code SPALLS has been written. Using the evaporation model the energy spectrum of the heavy recoil atom as well as that of the smaller emitted particles are calculated with a Monte Carlo procedure. Up to three emitted particles are covered, so that all partial cascade reactions can be treated, which are of importance up to 40 MeV, as shown for iron-56 in figure 1.

If the emitted particle is charged, then the Coulomb barrier suppresses the lowest energies and we get for the energy distribution of the evaporated particles in CMS

$$f(E_n \to T)dT = C(T-KT_c) \cdot \exp[-(T-KT_c)/\theta] \cdot dT$$

with E_n = energy of incoming neutron,
T = energy of emitted particle,

275

```
C  = normalisation constant,
T_C = Coulomb barrier,
θ  = fitted nuclear temperature,
K  = {0,47 protons }  effective constants for iron-56,
     {0,6  alphas  }  /5/.
```

The code SPALLS is operated within our modular system RSYST
/6/. The total system applied for processing of kerma and damage
data is shown in figure 2. It should be reminded, that the cross-
section input to SPALLS is a major source of uncertainty. By comparing two evaluations e.g. of the iron-56 (n,n'p) cross-section
in figure 3, large differences can be recognized. The higher data
values are based on the LANL code GNASH and the lower ones are
calculated by the Vienna code STAPRE /7, 2/. Similar relations are
found for the (n,n'α) cross-section and both effects are compensated by the reverse relation for the (n,2n) cross-section, where
the LANL data are lower. In all cases the largest deviations occur
just in the energy region, where the cross-section has its highest
values.

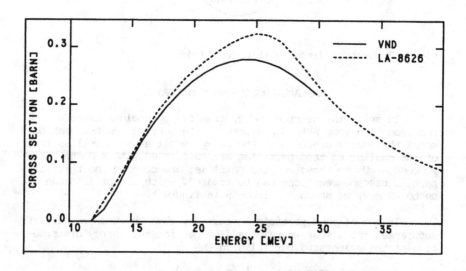

Figure 3: Cross-section for ^{56}Fe(n,n'p)

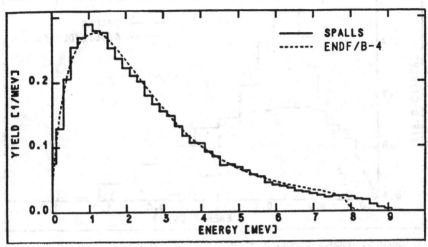

Figure 4: Neutron emission spectrum ^{56}Fe(n,2n) at 20 MeV

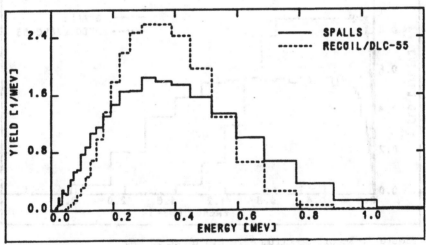

Figure 5: Recoil spectrum ^{56}Fe(n,2n) at 20 MeV

ENERGY TRANSFER BY CASCADE REACTIONS

To check, whether a simple evaporation model works to simulate the energy transfer in cascade reactions, we made detailed
intercomparisons with neutron emission spectra available below

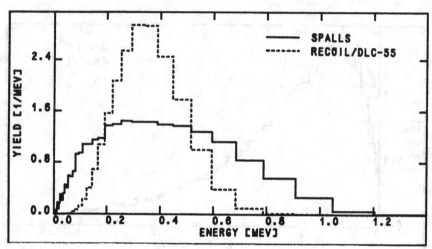

Figure 6: Recoil spectrum $^{56}Fe(n,n'p)$ at 20 MeV

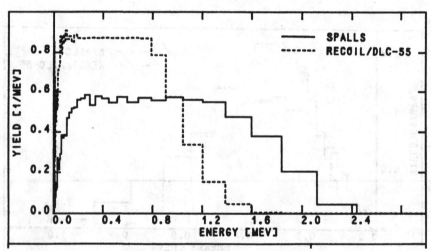

Figure 7: Recoil spectrum $^{56}Fe(n,n'\alpha)$ at 20 MeV

20 MeV. A perfect agreement between the Monte Carlo results of SPALLS and the ENDF/B-4 values are shown in figure 4 for the (n,2n) reaction at 20 MeV. Of course, we cannot conclude, that the same holds for neutron energies of 30 MeV and 40 MeV. But if the neutron emission spectrum can be fitted somehow by an evapora-

tion curve, we can conclude, that the respective spectrum of the heavy recoil atom is right, too. Therefore the recoil spectrum calculated by SPALLS for iron-56 (n,2n) reaction should be the right one compared to the data in DLC-55, as shown in figure 5 /8/. The narrow peak in the RECOIL data for the (n,2n) reaction indicates, that an approximation by inelastic scattering might have been applied. The same holds for the intercomparison of the iron-56 (n,n'p) recoil spectrum in figure 6. Finally the recoil spectrum is given in figure 7 for the iron-56 (n,n'α) reaction. Here the average recoil energy was underestimated in the RECOIL library strongly. It should be stated, that the data taken from the RECOIL library are for natural iron. Therefore especially in figure 6 and figure 7 small corrections have to be attributed to iron-54. We can conclude, a better processing of kerma and damage data is needed for high energetic neutrons. A special code for cascade reactions has been presented and results discussed.

REFERENCES

/1/ P.G. Young, E.D. Arthur, GNASH, A Preequilibrium, Statistical Nuclear Model Code for Calculation of Cross-Sections and Emission Spectra, LA-6947, 1977.

/2/ B. Strohmaier, M. Uhl, Nuclear Model Calculations of Neutron Induced Cross-Sections for ^{52}Cr, ^{55}Mn, ^{56}Fe, and $^{58+60}$Ni for Incident Energies up to 30 MeV, in Nuclear Data for Science and Technology, Ed., K.H. Böckhoff, Reidel Publishing, Dordrecht/Holland, 1983.

/3/ D.M. Parkin, A.N. Goland, Calculation of Radiation Effects as Function of Incident Neutron Spectrum, Radiation Effects, 28, 1976.

/4/ R.E. McFarlane, D.W. Muir, R.M. Boicourt, The NJOY Nuclear Processing System, LA-9303-M (ENDF-324), 1982.

/5/ R.E. McFarlane, D.W. Muir, F.M. Mann, Radiation Damage Calculation with NJOY, 3. Topical Meeting on Fusion Reactor Materials, Albuquerque, 1983.

/6/ R. Rühle, RSYST, ein integriertes Modulsystem mit Datenbasis zur automatisierten Berechnung von Kernreaktoren, IKE 4-12, 1973.

/7/ E.D. Arthur, P.G. Young, Evaluated Neutron-Induced Cross-Sections for ^{54,56}Fe to 40 MeV, LA-8626-MS(ENDF-304), 1980.

/8/ T.A. Gabriel, J.D. Amburgey, N.M. Greene, Radiation-Damage Calculations: Primary Recoil Spectra, Displacement Rates, and Gas-Production Rates; RECOIL, DLC-55, ORNL/TM-5160, 1976.

REMOVAL CROSS-SECTION TECHNIQUE FOR KERMA ESTIMATION OF 14 MeV NEUTRONS PASSED THROUGH SHIELDING LAYERS

J. Jordanova, K. Ilieva, V. Christov

G. Voykov

Institute for Nuclear Research and Nuclear
Energy, Bulgarian Academy of Sciences,
Sofia 1184, Bulgaria

On the basis of the removal cross-section
method and the averaged kerma-factors for polyethylene, tissue and carbon, an effective procedure for express estimation of kerma from
14-MeV neutrons passed through alluminium, iron,
copper and lead slabs is proposed.

The removal cross-section method gives the possibility for a fast and effective estimate of neutron fluxes behind shielding media and materials. The attenuated neutron flux behind a slab of the considered material is approximated by an exponential function

$$N(d, E_{trs}) = S \exp(-\sum_{rem}(E_{trs})d) \qquad (1)$$

where $\sum_{rem}$ is a removal cross-section, E_{trs} is the detection threshold energy (or the conditionally chosen
lower limit of the energy range, where the flux is numerically estimated); d is the barrier thickness; S =
= $S_o B$ represents the fast neutron source in the common
case where the accumulation of the slowing-down neutrons
is taken into account by means of B and in the case of
monodirectional source B=1; S_o is the initial source
intensity.

The removal cross-section method has come into being from the peculiarities of the neutron flux penetration through hydrogen containing materials, and later has been extended to neutron flux attenuation through
pure metals /1/.

J. P. Genthon and H. Röttger (eds.), Reactor Dosimetry, 281–286.
© 1985 ECSC, EEC, EAEC, Brussels and Luxembourg

Although, initially, this method has been used to describe the fast penetrating component down to 4 MeV only, the idea of the method could be spread over the whole energy range if the approximation (1) is sufficiently good.

For an accurate evaluation of the absorbed energy D(kerma), it is necessary to know the spectrum formed in the irradiation point along with the total flux. This can be done by numerical solution of the problem of neutron penetration through flat layers or by complicated and time-consuming experiments.

This paper presents the data obtained on the basis of numerical calculations (ANISN/2/, SUPERTOG/3/, L26P3S34/4/), for removal cross-sections Σ_{rem} of neutrons from a 14 MeV source and for basic construction and shielding materials - alluminium, iron, copper, lead. By means of the average kerma-factors obtained for polyethylene, tissue and carbon, the dose can be evaluated taking into account the spectrum influence which depends on the shielding-layer thickness. The geometry of the problem treated is as follows - the neutrons from an infinite flat monodirectional or isotropic 14 MeV source are passing through the shielding slab of 1-3 mfp thickness. The irradiation point is just behind the layer.

The flux numerically obtained for monodirectional and isotropic source which depends on d has been approximated by the empirical relation /1/ for different threshold energies E_{trs}. The removal cross-section as a function of E_{trs} down to 4 MeV can be easily defined using the linear dependence (5,6).

$$\Sigma_{rem}(E_{trs}) = \Sigma_{tot}(14\ MeV)\ (\alpha + \beta\ E(MeV)),\qquad (2)$$

where α and β are the coefficients presented in Table 1. The same table gives the removal cross-sections Σ_{rem}^{tot} obtained for the flux in the whole energy range as well as the average total macroscopic cross-section $\Sigma_{tot}(14\ MeV)$ in the first energy group (14,9-12,2 MeV). The accumulation coefficients B are shown in Table 2.

As it is well known, the absorbed energy D (kerma) can be expressed as a neutron flux functional

$$D(d) = \int K(E)N(E,d)dE\ ,\qquad (3)$$

where K(E) is the kerma-factor taking into account the first collision energy and the contribution of other possible reaction - (n,n'), (n,α) etc.

On the other hand

Table 1

material	alluminium	iron	copper	lead
$\sum_t$(14 MeV, cm^{-1}	0,1047	0,2264	0,2558	0,1772
ϱat, cm^{-3}	0,0603	0,0849	0,0847	0,0335
	monodirectional source			
α	0,5867 ± 0,0036	0,4712 ± 0,0042	0,4868 ± 0,0013	0,4505 ± 0,0028
β	0,0130 ± 0,0004	0,0104 ± 0,0005	0,0054 ± 0,0002	0,0068 ± 0,0003
$\sum_{rem}^{tot}$,cm^{-1}	0,0246 ± 0,0003	0,0457 ± 0,0002	0,0644 ± 0,0021	*
	isotropic source			
α	0,8342 ± 0,0048	0,7293 ± 0,0026	0,7174 ± 0,0006	0,6984 ± 0,0023
β	0,0196 ± 0,0006	0,0106 ± 0,0003	0,0045 ± 0,0001	0,0061 ± 0,0003
$\sum_{rem}^{tot}$, cm^{-1}	0,0527 ± 0,0047	0,1017 ± 0,0061	0,1034 ± 0,0028	0,0277 ± 0,0002

$$D(d) = \frac{\int K(E)N(E,d)dE}{\int N(E,d)dE} \cdot \int N(E,d)dE \cong$$

$$\bar{K}(d)S\exp\left(-\sum_{rem}(E_{trs})d\right) \tag{4}$$

where $\bar{K}(d)$ is the kerma-factor averaged over the spectrum.

The approximation (4) error is mainly due to the incorrectness in defining the $\sum_{rem}$ and the accumulation factor B in the case of isotropic source. The comparison between precise calculations of total doses $D_{tot}(d)$ using (3) and approximate evaluation by formula (4) exhibits maximum relative error below 3% for shielding layer thcikness up to 3 mfp. The kerma-factors $\bar{K}(d) = D_{tot}(d)/N_{tot}(d)$ averaged over the flux in the whole energy range are given in Table 3.

* The exponential approximation (1) is not good.

Table 2

Accumulation factors for an isotropic source

Energy interval (MeV)	alluminium	iron	copper	lead
14,9–12,2	0,705 ±0,012	0,715 ±0,0006	0,685 ±0,012	0,705 ±0,008
14,9–10,0	0,725 ±0,012	0,731 ±0,005	0,697 ±0,012	0,714 ±0,007
14,9–8,2	0,735 ±0,010	0,740 ±0,005	0,703 ±0,012	0,720 ±0,007
14,9–5,0	0,767 ±0,008	0,773 ±0,004	0,721 ±0,011	0,737 ±0,005
14,9–4,0	0,776 ±0,008	0,803 ±0,003	0,728 ±0,011	0,751 ±0,004
14,9–0,0	0,772 ±0,040	0,676 ±0,012	0,710 ±0,016	0,944 ±0,002

Throughout this work Caswell's data on $K(E)$ (7) were used.

The estimates of the doses due to neutrons of energy down to 4 MeV show that $D_{>4MeV}$ practically does not depend on the flux shape. The reason is that the 14 MeV neutrons contribute mostly in the fast component (more than 80%). Hence, the average kerma-factors $\bar{K}$ for the fast component down to 4 MeV can be replaced by the kerma-factor at 14 MeV

$$D_{>4MeV} \cong SK(14 \text{ MeV})\exp(-\sum_{rem}(>4 \text{ MeV})d) . \qquad (5)$$

The approximation error (5) reaches its maximum for the carbon kerma and decreases with the increase of the moderating-material atomic number: from 20% for alluminium to 10% for lead. The kerma in tissue and polyethylene has been estimated with maximum error of 3% for all the materials considered.

The offered data for the removal cross-sections (Table 1), the average kerma-factors (Table 3) and the accumulation factors (Table 2) open a possibility, by means of formulae (2-4), for an effective express estimation of the dose from a 14 MeV neutron source (fusion

Table 3

Spectrum-averaged kerma-factors (rad.cm^2 10^{-8})

d,cm	Monodirectional source			Isotropic source		
	carbon	poly-ethylene	tissue	carbon	poly-ethylene	tissue
Alluminium						
5	0,1722	0,4910	0,5722	0,1608	0,4823	0,5649
10	0,4153	0,4315	0,5126	0,1309	0,4217	0,5011
15	0,1163	0,3896	0,4580	0,1048	0,3637	0,4289
20	0,0977	0,3414	0,3904	0,0843	0,3051	0,3615
Iron						
4	0,1591	0,4784	0,5486	0,1472	0,4513	0,5435
6	0,1395	0,4505	0,5163	0,1315	0,4237	0,4948
8	0,1258	0,4082	0,4779	0,1108	0,3785	0,4463
10	0,1116	0,3763	0,4411	0,0966	0,3447	0,4047
12	0,0982	0,3364	0,4033	0,0827	0,3097	0,3653
Copper						
4	0,1584	0,5008	0,5538	0,1506	0,4511	0,5258
6	0,1427	0,4273	0,5098	0,1287	0,4025	0,4709
8	0,1395	0,3954	0,4613	0,1084	0,4415	0,4156
10	0,1359	0,3602	0,4339	0,0922	0,3157	0,3728
12	0,1295	0,3128	0,3778	0,0761	0,2793	0,3384
Lead						
5	0,1193	0,3821	0,4415	0,1123	0,3702	0,4328
7,5	0,1015	0,3445	0,4046	0,0911	0,3235	0,3797
10	0,0919	0,3105	0,3662	0,0754	0,2880	0,3391
12,5	0,0746	0,2831	0,3334	0,0623	0,2413	0,3150
15	0,0702	0,2592	0,3076	0,0521	0,2202	0,2663

systems, accelerators, etc.). The data for the average
kerma-factors can be successfully used in experimental
determination of doses by means of homogeneous ionisat-

ion chambers.

REFERENCES

1. Biological Protection of Transport Reactor Equipment, book, ed. Broder, Atomizdat, Moscow, 1960,(in Russian).

2. ANISN-W Code. One-Dimensional Discrete Ordinate Transport Technique. WANL-PR-(LL)-034, v.IV, 1970.

3. G. Voykov, V. Gadjokov, K. Ilieva, "Modification of the SUPERTOG Program Applied to Libraries with Tabulated Elastic Scattering Anisotropy Densities", IAEA INDC(BUL)-6/GV, Vienna, 1982.

4. G.Voykov, V. Gadjokov, S. Minchev, "L26P3S34-A 26-Group Library for the Computation of Neutron Transfer in Shielding Media", IAEA-INDC(BUL)-007/GV, Vienna, 1983.

5. J. Jordanova, K. Ilieva, V. Christov, G. Voykov, "Removal Cross-section Energy Dependence in Iron Shielding Slabs", Compt.rend.Bulg.Acad.Sci., 3,1984 (in print, in Russian).

6. J. Jordanova, K. Ilieva, N. Sokolinova, V. Christov, "Removal Cross-section in Dependence of Threshold Energy for Monodirectional and Isotropic 14 MeV Neutron Source", Yadrena Energia, Bulg.Acad.Sci., Sofia, 22, 1984 (in print, in Russian).

7. R.S. Caswell, J.J. Coyne, "Kerma-Factors for Neutron Energies below 30 MeV", Radiation Research 83, 217-254 (1980).

THE IMPORTANCE OF ANISOTROPIC SCATTERING IN HIGH ENERGY NEUTRON

TRANSPORT PROBLEMS

G. Prillinger, M. Mattes

Institut für Kernenergetik und Energiesysteme,
University of Stuttgart, Pfaffenwaldring 31,
7000 Stuttgart 80, Germany

ABSTRACT

Neutronic calcuations for high-energy fusion and spallation neutron sources are of increasing interest. The neutron distributions in reactor blankets or deep within shields is influenced by the elastic scattering which becomes highly anisotropic above 10 MeV expecially for heavy elements. Since common cross-section libraries are limited by Legendre expansion, or by their upper energy boundary, or exclude elastic scattering above 20 MeV a special library has been created.

To describe the highly anisotropic scattering of very fast neutrons adequately the transport code ANISN has been improved. Fokker-Planck terms have been introduced into the transport equation which accurately describe the small changes in energy and angle. The new code has been tested for a d(50)-Be neutron source in a deep penetration iron problem. The influence of the forward peaked elastic scattering on the fast neutron spectrum is shown to be significant and can be handled efficiently in the new ANISN version.

J. P. Genthon and H. Röttger (eds.), Reactor Dosimetry, 287–293.
© 1985 ECSC, EEC, EAEC, Brussels and Luxembourg

287

1. INTRODUCTION

Standard transport codes using adequate cross-section libraries have been successfully applied for fission neutrons in the past. For neutron sources above 10 MeV neutronic calculations are now of increasing interest. Beside the d-T fusion source around 14 MeV accelerator neutrons like d-Be up to 50 MeV and finally spallation sources up to 1100 MeV are used or will be constructed.

With growing energy the elastic scattering cross-section especially for heavy elements becomes highly anisotropic. Consequently treating the small changes in angle and energy in standard transport calculations correctly Legendre expansions of high-order are necessary. Since common cross-section libraries are limited by their number of Legendre scattering matrices, or by their upper energy boundary a special library has been created.

A more efficient treatment of the forward peaked scattering can be attained by incorporating Fokker-Planck terms into the Boltzmann equation which has been proposed recently also by Caro /1/.

The well proved transport code ANISN /2/ has been extended therefore to handle Fokker-Planck (FP) coefficients also. The new version of the code, ANISN-FP /3/, has been tested for a 50 MeV deuterium on beryllium neutron source in an iron deep penetration problem. Iron is considered to be an adequate shielding material for high energy neutrons sources.

For comparison standard transport calculations have been performed for the same problem with two available cross-section libraries above 20 MeV. Large differences in reaction rates for threshold detectors were obtained behind three meters of iron.

2. HIGH ENERGY MULTIGROUP CROSS-SECTION LIBRARIES

For neutron energies below 20 MeV a large number of multigroup cross-section libraries for different applications exist. To the contrary for energies above 20 MeV there is a lack of evaluated nuclear data files. Consequently there are only a few multigroup libraries available. These libraries have several limitations and can not be applied generally. Two libraries, HILO /4/ and MENSLIB /5/, have been used in this work. Their main features are listed in Table 1. It is a striking fact that MENSLIB does not contain elastic scattering except for hydrogen. This explains the great difference in the total neutron interaction cross-section which can be seen for iron in Figure 1. Elastic scattering becomes very anisotropic for higher energies and heavy elements, therefore it is sometimes argumented it has only a minor effect on neutron

288

1. HILO:	66 neutron groups, 21 gamma energy groups 0 - 400 MeV P_5, no elastic scattering for W, Pb H, B, C, N, O, Na, Mg, Al, Si, S, K, Ca, Cr, Fe, Ni, W, Pb
2. MENSLIB:	No elastic scattering, except H. 2 versions: A) 60 neutron groups, P_5, 0 - 60 MeV B) 41 neutron groups, P_3, 0 - 800 MeV
3. N108:	108 neutron groups, 0 - 40 MeV, elastic (P_{20}), non-elastic scattering separate iron (LA-8626, 1980) Fokker-Planck coefficients

Table 1: Transport cross-section libraries

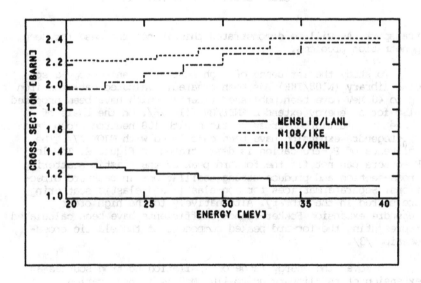

Figure 1: Total neutron interaction cross-section of different
multigroup libraries for iron

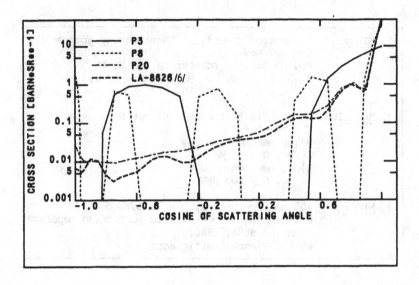

Figure 2: Truncated Legendre expansion of iron elastic
cross-section at 40 MeV

transport. As will be demonstrated this is not the case for deep
penetration problems.

To study the influence of high scattering anisotropy a special library (N108/IKE) has been created. Evaluated data for iron
up to 40 MeV have been published recently which have been proposed
also for an energy extended ENDF/BV file /6/. On the basis of
these data multigroup cross-sections with 108 neutron groups and
P_{20}-Legendre expansion have been calculated with NJOY /7/. The
influence of P_1-truncation is demonstrated in Figure 2. Too low
P_1-orders can not fit the forward peak of the elastic scattering
cross-section and produce strong oscillations in backward directions. Separate matrices for non-elastic and elastic scattering
are stored in the library. Alternatively to the high order
Legendre expansion Fokker-Planck coefficients have been calculated
representing the forward peaked component of the elastic cross-
section /3/.

To cover the energy range of spallation neutron sources an
extension of the library up to 1100 MeV is in preparation.

3. ANISN-FP FOR HIGH ENERGY NEUTRON TRANSPORT CALCULATIONS

In standard transport codes highly anisotropic scattering can not be handled efficiently. High Legendre expansions of the scattering cross-sections are necessary especially for deep penetration problems or if accurate angular fluxes are required. A more efficient treatment can be obtained if the high anisotropic component of the cross-section is separated in the Boltzmann equation. The scattering integral

$$\iint \Sigma(E' \to E, \ \overline{\Omega} \cdot \overline{\Omega}') \phi(E',\Omega') dE' d\overline{\Omega}'$$

is splitted into two parts, where the anisotropic component is approximated by a linear ansatz leading to Fokker-Planck terms of the form

$$\frac{\partial}{\partial E} S(E) \phi + T(E) \frac{\partial}{\partial \mu} (1-\mu^2) \frac{\partial \phi}{\partial \mu}.$$

The new equation, known also as Boltzmann-Fokker-Planck equation, can not be solved with transport codes like ANISN directly since the additional parabolic term in μ leads to a coupling of the angular fluxes in the S_N-equations. A new algorithm has therefore been written which is active in ANISN-FP for energy groups where FP-coefficients S and T are given /3/. S and T values can be determined from the angular distribution of the scattering cross-section or alternatively from the Legendre moments /1/.

4. FIRST RESULTS AND CONCLUSIONS

For testing the ANISN-FP code and comparison of the N108 library with HILO and MENSLIB the transport of d(50)-Be neutrons incident on a 3 meter slab iron shield has been analysed. The source neutron spectrum is given in Figure 3. To get a reference solution a standard ANISN calculation with S40, 1.5 cm mesh size and P_{20}-N108 data has been performed. The reaction rates for AU197(n, 3n) and AU197(n, 4n) with thresholds about 18 and 26 MeV have been calculated and compared with the HILO and MENSLIB results. The 4th curve shown in Figures 4 and 5 represents the solution of the ANISN-FP code where FP-coefficients for the highly anisotropic elastic component and a P_5-expansion for all other cross-sections have been used.

The following conclusions can be drawn:

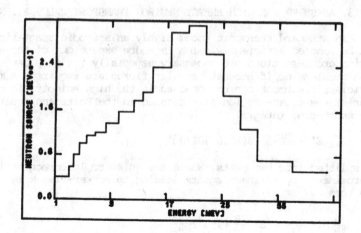

Figure 3: Neutron source spectrum used in the transport calculations

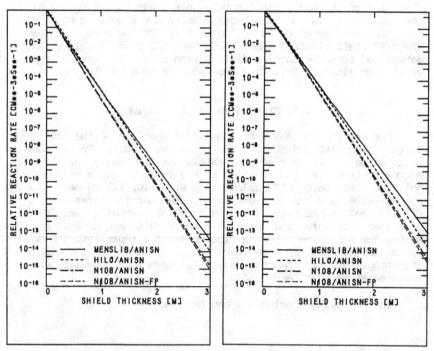

- Presently avaiblable cross-section libraries like HILO and MENSLIB overestimate appreciably the neutron spectrum above 20 MeV in thick iron shields. Differences to the results obtained with N108 are caused by different basic cross-section data and by the lack of elastic scattering in MENSLIB. Total reaction cross-section in HILO seems to be too low.

- Elastic scattering is important for the transport of high energy neutrons and must be included also for heavy elements in the scattering matrices.

- Scattering anisotropy can require orders of Legendre polynoms which are unreasonably high for standard transport codes. With the ANISN-FP code accurate solutions with reduced effort can be obtained.

- Cross-section libraries must be extended to higher energies for spallation neutrons. FP-coefficients should be stored in the libraries to account for scattering anisotropy increasing with energy. This can reduce the amount of data stored in high P_1-scattering matrices.

LITERATURE

/1/ M. Caro, J. Ligou, Treatment of Scattering Anisotropy of Neutrons through the Boltzmann-Fokker-Planck Equation, Nucl. Sc. Eng. 83, p. 242 (1983).

/2/ W.W. Engle, A User's Manual for ANISN, USAEC Report K-1693 (1967).

/3/ G. Prillinger, ANISN-FP, a ANISN Version for High Energy Neutron Transport, to be published.

/4/ R.G. Alsmiller, Jr., J. Barish, Neutron Photon Multigroup Cross Sections for Neutron Energies < 400 MeV, ORNL/TM-7818 (1981), RSIC, DLC-87.

/5/ W.B. Wilson, Nuclear Data Development and Shield Design for Neutrons Below 60 MeV, LA-7159-T (1978), RSIC, DLC-84.

/6/ E.D. Arthur, P.G. Young, Evaluated Neutron-Induced Cross Sections for ^{54,56}Fe to 40 MeV, LA-8626-MS (ENDF-304) (1980).

/7/ R.E. McFarlane, D.W. Muir, R.M. Boicourt, The NJOY Nuclear Processing System, Vol. 1, LA-9303-M (ENDF-324) (1982).

FUSION MATERIALS RESEARCH AND NEUTRON ACTIVATION

Theodore C. Reuther

U.S. Department of Energy

Washington, D.C. U.S.A.

INTRODUCTION

It is a pleasure to participate in the Fifth ASTM Euratom Conference
on Reactor Dosimetry. This is my first opportunity to participate in one
of the conferences, but I have been hearing most interesting stories
of these conferences for some time, especially of the first meeting that
took place in Europe in 1975. My colleagues in our laboratories brought
back documentary evidence of the very creative time they had in their
post-conference travels.

The good humor of that documentation was more than matched by the
outstanding accomplishments of the collaboration that went on in the U.S.
programs at that time. That collaboration among Norman Dudy of Argonne
National Laboratory (ANL), Bill McElroy and Don Doran of Hanford
Engineering Development Laboratory (HEDL), Harry Farrar of Atomic

J. P. Genthon and H. Röttger (eds.), Reactor Dosimetry, 295–308.
© 1985 ECSC, EEC, EAEC, Brussels and Luxembourg

International (now Rockwell International), and Jim Grundle of the
National Bureau of Standards in consultation with Alber Fabry of Mol,
Belgium, was extremely productive. It was trend-setting research which
has led to standards and practices that are used in dosimetry for many
different kinds of environments.

In the same general time period, the international collaboration of
Norgett, Robinson, and Torrens, followed by the general adoption of their
model to compute the now almost universally accepted DPA unit to describe
irradiation exposure level, set the stage for preparation, negotiation, and
in due course, acceptance of a most effective set of working standards for
dosimetry exposure units. That combination of research and negotiation to
establish documented standards for neutron dosimetry and to find an
acceptable if imperfect standard for the associated computations to turn
that dosimetry into an exposure unit that can, in part, account for
variation in both spectrum and the response of particular materials to
various spectra, is an accomplishment of lasting practical use. Those
accomplishments have their origins in the discussions both informal and
formal that have grown out of this series of biannual conferences. The
conferees are to be congratulated for that kind of an accomplishment.

The advent of fusion materials research in the mid-1970's found the
need to also account for helium and other transmutant products to a degree
not previously considered in light water reactor or LMFBR environments.
This need has led to the development of exposure units and dosimetry

296

measurements to represent displacement and helium (and other element) transmutation. The latter in turn led to development of the helium (stable product) accumulation monitors now widely used throughout dosimetry.

In 1980 when the International Energy Agency (IEA) Implementing Agreement on a Program of Research on Radiation Damage in Fusion Materials began, the initial tasks included the need to establish common exposure units. The existance of the ASTM/Euratom recommended practice made that task a very easy one. The working group that was responsible for that multinational Agreement simply adopted your recommended practice as a basis for representing the displacement component of irradiation damage in fusion materials. That practice continues to be in use internationally for this area of research.

In the United States (U.S.), the principle center for dosimetry for fusion materials research has been at ANL. You are well informed of the work which Dr. Larry Greenwood carries out at ANL as part of our Damage Analysis and Fundamental Studies (DAFS) program. A central subelement of that program continues to be dosimetry. The ever more efficient methods of physical measurements which he has organized and developed at ANL and the extensive library codes of evaluated nuclear reaction data which Dr. Greenwood has established in these recent years, are very important tools in the efficient and effective characterization of the neutron environment in our irradiation damage experiments. I am pleased that through the work of ANL, the Department of Energy (DOE) has been able to make these codes and libraries available to our international colleagues.

Fusion Materials Irradiation Test Facility (FMIT)

The FMIT has been a central element in the planning and in the activities of materials research in the fusion program of the U.S. for several years and it has been a considered factor in the program planning of both the European and Japanese fusion programs as well. The neutron spectrum of the Li(d, n) neutron source, broadly peaked about 14 MeV with a tail extending up to energies about three times that level has been a major consideration in the research and development effort in neutron dosimetry in the U.S. fusion materials program for the last several years. That research has been reported in technical details before your conferences and I am sure will be discussed at other times during the present conference. However, I would like to take a few moments to recount the activities that have been taken relative to the FMIT project since the time of your last conference and to acknowledge the very important role the dosimetry research has placed in the judgments and recommendations which have been made regarding the FMIT concept.

The concept of using the Li(d, n) reaction as a neutron source for irradiation damage studies was first raised by Norman Dudy (then of ANL) and Donald Parkin (then of Brookhaven National Laboratory) in the early

1970's. They hold a U.S. patent on their concept. In the mid-1970's as

fusion materials research was being initiated, damage analysis

calculations made the Li(d, n) source appear quite attractive as a high

fluence, high energy neutron irradiation concept. In 1975 at the

International Conference on Irradiation Sources for Research in Fusion

Materials that was hosted by ANL, there were several papers which

described and proposed the use of this reaction for a neutron source for

fusion. A recommendation of the workshop associated with that Conference

was that such a source be strongly considered for use in the fusion

materials research program. That recommendation of an international

scientific group was given full consideration within the then Energy

Research and Development Administration in the U.S.; and in 1978 the

Congress authorized the research, development, and design of the FMIT

project and appropriated funds for that work. In 1980 with the

establishment of the IEA Implementing Agreement on a Program of Research

and Development on Radiation Damage in Fusion Materials, international

collaboration on the FMIT became a formal activity as the first annex to

that Implementing Agreement. International participation in the agreement

included the assignment of personnel to the U.S. laboratories during the

research, development, and design work. It included dosimetry research.

Within a year after signing of the annex on the FMIT project, the

U.S. DOE informed its partners in that Agreement that the U.S. would not

proceed to construction on a unilaterial basis. It was our judgment that

the nature of the FMIT, in particular its unique qualities and almost certain recognition that only one such facility would be needed on an international basis, made it appropriate that the FMIT be constructed on the basis of cost sharing among international partners. In the intervening years research, development, and design of the FMIT has continued in the U.S. including a modest level of contributed personnel support from its European, Japanese, and Canadian partners. Over that same period of years, at levels ranging from that of the technical research level to the highest levels of government, there have been very extensive discussions, deliberations, and consideration of international support for the FMIT.

Perhaps the central activity in that discussion were the meetings and deliberations of a Senior Advisory Panel that was convened by the IEA to consider the role of materials research in the development of fusion energy. Included in that broad question, the Panel also considered specifics with respect to the FMIT concept. Professor Sir Allan Cottrell of Cambridge University, England, chaired that Panel and the Panel report was issued in the Summer of 1983. That report strongly endorsed the unique qualities of the FMIT concept and urged that there be an international collaboration to construct and operate the FMIT. The panel recommendations included a research program that would be strongly collaborative rather than simply allocation of space in which individual parties would place their particular experiments. Those recommendations have not led to a common agreement to undertake mutually funded constructed of the FMIT among the principle partners to that Agreement.

However, the concept, the need, and the programmatic recommendations have been enthusiastically endorsed at all levels of government. We are hopeful that such agreement can yet be developed, but we do not expect it to be in the immediate future. Therefore, the DOE is taking steps to complete and conclude the research, development, and design phase activities of the FMIT project in U.S. fiscal year 1985 (FY85).

Technically, the concluding work will include the testing of the accelerator through its 2 MeV stage, completion of the facility design, completion of the major developmental systems at HEDL, and placing those facilities in stand-by for later startup. Over the course of this year, staffing would be reduced to the level required to carry out international discussions. We anticipate that the FMIT project will continue to receive substantial discussion in international forums. This has been a much talked about project throughout the international materials research community which deals with radiation damage, and I hope that the time I have taken to summarize the background and current situation on the FMIT is informative and useful to you.

Before leaving the subject of FMIT, I would like to tell you a little about the relationship of the work of the technical area of this conference to the judgments of the Cottrell panel in evaluating the merit of FMIT. As you know, the broad spectrum and particularly the high energy tail of the Li(d, n) reaction was of substantial concern from the point of view of both dosimetry and damage analysis from the time of the earliest proposals concerning the FMIT. The initial evaluation of the potential effects of that tail were treated by individual efforts or drawn from the general scientific literature. As the potential for the FMIT was advanced, however, the DOE undertook a specific research and development program to develop dosimetry methodology and to examine the basic displacement damage phenomena in a variety of neutron spectra including that of the Li(d, n) and Be(d, n) neutron reactions. That research program of the Office of Fusion Energy (OFE) was strongly supported by basic energy science studies in the U.S. and elsewhere.

The dosimetry and damage analysis research has been amply described in past conferences of this series and I think will be touched on in this conference by Dr. Greenwood. However, it consisted essentially of intercomparison experiments using the 14 MeV RTNS-II, fission reactors and the Li(d, Be) source at the University of California. These experimental efforts in characterizing neutrons were backed up with a continuing program in the development and evaluation of nuclear data for the dosimetry range that was entailed, and that work continues. A dominant concern of the Cottrell panel was the question of helium generation at high energy and concern that differential cross section data were inadequate. Those concerns were resolved by direct measurements of spectral averaged cross sections by Farrar and co-workers at Rockwell International, again as has been reported before in this conference series. The simple expedient of direct measurement of helium production for a variety of elements and isotopes satisfied the Cottrell panel on this central concern. For those several contributions which came from the community represented at this conference, I would like to express the appreciation of the OFE, for those contributions led in a significant way to the very firm conclusions of the IEA Senior Advisory Panel.

Current Directions in the U.S. Fusion and Fusion Materials Programs

In response to the science policy of the U.S. Government, the goal and the program of magnetic fusion energy research in the U.S. has been defined to establish the scientific and technological knowledge base that is required for an attractive fusion energy option. A principle element of the technological objective is to establish knowledge of the behavior of materials in the fusion environment, and to do so in the context of conducting research and development on materials issues that are unique to or special to the concept of fusion as an energy resource.

In this regard we have given very substantial consideration to the question of neutron activation of the materials of construction of a fusion reactor system. As you know, the deuterium tritium fusion reaction itself does not produce radioactive byproducts. The interaction of the 80 percent of the energy released by that reaction which is contained as neutrons, does indeed activate the materials of the reactor system. However, we recognize mother nature provides us with some elements and isotopes which have the potential for reduced neutron activation, and our reactor studies and materials programs have given modest consideration to the issue. About two years ago we convened a panel to do a technical evaluation of the potential for the development of reduced activation materials for application to fusion energy systems. That panel was chaired by Professor Robert Conn of the University of California, Los Angeles. The report which they prepared for the DOE was summarized in a paper at the Albuquerque Topical Meeting on Fusion Materials in 1983. In brief summary, the panel considered the subject of reduced or low activation in the context of three general concerns: safety, maintenance, and long term waste management. The primary conclusion of the panel was that the concern of long term waste management could indeed be treated in an effective way by the development of materials with reduced activation.

The panel recommendation that we seek the development of materials which have reduced activation in the context of reducing the long term waste management burden has been accepted as a program goal of the fusion materials research program of the U.S. DOE. That goal has two levels to it. The first is a target or goal of activation reduced to the level of class B waste in the category of the current U.S. Nuclear Regulatory Commission criteria (which translates approximately to near surface (3 meter) burial with no further concern regarding intruder danger after 100 years.) The second level considered in the recommendation was for class C waste criteria (burial of at least 5 meters with engineered storage and controlled access for a period of 500 years). That criteria now is identified as a requirement for the program. Thus, the present U.S.

304

fusion materials program has a requirement that its materials development program produce materials which meet the 500 year goal or criteria but it has a higher goal to try to achieve materials which would meet the 100 year criteria.

U.S. FY84 has been a transition year from our previous program structure to a new distribution of effort that focuses on this reduced activation goal. During this time we have achieved a shift in research on reduced activation materials from about 25 percent to something in excess of 50 percent of our effort in our principle neutron irradiation effects program elements, Damage Analysis and Fundamental Studies and Alloy Development for Irradiation Performance. This corresponds to about 3.5 million dollars. About 60 percent of the program effort in this coming fiscal year is directed to the reduced activation materials research activities. Thus far those efforts have been directed primarily at the immediately obvious approaches of eliminating the particular noxious elements including particularly niobium, nickel, and molybenum in conventional alloys steels; and increasing the effort on intrinsically low activation materials such as vanadium and certain of its alloys. In the coming fiscal year and more strongly in the following years, we will initiate additional effort on ceramics materials which may have even lower activation potential.

Figure 1 (courtesy of D. G. Doran, HEDL) represents the principle elements that must be considered in thinking of this factor of neutron

activation from a waste management point of view. It is immediately
evident that the elements Nb, Mo, Ni, Cu, N and Al must be carefully
controlled or eliminated for one particular set of assumptions. The
reality of the situation, however, is far more complex than this bar graph
would illustrate, for we are talking about very high fluence irradiation
applications and, therefore, a proliferation of transmutation products
from the starting elements in alloy becomes significant. A few thousand
reactions must be considered. Dr. Fred Mann of HEDL has addressed this in
a recent paper and I refer to you to it for technical details. A central
issue which Dr. Mann has raised is that consideration of very long term
irradiation exposures and the associated transmutations and daughter
products and the need to consider extremely small quantities of materials
which have very long half-lives, has opened a major need for nuclear data
generation and analysis on a scale and range of isotopes that may not have
been considered before.

It is evident that a very similar set of considerations must be taken
into account by the dosimetry community when you analyze your foils and
especially when you undertake insitu dosimetry. Therefore, it would
appear to me that this dosimetry community may have a great deal to offer
in the development of the information base that would be required to guide
the materials research community in the development of alloys that will
indeed have the potential for reduced activation. It would do little good
to successfully pursue alloy development with a given assumed initial
composition and then find that a forgotten daughter product transmutes to
some very unacceptable isotope.

CONCLUSION

In concluding, let me emphasize that we in the U.S. are very enthused with the challenge that has been placed before us to develop materials for fusion which have low neutron activation and which might provide a very substantial benefit in terms of waste management burden. If we can achieve that goal it will give fusion an engineering and a technological quality that corresponds to the intrinsic characteristics of the deuterium-tritium reaction. Furthermore, if successful, I am certain that that kind of development would be adopted by the fission community as well. For while fission would still have its fission products, a reduced structural materials waste burden would most certainly be welcome. Then, of course, it is very exciting to the metallurgical engineers because there now is a whole array of alloy steel compositions that are at substantial variance to those which have been developed and researched over the past 50 or 60 years. Thus, while this new goal in fusion materials research has the opportunity of taking fusion to a higher potential, it most certainly will provide some exciting opportunities for materials research. It also has the need for substantial research and development in the area of nuclear data and nuclear reactions particularly at high levels of exposure and for multiple reactions at concentration levels that perhaps previously had not been given much consideration except perhaps by the dosimetry community. Therefore, I think there is a particular opportunity for this community to contribute to this general objective in materials research for fusion, indeed an opportunity in materials research for all nuclear energy options.

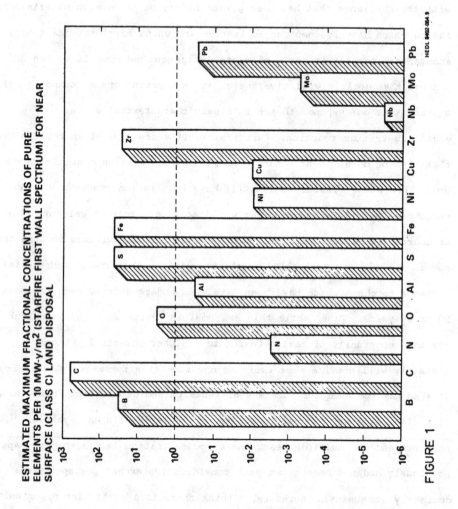

ESTIMATED MAXIMUM FRACTIONAL CONCENTRATIONS OF PURE ELEMENTS PER 10 MW-γ/m² (STARFIRE FIRST WALL SPECTRUM) FOR NEAR SURFACE (CLASS C) LAND DISPOSAL

FIGURE 1

HEDL 8402-054.9

DELAYED NEUTRON COUNTING SYSTEM FOR JET PLASMA
NEUTRON YIELDS DIAGNOSTICS

H. Tourwé*, A. Fabry and P. D'hondt

SCK/CEN, Mol, Belgium

ABSTRACT

Two fission reactions, $^{238}U(n,f)$ and $^{232}Th(n,f)$, have been identified as ideal fusion plasma neutron yield diagnostics monitors based on detection, within a time span of ~ 10 to 100 s, of their delayed-neutron signals; these responses, entirely decayed between two successive plasma pulses, are induced during exposure of the sensors in the vicinity of tokamak containment walls, using a fast pneumatic transfer system as vehicle between the machine and the remote counting stations.

The dynamic range of this method is very large and its sensitivity unchallengeable, reaching down to neutron fluences as low as a few $10^6 cm^{-2}$. Routine and automatized application is straightforward while encompassing a minimal inventory of recyclable and well-assayed targets.

At neutron fluences $\geq 10^7 cm^{-2}$, the accuracy reachable is estimated as $\pm$ 5 % in D,D operation provided a direct calibration be performed relative to absolute fission chambers in the MOL cavity uranium-235 fission spectrum standard neutron field.

INTRODUCTION

The objective of this work consisted in developing a

---

*At present : EURATOM, Luxemburg.

neutron yield diagnostics monitor for JET which is amenable to
automatic control with on-line data acquisition and analysis.
The principle of this diagnostic consist in irradiating a fissile target during the plasma pulse, transferring it to a
delayed neutron counting assembly, measuring the delayed neutron emission within a time span of ~ 10 to 100 s and recycling
the target for the next pulse. Since the measurements are done
between the discharges the technique is not sensitive to radiation, electrical or magnetic background.

FEASIBILITY OF THE DELAYED NEUTRON COUNTING TECHNIQUE

The two fission reac-
tions, $^{238}U(n,f)$ and
$^{232}Th(n,f)$ have been identified as ideal fusion
plasma neutron yield dia-
gnostics monitors since
both are little perturbed,
due to their effective
energy thresholds of
~ 1.5 MeV, by neutron
scattering effects on
structural materials under
the 2.45 MeV spectral peak
conditions prevalent in
D,D regime. Corrections
are more significant for
D,T operation and the use
of the $^{17}O(n,p)^{17}N$ reaction appears a satis-
factory alternative since
Nitrogen-17 is a delayed-
neutron precursor which
decays as follows :

$$^{17}_{7}N \xrightarrow[4.17s]{\beta^-} {}^{17}_{8}O \xrightarrow{prompt} {}^{16}_{8}O + n$$

$$Q = + 4.534 \text{ MeV}$$

Figure 1 gives a
schematic representation
of the delayed neutron
counter. The counter end
of the rabbit system is

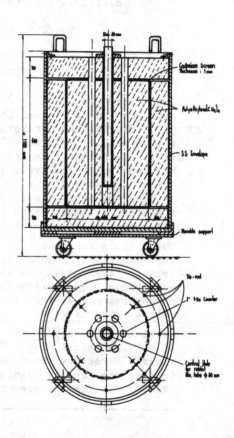

Fig. 1. Delayed neutron counter

mounted in the centre of a cylindrical polyethylene block of
40 cm diameter and is surrounded by 6 one inch diameter ^{3}He
counters, with a sensitive length of 15 cm, at a radial distance of 8 cm. The whole system is protected against background neutrons by means of a 1 mm Cd screen and a 10 cm thick
polyethylene shield.

In order to demonstrate the practical feasibility of the
technique, measurements were performed in the BR1 fission
spectrum with a rabbit system similar to the one that will be
installed at JET. The BR1 fission spectrum was chosen for this
purpose because the average neutron energy of the fission spectrum is comparable with the average neutron energy of the
fusion plasma neutron spectrum in the D,D phase. A second
reason is that the facility is well characterized and perfectly suited for the calibration of the final delayed neutron system prior to eventual installation at JET. The experimental set-up is shown in figure 2. The BR1 fission

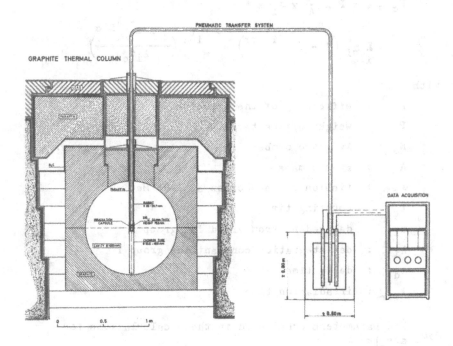

Fig. 2. Experimental set-up

spectrum is located in the 1 m diameter cavity of the vertical thermal column. A pure ^{235}U neutron fission spectrum is obtained inside the volume of a ^{235}U (93 % enrichment) cylinder in the centre of the spherical cavity. All thermal neutrons are absorbed by the Cd tube. A special set-up was designed (so-called Mk III fission spectrum) to accommodate the irradiation end of the pneumatic transfer system. The fission neutron flux inside the activation capsules was measured with In foils and is equal to $\sim 2.8 \ 10^8 n \ s^{-1} cm^{-2}$ at a reactor power of 1 MW. More detailed discussion of the BR1 fission spectrum is given in reference [1]. The irradiation target was made of a stack of 20 metallic natural uranium foils ($\emptyset$ 19 mm x 0.1 mm). The total target weight was 10.5 g.

The calculation of the number of delayed neutrons is based on the empirical KEEPIN six-group representation of the delayed neutron data. The total number of counts in the delayed neutron counter is given by :

$$N_c = \epsilon \times P \times \frac{N_o}{A} \times \sigma_f \times \Phi \times t_c \times$$

$$\sum_i \alpha_i \left(1 - e^{-\lambda_i t_{irr}}\right) e^{-\lambda_i t_d} \left(\frac{1 - e^{-\lambda_i t_c}}{\lambda_i t_c}\right)$$

with

ϵ	:	efficiency of the detector
P	:	weight of the target
N_o	:	Avogadro number
A	:	atomic mass
σ_f	:	fission cross-section at 2.45 MeV
t_c	:	counting time
α_i	:	delayed neutron yield for group i
λ_i	:	desintegration constant for group i
t_d	:	delay time
t_{irr}	:	irradiation time

The parameters considered in these calculations for ^{238}U are [2] :

Group	$T_{1/2}$ (s)	Yield	$\bar{E}_n$ (keV)
1	52.38	0.00054	200
2	21.58	0.00564	457
3	5.00	0.00667	381
4	1.93	0.01599	415
5	0.490	0.00927	406
6	0.172	0.00309	522

Since the delayed neutrons are decaying with short half-
lives it should be envisaged to start the counting as soon as
possible after irradiation. In normal JET operation conditions
the countings can only start about 10 s after the end of the
irradiation due to the rather important travel distance between
the torus hall and the counting laboratory. The maximum measuring time to be considered is of the order of 100 s since for
higher values one gets only small increases in the total number
of counts and background corrections would become more important.

To verify the decay of delayed neutrons, calculated with
the KEEPIN data, measurements were performed with an irradiation time of 10 s and a counting time of 5 s as a function of
the delay time.
The results are given in the next table :

T_d (s)	C (arb. units)	M	C/M
9	$2.451 \ 10^{-3}$	$4695 \pm 1\%$	$5.22 \ 10^{-7}$
16	$1.322 \ 10^{-3}$	2494	$5.30 \ 10^{-7}$
23	$8.810 \ 10^{-4}$	1669	$5.28 \ 10^{-7}$
30	$6.456 \ 10^{-4}$	1209	$5.34 \ 10^{-7}$
37	$4.970 \ 10^{-4}$	971	$5.12 \ 10^{-7}$
44	$3.926 \ 10^{-4}$	731	$5.37 \ 10^{-7}$
51	$3.143 \ 10^{-4}$	$600 \pm 3\%$	$5.24 \ 10^{-7}$
58	$2.535 \ 10^{-4}$	488	$5.19 \ 10^{-7}$
65	$2.054 \ 10^{-4}$	396	$5.19 \ 10^{-7}$
72	$1.669 \ 10^{-4}$	338	$4.94 \ 10^{-7}$
79	$1.361 \ 10^{-4}$	263	$5.18 \ 10^{-7}$
86	$1.113 \ 10^{-4}$	189	$5.25 \ 10^{-7}$
93	$9.123 \ 10^{-5}$	160	$5.72 \ 10^{-7}$
100	$7.500 \ 10^{-5}$	$140 \pm 7\%$	$5.37 \ 10^{-7}$

The calculations were made for the same irradiation and counting conditions. The C/M ratio can be considered constant taking into account the quoted uncertainties on the measurements.

Another check of the KEEPIN data was made by performing measurements for a counting time of 100 s and a delay of 10 s as a function of the irradiation time. Again calculations were performed for the same irradiation and counting conditions. The results are given in the next table.

T_{irr} (s)	C (arb. units)	M	C/M
5	$2.684 \ 10^{-4}$	11721	$2.29 \ 10^{-8}$
6	$3.129 \ 10^{-4}$	14003	$2.23 \ 10^{-8}$
7	$3.551 \ 10^{-4}$	16027	$2.21 \ 10^{-8}$
8	$3.953 \ 10^{-4}$	17165	$2.30 \ 10^{-8}$
9	$4.336 \ 10^{-4}$	18495	$2.34 \ 10^{-8}$
10	$4.702 \ 10^{-4}$	20817	$2.26 \ 10^{-8}$
11	$5.052 \ 10^{-4}$	21943	$2.30 \ 10^{-8}$
12	$5.388 \ 10^{-4}$	22283	$2.42 \ 10^{-8}$
14	$6.019 \ 10^{-4}$	25701	$2.34 \ 10^{-8}$
16	$6.603 \ 10^{-4}$	27743	$2.38 \ 10^{-8}$
18	$7.145 \ 10^{-4}$	29359	$2.43 \ 10^{-8}$

The standard deviation on C/M ratio is 3 %. The uncertainty of the measured data is also of the order of 3 of 4 %. The experimental uncertainty is larger than the statistical uncertainty on the counting data since the counting were started manually. For the final design one should use the information supplied by the photo-sensors at the irradiation end to start the countings.

CONCLUSION

From the above considerations one can conclude that a good consistency is obtained between measurements and calculations when using the KEEPIN data to describe the relative decay of delayed neutrons. The efficiency of the counter can be calculated from the above mentioned results since the flux in the irradiation position is known. The efficiency amounts to ~ 15%, this value is in very good agreement with the value obtained using a calibrated Am-Li neutron source from LMRI-France.

In case of a 50 g ^{238}U target, irradiated during 10 s in a D,D flux of $\emptyset = 2.10^5$n s^{-1}cm^{-2} and counted for 100 s after a delay of 10 s the total number of counts in the proposed system would be $\sim$ 100 counts. This figure demonstrates very clearly the sensitivity of the delayed neutron technique. The efficiency of the final system can even be slightly increased by taking ^{3}He tubes with a sensitive length of more than 15 cm.

REFERENCES

[1] A. Fabry et al., "The Mol Cavity Fission Spectrum Standard Neutron Field and Its Applications," Proc. Fourth ASTM-EURATOM symposium on reactor dosimetry, NUREG/CP0029, Vol. 2, 665 (1982).

[2] J.R. Lamarch, Introduction to Nuclear Reactor Theory (1966), p. 102, Addison-Wesley Publishing Company.

PART IV
TECHNIQUES

USE OF NIOBIUM FOR ACCURATE RELATIVE FAST NEUTRON FLUENCE

MEASUREMENTS AT THE PRESSURE VESSEL IN A VVER-440 NPP.

Bruno Bärs and Hari Karnani

Technical Research Centre of Finland, Reactor Laboratory

Otakaari 3A, SF-02150 Espoo, Finland

ABSTRACT

The reaction $^{93}Nb(n,n')^{93m}Nb$, with low threshold energy
(≈0.9 MeV) was used in a novel way to obtain both absolute
and accurate, relative experimental estimates for the fast
neutron (threshold) fluence at the inner wall of the pressure
vessel (PV) in a VVER-440 type PWR reactor. Niobium was
separated using chemical methods from steel samples containing
0.5 ... 1 % niobium, which had been scraped from the inner
wall of the Loviisa NPP (unit 1). Irradiated niobium foils from
the surveillance chains (SC) were also treated in a similar
way. Niobium from the liquid samples was then electrodeposited
in very thin layers on copper discs and the X-ray emission rate
measured with a Si(Li) detector. The threshold flux $\phi(>0.9$
MeV) was then obtained. The relative flux estimates (lead
factors) were obtained with 7-10 % accuracy.

1. INTRODUCTION

The fluence estimation at the pressure vessel (PV) in a nuclear
power plant (NPP) normally includes large uncertainties (up to
several tens of per cent), which mainly arise a) from the
limited amount or from the complete lack of experimental data at
the PV, b) from the poor response of conventional threshold
detectors in the most important energy region (≈0.5-2 MeV) and c)
from the large systematic uncertainties in the reaction cross
sections. The fluence estimates are needed for the estimation of
the fast-neutron induced pressure vessel embrittlement.

J. P. Genthon and H. Röttger (eds.), Reactor Dosimetry, 319–325.
© 1985 ECSC, EEC, EAEC, Brussels and Luxembourg

In order to reduce the uncertainty in the fluence estimates standardized relative (SC/PV) measurements have been developed and applied in a novel way /1/. The present work is a continuation of this development work including applications. The procedure is based on relative measurements of the fast neutron fluence at the SC and the PV and on standardization of the data and the measurements.

For practical purposes the fluence dependence of the embrittlement of the PV steel can be approximated by the relation

$$\Delta RT_{NDT,PV} = \Delta RT_{NDT,SC} \left(\frac{\Phi_{oPV}}{\Phi_{oSC}} \right)^b \qquad (1).$$

ΔRT_{NDT} is the shift in the nil-ductility transition temperature. $\Phi_o \equiv \Phi_o(>E_o)$ is the fast neutron fluence above a specified cut-off energy E_o, e.g. 1 MeV. $b \approx 1/3$. Thus only fluence ratios are needed to estimate the PV embrittlement from the corresponding measured SC embrittlement parameter.

In the 1) <u>initial stage of development</u> the reaction

$$^{54}Fe(n,p)^{54}Mn \qquad (2)$$

with a threshold energy $E_T = 2.8$ MeV was used /1/. The ratios ϕ_{TPV}/ϕ_{TSC} of the threshold fluxes above $E_T = 2.8$ MeV were experimentally determined by using steel samples from the inner surface of the PV and of the mechanical test specimens from the SC as activation detectors. The standardization includes the use of steel samples, the same fast neutron reaction (2), the same cross section data and other nuclear constants to determine the fluence ratio. In this manner one can almost completely eliminate the systematic errors, e.g. in the reaction cross sections. As to the application of this method on the WER-440 type PWR in Loviisa, the estimated uncertainty in the threshold fluence ratio was 4...6 %, in the fluence ratio at the cut-off energy $E_o = 1$ MeV was 12 % and in the measured absolute fluence at the PV 27 % /1/. The main uncertainty left in the flux ratio arises from the extrapolation of the threshold flux ratio ϕ_{TPV}/ϕ_{TSC} from 2.8 MeV to the cut-off energy $E_o = 1$ MeV of interest.

A <u>second stage of development</u> was started based on the reaction

$$^{93}Nb(n,n')^{93m}Nb \qquad (3)$$

with a low threshold energy (≈ 0.9 MeV). The Nb-foils in the SC, and the separated Nb in the steel samples scraped from the inner surface of the PV in the Soviet made reactor at Loviisa (unit 1) were used as neutron dosimeters in order to determine the flux ratio (SC/PV) at the threshold energy 0.9 MeV.

2. CHEMICAL TREATMENT AND ELECTRODEPOSITION OF NIOBIUM.

A scheme for separation of niobium from PV steel samples and impure niobium from SC, was worked out, based for the most part, on the published works by other authors /3/-/5/. Each sample was dissolved in 2 ml hydrofluoric acid and a drop of nitric acid. Anion exchange chromatography was used for the separation of Nb from other components and impurities. The niobium fraction was obtained in 3M HCl-0,1M HF. The concentration of Nb was determined by spectrophotometry /6/-/8/. The yields obtained for SC were above 98,5 %. For PV samples yields could not be determined because Nb content in PV surface samples varies from 0,5 to 1 %. The column purified Nb was electroplated on copper plates, as very thin layers (10-200 $\mu g/cm^2$). We describe here briefly, a method to electrodeposit Nb and to determine its mass.

A known amount (200-500 μg) of Nb (PV/SC) was taken and its pH raised to 4 by addition of ammonium hydroxide. The concentration of Nb in this bath solution was adjusted to 2-10 $\mu g/g$. A number (3-5) of samples, containing not more than 3 μg each, were taken out of the bath solution into 20 ml standard plastic vials for liquid scintillation counting (LSC). The rest of the bath solution was divided into several parts ranging from 20 to 200 μg Nb/part. From each part of the bath solution, Nb was electrodeposited at the cathode in an electrolytic cell. The cell consisted of a teflon vessel with a circular copper disc cathode at the bottom and a platinum wire as an anode. A current of 50-100 mA was passed through, for a cathode surface area of $\pi/4$ cm^2, for a period of 30-180 min. For more details, see /8/. At the end of the process, total weight of the bath solution was taken. A small part (1-5 ml) was taken out of this into a 20 ml vial for LSC. The deposited mass m_d, of Nb was determined from the equation

$$m_d = m - M_f A_f / m_f A_s \qquad (4)$$

where m is the net mass of Nb taken for an individual deposition process, M_f is the final mass of bath solution (before taking out of sample for LSC), m_f is the mass of solution taken out for LSC, A_f is the relative activity (cpm) corresponding to m_f, A_s is the mean relative specific activity (cpm/μg Nb) measured and calculated from the samples taken out before the process.

3. MASS DETERMINATION OF NIOBIUM

The main method of mass determination was to measure the Nb concentration of liquid solution samples by spectrophotometry (section 2). The activity to Nb mass ratio for the deposits could be determined with 3-7 % accuracy by using accurate spectrophotometric concentration values in an optimal absorbance range and by

321

using the radioactivity (^{93m}Nb + ^{94}Nb or ^{93m}Nb only) measured by LSC. A number of niobium deposits, for all PV and SC samples were made, and the mean values or the slopes of the absolute activity versus mass plot were used. The mass determination was checked by an additional method, which included the preweighing of some of the powedered SC samples and the gammaspectrometric (702.5 and 871.1 keV) determination of their longlived (20300 a) thermal neutron induced ^{94}Nb activity before dissolution. The absolute specific activity A^a_{94}/m_{Nb} corresponding to the total mass m_{Nb} of a speci__fic SC probe together with the measured ^{94}Nb-activities of the deposited Nb-masses were then utilized to determine the masses of deposited Nb for the samples of a specific origin. For the SC-2G 34 the spectrophotometric (+LSC) method gave a 1.5 % higher (total) mass estimate than the weighing (+A^a_{94}) method.

The third independent method to determine the amount of deposited Nb would be to take off the deposited Nb from the copper discs (after X-ray measurements), in some cases to purify it from Cu, to reactivate in the 250 kW TRIGA in our laboratory, to measure the short lived ($t_{1/2}$= 6.26 min) induced ^{94m}Nb activity (871.1 keV) and to compare with a similarly irradiated Nb standard of known weight. The feasibility of this method was tested and found acceptable but will still need some development for final application. It would be of special importance for PV samples where the weighing technique could not be applied.

Due to the low thermal neutron flux in the available TRIGA reactor a fourth possible method reactivation based on ^{94}Nb (reaction 6) could not be applied but is obviosly applicable in reactors with higher fluxes.

A LSC method combined with relative thermal neutron measurements may also be used based on the relation

$$\Phi_T(<0.9 \text{ MeV}) \ \sigma_T = R_s t \ = k_1 \ A^a_{93}/m = k_2 \ A^L_{93} \ \Phi_{th}/A^L_{94} \quad (5)$$

The upper indexes "a" and "L" refer to absolute and LSC measurements. The relative activity ratio A^L_{93}/A^L_{94} for each individual sample can be obtained in a simple and rapid way by one single LSC measurement on a very small ($\approx\mu$g) sample. They are then multiplied by the relative thermal neutron fluences or multiplied by the ratio A^L_{94} /m or alternatively A^a_{94} /m. For absolute activity measurements with LSC one must also determine the proportionality constant between the LSC and absolute activity ratios.

4. RESULTS

The reaction rates R_s per target nucleus and the threshold fluxes $\phi_T(>0.9$ MeV) were obtained from the relations

$$R_s = \frac{A_{93}^a}{m} \times \frac{1}{p_k \; q \; N_m \; \alpha_o} \quad (6), \quad \phi_T(>0.9 \text{ MeV}) = \frac{R_s}{\sigma_T} \quad (7)$$

A_{93}^a/m is the activity (converted to the shut down moment of the NPP, $t_{1/2}=16.11\pm0.19$ a, /9/) to mass ratio obtained as a mean or as a slope from the activity versus mass plot for at least three discs with niobium deposits. The absolute ^{93m}Nb-activity was obtained from X-ray measurements of the Nb-deposits on the Cu-discs with a calibrated Si(Li) X-ray detector and multichannel analyser. $q=1$ is the number of 30 keV gammas per disintegration, N_m the number of target atoms (Nb) per gram and $\alpha_o = A/A_{sat}$ from the irradiation history. $p_k = p_{k\alpha} + p_{k\beta} = 0.1066\pm3$ % is the probability that an emitted 30 keV gamma quantum from ^{93m}Nb is converted into the measured 16.58 keV and 18.62 keV X-rays /10/. The calculated threshold cross sections were 0.19016 b (SC) and 0.19078 b (PV) at the threshold energy $E_T=0.9$ MeV with $\sigma(E)$ taken from DOSCROS81 (same as IRDF-82) and normalized to agree with integral measurements by Alberts et al /10/ in a ^{252}Cf field. Table 1 contains measured threshold flux values for four SC foil samples and two PV samples, the flux ratios (in the range 0.873-0.997) from the present Nb-measurements and the previous Fe-measurements /1/ and the estimated uncertainties for the activity to mass ratio to be used at relative measurements.

Table 2 contains some measured lead factors ϕ_{SC}/ϕ_{PV} at the threshold energy 0.9 MeV, the estimated uncertainties and comparison with the same lead factors obtained from Fe-measurements /1/. The obtained flux ratio R_2 for the foil positions is converted to the wanted ratio R_4 for the center of the Fe-rod pairs by multiplying by an experimentally measured flux ratio R_3. The ratios R_4/R_5 (0.875 ... 1.065) of the lead factors from the Nb- and from the Fe-reaction (extrapolated) fall within the estimated error limits. The error limits 6.6...9.7 % for the measured lead factors R_5 (from the Nb-reaction) are lower than the previously estimated mean error limit ±12 % from the Fe-reaction. Thus the accuracy could and may further be improved by utilizing the Nb-reaction (3) instead of the Fe-reaction (2) in relative measurements. If we assume a $\pm$ 15 % uncertainty in the threshold cross section then the estimated uncertainties in the absolute flux estimates would be $\pm$ 17 % (SC) and $\pm$19 % (PV). The development work and application of the Nb-reaction has been quite laborious, all ideas have not yet been tested. Further development will be carried out in an attempt to improve the accuracy and to develop some more easily applicable routines.

ACKNOWLEDGEMENTS

The work was financially supported by Imatran Voima Oy (Imatra Power Co. Ltd.) and by the Ministry of Trade and Industry. The

323

cross section calculations carried out by Mr Tom Serén and the discussions with him are greatly acknowledged. Thanks are also due to Dr. Laina Salonen for help with the LSC measurements, to Dr Pekka Hiismäki for useful discussions and to Mr Tapio Saarenpää for assistance with some practical matters.

TABLE 1. Some measured threshold fluxes $\phi_T(>0.9$ MeV) using Nb (3) with estimated uncertainties, the ratio to the extrapolated cut off flux at the same energy obtained from the Fe-reaction (2) and the estimated uncertainty in the measured and used ^{93m}Nb-activity to mass ratio. P=1375 MW$_{th}$.

SAMPLE	$\dfrac{\phi_{Nb}(>0.9 \text{ MeV})}{10^{11}}$ (n/cm^2)	$\dfrac{\phi_{Fe}(>0.9)_{extr}}{\phi_{Nb}(>0.9)}$	$\Delta(A^a_{93}/m)$ ACT. TO MASS UNCERT. IN REL.MEAS.
A) PRESSURE VESSEL	Nb FROM SCRAPED PV	Fe/Nb SAMPLES	
PV-12	1.376	0.977	± 5 %
PV-17	1.865	0.873	± 7 %
UNCERT.	± 19 %		
B) SC 2G	Nb-FOIL	Fe-FOIL/Nb-FOIL	
2G-15	25.27	0.919	± 7 %
2G-34	23.29	0.917	± 3 %
2G-17	20.68	0.997	± 5 %
2G-19	13.82	0.964	± 5 %
UNCERT.	± 17 %		

TABLE 2. LEAD FACTORS ϕ_{SC}/ϕ_{PV} (R_4, R_5) ABOVE 0.9 MeV FROM THE Nb (3) AND THE Fe-REACTIONS (2).

R_1	R_2 $\left(\dfrac{\phi_{SCFOIL}}{\phi_{PV}}\right)_{Nb}$ X	R_3 $\dfrac{\text{Fe-ROD}}{\text{Fe-FOIL}}$ =	R_4 $\dfrac{\phi_{SC}}{\phi_{PVNb}}$ ± σ	R_5 $\left(\dfrac{\phi_{SC}}{\phi_{PV}}\right)_{Fe}$	R_4/R_5 Nb/Fe RATIO
$\dfrac{2G-19}{PV-12}$	10.10	0.902	9.11 ± 7.8 %	8.95	1.019
$\dfrac{2G-19}{PV-17}$	7.41	0.902	6.69 ± 8.4 %	7.38	0.906
$\dfrac{2G-34}{PV-12}$	16.93	0.869	14.71 ± 6.6 %	13.81	1.065
$\dfrac{2G-34}{PV-17}$	12.49	0.869	10.85 ± 7.4 %	11.40	0.951
$\dfrac{1G-17}{PV-17}$	11.08	0.859	9.53 ± 9.7 %	10.89	0.875
$\dfrac{2G-15}{PV-12}$	18.37	0.831	15.26 ± 9.1 %	14.35	1.063
MEAN UNCERT		±2-3 %		± 12 %	

REFERENCES

1. L.B.Bärs, P.A.Liuhto and T.O.Serén, "A method for improving
   the accuracy in neutron fluence measurements in power
   reactors and its application in a WWER-440 nuclear power
   plant". Kernenergie 27(1984)8, p.342-347.

2. F.Hegedüs et al, Measurement of the activation cross-section
   of the reaction ^{93}Nb(n,n')^{93m}Nb for 0-25 MeV neutrons.",
   Proc. the advisory group meeting on nuclear data for reactor
   dosimetry, Vienna (Nov. 13-17, 1978)

3. J.P.Faris,"Adsorption of the elements from hydrofluoric
   acid by anion exchange," Anal. Chem. 32(1960)4, p.520-522.

4. J.L.Hague et al, "Separation of Ti, W, Mo and Nb by ion
   exchange," J. of Research of the N.B.S 53(1954)4, p. 262.

5. M.J.Cabell & I.Milner, "The quantitative separation and
   determination of Ta and Nb using anion exchange
   chromatography," Anal. Chim. Acta 13(1955), p. 258-267.

6. A.D.Westland & J.Bezaire, "An improved spectrophotometric
   determination of Nb with thiocyanate.", Anal. Chim. Acta
   66(1973), p. 187-193.

7. C.S.P.Iyer & V.A.Kamath, "Determination of Nb in steels"
   Talanta 27(1980), p.537.

8. H.Karnani, "Chemical treatment and Electrodeposition of Nb
   for neutron dosimetry.", A forthcoming report, Technical
   Research Centre of Finland, Reactor Laboratory.

9. R.Lloret,"Complement à la mesure de la periode de
   decroissance radioactive du ^{93m}Nb." Radiochem. Radioanal.
   Lett. 50(1981)2, p.113-120.

10. W.G.Alberts et al, "Measurements with the Niobium neutron
    fluence detector at the PTB.", Proceedings of the Fourth ASTM
    -EURATOM Symposium on Reactor Dosimetry, Gaithersburg,
    Maryland, March 22-26, 1984. Vol. 1, p. 443-442.

NIOBIUM NEUTRON FLUENCE DOSIMETER MEASUREMENTS

R.J. Gehrke, JW Rogers and J.D. Baker

Idaho National Engineering Laboratory
EG&G Idaho, Inc.

Idaho Falls, Idaho, 83415 U.S.A.

ABSTRACT

The use of the ^{93}Nb(n,n')^{93m}Nb reaction to monitor fast neutron fluence is currently being applied at several laboratories around the world. The niobium dosimeter appears to have the desirable properties of providing a neutron fluence measure that corresponds to the changes in the mechanical properties of ferritic materials used in structural components of nuclear reactors and a reaction product with a suitably long half-life [$T_{1/2} = (16.13 \pm 0.15)$y]. Recently, an effort has been made to develop methods and procedures to routinely analyze niobium dosimeters at the Idaho National Engineering Laboratory (INEL). This effort includes the study of sample preparations, counting techniques, detector calibrations, calibration standards, data analyses and uncertainty evaluations.

---

1. INTRODUCTION

The use of the ^{93}Nb(n,n')^{93m}Nb reaction to monitor fast neutron damage fluences has received considerable attention over the last 10 years,[1-7] and as a result precise values for the pertinent decay parameters are becoming available and the cross section is being investigated. Because of the potential uses for this reaction, the Radiation Measurements Laboratory at the INEL has decided to develop the capability to routinely analyze niobium dosimeters. Expertise at the INEL in neutron dosimetry, detector efficiency measurements, radiation counting techniques, and spectral analysis

J. P. Genthon and H. Röttger (eds.), Reactor Dosimetry, 327–335.
© 1985 ECSC, EEC, EAEC, Brussels and Luxembourg

methods was applied to this dosimetry measurement problem and the
findings of this investigation are described below.

2. EXPERIMENTAL

2.1 Source Preparation

Because of the low energy of the radiations, special care is
required in the preparation of radionuclide x-ray sources. Serious
errors can be introduced by attenuation in the source and the
source mount and by any x-ray fluorescence of the carrier material. For this reason all sources described in this paper were
prepared in the same manner and the magnitudes of the various
corrections were determined. It was decided to use niobium wires,
but not to count the wires directly because this would maximize any
errors due to attenuation or to fluorescence caused by ^{182}Ta
photons.

Dissolution of the irradiated niobium dosimeters were performed
in essentially the manner described by Tourwe et al.[3] After
dissolution the final volume of solution was brought to 1 to 2 mL
by adding distilled water. The entire solution was then taken up
in a preweighed polyethylene pycnometer (weighing bottle).[8] The
niobium concentration in the solution was determined from a ratio
of the mass of the niobium wire to that of the solution. Mass
aliquots of this solution (determined to ± 0.05%) were deposited on
wafers of analytical paper (1.27 cm dia. x 0.076 cm thick). The
liquid deposits were air dried overnight.

It was decided to adopt our standard source mounting, as shown
in Figure 1, for γ-ray sources counted with NaI(Tl) and Ge
spectrometers. Dried sources were covered with polyester tape
(0.0076 cm thick). Four eyelet rivets hold the source mount
together. The effect of any warp or curvature in the entire source
was reduced by counting the source twice, once with each face
toward the detector.

The diameter of the analytical paper, used in the above source
mount, was small enough so that the data could be analyzed with the
point-source efficiencies of the NaI(Tl) or Ge(Li) spectrometers[9]
with negligible error.[10]. The thickness of the analytical paper
was chosen so that up to ~100 mg of solution could be absorbed by
the paper without becoming saturated. If too much solution was
placed on the analytical paper, a ring of activity formed around
the outer edge. This ring is of nonuniform thickness and, therefore, of unknown self-attenuation.

The techniques used to determine the x-ray peak areas allow
for some degradation in the peak shape due to the use of analytical

paper and a Kapton* film cover. The first technique consists of adding the counts under the K_α + K_β peaks and subtracting a sloping straight line background. The second technique consists of fitting the x-ray region of the spectrum in a linear least-squares manner with the various component spectra whose activities are known.[11]

2.2 Attenuation and Fluorescence Effects

Although the sources used in both the efficiency calibration and niobium dosimetry measurements were prepared in the same manner, it was decided that a measure of the self-attenuation due to the use of analytical paper would be worthwhile. Sources were made from mass aliquots of carrier-free ^{93m}Nb (separated from fission produced ^{93}Zr) as above but with no analytical paper present as well as with one, two and three wafers of analytical paper. Any effects due to the source asymmetry were reduced by the process of counting the sources from both sides. The attenuation of the carrier-free activity in one wafer was measured to be 1.04%.

The presence of carrier niobium in the ^{93m}Nb sources can cause self-attenuation of the emitted niobium K x rays. Although this effect is reduced (as compared to a wire) by the source preparation technique adopted above, an attempt was made to measure the residual effect. This was done by depositing a known amount of carrier-free ^{93m}Nb and known masses of niobium carrier. The results of these measurements are superimposed on a transmission curve (see Figure 2) determined with a mass attenuation coefficient of 18.219 cm^2/g[12] and show good agreement with it. The amount of niobium carrier present in a source was kept below ~500 µg to keep the self-attenuation below 0.4%.

Because Nb and Ta are both group VA elements, Ta is a common contaminant in Nb. These dosimeters contained ~200 ppm tantalum. Upon irradiation, 114.5-d ^{182}Ta is produced along with ^{93m}Nb. The possibility of x-ray fluorescence of the niobium material by the radiation emitted by the contaminant ^{182}Ta was investigated for the source geometry employed. This was done by preparing a set of sources containing a known amount of ^{182}Ta and various masses of niobium carrier. The number of niobium K fluorescence x rays produced per Becquerel of ^{182}Ta as a function of niobium carrier mass for our source geometry is given in Figure 3. For example, if 300 µg of niobium and 10^4 Bq of ^{182}Ta are present in a dosimeter source, 5.5 K x-rays per second will be produced by fluorescence. This correction can usually be ignored.

---

*Trademark of E.I. Du Pont de Nemours and Company, Inc.

2.3 Detector Efficiency Measurements

Although some methods of data analysis would not require a detector efficiency calibration, one was measured for other purposes. Because there is a paucity of radionuclides with which to measure the peak efficiency between 10 and 25 keV, it is common to use both L and K x-rays and γ-ray lines for calibration. Because the niobium K x rays are the only easily measurable radiation emitted from ^{93m}Nb, our preference is to use only K x-ray lines in the efficiency measurements. This avoids any problem of the difference in line shape for the L and K x rays and γ rays. Further, the L fluorescence yields are not as well known as the K fluorescence yields.

Based on the above criteria the following radionuclides have been used for measurement of the detector efficiency between 13 and 24 keV: ^{85}Sr, ^{88}Y, ^{92m}Nb, ^{99m}Tc, and ^{109}Cd. Table 1 lists the relevant decay data useful in determining their K x ray emission rates from the activity or the prominent γ ray emission rate. For consistency we used the decay data from Refs. 13, 16, and 17 to determine the K x-ray emission rates. The radionuclides ^{85}Sr, ^{88}Y, and ^{109}Cd are common efficiency standards and are available from several metrology laboratories. Some laboratories have used ^{99m}Tc or its parent ^{99}Mo, but due to their short half-lives, neither of these radionuclides have found general use. The use of ^{92m}Nb as a radionuclide standard is thought to be new. This source decays via electron capture with > 99% of the decays passing through the 934 keV transition which feeds the ^{92}Zr ground state. It is particularly well suited because its mean K x-ray energy (16.06 keV) is only 0.86 keV away from that of niobium. Both ^{99m}Tc and ^{92m}Nb can be produced simultaneously by irradiation of natural molybdenum in a reactor environment with both fast and thermal neutron components. Carrier-free sources are obtained by radiochemical separation of each of these elements (i.e., Nb and Tc) from the irradiated molybdenum.

The efficiency of a 1000 mm^2 area x 5 mm thick Ge detector was measured at 3- and 7-cm source-detector positions. The Ge detector has a resolution (FWHM) of 415 eV at 6.4 keV and 600 eV at 122 keV and a 0.005 cm thick beryllium window. The measured efficiencies for the Nb K x rays are 0.0492 ± 0.0014 and 0.0122 ± 0.0003 at 3 and 7 cm source-detector positions, respectively. The dominant contributions to the uncertainty, which cannot be reduced by averaging, were traced to that in the K fluorescence yield (±2%) and in the calculated K/L capture ratios (±1%).

TABLE 1

Radio-nuclide	Mean K X-ray Energy (keV)	Standardization γ Ray (keV)	γ Ray Emission Probability	K X-Ray to γ-Ray Intensity Ratio
^{85}Sr	13.62	514.0	98.0 ± 1.0 [a]	0.600 ± 0.010 [a]
			99.29 ± 0.04 [b]	0.595 ± 0.008 [b]
			98.4 ± 0.5 [c]	0.597 ± 0.009 [c]
^{88}Y	14.40	898.0	93.7 ± 0.3 [a]	0.644 ± 0.016 [a]
		898.0	94.6 ± 0.5 [c]	0.641 ± 0.018 [c]
		1836.1	99.3 ± 0.3 [a]	0.607 ± 0.015 [a]
		1836.1	99.24 ± 0.07 [c]	0.611 ± 0.017 [c]
^{92m}Nb	16.06	934.4	99.12 ± 0.04 [d]	0.641 ± 0.019 [e]
^{99m}Tc	18.72	140.5	88.97 ± 0.24 [a]	0.083 ± 0.003 [a]
			89.0 ± 0.2 [b]	0.081 ± 0.006 [b]
^{109}Cd	22.68	88.0	3.73 ± 0.06 [a]	27.3 ± 0.7 [a]
			3.65 ± 0.06 [b]	27.4 ± 1.2 [b]

a) Ref. 13 b) Ref. 14 c) Ref. 15 d) Ref. 16 e) Ref. 17

2.4 Counting Techniques

At least 50,000 counts were acquired in the net area of the $K_\alpha + K_\beta$ x ray peak. This resulted in ≲0.5% uncertainty (all quoted uncertainties are one estimated standard deviation) in the counting statistics. In all measurements, the gross count rate was kept below ~500 c/s so that no random summing correction was needed. As stated above, the amount of niobium carrier was kept below ~500 µg. After irradiation the dosimeters were not counted until the short half-life contaminants (e.g., ^{90}Y, ^{90m}Y, ^{92m}Nb, and ^{183}Ta) decayed away. Because of the low count rates employed, the background was monitored routinely and any interferring background peaks were subtracted.

2.5 Analysis Techniques

Two techniques can be applied to the analysis of the ^{93m}Nb spectral data (i.e., dosimeters). First, with a known efficiency and K x-ray emission probability (see Section 3), the ^{93m}Nb activity can be determined instrumentally with a Ge or Si(Li) spectrometer from the area of the $K_\alpha + K_\beta$ x-ray peaks. Second, if an activity standard is available, the activity of the ^{93m}Nb can be determined by direct comparison with this standard.

Measurements of several mass aliquots of a dissolved dosimeter under the same experimental conditions indicated that a reproducibility of $\pm$ 0.25% in the measured activity was achieved.

3. K X-RAY EMISSION PROBABILITY

Using the counting techniques described above, a measurement of the ^{93m}Nb K x-ray emission probability (P_{kx}) was undertaken. A solution of high specific activity ^{93m}Nb, from A.J. Fudge at AERE-Harwell, was standardized at a radioactivity concentration of (7.70 $\pm$ 0.06) x 10^5 Bq/g of solution at CBNM[18] by liquid scintillation. The K x-ray emission rate from four aliquots of this source material was measured to be (8.51 $\pm$ 0.21) x 10^4 K x rays per second per gram of solution. From the ratio of the K x-ray emission rate to the disintegration rate, the ^{93m}Nb K x-ray emission probability was determined to be:

11.04 $\pm$ 0.28   K x rays per 100 decays.

This value lies above the value of 10.7 $\pm$ 0.3 reported by Alberts et al.[5] and below that reported to Vaninbroukx et al.[7,19] of 11.5 $\pm$ 0.5. Intercomparison of ^{93m}Nb activity standards by these laboratories[19] indicated that the difference in these values was not due to the ^{93m}Nb activity measurements but rather due to differences traceable to the techniques used to measure the detector efficiency. Because our efficiency measurement technique avoids the attenuation corrections used by Alberts et al. and the use of standards with various peak shapes as used by Vaninbroukx, it is easy to use and yet does not sacrifice accuracy. It is our opinion that the difference between the previously reported P_{kx} values was caused by the lack of a suitable efficiency standard in the immediate energy region of the Nb K x-rays and the underestimation of the uncertainty in interpolating between the measured efficiency values.

4. SUMMARY

The techniques and measurements of this investigation have demonstrated a capability of assaying niobium dosimeters for activity with a Ge detector to an accuracy of ~0.8% by use of a ^{93m}Nb reference standard ($\pm$ 0.76%). A technique for measuring the detector efficiency for K x-rays (ε_{kx}) has been described and used to determine the ^{93m}Nb K x-ray emission probability (P_{kx}). The introduction of a standardized ^{92m}Nb K x-ray source provides better definition of the efficiency curve in the region of interest. The ε_{kx} and P_{kx} parameters can be used in place of a ^{93m}Nb standard to measure the activity of niobium dosimeters with a Ge detector to an accuracy of ~$\pm$1.4%. This accuracy is possible because the major component in both P_{kx} and ε_{kx} is the K fluorescence yield ($\pm$ 2%) which cancels in the activity determination.

Currently, the automation of the analysis routines using a linear least-squares spectral fitting technique is being implemented.

ACKNOWLEDGEMENTS

We appreciate the assistance of L. D. Koeppen in acquiring the data used in the measurements and R. G. Helmer for technical discussions. We are especially grateful to B. M. Coursey and R. Vaninbroukx for making the ^{93m}Nb activity standards available to us. This work was supported by the U.S. Department of Energy under DOE Contract No. DE-AC07-76ID01570.

REFERENCES

1. R. Lloret, 1st ASTM-EURATOM Symposium in Reactor Dosimetry, Petten 1975, EUR 5667 E/F, Part I (1977) 747.

2. F. Hegedus, ibid. p. 757.

3. H. Tourwe, N. Maene, 3rd ASTM-EURATOM Symposium on Reactor Dosimetry, Ispra 1979, EUR 6813 EN-FR (1980) 1245.

4. H. Tourwe, W.H. Taylor, D. Reher, R. Vaninbroukx, R. Lloret, H.J. Nolthenius, P. Wille, and R. Schweighofer, 4th ASTM-EURATOM Symposium on Reactor Dosimetry, Gaithersburg 1982, NUREG/CP-0029 (1982) 401.

5. W.G. Alberts, R. Hollnagel, K. Knauf, M. Matzke, and W. Pessara, ibid. p. 433.

6. R. Vaninbroukx, in Liquid Scintillation Counting, Recent Applications and Development, Vol I, Academic Press, Inc. New York and London (1980) p. 143.

7. R. Vaninbroukx, Int. J. Appl. Radiat. Isot. 34 (1983) 1211.

8. J.S. Merritt, Nucl. Instr. and Methods 112 (1973) 325.

9. R. G. Helmer, Nucl. Instr. and Methods 199 (1982) 521.

10. R.G. Helmer, Int. J. Appl. Radiat. Isot. 34 (1983) 1105.

11. A Handbook of Radioactivity Measurements Procedures, U.S. NCRP Report No. 58 (1978) p. 157.

12. W.H. McMaster, N. Kerr Del Grande, J.H. Mallett, and J.H. Hubbell, Compilation of X Ray Cross Sections, Section I, Revision I, UCRL-50174 (1969).

13. A Handbook of Radioactivity Measurements Procedures, U.S. NCRP Report No. 58 (1978) Appendix A.

14. Table of Radionuclides, F. Lagoutine, N. Coursol, and J. Legrand, Laboratory of Metrology of Ionizing Radiations, Vol 1 (1983).

15. Radioactivity Standard Cettificate, Physikalisch-Technische Bundesanstalt; H.M. Weiss (1984) private communication.

16. P. Luksch, Nuclear Data Sheets, 30 (1980) 573.

17. M.J. Martin, ORNL, private communication, August 1984.

18. B.M. Coursey, R. Vaninbroukx, and W. Bambynek, CBNM, private communication, January 1984 and May 1984.

19. R. Vaninbroukx, W. Zelmer, Annual Progress Report on Nuclear Data 1982, CBNM, Geel, Belgium (1983), INDC(EUR)017/G p. 46.

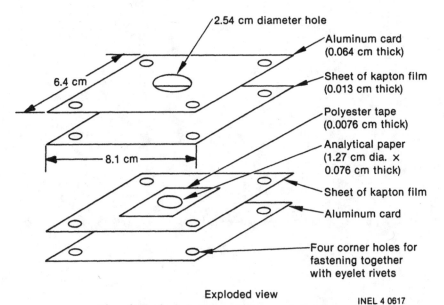

2.54 cm diameter hole

Aluminum card
(0.064 cm thick)

Sheet of kapton film
(0.013 cm thick)

Polyester tape
(0.0076 cm thick)

Analytical paper
(1.27 cm dia. ×
0.076 cm thick)

Sheet of kapton film

Aluminum card

Four corner holes for
fastening together
with eyelet rivets

6.4 cm

8.1 cm

Exploded view

INEL 4 0617

Fig. 1 Exploded view of source holder.

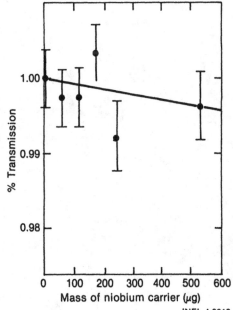

Fig. 2 Self-attenuation of ^{93}Nb K
x rays due to niobium.

INEL 4 0616

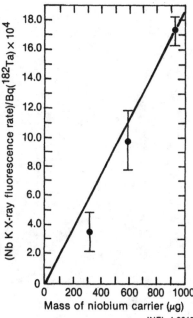

Fig. 3 K fluorescence produced
per Becquerel of ^{182}Ta.

INEL 4 0615

335

CALCULATION OF THE THERMAL NEUTRON FLUX PERTURBATION FACTOR
FOR FOIL DETECTORS IN VARIOUS MEDIA USING THE CODE PERTURB.D*

M. Carmo Freitas and Eduardo Martinho

LNETI-Instituto de Energia

2685 Sacavém, Portugal

ABSTRACT

The code PERTURB.D computes the thermal neutron flux perturbation factor, K, due to circular foils located in an isotropic neutron field. The calculation is based on the expression K = G.E.F, where G denotes the neutron self--shielding in the foil, E the edge correction factor and F the flux depression in the diffusing medium surrounding the foil. The code has been tested through an extensive comparison of calculated K-values with published experimental re-sults (87 values) for several kinds of detector of various radii and thicknesses, and in different media. It was found that, for 66% (85%) of the sample cases, our calculated K--values agree to better than 1% (2%) with the experimental ones. As an application of the code PERTURB.D, tables of K--values for different materials, including Mn, Co, In and Au, for diameters in the range 0.5 to 2 cm, and various prac_tical thicknesses, both in water and graphite, are calcu-lated and presented*.

---

*Abridged version.Full paper (3) - presenting the following further results: (i) tables of K-values for Sc, Cu and Dy; (ii) tables of G and G.E values for all the materials studied; and (iii) table of K-values for thermal neutron foil detectors from one of the current manufacturers, in water, paraffin, graphite and heavy water - avail_able from the authors.

J. P. Genthon and H. Röttger (eds.), Reactor Dosimetry, 337–343.
© 1985 ECSC, EEC, EAEC, Brussels and Luxembourg

INTRODUCTION

The activation technique can be used to determine the absolute thermal neutron flux in a nuclear reactor. Usually, the measurement is made with coin-shaped detectors (foil geometry). As is well-known, the original neutron flux is locally modified by the detector and this perturbation must be evaluated.

The neutron flux perturbation factor, K, is defined as

$$K = \bar{\Phi}/\Phi_0,$$

where Φ_0 represents the unperturbed neutron flux which was present before the foil being inserted and $\bar{\Phi}$ the average flux in the foil.

This factor can be expressed as the product of two other factors (BENSCH [1], for example):

(i) The global self-shielding factor, $(G.E) = \bar{\Phi}/\Phi_s$, which is the ratio between the average flux in the foil and the flux on the foil surface. Its value depends on the properties of the detector (material and dimensions). E designates the edge factor, which is a correction to G taking into account the neutrons incident through the edges of the foil.

(ii) The neutron flux depression factor, $F = \Phi_s/\Phi_0$, which is the ratio between the flux on the foil surface and the unperturbed neutron flux. Its value depends on both the properties of the diffusing medium and the properties of the detector.

Using these definitions, the neutron flux perturbation factor can be written as follows

$$K = (G.E).F.$$

The assumption concerning the separability of the corresponding effects appears to be acceptable provided that the error which is introduced remains minor as compared to other errors in the absolute measurement of the neutron flux.

The code PERTURB.D [2,3] computes the thermal neutron flux perturbation factor, K, due to foil detectors located in an isotropic neutron field. The calculated K-values were compared with published experimental results (87 values) for several kinds of detector (gold, indium, cobalt, dysprosium and copper) of various radii and thicknesses, and in different media (light water, paraffin and graphite). It was found that, for 66% (85%) of the sample cases, our calculated K-values agree to better than 1% (2%) with the experimental ones as shown on the figure. Thus the validity of the code PERTURB.D appears to be satisfactorily established.

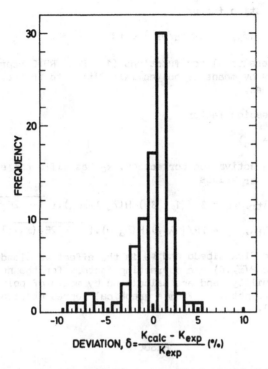

Frequency distribution of the deviation between calculated K-values and experimental ones.

PERTURB.D CODE

The expressions used for the calculation of G, E and F, for a foil of radius R and thickness d, are presented bellow*. They are based on the works of BECKURTS and WIRTZ (4) and BENSCH (1), which in turn use the results from theoretical and experimental investigations carried out by other authors.

(a) Self-shielding factor

$$G = \{ \overline{1 - 2E_3(x_t) + x_s\chi(x_t)} \} / (2\overline{x}_t)$$

where $x_i = d\Sigma_i$ (i accounting for t=total, s=scattering, a=absorption macroscopic cross section), E_3 is the exponential integral and χ is the intermediate scattering function (4).

The bar over the symbol of a function designates the average of that function taken over the Maxwellian distribution of neutron energies.

(b) Edge correction factor

$$E = 1 + \{ I(\overline{x}_a) - \Delta(2R\overline{x}_a/d) \}.d/\pi RG$$

where I and Δ are tabulated functions (1). PERTURB.D represents these functions by means of polynomials fitted to the corresponding calculated values.

(c) Flux depression factor

$$F = 1/(1 + \kappa_c)$$

where κ_c is the activation correction. κ_c has different expressions according to R/λ_{tr} values:

(i) when $R \geqslant \lambda_{tr}$, $\kappa_c = \beta_f/(1 - \beta_f).H(\overline{x}_a,\beta_f/\beta_e).\{ \overline{1 - 2E_3(x_a)} \}$;

(ii) when $R < \lambda_{tr}$, $\kappa_c = 2R/(\pi\lambda_{tr}).H(\overline{x}_a,0).\{ \overline{1 - 2E_3(x_a)} \}$

where β_e is the flux albedo and β_f is the effective albedo. $H(\overline{x}_a,\beta_f/\beta_e)$ and $H(\overline{x}_a,0)$ are correcting factors for the neutron absorption probability, and are calculated by means of polynomials fitted to values obtained from a graphical representation of the functions (1,4).

RESULTS

As an application of the code PERTURB.D, the thermal neutron flux perturbation factor, K, for different foil detectors in various diffusing media were calculated. The properties of the detector materials and the diffusing parameters of the media adopted in the calculation are given in Tables 1 and 2, respectively. K-values for manganese, cobalt, indium and gold foils, for diameters in the range 0.5 to 2 cm, and for various practical thicknesses, both in water and graphite, are shown in Tables 3 through 6.

REFERENCES

1. F. Bensch, "Flux Depression and the Absolute Measurement of the Thermal Neutron Flux Density", Atomkernenergie 25, 257 (1975).

2. M.C. Freitas e E. Martinho, PERTURB.D - Programa de Cálculo do Factor de Perturbação do Fluxo de Neutrões Térmicos devida a Detectores por Activação Circulares (in Portuguese), Report LNETI/ /DEEN-B 61α (1984).

3. M.C. Freitas and E. Martinho, Calculation of the Thermal Neutron Flux Perturbation Factor for Different Foil Detectors in Various

Diffusing Media using the Computer Code PERTURB.D, Portuguese
Report LNETI/DEEN-A 11 (1984), Paper distributed at this Symposium.

4. K.H. Beckurts and K. Wirtz, Neutron Physics, Springer-Verlag,
    Berlin, 1964.

5. Handbook of Chemistry and Physics, 62nd Edition, R.C. Weast,
    Editor, CRC Press, Inc., Boca Raton, Florida, 1981.

6. S.F. Mughabghab, M. Divadeenam, and N.E. Holden, Neutron Cross
    Sections, Volume 1, Academic Press, New York, 1981.

7. C.H. Westcott, Effective Cross Section Values for Well-Moder-
    ated Thermal Reactor Spectra, Canadian Report AECL-1101 (1960).

TABLE 1

Properties of the detector materials (5,6,7)

Detector material	Atom-gram (g/mol)	Density (g/cm^3)	(n,γ) cross section at 2200 m/s σ_γ^0 (barn)	Scattering cross section σ_s (barn)	Westcott factor g
Manganese	54.938	7.3	13.3	2.2	1
Cobalt	58.933	8.9	37.18	6.00	1
Indium	114.82	7.30	193.8	2.45	1.02
Gold	196.97	18.88	98.65	9.3*	1.006

*D.J. Hughes and R.B. Schwartz, Neutron Cross Sections, BNL-325
(1958)

TABLE 2

Diffusion parameters of the media (4)

Diffusing medium	Diffusion length L (cm)	Transport mean free path λ_{tr} (cm)
Water	2.755	0.431
Graphite	52.5	2.574

TABLE 3

Calculated values of the thermal neutron flux perturbation factor
for manganese foils in water and graphite

Thickness (cm)	WATER				GRAPHITE			
	Radius (cm)				Radius (cm)			
	0.25	0.50	0.75	1.0	0.25	0.50	0.75	1.0
0.0025	0.992	0.990	0.988	0.987	0.993	0.992	0.992	0.991
0.005	0.986	0.981	0.978	0.976	0.988	0.986	0.985	0.984
0.01	0.975	0.966	0.960	0.956	0.980	0.976	0.974	0.972
0.02	0.958	0.941	0.930	0.921	0.966	0.959	0.955	0.952
0.03	0.944	0.919	0.903	0.890	0.956	0.944	0.938	0.934
0.04	0.932	0.899	0.879	0.863	0.947	0.932	0.924	0.918
0.05	0.921	0.882	0.857	0.838	0.939	0.920	0.911	0.904
0.075	0.897	0.844	0.810	0.784	0.922	0.895	0.882	0.872
0.01	0.878	0.813	0.772	0.738	0.909	0.874	0.857	0.844

TABLE 4

Calculated values of the thermal neutron flux perturbation factor
for cobalt foils in water and graphite

Thickness (cm)	WATER				GRAPHITE			
	Radius (cm)				Radius (cm)			
	0.25	0.50	0.75	1.0	0.25	0.50	0.75	1.0
0.001	0.989	0.986	0.985	0.983	0.990	0.990	0.989	0.989
0.0025	0.976	0.970	0.966	0.962	0.979	0.977	0.976	0.975
0.005	0.958	0.946	0.938	0.932	0.964	0.961	0.958	0.957
0.0075	0.942	0.926	0.914	0.905	0.952	0.946	0.943	0.940
0.01	0.928	0.906	0.892	0.880	0.940	0.933	0.929	0.925
0.0125	0.915	0.889	0.871	0.857	0.930	0.921	0.916	0.912
0.015	0.903	0.873	0.852	0.836	0.920	0.910	0.904	0.899
0.02	0.881	0.844	0.818	0.797	0.902	0.889	0.881	0.875
0.025	0.861	0.818	0.788	0.762	0.886	0.870	0.860	0.853

TABLE 5

Calculated values of the thermal neutron flux perturbation factor
for indium foils in water and graphite

Thickness (cm)	WATER Radius (cm)				GRAPHITE Radius (cm)			
	0.25	0.50	0.75	1.0	0.25	0.50	0.75	1.0
0.0005	0.987	0.984	0.982	0.981	0.988	0.988	0.987	0.987
0.001	0.976	0.971	0.968	0.965	0.979	0.978	0.977	0.977
0.003	0.942	0.928	0.918	0.910	0.950	0.947	0.944	0.942
0.005	0.914	0.892	0.877	0.865	0.927	0.921	0.917	0.914
0.0075	0.883	0.854	0.832	0.816	0.901	0.893	0.888	0.884
0.01	0.856	0.821	0.795	0.773	0.879	0.868	0.862	0.856
0.015	0.810	0.765	0.732	0.702	0.840	0.826	0.817	0.809
0.02	0.772	0.719	0.679	0.644	0.808	0.789	0.778	0.769
0.025	0.738	0.679	0.634	0.596	0.779	0.757	0.744	0.734

TABLE 6

Calculated values of the thermal neutron flux perturbation factor
for gold foils in water and graphite

Thickness (cm)	WATER Radius (cm)				GRAPHITE Radius (cm)			
	0.25	0.50	0.75	1.0	0.25	0.50	0.75	1.0
0.0005	0.990	0.988	0.987	0.985	0.991	0.991	0.990	0.990
0.001	0.982	0.978	0.975	0.973	0.984	0.983	0.982	0.982
0.003	0.955	0.944	0.936	0.930	0.962	0.958	0.957	0.955
0.005	0.933	0.915	0.903	0.893	0.943	0.938	0.935	0.932
0.0075	0.908	0.884	0.867	0.853	0.923	0.916	0.911	0.908
0.01	0.887	0.856	0.834	0.817	0.905	0.896	0.890	0.886
0.015	0.849	0.809	0.780	0.755	0.874	0.861	0.853	0.846
0.02	0.817	0.769	0.734	0.704	0.847	0.830	0.820	0.812
0.025	0.788	0.734	0.694	0.660	0.824	0.803	0.791	0.782

NONDESTRUCTIVE DETERMINATION OF

REACTOR PRESSURE VESSEL NEUTRON EXPOSURE

BY CONTINUOUS GAMMA-RAY SPECTROMETRY

Raymond Gold, William N. McElroy, Bruce J. Kaiser*
and James P. McNeece

Westinghouse Hanford Company
Hanford Engineering Development Laboratory
Richland, Washington 99352

ABSTRACT

A nondestructive method for determinating reactor pressure vessel (RPV) neutron exposure is advanced. It is based on the observation of characteristic gamma-rays emitted by activation products in the RPV with a unique continuous gamma-ray spectrometer. This spectrometer views the RPV through appropriate collimators to determine the absolute emission rate of these characteristic gamma-rays, thereby ascertaining the absolute activity of given activation products in the RPV. These data can then be used to deduce the spatial and angular dependence of neutron exposure at regions of interest in the RPV. In addition, this method can be used to determine the concentrations of different constituents in the RPV by measuring the absolute flux of characteristic gamma-rays from radioactivity induced in these constituents through neutron exposure. Since copper concentration may be a crucial variable in radiation-induced embrittlement of RPVs, the ability of this method to measure copper concentrations in base metal and weldments is examined.

---

*General Electric Company, P.O. Box 780, Wilmington, NC.

J. P. Genthon and H. Röttger (eds.), Reactor Dosimetry, 345–356.
© 1985 ECSC, EEC, EAEC, Brussels and Luxembourg

INTRODUCTION

Neutron-induced radiation damage experienced by the pressure vessel of a power reactor can be a controlling factor in defining the effective life of plant operation. As a consequence, methods of quantifying the neutron exposure fluence of reactor pressure vessels are of worldwide interest. Therefore, a new nondestructive method of reactor pressure vessel (RPV) neutron dosimetry based on the observaton of characteristic gamma-rays emitted by activation products in the RPV with a unique Si(Li) Compton continuous gamma-ray spectrometer is advanced.

The ability to measure complex gamma-ray continua in reactor environments through Compton recoil gamma-ray spectrometry is well established (1-3). On this basis, the general applicability of continuous gamma-ray spectrometry for neutron dosimetry has already been described (4). This method is based upon the complementarity of the components of a mixed radiation field (5). Neutron and gamma-ray components possess a strong interrelationship, particularly for mixed radiation fields in reactor environments. This interrelationship is manifested through the existence of intense gamma-ray peaks that lie above the gamma continuum at characteristic and identifiable gamma-ray energies.

Actually, in-situ continua in reactor environments possess many peaks observed above the general level of the continuum. Furthermore, each of these peaks can be analyzed separately to determine absolute activity concentrations within the pressure vessel. Since these different peaks arise from neutron reactions with the constituent isotopes of the pressure vessel, a potential to produce considerably more information exists. For example, peaks in Si(Li)-observed gamma continua arise from different neutron reaction cross sections so that absolute Si(Li) gamma-ray data can be used the same way radiometric dosimetry data is analyzed with unfolding or least-squares adjustment codes to infer neutron energy spectral information. On the other hand, some of these peaks can be analyzed to determine the concentration levels of different RPV constituents. Of particular interest are those constituents which may play a significant role in the neutron induced embrittlement of RPVs, such as copper.

REACTOR PRESSURE VESSEL NEUTRON DOSIMETRY

Recent work for Three Mile Island Unit 2 (TMI-2) reactor recovery (6,7) has demonstrated that this unique Si(Li) Compton gamma-ray spectrometer can be operated in very intense gamma fields. In fact, these efforts demonstrated that fields of up to roughly 2000 R/h could be accommodated with shielded collimators of appropriate design. In RPV neutron dosimetry, measurements

346

could be conducted on both sides of the RPV, depending on accessibility. For example, on the core side of the RPV, the shielded Compton spectrometer could be placed in a corner fuel assembly location to measure the maximum exposure experienced by the RPV. Measurements on the other side, i.e., in the reactor cavity, would have the advantage of reduced background. Owing to count rate limitations, measurements on the core side of the RPV would have to be carried out with the reactor shutdown. In the reactor cavity, however, measurements may be possible at low reactor power depending on the collimator size that can be used within the spatial constraints of the cavity.

The general configuration for such reactor cavity measurements is shown in Figure 1. Here the Si(Li) Compton spectrometer views the RPV through a collimator shield, which possesses a gap of diameter d_1 and a length x_1. The absolute flux intensity of a characteristic gamma-ray observed at energy ε_0, $I_1(\varepsilon_0)$, is given by

$$I_1(\varepsilon_0) = \int_0^T A(s)e^{-\mu(\varepsilon_0)s}\Omega(s)ds \quad . \tag{1}$$

where $A(s)$ is the absolute activity per unit volume, including appropriate branching ratios, at a depth s in the RPV. The depth variable s is measured from the outer surface of the RPV as shown in Figure 1.

At a depth s, $\Omega(s)$ is the solid angle projected through the collimator and $\mu(\varepsilon_0)$ is the attenuation coefficient of the RPV for gamma-rays of energy ε_0. The solid angle $\Omega(s)$ is given by

$$\Omega(s) = \frac{1}{4\pi}\int \frac{dA}{r^2} = \frac{1}{4\pi}\int_0^{\alpha_1} rd\theta \int_0^{2\pi} \frac{r\sin\theta d\phi}{r^2},$$

where $\alpha_1 = \tan^{-1}(d/2x_1)$ is the half angle of the collimator, so that

$$\Omega(s) = \frac{1 - \cos \alpha_1}{2} \tag{2}$$

Since $\Omega(s)$ is the independent of s and is a function of only the collimator property α_1, it can be identified as $\Omega(\alpha_1)$. Use of the geometric solid angle is an assumption explored in greater detail in a companion paper presented at this symposium (8), which describes the very first application of this method in the BR-3 reactor at the CEN/SCK laboratory in Mol, Belgium.

The spatial dependence of the activity density $A(s)$ has been shown to possess exponential behavior (9), so that to a reasonable approximation one can write

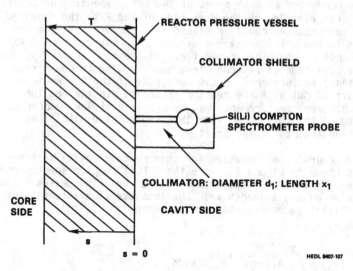

FIGURE 1. Overhead View of Reactor Cavity Measurements with
Continuous Gamma-Ray Spectrometer.

$$A(s) = Ce^{\lambda s} \quad . \tag{3}$$

where C is a constant, i.e., C = A(o), that represents the surface
activity density and λ is the neutron attenuation coefficient of
the RPV. Using Equations (2) and (3) in Equation (1), one finds
the quadrature result

$$I_1(\epsilon_0) = \frac{C\Omega(\alpha_1)}{(\mu-\lambda)} \left[1 - e^{-(\mu-\lambda)T} \right] \quad . \tag{4}$$

Using numerical estimates in Equation (4), one can show that
the exponential term is negligible for many applications, in which
case Equation (4) reduces simply to

$$I_1(\epsilon_0) = \frac{C\Omega(\alpha_1)}{\mu-\lambda} \quad . \tag{5}$$

The constants C and λ of the parametric representation of
activity density can be determined from Equations (4) or (5) in a
number of ways. The neutron attenuation coefficient λ can be
measured in separate RPV benchmark field calibration experiments,
such as the light water reactor-pressure vessel (LWR-PV) mockups
studied in the pool critical assembly (PCA) (9). Using this value
of λ, Equation (4) or (5) can be solved directly for C. On the
other hand, both parameters, C and λ, can be regarded as unknown,
in which event an additional measurement is required.

Consider, therefore, a second measurement with a different collimator of solid angle $\Omega(\alpha_2)$, which makes an angle θ with respect to the normal of the RPV surface. Using the above results, it can be shown for this case that

$$I_2(e_0) = \frac{C\Omega(\alpha_2)}{(\mu-\lambda_2)} \left[1 - e^{-(\mu-\lambda_2)T_2} \right] , \tag{6}$$

where $\lambda_2 = \lambda \cos\theta$, $\qquad$ (7a)

and $\quad T_2 = T/\cos\theta$. $\qquad$ (7b)

Since $\lambda_2 \leq \lambda$, the exponential term can be neglected for many applications so that

$$I_2(\epsilon_0) = \frac{C\Omega(\alpha_2)}{\mu-\lambda_2} . \tag{8}$$

Taking the ratio of Equation (5) by Equation (8), one can write

$$\frac{\mu-\lambda_2}{\mu-\lambda} = \delta , \tag{9a}$$

where $\delta = \left[\frac{I_1(\epsilon_0)}{I_2(\epsilon_0)}\right] \left[\frac{\Omega(\alpha_2)}{\Omega(\alpha_1)}\right] .$ $\qquad$ (9b)

The constant δ can be determined in terms of the Si(Li) gamma-ray spectrometer results, $I_1(\epsilon_0)$ and $I_2(\epsilon_0)$, obtained with the two different collimators. Consequently, use of Equation (7a) in Equation (9a) provides a relation that can be solved for λ. One finds

$$\lambda = \mu \left(\frac{\delta - 1}{\delta - \cos\theta} \right) . \tag{10}$$

This value of λ can then be used in Equation (5) or Equation (8) to determination C. The more general result, which follows from Equations (4) and (6), is

$$\left(\frac{\mu-\lambda_2}{\mu-\lambda}\right) \left[\frac{1 - e^{-(\mu-\lambda)T}}{1 - e^{-(\mu-\lambda_2)T_2}} \right] = \delta , \tag{11}$$

where δ is again the constant given in Equation (9b). Equation (11) is a transcendental relation that can be solved for λ iteratively. In fact, the iterative process would start with the approximate solution given by Equation (10). Having determined λ iteratively, Equations (4) or (6) can be used to find C.

An additional point that must be stressed is the advantage of reduced background that arises for measurements conducted at an angle θ with respect to the normal to the RPV surface. Here the angle θ can be chosen so that the collimated spectrometer no longer directly views leakage radiation from the core that penetrates through the RPV. Consequently, measurements can be carried out with two different collimators which make angles θ_1 and θ_2, respectively, with respect to the normal to the RPV surface. Under these conditions:

$$I_1(\epsilon_0) = \frac{C\tilde{u}(\alpha_1)}{\mu - \lambda_1}\left[1 - e^{-(\mu - \lambda_1)T_1}\right], \tag{12.}$$

$$\text{and} \quad I_2(\epsilon_0) = \frac{C\tilde{u}(\alpha_2)}{\mu - \lambda_2}\left[1 - e^{-(\mu - \lambda_2)T_2}\right], \tag{13}$$

Using these results, Equations (10) and (11) generalize to

$$\lambda = \mu\left(\frac{\delta - 1}{\delta \cos \theta_1 - \cos \theta_2}\right), \tag{14}$$

$$\text{and} \quad \left(\frac{\mu - \lambda_2}{\mu - \lambda_1}\right)\left[\frac{1 - e^{-(\mu - \lambda_1)T_1}}{1 - e^{-(\mu - \lambda_2)T_2}}\right] = \delta, \tag{15}$$

respectively. Here δ is again given by Equation (9b) and

$$\lambda_1 = \lambda \cos \theta_1$$

$$\lambda_2 = \lambda \cos \theta_2$$

$$T_1 = T/\cos \theta_1 \tag{16}$$

$$T_2 = T/\cos \theta_2$$

One can easily show that Equations (14) and (15) obey the correct limiting condition for $\theta_1 \rightarrow 0$, reducing to Equations (10) and (11), respectively. As before, the solution of λ given by Equation (14) can be used in the approximations and obtained from Equations (12) or (13); i.e., when the exponential term is neglected in these equations, to provide C. In an analogous manner, the more general result can be obtained by using the iterative solution of λ, found from Equation (15), in either Equation (12) or (13) to provide C.

It must be noted that limitations on accessibility do exist for the collimated Si(Li) spectrometer. For certain reactor designs, the reactor cavity is too small to permit insertion of the collimated spectrometer. On the core side, the thermal shield, pad, or barrel may lie between the collimated spectrometer and the RPV. In this case, the method is actually applied to the specific configuration viewed by the collimator. Often the collimated Si(Li) spectrometer can be inserted into reactor instrument tubes to allow a view of the RPV. In fact, for the very first application of this method in the BR-3 reactor (8), the Si(Li) spectrometer was located in an instrument tube. The advantage of viewing the bare RPV surface lies in the direct quantification of activity within the RPV, so that neutron dosimetry for the RPV can be performed without the need for extrapolation.

In addition to being nondestructive, this method possesses the advantage of providing exposure fluence data at very localized regions of the RPV. As a consequence, the exposure fluence of RPV welds can be mapped as a function of position.

MEASUREMENT OF RPV CONSTITUENT CONCENTRATIONS

The concentration of copper is a crucial variable governing radiation-induced embrittlement of RPV steels (10). Hence, copper concentration is a critical factor in end-of-life determinations for nuclear power RPVs. Copper concentration is not only important in RPV base metals, but is of particular significant in PV weldments. Consequencly, copper concentration is used to illustrate this method of measuring RPV constituent concentrations.

Owing to constraints that arise from the gamma-ray field intensity and limited spatial access, it will be assumed that measurements must be conducted with the reactor shutdown. To determine such RPV copper concentrations, measurements would have to begin soon after power reactor shutdown. Two radionuclides are produced by neutron capture on natural copper, namely copper-64 and copper-66. While the short half-life of copper-66, only 5.1 min., makes this radionuclide impractical to use in this application, copper-64 possesses a 12.7-h half-life and consequently can be used for RPV observations. With advance preparations made inside reactor containment, it should take only a few hours after shutdown to set up the collimated Si(Li) Compton spectrometer for measurement of RPV gamma spectra.

Two candidate gamma-rays in the copper-64 decay, namely the 1.346-MeV transition from the low intensity electron capture branch (0.6%) and annihilation radiation at 0.511 MeV from the positron decay branch (19%). The analysis given above for peak intensities above the general level of the gamma continuum is applicable for these two gamma-rays from copper-64. In this analysis, the absolute activity per unit volume A(s), at a given depth s in the RPV, is quantified in the exponential form given in Equation (3), where C and λ are determined by the measurements.

Gamma-ray peaks due to the decay of iron-59 will exist in the very same spectral measurements. The iron-59 radionuclide (45.5-day half-life) is produced by neutron capture on natural iron, wherein iron-58 exists at a level of 0.3%. Two candidate peaks from iron-59 exist, namely the transition at 1.292 MeV (45%) and the transition at 1.099-MeV (53%). Again following the same analysis given above, the absolute iron-59 activity per unit volume A(s) can be quantified. Consequently, copper-64 activity per unit volume can be written

$$A_1(s) = C_1 e^{\lambda s} \quad , \tag{17}$$

and for the iron-59 activity per unit volume

$$A_2(s) = C_2 e^{\lambda s} \quad , \tag{18}$$

The copper-64 and iron-59 activities per unit volume at a depth s can be simply expressed in terms of the thermal neutron flux $[n/cm^2 \cdot s)]$, $\phi_{th}(s)$, at depth s:

$$A_1(s) = \phi_{th}(s) \cdot \sigma_1 \cdot \rho_1 \cdot e^{-\lambda_1 t_d} \cdot \left(1 - e^{-\lambda_1 t_x}\right), \tag{19}$$

and

$$A_2(s) = \phi_{th}(s) \cdot \sigma_2 \cdot \rho_2 \cdot e^{-\lambda_2 t_d} \cdot \left(1 - e^{-\lambda_2 t_x}\right). \tag{20}$$

where:

λ_1 is the copper-64 decay constant

λ_2 is the iron-59 decay constant

σ_1 is the copper-63 thermal neutron capture cross section

σ_2 is the iron-58 thermal neutron capture cross section

ρ_1 is the copper-63 concentration (at./cm^3)

ρ_2 is the iron-58 concentration (at./cm^3)

t_x is the duration time of the irradiation

t_d is the elapsed time since reactor shutdown

All these parameters are known except for the copper-63 concentration ρ_1. Hence, taking the ratio of Equation (19) to Equation (20), one can write

$$\rho_1/\rho_2 = K \quad , \tag{21}$$

where K is expressed in terms of known parameters as

$$K = \frac{A_1(s)e^{\lambda_1 t_d}\left(1-e^{-\lambda_2 t_x}\right)\sigma_2}{A_2(s)e^{\lambda_2 t_d}\left(1-e^{-\lambda_1 t_x}\right)\sigma_1} . \tag{22}$$

Thus, the copper concentration can be simply obtained from the ρ_1/ρ_2 ratio by using the known percent abundances of copper-63 and iron-58 in natural copper and iron, respectively.

The copper concentration of the base metal can often be determined from archive RPV specimens, so that only the copper concentration of RPV weldments is desired for certain power reactors. In this case, the analysis given above can be used to show that only relative gamma spectra observations are necessary between RPV base metal and RPV weldments. Consequently, one can write

$$\left(\rho_1/\rho_2\right)_{weld} = \frac{\left(I_1/I_2\right)_{weld}}{\left(I_1/I_2\right)_{base\ metal}} \cdot \left(\rho_1/\rho_2\right)_{base\ metal}, \tag{23}$$

where I_1 and I_2 are the observed peak intensities of the copper-64 and iron-59 gamma-rays, respectively, so that (I_1/I_2) weld is the ratio of these intensities obtained observing the weld, whereas (I_1/I_2) base metal is the ratio of these intensities obtained observing the base metal.

Hence, if the copper concentration is known for the base metal, then only relative Si(Li) gamma spectra observations are needed between the base metal and weldment to determine the copper concentration of the weldment.

Interferences and background can arise in RPV gamma spectra observed with the collimated Si(Li) Compton spectrometer that could make the detection of copper-64 very difficult. Additional radionuclides are produced that possess gamma-ray transitions close to the gamma-ray energies emitted by either copper-64 or iron-59. For example, cobalt-58 can be produced by an (n,p) reaction on nickel-58. Since cobalt-58 is a positron emitter, annihilation radiation would be produced at 0.511 MeV from the cobalt-58 decay just as it is produced in the decay of copper-64. However, the $^{58}Ni(n,p)^{58}Co$ reaction cross section is very small relative to the $^{63}Cu(n,\gamma)$ cross section, and the cobalt-58 half-life is 70.8 days, which is considerably longer than the 12.7-h half-life of copper-64. Hence, the background annihilation component from cobalt-58 will be small relative to the copper-64 annihilation gamma peak.

In general, the time-dependent decay of the different radio-nuclides contributing to a given gamma peak can be used to separate signal from background. For example, that component of peak intensity at the annihilation energy possessing a 12.7-h half-life can be determined by measuring time-dependent RPV gamma spectra. Sequential gamma spectra measurements over a time period of a few days should serve to isolate the 12.7-h decay component uniquely attributed to copper-64.

Another example of background is the production of cobalt-60 in the RPV by neutron capture on trace concentrations of natural cobalt, i.e., cobalt-59. The cobalt-60 decay possesses gamma-ray transitions at 1.173 MeV (100%) and 1.332 MeV (100%). Since the energy resolution of the Si(Li) spectrometer is about 30 keV (FWHM), the 1.332-meV gamma-ray from cobalt-60 would interfere with the 1.346-MeV transition of copper-64, and to a lesser extent with the 1.292-MeV transition of iron-59. Fortunately, cobalt-60 has a half-life of 5.27 yr so that time-dependent measurements can be used to separate signal from background, should the need arise.

Another general method for isolating background contributions exists based on the observation of additional peaks in the gamma spectrum that are emitted by the very same background producing radionuclide. If such a peak can be identified, then the absolute activity of the background radionuclide can be quantified. Knowledge of the decay scheme of this background radionuclide together with the absolute activity of the background radionuclide provide the means to determine the background contribution to the peak intensity in question. For example, cobalt-58 possesses a gamma transition at 0.8108 MeV (99%). Consequently, if the cobalt-58 decay is contributing to the annihilation peak at 0.511 MeV, then a peak in the gamma spectrum should be observed at 0.8108 MeV. Hence, observation of this peak at 0.8108 MeV with the collimated Si(Li) spectrometer can be used to identify the absolute activity per unit volume of cobalt-58. This absolute cobalt-58 activity, together with a knowledge of the cobalt-58 decay scheme, will permit evaluation of the background component at the 0.511-MeV annihilation gamma-ray energy.

In spite of the much higher specific activity of copper-64, extracting copper-64 data from continuous gamma-ray spectra may still be difficult. In view of the importance attributed to copper in the radiation-induced embrittlement of RPVs, further investigation of this method is warranted. Realistic field tests should be conducted to evaluate the actual capabilities and limitations of this method. For some applications, particularly for measurements that might be made on the inside surface of a RPV, the differences between thermal and fast neutron-induced activations in Cu and Fe would have to be considered. Generally, this should involve only a small correction to the thermal neutron-induced events.

Advantages

1. This method is nondestructive.

2. Copper concentrations of base metal and weldments can be determined locally as a function of spatial position on the RPV surface. The exact location of a weld of interest need not be known, since the change in gamma spectra between the base metal and weldment can be used to locate the collimated spectrometer at the weld.

3. If the copper concentration of the base metal is known, then only relative gamma spectra measurements between base metal and weldment are needed to determine the copper concentration of the weldment.

4. Copper concentrations are determined without the need to quantify the thermal neutron exposure flux ϕ_{th}.

REFERENCES

1. R. Gold and B. J. Kaiser, "Reactor Gamma Spectrometry: Status," Proc. of the 3rd ASTM-EURATOM Symposium on Reactor Dosimetry, Ispra (Varese), Italy, October 1-5, 1979, Commission of the European Communities, EUR 6813, Vol. II, 1160, 1980 (invited paper).

2. R. Gold and B. J. Kaiser, "Gamma-Ray Spectrometry," LWR-PV-SDIP: PCA Experiments and Blind Test, NUREG/CR-1861, HEDL-TME 80-87, Section 5.3, NRC, Washington, DC (July 1981).

3. R. Gold, B. J. Kaiser and J. P. McNeece, "Gamma-Ray Spectrometry in Light Water Reactor Environments," Proc. of the 4th ASTM-EURATOM Symposium on Reactor Dosimetry, Gaithersburg, MD, March 22-26, 1982, NUREG/CP-0029, NRC, Washington, DC, July 1982 (invited paper).

4. R. Gold, "Estimates of High-Energy Gamma and Neutron Flux from Continuous Gamma-Ray Spectrometry," LWR-PV-SDIP Quarterly Progress Report - July-September 1983, NUREG/CR-0551, HEDL-TME 78-8, NRC, Washington DC (December 1978).

5. R. Gold, "Compton Recoil Gamma-Ray Spectroscopy," Nucl. Instr. Methods 84, 173 (1970).

6. J. P. McNeece, B. J. Kaiser, R. Gold and W. W. Jenkins, Fuel Content of the Three Mile Island Unit 2 Makeup Demineralizers, HEDL-7285, Hanford Engineering Development Laboratory, Richland, WA (February 1983).

7. R. Gold, J. H. Roberts, J. P. McNeece, B. J. Kaiser,
    F. H. Ruddy, C. C. Preston, J. A. Ulseth and W. N. McElroy,
    "Fuel Debris Assessment for TMI-2 Reactor Recovery by
    Gamma-Ray and Neutron Dosimetry," 6th International Conf.
    on Nondestructive Evaluation in the Nuclear Industry,
    Zurich, Federal Republic of Germany, November 27–
    December 2, 1983.

8. J. P. McNeece, R. Gold, A. Fabry, S. DeLeeuw and P. Gubel,
    "Nondestructive Measurement of Neutron Exposure in the BR-3
    Pressure Vessel by Continuous Gamma-Ray Spectrometry,"
    HEDL-SA-3124 and 5th International ASTM-EURATOM Symposium on
    Reactor Dosimetry, September 24-28, 1984, Geesthacht, Federal
    Republic of Germany.

9. W. N. McElroy, Ed., LWR-PV-SDIP: PCA Experiments and Blind
    Test, NUREG/CR-1861, HEDL-TME 80-87, 7.2-30, NRC,
    Washington, DC (July 1981).

10. W. N. McElroy, Ed., LWR-PV-SDIP - 1983 Annual Report,
    NUREG/CR-3391, Vol. 3, HEDL-TME 83-23, NRC, Washington, DC
    (June 1984).

ADVANCES IN CONTINUOUS GAMMA-RAY SPECTROMETRY

AND APPLICATIONS

Raymond Gold and James P McNeece

Westinghouse Hanford Company
Hanford Engineering Development Laboratory
Richland, WA USA

Bruce J. Kaiser
General Electric Company
P.O. Box 780
Wilmington, NC USA

ABSTRACT

Recent advances and applications in continuous Compton recoil gamma-ray spectrometry are described. Applications of continuous gamma-ray spectrometry are presented for:

1) Characterization of light water reactor (LWR) pressure vessel (PV) environments.

2) Assessment of fuel distributions for Three Mile Island Unit 2 (TMI-2) reactor recovery.

3) Measurement of LWR-PV-neutron exposure.

The latest improvements attained with the Janus probe, a special in-situ configuration of Si(Li) detectors, are presented. The status of current efforts to extend the domain of applicability of this method beyond 3 MeV is discussed with emphasis on recent work carried out with Si(Li) detectors of much larger volume.

J. P. Genthon and H. Röttger (eds.), Reactor Dosimetry, 357–371.
© 1985 ECSC, EEC, EAEC, Brussels and Luxembourg

INTRODUCTION

Since the inception of continuous Compton recoil gamma-ray spectrometry, (1-3) rather than being static, this method has evolved and improved. Earliest efforts were directed toward in-situ observation of gamma-ray continua in reactors (4,5). Almost simultaneously, the significance of this method for gamma-ray dosimetry was recognized (6). It was, therefore, not surprising that after these initial reactor experiments, applications arose in gamma-ray dosimetry, (7,8) health physics, (9) and environmental science (10-12). An environmental survey of the Experimental Breeder Reactor II (EBR-II) was conducted with these techniques (13). This method has been applied in reactor environments in Europe (14,15), and recognition of the general need for gamma-heating data (16) led to spectrometry measurements in fast breeder reactor environments. Compton recoil gamma-ray spectrometry was actually the first experiment performed in the Fast Flux Test Facility (FFTF) at startup (17). Efforts to characterize the gamma-ray field in light water reactor pressure vessel (LWR-PV) environments have already been reported at earlier meetings of this series of bi-annual symposia (18,19). Consequently, only the most recent applications and advances need be considered herein.

CHARACTERIZATION OF LWR-PV GAMMA-RAY FIELDS

To meet the needs of the LWR-PV Surveillance Dosimetry Improvement Program (SDIP), continuous gamma-ray spectrometry has been carried out in simulated LWR-PV environments. These in-situ observations provide gamma-ray spectra, dose, and heating rates needed to:

1) Benchmark industry-wide reactor physics computational tools, e.g., independently, the gamma-ray spectrometry measurements provide absolute data for comparison with calculations.

2) Assess radial, azimuthal, and axial contributions of gamma heating to the temperature attained within surveillance capsules, the PV wall, and other components of commercial LWR power reactors (20).

3) Design, control, and analyze high-power metallurgical irradiation tests.

4) Interpret fission neutron dosimetry in LWR-PV environments, where non-negligible photofission contributions can arise (21-24).

To meet these needs, gamma-ray spectrometry has been conducted in three LWR-PV-SDIP mockups, namely the Pool Critical

Assembly (PCA) in Oak Ridge National Laboratory (ORNL) (USA), VENUS in CEN/SCK (Belgium), and NESDIP in Atomic Energy Establishment Winfrith (AEEW) (UK). A significant outgrowth of these collaborative efforts was the recognition and subsequent quantification of the perturbation factor (PF) created by the Janus probe. It was conjectured that the PF arises from the void or semi-voided regions introduced by the Janus probe into the gamma-ray intensity gradient that exists in the PV block. Initial analysis of the 1981 work performed in the 4/12 SSC configuration at the PCA has already been presented and confirms the existence of such PF. Since the significance of this PF is now clearly established, recent followon PF measurements at NESDIP will be elaborated upon here.

Perturbation Factor (PF) Measurements at NESDIP

Two different gamma-ray dosimetry methods were used at NESDIP to measure Janus probe PFs, namely, ionization chambers (IC) and thermoluminescence dosimetry (TLD). Both techniques were implemented using a "dummy" Janus probe. Measurements are first carried out at a given location by incorporating the miniature IC or TLD in the "dummy" Janus probe. Measurements are then repeated at this location with the channel completely back-filled with appropriate material so as to eliminate voids. The PF is defined by the ratio

$$PF = \dot{D}_p / \dot{D}_u \tag{1}$$

where $\dot{D}_p$ is the perturbed dose rate observed in the presence of the "dummy" Janus probe and $\dot{D}_u$ is the unperturbed dose rate observed in the back-filled channel.

Special miniature ICs were developed at Hanford Engineering Development Laboratory (HEDL), specifically for PF measurements in the PV block, and these were employed in the earlier PF measurements conducted in the PCA (19). The PF measurements with TLD were carried out in collaboration with T. A. Lewis and colleagues of the Berkeley Nuclear Laboratories (UK), who used beryllium-oxide (BeO) TLD (25). The results of these PF measurements at NESDIP, which were performed only for the 12/13 configuration, are compared to the earlier PF observations obtained in the 4/12 SSC configuration at the PCA in Table 1.

Because of the limitations of the miniature IC design as well as the NESTOR power operation, IC measurements could be carried out only for the 1/4-T location of the 12/13 configuration at NESDIP. The gamma-ray intensity levels that could be attained at the 3/4-T and void box (VB) locations were too low to provide

TABLE 1

JANUS PROBE PERTURBATION FACTORS

Location	PCA - 4/12 SSC IC	NESDIP - 12/13 TLD	NESDIP - 12/13 IC
A2	--	1.12	--
1/4 T	1.16	1.30	1.27
1/2 T	1.14	1.24*	--
3/4 T	1.11	1.18	--
VB	--	0.90	--

*Since the 1/2-T location is not readily available
at NESDIP, this value was obtained by linear interpolation of the 1/4-T and 3/4-T results.

reliable readings. Moreover, it is well to note that the design
of these miniature ICs restricts applicability for PF measurements
to the PV block. In view of the restricted nature of the IC
results for the 12/13 configuration, the BeO TLD results, which
represent a consistent set of PF for the 12/13 configuration, are
recommended for use at this time. Nevertheless, it is important
to stress that the IC and TLD results agree within experimental
uncertainty at the 1/4-T location of the 12/13 configuration.

The PF results shown in Table 1 vary with both configuration
and location. In order to understand this behavior, it is instruc-
tive to examine the spatial dependence of dose rates within the PV
block. Figure 1 compares (uncorrected) finite-size dose rates for
the 4/12 SSC and 12/13 configurations. It is clear from Figure 1
that the 12/13 configuration gamma data possesses a larger gradient. In light of the results in Table 1, one finds that, when
the Janus probe is used in a field possessing a larger gradient,
the PFs are, in turn, larger.

This conclusion is also supported by the PF result for the A2
water position of the 12/13 configuration. This PF result, namely
1.12, is essentially as low as any result obtained within the PV
block for either the 12/13 or 4/12 SSC configuration. However, it
is well known that water is a rather poor attenuator of gamma
radiation compared with the iron medium of the PV. Hence, gamma-
ray intensity gradients at water locations are generally less than
those in the PV block, and the corresponding Janus probe PF is
indeed lower. Consequently, these overall PF results confirm the

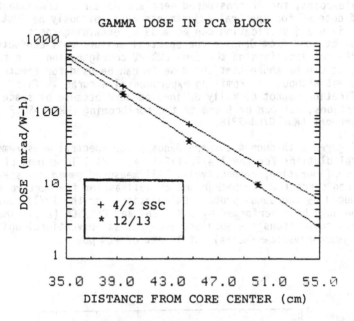

GAMMA DOSE IN PCA BLOCK

+ 4/2 SSC
* 12/13

FIGURE 1. Comparison of the Spatial Behavior of the Finite-Size
Dose Rate, $\dot{D}_{FS}$, for the 4/12 SSC and 12/13 Configurations. (The smooth lines are linear least-squares fits
of the logarithm of the experimental data.)

original conjecture that Janus probe PF stem from the introduction
of voids or semi-voided regions into a gamma field possessing an
intensity gradient.

The existence of a PF of less than unity for the VB location
of the 12/13 location can also be qualitatively explained.
Comparison of the Janus probe with a point detector for measurements in a void reveals that the probe must produce some attenuation of gamma radiation in the solid angle that the probe subtends
at the Si(Li) sensitive volume. Consequently in a void, one must
expect that the perturbed dose rate $\dot{D}_p$ would be less than the
unperturbed dose rate $\dot{D}_u$. Hence, an observed PF of less than
unity for the VB location of the 12/13 configuration is in accord
with very simple physical considerations.

Gamma-Ray Spectra and Dose Rates in the PCA

The PFs provided in Table 1 have been used to correct both gamma-ray spectra and dose rates measured with the Janus probe in LWR-PV configurations studied in the PCA. However, from a rigorous viewpoint, the PF considered here are dose PF. Consequently, use of dose PF for spectral adjustments must obviously be justified. Such a justification can be made by examining spectral ratios obtained from Janus probe spectral measurements conducted at different locations of the same LWR-PV configuration. On this basis, it can be shown that the dose PF can be used for spectral adjustment without compromising experimental accuracy. This justification cannot be fully delineated here because of space restrictions, but can be found in the forthcoming LWR-PV-SDIP NUREG report (NUREG/CR-3318).

Figures 2 through 4 compare Janus probe spectral measurements with calculations for the 1/4-T, 1/2-T, and 3/4-T locations of the 12/13 configuration, respectively. All measured gamma-ray spectra have been corrected for background as well as for the perturbation introduced by the Janus probe. Calculations for the 12/13 configuration have been performed by ORNL (26) and CEN/SCK (27). These spectral comparisons are absolute and possess conventional units, i.e., gamma-rays/(cm^2·MeV·s), at 1 watt of PCA power.

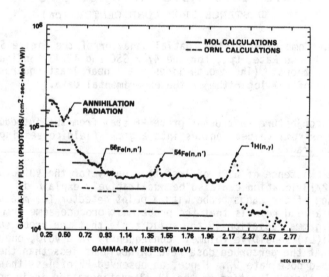

FIGURE 2. Low-Energy Gamma-Ray Continuum for the 1/4-T Location of the 12/13 Configuration as Compared with CEN/SCK and ORNL Calculational Results.

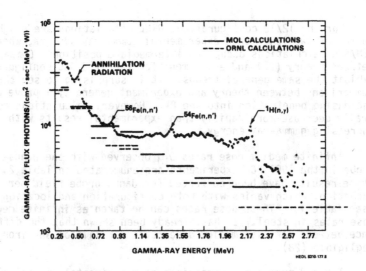

FIGURE 3. Low-Energy Gamma-Ray Continuum for the 1/2-T Location of the 12/13 Configuration as Compared with CEN/SCK and ORNL Calculational Results.

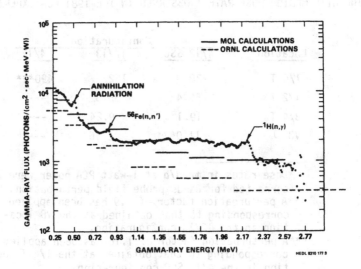

FIGURE 4. Low-Energy Gamma-Ray Continuum for the 3/4-T Location of the 12/13 Configuration as Compared with CEN/SCK and ORNL Calculational Results.

For the 12/13 configuration, ORNL calculations are roughly a factor of two lower than experimental gamma-ray spectra, whereas CEN/SCK calculations occupy an intermediate position. Comparisons between theory (27) and experiment for the 4/12 SSC configuration exhibit the same general trends. It is surprising to see that comparisons between theory and experiment generally improve with increasing penetration into the PV. However, calculations generally decrease more rapidly than experimental results with increasing gamma-ray energy.

Infinite medium dose rates $\dot{D}_{IM}$ observed with the Janus probe in the 1981 PCA experiments are enumerated in Table 2. These results have been corrected for Janus probe field perturbation, which varies with both configuration and location (see Table 1). These dose rates can be taken as infinite medium dose rates in steel. It has already been shown that the difference between infinite medium dose rates for silicon and iron is negligible (28).

Table 3 presents a comparison of experimental and calculated gamma-ray dose rates for the 4/12 SSC configuration. In addition to the $\dot{D}_{IM}$ results from the 1981 Janus probe experiments, this table presents results obtained by the CEN/SCK group (27,29), who performed both TLD measurements and calculations.

TABLE 2

INFINITE MEDIUM DOSE RATES* OBSERVED IN THE 1981 PCA EXPERIMENTS

| | Configuration | | |
Location	4/12 SSC	12/13	4/12
1/4 T	220	152	490***
1/2 T	65.4	35.6	---
3/4 T	19.1	9.24	---
VB	11.0**	2.56	---

*Dose rates in mrad/h at 1-watt PCA power were corrected for Janus probe field perturbation.
**A perturbation factor of 0.9 has been applied corresponding to that obtained at the VB location in the 12/13 configuration.
***A perturbation factor of 1.16 has been applied corresponding to that obtained at the 1/4-T location in the 4/12 SCC configuration.

TABLE 3

GAMMA-RAY DOSE RATES* FOR THE 4/12 SSC CONFIGURATION

| Location | Experiment | | Calculation | | |
	TLD	$\dot{D}_{IM}$	CEN/SCK	$\dot{D}_{IM}$/TLD	$\dot{D}_{IM}$/CAL
1/4 T	255	220	210	0.86	1.05
1/2 T	68	65.4	52	0.96	1.26
3/4 T	21.5	19.1	19.1	0.89	1.00
VB	11.5	11.0	2.2	0.96	5.05

*Dose rates in mrad/h at 1-watt PCA power.

Using the results from all four locations of the 4/12 SSC configuration given in Table 3, one finds a $\dot{D}_{IM}$/TLD average ratio of 0.92. Consequently, the Si(Li) and TLD methods agree within experimental uncertainty. Comparison of these experimental results with calculations does not show consistent agreement. The extremely low calculational result at the VB location might be due to inadequate modeling of the actual geometric configuration used in the PCA.

RECENT DEVELOPMENTS

LWR-PV Neutron Dosimetry

The possibility that neutron dosimetry could be performed by appropriate analysis of gamma-ray continua was advanced some time ago (23). It is based on the existence of intense gamma-ray peaks that lie above the continuum at characteristic and identifiable gamma-ray energies. The first important realization of this concept came from the recognition that this method could be applied for LWR-PV neutron dosimetry. Further details on this application can be found in two companion papers presented at this symposium, the first of which provides a general exposition of the method (30) and the second describes results obtained from the very first field test of this method at BR-3 (31).

TMI-2 Fuel Distribution Assessments

Recent data at TMI-2 indicates that the void in the core geometry constitutes about 26% of the volume, and much of the displaced fuel appears to have been reduced to rubble. It is possible that significant amounts of this fuel debris have been transported out of the core into off-normal locations. For planning and recovery operations, it is important to identify the location of the fuel debris.

TMI-2 fuel distribution assessments have been carried out nondestructively with two highly specialized dosimetry methods, namely solid state track recorder (SSTR) neutron dosimetry and continuous gamma-ray spectrometry with a Si(Li) Compton spectrometer. In neutron dosimetry, neutrons generated from a combination of spontaneous fission, (α,n) reactions, and subcritical multiplication are measured. In gamma-ray spectrometry, the absolute flux of gamma-rays emitted by specific fission products is measured. The overall status of these TMI-2 efforts is described in a companion paper presented at this symposium (32).

Assessment of the fuel debris content in TMI-2 demineralizer A with continuous gamma-ray spectrometry is particularly worthy of note. To our knowledge, gamma-ray spectrometry has never before been conducted under such adverse conditions, in which the general radiation field intensity exceeded 2000 R/h. The Si(Li) gamma-ray spectrometry result for TMI-2 demineralizer A of 1.3 + 0.6 kg is in excellent agreement with the SSTR result of 1.7 + 0.6 kg.

The complementary nature of these two independent methods must be stressed. Indeed, source spatial distribution data obtained with the collimated Si(Li) spectrometer were used to guide the SSTR calibration experiments. On the other hand, SSTR evidence from the vertical stringer data implied that the demineralizer A tank was dry. This information, in turn, provided useful guidance for followon gamma-ray calibration experiments used to analyze the Si(Li) spectrometer data. Hence, these two independent nondestructive dosimetry methods provided concordant and complementary results. Some six months later, samples taken from the demineralizer A tank substantiated the conclusion that this tank was dry.

High Energy Measurements at the Little-Boy Replica

The US Department of Energy (DOE) has established a program for the replication of Little-Boy, the atomic bomb exploded over Hiroshima. This program is motivated by the crucial role in setting radiation protection standards that is played by the medical data base of A-bomb survivors at Hiroshima and Nagasaki. A key element in the generation of such radiation protection standards is the absolute accuracy of the exposure dose experienced by A-bomb survivors. To resolve discrepancies concerning these exposure doses, (33,34) a replica of Little-Boy has been constructed at the Los Alamos National Laboratory (LANL).

A Janus probe was constructed specifically for continuous gamma-ray spectrometry at Little-Boy using Si(Li) detectors of much larger volume. To extend high energy capabilities, this probe incorporated two Si(Li) detectors of 3 cm^3 volume each, as opposed to the 1 cm^3 volume of each of the two Si(Li) detectors used in the earlier versions of the Janus probe (18,19). Gamma-ray spectrometry was carried out at the Little-Boy replica with this special spectrometer in late July of 1984.

Although these data are still undergoing analysis, the raw electron spectra obtained with this special Janus probe show an enormous enhancement in sensitivity, especially at high energy. Figure 5a displays a typical electron spectrum observed at Little-Boy. The pronounced double escape peak at about 6.6 MeV in this electron spectrum is due to the intense neutron capture gamma-ray doublet in iron at ~7.64 MeV. The high energy portion of the electron spectrum is displayed in Figure 5b using an expanded scale to reveal more detail. Of lower intensity, but still evident in this spectrum, is a double escape peak at about 8.3 MeV, which in this case is due to a 9.3-MeV neutron capture gamma-ray produced again in iron. These results hold much promise for the extension of continuous gamma-ray spectrometry to considerably higher energy.

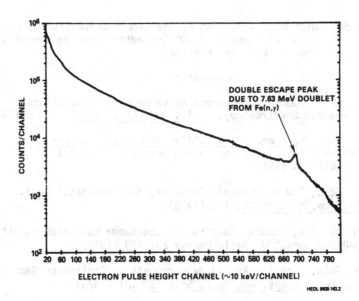

FIGURE 5a. Electron Spectrum Observed with the Janus Probe at a Polar Angle of 30° One Meter from the Center of the "Little-Boy" Replica.

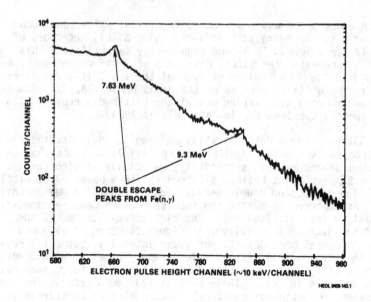

FIGURE 5b. High Energy Portion of the Electron Spectrum Observed at a Polar Angel of 30° One Meter from the Center of the "Little-Boy" Replica.

REFERENCES

1. R. Gold, "Compton Continuum Measurements for Continuous Gamma-Ray Spectroscopy," Bull. Am. Phys. Soc. 13, 1405 (1968).

2. M. G. Silk, Iterative Unfolding of Compton Spectra, AERE-R-5653, Atomic Energy Research Establishment, Harwell, UK (1968).

3. R. Gold, "Compton Recoil Gamma-Ray Spectroscopy," Nucl. Instrum. Methods 84, 173 (1970).

4. M. G. Silk, "Energy Spectrum of the Gamma Radiation in the DAPHINE Core," J. Nucl. Energy 23, 308 (1969).

5. R. Gold, "Compton Recoil Measurements of Continuous Gamma-Ray Spectra," Trans. Am. Nucl. Soc. 13, 421 (1970).

6. R. Gold and I. K. Olson, Analysis of Compton Continuum Measurements, ANL-7611, Argonne National Laboratory, Argonne, IL (1970).

7. R. Gold and A. M. Strash, "Gamma-Ray Dosimetry for the Natural Environment," Trans. Am. Nucl. Soc. 13, 502 (1970).

8. A. M. Strash and R. Gold, "Absolute Gamma-Ray Dosimetry by Recoil Electron Spectroscopy," Nature 234, 260 (1971).

9. R. Gold, "Gamma-Continuum at the Air-Land Interface," Health Phys. 21, 79 (1971).

10. R. Gold and A. M. Strash, "Pre-Operational Measurement of Environmental Gamma-rays at the Monticello Site with a Continuum Spectrometer," Trans. Am. Nucl. Soc. 14, 31 (1971).

11. R. Gold, F. J. Congel, J. H. Roberts and A. M. Strash, "Gamma Continuum at a Fossile-Fuel Plant Site," Trans. Am. Nucl. Soc. 15, 639 (1972).

12. R. Gold, A. M. Strash, F. J. Congel and J. H. Roberts, "Continuous Gamma-Ray Spectroscopy in the Natural Environment," Trans. IEEE NS-20, (1), 48 (1973).

13. R. Gold, B. G. Oltman, K. F. Eckerman and A. M. Strash, "Environmental Radiation at the EBR-II Site," Trans. IEEE NS-21, (1), 596-601 (February 1974).

14. H. E. Korn, Measurement of the Energy Distribution of the Gamma Field in a Fast Reactor, EUR-FNR-1316, Karlsruhe Nuclear Research Center, Karlsruhe, Federal Republic of Germany (1975).

15. S. H. Jiang and H. Werle, "Fission Neutron-Induced Gamma Fields in Iron," Nucl. Sci. Eng. 66, 354 (1978).

16. R. Gold, "Overview of Gamma-Ray Energy Deposition and Spectra in Fast Reactor Environments," Proc. of the 2nd ASTM-EURATOM Symposium on Reactor Dosimetry, Palo Alto, CA, October 3-7, 1977, NUREG/CP-0004, NRC, Washington, DC, Vol. 1, 101-140 (1978).

17. R. Gold, R. B. Kaiser, F. S. Moore Jr, W. L. Bunch, W. N. McElroy and E. M. Sheen, "Continuous Gamma-Ray Spectrometry in the Fast Flux Test Facility," HEDL-SA-2166 and Proc. of the ANS Topical Meeting on 1980 Advances in Reactor Physics and Shielding, Sun Valley, ID, September 14-17, 1980, ISDN 0-89448-107-X, 803 (1980).

18. R. Gold and B. J. Kaiser, "Reactor Gamma-Ray Spectrometry: Status," Proc. of the 3rd ASTM-EURATOM Symposium on Reactor Dosimetry, Ispra, Italy, October 1-5, 1979, EUR 6813, Commission of the European Communities, Vol. II, 1160-1171 (1980).

19. R. Gold, B. J. Kaiser and J. P. McNeece, "Gamma-Ray Spectrometry in LWR Environments," Proc. of the 4th ASTM-EURATOM Symposium on Reactor Dosimetry, Gaithersburg, MD, March 22-26, 1982, NUREG/CP-0029, NRC, Washington, DC, Vol. 1, 267-279 (July 1982).

20. P. N. Randall, "Status of Regulatory Demands in the US on the Application of PV Dosimetry," Proc. of the 4th ASTM-EURATOM Symposium on Reactor Dosimetry, Gaithersburg, MD, March 22-26, 1982, NUREG/CP-0029, NRC, Washington, DC, Vol. 2, 1011-1022 (July 1982).

21. C. D. Bowman, C. M. Eisenhauer and D. M. Gilliam, "Photofission Effects in Reactor PV Dosimetry," Proc. of the 2nd ASTM-EURATOM Symposium on Reactor Dosimetry, Palo Alto, CA, October 3-7, 1977, NUREG/CP-0004, NRC, Washington, DC, Vol. 2, 575-582 (1978).

22. G. L. Simmons, V. V. Verbinski, W. K. Hagan and C. G. Cassapekis, Measurement and Analysis of Gamma-Ray Induced Contamination of Neutron Dosimetry Procedures Used for Reactor PV Applications, EPRI NP-1056, Electric Power Research Institute, Palo Alto, CA (1979).

23. R. Gold, "Estimates of High-Energy Gamma and Neutron Flux from Continuous Gamma-Ray Spectrometry," LWR-PV-SDIP Quarterly Progress Report - July-September 1978, NUREG/CR-0551, HEDL-TME 78-8, NRC, Washington, DC (1979).

24. V. V. Verbinski, C. G. Cassapakis, W. K. Hagen and G. L. Simmons, "Photointerference Corrections in Neutron Dosimetry for Reactor PV Lifetime Studies," Nucl. Sci. & Eng. 75, 159 (1980).

25. T. A. Lewis, "Use of Beryllium-Oxide Thermoluminescence Dosimeters for Measuring Gamma Exposure Rates in Graphite Moderated Reactors," TPRD/B/0349/N83 (October 1983).

26. R. E. Maerker et al., "Neutron and Gamma Transport Results for 12/13 Configurations," PCA Dosimetry in Support of the PSF Physics-Dosimetry-Metallurgy Experiments, NUREG/CR-3318, HEDL-TME 84-1, NRC, Washington, DC, January 1984.

27. N. Maene, R. Menil, G. Minsart and L. Ghoos, "Gamma Dosimetry and Calculations," Proc. of the 4th ASTM-EURATOM Symposium on Reactor Dosimetry, Gaithersburg, MD, March 22-26, 1982, NUREG/CP-0029, NRC, Washington, DC, Vol. 1, 355-363 (July 1982).

28. B. J. Kaiser, R. Gold and J. P. McNeece, "Si(Li) Gamma-Ray Dosimetry," LWR-PV-SDIP: PCA Experiments and Blind Test, NUREG/CR-1861, HEDL-TME 80-87, NRC, Washington, DC, pp. 5.3-1 - 5.3-8 (July 1981).

29. A. Fabry, N. Maene, R. Menil, G. Minsart and J. Tissot, "Validation of Gamma-Ray Field Calculations by TLD Measurements," LWR-PV-SDIP: PCA Experiments and Blind Test, NUREG/CR-1861, HEDL-TME 80-87, Sec. 5.4, NRC, Washington, DC, (July 1981).

30. R. Gold, W. N. McElroy, B. J. Kaiser and J. P. McNeece, "Non-Destructive Determination of Reactor PV Neutron Exposure by Continuous Gamma-Ray Spectrometry," HEDL-SA-3064 and 5th International ASTM-EURATOM Symposium on Reactor Dosimetry, September 24-28, 1984, Geesthacht, Federal Republic of Germany.

31. J. P. McNeece, R. Gold, A. Fabry, S. DeLeeuw and P. Gubel, "Nondestructive Measurement of Neutron Exposure in the BR-3 Pressure Vessel by Continuous Gamma-Ray Spectrometry," HEDL-SA-3124 and 5th International ASTM-EURATOM Symposium on Reactor Dosimetry, September 24-28, 1984, Geesthacht, Federal Republic of Germany.

32. R. Gold, J. H. Roberts, F. H. Ruddy, C. C. Preston, J. P. McNeece, B. J. Kaiser and W. N. McElroy, "Characterization of Fuel Distributions in the TMI-2 Reactor System by Neutron and Gamma-Ray Dosimetry," HEDL-SA-3063 and 5th International ASTM-EURATOM Symposium on Reactor Dosimetry, September 24-28, 1984, Geesthacht, Federal Republic of Germany.

33. R. E. Malenfant, "Little-Boy Replication: Justification and Construction," Proc. of the 17th Midyear Topical Symposium of the Health Physics Society, February 5-9, 1984, Pasco, WA.

34. W. E. Loewe, "Why Replicate Little-Boy After Nearly Forty Years?," 29th Annual Meeting of the Health Physics Society, June 3-8, 1984, New Orleans, LA.

THE USE OF BERYLLIUM OXIDE THERMOLUMINESCENCE DOSEMETERS

FOR MEASURING GAMMA EXPOSURE RATES

P J H Heffer and T A Lewis

Central Electricity Generating Board (CEGB)

Berkeley Nuclear Laboratories, Berkeley, GL13 9PB, England

ABSTRACT

Beryllium Oxide(BeO) is shown to be the best thermoluminescence dosemeter(TLD) for measuring gamma exposure rates in graphite moderated reactors. It has an almost negligible response to thermal neutrons and the correction for the unwanted fast neutron response is small. The experimental procedure is described and is based on TL from a glowpeak at $300\,^{0}C$ rather than that at $190\,^{0}C$, which is usually used.

The variation of response of BeO as a function of photon energy has been measured for TLDs built up in graphite, a medium to which they are well matched dosimetrically, and within aluminium and steel. The results are compared with calculated cavity effects which are shown to be insufficient to describe the measured energy dependence. TL effects in the material appear to dominate.

INTRODUCTION

The radiation field within the core of Commerical Advanced Gas-Cooled Reactors (CAGRs) induces a chemical reaction between the graphite moderator and the coolant gas which reduces the weight and hence the strength of the core (1). The effect is dominated by the gamma radiation dose, though fast neutrons also enhance corrosion. A detailed knowledge of the radiation dose distribution in the moderator is fundamental to the prediction and understanding of radiolytic corrosion, and hence it forms an important input into reactor lifetime assessments.

J. P. Genthon and H. Röttger (eds.), Reactor Dosimetry, 373–379.
© *1985 ECSC, EEC, EAEC, Brussels and Luxembourg*

This paper explains why BeO was chosen to measure the gamma energy deposition even though it was less well developed than other TLD materials and describes the experimental technique employed. In addition the supporting measurements of the thermal and fast neutron sensitivities of the TLDs are outlined; these being necessary to derive measurements of the signal from gamma rays only.

The photon energy response of the TLDs is necessary for interpretation of any results in terms of dose to a particular material. Its measurement is described and in an attempt to understand the response, the measurements were made with TLDs built up in several different materials and the results compared with cavity theory calculations. This comparison indicated that TL effects were dominating the energy response, rather than cavity effects; a fact supported by the observation that different glowpeaks were shown to have different responses.

CHOICE OF DOSEMETER

A meaningful measurement of gamma exposure can only be made in a mixed radiation field if neutron effects do not dominate the dosemeter response. Traditionally TLD700 has been regarded as insensitive to thermal neutrons. However, in a typical CAGR moderator almost 60% of the response of TLD700 derives from thermal neutrons. This is clearly unacceptable. Of dosimetry materials only three had an acceptably low thermal neutron sensitivity; BeO, calcium fluoride and aluminium oxide. The variation of mass stopping power, relative to that of graphite, for these three materials is shown in fig 1. BeO was chosen as the material best matched to graphite and its neutron sensitivity in TLD form was measured.

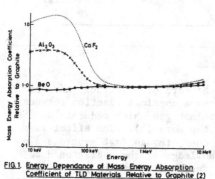

FIG 1. Energy Dependance of Mass Energy Absorption Coefficient of TLD Materials Relative to Graphite (2)

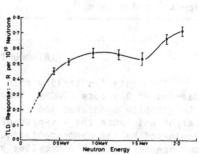

FIG 2. Fast Neutron Response of Beryllium Oxide

The thermal neutron sensitivity was measured in a graphite thermal column; a mixed field of neutron and gamma radiation. Two methods were used: the first was to absorb the neutrons with a small sphere of Lithium6 carbonate: the second was to measure the gamma field with a neutron insensitive ionisation chamber (3). Both techniques agreed to within the experimental uncertainty that the thermal neutron response of the TLDs was unmeasurable. A value of 0.000 ± 0.005 R per 10^{16} fluence in the Westcott notation has been adopted. This value only applies to well separated TLDs of the geometry used. The value is lower than kerma calculations suggested because of the high escape probability of the gamma rays produced in (n,γ) events in the TLDs; most of the energy is dissipated in the surrounding medium.

The fast neutron sensitivity has been measured in a calibrated beam of neutrons at Birmingham University (4). The results are shown in fig 2. These values are much less than would be expected from kerma calculations and probably arises from reduced TL efficiency for the high LET of the recoil nuclei.

EXPERIMENTAL TECHNIQUE

Ceramic discs measuring 6mm diameter and 0.5mm thick have been adopted as the most convenient size and have been supplied by Consolidated Beryllium (5). A typical glowcurve is shown in fig 3. During readout the signal from the two glowpeaks is separated by holding the temperature at $210\,^{0}$C for a few seconds.

All of the readouts have been performed on a TOLEDO 654 (6) with glowcurves recorded onto a micro-computer. The lower temperature glowpeak (peak 1) was found to be supralinear in the measurement range 5-100 Roentgens(R) and generally less reproducible. In contrast the higher temperature peak (peak 2) was linear to within a few percent and highly reproducible in

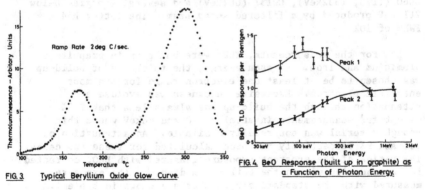

FIG. 3   Typical Beryllium Oxide Glow Curve.

FIG. 4.   BeO Response (built up in graphite) as a Function of Photon Energy.

the range 5-400R. The exact origin of peak 2 is not known but
at least half of the white beryllia offered by the manufacturer
has had a sufficiently high content of peak 2 traps. Higher
purity hot pressed material had a completely different
glow-curve and very erratic characteristics.

As delivered the sensitivities of peaks 1 and 2 have
standard deviations of 15% and 9% respectively and all TLDs must
be identified and calibrated to achieve accurate results. These
calibrations can be applied over as many as 50 uses (read,
anneal and irradiate) over periods of up to two years. After
use TLDs are annealed between gold foils at $800\,^0C$ for 5 minutes
and cooled quickly. This high temperature keeps the discs clean
and eliminates charge transfer from deeper traps.

Before irradiation the TLDs are handled in dull
laboratory lighting: after irradiation red light is used
because the TL signal is annealed by UV light. Peak 1 is much
more strongly effected than peak 2.

TLDs are secondary dosemeters in that they do not provide
an absolute measure of exposure. A calibration procedure
closely analogous to that of ionisation chambers has been
adopted. TLDs were exposed to Co60 radiation, built-up in a
dummy graphite ionisation chamber, the attenuation of whose wall
was well defined (7). This approach yielded a calibration
factor in terms of dose in air (ie. Roentgens) at an average
energy of 1250 keV. All measurements are quoted in these units
of Co60 equivalent R.

ENERGY DEPENDENCE

The dependence of the BeO response as a function of
photon energy and build-up material was studied in well defined
fields calibrated with a standard Farmer Dosemeter NE 2570 (8)
and 0.6 cc graphite walled chamber. Measurements were made at
Co60 (1173, 1332keV), Cs137 (662keV) and several energies below
211 keV produced by a filtered X-ray beam. The latter had a
FWHM of 20%.

For the measurements, TLDs were built up in graphite,
aluminium and iron. At each energy, the thickness of build-up
was chosen to be at least one electron range for the most
energetic electron. Excessive thickness was avoided and
attenuation through the build-up was always less that 2.5%
except for measurements in steel at 30 and 48keV where thin
enough material was not readily available. An attenuation of
38% and 9% respectively has been calculated for these two cases.
The ratios of the apparent exposure measured with BeO, corrected
for attenuation within the build-up medium, to the exposure
measured with the standard air chamber are shown in table 1.

These results are compared with predicted ratios which
have been calculated using Monte Carlo methods. The
calculations have taken account of both photon and electron
processes, with the primary photon transport and electron
production data being obtained from the 3-dimensional particle
transport code SPARTAN(9) and subsequent electron transport
being simulated with the condensed history technique code BETA
II/B(10). Considerable care has been taken to model all the
particle transport processes as closely as possible. In
particular, special attention has been paid to the accurate
representation of the highly anisotropic angular distributions
of electrons generated from both Compton and photoelectric
processes. The results from these calculations are presented in
table 1 and their implications are considered in the following
section.

DISCUSSION

In graphite the measured energy response reduces more
rapidly with photon energy than would have been expected from
the mass energy absorption coefficients in Figure 1. However
averaged over realistic spectra in CAGR moderators the effect is
only to change the deduced graphite dose by a few percent. If a
response closer to that of graphite were essential then thin
foils of a heavier material could be used to sandwich the TLDs
and engineer the desired response function.

The reduction in TLD response with reducing photon energy
is more than can be explained by mismatch between TLD and its
surrounding medium. The implication of this result is that the
basic TL mechanism introduces an effect which controls the
response function of the TLD.   To demonstrate this more
clearly, the experimental data has been reanalysed using the
peak 1 response rather than peak 2. The results are compared in
Figure 4, which indicates the markedly different behaviour of
the TLD response when different trap types are examined.

Table 1. Photon Energy Response of BeO Relative to Air

Photon Energy (keV)	BeO Exposure/Ion Chamber Exposure GRAPHITE BUILD-UP			BeO Exposure/Ion Chamber Exposure ALUMINIUM BUILD-UP			BeO Exposure/Ion Chamber Exposure IRON BUILD-UP		
	Experiment *	Calculation	Calculation Experiment	Experiment *	Calculation	Calculation Experiment	Experiment *	Calculation	Calculation Experiment
1252	1.01 ± .03	0.98 ± .02	0.97 ± .03	1.01 ± .03	0.99 ± .02	0.98 ± .04	0.98 ± .03	0.99 ± .02	1.01 ± .04
662	0.96 ± .03	1.00 ± .02	1.04 ± .03	0.96 ± .03	1.04 ± .03	1.08 ± .04	data lost	1.01 ± .03	---
211	0.88 ± .03	0.99 ± .02	1.12 ± .03	0.87 ± .03	1.02 ± .02	1.15 ± .04	0.88 ± .03	1.01 ± .04	1.15 ± .05
109	0.75 ± .02	1.01 ± .02	1.35 ± .03	0.77 ± .02	1.03 ± .04	1.34 ± .03	0.99 ± .04	1.27 ± .03	1.28 ± .05
48	0.66 ± .02	0.82 ± .02	1.24 ± .03	0.69 ± .02	0.81 ± .03	1.18 ± .04	0.81 ± .03	1.10 ± .03	1.24 ± .04
30	0.62 ± .03	0.81 ± .03	1.31 ± .04	0.64 ± .03	0.85 ± .04	1.30 ± .05	0.53 ± .02	1.05 ± .04	1.43 ± .04

* Experimental data is corrected for attenuation using calculated data

Examination of Table 1 shows that there is not an apparent simple trend of C/E ratio with photon energy and nor is there consistency between the sets for different build up material, with the lower energy iron results showing a different variation from the other two comparisons. More detailed consideration of the physical processes occurring in the TLD and build up material, however, produces an explanation for the observed behaviour. For the data of the carbon and aluminium systems, for which the nett electron transfer from build-up to TLD material is always small, it is found that the electron recoil spectrum reaches its lowest mean value at a photon energy of 109keV since the photoelectric cross-section is still small and Compton recoil electrons cannot be generated above an energy of 33keV. For the 48 and 30keV energy points, the photoelectric interaction is the most important energy depositing process and gives rise to a harder electron spectrum. Correlating the C/E ratios against electron energy rather than incident photon energy produces a consistent and understandable trend.

For the iron build-up case, the theoretical data appears to underpredict the dose enhancement at 109keV while overpredicting the 30keV measurement. Examination of the electron spectra shows that at 109keV the electrons entering the TLD from the iron build-up material have a mean energy well in excess of that generated within the TLD itself. This gives rise to an enhanced TL efficiency and hence an apparent over-meaurement of the energy deposition from this source. At 30keV the electron spectrum entering the TLD from the iron is significantly softer than the TLD spectrum and so the energy deposited by these electrons is recorded with a lower TL efficiency.

By taking account of the electron spectrum effects in the TLD material a consistent description of the experimental results can be made. Examination of the peak 1 data, though not of direct importance to the current work, also shows that the deviations from expected behaviour also occur at energies just in excess of 100keV, supporting a model of response which is a function of the TL mechanism and the electron energy or LET. A full description of such a model is being prepared (11). It is of interest to note that such a mechanism of reducing TL efficiency of peak 2 with LET is entirely consistent with the observed fast neutron behaviour of BeO described above.

CONCLUSIONS

Beryllium Oxide is the best TLD for measuring gamma exposure rates in graphite moderated reactors. It is dosimetrically well matched to graphite although some of this advantage is lost because the TL efficiency reduces with photon energy. However, the overall response is still adequately well matched and could be improved/engineered with suitable sandwiching by thin foils of higher Z materials.

The photon energy response has been measured over an energy range through which the TLD changes from a small to a very large cavity. The cavity corrections have been adequately modelled for build-up in graphite, aluminium and iron, provided that full account is taken of the TL efficiency and electron spectrum effects.

ACKNOWLEDGEMENT

This paper is published by permission of the Central Electricity Generating Board.

REFERENCES

1. C J Wood and A J Wickham, Nucl. Energy, Vol. 19 No. 4 277-282, 1980
2. P Bassi, G Busuoli and O Rimondi, Calculated Energy Dependence of some RTL and RPL Detectors, Int. J. Appl. Rad. Isotopes 27-291, 1976
3. S J Kitching and T A Lewis, An Ionisation Chamber for Measuring Gamma Exposure Rates in Mixed Radiation Fields , CEGB Report TPRD/B/0348/N83
4. Birmingham Radiation Centre, University of Birmingham, England
5. Consolidated Beryllium, Marble hall Road, Milford haven, Dyfed, Wales
6. Vinten Instruments, Mill Works, Jessamy Road, Weybridge, Surrey, England
7. G P Barnard, A R S Marsh and D G I Hitchman, Studies of Cavity Ionization Chambers with Megavoltage X-rays, Phys. Med. Biol. Vol. 13 No. 3 p295, 1964
8. Nuclear Enterprises Ltd., Bath Road, Beenham, Reading, England
9. Bending R C and Heffer P J H, The Generalised 3D Monte Carlo Particle Transport Code SPARTAN. CEGB Report RDB/N3178
10. Jordan T M BETA II/B Time Dependent Monte Carlo Bremsstrahlung and Electron Analysis Code RSIC CCC 117, Oak Ridge National Laboratory
11. Heffer P J H and Lewis T A. To Be Published.

FAST NEUTRON DOSIMETRY BY MEANS OF THE

SCRAPING SAMPLING METHOD

F. Hegedues

Swiss Federal Institute for Reactor Research

CH-5303 Wuerenlingen

Abstract

During the shut-down period of a nuclear plant, scraping samples could be taken from the surface of structural materials. From the measured 93m-Nb (16 y) and 54-Mn (315 d) activities the fast neutron fluence and flux density could be determined. Furthermore, by means of the 93m-Nb to 54-Mn activity ratio the calculated neutron spectrum could be adjusted.

From given steel samples the 835 keV gamma activity of 54-Mn was measured directly. In order to measure the low energy (16.6 keV) x-ray activity of 93m-Nb, it was necessary to separate the niobium from the steel. The yield of the chemical separation (ionic precipitation by HAP) was higher than 95%.

The scraping sampling method is useful for measuring the azimuthal neutron flux distribution on the PV inner wall and for determining experimentally the lead factor of the material samples located in the surveillance capsule.

1.0 INTRODUCTION

The objective of the fast neutron dosimetry for the reactor pressure vessel (RPV) surveillance program is to determine the fast neutron spectrum, fluence, fluence-rate, dpa and dpa-rate both at the specimen location and on the RPV inner wall.

At the specimen location, between the thermal shield and the RPV, these parameters are determined by measurement (activation detectors) and by calculation (e.g. I-D transport code). In case of a discrepancy between the results of the two methods, the method of calculation is adjusted to get agreement between the two results. By that means the method of calculation is validitated.

On the RPV inner wall, these parameters are usually not measured but only calculated using the same adjusted method as for the specimen location. However, since the neutron spectrum at the RPV inner wall differs from the spectrum at the specimen location, the calculated results would have higher uncertainities than at the specimen location. Another more important source of uncertainity is the azimuthal neutron fluence distribution. The neutron fluence, due to the core edge structure and the time dependent fuel burn-up, has an important azimuthal variation (up to a factor of 2). In order to calculate the azimuthal distribution, 2-D transport code is required. In this calculation, many not well known input data are involved.

In order to improve the reliability of the RPV inner wall fluence values, neutron fluence measurements were done by means of the scraping sampling method.

2.0 DESCRIPTION OF THE SCRAPING SAMPLING METHOD

2.1 PRINCIPLE OF THE METHOD

In the past the principles of this method were very often applied at many places. By taking few miligrams of splinter from the surface of an irradiated steel specimen and measuring its 54-Mn and 58-Co activity, the neutron fluence could be determined. To properly interpret the result, the weight as well as the Fe and Ni content of the steel splinter should be accurately known.

This method could be applied to the RPV inner wall of a power plant (Ref.1). During the shut down period, with a special boring machine, splinter samples (100mg on 1 cm surface) were taken. As the inner plating of the RPV contained niobium, three activities (93m-Nb, 54-Mn, 58-Co) were measured in the splinter.

2.2 SAMPLING POSITIONS

Fig. 1 shows the core structure and the sampling positions. It is assumed that the azimuthal fluence distribution has the same shape in each quadrant (0° -90°, 90° - 180°, 180° -270°, 270° - 360°) and furthermore, that each quadrant is symmetric about its median (45°) line. Therefore it was enough to measure the fluence distribution in one eight of the core (90° - 135°).

All positions, with the exception of Pos.13 (+ 50 cm) and Pos.14 (- 50 cm) were at the same vertical level (mid-plane of the core hight).

By using positions 1 to 10 the azimuthal fluence distribution was measured.

By using positions 2 (93°), 11 (210°) and 122 (330°) the core symmetry was measured.

By using positions 13, (+ 50 cm), 2 (0 cm), 14 (- 50 cm) the vertical fluence distribution was measured.

3.0 DESCRIPTION OF THE MEASUREMENT

3.1 MEASUREMENT OF THE Nb, Fe AND Ni CONTENT OF
THE SCRAPING SAMPLES

3.1.1 ABSOLUTE MEASUREMENT WITH ATOMIC ABSORPTION.

In three samples, the Nb, Fe and Ni contents were determined by means of atomic absorption. The NBS standard steel Cr18 - Ni9 was used as the reference material. The composition of this standard is very similar to the RPV inner plating. The composition of the RPV inner plating was: Nb: 0.55%, Fe: 67.0%, Ni: 10.6%.

3.1.2 Relative measurement with XRF

The relative variation of the Nb, Fe and Ni contents of the samples were measured by means of the XRF (x-Ray-Fluorescence) method. The Nb/Fe and Ni/Fe weight fractions were found constant with a fluctuation of $\pm$ 8% (over 14 samples).

3.2 CHEMICAL SEPARATION OF NIOBIUM AND PREPARATION OF

THIN NIOBIUM SAMPLES

In order to count the low energy x-ray (16.6 keV) of 93m-Nb it is necessary to separate the niobium from the steel splinter. The steel was dissolved in an $HF-HCl-HNO_3$ acid mixture, evaporated and solved again in HNO_3. This solution was passed through an HAP (Hydrated Antimony Pentoxide) column, then evaporated, eluted with HF and dried on a thin paper filter. The chemical separation procedure was optimized by means of 94-Nb labeled specimens. The Nb yield was as high as 97 ± 2%.

The characteristics of the thin Nb sample are:

Surface: 1.77 cm
Paper filter thickness: 8.3 mg/cm^2
Niobium thickness: 0.2 - 0.3 mg/cm^2
Mylar protection foil: 1.4 mg/cm^2

The total absorption and self-absorption correction for the 93m-Nb $XK_{\alpha,\beta}$ -ray is practically negligeable: 0.994 ± 0.001.

3.3 COUNTING OF THE 54-Mn and 58-Co ACTIVITIES

In the steel splinter the 54-Mn and 58-Co activities were counted using a Ge(Li) counter. The 835 keV (54-Mn) and the 811 keV (58-Co) gamma lines are not perturbed by other gamma rays.

3.4 COUNTING OF THE 93m-Nb XK_-RAYS

A pure Ge counter was used with a resolution of 180 eV for the 16.6 keV Nb XK_α line (Ref.2). The absolute calibration of the counter was done by a set of standard sources (5 - 80keV). This calibration was checked by means of a standard 93m-Nb source of PTB, the agreement was very good (± 2%).

The corrections due to the 182-Ta and 94-Nb activities were experimentally determined. The Nb $XK_{\alpha,\beta}$ emission due to the 182-Ta activity could be calculated by the following emprical formula:

$$A(NbX) = 2.2 \cdot 10^{-3} \cdot A(182\ Ta) \cdot t$$

where

$A(NbX)$: $XK_{\alpha,\beta}$ (16.6 and 18.6 keV)-ray emission (s^{-1})

$A(182\ Ta)$: activity of 182-Ta (Bq)

t : sample thickness (only Nb) $(mg \cdot cm^{-2})$

The constant of $2.2 \cdot 10^{-3}$ was determined with a series of samples ranging from 0.1 to 10 mg/cm². The samples consisted of inactive Nb mixed with 182-Ta.

The NB $XK_{\alpha,\beta}$ emission due to 94-Nb was measured by means of a series of samples (niobium – 94-Nb mixture). The following result was obtained:

$$A(NbX) = C \cdot A(94-Nb) \cdot t$$

the constant c was found to be unmeasurably small: $< 4,5 \cdot 10^{-6}$.

The self absorption correction was calculated by means of the following formula:

$$I(Measured) = I\ (emitted) \cdot \frac{1-\exp(-\mu t)}{\mu \cdot t}$$

$(I : XK_{\alpha,\beta}$ -ray intensity)

The absorption coefficient was measured in Nb metal: $2.55 \cdot 10^{-2}$ cm² mg⁻¹ which is slightly higher than the value: $1.80 \cdot 10^{-2}$ (Ref.3).

In case of the thin 0.2 - 0.3 mg/cm^2 Nb samples on filter paper, the total acitivity correction, including the 182-Ta 94-Nb fluorescence effect and the self absorption was less than 1%.

4.0 EVALUATION OF THE FAST NEUTRON FLUENCE

The DIN norm (Ref.4, 5) was applied to evaluate the fast neutron fluence from the measured 93m-Nb, 54-Mn and 58-Co activities. The cross sections $\sigma(E)$ were taken from ENDF/B-V File with the exception of the 93-Nb(n,n') reaction (Ref.6). Other nuclear data were taken from the EURATOM-Guide (Ref.7) .

The relation between the fast neutron fluence: $\phi(E> 1.0$ MeV) and the detector activity: A (per atom at EOI, S^{-1}) is:

$$\phi(1.0\ MeV) = \frac{A}{\sigma_{eff}(1.0\ MeV)} \cdot \frac{\sum f_i \cdot \Delta t_i}{\sum f_i \cdot (1 - \exp(-\lambda \Delta t_i)) \cdot \exp(-\lambda(t_n - t_i))}$$

where

$$\sigma_{eff}(1.0\ MeV) = \frac{\int_0^\infty \sigma(E)\ \varphi(E)\ dE}{\int_{1.0\ MeV}^\infty \varphi(E)\ dE}$$

f_i : average reactor thermal power in the interval of time $t_i - t_{i-1}$

t_n : time between the beginning and the end of irradiation (EOI)

$\varphi(E)$: calculated neutron spectrum (arbitrary units)

5.0 DISCUSSION OF THE RESULTS

There is an agreement (± 10%) between the neutron fluences obtained from the 93m-Nb and 54-Mn measurements. The 58-Co measurement is not comparable to the 93m-Nb and 54-Mn due to its short half life. Nevertheless it is useful for assessing the fluence-rate for the last 5 months.

These results, especially the azimuthal and vertical fluence distribution and the core symmetry, are very useful. It was possible to estimate the EOL fluence at the most irradiated point on the RPV inner wall. The so called "lead factor" was also determined.

By comparing the measured fluence values to the calculation, the 2-D transport code has been validitated.

REFERENCES

1. W. Schweighofer (KWU-Erlangen)
   27, PTB Seminar, Braunschweig Oct. 20 -21, 1980
   (Proceedings not available)

2. F. Hegedues
   Fast neutron dosimetry by the reaction 93-Nb
   (n,n'). Counting Technique for 93m-Nb Activity.
   Proceeding of the Second ASTM-EURATOM Symposium
   on Reactor Dosimetry (Palo Alto, 1977)
   NUREG/CP-0004.

3. H. Joffre, L. Pages: Rapport CEA-R-3655 (1968)

4. Bestimmung der Fluenz schneller Neutronen mit
   Aktivierungs- und Spaltungsdetektoren. DIN
   25456, Teil 1 (1982)

5. Bestimmung der Fluenz schneller Neutronen mit
   Eisen-Aktivierungs detektoren. DIN 25 456, Teil
   2 (1982)

6. F. Hegedues
   Detector of fast neutron flux based on the
   reaction 93-Nb (n,n') 93m-Nb. EIR-Bericht 195
   (1971)

7. Nuclear data guide for reactor neutron metrology.
   EUR 7164 EN (1982)

Fig. 1   Location of the Scarping Positions

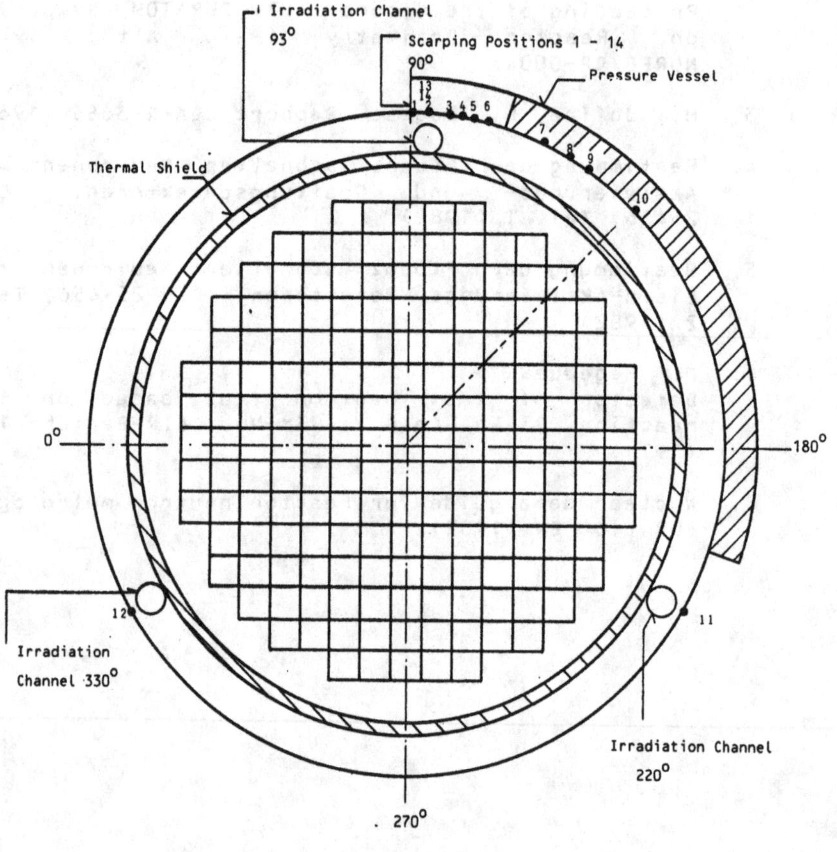

390

MESURE DES DEBITS DE DOSE D'IRRADIATION IONISANTE

EN ATMOSPHERE SURCHAUFFEE

Serge LORRAIN, Guy PORTAL et Georges VALLADAS(*)

CEA/IPSN/DPT/SIDR - BP n°6 - 92260 Fontenay-aux-Roses, FRANCE

CNRS - LFR - 91190 - Gif sur Yvette, FRANCE (*).

Nous présentons une dosimétrie des rayonnements ionisants par Radiothermoluminescence (RTL), adaptée aux irradiations à hautes températures; celle-ci utilise l'ALUMINE (Al2O3:Si,Ti) ou le SULFATE DE CALCIUM dopé au dysprosium (SO4Ca:Dy) comme détecteurs solides réfractaires.

Nous indiquons ensuite un certains nombre d'applications du domaine nucléaire, motivées par les nouvelles performances.

1 INTRODUCTION

La mesure "in situ" des doses d'irradiation ionisante délivrées à chaud exige une grande stabilité thermique du dosimètre. La théorie de J.T. RANDALL et M.H.F. WILKINS (1) sur le mécanisme de la RTL, permet de définir des paramètres caractéristiques de la stabilité thermique des pièges responsables de l'information dosimétrique pris en compte. Il s'agit principalement de l'énergie d'activation et du facteur de fréquence. Des méthodes qui conduisent à une détermination expérimentale de ces paramètres ont été proposées; nous les avons appliquées aux produits RTL réfractaires étudiés. Les résultats ont permis de conclure qu'une dosimétrie à chaud employant l'ALUMINE (Al2O3: Si,Ti) ou le SULFATE DE CALCIUM dopé au dysprosium (SO4Ca:Dy) est concevable (2,3). Ces recherches ont souligné l'intérêt des pièges profonds qui donnent une thermoluminescence (TL) stimulable dans l'intervalle 400-800°C, pièges doués d'une stabilité thermique remarquable, pouvant assurer la mesure des doses reçues dans des environnements radioactifs chauffés. L'application dosimétrique

J. P. Genthon and H. Röttger (eds.), Reactor Dosimetry, 391–397.
© 1985 ECSC, EEC, EAEC, Brussels and Luxembourg

de telle TL se heurte à de grandes difficultés; peu de chercheurs ont suivi cette voie. La pratique usuelle en RTL se limite à des températures de dépouillement voisines de 350°C; il y a deux raisons à cela: on évite d'une part les inconvénients causés par l'émission thermique infrarouge du four qui devient vite importante aux températures supérieures et d'autre part, de nombreux matériaux RTL utilisés ne possèdent pas de tels pièges profonds. On doit beaucoup, par contre, aux recherches en radiodatation par RTL qui détermine l'âge de poteries et de céramiques anciennes (2). Dans ce domaine, il est indispensable de recourir à l'exploitation d'une TL de haute température mettant en jeu des pièges profonds, découverts en particulier dans certains minéraux isolables (variétés de SiO2). Nos travaux se sont déroulés en trois étapes, guidés par les impératifs des programmes d'applications nucléaires. Il faut satisfaire d'abord aux besoins d'une dosimétrie s'adaptant aux mesures en atmosphères chauffées à des températures voisines de 100°C, ensuite de 240°C, enfin de 450°C. La dosimétrie appliquée dans la première étape au voisinage de 100°C, considère encore l'information donnée par la RTL du pic classique, non profond (dénommé pic III): celle-ci est corrigée du "fading" thermique inévitable. Le calcul peut emprunter les valeurs des constantes caractéristiques publiées (E et S), si on connaît la température d'irradiation; cependant, il est préférable de se baser sur une détermination expérimentale directe du déclin de la RTL qu'on mesure "in situ", à partir des lectures de prédoses irradiées avec les dosimètres vierges. Cette technique donne des résultats satisfaisants vérifiés à diverses reprises par une seconde lecture des mêmes dosimètres en considérant les TL de hautes températures; on valide ainsi le principe simple appliqué. Nous insistons maintenant sur la mise en valeur des TL de hautes températures de l'(Al2O3:Si,Ti) et du (SO4Ca:Dy) lors des dosimétries au-dessus de 100°C. Nous traitons en détails le cas de l'ALUMINE.

Les matériaux RTL ont la particularité de fixer à l'état métastable dans des sites privilégiés appelées "pièges", une partie des charges mobiles arrachées par interaction des rayonnements ionisants sur les atomes du cristal. Un apport de chaleur chasse les porteurs bloqués vers des centres de recombinaison luminogènes préexistants dans le matériau sous forme d'activateurs et responsable de la luminescence. Chaque type de pièges possède sa stabilité propre, définie par sa profondeur (ou énergie d'activation) et son facteur de fréquence. Les transitions radiatives émettent une luminescence de longueur d'onde caractéristique de l'élément activateur présent dans la substance. Les courbes RTL comportent plusieurs pics successifs dûs à la libération progressive des porteurs piégés puis chassés. Les plus énergétiquement captés apparaîssent les derniers donc aux températures les plus élevées. L'intensité lumineuse correspondant aux pics de bas niveaux est beaucoup plus importante que celle due aux pièges profonds, moins denses en population de porteurs à expulser.

En conséquence, l'emploi en dosimétrie d'une TL de haute température présente trois difficultés majeures qu'il convient de surmonter.

1° Il faut isoler une TL de courte longueur d'onde, proche de l'ultraviolet, pour éviter une interférence gênante de l'émission infrarouge provenant des parties métalliques du four de chauffage.

2° On doit transférer une énergie calorifique capable de stimuler les pièges profonds activés à l'irradiation, par une élévation régulière de la température de l'échantillon au-delà de 700°C.

3° Enfin, il faut détecter puis mesurer avec précision la fraction du flux lumineux, relativement faible, émise uniquement par les pièges profonds concernés.

Le tableau A rassemble les données utiles concernant les propriétés RTL du (SO4Ca:Dy) et de l'(Al2O3:Si,Ti); elles proviennent d'études antérieures publiées (3,4). Les thermogrammes de ces deux matériaux fondamentaux en dosimétrie par RTL à hautes températures, sont représentés figure 1 et 2.

2 PARTIE EXPERIMENTALE

Nous travaillons sur un prototype conçu au laboratoire et considéré comme une version améliorée de celui décrit dans (5). Ses caractéristiques reposent sur:

1° un programmateur de températures performant de grande souplesse qui permet un chauffage linéaire jusqu'au-delà de 700°C; la vitesse de chauffe fixe peut varier de 0,2 à 15°/s. Il y a possibilité d'arrêt en palier à une température constante au choix. Il assure surtout une bonne répétivité des cycles thermiques...

2° un dispositif d'adaptation de filtres optiques épais (10 mm) qui est prévu sur le parcours de la luminescence...

3° un détecteur de flux lumineux sensible qui est constitué d'un photomultiplicateur (PM) à haut rendement quantique qu'on associe à une électronique sophistiquée de comptage des photoélectrons provenant de la cathode refroidie à basse température, pour réduire le courant noir.

Le (CaSO4:Dy) utilisé provient d'une fabrication mise au point au laboratoire, selon une variante de la méthode proposée par T.YAMASHITA et al.(6); ce (SO4Ca:Dy) a fait l'objet d'études approfondies (7,8). Ce matériau RTL est actuellement considéré comme le plus sensible et nous avons montré que sa stabilité thermique est appréciable. Nous l'employons avec intérêt dans la technique des prédoses, en dosimétrie "tiède". Malgré la potentialité de ses pics profonds, il n'est pas encore entré dans les applications courantes, pour la dosimétrie à haute température.

L'alumine RTL CEC-DESMARQUEST constitue le matériau dosimètre RTL le plus apprécié dans cette étude. Son originalité tient au thermogramme qui montre des pics d'émission TL à haute température. Nous signalons la similitude de ses caractéristiques avec celles publiés par SUDERSHAN K.MEHTA (9) concernant des Al2O3 d'origine indienne. La courbe de TL de l'Al2O3 comprend une succession de trois groupes principaux de pics, culminant à 270°, 460° et 640°C. Nous nous intéressons aux deux derniers, dénommés pic V et pic VI. Lorsqu'on développe le thermogramme complet jusqu'au-delà de 750°C par lecture directe, même en réduisant au maximum la vitesse de chauffage (à 0,5°C/s par ex.) on ne peut pas empêcher le recouvrement des pics voisins. Par le jeu de deux préchauffages initiaux successifs (suivis de refroidissements) strictement définis, puis par une lecture terminale sélective du pic profond retenu, on fractionne les émissions TL successives.

Le premier objectif visé, lorsque les principes sont acquis et la technologie raffinée, est la mise au point d'une méthode de lecture de ces TL de hautes températures qui soit à la fois fiable et précise. Elle est ensuite longuement expérimentée et utilisée pour tracer des courbes de réponses en fonction de la dose. Les figures 3 et 4 montrent les courbes d'étalonnage de (Al2O3:Ti,Si) obtenues à partir de la TL des pics V & VI. On voit d'une part que le pic V permet la mesure des doses comprises entre 15 et 5000 rad avec une bonne précision et que le pic VI peut être utilisé dans le domaine plus large, des doses comprises entre 500 et 50000 rad. D'une part, nous avons vérifié par l'expérience directe, la grande stabilité thermique de ces TL de hautes températures. On a conservé plus de 200 heures, à des températures de 250°C puis de 450°C, des poudres d'Al2O3 prédosées, sans pouvoir noter dans les mesures, le moindre fading de la RTL. D'autre part, l'étude comparée à chaud et à froid des réponses données par des dosimètres chauffés (ou non chauffés) simultanément irradiés dans un faisceau gamma étalonné délivrant des doses connues identiques, a été poursuivie à divers débits de dose. Il ressort qu'à la question de la détermination des doses en enceinte surchauffée, il se greffe le problème connexe de l'interférence physique du container sur le matériau RTL intérieur. Nous avons mis en évidence, pour des doses connues délivrées à chaud (240°C), des cas d'irradiation au travers les containers métalliques où il faut parfois corriger la dose lue dans l'Al2O3 de $\pm$15% pour trouver la dose effective reçue par le dosimètre lui-même et provenant du faisceau extérieur.

3 APPLICATIONS

Pour des raisons diverses de radioprotection du matériel ou

du personnel, il existe un besoin pressant du contrôle des doses
règnant dans les zônes actives surchauffées qu'on rencontre soit
au voisinage du coeur dans les bâtiments réacteurs des centrales
nucléaires, soit dans certaines enceintes blindées radioactives
de l'industrie nucléaire. La stabilité de l'information permet de
réaliser des temps d'exposition très longs et d'effectuer en
conséquence la mesure des faibles débits de dose à chaud. Nous
allons passer en revue les principales applications de cette
méthode dosimétrique nouvelle.

1°) Le flux gamma est rendu responsable d'une fraction non
négligeable de la puissance électrique fournie par un réacteur
surrégénérateur; il est intéressant de chercher à l'explorer
expérimentalement. Des essais de mesure directe de débit de dose
en des points d'accès particuliers très chaud (450°C) sont à
l'étude. L'emploi d'(Al2O3:Si,Ti) permet une dosimétrie à cette
température à l'aide du pic RTL VI. Les mesures de dose projetées
à PHENIX (et SUPERPHENIX), ont pour but de vérifier et de mettre
au point les codes de calculs utilisés.

2°) Dans le caisson du réacteur UNGG dépanné SLA2 de SAINT
LAURENT DES EAUX, afin de suivre en continu la montée de l'activité des filtres placés sur le circuit du CO2 caloporteur, à
240°C, nous déterminons les débits de dose en-dessous de la
filtration à l'aide des dosimètres d'(Al2O3:Si,Ti) dont on lit la
réponse TL sur le picV.

3°) Dans le but d'un contrôle de l'environnement et de radioprotection pour le personnel vacant dans le voisinage, nous avons
été amenés à effectuer des mesures de débit de dose à l'usine
EURODIF de PIERRELATTE autour de batteries d'étuves maintenus à
125°C, servant à des opérations chimiques de traitement d'enri-
chissement de l'Uranium 235. Nous avons appliqué la technique de
correction du fading de la RTL des pics dosimétriques classiques
à bas niveaux de l'(Al2O3:Si,Ti) et du (SO4Ca:Dy). Le fait de
mettre ensemble ces deux matériaux aux mêmes points de mesure
autorise alors une double vérification des résultats.

4°) L'étude de l'évolution des débits de dose à proximité
de matériels électromécaniques importants (comme les capteurs,
vannes, moteurs etc...)placés à l'intérieur du bâtiment réacteur
à FESSENHEIM 2, où la température ambiante peut atteindre parfois
50°C, se poursuit. Les résultats réguliers tirés des différentes
campagnes successives montrent que le protocole proposé, basé sur
les principes décrits, convient tout à fait. Nanti de l'expérience ainsi acquise en pareil domaine, il nous a été demandé de
déterminer la distribution des doses présentes le long de deux
boucles en fibres optiques à l'essai, placées dans le bâtiment du
réacteur PALUEL 1, et qui sont en cours d'irradiation depuis le
démarrage.

5°) Une dernière application importante a pour but un contrôle d'homogénéité du flux autour du réacteur PALUEL 2. On
dispose à des endroits stratégiques de la périphérie, une série
de dosimètres RTL dans une ambiance tiède. Le dépouillement des

résultats doit aboutir à une cartographie radiologique de la zône périphérique du réacteur, riche d'enseignement.

Cette liste d'applications plus ou moins fortuites, n'a rien d'exhaustive. Nous avons cherché à faire ressortir ici le caractère nouveau de ces investigations.

Partant d'une étude fondamentale théorique antérieure (3,4) sur la RTL de l'ALUMINE et du SULFATE DE CALCIUM dopé au dysprosium, nous appliquons à des fins dosimétriques, les résultats d'une recherche technologique, ayant conduit à circonscrire le nouveau domaine d'utilisation, de la TL à hautes température, présente dans ces matériaux. Le but pratique a été la détermination, dans les meilleures conditions, des débits de dose de rayonnements ionisants délivrés en atmosphère surchauffée.

Les exemples d'applications mentionnés, viennent illustrer l'intérêt technologique croissant suscité par cette méthode; ils projettent en sorte l'éventail largement ouvert des nouvelles expérimentations offertes, compte tenues de la commodité de mise en oeuvre (quelques milligrammes de détecteurs en place suffisent) et de la bonne précision des mesures.

REFERENCES

1   J.T. RANDALL and M.H.F. WILKINS, Phosphorescence and electron traps. Proc.Roy.Soc.(London) A 184, 366 (1945).
2   G. and H. VALLADAS, High temperature thermoluminescence, Proc. 18th Int. Symp. on Archaeometry and Archaeological Prospection, Bonn, pp 506 (1978).
3   G.PORTAL, Etude et développement de la dosimétrie par thermoluminescence, Rapport CEA-R-4943 (1978).
4   G.PORTAL, S.LORRAIN and G.VALLADAS, Very deep traps in Al2O3 and CaSO4:Dy. Nucl.Instr. and Meth. 175, 12 (1980).
5   R.BROU and G.VALLADAS, Appareil pour la mesure de la TL de petits échantillons, Nucl.Inst. and Meth. 127,109 (1975).
6   T.YAMASHITA, N.NADA, H.ONISHI and S.KITAMURA, Calcium sulfate activated by thulium or dysprosium for thermoluminescence dosimetry, Health Physics, 21,295 (1971).
7   R.PIAGGIO-BONSI, S.LORRAIN, G.PORTAL, Etude de la stabilité des pièges dans le CaSO4:Dy RTL pour les mesures d'environnement, VIIIe Congrès international de la SFRP, Saclay (1976)
8   S.LORRAIN, R.PIAGGIO-BONSI, G.PORTAL, Stabilité de divers sulfates de calcium RTL destinés aux mesures d'environnement, IVth. Intern. Congress of IRPA, Paris (1977).
9   SUDERSHAN K.MEHTA and S.SENGUPTA, Gamma Dosimetry with Al2O3 Thermoluminescent Phosphor, Phys. Med. Biol. Vol. 21 (1976).

TABLEAU A   (Al$_2$O$_3$:Si,Ti)   q=1°C/s.

PIC	IV	V	VI
E(eV)	1,7 + 0,20	2,4 + 0,25	2,84 + 0,10
S(/s)	3,0 E14 + 1d	2,0 E15 + 1d	5,0 E13 + 1d
T max	270 °C	460 °C	640 °c

long.d'onde	(405 - 445) -- 600 (nm)
émission TL	bleu - violet rouge

(CaSO4:Dy)   q=0,5 °C/s.

PIC	III	IV	V	VI
E(eV)	(1,4 ----- 1,5)		3,0 + 0,3	2,8 + 0,3
S(/s)	(2.E14 - 6.E13)		2,5.E17 + 1d	3,0.E14 + 1,5d
T max	178 °C	230 °C	540 °C	605 °C

long.d'onde	(470 et 570) (nm)
émission TL	bleu-vert vert-jaune

FIG. 2   Al$_2$O$_3$:Si;Ti

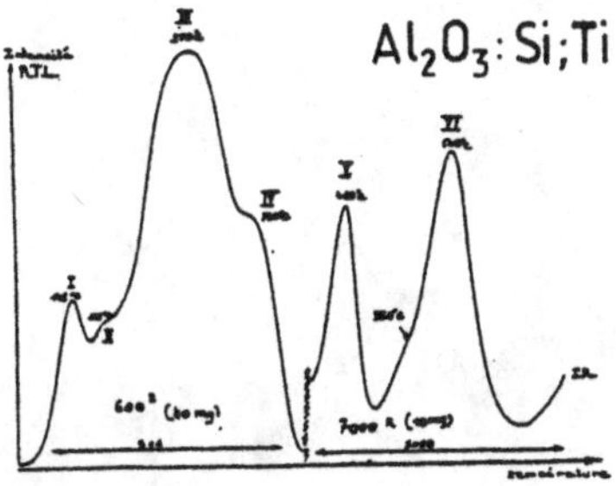

FIG.1   SO$_4$Ca:Dy

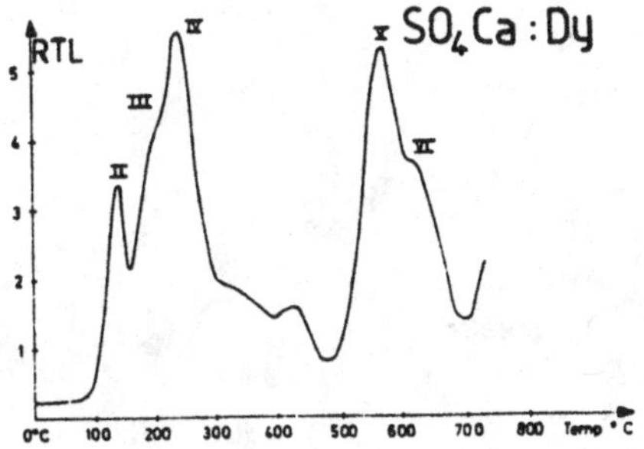

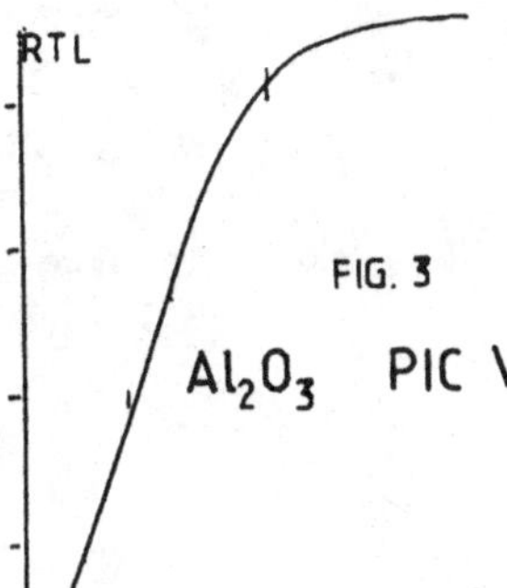

REPONSE A LA DOSE

FIG. 3   Al$_2$O$_3$ PIC V

FIG. 4   Al$_2$O$_3$ PIC VI   RTL = F (D)

DOSE (rad)

397

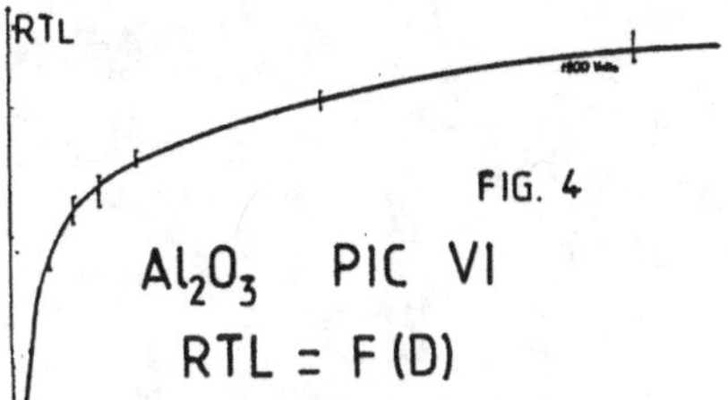

SPECIAL CONSIDERATIONS FOR LWR NEUTRON DOSIMETRY EXPERIMENTS

G. C. Martin, Jr.   and   C. O. Cogburn

General Electric Company   University of Arkansas
Vallecitos Nuclear Center   Mechanical Engineering Department
Pleasanton, CA 94566   Fayetteville, AR 72701

ABSTRACT

Characterization of the irradiation environments for operating
boiling water reactor (BWR) and pressurized water re ctor (PWR)
power plants is essential for the proper analysis and application
of research and power reactor test and surveillance program results.

As experience is gained, the experimenter understands the
general considerations for a successful dosimetry test; e.g., the
proper choice of detector materials, thermal neutron shields, and
capsule types for different flux levels and spectra based on cross
sections, half lives, purity, interfering reactions, space availability, etc.

Neutron dosimetry programs conducted by the Vallecitos Nuclear
Center for BWRs (1,2,3,4) and by the University of Arkansas for PWRs
(5,6), plus cooperative efforts in ASTM standards committees, have
given valuable experience to the authors. There are special considerations which have not been obvious to many experimenters.
These subtle phenomena have caused inaccuracies in results, contamination, frustration, and loss of experiment. Several of these
are described.

Research sponsored by the Electric Power Research Institute, the
Tennessee Valley Authority, Arkansas Power and Light, and the
General Electric Company.

SPECIAL CONSIDERATIONS

I. Local Power History vs. Total Reactor Power History

A significant problem appears to be the inability to use the normally provided reactor total power-time history for accurate spectrum and flux density determinations at all in-vessel locations. The reactor total power does not necessarily reflect local power at an experimental location due to movement of control rods and core burnup. This effect has been tested at a BWR at a major U.S. utility.

Three fission product nuclides (30-y) Cs-137, (64-d) Zr-95, and (39-d) Ru-103 were measured in the U-235, U-238, Np-237, and Th-232 dosimeters which were contained in three assemblies. These assemblies were located near the shroud, near the pressure vessel, and at a center location between the shroud and vessel at azimuthal angles 3°45' (core flat) and 45° (core corner). The irradiation duration was 750 days. Reaction rate ratios from these irradiations, designated Cs/Zr and Cs/Ru, were measured at the General Electric Vallecitos Nuclear Center. For the 3°45' midplane (G3) assembly, the ratios are close to unity (1.027 av.). For the 45° midplane (G4) and core top (G1) assemblies, however, the ratios are approximately 1.16 and 0.86, respectively. Since the experimental uncertainties in the activity measurements/fission yields are ±7%, the ratios for G4 and G1 are not attributed to experimental error.

Anomalies in the full-power fluxes derived from the measurements have resulted in an effort by Science Applications, Inc. (SAI) to make independent predictions of these measurements and in so doing understand these anomalies (7). An example (for U-235) is given in Table 1.

TABLE 1. Calculated and Measured Reaction Rate Ratios (Based on Reactor Total Power History) [Example: U-235 (n,f)]

	G1		G3		G4	
	Calc	Meas	Calc	Meas	Calc	Meas
$\frac{(Cs)}{(Zr)}$	0.85	0.87	1.07	1.05	1.15	1.24
$\frac{(Cs)}{(Ru)}$	0.85	0.89	1.08	1.03	1.18	1.19

Table 2 lists the adjustments (multiplicative factors) to all measured reaction rates which were made as a result of utilization of localized power histories (i.e., for a small region of the core near the dosimeters). For the G3 (3°45') assembly, adjustments of only 2% or less were made; for the G4 (45°) assembly, up to 9 to 10% (x1.1) adjustments were needed, particularly for the shorter-lived (40-85d) nuclides. For the G1 assembly (45°, top of core), multiplicative factors of approximately 0.80 were required to correct calculated reaction rates of the short-lived nuclides for local power histories.

TABLE 2. Adjustments[a] To Reaction Rates for All Dosimeter Radionuclides Using Localized Power Histories — Reactor Cycle 1

| Assembly | U-235, U-238, Np-237, Th-232 | | | Co | Fe | Ag | Sc | Ni | Fe |
	Cs-137	Zr-95	Ru-103	Cu Co-60	Mn-54	Ag-110m	Ti Sc-46	Co-58	Fe-59
G3	1.00	1.017	1.006	1.001	1.015	1.009	1.020	1.018	1.010
G4	1.00	1.096	1.104	1.001	1.034	1.034	1.087	1.093	1.103
G1	0.998	0.799	0.792	0.980	0.902	0.877	0.805	0.801	0.794

a. Local reaction rate equals adjustment factor times calculated rate based on total power.

Two important conclusions of SAI: a. "errors of up to ±20% may be introduced by the use of reactor total power history instead of local power history for the calculation of flux from activity measurements;" b. "specimens should be located in positions where flux gradients and sensitivity to small changes in control rod configurations are minimized."

II. Photonuclear Reactions

The concern of photo contamination with the application of pressure vessel dosimetry is relatively recent. Three practical dosimeters which activate in the important 1 MeV to 3 MeV neutron energy region are the fission dosimeters Np-237, U-238, and Th-232. These had been considered quite ideal since they could be used for a large range of irradiation durations (seconds to years). Any one of several measurable fission products induced with half lives up to 30 years could be utilized. However, recent calculations and measurements performed have shown that photofission (γ,f) reactions can significantly compete with neutron-fission (n,f) reactions, resulting in erroneous reaction rate data. When such fission dosimeters are applied without any adjustment for the photofission effect, an overprediction of the neutron flux density can result at locations where the relative gamma flux is high (i.e., as the distance from the reactor core is increased and the ratio of high-energy gamma rays to fast neutrons increases, the gamma-ray effects begin to become important). Although photocontamination at any particular point is unique for each reactor, the vicinity of the pressure vessel experiences the highest ratio of gamma rays to neutrons. This effect at pressure vessels appears to be more pronounced in BWR's than in PWR's.

Table 3 gives photofraction data from SAI for U-238, Np-237, and Th-232 at several capsule locations (7). These are one-dimensional calculations using the ANISN transport code and incorporating the effects of void fractions, localized power distributions, and equivalent core radii. The data are expressed as percent photofraction (pf) for each reaction, i.e., (1.0-% pf/100) is the correction factor to be applied to a measured reaction rate. The photofraction corrections are shown to be moderate (5 to 20%) at center locations (between shroud and pressure vessel) and in the reactor cavity, and large (20 to 54%) at near-vessel locations. The near-vessel numbers were reported with an uncertainty of ±25%.

TABLE 3. Photofraction Calculations by SAI

Reactor Location	U-238	Np-237	Th-232
PV-30 cm 3° 45' Midplane	7.6%	4.7%	14.2%
PV-7 cm 3° 45' Midplane	28.2	20.6	43.6
PV-30 cm 45° Midplane	10.5	6.1	18.8
PV-7 cm 45° Midplane	37.7	27.8	53.8
PV (Inner Diameter)	42.5	32.1	59.0
Cavity 30° Midplane	10.7	3.7	20.6

III. Two-Stage Competing Reactions

Long irradiations allow the possibility of two-stage competing reactions, the best examples being U-238, Np-237, and Th-232. At moderately high fluences, fission products from buildup of interfering fissionable nuclides (e.g., Pu-239 in the U-238 dosimeter, Np-238 and Pu-238 in the Np-237 dosimeter, U-233 in the Th-232 dosimeter) can dominate those produced directly by U-238, Np-237, or Th-232 fission. This complicates and limits the usefulness of these dosimeters.

Table 4 shows example calculations of Cs-137 produced from the irradiation of bare and cadmium-covered U-238, Np-237, and Th-232.

TABLE 4. Cs-137 (Interfering Reaction/Primary Reaction)

	Total Fluence (Bare)		Epithermal Fluence (Cd-covered)	
	1×10^{19}	3×10^{19}	1×10^{20}	4×10^{20}
U-238	0.23	0.69	0.07	0.27
Np-237	0.03	0.08	0.004	0.014
Th-232	0.51	1.5	0.13	0.50

IV. Shield Materials

Shield materials are frequently used to eliminate interference from thermal neutron reactions when resonance and fast neutron reactions are being studied. Cadmium is commonly used. However, elemental cadmium (m.p. = 320°C) will melt if placed within the vessel of an operating water reactor. High temperature filters include cadmium oxide (or other cadmium compounds or mixtures), boron (enriched in the B-10 isotope), and gadolinium.

The thickness of the shield materials must be selected to account for burnout from high fluences. Table 5 shows calculation variations of thermal neutron transmission with exposure for given cadmium thicknesses (8).

TABLE 5. Thermal Neutron Transmission of Cadmium versus Exposure

Thermal Neutron Exposure (nvt)	Thermal Neutron Transmission		
	20-mil	30-mil	40-mil Cd
5×10^{20}	0.06	0.004	0.0002
7×10^{20}	0.2	0.01	0.0006
1×10^{21}	0.7	0.07	0.003

Other precautions to consider: cadmium oxide is a powder and is toxic; high temperatures from (n,γ) gamma heating can cause melting and diffusion bonding between shield material and contents (this is a particular problem for in-vessel and in-core dosimetry capsules).

V. Selecting Materials and Planning Experiments

1. Shield and capsule material should be carefully chosen to avoid altering the neutron spectral shape. For example, stainless steel capsules contain elements with high thermal neutron cross sections that can alter the shape of the thermal neutron spectrum. The presence of 1.3-cm-diameter capsule assembly tubes in the BF3 reactor which consisted of 5-mm wall carbon steel and solid aluminum spacers, led to a calculational study of the possibility of spectral shape changes and perturbation effects affecting the fast neutron energy region. Reported SAI perturbation data expressed as correction factors to be applied to (divided into) measured reaction rates are given in Table 6 (9).

TABLE 6. Capsule Correction Factors (With Assembly/No Assembly)

Location	U-238	Np-237	Th-232	ϕ(>0.1 MeV)	ϕ(>1 MeV)
Near-Shroud	0.95	0.94	0.93	0.96	0.93
Near-Vessel	0.96	0.99	0.95	1.01	1.00

2. Nickel must be shielded to prevent burnout of Co-58 and Co-58m by thermal and epithermal neutrons. For example, a 5% loss of Co-58 occurs at 60 days at a thermal flux density of approximately 8×10^{12} n·cm^{-2}·s^{-1} (10). Burnout also occurs for several common dosimeters (and often their activation products) that exhibit high thermal or resonance cross sections (e.g., Co, Au, Ag, Sc, In, U-235, B, Li). A 5% effective loss of Co-60, Au-198, and Ag-110m occurs at thermal fluences of approximately 2×10^{21}, 5×10^{18}, and 1×10^{21} nvt, respectively.

3. The neutron self-shielding phenomenon occurs for dosimeters with high thermal cross sections, and occurs with essentially all resonance detectors. This can be minimized by using low weight percentage alloys, for example, Co-Al, Co-Zr, Ag-Al, In-Al, B-Al, Li-Al.

4. Fission fragment loss has been observed (7%) for thin (0.03-mm) fission foils. Increasing the thickness to 0.13 mm will reduce this loss to about 1%.

5. There are certain elements that should be avoided as impurities, particularly when thermal or intermediate flux analysis is desired, because of their high thermal or resonance cross sections and the resulting interfering activities. These include

Li, B, Cd, In, Hg, Au, Mn, Ta, W, Th, U, Bi, Co, Hf, K, Sn, and rare earths.

6. Thin-wall capsules of aluminum (such as A9-6061), with a provision for sealing out moisture, have been found satisfactory as foil holders for pressure vessel cavity dosimetry. Sealing prevents corrosion and possible contamination of the dosimeters, and allows thorough decontamination of outside surfaces before the experiment is removed from the power plant.

7. Spare ex-vessel detector wells, or other provisions for defined geometry, are recommended for cavity experiments because of air currents in the cavity.

8. Dosimetry capsules may be suspended at different elevations in the cavity on stainless steel beaded chains. This makes handling during entry and placement easy, and also provides a "flux wire" with five reactions: $Co-59(n,\gamma)$, $Fe-58(n,\gamma)$, $Cr-51(n,\gamma)$, $Fe-54(n,p)$, and $Ni-58(n,p)$.

9. Cavity dosimetry is less affected by local power peaks than in-vessel dosimetry. However, local support structures, or shielding, do have an appreciable effect. For example, a lead-shielded detector well (5.6-cm thickness of lead on a 15-cm steel pipe) as compared to a bare well (15-cm steel pipe) at symmetrical core positions, has reduced measured reaction rates by a factor of approximately two. Also, neutron shields at the top of the pressure vessel produce a large depression in flux density.

10. Where possible, a comparison is desirable of in-vessel and ex-vessel experiments along the same core radius and for the same irradiation period. This, when coupled with a transport calculation that is in agreement at the two experimental locations, gives the best prediction of exposure to the vessel wall.

CONCLUSION

Only when all considerations are addressed will consistent and accurate measurements and data from light water reactor neutron dosimetry experiments provide the means to a successful surveillance program.

REFERENCES

1. G. C. Martin, <u>Neutron Spectra and Dosimetry Determinations:</u> <u>GETR Pool, HBPS Pressure Vessel and Outer Core, BRP Outer</u> <u>Core</u>, NEDE-12613 (February 1976).

2. G. C. Martin, <u>Browns Ferry Unit 3 In-Vessel Neutron Spectral</u> <u>Analysis</u>, NEDO-24793 (August 1980).

3. G. C. Martin, <u>Browns Ferry Unit 3 Cavity Neutron Spectral</u> <u>Analysis</u>, NP-1997 (August 1981).

4. G. C. Martin, <u>Browns Ferry Unit 3 Drywell Neutron</u> <u>Measurements</u>, NEDO-24946 (June 1981).

5. C. O. Cogburn and J. G. Williams, "Pressure Vessel Dosimetry at the Arkansas Nuclear Plants," ANS Transactions (November 1983).

6. C. O. Cogburn and J. G. Williams, "Pressure Vessel Dosimetry at U.S. PWR Nuclear Plants," (this Symposium).

7. T. F. Albert, W. K. Hagan, and G. L. Simmons, <u>Analysis of</u> <u>Browns Ferry Unit 3 Irradiation Experiments — Final Report</u>, NP-3719 (August 1984).

8. B. F. Rider, J. P. Peterson, C. P. Ruiz, and F. R. Smith, <u>Accurate Nuclear Fuel Burnup Analyses Ninth Quarterly Progress</u> <u>Report December 1963–February 1964</u>, GEAP-4503 (March 1964).

9. G. L. Simmons and W. K. Hagan, Science Applications, Inc., Private Communication (June 1980).

10. C. H. Hogg, L. D. Weber and E. C. Yates, <u>Thermal Neutron Cross</u> <u>Sections of the Co-58 Isomers and the Effect on Fast Flux</u> <u>Measurements Using Nickel</u>, IDO-16744 (June 1962).

CHARACTERIZATION OF THE IMPERIAL COLLEGE

REFERENCE GAMMA RAY FIELD

J.A. Mason, A.N. Asfar and T.C. Jones

Reactor Centre, Imperial College, Silwood Park,
Ascot, Berkshire, England.

A.M. Fabry and M.R. Menil

SCK/CEN, Boeretang 200, B-2400 Mol, Belgium.

ABSTRACT

This paper describes the implementation and characterization of a reference gamma ray field in the vertical thermal column of the Consort II reactor at Silwood Park. The facility, which is based on the design of the reference gamma ray field at SCK/CEN Mol employs thermal neutrons to generate an internal prompt capture gamma ray field. A cylindrical cadmium radiation, which is black to thermal neutrons, is used as the thermal neutron to gamma ray converter. The driving thermal neutron flux has been calibrated using an NBS type double fission chamber and an installed thermal neutron sensitive borated ionization chamber. The gamma ray field has been characterized using calibrated ionization chambers.

INTRODUCTION

The importance of the gamma radiation component in reactor mixed radiation fields has long been realised in the study of reactor dosimetry. Unfortunately, in the absence of gamma ray fields characteristic of reactor gamma spectra, dosimeter calibration has been performed using low energy systems such as ^{60}Co. It has been necessary to make assumptions about the response of detectors in the higher energy spectra experienced in real reactor systems.

J. P. Genthon and H. Röttger (eds.), Reactor Dosimetry, 407–413.
© 1985 ECSC, EEC, EAEC, Brussels and Luxembourg

A significant step towards gamma ray metrology standardization
has been taken with the introduction of the reference gamma ray
field concept pioneered at SCK-CEN Mol, Belgium (1).   The Mol
system employs capture gamma ray black absorbers that are suspended
in the one meter cavity of the vertical thermal column of the BR1
reactor.   The cylindrical sources are attached to a borated pyrex
thimble and an intense gamma ray field, characteristic of the source
material, is generated inside the thimble.   Including the pyrex
thimble alone, (producing a gamma spectrum dominated by the 480 keV
line), sources of cadmium and cobalt are available and others including indium and samarium are under consideration.   A recent up-
grading involves the replacement of the boron loaded pyrex glass by
hot-pressed boron nitride tubing.   The gamma ray background in the
system which is less than 10% consists principally of carbon prompt
capture gamma rays and the associated Compton scattering continuum.

The reference gamma field concept has a variety of uses.   It
can play a central role in reactor gamma ray dosimetry standardisation.   Applications include instrument calibration and the
measurement of spectral sensitivity using different capture gamma
ray sources.   The measurement and unfolding of photofission cross-
sections is another area of application.   Perhaps the most important contribution of such systems concerns their application in
studying the response of detectors in gamma ray fields typical of
reactors.

The concept of the Mol reference gamma ray field has been
transferred to the Imperial College CONSORT II reactor and adapted
for use in the vertical thermal column.   This paper describes the
implementation and characterisation of the system.

DESIGN OF THE REFERENCE GAMMA RAY FIELD

The graphite column into which the reference field is inserted
forms a vertical branch to the 90° thermal column and it is known
as the vertical thermal column (VTC).   It is in the shape of a
rectangular block and it is penetrated by six accessible holes to a
depth of 1.424 m.   The holes are eccentric with maximum and minimum
dimensons of 99 mm and 93 mm respectively.   A cross section through
the VTC showing the location of the reference gamma ray field is
illustrated in Fig.1.

The design of the reference gamma ray system is based closely
on its precursor at SCK/CEN Mol, Belgium.   Cylindrical gamma ray
radiators (black absorber sources) are attached on a pyrex thimble
which serves as a central structural support.   The thimble is
77.5 cm long, 2 mm thick and has an outside diameter of 52 mm.   It
is also useful as an absorber of electrons and X-rays originating
in the sources.   The pyrex thimble, which is closed at the bottom

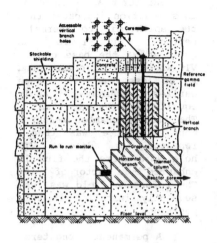

Fig.1 VTC with Reference
Field

end, is connected to a PTFE extension tube which in turn is
connected to an aluminium tube.
The aluminium tube locates in a
steel shielding plug in the concrete shielding block above the
VTC. Detectors and samples may
be fixed to a support rod which
passes through the shielding plug
and down the central axis of the
system.

In principle the borated
pyrex thimble may be employed as
a radiator producing a low
energy gamma ray spectrum dominated by the 480 keV line from
the $^{10}B(n,\alpha\gamma)^{7}Li$ reaction. However it contains only 13% B_2O_3
and is not black to thermal
neutrons: it is therefore not
an efficient gamma ray source.

The cylindrical cadmium radiator which has been used with the
system is 1 mm thick, 55 mm in diameter, 24 cm long and is black
to thermal neutrons. Because of the exponential thermal neutron
flux shape in the VTC, a black cadmium end cap has been added at
the base of the system in order to prevent thermal neutron streaming and to enhance the local gamma dose rate in the higher flux
region. The configuration of the system is illustrated in Fig.2.
where the detail of the radiator and pyrex thimble have been enlarged. A cobalt radiator is also available for the system
although it has not been used.

THERMAL NEUTRON CALIBRATION

Before installing the reference gamma ray field it was
necessary to establish a thermal neutron flux calibration for the
90° thermal column vertical branch or vertical thermal column,
(VTC). A run to run monitor was also calibrated against both
thermal neutron flux in a reference position and reactor power.

The thermal neutron calibration of the VTC was undertaken with
an NBS type double fission chamber employing two ^{235}U deposits of
different densities. The fission rates from the two deposits were
compared for consistency. The chamber was mounted in an aluminium
cylinder positioned on the central axis of VTC hole No.9 which had
previously been established as a reference position for thermal
neutron measurements. The axial variation of the thermal neutron
flux was then measured.

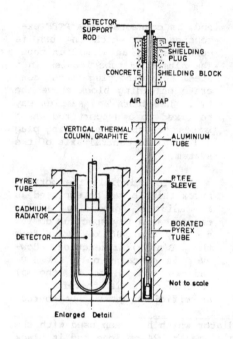

DETECTOR SUPPORT ROD

STEEL SHIELDING PLUG

CONCRETE SHIELDING BLOCK

AIR GAP

VERTICAL THERMAL COLUMN, GRAPHITE

ALUMINIUM TUBE

PYREX TUBE

P.T.F.E. SLEEVE

CADMIUM RADIATOR

DETECTOR

BORATED PYREX TUBE

Not to scale

Enlarged Detail

Fig.2   Reference Gamma Ray Field

Earlier measurements in this position indicated that the epithermal to thermal neutron flux ratio was greatest at the base of the VTC where the ratio was found to be less than 1%. These earlier measurements apply equally to adjacent holes. The fission chamber measurement procedure was then repeated for adjacent VTC hole No.8 where the flux is enhanced by a factor of 1.16. The reference field was subsequently installed in this position.

A permanent, long term, accurate and consistent form of thermal neutron flux monitor is required for use with the reference gamma field. It should be possible to normalize the monitor to reactor power and the monitor should not be sensitive to reactor fine control rod adjustments or flux perturbations due to the presence of the reference field. It was apparent that the fission chamber could not be used for this purpose as the thermal neutron absorbing gamma ray radiator significantly perturbed the thermal flux in hole No.9. A boron lined current type thermal neutron sensitive ionization chamber (Twentieth Centry Type RC7/CPBE/H150) was therefore obtained for use as a run to run monitor. In order to overcome the perturbations in the VTC, access was obtained to the end face of the horizontal branch of the 90° thermal column which drives the VTC. The chamber was installed in the graphite at the end face beyond the point at which thermal neutrons gain access to the VTC. No thermal flux perturbations have been observed with the ionization chamber installed in this position so that the monitor and reference system have been effectively decoupled. The run to run monitor has been calibrated against both thermal neutron flux in the reference position and reactor power.

NEUTRON AND GAMMA RAY MEASUREMENTS

Once the reference gamma field was installed in hole No.8 of the VTC, its characteristics were measured. The measurements reported in this section concern the full system with pyrex thimble and black absorber cadmium radiator with cadmium end cap. The system was position such that the end of the cadmium radiator was 65 mm from the base of the VTC.

Initially measurements were made of the thermal, epithermal and fast neutron flux within the reference field. The epithermal neutron flux was measured using CR-39 solid state nuclear track detectors (SSNTD). The CR-39 materials using natural boron deposits as radiators were enclosed in cadmium boxes 0.5 mm thick. A number of SSNTD's were placed at different positions along the axis of the system within and above the cadmium gamma ray radiator region. The measurements were made at 20 kW reactor power and each SSNTD was exposed for 15 minutes. Epithermal neutrons were detected through the reaction $^{10}B(n,\alpha\gamma)^{7}Li$ where both the alpha particle and ^{7}Li particle register in the CR-39 with a known efficiency for specific etching conditions. A solution of 6.25 N NaOH was used at a temperature of 70 ± 0.2°C for three hours. Track densities were counted manually for each SSNTD using an optical microscope and a magnification of 400.

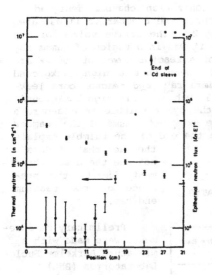

Fig.3 Reference Field Axial Neutron Flux

From the measured track densities and a knowledge of the boron resonance integral, CR-39 efficiency and particle ranges in boron, the epithermal neutron flux parameter per unit lethargy has been determined assuming a 1/E dependence. The epithermal neutron flux as a function of position from the base of the pyrex thimble is plotted as part of Fig.3.

The NBS type double fission chamber has been employed to measure the axial neutron flux along the centre of the reference field within and above the cadmium gamma ray radiator region. Using two ^{235}U deposits, again for consistency, the chamber measured a fission rate due to both thermal and epithermal neutrons. From a knowledge of the ^{235}U resonance integral and the epithermal neutron flux measured by SSNTD's it was possible to determine the thermal neutron flux in the system. This axial thermal neutron flux is plotted as part of Fig.3. It can be seen that the higher thermal neutron background in the upper part of the cadmium radiator region is due to the streaming of thermal neutrons from above.

Measurements were also made inside the pyrex thimble with the cadmium radiator removed. In this case the thermal neutron flux on the axis is about 35% of the value with no thimble present. This is due to the fact that the boron density is not sufficient to make the thimble black to thermal neutrons.

The fast neutron component of the flux was measured at a point 0.7 cm from the base of the pyrex thimble inside the radiator region. The NBS type double fission chamber was used with two ^{238}U deposits. The fast flux component above 2 MeV was measured as 2.12×10^5 n $cm^{-2} s^{-1}$. All of the fission chamber measurements have an uncertainty of about 2%.

Measurements of the gamma ray exposure in the reference field have been made with calibrated neutron insensitive graphite ionization chambers.(2) In order to increase the gamma ray intensity the reference gamma ray field system was moved to a lower position such that the bottom of the cadmium lined pyrex thimble was resting just above the base of the VTC.

Gamma ray measurements were made with the system in different configurations using a graphite ionization chamber designed to resemble a graphite microcalorimeter. The chamber is illustrated inside the reference field in Fig.2. The active volume of the chamber extended over 5 cm so that it was in a region of gamma ray exposure gradient. The results for a reactor power of 100 kW are tabulated in Table 1. It can be seen that a high background exists of carbon prompt capture gamma rays and reactor core leakage gamma rays amounting to about 83 R h^{-1}. This high background effectively precludes the use of the pyrex thimble as a separate low energy gamma ray radiator. The extended volume of the chamber and its measurement location above the end of the thimble explain the fact that it does not see the dose enhancement due to the presence of the cadmium end cap.

Preliminary measurements were made with a miniature Berkeley Nuclear Laboratories (BNL) graphite ionization chamber (2) of the exposure rate profile with cadmium radiator and end cap fitted. With the reactor at 100 kW measurements were made at 2 cm intervals from the base of

TABLE 1

GAMMA REFERENCE FIELD EXPOSURE RATES

Configuration	Exposure Rate (R h^{-1})
Pyrex tube, Cadmium sleeve with end cap	716
Pyrex tube only	114
Pyrex tube and Cadmium removed	83
Pyrex tube, Cadmium sleeve, Cadmium end cap removed	715

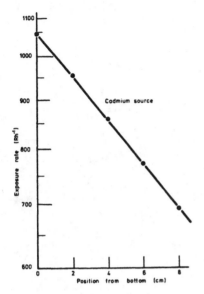

Fig.4 Exposure Rate in
Reference Field

the pyrex thimble. The results are displayed as Fig.4. The exponential shape of the gamma ray exposure rate profile is consistent with that of the thermal neutron field which drives the reference gamma ray system.

CONCLUSIONS

The reference gamma ray field concept provides a useful approach to gamma ray metrology problems. Most important, reference fields may be used to simulate reactor gamma rays and thus provide a useful environment for reactor dosimetry measurements.

Future extension of the system include the use of a cobalt radiator and the addition of shielding to reduce background gamma rays from the reactor core. It may also be possible to implement the system in the 50 cm spherical cavity in the 270°C thermal column.

The authors wish to acknowledge the support and assistance of Mr. J. Debrue of SCK-CEN. The assistance of Mr. T.A. Lewis of the CEGB, who performed the ionization chamber measurements and of Mr. G.D. Burholt, Reactor Manager, is acknowledged.

REFERENCES

1. A. Fabry, G. Minsart. F. Cops and S. De Leeuw, "The Mol Cavity Fission Spectrum Standard Neutron Field and its Applications", Proceedings of the Fourth ASTM-EURATOM Symposium on Reactor Dosimetry, Gaithersburg, Maryland, USA, 1982.

2. S.J. Kitching and T.A. Lewis, "An Ionisation Chamber for Measuring Gamma Exposure Rates in Mixed Radiation Fields", report TPRD/B/0348/N83, Central Electricity Generating Board, Berkeley Nuclear Laboratories, 1983.

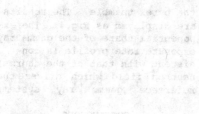

IMPROVED MICROCALORIMETRY FOR RADIATION

ABSORBED DOSE MEASUREMENTS

J.A. Mason, A.N. Asfar and P.J. Grant

Reactor Centre, Imperial College, Silwood Park, Ascot,

Berkshire, England.

ABSTRACT

The status of the Imperial College adiabatic microcalorimetry technique is described in the present paper. The
implementation and use of an improved computer based control and
logging system is described in conjunction with the MKII adiabatic
microcalorimeter. Results are reported of measurements of a few
tens of milligray per second in the Imperial College ^{60}Co facility.
These measurements are compared with calibrated ionization chamber
measurements. Areas of future development in the technique are
identified.

INTRODUCTION

A knowledge of the radiation absorbed dose or energy deposited
by radiation in reactor materials is of great importance in the
design of reactor systems. Predictions of the structural integrity
of reactor components and indeed reactor life itself are dependent
on dosimetry information. In mixed radiation fields this information is usually obtained by separately measuring the neutron
and gamma ray components of the field. Confidence in the calculation process is obtained using this experimental information.

Unfortunately, the performance of detailed dosimetry calculations for realistic high powered reactor facilities is seldom
feasible. Geometric complication and the mixture of materials are
the principal limiting factors. The solution that is generally
adopted is to perform measurements and calculations using a low
powered, benchmark system where the geometric structures are

J. P. Genthon and H. Röttger (eds.), Reactor Dosimetry, 415–423.
© 1985 ECSC, EEC, EAEC, Brussels and Luxembourg

simplified and direct comparisons may be made between measurements and calculations.

In practice neutron and gamma ray component measurements in mixed radiation fields are not simple. Neutron spectral information obtained from activation measurements is generally subject to large uncertainties. Gamma ray component measurements using ionization chambers or thermoluminescent dosimeters (TLD) often suffer from neutron interference and a non-linearity of gamma ray response.

An absolute technique which may be used to measure both components of a mixed radiation field is calorimetry. It provides a direct measure of the total energy deposition rate in materials. In the past, the technique has been limited by a lack of sensitivity and low power, benchmark field measurements have not been possible. Recent work at Imperial College has been directed towards developing the techniques for such benchmark measurements (1), and the approach has been applied to a shielding benchmark experiment (2).

The purpose of the present work is to improve the performance of the calorimeter and to identify and eliminate systematic errors in the measurement technique. As a first stage in this process the control and data logging instrumentation has been significantly upgraded and new control algorithms have been implemented. This paper reports on these developments and the improvement in calorimeter performance which has resulted. Gamma ray measurements for graphite and aluminium calorimeters using the Imperial College ^{60}Co irradiation facility are described. As a result of the work improvements in the calorimeter design have been identified.

CALORIMETER DESIGN

The calorimeter operates in adiabatic mode in that a central sample of absorbing material is maintained in adiabatic equilibrium with its surroundings. This is achieved by enclosing the cylindrical sample in cylindrical jackets of the same material. The purpose of the jackets is twofold. First they serve to prevent convective heat transfer between the sample and the environment and second they ensure electronic equilibrium between the sample and its surroundings. The temperature on the outer surface of the inner jacket or baffle is controlled by means of an electrical heater in such a way that it closely matches that of the sample. In this manner heat transfer between the sample and its surroundings is virtually eliminated and, in a radiation field, the temperature rise of the sample will be due solely to radiation energy deposition. This energy deposition rate is proportional to the rate of rise of sample temperature where the sample mass (m) and heat capacity (C) are constants of proportionality as illustrated by the following equation:

$$\frac{dQ}{dt} = mC\frac{dT}{dt} \qquad (1)$$

The geometry of the calorimeter is very similar to that reported earlier (1), with the exception that a new connection block has been mounted on a stem 23.5 cm above the calorimeter outer can. The component parts of the device as well as an assembled calorimetry may be seen in Fig.1. The sample is 10 mm in diameter and 58 mm high. The jacket walls are 1.5 mm thick and the outer aluminium can is 42 mm in diameter and 93 mm long ignoring the mounting stem. The structure of an assembled calorimeter is illustrated in Fig.2. The supports are made of PTFE and electrical heating coils of 40 SWG nickel-chromium wire are wound around both the sample and the inner jacket.

Temperature measurement is accomplished using 46 SWG chromel-constantan thermocouples. One measures the sample temperature and two others form a differential thermocouple between the sample and inner jacket. The signal from the differential thermocouple is used to control the jacket electrical heater and thus maintain adiabatic equilibrium. The cold junction consists of a dewer flask containing demineralised water at room temperature.

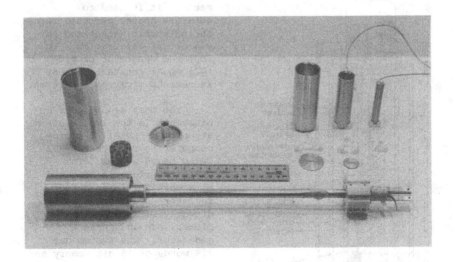

Fig.1 Modified MKII Microcalorimeter and its Component Parts

CONTROL AND DATA ACQUISITION SYSTEM

The original CAMAC based calorimeter control and data logging system (1) suffered from several significant limitations. The Harwell 7025 controller which it employed had a very limited memory and as a result the control algorithm was relatively crude. In particular the response time of the calorimeter to external temperature changes was long and the result was imperfect sample temperature control. Reliability was also a problem when the system was transported. In order to improve the performance of the calorimeter and eliminate systematic errors, the first phase of the development programme involved the implementation of an enhanced computer based control and data logging system.

Enhanced Electronic Hardware

The components of the enhanced electronic hardware are represented diagrammatically in Fig.3. The cold junction, amplifiers, multiplexer and digital voltmeter (DVM) remain as reported previously (1). All other components of the system are new.

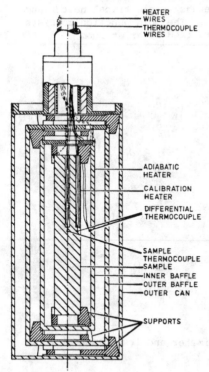

HEATER WIRES
THERMOCOUPLE WIRES

ADIABATIC HEATER

CALIBRATION HEATER

DIFFERENTIAL THERMOCOUPLE

SAMPLE THERMOCOUPLE
SAMPLE
INNER BAFFLE
OUTER BAFFLE
OUTER CAN

SUPPORTS

Fig.2   Microcalorimeter in Cross Section

The computer controlled constant current generator replaces a current source with 256 unequal current increments. It is used to generate heater currents for the adiabatic jacket and for electrical calibration. The improved generator provides 1024 equal constant current increments from zero to 20 mA.

The most significant enhancement is the implementation of an LSI-11/2 microcomputer to replace the 7025 CAMAC based system. The CAMAC interface is replaced by a parallel input/output register which communicates directly with the multiplexer, DVM and current generator. The system has 32K words of 16 bit memory and is programmed in FORTRAN and assembly language. The processor is capable of floating point arithmetic and sophisticated interrupts are available. These two features greatly facilitate real time control of microcalorimeters.

Software and the Adiabatic Heater Control Algorithm

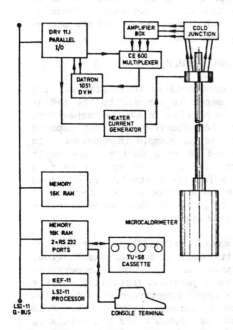

Fig.3 Microcalorimeter Control System Hardware

The increased sophistication of the software is the most important factor in the first stage of improvement in the calorimetric technique. Most of the routines have been written in the FORTRAN language and the communication routines for the peripheral devices, multiplexer, DVM and current generator have been written in assembly language. The system of commands is menu driven and all system functions are accessible to the software program. The main program contains a main control loop and clock driven interrupts command the logging of data from the calorimeter and the control action for the adiabatic heater. The software is easy to use, may be readily modified and upgraded and is effectively portable. Many more commands are incorporated into the system in order to facilitate operator intervention. The program also makes much more information available to the experimenter.

A prime feature of the software is the four term proportional control algorithm, which is used to control the jacket temperature and maintain adiabatic equilibrium. The previous system employed a crude control algorithm in which the jacket heater current was switched on and off and its magnitude was varied in relatively large steps. The response of the algorithm to significant events such as the onset of nuclear heating or a change in the calorimeter environment temperature was slow. The new system employs a fast response digital control algorithm (3) consisting of a component proportional to the error signal, a component which varies as the time derivative of the error signal, an integral component and a double integral component. The error signal in this case is the differential thermocouple voltage which measures the sample - inner baffle temperature difference. The first three terms correspond to the three terms of a conventional proportional controller. The fourth term is necessary to correct for the fact that the sample-jacket equilibrium temperature is itself changing

419

during a nuclear heating measurement (4). The controller gains have been determined empirically and the generation of excessive integral values, the so-called integral windup problem known to control engineers, has been prevented in the software. Measurements on the system indicate that both the response and the degree of control have been significantly improved.

MEASUREMENTS AND RESULTS

A calorimetry measurement is conducted in three stages. The main stage during which the calorimeter is exposed to the radiation field, results in a linear rate of rise in sample temperature Before and after this stage, when the calorimeter is not exposed to radiation, the temperature drift of the sample is observed. In theory the sample temperature should be zero during these drift periods. In practice this is not the case, either due to cold junction instability or imperfect adiabatic control. In order to obtainthe true time rate of sample temperature rise an average of the observed sample temperature drift rates is subtracted from the measured sample radiation induced heating rate.

As the drift rate correction makes a significant contribution as a source of measurement error it required further examination. A first question concerned the magnitude of sample temperature drift that could be attributed to cold junction instability. This was accomplished by monitoring the room temperature cold junction with a thermocouple against an ice junction. In most cases cold junction drift was found to be less than 0.003°C per hour: this is a small fraction of the typical sample temperature drift rate.

In previous measurements it was found that a positive bias voltage applied to the differential thermocouple chopper amplifier reduced the sample temperature drift rate. It had also become apparent that the significant component of the drift resulted from the action of the adiabatic jacket heater: heat in excess of that required to maintain a constant sample temperature was available and resulted in a gradual sample temperature rise. Further investigation revealed that a small offset or bias voltage could be measured on the differential thermocouples of calorimeters which were in thermal equilibrium. This spurious voltage with a magnitude of about 4 μV, did not represent a sample-inner baffle temperature difference. The resulting amplified signal however was interpreted by the control system as representing a temperature difference and excess jacket heating therefore resulted. The voltage is thought to arise as a result of a temperature gradient imbalance between the two arms of the differential thermocouple.

In order to study the sample temperature drift effect a bias facility was built into the software, equivalent to the differential thermocouple amplifier bias adjustment, but more reproducible. Also a thermocouple was fitted to the calorimeter outer

420

can to monitor the external environment temperature seen by the calorimeter. Measurements were then performed using the aluminium calorimeter to measure the sample temperature drift rate as a function of equivalent amplifier bias at constant sample temperature. Then the sample temperature drift rate was measured as a function of sample temperature with a constant external temperature. The results of the former measurement are displayed as Fig.4. In the measurements the sample temperature drift rate appears to be inversely proportional to both the differential thermocouple amplifier bias and the sample-external temperature difference.

A measurement was then performed to determine if a heat loss process was involved in the drift phenomenon. A calorimeter was left to reach thermal equilibrium with the environmental temperature. With no adiabatic control a negative (by convention) differential thermocouple voltage was measured. It was identified as the spurious voltage mentioned earlier. The equivalent amplifier bias was then set to zero the differential thermocouple and the calorimeter sample temperature was raised slightly. The adiabatic jacket heater was then enabled and the calorimeter came under control with virtually zero sample temperature drift. At this point the environmental temperature was lowered and a negative sample drift (temperature decrease) resulted. This effect is interpreted as a heat loss, probably from the ends of the sample where the adiabatic jacket is less effective.

As a result of these measurements it has become apparent that the two effects observed operate in competition in causing sample temperature drifts. The procedure of adjusting the equivalent amplifier bias has the effect of compensating to reduce the resulting observable sample temperature drift to zero.

Gamma ray measurements were undertaken using the Imperial College ^{60}Co facility in order to assess the performance of the calorimetry system. The calorimeters were suspended both in air and in a graphite sleeve at fixed distances from a gamma ray source. A knowledge of the sample temperature drift effects was used to adjust the equivalent amplifier bias to reduce the drift. The calorimetry heating measurements were converted to equivalent exposure rates and they are compared with calibrated ionization chamber measurements (5). The results are given in Table 1.

The results represent an improvement over previous results where the systematic error was in reality 12%-15%. The scatter in the data is probably a result of the inability to control the environmental temperature which significantly influences the sample temperature drift rate and hence the error.

Table 1. ^{60}Co Gamma Ray Energy Deposition Measurements Expressed as Exposure Rates (R.s^{-1})

Calor- meter	Con- figuration	Position from Source No.1 (cm)	Calori- meter Measure- ment	Ioni- zation Chamber Measure- ment	Differ- ence(%) (Calori- meter − Ioni- zation Chamber)
Graphite	Graphite Sleeve	30	10.38	11.15	− 6.9
			10.03	11.15	− 10.1
		60	2.95	3.09	− 4.6
			2.90	3.09	− 6.2
Graphite	Air	60	2.84	3.166	− 10.2
			3.04	3.166	− 4.0
			3.03	3.166	− 4.5
			2.84	3.166	− 10.2
			3.00	3.166	− 5.2
Alumin- ium	Graphite Sleeve	30	10.65	11.15	− 4.5
			10.62	11.15	− 4.8
			10.72	11.15	− 3.9
Alumin- ium	Air	60	2.90	3.166	− 8.5
			3.07	3.166	− 3.1
			2.81	3.166	− 11.1

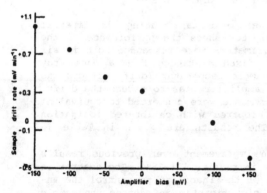

Fig.4 Sample Temperature Drift Rate versus Amplifier Bias

CONCLUSION

The improved calori- meter performance is attributed to the improve- ment in adiabatic control and an increased under- standing of the mechanism of sample temperature drift. Further improve- ments in the technique which have been identi- fied include a controlled outer baffle heater to isolate the calorimeter from the environmental temperature, a reduction in sample length to reduce end heat loss and further improvements in the adiabatic control algorithm.

The authors wish to acknowledge the assistance of Mr. T.A. Lewis of the CEGB Berkeley Nuclear Laboratories who performed the ionization chamber measurements, and of Mr. A. Packwood and Dr. I.J Curl of AEE Winfrith for useful discussions. The support and encouragement of Dr. J. Butler of AEE Winfrith is gratefully acknowledged. The work is supported by the United Kingdom Atomic Energy Authority.

REFERENCES

1. J.A. Mason, "Development of Sensitive Microcalorimeters for Absorbed Dose Measurements in Benchmark Radiation Fields," Proceedings of the 4th ASTM-EURATOM Symposium on Reactor Dosimetry, Gaithersburg, Maryland, U.S.A., 22-26 March 1982.

2. I.J. Curl, J.A. Mason, A. Packwood and P.C. Miller, "Adiabatic Microcalorimetry in Shielding Benchmark Experiments," Proceedings of the 6th International Conference on Radiation Shielding, Tokyo, 1983.

3. R.J. Bibbero, Microprocessors in Instruments and Control, J. Wiley & Sons Inc., New York, 1977.

4. R.W. Wilde, private communication on control, Electrical Engineering Department, Imperial College, 1984.

5. T.A. Lewis, private communication concerning [60]Co Ionization Chamber Measurements, Berkeley, 1984.

APPLICATION OF THE MULTI-COMPONENT WIRE ACTIVATION

DETECTOR SYSTEM FOR IN-CORE NEUTRON SPECTROMETRY

AT A NPS

H.-C. Mehner, S. Nagel, M. Schöne,
I. Stephan, U. Hagemann

Zentralinstitut für Kernforschung
Rossendorf, GDR

U. Pieper, W. Gehrig

VEB Kombinat Kraftwerksanlagenbau
Berlin, GDR

ABSTRACT

A new-developed sensor technique based on
a multi-component wire activation detector
(MWAD) together with an automated measuring
system has been utilized for measuring the
axial spectral neutron flux distribution inside
reactor core of a WWER-440 of the Greifswald
NPS. The improved MWAD consisting of Au, Mn,
Mo, Ni, and W has been inserted into a fresh
fuel assembly enriched with 3.6 % ^{235}U. The
measurements were carried out during one cycle
in monthly intervals. By using simple on-line
evaluation models the thermal and fast neutron
flux as well as the spectrum hardness were
calculated on the basis of the measured reaction rates.

INTRODUCTION

To study the axial spectral neutron flux distribution inside WWER power reactors an in-core measuring
system based on a multi-component wire activation
detector (MWAD) has been developed (1). The system has
been applied at the Greifswald NPS in order to investigate the changes of neutron flux and neutron spectrum
caused by fuel burnup and fission product poisoning

during reactor operation. The results should provide
reactor physics core calculation codes with experimentally adjusted data as well as generate input data
for reactor shielding codes, which may be used for
estimating the radiation damage of the reactor pressure vessel.

MULTICOMPONENT WIRE ACTIVATION DETECTOR

The new-developed MWAD represents an alloy of different metals appropriate to measure neutrons at a
broad energy region. Comparing with commonly used activation foils the MWAD offers the advantage to determine simultaneously several reaction rates with one
probe exposition and one gamma-ray spectrum measurement. Designing the detector in wire form the axial
distribution of neutron flux and neutron spectrum can
be determined with a single irradiation process.
This wire with a diameter of 0.9 mm consisting mainly
of the metals Au, Mn, Mo, Ni, and W is the basis of
the in-core measuring system (1). The quantitative
composition and the reactions used are summarized in
Table 1.

Especially by adding molybdenum to the alloy and
by improving the fabrication technique the tensile
strength and the hardness of the detector was increased to meet the requirements at the NPS. The
manganese content was reduced to about 30 % compared
with the earlier version (1) in order to optimize the
composition and the gamma-ray acquisition.

Table 1. Composition and reactions of the MWAD

Element	Mass fraction (%)	Reaction
Gold	0.30 ± 0.01	$^{197}Au(n,\gamma)^{198}Au$
Manganese	0.34 ± 0.01	$^{55}Mn(n,\gamma)^{56}Mn$
Molybdenum	14.68 ± 0.08	$^{98}Mo(n,\gamma)^{99}Mo$
		$^{100}Mo(n,\gamma)^{101}Mo \rightarrow {}^{101}Tc$
Nickel	81.96 ± 0.10	$^{58}Ni(n,p)^{58}Co$
Tungsten	2.44 ± 0.05	$^{186}W(n,\gamma)^{187}W$

APPLICATION AND EVALUATION AT THE NPS

The measuring system was installed at the fourth unit of the Greifswald NPS on the top of the reactor shielding. From here, a dry channel led to the core into a 3.6 % enriched standard fuel assembly situated in a nearly undisturbed core region.

The measurements were performed during the fourth cycle of the unit in monthly intervals. In course of these experiments the single components of the measuring system proved reliable and practicable.

The evaluation of the measurements was carried out on the spot by using a EMG—666/B mini-computer coupled with a multichannel analyser. Programmes were elaborated to calculate the gamma-ray intensities, the reaction rates as well as some parameters of the neutron field like thermal and fast neutron flux and neutron spectrum hardness.

The neutron spectrum hardness α defined by

$$\alpha = \frac{\gamma_e}{\phi_t}$$

where γ_e – epithermal neutron flux per unit lethargy,
ϕ_t – thermal neutron flux,

is the mean value between $\alpha(Au/Mn)$ and $\alpha(W/Mn)$, where $\alpha(Au/Mn)$ and $\alpha(W/Mn)$ are calculated from the ratios of the reaction rates $R(Au)/R(Mn)$ and $R(W)/R(Mn)$, respectively:

$$\alpha(Au/Mn) = \frac{\frac{\sqrt{\pi}}{2}\sqrt{\frac{T_0}{T}}\left(1 - \frac{R(Au)}{R(Mn)}\bigg/\frac{\sigma_0(Au)}{\sigma_0(Mn)}\right)}{\left(\frac{R(Au)}{R(Mn)}\bigg/\frac{\sigma_0(Au)}{\sigma_0(Mn)}\right)\frac{I(Mn)}{\sigma_0(Mn)} - \frac{I(Au)}{\sigma_0(Au)}}$$

with T – neutron temperature (T_0 = 293 K)
σ_0 – thermal activation cross section
I – resonance integral.

By using the effective thermal activation cross section

$$\sigma_{eff,t} = \frac{\sqrt{\pi}}{2} \sqrt{\frac{T_0}{T}} \sigma_0 + \alpha I$$

the thermal neutron flux is calculated from the reaction rates of Au, W, Mn, ^{98}Mo, and ^{100}Mo:

$$\phi_t = \frac{R}{\sigma_{eff,t}}$$

The fast neutron flux is derived from the reaction rate of ^{58}Co

$$\phi_f = \frac{R(^{58}Co)}{\langle \sigma \rangle \cdot C}$$

with $\langle \sigma \rangle$ — cross section averaged over a fission
                   neutron spectrum
     C   — correction factor for the decay of ^{58}Co
            (3).

EXPERIMENTAL RESULTS

A typical gamma-ray spectrum obtained from the MWAD is shown in Fig. 1.

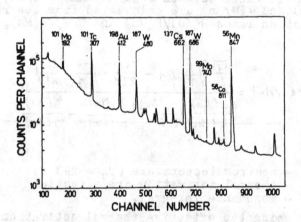

Fig. 1.   Gamma-ray spectrum of the MWAD
          (irradiation time 6 min, cooling
          time 4 min, counting time 75 min)

The single peaks at 307 keV (^{101}To), 412 keV (^{198}Au), 480 keV and 686 keV (^{187}W), 740 keV (^{99}Mo), 811 keV (^{58}Co), and 847 keV (^{56}Mn) were used for calculating the reaction rates R(Au), R(Mn), R(^{98}Mo), R(^{100}Mo), R(Ni), and R(W). The gamma-ray peak at 662 keV is caused by the monitor and calibration source ^{137}Cs. Table 2 contains e.g. the reaction rates measured at the end of the cycle (267 effective days of operation) at 10 axial core heights. The nuclear data were taken from ref. (2).

Fig. 2 shows the axial distribution of the thermal neutron flux, the spectrum hardness and the fast neutron flux at full power. The plotted lines should guide the eye. The errors are statistical ones. The thermal neutron flux $\emptyset_t$ is the mean value from the neutron fluxes $\emptyset_t$(Au), $\emptyset_t$(Mn) and $\emptyset_t$(W).

The thermal neutron flux shows the typical distribution in the core of pressurized water reactors: at the beginning of the cycle a cos-distribution which is shifted towards the bottom due to the in-

Table 2. Reaction rates per nucleus at the end (267 effective days) of the fourth cycle of unit 4

Measuring position	^{55}Mn $[10^{-10}s^{-1}]$	^{98}Mo $[10^{-11}s^{-1}]$	^{100}Mo $[10^{-11}s^{-1}]$	^{186}W $[10^{-9}s^{-1}]$	^{197}Au $[10^{-8}s^{-1}]$	^{58}Ni $[10^{-11}s^{-1}]$
1	2.34	4.22	2.81	2.45	0.766	C.546
2	3.40	6.11	3.97	3.63	1.13	0.834
3	3.62	6.22	4.70	3.94	1.22	0.800
4	3.90	6.63	4.95	4.19	1.29	0.877
5	3.99	6.87	4.91	4.19	1.34	0.895
6	4.03	7.98	5.20	4.34	1.39	0.937
7	4.18		5.41	4.61	1.49	
8	4.39	8.30	5.58	4.64	1.47	0.881
9	3.81	7.86	4.75	4.16	1.39	1.09
10	2.68	4.90	3.33	2.80	0.895	0.669
Statistical error (%)	2	8	4	2	2	5
Total error (%)	5	9	7	5	5	7

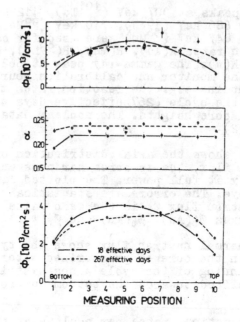

Fig. 2. Thermal neutron flux
ϕ_t, spectrum hardness
α and fast neutron
flux ϕ_f at the beginning and the end of
the fourth cycle of
unit 4.

fluence of the control rods and at the end of the
cycle a nearly constant distribution. The shifting of
the neutron flux maximum into the upper core region is
caused by extracting the control rods. The mean axial
thermal neutron flux in the assembly is nearly constant: $3.25 \cdot 10^{13}/cm^2s$ at the beginning and
$3.20 \cdot 10^{13}/cm^2s$ at the end. The temporal axial mean
value is characterized by $3.25 \cdot 10^{13}/cm^2s$ with a
standard deviation of 0.9 %.

The fast neutron flux ϕ_f follows mainly the thermal one. The statistical error of the fast neutron
flux is greater than that of the thermal one due to
the lower accuracy of the 811 keV peak analysis of
^{58}Co. The temporal axial mean value of the fast
neutron flux is $7.93 \cdot 10^{13}/cm^2s$ with a standard deviation of 7 %.

The spectrum hardness α is in axial direction nearly constant. From the beginning to the end of the cycle it is increased at about 10 %. Moreover, the greatest ascending happens just at the beginning of the cycle during the first 40 days of operation. The hardening of the spectrum in the epithermal energy region is caused by increasing the partial burnup of Pu and by growing fission product poisoning.

The strongest radiation damages of the reactor pressure vessel have to be expected in the axial neutron flux maximum, about in the core middle plane. Here, the greatest differences of the thermal as well fast neutron flux during the cycle only amount to 10 - 20 % (Fig. 2). Therefore, rough estimations of neutron flux and radiation damage of the reactor pressure vessel can be carried out assuming temporal mean values over the cycle. However, for detailed studies the temporal axial neutron flux and spectrum fluctuations have to take into account. Moreover, for analysing and evaluating the accelerated metallurgical test specimen which are irradiated in the lower and upper core region not far from the welded joints of the vessel, the temporal axial fluctuations in the order of 70 % have carefully to be taken into consideration.

FINAL REMARKS

For the present purpose, only some results from our measurements which have a bearing on the topics here discussed, have been   quoted. To investigate the neutron spectrum at reduced core coolant flow rates we mountered the MWAD measuring system at the Rheinsberg NPS and measured within a experimental fuel assembly (4). Further efforts in our work should be directed to study the neutron spectrum changes at higher fuel burnup during some operation cycles and to measure immediately at or near the reactor pressure vessel inner wall. The measuring system gives here the advantage to investigate directly the temporal changes of the neutron spectrum, a work which can be performed hardly with the conventionally used multiple foil technique.

REFERENCES

1. H.-C. Mehner, I. Dennstädt, U. Hagemann, S. Nagel, and M. Schöne, Multicomponent Wire Activation Detector System for Neutron Spectrometry on Power Reactors, Proc. of the Fourth ASTM-EURATOM Symposium on Reactor Dosimetry, Gaithersburg, USA, March 22-26, 1982, NUREG/CP-0029, Vol. 1, 379.

2. W.L. Zijp, and H.J. Baar, Nuclear Data Guide for Reactor Neutron Metrology, Part I: Activation Reactions, Netherlands Report ECN-70 (August, 1979).

3. I. Dennstädt, and H.-C. Mehner, Messung der Neutronenflußdichte mit einer Mehrkomponentensonde an Leistungsreaktoren des KKW "Bruno Leuschner" Greifswald, Report ZfK-449 (July, 1981).

4. U. Hagemann, and H.-C. Mehner, In-core Neutron Spectrometry at Reactors of WWER Types, IAEA Seminar on Diagnosis and Response to Abnormal Occurences at Nuclear Power Plants, Dresden, GDR, 12-15 June 1984.

DATA ACQUISITION AND CONTROL SYSTEM FOR

THE K_{1C}-HSST EXPERIMENTS AT THE ORR

L.F. Miller and R.W. Hobbs

The University of Tennessee and Oak Ridge National Laboratory

Knoxville and Oak Ridge, Tennessee, USA

ABSTRACT

Major components and primary functions of the process
control system for the K_{1C}-HSST irradiation experiments at
the Oak Ridge Research Reactor (ORR) are described. Information relative to methodology for integrating unique
features of the Digital Equipment Corporation's RSX-11M
Operating System with analog-to-digital and digital-to-
analog hardware is presented. In particular, data flow
among various real-time applications programs relative to
system hardware is presented. General features of the
temperature control algorithm are presented, and results
that illustrate the spatial temperature distribution in
the capsule achieved by the control system are included.

INTRODUCTION

This paper describes the computer system hardware, software,
and methodology relative to data acquisition and control for the
K_{1C}-HSST metallurgical capsule irradiation program at the Oak
Ridge Research Reactor (ORR). Previous publications (1,2,3,4)
document data acquisition and control for the Light Water Reactor
Pressure Vessel Surveillance Dosimetry Improvement Program (LWR-
PVS) irradiations performed at the ORR. Since the completion of
the LWR-PVS irradiations, significant improvements have been made
in the computer system hardware and software; thus, functional
objectives relative to the upgraded system are achieved by significantly different methods than those utilized for the LWR-PVS
capsule irradiations.

J. P. Genthon and H. Röttger (eds.), Reactor Dosimetry, 433–439.
© 1985 ECSC, EEC, EAEC, Brussels and Luxembourg

HARDWARE, OPERATING SYSTEM, AND APPLICATION SOFTWARE

Major components of the data acquisition and control system include: (1) a Digital Equipment Corporation (DEC) LSI-11/23 computer with 256 kilobytes of random access memory, (2) a DEC Industrial Products I/O Subsystem housed in an IP-300 system unit, and (3) a Data Systems Design (DSD) 30-megabyte Winchester/Floppy (DSD-880) disk subsystem. Functional relationships among these components are illustrated by Fig. 1. Note that details relative to components and to integration of this equipment into a system configuration are provided elsewhere (5,6).

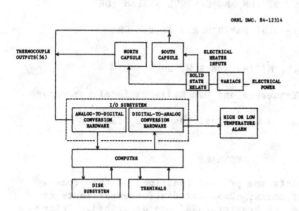

ORNL DWG. 84-12314

Fig. 1. Functional Illustration of the Data Acquisition and Control System for the K_{1C}-HSST Irradiation Experiments.

Computer system hardware control and data management are accomplished under the RSX-11M, Version 4.0, Real-Time Operating System supplied by DEC. This system allows for multiple programs to run concurrently at various levels of priority. These programs may be automatically rescheduled at fixed-time intervals or be designated to execute a single time. In addition, the rescheduled programs may be fixed in a core memory partition or be loaded from disk each time they are executed.

Data acquisition requirements for the K_{1C}-HSST irradiation experiments consist of acquiring capsule temperature and reactor power data for permanent records and for process control. The temperature and reactor power histories are needed for analysis of metallurgical data, real-time control of power to 36 electrical heaters (18 per capsule), and annunciating alarms. The data are processed by the control algorithm to determine the fraction of maximum heater power to which each heater is energized. The maximum power is determined by the physical setting of the variac.

Application software for support of the K_{1C}-HSST experiment consists of several programs that execute on a specified schedule and communicate through system-controlled common areas. Communication with data acquisition and control hardware is accomplished through a non-executable assembly language program that overlays the I/O page (IPXCOM), and communication among various applications programs is accomplished through a similar program that overlays a system controlled area (K1CCOM). Table 1 lists independent K_{1C} support programs along with a brief functional description of each, and Fig. 2 illustrates data flow among these programs.

Table 1. Brief Functional Descriptions of K_{1C}-HSST Support Programs

Program Name	Function
K1CSTR	Initializes data for intertask communication
FSSR	Fires solid state relays
ADEUSA	Reads analog data, converts it to engineering units, and performs a statistical analysis of the digitized data
K1CTCL	Calculates heater demand for control of K_{1C}-HSST experiment temperatures
K1CDSV	Saves K_{1C} data on disk
K1CPRT	Prints K_{1C} data on terminal TTO:
K1CDPR	Prints K_{1C} data stored on disk at terminal TTO:
K1CCOM	Provides linkage for intertask data communication
IPXCOM	Provides linkage for I/O page communication

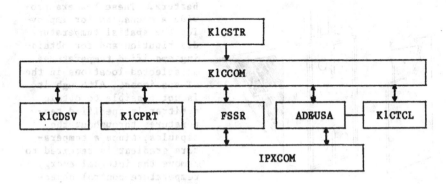

Fig. 2. Data Flow Among K_{1C}-HSST Irradiation Experiment Support Computer Programs.

The computer program that fires the solid-state relays (SSRs) is scheduled on the same interval as the SSR hardware timeout interval of 0.2 seconds and consists of a main routine written in FORTRAN that sets up binary masks for controlling the digital output hardware and of an assembly language subroutine that communicates directly with the I/O subsystem circuit boards.

Analog data are read and processed by a FORTRAN main routine (ADEUSA) which utilizes an assembly language subroutine that controls the analog-to-digital converter hardware. This routine is scheduled for execution every 60 seconds based on requirements for control of the irradiation capsule temperature. The program which calculates heater demand for controlling irradiation capsule temperature (K1CTCL) and ADEUSA are ensured of sequential execution through the use of global event flag 41. In particular, the execution of K1CTCL is delayed until ADEUSA completes data acquisition to ensure that K1CTCL obtains new data each periodic interval. K1CTCL calculates heater demand based on a proportional-integral-differential (PID) control algorithm.

Data are saved on a disk file by program K1CDSV at one-hour intervals during steady state conditions and at 15-minute intervals during transient conditions. Data are printed on terminal TT0: at one-hour intervals by K1CPRT.

CONTROL METHODOLOGY

General features of the K_{1C}-HSST irradiation capsules are illustrated by Figs. 3 and 4. Note that the 4-in. thick tensile (4T) metallurgical test specimens are sandwiched between two steel plates and that each plate contains nine independently controllable electrical heaters. These heaters provide a mechanism for improving the spatial temperature distribution and for obtaining specified temperatures at selected locations in the 4T specimens. Although it is not possible to obtain arbitrarily specified temperatures throughout the capsules, since a temperature gradient is required to remove the internal energy, temperature control objectives of improving the peak-to-average temperature ratio and of obtaining

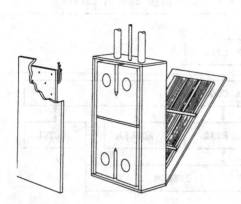

Fig. 3. Simplified Layout of
4T Capsule Design.

specified temperatures at selected locations can be readily achieved. These objectives are accomplished by: (1) control of the helium-neon cover gas composition, (2) manual adjustment of variacs that supply the electrical heaters, and (3) computer control of the solid-state relays. The helium-to-neon ratio significantly impacts overall capsule heat transfer characteristics and, hence, provides a coarse control of capsule temperature. Variac positions fix the maximum power that can be delivered to a particular heater, and computer control of the SSRs determines the fraction of time that voltage is applied across the heater wires.

The control algorithm operates in two modes: (1) proportional-integral-differential (PID) when capsule temperatures are near their desired values and (2) on-off when capsule temperatures are outside of the PID control band. The algorithm transfers from on-off to PID for each heater when applicable thermocouples obtain a $\pm10°C$ control band relative to the set-point temperature. Actual heater demand is determined from proportional and differential errors (multiplied by their associated gains) relative to a reference demand. The reference demand is updated based on the integral gains applied to each heater demand. During steady-state conditions, the proportional and differential error signals go to zero, and the heaters are energized at the reference set point condition. More detail on the algorithm is provided by Ref. 6.

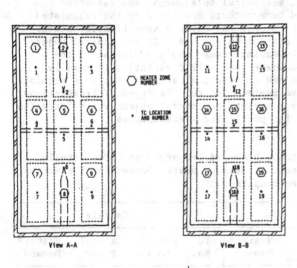

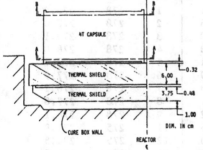

Fig. 4. Plan View of HSST 4T Capsule in Poolside Facility and Location of Heater Zones and Specimen Surface Temperature.

437

RESULTS AND CONCLUSIONS

Operating experience with the capsule indicates that, in the zones where electrical heat can be applied, the control system is capable of maintaining surface temperatures within approximately 2°C of specified values.

An illustration of the control system performance, subject to present surface temperature demands, is provided by Table 2. Note that the actual thermocouple readings are very near to the demanded value. The several thermocouples that differ significantly from the demanded values are in regions that cannot be precisely controlled because of limitations in heat transfer capabilities relative to the cover gas, mechanical tolerances, and radiation heating. The demanded temperatures are based on the calculated spatial energy deposition rate and capsule heat transfer characteristics relative to control objectives. If the control objective is to minimize the maximum temperature variation in the capsule, test cases indicate that a 12°C variation is readily obtainable. Data obtained to date indicate that the data acquisition and control system implemented for the K_{1C}-HSST irradiations at the ORR can satisfy a variety of control objectives relative to demanded temperatures for particular thermocouples.

Table 2. Comparison Among Demanded Temperatures (°C), Actual Temperatures, and Percent of Maximum Heater Demand for the North and South Capsule Thermocouples

North Capsule				South Capsule			
TC No.	Demand Temp.	Actual Temp.	Heater* Demand	TC No.	Demand Temp.	Actual Temp.	Heater* Demand
1	283	283.0	91	1	278	278.0	86
2	283	283.1	86	2	278	278.0	74
3	283	283.1	90	3	278	278.1	91
4	283	282.7	63	4	278	278.0	68
5	283	282.8	81	5	278	277.7	95
6	283	282.9	79	6	278	277.9	77
7	283	283.1	93	7	278	278.2	91
8	283	282.7	97	8	278	277.8	81
9	283	282.8	92	9	278	278.1	94
11	278	277.7	76	11	275	274.9	18
12	278	279.3	0	12	275	280.5	0
13	278	278.0	83	13	275	274.8	73
14	278	277.8	69	14	275	285.0	0
15	278	278.3	0	15	275	284.4	0
16	278	277.5	0	16	275	275.3	0

Table 2. (Continued)

North Capsule				South Capsule			
TC No.	Demand Temp.	Actual Temp.	Heater* Demand	TC No.	Demand Temp.	Actual Temp.	Heater* Demand
17	278	277.9	93	17	275	274.5	100
18	278	277.3	100	18	275	282.0	0
19	278	276.8	100	19	275	274.9	100
20		288.4	--	20		288.4	--
21		291.1	--	21		291.9	--
22		288.4	--	22		289.1	--
23		288.2	--	23		286.3	--
24		290.8	--	24		290.3	--
25		287.8	--	25		289.5	--
26		289.0	--	26		289.0	--
27		293.5	--	27		294.7	--
28		289.8	--	28		293.6	--

*Thermocouples 20 through 28 are interior to the test specimens and are indirectly associated with surface energy input.

REFERENCES

1. S.H. Merriman, A Computerized Process Control System for the ORR-PSF Irradiation Experiment Part 1: Overall View of the Control System, NUREG/CR-1710, ORNL/NUREG/TM-405/P1, Nuclear Regulatory Commission, Washington, DC, Vol. 1, July 1982.

2. L.F. Miller, A Computerized Process Control System for the ORR-PSF Irradiation Experiment Part 2: Mathematical Basis and Computer Implementation of the Temperature Control Algorithm, NUREG/CR-1710, ORNL/NUREG/TM-405/P2, Nuclear Regulatory Commission, Washington, DC, Vol. 2, November 1980.

3. L.F. Miller, "Control of the ORR-PSF Pressure Vessel Surveillance Irradiation Experiment Temperature," Proc. of the Fourth ASTM-EURATOM Symposium on Reactor Dosimetry, NUREG/CR-0029, Nuclear Regulatory Commission, Washington, DC, 1982.

4. L.F. Miller, Analysis of Temperature Data from the ORR-PSF Irradiation Experiment: Methodology and Computer Software, NUREG/CR-2273, ORNL/TM-7766, Nuclear Regulatory Commission, Washington, DC, 1981.

5. Industrial Products, I/O Subsystem Users Guide, EK-OPIOS-UG-003, Digital Equipment Corporation, Maynard, MA, 1979.

6. L.F. Miller and R.W. Hobbs, Data Acquisition and Control of the K_{1C}-HSST Experiments at the ORR: Description of the Computer System Hardware, Real-Time Application Support Software, and Control Methodology, NUREG/CR-3872, ORNL/TM-9253, Nuclear Regulatory Commission, Washington, DC (to be published).

USE OF STAINLESS STEEL FLUX MONITORS

IN PRESSURE VESSEL SURVEILLANCE

T. H. Newton, Jr., C. O. Cogburn, J. G. Williams

University of Arkansas, Mechanical Engineering Dept.

Fayetteville, Arkansas

ABSTRACT

Stainless steel beaded chains have been suspended in detector wells adjacent to the outside of the pressure vessels of the two Arkansas Power & Light Company reactors. The chain provides more ease in handling than wire in supporting activation foil capsules and also provides axial flux profiles by use of two fast and three thermal neutron induced reactions in the beads from the chain: $^{58}Ni(n,p)^{58}Co$, $^{54}Fe(n,p)^{54}Mn$, $^{58}Fe(n,g)^{59}Fe$, $^{59}Co(n,g)^{60}Co$ and $^{50}Cr(n,g)^{51}Cr$.

Activation analysis using ^{252}Cf sources was performed on samples of the beads from the chain and revealed good uniformity in composition among the beads. The composition was determined to be: Fe:63.25%, Cr:17.08%, Ni:8.85%, Co:0.12%.

Results from the vessel dosimetry experiments showed no difference in axial flux shape between different parts of a fuel cycle, as well as consistency with activation foil experiment results and transport calculation results.

Fluence rate values from the beads were extrapolated to the inside surface of the vessel of Unit I to determine the mid-plane fluence rate.

J. P. Genthon and H. Röttger (eds.), Reactor Dosimetry, 441–448.
© 1985 ECSC, EEC, EAEC, Brussels and Luxembourg

INTRODUCTION

Stainless steel beaded chains have been used in reactor pressure vessel surveillance experiments of the two Arkansas Power & Light Company reactors. The chain provides a convenient method of supporting activation foil capsules in detector wells adjacent to the two reactor pressure vessels, as well as providing axial flux profiles by using the beads as flux monitors.

The chains consist of beads which have a diameter of about 4.5mm, weight about 210mg and contain sufficient quantities of iron, chromium, nickel and cobalt to allow monitoring of two fast and three thermal neutron induced reactions: $^{58}Ni(n,p)^{58}Co$, $^{54}Fe(n,p)^{54}Mn$, $^{58}Fe(n,g)^{59}Fe$, $^{59}Co(n,g)^{60}Co$ and $^{50}Cr(n,g)^{51}Cr$.

ACTIVATION ANALYSIS

Activation analysis was performed on the beads (1) to determine the elemental composition of the four elements used in the flux analysis by taking twelve beads from various positions along the length of rolls of unirradiated chain which is used for the experiments. The specimens were taken from four 150 meter rolls, designated as TN1, TN2, TN3 and TN4, obtained in a single purchase. A single roll from an earlier batch was designated TN0. Standard materials for comparative neutron activation analysis were pure foils of iron and nickel, a dilute cobalt-aluminum foil and 37 mg of pure chromium sesquioxide, Cr_2O_3, in powder form.

The analysis was performed using ^{252}Cf fission neutron sources in a H_2O moderator tank at SEFOR Calibration Center. The irradiation facility contains positions for up to three intense californium sources and a spinner with five sample positions, each containing up to 1 gram of material. Two sources were used, having a total neutron emission rate at the time of the 118 hour irradiation of 3.97×10^{10}/s. Specimens were counted using semiconductor gamma spectrometers for the five reaction products mentioned.

Comparison of the results for the twelve specimens showed that the cobalt content of the bead from roll TN0 was significantly different from the other eleven values, but that the twelve values for iron, chromium and nickel were similar. Standard deviations for the twelve values for iron and chromium, and eleven values for cobalt (excluding TN0) deduced from the (n,g) reaction products were 1.5%, 1.5% and 1.9%, respectively. The twelve values for iron and nickel determined from the (n,p) reaction products had larger variations between the samples, which was attributed to the larger fast flux gradient within the irradiation volume than that for thermal neutrons. This effect was

442

compensated by adjusting the $^{58}Ni(n,p)$ values by a bias factor equal to the ratio of the $^{58}Fe(n,g)$ results to the $^{54}Fe(n,p)$ results. This adjustment reduced the standard deviation of the twelve nickel values to 2.7%. Final values of the sample compositions are given in Table 1. The variance of the results shows no significant difference between the three sets when the value for cobalt in the TNO sample is excluded.

DOSIMETRY EXPERIMENTS

The chains were lowered into detector wells in the cavities next to the pressure vessels, such as the one shown in Figure 1 from Unit I. The experiments were irradiated for periods up to a full fuel cycle, and retrieved when an outage of sufficient duration occured to facilitate removal of experiments for analysis. The chain has been used during three fuel cycles for each reactor.

Table 1

Composition of Steel Beads % by Weight

Roll	Set	Iron	Chromium	Nickel	Cobalt
TN1	1	65.1	17.6	9.39	0.121
TN1	1	63.4	17.0	8.90	0.121
TN1	1	63.7	17.3	8.85	0.119
TN1	1	63.5	17.2	8.92	0.119
TN1	2	63.4	17.1	9.83	0.118
TN1	2	61.7	16.6	8.42	0.117
TN1	2	62.1	16.9	8.83	0.117
TN1	2	62.5	17.0	8.70	0.121
TN2	3	63.6	16.8	8.68	0.122
TN3	3	62.5	17.0	8.85	0.118
TN4	3	64.3	17.1	8.60	0.123
TNO	3	63.2	17.3	9.63	0.100
mean		63.25	17.08	8.85	0.120*
s.d.		0.94	0.26	0.25	0.002*

* excluding the TNO value for set 3

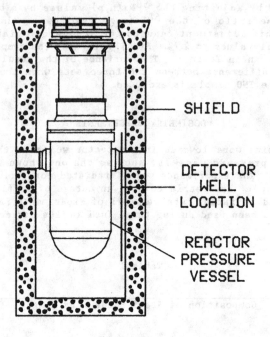

SHIELD

DETECTOR
WELL
LOCATION

REACTOR
PRESSURE
VESSEL

Figure 1   Cavity Dosimetry Location for ANO, Unit I

Arkansas Nuclear One (ANO) Unit I has two diametrically opposite wells available for the experiments. Well 2 is a bare 6-inch schedule-40 steel pipe, while well 1 is covered with a radial thickness of 55 mm inches of lead up to about 0.5 m above the top of the active fuel.

Unit II has four wells available for the experiments. Two of the wells contain obstructions, preventing measurements below a point 1.2 m below the top of the fuel.

Beads from the irradiated chain were cut at various axial intervals and counted using semiconductor gamma spectrometers. Axial flux profiles were determined from the reaction rates for the reactions mentioned.

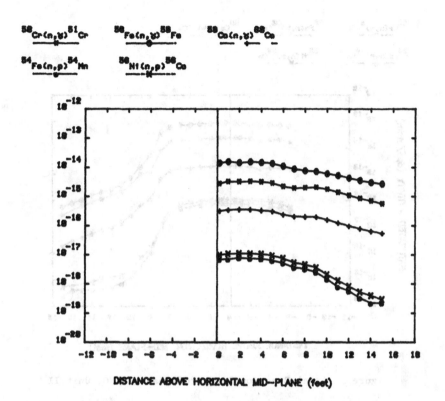

Figure 2   Axial Reaction-Rate Profiles for ANO, Unit I

RESULTS

Figure 2 shows a typical profile from Unit I. This profile was taken in well 1, the lead shielded well. The effects of the lead shielding are apparent in the slight rise in the reaction rates at about 2.4 m above mid-plane, where the lead shielding ends.

A typical profile from Unit II is shown in Figure 3. The large drop in activities beginning above 6 feet above mid-plane is due to the presence of a neutron shield which begins just above the top of the active fuel. The slight dip in the thermal activities at about 3 feet below mid-plane is due to the presence of a support structure at that point.

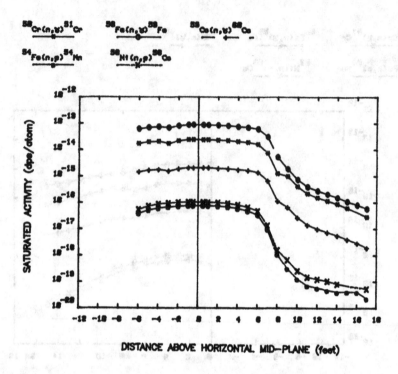

Figure 3   Axial Reaction-Rate Profiles for ANO, Unit II

Comparisons of the $^{58}Ni(n,p)^{58}Co$ activities around the fuel region of four different experiments in Unit II are shown in Figure 4. Experiment II-2 was irradiated in the middle of fuel cycle 1, II-4 at the beginning of fuel cycle 2, II-6 at the end of fuel cycle 2 and II-7 over the entirety of cycle 3. No significant differences in the shape of the axial distribution are apparent in these results.

The chain experiment results have shown excellent agreement with foil experiments which have been suspended by the chain. Results also show excellent agreement with the results of transport calculations performed by the University of Missouri - Rolla (UMR), as shown in Figure 5, which shows the results of the $^{58}Ni(n,p)$ reactions of experiment I-7 from Unit I. The larger gradient away from the mid-plane in the transport calculation may be partly due to the fact that the experimental data was averaged over the entire irradiation, whereas the transport calculation is an instantaneous value. Fluence rates at the inside surface of the pressure vessel have been determined by dividing the saturated

446

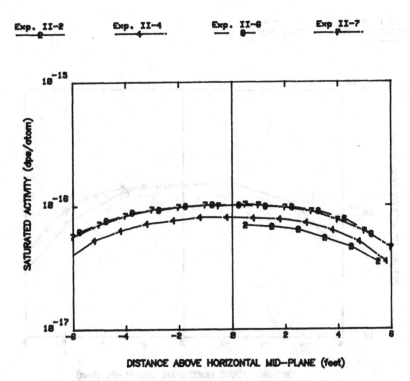

Figure 4   Axial Reaction Rate Profiles
for Four Experiments in Unit II

activities of the fast reactions from the beads by their cross
sections, furnished by the University of Missouri, Rolla (UMR) and
multiplying by the ratio between the calculated ratio of the
reaction rates at the vessel and in the detector well.   The midplane fluence rate above 1 MeV at the inside surface of the
pressure vessel of Unit I was determined from experiment I-7 to be
8.79×10^9 n/s-cm^2.

CONCLUSION

The stainless steel beaded chains offer considerable ease of
handling when used to support activation foil capsules in detector
wells adjacent to the pressure vessel. This handling ease, as
well as the availability of two fast and three thermal neutron
induced reactions, makes the chains ideal for axial dosimetry in
the wells.

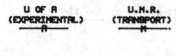

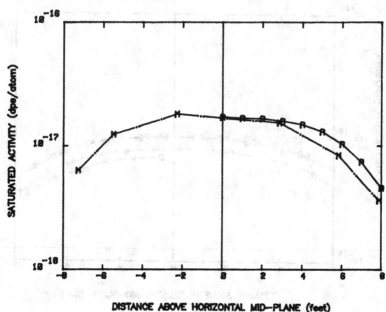

Figure 5   Comparison Between Measurement and Calculation
for the ^{58}Ni(n,p) Reaction in Unit II

Activation analysis has shown uniformity in composition among
samples of the beads, and the bead results have been consistent
when compared with both activation foil results and transport
calculations. These consistencies considerably enhance the value
of the beaded chain as a neutron flux monitor.

REFERENCES

1. J. G. Williams, T. H. Newton, Jr., and C. O. Cogburn, "Activa-
    tion Analysis of Stainless Steel Flux Monitors Using ^{252}Cf
    Sources", 5th International Conference on Nuclear Methods in
    Environmental and Energy Research, Mayaguez, Puerto Rico,
    April 2-6, 1984. ANS Topical Conference Proceedings in Press.

SIMULTANEOUS APPLICATION OF ACTIVATION AND DAMAGE DETECTORS

H.J. Nolthenius[1], W.P. Voorbraak[1], W.L. Zijp[1],

A. Alberman[2], M. Benoist[2], M. Thierry[2]

[1]Netherlands Energy Research Foundation ECN, Petten, The Netherlands

[2]Centre d'Etudes Nucléaires de Saclay, Gif-sur-Yvette, France.

ABSTRACT

Neutron fluence rate spectrum information derived from activation and fission detectors and from two types of damage detectors irradiated in an HFR core is summarized. The spectrum information refers to various experiment positions in the HFR. Damage to Activation Ratios (DAR), derived from the results of the damage detectors and of the adjusted neutron spectra, are in good agreement with each other.

INTRODUCTION

The High Flux Reactor (HFR) at Petten is a materials testing reactor (MTR) in which specimens of fissile materials, graphite, steel and other materials are tested. Many of the irradiations are carried out in reloadable facilities (fig. 1).

The study on the damage effect of radiation incident on materials requires detailed information on the neutron fluence rate spectrum and damage characteristics. Therefore spectrum characteristics measurements have been carried out in mock-up assemblies of experiments which are frequently irradiated in core positions which are used regularly for that type of measurements. These positions are shown in fig. 1.

J. P. Genthon and H. Röttger (eds.), Reactor Dosimetry, 449–456.
© *1985 ECSC, EEC, EAEC, Brussels and Luxembourg*

The S-assemblies (aluminium filler assemblies with an inner aluminium plug surrounded by a stainless steel layer, see fig. 1) have been used also in previous irradiations to determine the neutron characteristics in experiment assemblies. The measurements have been carried out with 28 activation detector sets, 30 TUNGSTEN and 25 GAMIN damage detectors. The GAMIN detector gives a damage rate representative for graphite. The TUNGSTEN detector determines directly the damage rate in tungsten, which can be converted into the damage rate for steel.

In general, a set of activation detectors consists of a series of activation and fission detectors, covered by cadmium or boron carbide boxes, together with 2 series of uncovered detectors above and below the covered set (in an aluminium, graphite or steel box, depending on the type of experiment (see fig. 2). A set of GAMIN detectors consisted of 5 detectors positioned above each other along the centre line of an experiment. The sets of TUNGSTEN detectors consist of 2 groups of 3 detectors positioned near the centre line. Both types of damage detectors can only be used within certain limits for temperature and fluence. In the neighbourhood of the TUNGSTEN detectors cobalt wire pieces have been irradiated in order to obtain information which was required to apply a correction to the responses for the recoil of the TUNGSTEN nucleus due to the (n,γ) reaction. The reported measurements are a collaboration of CEA Saclay, ECN and JRC-Petten.

The irradiation was performed during an irradiation of 5 hours at a reactor power of 300 kW. The temperature at the positions of the GAMIN and TUNGSTEN detectors was about 33°C. Reference TUNGSTEN and GAMIN detectors accompanied the mock-up experiments during all treatments (inclusive of transports), except the irradiation. The activation detectors have been counted in Petten and the damage detectors have been measured in Saclay.

With the ECN version of the SAND-II program, the neutron fluence spectra have been adjusted. As input spectrum the result of a multigroup diffusion calculation for core position E5 has been used. The activation cross-section values for the adjustment were taken from the DOSCROS81 library (1). No self-shielding corrections have been applied, because for all sensitive materials strongly diluted materials have been used.

A few adjustment runs have been performed in which also the response of a TUNGSTEN detector was applied as one of the experimentally derived reaction rates. The main advantage is, that now response is available for the energy range important for damage. This energy range has negligible response from the activation and fission detectors (see fig. 3). The cross-section for this TUNGSTEN detector was obtained with an adjustment procedure also. In this latter adjustment the responses of TUNGSTEN detectors in 8 different neutron spectra were considered (2).

DAMAGE PARAMETER

The neutron damage can be characterized by various parameters. In the case that the displacement rate is considered (R_{dpa}) it can have advantages to compare it with for instance the activation rate of nickel (R_{Ni}). A special normalization of this ratio yields the so called Damage to Activation Ratio (DAR). The normalization realizes a value of 1 if the neutron spectrum is a pure fission neutron spectrum. The definition formula for the DAR is:

$$(DAR)^a = \frac{<\sigma_{dam}>^a / <\sigma_{dam}>^f}{<\sigma_{act}>^a / <\sigma_{act}>^f} = \frac{\phi_{dam}}{\phi_{act}}$$

When applied to displacements in materials and activation of nickel one has

$$(DAR)^a = \frac{R_{dpa} / <\sigma_{dpa}>^f}{R_{Ni} / <\sigma_{Ni}>^f}$$

where $<\sigma_{dam}>^a$ = average cross-section for damage over the total energy region for the actual neutron spectrum;

$<\sigma_{act}>^f$ = average cross-section for activation over the total energy region for a fission neutron spectrum (Watt);

ϕ = equivalent fission fluence rate.

The DAR value for the materials steel and graphite with nickel as activation reaction was calculated for the output spectra of the adjustment procedure (with and without the reaction rate for the TUNGSTEN damage detector). These data are compared in table 1 with the DAR values obtained in a more direct way for steel and graphite with the damage detectors TUNGSTEN and GAMIN.

CONCLUSION

The results in table 1 show, that in general a good agreement is obtained for the results for the DAR values which are obtained with the direct method with TUNGSTEN or GAMIN and those obtained with the indirect method, using a number of reaction rates in an adjustment procedure, followed by the calculation of the damage characteristics with appropriate cross-sections. Except for C7 the DAR values derived from the TUNGSTEN detectors have large standard deviations; the DAR value for E7 is roughly two standard deviations larger than for the indirect method.

The DAR values which are obtained using the response of TUNGSTEN also in the input of the adjustment are for the positions C7 and E3

451

in good agreement with the other two methods. For the positions F8 and H2 differences in the order of 6% are found. At this moment no explanation is available for this effect in the neutron spectra, which yield large DAR values. A detailed uncertainty analysis is required to obtain a good understanding of this effect.

ACKNOWLEDGEMENT

The work described in this document has been carried out under contract to the European Commission and has been financed by the JRC budget.

REFERENCES

1. W.L. Zijp, H.J. Nolthenius, H.Ch. Rieffe: "Cross-section library DOSCROS81 (in a 640 group structure of the SAND-II type)", Report ECN-111 (Netherlands Energy Research Foundation ECN, Petten, 1981).

2. A. Alberman, J.P. Genthon, H.J. Nolthenius, W.L. Zijp: "Nouveaux developpements de la dosimétrie des dommages par technique tungsten (W)", Proc. 4th ASTM-Euratom symposium on reactor dosimetry, Gaithersburg, 22-26 March 1982, Report NUREG/CP-0029, Vol. 1 (also CONF-820321, Vol. 1), p. 321 (USA Nuclear Regulatory Commission, Washington D.C., 1982).

Table 1. Comparison of DAR values obtained with GAMIN and TUNGSTEN damage detectors and values obtained with activation spectrometry.

The DAR values refer to the reaction $^{58}Ni(n,p)$.

core position	assembly	TRIO channel	environment	DAR for graphite		DAR for steel		
				GAMIN technique	activation spectrometry	TUNGSTEN technique	activation spectrometry	activation spectrometry and TUNGSTEN reaction rate
C3	S(tandard)	-		1.86 (0.8%)	1.84		1.40	
C5	TRIO	II	graphite	1.81 (0.8%)	1.82		1.39	
C7	TRIO	I	graphite	1.85 (0.4%)	1.87		1.42	
		II	graphite		1.85		1.41	
		III	graphite		1.95	1.49 (3.9%)	1.46	1.45
D2	TRIO	I	graphite		2.13		1.54	
		II	graphite		2.58		1.72	
		III	graphite		2.15		1.54	
E3	TRIO	I	steel	2.04 (0.5%)	2.07	1.53 (13.1%)	1.53	1.49
		II	steel		1.98		1.49	
E7	S(tandard)	-			1.88	1.77 (9.7%)	1.43	1.53
F2	REFA	-	graphite	2.18 (0.8%)	2.40		1.64	
F8	TRIO	II	steel		2.35	1.58 (11.2%)	1.63	1.54
H2	REFA	-	steel/aluminium		2.51	1.64 (20.5%)	1.69	1.59

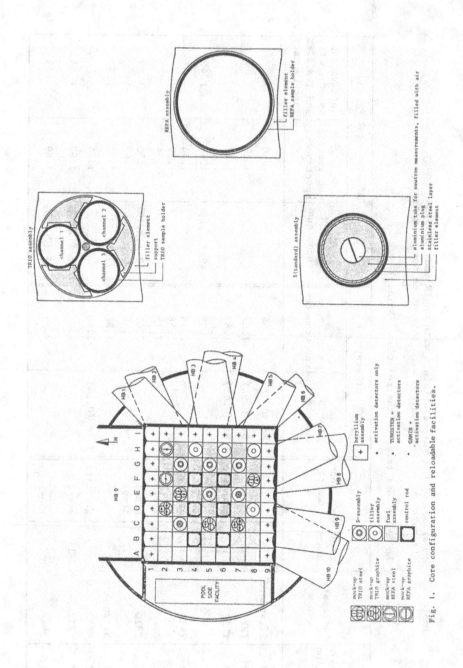

Fig. 1. Core configuration and reloadable facilities.

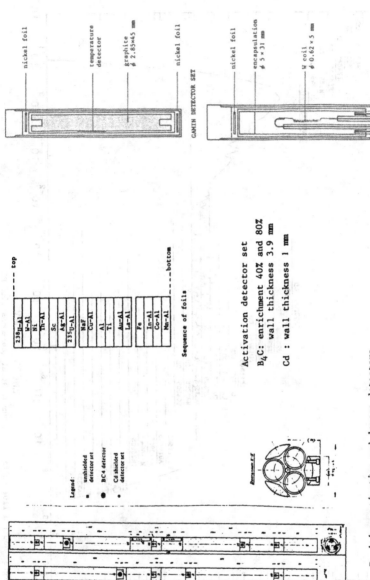

nickel foil

temperature detector

graphite
⌀ 2.85×45 mm

nickel foil

GAMIN DETECTOR SET

nickel foil

encapsulation
⌀ 5 × 31 mm

W coil
⌀ 0.62 × 5 mm

socle

nickel foil

TUNGSTEN DETECTOR SET

---- top

| 238U-Al |
| W-Al |
| Ni |
| Th-Al |
| Sc |
| Ag-Al |
| 235U-Al |

| NaF |
| Cu-Al |
| Al |
| Ti |
| Au-Al |
| La-Al |

| Fe |

| In-Al |
| Co-Al |
| Mo-Al |

----bottom

Sequence of foils

Activation detector set

B_4C: enrichment 40% and 80%
wall thickness 3.9 mm

Cd : wall thickness 1 mm

Legend:

■ unshielded detector set

● BC4 detector

● Cd shielded detector set

Fig. 2. Position activation and damage detectors.

455

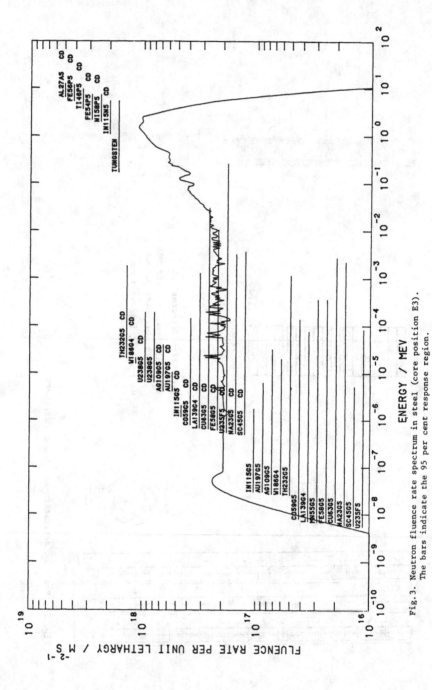

Fig.3. Neutron fluence rate spectrum in steel (core position E3). The bars indicate the 95 per cent response region.

THE S^{32} (n,p)P^{32} THRESHOLD DETECTOR AND ITS APPLICATION FOR FAST NEUTRON DOSIMETRY (FAST REACTORS AND FUSION REACTORS)

G.Perlini, H.Rief
Joint Research Centre
Ispra (Va) Italy

M.D. Carter, M.F. Murphy
U.K.A.E.A.
AEE-Winfrith (England)

ABSTRACT

The knowledge of fast neutron attenuation at great penetration depth is very important in reactor shielding analysis. To validate cross section libraries and radiation transport codes several clean, single material benchmarks have been performed. For two of the most important materials, namely iron and sodium, neutron deep penetration measurements were carried-out at the Pavia EURACOS facility. Using a plane fission disk source the fast neutron propagation was measured by means of the S^{32} (n,p)P^{32} threshold reaction in a large block of iron (145 x 145 x 130 cm^3) and sodium (180 x 200 x 390 cm^3). The quantity of the irradiated sulphur detectors ranged from 1 to 500 grams.

J. P. Genthon and H. Röttger (eds.), Reactor Dosimetry, 457–463.
© *1985 ECSC, EEC, EAEC, Brussels and Luxembourg*

IRRADIATION INSTALLATION

The EURACOS II facility is installed in front of the thermal column of the LENA reactor operated by the University of Pavia (Fig. 1). Its neutron source is made up of six trapezoidal U-Al alloy plates containing a total of 4.2 Kg. of U^{235}. They form an almost circular fission plate with a diameter of 80 cm.

With the TRIGA Mark II reactor at full power (250 KW) on the axis at 6 cm from the converter plate the $S^{32}(n,p)P^{32}$ reaction rate was measured. If the neutron spectrum is assumed, to be close to a U^{235} fission spectrum, this leads to a fission neutron flux of $1,2.10^8$ $n/cm^2.sec$.

In front of the converter disc there is a large irradiation chamber in which models of the shield under study can be installed.

Fig. 1 - EURACOS II Facility

In the first benchmark experiment worked plates (145 x 145 x 4 cm^3) were assembled to study <u>neutron penetration in iron</u>. The resulting block with dimensions 145 x 145 x 130 cm^3 was placed on a borated concrete and aluminium trolley and put in front of the converter.

Despite the care taken in the working of the iron plates they could not be made perfectly plane. It was found that in structures made up of groups of individual plates the thickness exceeded by about 1% the nominal value.

As the fast flux penetrating the iron suffers attenuations up to 10^{-7} ÷10^{-8} after one metre, it is not possible to use a single type of detector. For this reason sulphur pastilles, 2 cm in diameter and of thickness 2 mm, were used for the first part, and sulphur sheets of increasing weight were used for the intermediate and final parts of structure.

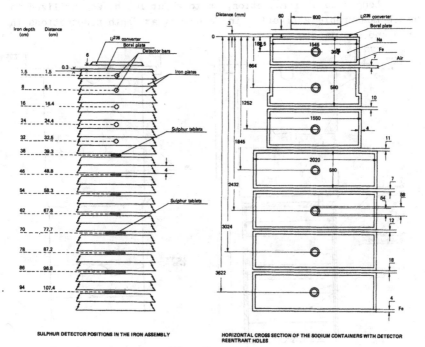

SULPHUR DETECTOR POSITIONS IN THE IRON ASSEMBLY   HORIZONTAL CROSS SECTION OF THE SODIUM CONTAINERS WITH DETECTOR REENTRANT HOLES

Fig. 2

The small pastilles could be inserted into the centre of the iron plates, while the 20x20x1 cm^3 sulphur sheets had to be placed between two iron plates. All the detectors were enclosed in Cd boxes with a thickness of 1 mm. The irradiation geometry is that of Fig.2.

For the study of <u>penetration in sodium</u>, 7 iron boxes (wall thickness 4 mm) were filled with a total of about 10 tonnes of sodium. The boxes mounted on three aluminium concrete trolleys occupied all the available space in the irradiation chamber for an overall length of 390 cm with a section of 150 x 170 cm^2 between 0 and 100 cm and 180x200 cm^2 between 100 and 390 cm.

Pastille detectors with diameter 2 cm and thickness 2 mm were also used for this structure in the first layers while in the other layers cylindrical detectors with diameter 8 cm and a length of about 10 cm were used. Each detector was clad with 1 mm of Cd and was positioned at the centre of a box (Fig. 2 and 3).

Long time of irradiation, up to about 100 h, was utilized to reach measurable activation in detectors at depth penetrations in iron and sodium.

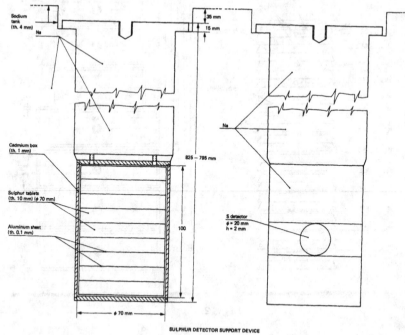

SULPHUR DETECTOR SUPPORT DEVICE

Fig. 3

MEASUREMENT OF INDUCED ACTIVATION

The threshold reaction $S^{32}(n,p)P^{32}$ can be conveniently used to measure fast neutron fluxes even at large penetration depths because of the great sensitivity of these detectors.

The sulphur samples are covered with Cd to minimize activities other than P^{32}. An aluminium beta filter on the counter stops beta-rays of lower energies. Any short lived activities are allowed to decay before the samples are counted. With these precautions, the beta-activity measured is found to decay with the half-life appropriate to P^{32}.

To make a thorough analysis of the background, in the case of iron the samples were divided and counted indipendently both at Ispra and at Winfrith. The Ispra detector was a gasflow with a guard anticoincidence counter. The counter was placed inside an old-lead shielding 10 cm thick. The Winfrith counter was a set of three scintillators. Their radioactive decay was monitored at both laboratories over a period of three months.

The absolute induced activity in the sulphur is determined by counting the 1.76 MeV rays emitted from the decay of the phosphorus 32. By slow combustion of the irradiated sample it is possible to remove the sulphur so that the ash residues contain the phosphorus 32 produced by neutron capture.
Its activity can be counted practically without selfabsorption and by collecting it in a small volume.
This method allows the analysis of large quantities of sulphur (up to several hundred gr.) used as fast neutron detectors.

As the detectors far from the source have very weak activations an accurate study was made of the background contributions. First of all anticoincidence counters were used to reduce drastically the background of environmental activity. Second, by analysing the count behaviour during a long decay time one notes the presence of a weak residual activity which if it is not constant certainly has a very long half-life and may therefore be considered, at least in a first approximation, as an additional background.

It is also observed that the material of support for the ashes (Al) and the sulphur ashes themselves which are not activated initially, contribute to the background.

461

Tab. I - Saturation Activity of P^{32} per gr./sec. at Different Depths in Iron Block

Iron depth cm	Ispra counting dis/gr.sec.		S%	R%	Winfrith counting dis/gr.sec.		S%	R%
1.5	8.5733	E+4	5.4	0.2				
8.0	1.7975	E+4	5.4	0.7				
16.	3.7888	E+3	5.4	1				
24.	8.4142	E+2	5.4	2				
32.	1.8627	E+2	5.4	3				
38.	5.9010	E+1	5.4	0.2	5.5121	E+1	5.3	0.03
46.	1.286	E+1	5.4	0.3	1.2198	E+1	5.0	0.06
54.	3.052	E+0	5.4	1.	2.872	E+0	5.1	0.1
62.					6.398	E-1	5.0	0.3
70.	1.420	E-1	5.4	2.6	1.338	E-1	5.3	0.3
78.	3.165	E-2	5.4	3.4	3.045	E-2	6.2	0.7
86.			5.4		7.025	E-3	5.3	4.8
94.			5.4		2.372	E-3	5.5	10.5

Tab. II - Saturation Activity of P^{32} per gr./sec. at Different Depth in Sodium Block

Distance cm	Winfrith counting dis/gr.sec.		S%.	R%.	Inter. air cm	Plating of the container (Fe) cm
183.5	3.237	E+4	5	0.7	--	0.4
66.4	1.971	E+3	5	0.7	0.7	1.2
125.2	1.036	E+2	5	0.7	1.7	2.0
184.5	6.27	E+0	5	1.0	2.8	2.8
243.2	4.20	E-1	5	1.0	3.5	3.6
302.4	3.03	E-2	5	1.0	4.7	4.4
362.2	1.99	E-3	5	10.0	6.5	5.2

S = Systematic Error
R = Random Error

As the decay constant for phosphorus 32 is known a fit was made for every series of measurements of the decay of each sample with a function of the type $C = Ae^{-\lambda t} + B$ where λ is the decay constant of phosphorus 32, t the elapsed time after irradiation and A and B are parameters to be determined. They represent the activity at t=0 (end of irradiation) and the residual activity of the background while C is the count rate at time t.

Tables I and II and the graph of Fig. 4 represent the absolute disintegration per second measured per gram of infinitly long irradiated sulphur in iron and in sodium.
In the case of sodium the geometry is a little more complicated since the boxes made of iron sheeting of only thickness 4 mm were filled with sodium at 200 C°.

For this reason the mating surfaces of the cases show slight swelling. When the cases were stacked the total thickness of the intervening gaps was measured. This value is reported in Table II together with the sum of the iron walls of the containers lying between the detector and the source

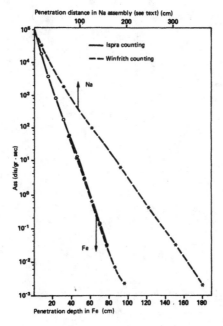

SATURATION ACTIVITY FOR THE S³² (n, p) P³² REACTION
AT DIFFERENT DETECTOR POSITIONS IN Fe AND Na

Fig. 4

PUISSANCE DEPOSEE DANS UN DISPOSITIF COMPLEXE

PAR LES GAMMAS DE REACTEUR D'ESSAIS

EXPERIENCES ET CALCULS REALISES A SILOE

H.PETITCOLAS[1]   (1)CEN-CENG - Service des Piles
G.COSOLI[2]   85X - 38041 GRENOBLE CEDEX - France

A.BESSON[1]   (2) ENEA VEL TECN Casaccia
A.BEVILACQUA[1]   BP 2400 - ROME - Italie.

ABSTRACT

Eight samples, which represent different materials used in testing reactors, were irradiated in the device "CYRANO" placed in the water reflector at different distances from the reactor core. The power dissipated in the device was measured by the "CYRANO" equipement itself, whereas the calorimeter juxtaposed served to monitor the gamma flux.

Parallel to each experiment, the power deposited in the samples, the device materials and the calorimeter were calculated by the code MERCURE 4.

The measured values were compared with the calculated ones, both in relative and in absolute values, for each sample and for each distance in the reflector. The comparison shows very good agreement.

NATURE DU PROBLEME ET OBJECTIFS.

Dans une géométrie simple, on sait déterminer/1/ la puissance déposée par les gammas de réacteur dans un échantillon, si on connaît sa nature, sa densité et ses dimensions - si toutefois elles sont petites - à partir d'une mesure de la puissance dans un calorimètre moniteur.

Par contre, cette méthode est insuffisante si les dimensions sont grandes et surtout si la géométrie est complexe, avec juxtaposition de matériaux différents.

J. P. Genthon and H. Röttger (eds.), Reactor Dosimetry, 465–471.
© *1985 ECSC, EEC, EAEC, Brussels and Luxembourg*

Or, la puissance déposée par les gammas dans un dispositif d'irradiation est souvent considérable. Seul un calcul performant peut la prévoir, en raison de la complexité de la géométrie, de la multiplicité des sources gamma à spectres différents et de la diversité des matériaux plus ou moins absorbants.

Objectif général. Qualifier, par un programme expérimental, le code MERCURE 4, qui calcule, dans une géométrie complexe, la distribution spatiale et spectrale du flux gamma et les dépôts de puissance.

Objectifs particuliers .
1) déterminer les valeurs relatives des puissances déposées dans chaque matériau en fonction du numéro atomique et de la densité.
2) corréler la puissance déposée dans les matériaux du dispositif à celle mesurée par le calorimètre moniteur au graphite.

MOYENS EXPERIMENTAUX

Le programme expérimental a été réalisé dans un four CYRANO, dispositif qui peut, en particulier, mesurer la puissance totale.(fig.2).

1-Principe de fonctionnement.

Le four CYRANO est un dispositif d'irradiation servant à tester les matériaux. Il permet de connaître à tout instant la puissance nucléaire totale dissipée.

Pour mesurer la puissance sortant du dispositif, on utilise une résistance de mesure qui fonctionne en thermomètre à résistance. En effet, sa température dépend de la puissance traversant le dispositif, que celle-ci soit nucléaire ou électrique.

L'étalonnage préalable est effectué en injectant une puissance électrique connue dans la résistance de chauffage. (Fig.1).

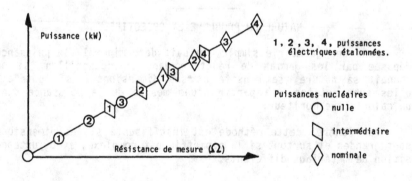

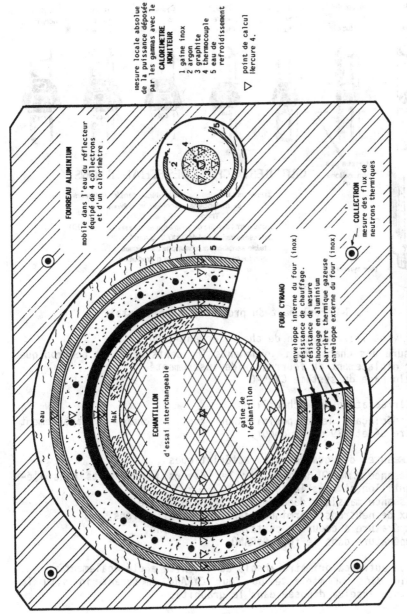

mesure locale absolue
de la puissance déposée
par les gammas avec le
**CALORIMETRE
MONITEUR**

1 gaine inox
2 argon
3 graphite
4 thermocouple
5 eau de
  refroidissement

▽ point de calcul
Mercure 4.

FOURREAU ALUMINIUM

mobile dans l'eau du réflecteur
équipé de 4 collectrons
et d'un calorimètre.

FOUR CYRANO

enveloppe interne du four (inox)
résistance de chauffage.
résistance de mesure
shoopage en aluminium
barrière thermique gazeuse
enveloppe externe du four (inox)

COLLECTRON
mesure des flux de
neutrons thermiques

ECHANTILLON
d'essai interchangeable

gaine de
l'échantillon

eau

NaK

Fig. 2 - SCHEMA D'ENSEMBLE DU DISPOSITIF EXPERIMENTAL.

2-Echantillons irradiés (fig.3)

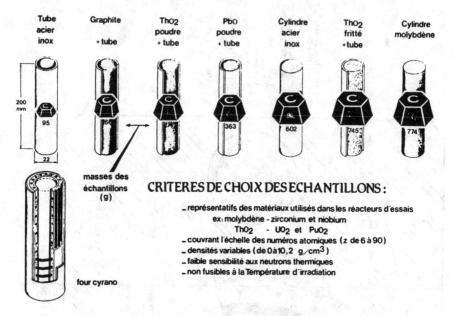

| Tube acier inox | Graphite + tube | ThO₂ poudre + tube | PbO poudre + tube | Cylindre acier inox | ThO₂ fritté + tube | Cylindre molybdène |

masses des échantillons (g)

four cyrano

CRITERES DE CHOIX DES ECHANTILLONS :

_ représentatifs des matériaux utilisés dans les réacteurs d'essais
ex: molybdène - zirconium et niobium
ThO_2 - UO_2 et PuO_2
_ couvrant l'échelle des numéros atomiques (z de 6 à 90)
_ densités variables (de 0 à 10,2 g/cm^3)
_ faible sensibilité aux neutrons thermiques
_ non fusibles à la Température d'irradiation

3-Déroulement du programme expérimental.

Avant l'irradiation de chaque échantillon :
- mesure des champs de rayonnements incidents gamma et neutronique par déplacement du fourreau vide instrumenté.
- étalonnage du système CYRANO.

Irradiation des échantillons
- en déplaçant le dispositif dans tout le champ de rayonnement entre 34 et 179 mm du coeur.

Mesures des grandeurs caractéristiques :
- puissance CYRANO déposée par les divers rayonnements dans l'échantillon et dans le four,
- puissance massique déposée dans le calorimètre,
- flux de neutrons thermiques dans les 4 collectrons,
- température à coeur de certains échantillons,
- température dans le four CYRANO.

Un autre calorimètre absolu au graphite a été également irradié dans la maquette du four CYRANO, mesurant in situ la puissance absolue déposée dans le graphite.

MOYEN DE CALCUL : CODE MERCURE 4

1-Principe de calcul de la puissance gamma en un point r_0

- Intégration de noyaux ponctuels d'atténuation en ligne droite.
- Calcul de build-up.
- Géométrie à 3 dimensions.

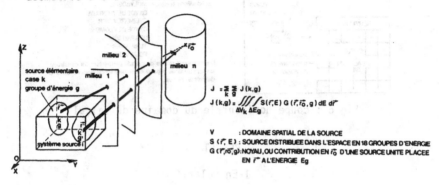

$$J = \sum_k \sum_g J(k,g)$$

$$J(k,g) = \iiint_{\Delta V_k} \int_{\Delta E_g} S(\vec{r},E) \, G(\vec{r},\vec{r_0},g) \, dE \, d\vec{r}$$

V : DOMAINE SPATIAL DE LA SOURCE
S ($\vec{r}$, E) : SOURCE DISTRIBUEE DANS L'ESPACE EN 18 GROUPES D'ENERGIE
G ($\vec{r},\vec{r_0}$,g) : NOYAU, OU CONTRIBUTION EN $\vec{r_0}$ D'UNE SOURCE UNITE PLACEE EN $\vec{r}$ A L'ENERGIE Eg

2-Synoptique du code (fig 5)

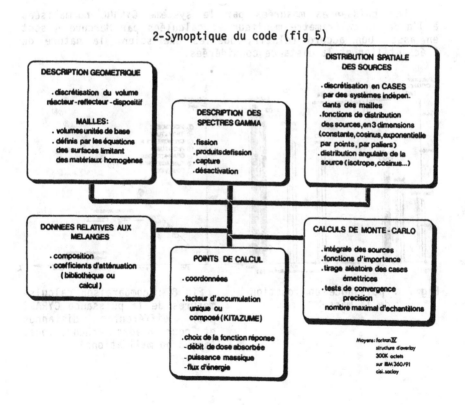

3-Applications :

puissances gamma déposées dans le four et dans les échantillons.

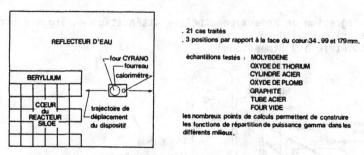

REFLECTEUR D'EAU
four CYRANO
fourreau
calorimètre
BERYLLIUM
CŒUR du REACTEUR SILOE
trajectoire de déplacement du dispositif
. 21 cas traités
. 3 positions par rapport à la face du cœur:34 . 99 et 179mm.
échantillons testés : MOLYBDENE
OXYDE DE THORIUM
CYLINDRE ACIER
OXYDE DE PLOMB
GRAPHITE
TUBE ACIER
FOUR VIDE
les nombreux points de calculs permettent de construire les fonctions de répartition de puissance gamma dans les différents milieux.

Fig.6 : coupe horizontale du domaine du calcul.

RESULTATS

1-En relatif

Les puissances mesurées par le système CYRANO normalisées à l'aide du calorimètre moniteur et calculées par Mercure 4 sont en assez bon accord relatif, quelles que soient la nature de l'échantillon et la distance considérées.

Puissance CYRANO (W)
mesure
Calculs Mercure 4 après normalisation
Four vide
tube acier
Graphite + tube
PbO + tube
Cylindre acier
ThO2 + tube
Molybdène
masse intérieure au four (g)

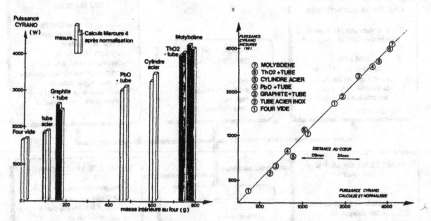

Fig 7 : puissance en fonction de la masse

PUISSANCE CYRANO MESUREE (W)
⑦ MOLYBDENE
⑥ ThO2 +TUBE
⑤ CYLINDRE ACIER
④ PbO +TUBE
③ GRAPHITE+TUBE
② TUBE ACIER INOX
① FOUR VIDE
DISTANCE AU CŒUR
179mm
34mm
PUISSANCE CYRANO CALCULEE ET NORMALISEE

Fig 8 : comparaison calculs-mesures de la puissance CYRANO pour différentes distances et pour divers échantillons. (après normalisations).

2-En absolu.

Les mesures, effectuées avec les calorimètres dans la position de monitorage et dans la maquette CYRANO à diverses distances du coeur, donnent la puissance gamma déposée dans le graphite à mieux que 6% en absolu /2/.

Les calculs Mercure 4 correspondants ont été comparés :

1) les gradients mesurés et calculés sont identiques au pour cent près sur une distance de 20 cm.

2) les valeurs absolues mesurées et calculées sont très proches :
mes/calc environ 1.08

3) si $R = \dfrac{\text{W/g dans le calorimètre intérieur à la maquette}}{\text{W/g dans le calorimètre moniteur}}$

R calculé / R mesuré = 1,02 $\pm$ 0,02 (à 1 σ)

CONCLUSION.

Les calculs **"Mercure 4"** et les mesures par le **"système Cyrano"** de la puissance déposée dans le four et ses échantillons sont en assez bons acccords relatifs lorsque varient la distance au coeur, le numéro atomique et la densité de l'échantillon.

Les calculs **"Mercure 4"** et les mesures avec les **"calorimètres absolus"** situés dans le dispositif complexe Cyrano ou dans l'emplacement de monitorage présentent un écart systématique absolu assez faible (environ 8%) pour diverses distances dans le réflecteur.

Le **code Mercure 4 permet de corréler** de manière très satisfaisante la puissance gamma déposée dans les matériaux d'un dispositif complexe et volumineux à celle mesurée à proximité par un calorimètre moniteur absolu.

REFERENCES

/1/ Nouveaux développements en calorimétrie au CEN-G.
H.PETITCOLAS - 3ème Symposium ASTM-Euratom - Ispra - 1979.

/2/ Calorimètre à grande sensibilité pour la mesure de rayonnement auprès des réacteurs.
H.PETITCOLAS, J-J.BONNIN - Euratom - Petten - 1980.

SOLID-STATE TRACK RECORDER NEUTRON DOSIMETRY IN
LIGHT WATER REACTOR PRESSURE VESSEL SURVEILLANCE MOCKUPS

F. H. Ruddy, J. H. Roberts, R. Gold
and C. C. Preston

Westinghouse Hanford Company
Hanford Engineering Development Laboratory

Richland, WA USA

ABSTRACT

Solid-State Track Recorder (SSTR) measurements
of neutron-induced fission rates have been made in
several pressure vessel mockup facilities as part of
the U.S. Nuclear Regulatory Commission's (NRC) Light
Water Reactor Pressure Vessel Surveillance Dosimetry
Improvement Program (LWR-PV-SDIP).

The results of extensive physics-dosimetry mea-
surements made at the Pool Critical Assembly (PCA) at
Oak Ridge National Laboratory (ORNL) in Oak Ridge, TN
are summarized. Included are ^{235}U, ^{238}U, ^{237}Np and
^{232}Th fission rates in the PCA 12/13, 8/7, and 4/12
SSC configurations.

Additional low power measurements have been made
in an engineering mockup at the VENUS critical assembly at CEN-SCK, Mol, Belgium. ^{237}Np and ^{238}U fission
rates were made at selected locations in the VENUS
mockup, which models the in-core and near-core regions
of a pressurized water reactor (PWR). Absolute core
power measurements were made at VENUS by exposing
solid-state track recorders (SSTRs) to polished fuel
pellets within in-core fuel pins.

J. P. Genthon and H. Röttger (eds.), Reactor Dosimetry, 473–486.
© 1985 ECSC, EEC, EAEC, Brussels and Luxembourg

Low power measurements have also been made at a
PV simulator similar to PCA at the NESTOR reactor at
Winfrith, England as part of the NESTOR Shielding and
Dosimetry Improvement Program (NESDIP). The results
of absolute ^{235}U, ^{238}U and ^{237}Np fission rate measurements made at selected locations in NESDIP are
reported.

SSTR dosimetric measurements for high fluence
LWR-PVS metallurgical studies have been carried out
at the Poolside Facility (PSF) and subsequent Surveillance Dosimetry Measurement Facility (SDMF) at
the Oak Ridge Research (ORR) reactor.

General data trends are discussed and comparisons
are made with independent experimental measurements.

INTRODUCTION

Solid-State Track Recorder (SSTR) fission rate measurements
have continued at light water reactor pressure vessel (LWR-PV)
mockup facilities. In addition to the previously reported (1-3)
fission rate measurements carried out at the Pool Critical
Assembly (PCA) (4) at Oak Ridge National Laboratory (ORNL) in Oak
Ridge, TN, further measurements have been made at PCA, at the
NESTOR Dosimetry Improvement Program (NESDIP) facility at
Winfrith, England, and in the VENUS critical facility in Mol,
Belgium. High fluence tests in LWR mockups have been carried out
in the Poolside Facility (PSF) and subsequent Surveillance
Dosimetry Measurement Facility (SDMF) at the Oak Ridge Research
(ORR) reactor.

PCA MEASUREMENTS AND RESULTS

SSTR fission rate measurements for ^{235}U (bare and cadmium
covered), ^{238}U, ^{237}Np, and ^{232}Th in the PCA 8/7 and 12/13 configurations have been reported previously (1-3). Recent work has
concentrated on a repeat of selected PCA 12/13 measurements and on
measurements in the PCA 4/12 SSC configuration.*

---

*Nomenclature and acronyms as introduced in Reference 1.

PCA 12/13 Configuration

A comparison of measurements made in the PCA 12/13 configuration in separate sets of experiments carried out in October and November of 1981 is shown in Table 1. In general, the agreement between the two sets of data is excellent, indicating that the measurements are reproducible within the quoted experimental uncertainties.

PCA 4/12 SSC Configuration

^{237}Np and ^{238}U fission rates were measured in the SSC, 1/4-T, 1/2-T and 3/4-T locations in the PCA 4/12 SSC configuration during January 1981. These data are summarized in Tables 2 and 3. The relative fission rates are plotted as a function of axial location in Figure 1. All data were normalized to the midplane location. The solid line plotted for comparison is the result of Mol fission chamber traverses (5). The agreement of the relative SSTR fission rates with the shape of the axial distribution indicated by the fission chamber is consistent with the experimental uncertainties of the data. Fission rates as a function of radial location are plotted for ^{237}Np in Figure 2 and for ^{238}U in Figure 3. Data from the 8/7 and 12/13 configurations are also plotted for comparison. Relative uncertainties are indicated for the data in Tables 2 and 3. To obtain the absolute uncertainties from these relative uncertainties, the 4.1% uncertainty in the absolute power normalization must be combined in quadrature with the tabulated values. The absolute uncertainties in these data are generally <5% (1σ).

General Data Trends

The data plotted in Figures 2 and 3 show that the slopes of the attenuation in the PV block appear to be independent of configuration. This fact, which was first noted in Ref. 5, is further substantiated by the data in Table 4, where all fission rates have been normalized to one at the 1/4-T location, and the 1/2-T and 3/4-T relative fission rate values are seen to be independent of configuration. The small standard deviations of the means of the relative reaction rates for each location indicate that the precision of the SSTR results is within the quoted uncertainties.

As a further check on the consistency of the SSTR reaction rates, ratios were taken for equivalent locations in the different configurations. These data are contained in Table 5. For the PV block, the reaction rate ratios are independent of location. Again, the standard deviations of the means are consistent with the experimental uncertainties of the data.

TABLE 1

RATIOS OF DUPLICATE PCA 12/13 SSTR FISSION RATE MEASUREMENTS

Isotope	Location	Fission Rate [(fissions per atom per core neutron) x 10^{33}]*		
		October 1981	November 1981	Oct 1981/Nov 1981
^{238}U	1/4 T	17.4 (±2.7%)	17.5 (±2.7%)	0.992 (±3.6%)
	1/2 T	7.44 (±2.7%)	7.50 (±2.7%)	0.992 (±3.6%)
	3/4 T	3.24 (±2.7%)	3.23 (±2.7%)	1.01 (±3.6%)
	VB	0.970 (±2.7%)	0.970 (±2.7%)	1.00 (±3.6%)
^{237}Np	1/4 T	116.0 (±3.3%)	118.0 (±3.6%)	0.983 (±4.7%)
	3/4 T	32.3 (±3.3%)	33.2 (±3.3%)	0.970 (±4.5%)
	VB	9.79 (±3.3%)	9.70 (±3.4%)	1.01 (±4.6%)
			Average	0.994 (±1.45%)

*The uncertainties on individual ratios were obtained by combining the uncertainties on the two measurements in quadrature. The uncertainties on the averages are the standard deviations of the mean of the individual ratios.

TABLE 2

^{237}Np FISSION RATES IN PCA 4/12 SSC CONFIGURATION

Axial Location (mm)	Fission Rate (fissions per atom per core neutron)		
	SSC Position	1/4-T Position	1/2-T Position
+150	---	5.08 E-31 (±2.6%)	---
+75	8.16 E-30 (±2.6%)	---	---
0	8.48 E-30 (±2.6%)	6.25 E-31 (±2.6%)	3.44 E-31 (±2.6%)
-75	8.42 E-30 (±2.6%)	---	---
-130	---	5.52 E-31 (±2.6%)	---

TABLE 3

^{238}U FISSION RATES IN PCA 4/12 SSC CONFIGURATION

Axial Location (mm)	Fission Rate (fissions per atom per core neutron)			
	SSC Position	T/4-T Position	T/2-T Position	3/4-T Position
+75	1.04 E-30 (±2.6%)	5.89 E-32 (±2.4%)	---	---
0	1.15 E-30 (±2.6%)	6.60 E-32 (±2.4%)	2.99 E-32 (±2.4%)	1.29 E-32 (±2.4%)
-75	1.12 E-30 (±2.6%)	6.51 E-32 (±2.4%)	---	---

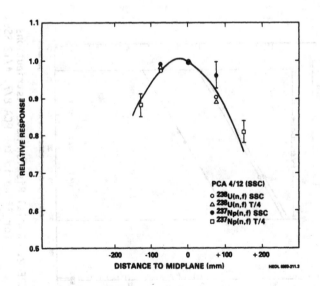

FIGURE 1. Axial Distributions of the ^{237}Np and ^{238}U Fission Rates in the PCA 4/12 SSC Configuration

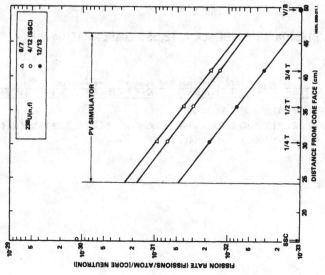

FIGURE 2. Radial Fission Rate Distributions for ^{237}Np in the PCA 8/7, 4/12 SSC, and 12/13 Configurations. The distances from core face apply to the 4/12 SSC data only.

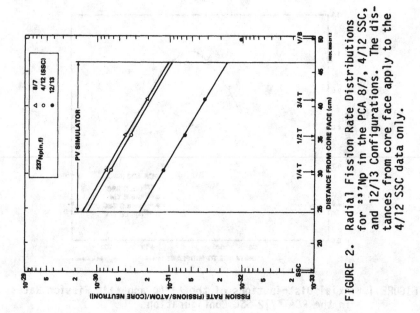

FIGURE 3. Radial Fission Rate Distributions for ^{238}U in the PCA 8/7, 4/12 SSC, and 12/13 Configurations. The distances from core face apply to the 4/12 SSC data only.

478

The relative reaction rate data of Tables 4 and 5, as well as the data of Table 1 indicate that all the PCA SSTR reaction rate measurements are self-consistent on a relative basis and that the measurements are reproducible within the stated experimental uncertainties on an absolute basis.

Discrepancies have been found with the results of fission chamber measurements, however, and these discrepancies are described in the following section.

Comparison of SSTR and Fission Chamber PCA Fission Rates

At present, pending the result of benchmark irradiations of the SSTR fissionable deposits, the SSTR results are reported as absolute fission rates. The fission chamber results, on the other hand, have been benchmark referenced, and the fission chamber results are reported as fission equivalent fluxes. In order to make direct comparisons between the SSTR and the National Bureau of Standards (NBS) fission chamber results, the fission chamber data from Reference 5 were converted into the corresponding reaction rates. These comparisons are contained in Table 6.

The data of Table 6 indicate that the NBS fission chamber results are consistently about 10% higher than the SSTR results. The mean of all of the ^{237}Np ratios in Table 6 is 0.906 ± 0.026, the mean for all of the ^{238}U ratios is 0.890 ± 0.023, and the overall mean of all 22 ratios is 0.897 ± 0.025. A possible explanation for the 10% bias is the fact that the void introduced by the NBS fission chamber causes a perturbation. Measurements have been conducted at PCA to compare SSTRs, Radiometric Monitors (RMs), and the NBS fission chamber directly. The results of these comparisons are presently being analyzed.

Direct comparisons between the SSTRs and CEN/SCK chamber can be made only at two locations in the 12/13 configuration. As can be seen from the data in Table 7, no detectable bias exists between the SSTR and CEN/SCK chamber measurements at these two locations, which are both external to the PV block.

NESDIP MEASUREMENTS AND RESULTS

SSTRs were exposed at NESDIP during Phase I (PCA 12/13 Replica Experiments) of the program. ^{237}Np, ^{235}U and ^{238}U were measured in the pressure vessel front location, in the PVS block and in the void box location. The results of these measurements are summarized in Table 8. Since the source of neutrons at the NESDIP facility (6) is a uranium driver plate driven by neutrons from the NESSUS reactor core, measurements must be made with

TABLE 4
RELATIVE REACTION RATES IN THE PCA PRESSURE VESSEL SIMULATOR

Isotope	Configuration	Location		
		1/4 T	1/2 T	3/4 T
^{238}U	8/7 (Oct 1981)	1.00	0.442 (±3.7%)*	0.193 (±3.7%)*
	8/7 (Oct 1978)	1.00	0.443 (±3.8%)	0.187 (±3.6%)
	4/12 SSC	1.00	0.454 (±3.2%)	0.196 (±3.2%)
	12/13 (Oct 1981)	1.00	0.427 (±3.7%)	0.187 (±3.7%)
	12/13 (Nov 1981)	1.00	0.429 (±3.7%)	0.185 (±3.7%)
	Average	--	0.439 (±2.5%)	0.190 (±2.5%)
^{237}Np	8/7 (Oct 1981)	1.00	0.593 (±3.7%)	0.275 (±3.7%)
	8/7 (Oct 1978)	1.00	0.564 (±3.5%)	0.275 (±5.1%)
	4/12 SSC	1.00	0.549 (±3.5%)	---
	12/13 (Oct 1981)	1.00	---	0.278 (±3.7%)
	12/13 (Nov 1981)	1.00	0.524 (±6.4%)	0.281 (±4.7%)
	Average	--	0.558 (±5.2%)	0.277 (±1.0%)

*Reaction rates normalized to the 1/4-T position. Uncertainties in relative reaction rates were obtained by combining the uncertainties of reaction rates in quadrature. The uncertainty of the average is the standard deviation of the mean of the values averaged.

TABLE 5

Isotope	Ratio	Location			Average
		1/4 T	1/2 T	3/4 T	
^{238}U	(8/7)/(4/12 SSC)	1.38 (±3.4%)*	1.35 (±3.6%)*	1.32 (±3.4%)*	1.35 (±2.2%)*
	(4/12 SSC)/(12/13)	3.78 (±3.5%)	3.99 (±3.5%)	4.01 (±3.5%)	3.93 (±3.2%)
	(8/7)/(12/13)	5.22 (±3.6%)	5.39 (±3.8%)	5.30 (±3.6%)	5.30 (±1.6%)
^{237}Np	(8/7)/(4/12 SSC)	1.14 (±5.3%)	1.17 (±5.9%)	---	1.16 (±1.8%)
	(4/12 SSC)/(12/13)	5.30 (±4.3%)	5.55 (±6.0%)	---	5.42 (±3.3%)
	(8/7)/(12/13)	6.06 (±5.8%)	6.51 (±7.6%)	5.92 (±5.5%)	6.16 (±5.0%)

*Uncertainties on the reaction rate ratios were obtained by combining the uncertainties of the reaction rates in quadrature. The uncertainty of the average is the standard deviation of the mean of the ratios averaged.

TABLE 6

COMPARISON OF SSTR AND FISSION CHAMBER MEASURED PCA FISSION RATES

Configuration	Isotope	Location	Fission Rate [(fissions per atom per core neutron) x 10¹¹] Fission Chamber	SSTR	SSTR/ Fission Chamber Ratio*
12/13 (Oct. 81)	²³⁷Np	1/4 T	12.55 (±2.9%)	11.6 (±3.3%)	0.924 ± 0.037
		3/4 T	3.690 (±3.1%)	3.23 (±3.3%)	0.875 ± 0.040
				Average	0.899 ± 0.035
	²³⁸U	1/4 T	1.943 (±3.0%)	1.74 (±2.7%)	0.896 ± 0.037
		1/2 T	0.8536 (±3.1%)	0.744 (±2.7%)	0.872 ± 0.036
		3/4 T	0.3546 (±3.1%)	0.324 (±2.7%)	0.914 ± 0.021
				Average	0.894 ± 0.021
12/13 (Nov. 81)	²³⁷Np	1/4 T	12.55 (±2.9%)	11.8 (±3.6%)	0.940 ± 0.043
		1/2 T	7.045 (±3.1%)	6.19 (±5.4%)	0.879 ± 0.055
		3/4 T	3.690 (±3.1%)	3.32 (±3.3%)	0.900 ± 0.041
				Average	0.906 ± 0.031
	²³⁸U	1/4 T	1.943 (±3.0%)	1.75 (±2.7%)	0.901 ± 0.036
		1/2 T	0.8536 (±3.1%)	0.750 (±2.7%)	0.879 ± 0.036
		3/4 T	0.3546 (±3.1%)	0.323 (±2.7%)	0.911 ± 0.037
				Average	0.897 ± 0.016
8/7	²³⁷Np	1/4 T	7.789 (±2.9%)	7.15 (±4.6%)	0.918 ± 0.050
		1/2 T	4.321 (±2.9%)	4.03 (±5.4%)	0.933 ± 0.057
		3/4 T	2.282 (±2.9%)	1.97 (±4.4%)	0.863 ± 0.045
				Average	0.905 ± 0.037
	²³⁸U	1/4 T	1.050 (±2.8%)	0.913 (±2.6%)	0.870 ± 0.033
		1/2 T	0.4575 (±3.0%)	0.404 (±2.9%)	0.883 ± 0.037
		3/4 T	0.1899 (±3.0%)	0.171 (±2.7%)	0.900 ± 0.036
				Average	0.884 ± 0.016
4/12 SSC	²³⁷Np	1/4 T	6.826 (±1.7%)	6.25 (±2.6%)	0.916 ± 0.031
		1/2 T	3.765 (±1.9%)	3.44 (±2.6%)	0.914 ± 0.029
				Average	0.915 ± 0.001
	²³⁸U	1/4 T	0.7845 (±1.8%)	0.660 (±2.4%)	0.841 ± 0.025
		1/2 T	0.3392 (±2.3%)	0.299 (±2.4%)	0.882 ± 0.029
		3/4 T	0.1409 (±2.6%)	0.129 (±2.4%)	0.917 ± 0.033
				Average	0.880 ± 0.038

*The uncertainties on individual ratios were obtained by combining the uncertainties on the SSTR and fission chamber measurements in quadrature. The uncertainty on the average is the standard deriva-tion of the mean of the three ratios.

TABLE 7

COMPARISON OF SSTR AND CEN/SCK FISSION CHAMBER ^{237}Np FISSION RATES IN THE PCA 12/13 CONFIGURATION

Location	Equivalent Fission Fluxes (x 10^{8}) CEN/SCK	SSTR	Fission Chamber/SSTR Ratio
PVF	22.7 (±6.3%)	23.1 (±3.3%)	0.983 ± 0.071
VB	0.73 (±9.2%)	0.711 (±3.4%)	1.027 + 0.101
		Average	1.005 ± 0.031

TABLE 8

PRELIMINARY PCA-NESDIP FISSION RATE COMPARISON

Isotope	Location	Fission Rate [f/at./(kW-h)] NESDIP*	PCA	PCA/NESDIP
^{237}Np	PVF	3.78 x 10^{-16} (±3.3%)	8.67 x 10^{-14} (±3.3%)	229.6 (±4.6%)
	1/4 T	1.40 x 10^{-16} (±3.8%)	3.21 x 10^{-14} (±3.6%)	228.8 (±5.3%)
	3/4 T	4.11 x 10^{-17} (±2.1%)	9.02 x 10^{-15} (±3.3%)	219.5 (±3.9%)
	VB	1.27 x 10^{-17} (±3.4%)	2.63 x 10^{-15} (±3.4%)	207.5 (±4.8%)
			Average	221.4 (±4.7%)
^{235}U	PVF	1.13 x 10^{-13} (±3.4%)	---	---
	1/4 T	2.54 x 10^{-15} (±3.1%)	3.42 x 10^{-13} (±2.8%)	134.8 (±4.2%)
^{238}U	1/4 T	2.07 x 10^{-17} (±2.9%)	4.76 x 10^{-15} (±2.7%)	230.2 (±4.0%)

*Preliminary fission rates have been calculated using power levels as indicated by the NESDIP instruments. Final more accurate fission rates will be reported when NESDIP run-to-run monitoring data for these runs become available. Relative uncertainties are tabulated in parentheses. The absolute uncertainty in the core power is 4.1% for PCA and is unknown for NESDIP at this time.

both the plate in and with the plate out to differentiate between
neutrons coming from the driver plate and neutrons coming directly
from the NESSUS core.

This correction, which was measured experimentally, can be as
large as 8% as shown in Table 9. Also shown in Table 9 are the
corrections made for the presence of impurities in the fission
deposits. Most notable is the correction necessary for the 435 ppm
of ^{235}U in the ^{238}U foils. Plate-out measurements were not made
for ^{238}U, and the ^{238}U fissions rates shown in Table 8 assume that
the correction is the same for ^{238}U as it is for ^{237}Np (3.42%).
The ^{238}U fission rate is not reported at the PVF location because of
the large contribution from ^{235}U in the ^{238}U foils. 1/2-T and VB
results are not reported for ^{238}U and ^{235}U because of foil placement problems. Shown for comparison in Table 8 are the corresponding PCA 12/13 measurements. The agreement of the PCA/ NESDIP
reaction rate ratio for ^{237}Np is consistent with experimental
uncertainties. Also, the ^{238}U reaction rate ratio is in agreement
with ^{237}Np. The reaction rate ratio for ^{235}U is somewhat smaller,
indicating that the epithermal component of the NESDIP neutron
energy spectrum is proportionatly about 70% greater than PCA. The
^{237}Np/^{238}U spectral indices at the 1/4-T location are contained in
Table 10. The agreement between PCA and NESDIP is well within the
experimental uncertainty.

TABLE 9

NESDIP CORRECTION FACTORS FOR FISSION CAUSED BY IMPURITIES
AND FOR NEUTRON LEAKAGE FROM THE NESSUS CORE

Isotope	Location	Plate in/Plate Out (%)	Impurities (%)
^{237}Np	PVF	2.16	<0.1
	1/4 T	3.42	<0.1
	3/4 T	6.99	<0.1
	VB	8.35	<0.1
^{238}U	PVF	*	48
	1/4 T	*	4.92
	VB	*	20
^{235}U	PVF	0.71	<0.1
	1/4 T	0.12	<0.1

*Plate out measurements were not made at NESDIP.

TABLE 10

$^{237}Np/^{238}U$ SPECTRAL INDICES AT THE 1/4-T LOCATION

$$(^{237}Np/^{238}U)_{PCA} = 6.74 \ (\pm 4.5\%)$$

$$(^{237}Np/^{238}U)_{NESDIP} = 6.76 \ (\pm 4.8\%)$$

$$\frac{(^{237}Np/^{238}U)_{NESDIP}}{(^{237}Np/^{238}U)_{PCA}} = 1.003 \pm 0.066$$

The NESDIP ^{237}Np fission rates are plotted as a function of distance from the driver plate in Figure 4. Included for comparison are the corresponding PCA fission rates. The slope of the attenuation in the PV block is the same for the NESDIP and the PCA with a difference in absolute intensity of about a factor of 220. The absolute values of the NESDIP reaction rates are based on power levels derived from the NESDIP operating instruments. The results of run-to-run monitoring performed during the SSTR runs are as yet unavailable. Final absolute results will be reported when these run-to-run monitoring results are made available.

VENUS MEASUREMENTS

SSTR fission rates were measured for ^{237}Np, ^{238}U, and ^{235}U in the VENUS (6) mockup, which models the in-core and near-core regions of a PWR. Also, the SSTR method was used to establish the absolute power level of VENUS by exposing mica SSTRs, under well-moderated conditions, between polished pellets within fuel pins in the VENUS core. Six SSTRs were exposed in VENUS fuel and 6 SSTRs were exposed in fuel in BR-1 during a reference irradiation. SSTR measurements were made in a total of 17 locations for ^{238}U (Cd covered) and ^{237}Np (bare) in the VENUS water gap, barrel, and pad. These SSTRs are presently being processed.

HIGH FLUENCE MEASUREMENTS

Mica and quartz crystal SSTRs have been exposed with ultra-low mass (10^{-12} - 10^{-19} gram) fissionable deposits in high fluence irradiations in PSF and in the third and fourth SDMF tests (6). In most cases, very high track densities (10^6 - 10^7 tracks/cm^2) result, requiring specialized scanning techniques (8), which are currently under development and testing.

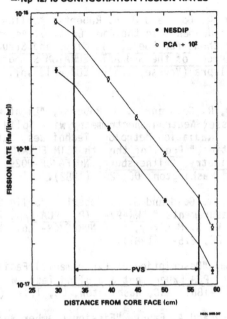

FIGURE 4. Radial Fission Rate Distribution for ²³⁷Np in NESDIP.
PCA 12/13 ²³⁷Np fission rates are shown for comparison.

CONCLUSIONS

The low fluence SSTR fission rates reported here and previously (1-3) for PCA and NESDIP represent a data set that is
internally consistent within the bounds set by the experimental
uncertainties. The absolute accuracies of these measurements are
in the range 2-5%. When compared to absolute NBS fission chamber
results with similar absolute accuracies and internal consistency,
a 10% bias results. Measurements have been performed to directly
compare SSTR and fission chamber measured fission rates and to
directly compare both to radiometric fission rates. These measurements are currently being analyzed and will hopefully resolve
the discrepancy between these two techniques.

SSTR measurements have been extended to high fluences by
developing ultra-low mass fissionable deposit fabrication techniques. The impending development of higher track density scanning methods will complete a useful extension of this technique.

REFERENCES

1. F. H. Ruddy, R. Gold and J. H. Roberts, "Solid-State Track Recorder Measurements in the Pool Critical Assembly," Dosimetry Methods for Fuels, Cladding, and Structural Materials, Proc. of the 3rd ASTM-EURATOM Symposium on Reactor Dosimetry, Ispra (Varese) Italy, EUR 6813 Vol. II, 1069 (1980).

2. F. H. Ruddy, R. Gold and J. H. Roberts, "Light Water Reactor Pressure Vessel Neutron Spectrometry with Solid State Track Recorders," "Radiation Metrology Techniques, Data Bases, and Standardization," Proc. of the 4th ASTM-EURATOM Symposium on Reactor Dosimetry, Gaithersburg, NUREG/CP-0029, CONF-820321, Vol. I, NRC, Washington, DC, 293 (1982).

3. F. H. Ruddy, R. Gold and J. H. Roberts, "Solid State Track Recorder Measurements," LWR-PV-SDIP: PCA Experiment and Blind Test, W. N. McElroy, Ed., NUREG/CR-1861, NRC, Washington, DC, 2.5-1 (1981).

4. F. B. K. Kam, "Description of Experimental Facility," LWR-PV-SDIP: PCA Experiment and Blind Test, W. N. McElroy, Ed., NUREG/CR-1861, NRC, Washington, DC, 1.1-1 (1981).

5. E. D. McGarry and A. Fabry, "Fission Chamber Measurements," LWR-PV-SDIP: PCA Experiment and Blind Test, W. N. McElroy, Ed., NUREG/CR-1861, HEDL-TME 80-87, NRC, Washington, DC, 2.3-1 (1981).

6. W. N. McElroy et al., LWR-PV-SDIP: 1983 Annual Report, NUREG/CR-3391, Vol. 3, HEDL-TME 83-23, NRC, Washington, DC, (June 1984).

7. J. Butler, M. D. Carter, I. J. Curl, M. R. March, A. K. McCracken, M. F. Murphy and A. Packwood, The PCA Replica Experiment. Part I: Winfrith Measurements and Calculations, NUREG/CR-324, Part I, AEEW R1736, NRC, Washington, DC, (January 1984)

8. C. C. Preston, R. Gold, J. P. McNeece, J. H. Roberts and F. H. Ruddy, "Progress in Automated Scanning Electron Microscopy for Track Counting," Nucl. Tracks 7, 53, (1983).

A EUROPEAN COMMUNITY SOURCE OF REFERENCE MATERIALS FOR

NEUTRON METROLOGY REQUIREMENTS

J. Van Audenhove and J. Pauwels

Central Bureau for Nuclear Measurements

Geel, Belgium.

A.J. Fudge

Atomic Energy Research Establishment

Harwell, United Kingdom

ABSTRACT

Recognising the importance of an assured supply of reference materials of high purity for use as neutron dosimeters, the Monitor Materials sub-group of the European Working Group on Reactor Dosimetry collaborated with the Central Bureau for Nuclear Measurements, Geel, Belgium to establish a list of requirements. C.B.N.M. is organising the availability and characterisation of these materials in collaboration with specialist laboratories within the European Community. At present 12 activation and 6 fission dosimetry materials are being examined prior to certification and issue.

This source of material is expected to be of great use for reactor surveillance, interlaboratory comparisons, and other neutron metrology requirements.

1. INTRODUCTION

The principal current method used for the measurement of neutron flux and fluence associated with nuclear reactors involves a range of nuclear reactions with different materials. These materials, or dosimeters, produce radioactive or stable products which after measurement enable the neutron flux and/or fluence to be calculated from the reaction probability (cross section) and the power-time history of the irradiation. This neutron fluence can be converted into a damage dose using a postulated relationship for particular energy ranges (such as flux of energy greater than 0.1 or 1.0 MeV) or damage parameters (such as displacements per atom). A particularly important application of this technique of activation dosimetry is the safety surveillance of pressurised water reactor pressure vessels. Test specimens of steel are irradiated at enhanced flux positions in the reactor with a series of dosimetry materials to measure the property changes of the steel over the neutron energy spectrum encountered in the reactor. From this response, assessment of the duration of the safe operation of the reactor and hence its operational lifetime is made.

In view of the importance of these dosimetry materials to such programmes the European Working Group on Reactor Dosimetry (EWGRD) set up a sub-group on monitor materials (MMSG) with the following objectives:-

-    To identify the uses, needs, and supply problems of reactor dosimetry materials for materials research studies and reactor pressure vessel surveillance requirements.

-    To advise and assist in the establishment of stocks of reference dosimetry materials for the European Community.

-    To discuss and exchange information on the use of dosimetry materials and establish further requirements.

These objectives have formed the basis for the work programme which is being undertaken by the Central Bureau for Nuclear Measurements, (C.B.N.M.) of the Joint Research Centre of the Commission of the European Communities.

The wide range of neutron energies which exist in a reactor necessitates a range of nuclear reactions of differing energy response to cover it. In general these reactions can be divided into 3 categories:-

-    These covering thermal neutron energies and which react with neutrons with energies below 1 eV. These are typically (n,γ) reactions such as $^{59}Co(n,\gamma)^{60}Co$.

-    Intermediate neutron energy reactions which occur between 1 eV and a few keV. In this energy region the reaction probability can be very large but localised at precise resonance energies. A typical reaction of this type is $^{197}Au(n,\gamma)^{198}Au$ with a main resonance energy of 4.9 eV.

- Fast neutron energy reactions covering neutron energies above 0.1 MeV. The reactions occurring in this region are of the following types:-

$$(n,p), \quad (n,2n), \quad (n,\alpha), \quad (n,f), \quad (n,n')$$

and occur above a threshold energy.

The materials acquired for this stock should conform to a number of requirements. They should meet the quality assurance needs of all uses and hence be characterised chemically to a high degree so that they can be a referencable stock. They should be of the highest purity achievable in the physical form best suited for dosimetry usage. They should have little or no spectral or nuclear interferences and in the form used the nuclide of interest for the dosimetry reaction should be known with an uncertainty of not more than ±1%. It is intended that these well characterised, high purity materials will be particularly useful for interlaboratory reaction rate comparisons.

2. THE CENTRAL BUREAU FOR NUCLEAR MEASUREMENTS PROGRAMME

The EWGRD sub-group on monitor materials decided to ask C.B.N.M. Geel to first carry out a survey of the needs of monitor materials in European Community member states. The results of this survey indicated very diverse requirements for types and forms of materials intended for both reactor pressure vessel surveillance and spectrum analysis, and research programmes on neutron dosimetry. The sub-group decided to request a list of priority materials to be established in a usable form for surveillance purposes and for a continuing programme to meet the remainder and any subsequent needs at a later date.

As a result of this request, C.B.N.M. has started a programme of work on the preparation and characterisation of certified reference materials for reactor neutron dosimetry. This programme has been carried out since 1982 in cooperation with MMSG which acts as an advisory group to C.B.N.M. As a large part of the work is of an analytical and radiochemical nature which cannot be covered completely by C.B.N.M., assistance from selected specialised laboratories has been enlisted.

The following types of dosimeters are being considered:-

- High purity metals certified for impurities which are activated by thermal neutrons yielding capture products which

489

interfere with the measurement of the fast neutron reaction of interest.

-    Alloys and composite materials certified for homogeneity and absolute amounts of the nuclide of interest; in the case of actinide materials also traces of other actinide nuclides with high fission cross section relative to the nuclide of interest have to be certified.

Up to the present time 12 activation and 6 fission dosimeter materials are being examined as acceptable sources of materials (Table 1).

3. WORKING PROCEDURE

Specifications for the reference materials have been established in order to decide initially whether or not the materials are suitable for use as dosimeters in the wide range of applications for which they may be used. If the material available commercially was not adequate for such use then further purification was required. In the case of alloys and mixtures, the specifications were set as a maximum uncertainty that could be tolerated for both mass and homogeneity of the nuclide of interest. In respect of the nuclear purity it was assumed that fast neutron dosimeters could be used in a neutron spectrum containing at least two orders of magnitude greater thermal than fast neutron flux. The resulting interference from induced radiation, was calculated as to not lead to a greater than 10% correction to the activity measurement of the dosimeter nuclide. These impurity specifications are only guides to the selection on economic grounds and as far as possible the highest achievable purity material will be characterised.

On the basis of the above specifications, base materials are purchased or prepared and transformed into their final chemical and physical forms before characterisation. In the case of pure metallic reference materials the preparation is in general limited to mechanical transformation into foil and wire. For alloys prepared by high frequency levitation melting[1] extensive homogenisation checks are carried out on small amounts of the material, comparable to those used for dosimetry purposes. Pure or mixed oxides are prepared by sol-gel or gel-precipitation techniques which results in easily dispensed and counted forms with a high degree of uniformity[2]. Niobium metal with a very low tantalum content has been manufactured[3] specifically for these dosimetry requirements.

The certification will be done in steps and will finish with the issue of an EC Certificate by C.B.N.M. after approval by the

TABLE 1   REFERENCE MATERIALS FOR NEUTRON DOSIMETRY

Element	Chemical Form	Physical Form (a)	Major Impurity Specification (b)	Accuracy Specification (c)
Al	Metal	F0.1; F1; W1	<0.5 Na	
Co	Al - 0.1% Co	F0.1; W0.5		±1% (20mg)
	Al - 1.0% Co	F0.1; W0.5		±1% (20mg)
	V - 0.1% Co	F0.1; W0.5		±1% (5mg)
	V - 1.0% Co	F0.1; W0.5		±1% (5mg)
Cu	Metal	F0.1; 1.0 W0.5; 1.0	<0.1 Co	
Ni	Metal	F0.1; W0.5	<1 Co (b)	
Fe	Metal	F0.1; W0.5	<0.1 Co, Mn	
Nb	Metal	F0.02; F0.1 W0.5	<1 Ta	
Ti	Metal	F0.1; F0.5; W0.5	<0.1 Sc	
In	Mixed oxide	S0.5		±1% (1 sphere)
Rh	Metal	F0.05	<15 Ir, Pt	
^{235}U	V - 0.2% U	W0.5		±1% (15mg)
	V - 5% U	W0.5		±1% (15mg)
^{238}U	Oxide	S0.5; 1.0		±1% (1 sphere)
^{237}Np	Oxide	S0.5; 0.8		±1% (1 sphere)
^{239}Pu	V - 3% Pu	W0.5		±1% (1mg)
^{232}Th	Oxide	S0.5; 0.8		±1% (1mg)
V	Metal	Encapsulation	<10 Ta	

(a) F = foil; W = wire; S = sphere; Figure diameter or thickness in mm

(b) Concentration in µg/g

(c) Accuracy on nuclide of interest (sample size)

Nuclear Certification Group and endorsement by a Commission
Advisory Committee. However, in order to make the reference
materials available to the customers as quickly as possible, first
deliveries are planned to be performed together with the draft
Certification Report and the draft Certificate, prepared by
C.B.N.M. after discussion at an in agreement with the sub-group.

4. PROGRESS OF THE PROGRAMME

A target of 2 to 3 reference materials per year, beginning in
1985, has been set for the programme. In 1984, three materials
have been purchased, prepared and distributed for certification
analysis. These were aluminium, nickel and uranium-238 oxide.
The detailed progress on each of these three materials is
described in section 4.1 to 4.3.

Further materials have been purchased and are undergoing initial
examination to establish their suitability for full characterisation.
These are copper, iron, neptunium-237, niobium and vanadium. The
progress on these materials is described in sections 4.4 to 4.8.
The establishments with specialist experience in chemical and
radiochemical analysis are collaborating on the certification of
these materials are: A.E.R.E. Harwell (U.K.), Bundesanstalt für
Materialprüfung, Berlin (F.R.G.), C.B.N.M. Geel (B), C.N.R.S.
Orleans (F), Institute for Nuclear Sciences of the State University
of Gent (B), Max Planck Institut für Metallforschung, Stuttgart
(F.R.G.) and SCK-CEN Mol (B).

4.1 Aluminium

Aluminium metal of a stated purity of >99.9995% was purchased
from Vereinigte Aluminiumweike A.G., Bonn F.R.G. and delivered in
rods of 50mm diameter. These were cut into slices 12mm thick,
deburred and cleaned by etching in a acid mixture of 90% phosphoric,
5% sulphuric and 5% nitric acids at 90°C. After water rinsing and
drying the slices were rolled into foil of 1mm and 0.1mm thickness.
The foil was washed with alcohol, dried, and stored in an argon
atmosphere. Some of the slices were rolled to 4mm thickness and
cut into strips 4mm wide. These strips were rolled into a wire
form 1.5mm diameter then swaged to 1.2mm and finally drawn into
wire of 1.0mm diameter through diamond dies. Pieces 1m in length
were cut, cleaned in alcohol, dried and stored in an argon
atmosphere.

Rolling and wire drawing tools were polished before use with
aluminium oxide powder, and lubricated with petroleum spirit during
use. Panoramic analyses using spark source mass spectrumetry
(SSMS) and thermal neutron activation analysis (NAA), as well as

detailed certification analyses using atomic absorption (AAS), emission spectrometry with inductively coupled plasma source excitation (ICP) and NAA are planned for the latter part of 1984. Initial indications from reception analyses carried out by NAA indicate a sodium content below 0.5 µg/g[5].

4.2 Nickel

Nickel of a stated purity of better than 99.995% was purchased as rods of 12mm diameter from Materials Research B.V. Utrecht, Holland. Foil of 0.1mm thickness was prepared by rolling of 10cm lengths of rod after etching in a mixture of 50% hydrofluoric and 50% nitric acids, washing and drying. The foils were annealed by heating to 700°C for 3 hours in vacuum at a thickness of 0.2mm and again on completion of the rolling.

Wire of 0.5mm diameter was prepared by reducing the original rod intially to 3mm diameter by wire rolling. A further reduction to 1mm was achieved by swaging in steps of 0.2mm; then to 0.7mm in steps of 0.1mm. Further reduction in diameter was achieved by wire drawing in steps through diamond dies down to 0.63mm, 0.58mm and 0.52mm respectively. The wire was finally reduced to 0.50mm by swaging again. After being cut into 1m lengths the wire was cleaned with ethanol, dried, and stored in an argon atmosphere. Rolling and wire drawing tools were polished before use as described in section 4.1.

Panoramic analyses by SSMS and certification by NAA, ICP and AAS are planned for the latter part of 1984. Preliminary analyses show a cobalt content of below 0.1µg/g[6,7,8].

4.3 Uranium-238

Uranium-238 with a stated isotopic purity of >99.999% and a chemical purity >99.98% was purchased from Oak Ridge National Laboratory, U.S.A. and delivered in the form of oxide powder. This powder was converted into 0.5 and 1.0mm uranium dioxide spheres at A.E.R.E., Harwell using the sol-gel technique[2]. Homogeneity of 1mm spheres was confirmed by gamma counting[9]. Panoramic analyses by SSMS and certification by mass spectrometry (isotopic analyses and isotope dilution), coulometry and O:U equilibrium are planned for the latter part of 1984.

4.4 Copper

Copper metal of a stated purity of >99.999% has been purchased from ASARCO Inc., Denver, U.S.A. The material was supplied in four pieces and checked by NAA for homogeneity. Although nominally from the same batch small differences were noted between the pieces. The pieces have now been melted together and further checks are being carried out on this material by NAA. There should be no difficulty in meeting the specification of <0.1µg/g of cobalt in the material.

4.5 Iron

Iron metal with a stated purity of >99.99% has been purchased from Ecole National Supérievre des Mines, St. Etienne, France. The reception analysis is being carried out by NAA and the specification of <0.1µg/g for cobalt and manganese should easily be met.

4.6 Neptunium-237

Neptunium-237 in the form of oxide has been purchased from the Oak Ridge National Laboratory, U.S.A. When the preliminary reception analyses by alpha spectrometry have been carried out for plutonium the material will be converted into 0.5mm and 0.8mm diameter oxide spheres using the sol-gel technique.

4.7 Niobium

A batch of niobium metal of very high purity has been prepared by Max-Planck Institut, Stuttgart, F.R.G. by electrolytic purification and zone refining. The material has been supplied in the form of single crystals of 0.8cm diam. and 15cm long. Initial reception analysis by N.A.A. indicate that the tantalum content is about 0.3µg/g[3].

4.8 Vanadium

Vanadium is required for alloy purposes and the manufacture of dosimeter capsules. A source of high purity material has been located for purchase from Gesellschaft für Electrometallurgie, Nürnberg, F.R.G.[4]. Analysis by N.A.A. of preliminary samples indicate a tantalum content of less than 4µg/g[5,6].

5. CONCLUSION

The setting up of a source of certified reference materials for use as neutron dosimeters is now in progress at the Central Bureau for Nuclear Measurements, Geel, Belgium. The monitor

materials sub-group of the European Working Group on Reactor
Dosimetry has collaborated by suggesting the materials, the form,
and the purity specification for these materials. A number of
specialist laboratories are assisting with the initial investigatory
analyses and with the final certification analyses. Three
materials, Al, Ni, and ^{238}U will be issued in 1985 and 4 others are
under investigation.

These materials will assist with the establishment of
quality assurance, interlaboratory comparisons, and overall safety
improvements to the field of neutron metrology.

6 REFERENCES

1. J. Van Audenhove, J. Joyeux, Journal of Nucl. Materials, 19
    (1966) p 97-102.

2. C.E. Lyon, 12th International Nuclear Target Development
    Society Conference, 25-28 September 1984, NIM in press.

3. K. Schulze, M. Krehl, International Nuclear Target
    Development Society Conference, 25-28 September 1984, NIM
    in press.

4. R. Hähn, International Nuclear Target Development
    Society Conference, 25-28 September 1984, NIM in press.

5. SCK-CEN Mol, private communication.

6. Inst. for Nucl. Sc. (State University Gent) private
    communication.

7. A.E.R.E., Harwell, private communication.

8. B.A.M., Berlin, private communication.

9. C.B.N.M., Geel, internal note.

NIOBIUM-ALUMINIUM FOILS AS MONITORS FOR ROUTINE

MEASUREMENT OF FAST NEUTRONS

P. Wille

GKSS-Forschungszentrum Geesthacht, Institut fur Physik

Box 1160, D-2054 Geesthacht, Germany

ABSTRACT

Niobium is an excellent monitor for fast neutrons.
The reasons are the long half-life of about 16
years of the Nb93m and the low neutron energy
threshold of about 0.1 MeV. Nb93m is produced by
inelastic neutron scattering. The low energy of the
emitted KX-radiation and conversion electrons
between 15 and 27 keV does not call for great care
when handling. On the other hand the monitors have
to be very thin in order to limit self-absorption,
and if the content of tantalum is small -less than
50 ppm- the influence of Ta182 can   also be
neglected. Thin monitors have an additional
advantage: small amounts of radioactivity compared
with monitors produced from parts of a large piece
of irradiated niobium. Thin niobium foils however
suffer at least some loss of mass, or may be
completely powdered by radiation embrittlement. To
overcome this problem, a thin (3 micron) Nb layer
was stabilized using 0.1mm Al foil. The niobium
layer did not show any loss of mass due to
irradiation up to $10^{27}/cm^2$. The problem of Nb-mass
determination in Nb/Al foils can be solved by
exact weighing before and after Nb deposition, or
by comparing Nb95 activities of thick enough Nb
plates (to determine the mass) and Nb/Al monitors
irradiated in direct contact with each other to
"see" the same thermal neutron flux.

J. P. Genthon and H. Röttger (eds.), Reactor Dosimetry, 497–504.
© *1985 ECSC, EEC, EAEC, Brussels and Luxembourg*

INTRODUCTION

Niobium is transformed into Nb93m, when neutrons with energies
more than 0.1Mev are scattered inelastically by Nb-nuclei. Nb93m
decays with a half-life of about 16 years by internal conversion
into its ground state. It emits conversion electrons of 26.5 keV
and the accompanying radiations: X-radiation and Augerelectrons.
Because of its simplicity the KX-radiation is measured to detect
Nb93m. The KX-radiation is emitted when electrons from the K-
shell are "internally converted". Unfortunately all radiations
with energies high enough, excite Nb-atoms to emit KX-radiation.
One of the sources of such radiation is Ta182. Ta182 is formed
by low energy neutrons from tantalum, which is a major impurity
in niobium. This influence can eliminated or calculated. Because of its long half-life and low neutron energy threshold,
which indicates a low sensitivity against neutron spectra variations, niobium is of interest for long term reactor surveillance (1,2). In most laboratories the niobium monitors are
chemically prepared after irradiation. This is an inconvenient
method involving the use of nitric and hydrofluoric acids. With
our monitors, stabilized on Al foil before irradiation, no chemical processing is necessary. However attention must be given
to careful mass determination. Pure Nb foils cannot be used
directly for irradiation since they are destroyed by radiation
embrittlement (3).

The requirements for the determination of fast neutron fluences via the measurement of Nb93m, are:
1. thin samples
2. mechanical stability after irradiation
3. known amount of niobium in the monitor
4. an additional but not essential feature concerns the ease of
   measurement if no chemical processing is necessary.

1. Thin samples.
   Thin samples of Nb -some mg Nb/cm^2- should be used in order
   to keep low or constant (c):
   a) self-absorption
   b) fluorescence of Nb by activated isotopes in the sample
   c) fluorescence of Nb by background radiation.

1a) Self-absorption for the KX-radiation (16.6 and 18.7 keV) in
niobium is high. One half of the 16.6 keV photons emitted in a
whole sample are absorbed in 0.093 mm thick foils. The self-
absorption coefficient depends on the sample-detector distance.
For samples with low activity the distance to the detector has
to be small. The self-absorption for 3 micron niobium is only
2.5%. The additional correction for the nearest possible distance, about 5 mm, is less than 1%. Any other geometrical
effects can be eliminated, if one uses standards with the same

geometrical cross sections.

1b) The radiation of any radioactive isotope present in the niobium excites the atoms to emit their x-radiation, which is i-
dentical with the radiation to be measured in the case of Nb93m.
The escape of interfering radiation is more probable the thinner
the sample. Ta182 is one of the most interfering iso- topes. It
is produced by slow neutrons from tantalum normally present in
niobium. It decays with a half-life of 115 days emitting betas
and gammas, and also its influence, $I_K(Ta182, t)$, on niobium
decays with that half-life. This influence can be eliminated, if
the monitor is measured at times t0 and t1 to get the KX-inten-
sities $I_K(t0)$ and $I_K(t1)$ still containing the contribution of
Ta182. The intensity from the Nb93m decay at the time t0 is then

$$I_K(Nb93m, t0) = \frac{I_K(t1) - I_K(t0)*\exp(-\lambda_{Ta}*\Delta t)}{\exp(-\lambda_{Nb}*\Delta t) - \exp(-\lambda_{Ta}*\Delta t)}$$

where λ_x is the decay constant of the isotope x. For 3 micron
Nb-layers, and a ten times higher activity of Nb93m over Ta182,
1.3% of the measured KX-radiation stems from the influence of
Ta182. The influence of other radioisotopes Nb94 e.g. can be
neglected.

1c) Niobium like other
elements is also excited by
background radiation to emit
its KX-radiation. In a
background of 6 micro-
roentgen/hour, 0.6 KX-photons
were measured leaving a 0.2 mm
thick Nb foil of 1 cm surface
area. The KX-count rates of
thick yttrium-, zirconium-,
cadmium-, and indiumplates
were found to be in that
range (fig. 1).

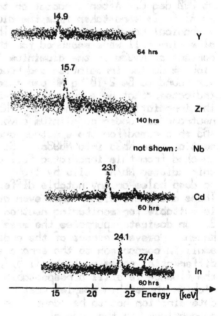

Fig. 1: "Activity" of some
nonactive metals under the
influence of background
radiation measured with a
Si (Li).

This low value does not limit the thickness of the monitor. It is a limit, however, for the lowest detectable neutron fluence. Corrections are only necessary, if very low neutron fluences(6) are to be measured; here the thickness of the monitor is supposed to change that fluorescence induced by background radiation. The fluorescence, which always occurs, is not proportional to the background level due to spectral variations and/or due to natural radioactivity in the sample. With heavy iron-shielding the value for the fluorescence went down by a factor of 6. For 3 micron Nb-layers with 3mm foil diameter this always existing KX-emission is about 0.03 photons/s/cm^2. ——

2. Niobium foils are known to become damaged under reactor irradiation. At least some loss of material might happen during irradiation or postirradiation handling or, in some cases, foils were transformed into powder (3). This is the reason why Nb foils cannot be used as monitors for neutron fluence measurements. It is now of interest, if Nb layers of some microns backed by suitable material such as aluminium, could withstand reactor irradiation. If so, an ideal Nb-monitor could be produced, which is thin enough to meet all the requirements, provided the mass is known. To investigate this, a layer of about 3 microns of niobium was produced by evaporation of Nb onto one surface of a 0.1 mm Al foil. From the Nb/Al foil six discs of 3mm diameter were punched, wrapped in aluminium foil and irradiated in FRG 2 at 290 deg C. After irradiation to $3*10^{19}$ neutrons/cm^2 the six Nb/Al foils were taken from the aluminium envelope after some mechanical handling procedures. The six Nb/Al foils and the aluminium foil were measured for their Nb93m-activities. The content of Nb93m in the aluminium foil indicates the loss of niobium during irradiation and handling. The loss of niobium was found to be $2*10^{-4}$ g Nb per inserted gram Nb. This low value indicates, that no substantial loss of monitor material occurred. The irradiation of another large Nb/Al foil, 7*10cm^2 to about 10^{21} neutrons/cm^2 under gadolinium cover was performed at KFA Juelich (63 mb assumed for the spectrum averaged activation cross section from Nb93 into Nb93m). Some discs 3mm diameter were punched from this irradiated foil and compared visually with unirradiated Nb/Al foils by light microscopy and SEM-techniques: no deep holes or any notable difference could be observed. These are indications, that even at high neutron fluences, Nb/Al is suitable for monitoring neutron fluences. ——

3. For dosimetry purposes the mass of the monitor must be known. However, errors of the order of 1% in the mass are still small in comparison to the errors in the neutron spectra and the differential cross sections (1,5,7,8,9,). An error of 10% for the fluence is still a respectable value. In the case of niobium where thin samples must be used, the Nb-mass determination has to be adapted to the method of thin monitor production: in the case of

a) chemical processing an aqueous solution of a Nb93m-containing niobium-compound can be brought on filter paper tissue, or on copper foils by a electrochemical process(4). The amount of Nb can be determined by measuring the activity of any radioisotope in the irradiated bulk niobium. The activity and the mass of the large niobium piece can be measured easily. If one compares the activity of the same Nb-isotope in both samples one gets the amount of niobium on the filter paper tissue. The accuracy of the method can be better than 1%. It depends on the counting statistics and might be time consuming, if one aims for some 0.1% error.

b) Another method of determining the amount of Nb in a thin monitor still to be produced, is to dissolve the bulk metallic niobium monitor (1 g Nb in 4mL 67%HNO3 and 4.5 mL 40%HF) after irradiation. Its mass has to be determined. The missing mass -some milligrams of the solution has been taken to produce the thin sample on filter paper tissue- divided by the whole mass of the solution is exactly the niobium mass fraction contained on the filter paper tissue. The error in Nb mass is reported to be 0.03% (2). We introduced a different kind of niobium monitors, which fulfills the fourth requirement mentioned above. For the Nb/Al foil the Nb mass determination was done in three ways:
(i) The method described in 3a) was used with the difference that instead of preparing a thin sample by chemical processing the Nb/Al foil was used.
(ii) From the Nb-deposited Al foil and another pure Al foil taken from the same stock about twenty 3 mm diameter disc-shaped foils were punched. Both sets of foils were weighed on a microbalance to determine the mean weights and the standard deviations for both sets of discs. We found an error of 1.3% Nb mass on the 3mm Nb/Al foils. It is advisable to use 0.03 mm Al foils instead of 0.1 mm Al for the evaporation of Nb. Then the error in the mass determination was calculated to be less than 0.3%.
(iii) Six Nb/Al foils 3 mm diameter of unknown Nb mass were irradiated each in between two 0.1 mm Nb foils of known mass representing 6 stacks. Each stack (3 foils) was positioned vertically in a slit of an aluminium disc. The irradiation was performed in the 15 MW swimming-pool FRG 2 in a rotating irradiation device (0.17 rpm in core) in six positions 16 mm apart (fig. 2). The rotation during irradiation eliminated local neutron flux variations to some extent. The resulting 35 day Nb95-activity, which emits 766 keV gammaradiation (fig. 3) was repeatedly measured with a Ge(Li) 117 times over a distance of 100 cm. The comparison of the 766 keV count rates of the foils, which were irradiated in direct contact (0.1 mm apart), gives the Nb mass of the Nb/Al foil. The standard deviations between the six mean masses (0.25 mg) were between 0.4% and 0.5% for each of the foils in the six positions of the rotatable irradiation device. In order to find the thermal neutron flux variations in the six positions, one has to compare the specific count rates (counts/second/gram Nb) of the six pairs of 0.1 mm Nb foils. Maximum

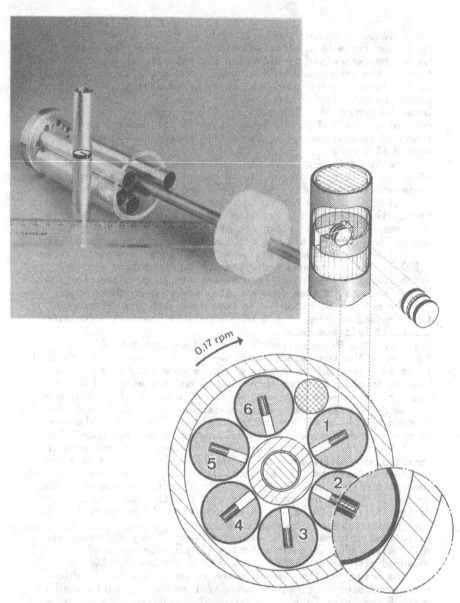

Fig. 2: Rotatable irradiation device for mass determination of
Nb in Nb/Al foils via Nb95-production. It shows the
configuration of six stacks consisting of 3 foils: two
Nb discs with one Nb/Al disc between. They are positioned verically in a slit of an aluminium plate.

differences of 2.8% were found. These flux variations occur
over 16 mm distances whereas for 0.1 mm, which was the relevant
distance for the mass determination, the influence of variable
flux was neglected in the error determination for the Nb mass.

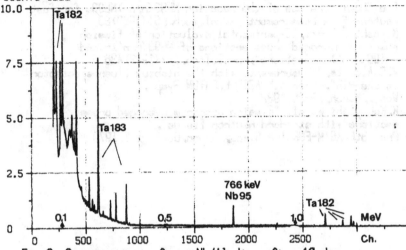

Fig. 3: Gammaspectrum of one Nb/Al disc after 10 days
irradiation, measured with a Ge(Li) over 100 cm
distance and 5400 sec. for mass determination.

4) The advantage of no chemical processing after irradiation is
obvious. The radioactivities produced in niobium under reactor
radiation require special handling in the case of chemical
proces- sing in order to prevent loss of radioactivity.
Especially when hydrofluoric acid is used to dissolve
radioactive material the working conditions are regulated. This
involves a high standard for the equipment of laboratories. It
is thus convenient to shift problems from the postirradiation
stage to the preirradiaton stage, whereever possible. How this
could be done has been shown with our foils.

REFERENCES

1. F. Hegedues, "Detecteur de fluence de neutrons rapides base
   sur la reaction Nb93(n, n')Nb93m", Report EIR 195,
   Switzerland 1971
2. P. Wille, "Messung der Aktivität des Niob93m zur Bestimmung
   der Fluenz schneller Neutronen in Leistungsreaktoren",
   Atke, 29 (1977) 166
3. H. Tourwe et al., "Niobium dosimetry intercomparison in EBR
   II and BR2", Proc. 4th ASTM-EURATOM Symp.,
   Wash., March, 22-26, 1982

4. W.H.Taylor, Proc. 2th ASTM-EURATOM Symp., Palo Alto, 1977, 3-7, 1977
5. B.Strohmaier et al., "Evaluation of the cross sections for the reactions..., Nb93(n,n')Nb93m,...", Physik Daten, FIZ, 13-2, 1980
6. Gad Shani, "The use of the reaction Nb93(n,n')Nb93m for fast neutron flux measurement", Nucl. Tech., 51 (1980) 83
7. Kiyoshi Sakurai, "Experimental evaluation of fission spectrum averaged cross sections of Nb93(n,n')- and Hg199(n,n')- reactions", Nucl. Tech., 57 (1982) 436
8. W.G.Alberts, "Measurements with the niobium fluence detector at the PTB", Proc. 4th ASTM-EURATOM Symp., Wash., March, 22-26, 1982
9. K.Kobayashi et al., "Fission spectrum averaged cross sections with standard neutron fields", Proc. 3th ASTM-EURATOM Symp., Ispra, Oct. 1-5, 1979

RECENT DEVELOPMENTS IN VERY HIGH-COUNT-RATE GAMMA SPECTROSCOPY

William H. Zimmer

EG&G ORTEC

100 Midland Road, Oak Ridge, Tennessee 37830 U.S.A.

ABSTRACT

Recent improvements in the electronics that amplify
and analyze gamma photon-induced pulses have made it possible for HPGe coaxial detectors to accept input rates of
one-million, one-MeV gamma photons-per-second and still
provide the spectroscopist with spectra that can be analyzed. Data are presented that illustrate gain shift, shape
variation, and changes in counting uncertainty statistics
due to the greatly extended count-rate range. Software
algorithms are presented that allow gain shift and peak
resolution to be adjusted automatically on a sample-by-
sample basis. Relationships are developed between integrated count rate and the variances of full-energy photon
peak area and counting uncertainty when using the real-
time-correction mode of pulse processing. When hardware
and software components that have been designed for both
wide-range and high-count-rate spectroscopy are integrated
into a mutually supportive system, qualitative and quantitative spectroscopy over counting rates of one to one-
million counts-per-second become achievable.

INTRODUCTION

A high-count-rate, high-resolution gamma spectrometer system is
described in ref. (1). The principal components of the system are a
7.6% relative efficiency, reverse-electrode HPGe coaxial detector,
a transistor reset preamplifier, a time-variant shaping amplifier,
a multichannel buffer (MCB), and a computer system. All data were
taken using this system with an amplifier shaping time of 0.25 μs

J. P. Genthon and H. Röttger (eds.), Reactor Dosimetry, 505–511.
© 1985 ECSC, EEC, EAEC, Brussels and Luxembourg

and a 2.1-mCi Ra-226 source.

The high purity germanium resistive-feedback preamplifier energy rate product is 160 000 MeV/second; that is, at rates higher than 160 000 one-MeV pulses-per-second (p/s), the preamplifier saturates and no longer transmits pulses. The recently developed transistor reset preamplifier (TRP) increases the throughput capabilities by a factor of 5 so that the preamplifier is no longer the limiting device. The TRP also eliminates the need for amplifier pulse-shaping, time-related pole-zero adjustment by removing the need for a feedback resistor.

A combination amplifier and gated integrator (1) integrates the area under the shaped photon pulse rather than measuring its voltage height. This change permits the choice of shaping time constants as short as 0.25 µs with excellent Full-Width-at-Half-Maximum (FWHM) resolution characteristics. The semigaussian amplifier limit to throughput for a Co-60 signal with 2-µs shaping into a 100-MHz Wilkinson Analog-to-Digital Converter (ADC) is 20 000 p/s. An amplifier/gated integrator using 0.25-µs shaping into a 10-µs fixed-conversion-time ADC improves throughput of the same signal by 3.5 times.

The fixed-conversion-time ADC approximately doubles the system throughput rate of a Wilkinson 100-MHz ADC. More importantly, the fixed-conversion-time ADC removes the energy dependence from the time of conversion. This change alone improves the accuracy of live-time-correction circuitry by at least an order of magnitude. It also allows implementation of recently improved real-time correction of counting losses techniques (2,3) to the analysis of highly variable count-rate samples.

This presentation is concerned with the output of the hardware described and with the integration of software preconditioning and correction algorithms to form a quantitative and qualitative system that will operate over an extended count-rate range. The system is capable of operating using traditional live timing and real-time-correction timing. In the former, while one pulse is being analyzed, the timing clock is shut off. One pulse is stored for each pulse analyzed, and live time is counted as the time available to analyze pulses; that is, dead time is the time utilized to analyze and store pulses. Real-time-correction mode of analysis counts in real- or clock-time. However, for each pulse analyzed, a variable number of counts are stored in the MCB. The number of counts stored equals the number of pulses detected while one pulse is being analyzed, plus one for the pulse that is analyzed. The real-time-correction mode of analysis is only possible in a constant-time ADC and is useful in correctly analyzing samples with variable counting rates.

506

EXPERIMENTAL

In traditional methods of high-count-rate gamma spectroscopy the amplifier shaping time was set at a value close to 2 μs in order to preserve detector resolution over as large a portion of the count-rate range as possible. The resulting maximum input count-rate range varied from about 40 000 counts-per-second (c/s) to about 200 000 c/s depending on the hardware and software integrated in the system.

There were always substantial peak-area-per-second, random summing, and losses due to pulse pileup, even with a pulse pileup rejector (4). The losses were about 8% at input rates of 100 000 c/s and were due to the fact that every time a pulse was processed through the amplifier, voltage did not return to the baseline until approximately six times the shaping time. For 2-μs shaping time there were many blocks of 12 μs in which pulse pileup and random summing occurred.

Use of the time-variant shaping amplifier allows shaping times as low as 0.25 μs while preserving good resolution over the usable count-rate range. The block of time in which random summing can occur is now reduced to approximately 1½ μs. Even at 10 x 100 000 c/s input rates, the losses due to random summing are substantially below 8%. Several independent systems of this type with 0.25-μs shaping time have produced random summing losses of <1% at input rates of 700k to 800k c/s. With such minor losses, the random summing correction needs no longer be applied. However, this correction remains available in the software when needed.

Resolution or shape varies rapidly as a function of input count rate, shaping time, and energy. Figure 1 illustrates the variability of resolution with input count rate and three shaping times at 1.33 MeV. Table 1 illustrates energy as an independent third variable.

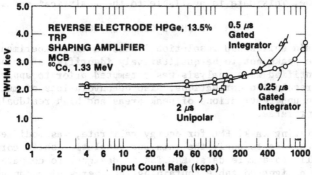

Fig. 1.   Resolution as a Function of Input Count Rate and Shaping Time.

Table 1.   Resolution as a Function of Input Count Rate and Photon
Energy Using 0.25-μs Shaping Time and a Gated Integrator.

Energy (keV)	Count Rate (counts-per-second)		
	5k	200k	500k
	FWHM Resolution (keV)		
295.21	1.854	2.326	3.107
351.92	1.853	2.396	3.146
609.31	1.966	2.445	3.204
1120.29	2.204	2.707	3.416
1238.11	2.564	2.769	3.456
1764.49	2.558	3.027	3.837
2204.22	2.808	3.203	4.023
2447.71	2.983	3.405	3.994

The algorithms in EG&G ORTEC's analytical software for germa-
nium-derived gamma spectra, GELIGAM^{TM}, use a channel vs shape calibration to establish integration range limits for single energy peaks,
the presence of multicentroid photon peaks, and the shapes to be fit
in multiplet deconvolution. It is essential to quality spectroscopy
that the appropriate shape calibration for the analysis of each
sample spectrum be used.

In GELIGAM a peak must be within 0.5 x FWHM (from the shape
calibration) of a userlibrary photon-centroid energy before the peak
is qualitatively identified with a nuclide.

Gain shift due to count rate is relatively small, but not negligible or unimportant. A nuclide may easily be missed from a sample
spectrum analysis due to gain shift.

Figure 2 is a plot of the number of counts stored per pulse
analyzed in real-time-correction mode vs integrated count rate recorded in the MCB. It was determined by taking a ratio of integrated,
real-time-correction rate to the integrated live-time count divided
by real time. All input count rates referenced were determined from
integrated spectral data, divided by live time when acquired in that
mode, and by real time when the data was acquired in real-time-correction mode. This data is available to the analytical software.

GAIN AND SHAPE COMPENSATION

The gain shifts and resolution losses in this experiment caused
many sample peaks not to be qualitatively identified or to be erron-
eously identified when analysis was attempted prior to applying gain
shift and resolution corrections. Also deconvolutions of multiplets
resulted in underestimations of peak areas and high residuals at
higher count rates.

An existing task, ENC for energy calibrate, was modified and
run in batch mode to apply the gain and resolution shift corrections
automatically to sample data. ENC.TSK is designed to compare the contents of a designated table containing the energy of clean photon
peaks and an estimate of their channel centroids to a real gamma spectrum. The results of the comparison are channel vs energy and channel

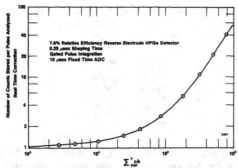

Fig. 2.   Number of Counts Stored per Pulse Analyzed in Real-Time-
Correction Mode vs Integrated Count Rate Recorded in the MCB.

vs FWHM quadratic equations. The equations replace the previous equations in the calibration file as long as the % difference between any
individual photon energy centroid and the calculated function does
not exceed 1%, or the % difference between any measured FWHM and the
calculated FWHM does exceed 10%.

Two modifications were made. A trigger level was entered in the
germanium parameter file together with the ENC table file specification. The units of the trigger level were integrated c/s from the
sample spectrum file. No correction was attempted until the trigger
level had been exceeded. Secondly, the corrected calibration functions
were applied to the calibration data in the sample Unformatted Output
(UFO) file. As a result, only the sample data were changed. The results of the changes were automatically reflected in the sample calculations and in the activity or concentration report. Each time the
gain and shape modifications were successfully applied, an ENC report
was generated on the printing device (Table 2). Each time the correction was not applied successfully, a message appeared on the terminal
that the attempt failed and the original calibration was used. The
same message was printed together with the attempted ENC report and
the original ENC report. The sample data were still usable by utiliz-

Table 2.   ENC Report Resulting From Successful Completion of Automatic Gain and Shift Adjustments to a 500k c/s Sample Spectrum.

CALIBRATION FILE SY:HICRTC.CLB
ENERGY=6.84111 + 0.332540 CHAN −4.518157E−08 CHAN**2
FWHM =8.92360 + 4.299594E−04 CHAN + 2.895942E−09 CHAN**2

	True Energy (keV)	Calculated Energy (keV)	Difference (%)	Measured FWHM (keV)	Calculated FWHM (keV)	Difference (%)
1	295.21	295.21	−0.00	3.107	3.092	−0.48
2	351.92	351.88	−0.01	3.146	3.117	−0.92
3	609.31	609.38	0.01	3.204	3.230	0.81
4	1120.29	1120.31	0.00	3.416	3.457	1.19
5	1238.11	1238.06	−0.00	3.456	3.510	1.57
6	1764.49	1764.52	0.00	3.837	3.751	−2.25
7	2204.22	2204.14	−0.00	4.023	3.955	−1.68
8	2447.71	2447.76	0.00	3.994	4.070	1.92

Table 3. Example of ENC Table Used to Implement
Gains and Shape Compensation.

Nuclide	Channel	Energy (keV)
A	871.000	295.213
B	1042.00	351.921
C	1819.00	609.312
D	3360.00	1120.29
E	3715.00	1238.11
F	5303.00	1764.49
G	6630.00	2204.22
H	7365.00	2447.71

ing the standard ENC task manually after reselecting the peaks in the
ENC table. There must be between three and twenty channel/energy pairs
in the ENC table specified that correspond to clean single peaks in
the sample spectrum (Table 3).

Use of the gain- and shape-shifting function is not restricted
to high-count-related applications. If gain and shape shifting is
being introduced to the sample spectra for any cause, including temperature variation, this function can compensate for it automatically.
It may be necessary to set the trigger level very low but there is
no restriction on how low it may be set.

REAL-TIME-CORRECTION MODE COUNTING UNCERTAINTY

The method of calculating counting uncertainty in GELIGAM was:

$$U_c = \frac{\sqrt{G + [N_c^2/(2n)^2 \cdot C] + B}}{\text{Net Counts}} \tag{1}$$

where

G = gross counts

N_c = number of channels integrated to determine continuum background (or net area) for a peak

n = number of consecutive channels used to determine continuum
background level higher and lower in energy than the peak;
for example, 3 or 5 channels

C = continuum background counts in the $2n$ channels above

B = net peak counts and gross counts from all other sources
that act as background to the peak being calculated.

Equation (1) assumes one count stored in the spectrum per pulse
analyzed. As previously explained, there is a variable number of
counts stored per pulse analyzed in real-time-correction mode. Therefore, the unaltered use of Eq. (1) for the calculation of counting
uncertainty in real-time-correction mode acquisitions is in error. A
better approximation is to correct for the average number of counts
stored per pulse analyzed as a function of the integrated count rate
(Fig. 2).

$$U_c = \frac{\{[G + (N_c^2/(2n)^2 \cdot C) + B] \cdot [R]\}^{1/2}}{(\text{Net Count}) \cdot (R)} \tag{2}$$

where R = the average counts stored per pulse analyzed at a determined integrated sample count rate.

The data in Fig. 2 become a calibration for the determination
of counting uncertainty in real-time-correction mode. In actual
application, the data in Fig. 2 are converted to a table file and
applied to all spectral data in which the real time equals the live
time. If a very low count-rate spectrum acquired in live-time mode
meets this criterion, the calculation of counting uncertainty remains correct because at low rates only one pulse is stored per
pulse analyzed and "R" is set to 1.

CONCLUSIONS

Recent improvements in hardware have made it possible to analyze
gamma-emitting samples with detector input rates of up to one-million
one-MeV p/s, but the resultant spectra often cannot be analyzed with
existing algorithms and exhibit biases and can understate counting
uncertainties. However, when software algorithms are incorporated
to adjust gain shift and shape from sample data and to correct for
biases and counting uncertainty as a function of integrated count
rate, then all peaks are identified with the appropriate nuclide and
all analytical results are within predictable total uncertainties.

REFERENCES

1. C. Britton, T. Becker, T.J. Paulus, and R. Trammell, Character-
    istics of High-Rate Energy Spectroscopy Systems Using HPGe
    Coaxial Detectors, presented at Nuclear Science and Nuclear
    Power Systems Symposia, San Francisco, October 1983.

2. G.P. Westphal, Real-Time Correction of Counting Losses in
    Nuclear Pulse Spectroscopy, presented at the 6th Modern Trends
    in Activation Analysis Conference, Toronto, 1981.

3. J. Harms, "Automatic Dead-Time Correction for Multichannel
    Pulse-Height Analyzers at Variable Counting Rates," Nucl.
    Instrum. Methods, 53, pp. 192-196 (1967).

4. W.H. Zimmer, "Analytical Software — Random Summing Corrections,"
    EG&G ORTEC Systems Application Studies, PSD No. 3 (1977).

INTERNATIONAL COMPARISON OF INTERPOLATION PROCEDURES FOR THE EFFICIENCY OF GERMANIUM GAMMA-RAY SPECTROMETERS
(Interim report on the GAM-83 exercise)

W.L. Zijp, A.N. Polle, H.J. Nolthenius

Netherlands Energy Research Foundation   ECN

Westerduinweg 3, 1755 ZG   PETTEN, The Netherlands

K. Debertin

Physikalisch-Technische Bundesanstalt

Bundesallee 100, D 3300 Braunschweig,
B-R Deutschland

ABSTRACT

Some first results are presented for the outcome of an international intercomparison of laboratory performances in determing photo-peak efficiencies and activities by means of gamma-ray spectrometric procedures and calibrated radionuclide sources.
The intercomparison (coded GAM83) was organized and analyzed under auspices of the International Committee for Radionuclide Metrology (ICRM).
The evaluation comprises the analysis of the contributions of 41 participants, and is not yet finished

INTRODUCTION

During a meeting of the International Committee for Radionuclide Metrology, ICRM, in Geel (May, 1983) the ICRM action 16/83 was planned. This action comprised an international intercomparison of laboratory performances in determing photo-peak efficiencies and activity values by means of gamma-ray spectrometric procedures.
The code name for this action, GAM83, is a short indication for the state-of-the-art of Gamma-ray spectrometric Activity Measure-

J. P. Genthon and H. Röttger (eds.), Reactor Dosimetry, 513–528.
© 1985 ECSC, EEC, EAEC, Brussels and Luxembourg

ments in the year 1983. The coordinator of the action is Dr. K. Debertin (PTB), chairman of the I.C.R.M. gamma-ray spectrometry working group. In July 1983 an information sheet was distributed to prospective participants. Candidate participants reacted while some of them provided also good quality input data for the intercomparison. A selection was made from these input data by K. Debertin in cooperation with W.L. Zijp of ECN Petten. As result of this preparatory work four input-data sets for the participants could be prepared at ECN.

The input-data sets contained a series of measured peak-efficiency values obtained with aid of germanium semi-conductor detectors. The uncertainties of the peak-efficiency values were also given. For one set a detailed specification of the uncertainty components was also given.

The task of the participants was to calculate with their own laboratory procedure the following three types of numerical data:

1. the adjusted efficiency values for the measured gamma-ray transitions of the calibrated sources, in case that an analytic fit is made with a smooth curve to the efficiency data;
2. the efficiency values for a few gamma radiations having other energies to be calculated on the basis of the experimental input-efficiency data and also the corresponding standard deviations;
3. the activity values for two radionuclide sources to be calculated on the basis of the experimental input-efficiency data and the photo-peak counting rates corresponding to the gamma-ray transitions of the radio-nuclides involved. Also the standard deviations should be calculated.

In addition to the three types of numerical data, the participants should give a <u>description of the model</u> used for the determination of the efficiency curve. The participants were requested to describe in some detail the specific calculation procedure (e.g. number of energies considered, any statistical weights used, model for the efficiency curve or for the interpolation, information on the quality-of-fit).

In case an analytic expression was used for fitting a smooth curve to the efficiency data, this function should be specified in terms of the parameters (and, if possible, their uncertainties and correlations).

Each participant was allowed to select from the complete list of energy values and efficiency values his own best set of efficiency data for his preferred analysis, but he should in any case also perform the analysis if possible for the complete input-data set.

The participants were requested to prepare their output data with

514

a prescribed set of nuclear data (of special importance are here the gamma-ray emission probabilities). They had the freedom to prepare, in addition to this output-data set, also output-data sets with preferred nuclear-data sets, different from the pre-scribed data set.

The input data sets were dispatched to some 60 candidate participants in November 1983. The participants were requested to return their contribution to the action before the 15th of February 1984.

The 41 returned contributions of the participants were received and analyzed at ECN in Petten.
Some preliminary results of this analysis are presented in this report.

INFORMATION ON THE INPUT-DATA SETS

Numerical data

The input data were selected from data sets which had kindly been supplied by a few standards laboratories. These data refer to actual calibrations with gamma-ray spectrometers using Ge(Li) and HPGe detectors.

The four selected input-data sets shown in figures 1...4 had the following properties:
- a high accuracy- (uncertainty < 1%) data set with a relatively large number of measured data (SET 1);
- a low accuracy- (uncertainty 3-5%) data set with a relatively small number of measured data (SET 2);
- a low energy-data set (SET 3);
- a high accuracy-data set with a relatively small number of measured data (SET 4).

Additional to the photo-peak efficiency values also data for the counting rate of an uncalibrated ^{134}Cs and ^{152}Eu source were given so that the participant could calculate the activity and its uncertainty.
It is assumed that, where necessary, appropriate corrections (for effects due to pulse pile-up, insensitive time, differences in source geometry) were taken into account in the input data.
The input-data sets have been carefully choosen with respect to the range in the gamma-ray energies, and the uncertainties and values of the gamma-ray emmision probabilities.

REVIEW OF PARTICIPANTS CONTRIBUTIONS

Each participant was assigned a serial number just for identification. The key for the relation between these serial numbers and the laboratories involved is not published. An overview of the participating countries and laboratories is shown in table 1. A survey of the submitted numerical data is given in figure 5.

CLASSIFICATION OF FITTING PROCEDURES

A large variety of interpolation procedures for the efficiency curve is in use. The 41 participants used some 20 different models.
Several types of numerical procedures can be distinguished in the contributed solutions:
- procedures using an algebraic expression for which all derivatives are continuous;
- procedures using spline functions, either passing exactly through the data points or fitting best to them, taking into account the uncertainties in the data points;
- procedures using an interpolation formula for a few successive data points;
- other procedures (comprising also hand-method procedures).

The actual mathematical treatment by the participants of the efficiency points showed differences in a number of aspects:
- different mathematical model;
- application of statistical weights;
- application of covariance data;
- different numerical precision, due to the applied computer.

Widely used is a model in which the logarithm of the efficiency is expressed as a polynomial function of the logarithm of the gamma-ray energy.
Only 4 participants have composed and used a covariance matrix for the input efficiencies by considering the components of the unertainty data as supplied for SET 1.

It is expected that especially in the cases where a matrix inversion is part of the calculation procedure the word length of the applied computer will determine the numerical precision.

EVALUATION METHOD

Software was developed for data treatment and for plotting tasks. The evaluation of the participants data involved the following actions:

- rough visual check of the supplied numerical data with respect to the presence of clerical errors;
- comparison of the numerical data supplied by the participants for the different data sets;
- overview of the various methods for efficiency determinations.

The computerized treatment of the submitted data yields, among others, the following numerical data:
- the mean value of the observations;
- median value of the observations (denoted with a tilde, e.g. $\tilde{x}$);
- the standard deviation of these observations;
- the estimate for this standard deviation obtained using the method of the Distribution of Differences (see below);
- the sum of squares of deviations (taken absolute, relative or "studentized");
- a ranking parameter obtained from the interpolated-efficiency values.

The main problems encountered in the data treatment of the contributed solutions were:
- incompleteness of the contributed data;
- obvious mistakes in numerical data supplied by the participants;
- treatment on outlying results;
- inaptitude for adequate comparison of different fitting procedures;
- influence of the length of the energy range in comparing fitting procedures (i.e. some fitting functions gave good result in a certain energy range while outside the range important systematic deviations could be observed.

In the program a number of mean values as well as standard deviations are calculated. A draw back of the straight-forward application of the normal definitions of these quantities is that possible discrepant values can influence the output values of these calculations in a serious way. For this reason the standard deviation is also calculated with the new method of the Distribution of Differences [1,2]. This method is based on the absolute differences between individual observations in the same series. These differences are sorted as a function of their absolute value after which their frequency is determined. From the obtained distribution function a new standard deviation can be derived (denoted as $\tilde{s}$) by determing that difference value for which 52 per cent of the total number of all possible difference values yields smaller differences. This method is quite insensitive for outlying results.

RESULTS FOR INTERPOLATED EFFICIENCIES AND ACTIVITIES

An example of the results from the participants is shown in figures 6...9.

For each interpolated efficiency and each calculated activity the following characteristic quantities were considered:
- the extremes and the average,
- the quartiles,
- the standard deviation according to the ordinary definition (s), or to the DoD method ($\tilde{s}$).

The main results are shown in figure 10. Presence of discrepant values will be clear from the position of the extremes and from the difference between the two values s and $\tilde{s}$ for the standard deviation. The positioning of the quartiles can show roughly a possible presence of a clear skewness in the data. There seems to be a tendency that the interpolated efficiency at the lowest energy in a set has somewhat larger spread than at the other energies. The larger spread is also found for the extrapolated efficiency at an energy of 2200 keV for SET 2.

The spread in the interpolated efficiencies is of the same order as the spread in the calculated efficiency data for the measurement points.

The 41 participants submitted for each of the two sources (^{134}Cs and ^{152}Eu) 34 values for the activity and for the corresponding standard deviation. All 34 values show consistency; no statistical significant deviations from the overall mean are present.

The interlab spread $\tilde{s}$ in the activity values is also of the same order as the interlab spread for the efficiencies. The estimated interlab-standard deviation is 0.35%.

The calculated uncertainties of the activity values submitted by the various participants have a mean value of about 0.78% with an interlab-standard deviation of 40%.

The methods which have the smallest standard deviation in the $\varepsilon_c/\varepsilon_m$ ratio for SET 1 are (in arbitrary order):
- a fit with three polynomial functions with a total of 14 parameters;
- a graphical method;
- an unspecified method;
- a least squares-spline regression.

The four solutions with the largest standard deviation in the $\varepsilon_c/\varepsilon_m$ ratio were obtained for the following functions:
- polynomial function with 4 parameters;
- polynomial function with 2 parameters;
- polynomial function with 10 parameters;

- exponential function with 5 parameters.

   From these results it follows that the applied model is not a guarantee for a good fit, since it was observed that the same type of fitting polynomial has been used by participants with "very good", "good" and "moderate" overall performance. No sharp conclusion can be made with respect to the actual reason of the large spread in fit quality. The following effects could contribute:
- the number of parameters in the fitting function;
- complete or partial coverage of the energy range (sometimes two fittings are made to cover the lower and higher energy range respectively);
- the number of input-data points used in the fitting procedure;
- numerical precision;
- the definition of the statistical weight applied in the calculations;
- the mathematical procedures;
- the units in which the energy is expressed (MeV, keV or eV; this has influence on the correlation coefficients between the fitting parameters).

   It should be repeated that not all models are capable to describe the real efficiency curve accurately enough, at least when the complete energy region is considered. With a simple calculation procedure a comparison is made for the output of the various participants. A ranking is obtained from these results (table 2). The ranking parameter is based upon comparison of interpolated efficiency values and calculated activity values with the participants' median value as reference value, which probably does not coincide with the best solution. The mathematical expression is given in table 2. The ranking is a rather arbitrary procedure in which the majority has much influence, so that very fine detailed information for a particular participant can have a reduced influence.

GENERAL OBSERVATIONS

1. Frequently used is a fitting function based on a polynomial model with logarithms of the variables ε and E. Such a function type is not a guarantee to obtain good results. The number of parameters in the fitting function need not to exceed 6 or 7.
2. Sometimes one applies two (or more) independent fits, each to cover a part of the energy range, instead of one function for the complete energy range. By this approach one can arrive more easily at better local fits, but one neglects possible correlations between efficiencies in the separate energy ranges.
3. When a fitting function is applied in a later stage as a calibration curve, one should realize that for the calculation

of uncertainties in the efficiency values also the role of
covariances between parameters of the fitting function may
become important. Numerical difficulties in the propagation of
uncertainties when highly correlated parameters are involved
can often be avoided by transformation of the variables.

For example, when an expression of the type $\ln \varepsilon = \sum\limits_{j=0}^{n} a_j (\ln E)^j$
is used, one can better express the energy in units of MeV than
in keV or eV.

Preferably one should use a system of coordinates in which the
centre of gravity is situated in the origin. In that case the
correlation coefficients between the fitting parameters are
close to zero.

When strong correlation coefficients for the parameters of a
polynomial model offers problems (e.g. with respect to comput-
er-word length), one should consider the application of
orthogonal polynomials, the coefficients of which are uncorrelated.

4. In order to judge the quality of fit of the efficiency-inter-
   polation function, one should also consider the differences
   between calculated and measured efficiency values, both with
   respect to sign and size. Preferably one should make a plot of
   the "studentized" residuals $(\varepsilon_i - \varepsilon_m)/s(\varepsilon_m)$ as a function of
   energy.

5. As an overall measure of inaccuracy for the method of calculating efficiency values with a function fitted to a series of
   experimental points one could report the value for $\omega^2 =$
   $[\Sigma(\varepsilon_c - \varepsilon_m)^2/s^2(\varepsilon_m)]/\nu$, where ε_c and ε_m denote the calculated and
   the measured efficiency values respectively.

6. The quantitative approach mentioned under points 4 and 5 fails
   for some spline functions if these by definition pass exactly
   through the measured-efficiency points. In such cases one may
   fit those spline functions through adjacent measurement points,
   while systematically taking one of these points as interpolation point.

7. Even in the absence of a good software-hardware combination one
   can very well perform interpolations between experimental
   points. One participant obtained very good results using a
   graphical (manual) interpolation by means of plots of ε versus
   E at low energy, and $\varepsilon \cdot E^{1.06}$ versus E at high energy. Such a
   procedure however cannot yield objective values for the
   standard deviations in the interpolated values. Only on the
   basis of long experience in the field of gamma-ray spectrometry, knowledge of measurement procedures, and a good ability to
   judge the input data, one can make reliable (but subjective)
   estimates for these standard deviations.

CONCLUSIONS

This GAM83 exercise gave much information on the state of art in 1983 to determine efficiency values with aid of a number of well chosen gamma rays of calibrated radio nuclide sources.
The following conclusions were drawn from a first analysis of the contributions of the 41 participants:
- A great variety of methods and fitting functions was applied in this exercise;
- The interlaboratory standard deviation $\tilde{s}$ in the calculated results (both for efficiencies and activities) is more or less of the same order of magnitude as the standard deviation of the corresponding input data set;
- The standard deviations calculated by the participants show a large spread without obvious outlying results (the standard deviation in these quoted standard deviation values for a set is in the order of 70%);
- The propagation of uncertainties is in most cases not completely considered in the fitting procedure, and in the activity calculation;
- The great variety of fitting functions in combination with the calculated large standard deviation of the parameters and their strong correlations seems to indicate that at present no single simple smooth function can be recommended as a general purpose model for describing the efficiency curve of a semiconductor gamma-ray spectrometer.

FINAL REMARKS

It might be useful to investigate the influence on the results if a particular good model is applied with the same input data by various participants.
In such a numerical exercise an improved idea can be obtained for the reason of the differences which are observed for obvious similar procedures.

The reader should consider that these results of the GAM83 exercise had to be achieved with a limited budget and with limited manpower. For that reason a number of interesting topics could not be studied in detail. But we hope that a good overview is established of the state of art.

REFERENCES

[1] Beyrich, W; Golly, W.; Spannagel, G.; "The DoD method; an
     empirical approach to the treatment of measurement data
     comprising extreme values"; Contribution to Stanchi, L.
     (editor); Proceedings Third Annual Symposium on Safeguards
     and Nuclear Material Management; Karlsruhe, May 1981, Report
     ESARDA-13 (CEC/JRC, Ispra, 1981) p 289-294.

[2] Beedgen, R.; "Statistical Analysis of the DoD Method";
     Contribution to Stanchi, L. (editor); Proceedings Sixth Annual
     Symposium on Safeguards and Nuclear Material Management;
     Venice, May 1984, Report ESARDA-17 (CEC/JRC, Ispra, 1984)
     p 533-538.

PARTICIPATION TO GAM83

Country code	Organization	Location
AT	Inst. Rad. Kernphys.	Wien
BE	S.C.K./C.E.N.	Mol (3x)
CA	A.E.C.L.	Chalk River
CA	A.E.C.L.	Pinawa
CH	I.E.R.-E.P.F.L.	Lausanne
CH	Kantonales Lab.	Basel
CH	KUeR	Fribourg
CN	Nat. Inst. Metrology	Beying
CN	Univ. Nucl. Phys. Lab.	Jilin
CS	Inst. of Physics	Bratislava
CS	Nucl. Res. Inst.	Řež
DE	G.K.S.S.	Geesthacht (2x)
DE	P.T.B.	Braunschweig
DE	Radiochemie	Garching
DK	National Lab.	Risø
ES	J.E.N.	Madrid
FR	L.M.R.I.	Gif-sur-Yvette
GB	A.E.R.E.	Harwell (2x)
GB	A.W.R.E.	Aldermaston
GB	C.E.G.B	Berkeley
GB	Imperial College	Ascot
GB	N.P.L.	Teddington
HU	B.M.E.	Budapest
HU	O.M.H.	Budapest
ID	B.T.A.N.	Jakarta
JP	J.A.E.R.I.	Tokai-Mura (2x)
JP	University	Hiroshima
NL	ECN	Petten
PL	I.E.A.	Swierk
RO	I.F.I.N.	Bucharest
TR	I.T.U.	Istanbul
US	Monsanto	Miamisburg
US	I.N.E.L.	Idaho
US	G.E.C.	Vallecitos
US	N.B.S.	Gaithersburg
ZA	NUCOR	Pretoria

Ranking parameter is $R = \sum\limits_{}^{n} (\frac{\epsilon_i - \tilde{\epsilon}_i}{\tilde{s}(\epsilon_i)})^2/n$ (for SET1, SET2 and SET3)

and $R = \sum\limits_{}^{n} (\frac{A_i - \tilde{A}_i}{\tilde{s}(A_i)})^2/n$ (for SET4).

where n is the number of data in the set.

$$R_{tot} = R_1 + R_2 + R_3 + R_4$$

	SET1		SET2		SET3		SET4		COMBINED SET	
	R_1	participant	R_2	participant	R_3	participant	R_4	participant	R_{tot}	participant
1	.026	60	*	30	*	30	*	30	.027	30
2	.027	30	*	41	*	40	*	43	.224	48
4	.053	32	.069	28	*	58	*	57	.291	28
5	.080	43	.133	33	*	57	*	14	.310	59
6	.138	48	.144	47	*	12	*	6	.316	52
8	.138	2	.151	4	*	45	.001	44	.319	54
9	.156	50	.168	9	*	31	.002	59	.328	44
10	.181	33	.192	60	*	44	.002	34	.336	60
12	.211	57	.265	19	*	41	.005	52	.377	38
14	.219	23	.320	36	.010	28	.011	48	.436	36
16	.238	44	.349	12	.082	38	.012	5	.502	50
19	.269	31	.418	50	.098	54	.025	33	.571	23
20	.282	1	.440	23	.121	23	.033	9	.578	32
23	.300	61	.470	54	.148	59	.057	38	.594	16
26	.320	8	.472	52	.148	48	.060	16	.616	47
28	.365	54	.472	2	.148	52	.160	28	.635	2
30	.420	36	.472	38	.151	34	.228	20	.639	34
31	.430	47	.473	48	.171	1	.249	8	.678	57
32	.433	59	.473	59	.207	5	.277	1	.738	56
33	.451	52	.548	44	.385	32	.303	46	.833	19
34	.456	42	.717	56	.490	16	.380	56	.910	42
36	.488	16	.764	42	.499	61	.380	54	.919	20
38	.663	6	.949	43	.551	60	.388	36	.971	1
40	.663	26	.962	40	.627	36	.405	32	1.081	8
41	.677	19	1.017	16	.636	50	.425	2	1.097	61
42	.704	38	1.071	34	.664	47	.522	19	1.189	9
43	.759	56	1.144	57	.694	46	.606	31	1.571	5
44	.780	20	1.300	61	.881	56	.864	12	1.779	41
45	.847	28	1.322	58	.882	20	.882	41	1.784	31
46	.950	34	1.326	32	.968	42	.933	60	2.158	4
47	1.417	5	1.372	20	1.139	26	1.243	50	2.577	6
48	1.584	9	1.905	8	1.156	6	1.499	4	3.065	12
50	2.138	41	2.736	1	1.351	8	2.145	47	5.269	26
52	4.004	4	3.027	46	1.377	2	2.320	42	8.030	33
54	6.661	12	3.420	14	1.683	19	2.899	23	9.320	40
56	15.177	40	3.713	5	2.278	9	4.079	61	15.166	10
57	23.417	10	3.769	31	2.581	4	7.493	10	33.140	46
58	61.812	58	5.498	45	10.429	10	15.571	40	40.273	58
59	108.835	46	6.746	6	26.978	33	24.146	26	258.241	14
60	723.671	14	6.751	26	47.631	14	25.033	45	311.981	43
61	1378.723	45	14.608	10	934.912	43	83.805	58	580.931	45

* These participants, listed in arbitrary order, have not submitted data for these sets.

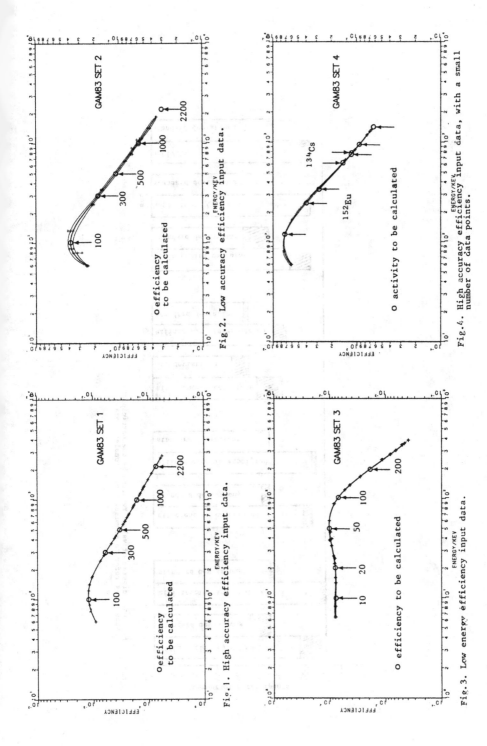

Fig.1. High accuracy efficiency input data.

Fig.2. Low accuracy efficiency input data.

Fig.3. Low energy efficiency input data.

Fig.4. High accuracy efficiency input data, with a small number of data points.

525

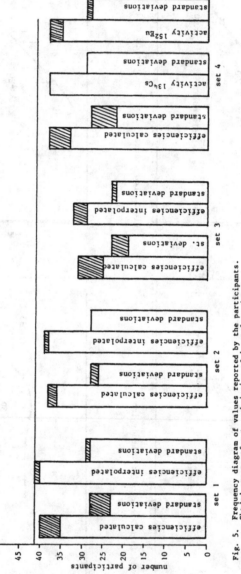

Fig. 5. Frequency diagram of values reported by the participants. Shaded areas refer to submission of incomplete data.

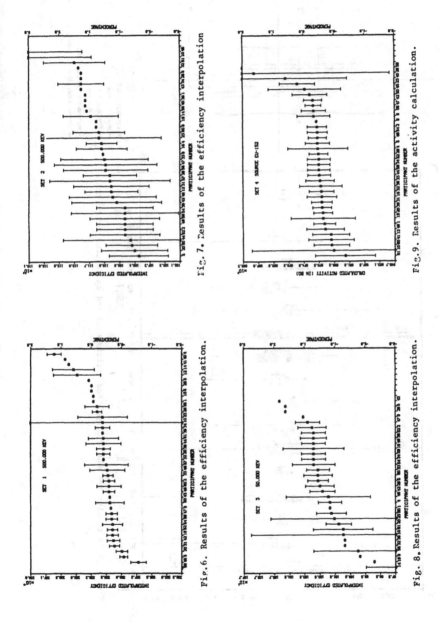

Fig.6. Results of the efficiency interpolation.

Fig.7. Results of the efficiency interpolation

Fig.8. Results of the efficiency interpolation.

Fig.9. Results of the activity calculation.

527

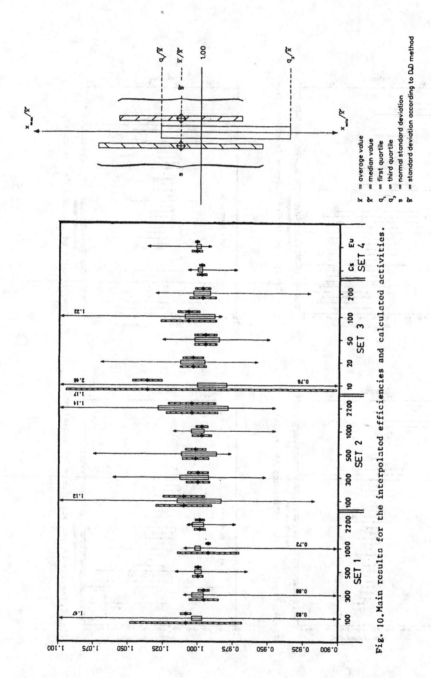

Fig. 10. Main results for the interpolated efficiencies and calculated activities.

x̄   = average value
x̃   = median value
q_1 = first quartile
q_3 = third quartile
s   = normal standard deviation
$\hat{s}$ = standard deviation according to DJD method

PART V
DAMAGE MODELS, PHYSICS, DOSIMETRY AND MATERIALS

EMBRITTLEMENT PROFILE IN THE FRACTURE PLANE OF IRRADIATED

CT100 SPECIMENS

J. Ahlf, D. Bellmann, G. Prillinger*
GKSS-Forschungszentrum Geesthacht GmbH
D-2054 Geesthacht, FRG
*Universität Stuttgart, IKE
D-7000 Stuttgart 80, FRG

ABSTRACT

In order to get more informations about the embrittlement profile induced by fast neutrons in the fracture planes of CT100 specimens, which were orientated parallel to the irradiation direction, the relevant specimen volume was replaced by charpy-V notch specimens. The irradiation was performed behind a gamma shield of tungsten in a reflector position of FRG-2, a swimming pool type reactor operating at 15 MW.

First evaluation show no significant embrittlement profile for the "Low Shelf Test Melt KS 07B", similar to 22NiMoCr37 but out of the specification. Additional investigation of the steel KS 12 (20MnMoNi55) are in progress.

INTRODUCTION

Irradiation experiments with compact tension specimens up to 100 mm thickness are planned in reflector positions of FRG-2, a swimming pool type research reactor operating at 15 MW /1/. In most cases the fracture planes of the specimens are arranged in a way that flux gradients in irradiation direction have no influence on the embrittlement, except CT100 specimens. In order to avoid large temperature gradients and high fast neutron self absorption in these specimens the fracture planes have to be orientated in irradiation direction. The neutron fluence gradient will be reduced by revolving the rig several times by 180 degree along its longitudinal axis during the irradiation period. By this practice the fluence distribution across the fracture plane has a parabolical shape which may induce gradients of embrittlement in the fracture planes of the CT100 specimens.

To get more information of these embrittlement gradients a special rig was irradiated behing a shield of tungsten up to a fluence of round about $2.5 \cdot 10^{23}$ m^{-2} (E > 1 MeV). The relevant volume of two CT100 specimens was replaced by a lot of charpy-V notch specimens /2/. By the way the embrittlement profile across the fracture plane could be determined from impact tests.

DOSIMETRY

The fluence distribution inside the rig was determined by evaluation of activation wires, an alloy of 63.8 wt% Fe and 36.2 wt% Ni. The Mn54 and Co58 activity was measured by scanning the wires along a Ge(Li)-detector.

2-dimensional multigroup neutron transport calculations were performed with DOT-IV code, using the EURLIB-library condensed to 51 energy groups > 0.1 MeV. The dosimetry cross sections were taken from ENDF/B V. The definitions of the cross sections are given in reference /3/.

EMBRITTLEMENT PROFILE ACROSS A STEEL PLATE OF 100 MM

The embrittlement is measured by the transition temperature shift at 41 Joule by charpy impact tests. For a specific steel heat this transition temperature shift depends mainly on irradiation temperature and neutron exposure.

By the mentioned irradiation techniques temperature and neutron exposure gradients cannot be avoided. Their influence is discussed in the following.

Impact tests

As can be seen from fig. 1 the relevant CT100 specimen volume is replaced by two columns of 20 plates of two different materials. From each plate 8 charpy specimens were machined after irradiation. This practice avoids gas gaps leading to additional temperature gradients during irradiation.

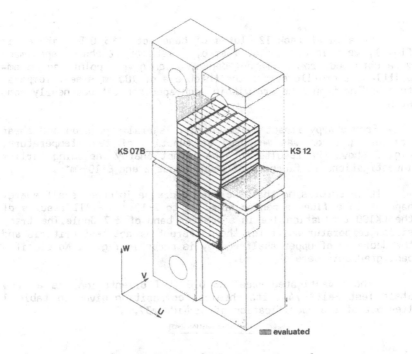

KS 07B — KS 12

▓ evaluated

Fig. 1 Specimen setup for temperature and
flux gradient evaluation experiment

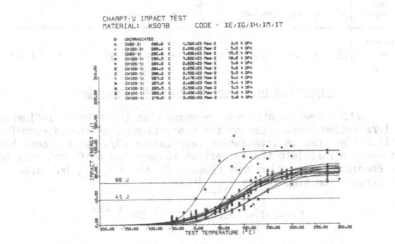

CHARPY-V IMPACT TEST
MATERIAL: KS07B   CODE - XE/XG/XH/XM/XT

⊙	UNIRRADIATED				
▲	CK60-2)	286.6	C	1.56E+23 flex-2	3.0 % DPA
+	CK120-2)	285.4	C	2.09E+23 flex-2	5.0 % DPA
X	CK80-1)	285.8	C	7.09E+23 flex-2	13.3 % DPA
◆	CK120-1)	290.5	C	7.82E+23 flex-2	18.6 % DPA
⊕	CK100-1)	280.5	C	2.83E+23 flex-2	5.0 % DPA
✕	CK100-1)	284.0	C	2.65E+23 flex-2	5.6 % DPA
Z	CK100-1)	286.2	C	2.55E+23 flex-2	5.5 % DPA
Y	CK100-1)	287.2	C	2.47E+23 flex-2	5.4 % DPA
M	CK100-1)	287.0	C	2.68E+23 flex-2	5.4 % DPA
M	CK100-1)	283.5	C	2.55E+23 flex-2	5.5 % DPA
M	CK100-1)	282.8	C	2.65E+23 flex-2	5.6 % DPA
I	CK100-1)	275.0	C	2.50E+23 flex-2	5.6 % DPA

68 J

41 J

IMPACT ENERGY (J)

TEST TEMPERATURE (°C)

Fig. 2 Charpy - V notch ductility of steel KS 07B before and after irradiation

As a first task 12 plates of base metal KS 07B, hatched in fig. 1, were investigated. Thereby 8 sets of 12 charpy specimens were obtained from different depth, which give 8 points on an embrittlement profile across the thickness of 100 mm steel. Temperature and neutron exposure within each specimen set was nearly constant.

From charpy impact tests energy, lateral expansion and shear fracture portion was measured as function of test temperature. Fig. 2 shows the results for the impact energy including earlier investigations at fluences of round about 2 and $8 \cdot 10^{23} m^{-2}$.

It is surprising that the mean decrease in upper shelf energy happens in a fluence region from 1.5 to $2 \cdot 10^{23} m^{-2}$. All results of the CK100 evaluation lie in a scatter band of ± 7 Joule. The transition temperatur shift for the different impact test criteria and the decrease of upper shelf energy is shown in fig. 3. No significant gradients were obtained.

The investigated steel KS 07B (L-T orientation) is a "Low Shelf Test Melt" /4/. Its chemical composition given in table 1 lies out of the specification of 22 NiMoCr37.

Chemical composition (wt%)

22 NiMoCr 37	C	Si	Mn	P	S	Cr	Mo	Ni	Al	Cu	V	Sn	Co	As	Sb
Specification															
VdTÜV	.17		.50			.30	.50	.60							
365(4.72)	.25	.35	1.00	.025	.025	.50	.80	1.00	.05	.20	.05	-	-	-	-
Base metal															
KS07B	.27	.29	.62	.022	.034	.49	1.03	.74	<.003	.26	.05	.012	.016	.026	<.005

Dependence of ΔT on irradiation temperature

Earlier investigations have shown that there is an influence on irradiation temperature on the embrittlement of steels even for small differences in irradiation temperature /3/. Van Asbroeck has proposed to correlate the transition temperature shift not only to the chemical composition and fluence $\emptyset t$ $[10^{23} m^{-2}]$, but also to the irradiation temperature T_{irr} [°C]:

$$\Delta T(41J) \sim (350 - T_{irr}) \cdot (\emptyset t)^{0.423}$$

Irradiations of steel KS 07B in different capsules behind gamma shields of stainless steel at nearly the same irradiation

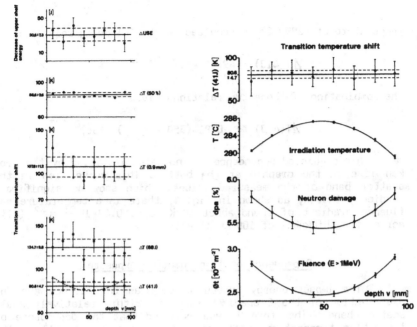

Fig. 3 Impact test results for steel KS 07B

Fig. 4 ΔT(41J), dpa, Øt versus depth of the specimen volume

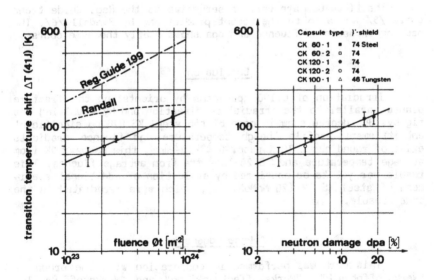

Fig.5 Transition temperature shift at 41J versus fluence
(E>1MeV) and dpa for KS 07B

temperature of (289 ± 2) °C resulted in the following relation:

$$\Delta T(41J) = 49,60 \cdot (\emptyset t)^{0.423}.$$

The combination of these two relations gives

$$\Delta T(41J) = 0,814(350 - T_{irr}) \cdot (\emptyset t)^{0.423}.$$

This predicted dependence is indicated in fig. 3 by the broken line in the graphic at the bottom. This curve lies in the scatter band of the measured values, which show no significant gradient, though, as shown in fig. 4, there is a temperature and fluence gradient of round about 10 K resp. $0.47 \cdot 10^{23}$ m^{-2} (17%) across the thickness of 100 mm steel.

Dependence of ΔT on neutron exposure

The impact energy curves as well as the fluence and dpa distribution of the 8 specimen sets lie within relatively small scatter bands. Therefore it was expected that the dependence of transition temperature shift versus neutron exposure will be the same as determined by earlier investigation, which is confirmed by fig. 5:

The ΔT values are very conservative to the Reg. Guide trend curve /5/ and also to the recent predictions by Randall /6/. The neutron exposure of fluence and dpa shows nearly the same course.

Conclusion

Irradiation of CT100 specimens by orientating the fracture planes parallel to the irradiation direction and by revolving the rig by 180 degree show in case of the steel KS 07B no significant embrittlement profile though temperature and fluence gradients exist of round about 10 K resp. $0.47 \cdot 10^{23}$ m^{-2}, this means $\pm$ 5K from average temperature and $\pm 0.24 \cdot 10^{23}$ m^{-2} from average fluence. This result has yet to be confirmed by evaluation of additional specimens of steel KS 12 (20 MnMoNi 55), which were irradiated in the same capsule.

Acknowledgement

This work was performed in cooperation with the group of "Werkstoffphysik", "Werkstoffentwicklung" and "Werkstoff-Prüflabor" of the GKSS-Forschungszentrum Geesthacht GmbH. The contributions of these groups are gratefully acknowledged.

REFERENCES

/1/ D. Bellmann, J. Ahlf, P. Wille, G. Prillinger, "Neutron Dosimetry in Irradiation Capsules for Large RPV Steel Specimens". Forth ASTM-EURATOM Symposium on Reactor Dosimetry, March 22-26, 1982 (NUREG/CP-0029, Vol. 1)

/2/ J. Ahlf, D. Bellmann. "Irradiation Techniques for Large Reactor Pressure Vessel Steel Specimen", p. 483 in Aspects of Nuclear Reactor Safety, P. von der Hardt, H. Röttger, Editor, EUR 6612, 1980

/3/ J. Ahlf, et. al., "Investigation on the dependence of RPV steel embrittlement on irradiation temperature and neutron exposure", Forth ASTM-EURATOM Symposium on Reactor Dosimetry, March 22 - 26, 1982 (NUREG/CP-0029, Vol. 2)

/4/ J. Föhl et al., "Forschungsvorhaben Komponentensicherheit", Band Bestrahlung, TWB 5/1, Dezember 1981

/5/ U.S. Nuclear Regulatory Commission Regulatory Guide 1.99 Rev. 1 (April 1977), "Effects of Residual Elements on Predicted Radiation Damage to Reactor Vessel Materials"

/6/ P.N. Randall, "NRC Perspective of Safety and Licensing Issues Regarding Reactor Vessel Steel Embrittlement - Criteria for Trend Curve Development", ANS Annual Meeting Detroit, Michigan, USA, June 14, 1983.

THE EFFECT OF DAMAGE RATES AND DIFFERENT KINDS OF IRRADIATION

ON THE DUCTILITY OF AMORPHOUS $Fe_{40}Ni_{40}B_{20}$

R. Gerling, F.P. Schimansky, P. Wille and R. Wagner

GKSS-Forschungszentrum Geesthacht, Institut für Physik

Box 1160, D-2054 Geesthacht, FRG

ABSTRACT

Thermally embrittled amorphous $Fe_{40}Ni_{40}B_{20}$ reveals a complete restoration of the ductility upon low dose neutron irradiation. The improvement of the ductility is studied both as a function of the damage type as caused either by thermal or by fast neutrons, and of the damage rate. While the damage rate has no influence, the type of irradiation strongly influences the damage level at which the ductility of amorphous $Fe_{40}Ni_{40}B_{20}$ is completely restored.

INTRODUCTION

The results of irradiation experiments are often compared on the basis of the damage level as given by the displacements per atom (dpa). However, the dpa-level does neither involve possible recombination effects during the irradiation nor a likely influence of the specific damage of the particle used for the irradiation experiment. It has been the objective of this work to study the irradiation induced changes of a physical property under various irradiation conditions and thus to investigate whether there exists a well defined relationship between the dpa-level and a certain physical property. During n-irradiation thermally embrittled amorphous $Fe_{40}Ni_{40}B_{20}$ experiences a unique property change: the complete restoration of the ductility (1). Since this drastic effect occurs within a rather narrow range of fluences this alloy appeared quite suitable for the intended experiments.

J. P. Genthon and H. Röttger (eds.), Reactor Dosimetry, 539–547.

During n-irradiation of $Fe_{40}Ni_{40}B_{20}$ the damage results from two different processes: i) displacements caused by high energy α- and Li-particles which result from the capture of thermal neutrons by ^{10}B and ii) displacements caused by fast neutrons. In order to separate these processes, two reels of amorphous $Fe_{40}Ni_{40}B_{20}$ have been prepared using either ^{10}B- or ^{11}B-enriched boron. With the aid of an optional Cd-shielding, the irradiation conditions can be chosen such that on one hand they vary with respect to the damage rate while the kind of irradiation remains unchanged, and on the other hand they vary with respect to the two types of irradiation.

EXPERIMENTAL

For the preparation of the two amorphous $Fe_{40}Ni_{40}B_{20}$ ribbons (Vakuumschmelze Hanau, FRG), for one alloy boron enriched to 86 % ^{10}B (natural abundance 19.8 %) was used, while for the other alloy ^{11}B enriched boron was taken: 99.2 % ^{11}B; 0.8 % ^{10}B (natural abundance ^{11}B: 80.2 %). Both alloys were about 3 mm wide and 40 μm thick. The ductility of the specimens was determined by measuring the relative strain at fracture, ε_f. Specimens are completely ductile if $\varepsilon_f=1$. The loss of ductility is characterized by $\varepsilon_f < 1$; the embrittlement is the more advanced the smaller ε_f is. In the as-quenched state, both alloys are completely ductile. Prior to the neutron-irradiation both, the $Fe_{40}Ni_{40}{}^{10}B_{20}$- and the $Fe_{40}Ni_{40}{}^{11}B_{20}$-alloy were embrittled by thermal annealing for 43 h at 280 and 304 °C, respectively. In spite of this treatment the specimens remained completely amorphous. The fluxes of thermal ($E_n < 0.1$ eV) and fast ($E_n > 0.1$ MeV) neutrons in the chosen incore position were $3.5 \cdot 10^{13}$ n_{th}/cm^2 sec and $2.2 \cdot 10^{13}$ n_f/cm^2 sec. Prior to the irradiation, the specimens were sealed in Al-tubes; the temperature during the irradiation never exceeded 70°C. For the optional shielding, sheets of 1 mm Cd were used.

DAMAGE CALCULATIONS

Radiation Damage from Thermal Neutrons

Due to their small kinetic energy, thermal neutrons are not able to displace target atoms by elastic or inelastic collisions. Nevertheless, there is a significant contribution from thermal neutrons to the damage of $Fe_{40}Ni_{40}B_{20}$: According to the capture of thermal neutrons by ^{10}B high energy α- and Li-particles are generated causing radiation damage. For 93.9 % of the nuclear $^{10}B(n,\alpha)^{7}Li$ reactions the total kinetic energy of the reaction products is 2.31 MeV. This corresponds to kinetic energies of $E_\alpha = 1.47$ MeV and $E_{Li} = 0.84$ MeV. The range of the α- and Li-particles in $Fe_{40}Ni_{40}B_{20}$ is $R_\alpha = 2.67$ μm and $R_{Li} = 1.29$ μm. To a large extent their initial kinetic energy is dissipated by electron excitation; only a small amount is lost by elastic collisions with target atoms. If the energy transfered during the

540

collision to an atom, exceeds the displacement energy E_d, it leaves
its original position as a primary knock-on atom (PKA). For amorphous
$Fe_{40}Ni_{40}B_{20}$, we have estimated a mean displacement energy of
$E_d = 33$ eV (2). If a PKA possesses a sufficient kinetic energy, it
is capable of creating additional displacements, i.e. a displacement
cascade is formed. The investigated amorphous $Fe_{40}Ni_{40}B_{20}$ ribbons
have been thin enough (d = 40 μm) for the flux of thermal neutrons to
be nearly constant within the ribbon volume. On the other hand, they
are thick enough for nearly all particles which contribute to the
radiation damage to remain inside the material. With these considerations the number of displacements per atom (dpa) is given as:

$$dpa = n_\alpha \cdot Z_\alpha \cdot \bar{v}_\alpha \cdot N^{-1} + n_{Li} \cdot Z_{Li} \cdot \bar{v}_{Li} \cdot N^{-1} \tag{1}$$

$n_{\alpha,Li}$ = the number of α- or Li-particles created during irradiation
$Z_{\alpha,Li}$ = the number of PKA's created by a single α- or Li-particle
$\bar{v}_{\alpha,Li}$ = the average number of displaced atoms within a displacement
       cascade
N   = the number of target atoms exposed to irradiation

In order to calculate the damage by virtue of eq. 1, $Z_{\alpha,Li}$ is obtained from the energy dependent cross section $\sigma_{\alpha,Li}(E)$ accounting
for the generation of a PKA by α- or Li-particles with energy E. This
cross section is calculated from the potential energy of the α- or
Li-particle and the PKA resulting from their mutual Coulomb repulsion
during the collision. The number v of atoms displaced in a single
cascade depends on the initial PKA-energy T. The calculation of v(T)
is based on a model of Kinchin and Pease (3). However, unlike in this
model we consider an energy loss of the PKA in the cascade by electron excitation and rather than a hard sphere potential we use a more
realistic inverse-power-potential for the description of the collisions within a cascade. These alterations lead to two correction factors
c_1 and c_2 in the simple v(T)-formula obtained by Kinchin and Pease:

$$v(T) = c_1(T) \cdot c_2(T) \cdot \frac{T}{2E_d} \tag{2}$$

The differential cross section $d\sigma/dT(E,T)$ for the generation of a PKA
with initial energy T by an α- or Li-particle with energy E depends
on E and T as:

$$d\sigma/dT \propto E^{-s} \cdot T^{-(s+1)} \tag{3}$$

The exponent s in eq. 3 depends on the kinetic energy of the incident
particle. For α- and Li-particles with E close to the initial energies, the collisions can be described by the Coulomb law: s = 1.
During the slowing down of the particles, s decreases and becomes
s = 0.6 for particle energies which are just high enough to transfer
E_d to a target atom. As a consequence of eq. 3, mainly PKA's with
small energies T are generated. According to eq. 2 this yields small
displacement cascades: $v < 10$. In Table 1 the probabilities for cascades within certain intervals of magnitude v are listed. For thermal

Table 1: The probabilities for cascades with a number v of displaced atoms generated by thermal and fast neutrons.

v [displaced atoms]	Thermal neutron induced damage[%]	Fast neutron induced damage[%]
1 - 10	97,8	24,9
10 - 50	1,9	19,9
50 - 100	0,2	16,7
> 100	0,1	38,5

neutron fluences for which the burn up of ^{10}B-atoms can be neglected ($\emptyset_{th} \cdot t < 1 \cdot 10^{19} \ n_{th}/cm^2$), the number of α- or Li-particles generated by irradiating n_{10_B} ^{10}B-atoms for a time t, is:

$$n_\alpha = n_{Li} = n_{10_B} \cdot \int \emptyset_{th}(E_n) \cdot t \cdot \sigma^{10_B}(E_n) \ dE_n \qquad (4)$$

The thermal neutron spectrum for the specific irradiation position of our specimens is shown in Fig. 1. It has been deduced from the neutron spectrum reported by Poole (4) with special regard to the water gap between the source and the specimens. Additionally in Fig. 1 the variation of the absorption cross section σ_a of ^{10}B with neutron energy is shown. By knowing $\emptyset_{th}(E_n)$ and $\sigma_a(^{10}B)$ n_α and n_{Li} can be calculated for any given fluence from eq. 4. Damage calculations for the $Fe_{40}Ni_{40}$ $^{10}B_{20}$-ribbon revealed a thermal neutron fluence of $1 \cdot 10^{19} n_{th}/cm^2$ to correspond to an atomic displacement level of 2.8 dpa.

The radiation damage by thermal neutrons can be lowered drastically by shielding the ribbons with Cd, since cadmium has a high absorption cross section σ_a for thermal neutrons (Fig. 1). However, due to the drastic decrease of $\sigma_a(Cd)$ for neutron energies ranging from 0.2 to 2 eV, some $^{10}B(n,\alpha)^7Li$ reactions can still be initiated by epithermic neutrons. A Cd-shielding of 1 mm reduces the damage generated by thermal neutrons by a factor of 40. In this case, irradiation to a fluence of $1 \cdot 10^{19} \ n_{th}/cm^2$ yields a displacement level of only about 0.07 dpa. With respect to the $Fe_{40}Ni_{40}$ $^{11}B_{20}$ alloy which contains a drastically reduced amount of ^{10}B, the damage caused by thermal neutrons is only 0.026 dpa after a fluence of $1 \cdot 10^{19} n_{th}/cm^2$. A Cd-shielding of 1 mm reduces the damage again by a factor 40. The results of these damage calculations are summarized in Table 2.

Radiation Damage by Fast Neutrons

Fast neutrons can generate PKA's by elastic or inelastic collisions with the nuclei of target atoms. The contributions of these two different scattering mechanisms to radiation damage will be computed

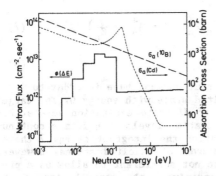

Fig. 1: The neutron flux $\emptyset(\Delta E)$ together with the absorptions cross sections σ_a for ^{10}B and Cd as a function of neutron energies.

separately. The number Z of PKA's generated by <u>elastic</u> neutron scattering after an irradiation time t in a target consisting of N atoms is:

$$Z = N \cdot \int_{E_d/\Lambda}^{\infty} \emptyset_f(E_n) \cdot t \int_{E_d}^{\Lambda E_n} \frac{d\sigma_{el}}{dT} (E_n, T) \, dT \, dE_n \qquad (5)$$

$E_n = E_d/\Lambda$ is the smallest energy within the fast neutron spectrum $\emptyset_f(E_n)$ which enables a neutron to displace a target atom. On the other hand, the maximum energy transfer to a PKA by a neutron with energy E_n is $T = \Lambda \cdot E_n$. The magnitude of Λ depends on the atomic mass number A of the target atom. For $Fe_{40}Ni_{40}B_{20}$ the mean value is $\Lambda = 0.08$. The number of displaced atoms can be obtained by counting the total number of atoms displaced within all Z displacement cascades. This is done by inserting the number v of displaced atoms per displacement cascade (eq. 2) into the inner integral of eq. 5. Dividing by N yields the displacements per atom caused by elastic scattering of fast neutrons:

$$dpa_{el} = \int_{E_d/\Lambda}^{\infty} \emptyset_f(E_n) \cdot t \int_{E_d}^{\Lambda E_n} c_1(T) \, c_2(T) \, \frac{T}{2E_d} \frac{d\sigma_{el}}{dT} (E_n, T) \, dT \, dE_n \qquad (6)$$

Elastic neutron scattering is isotropic below about 0.1 MeV. At higher energies fast neutrons are scattered more frequently in forward directions (small scattering angles in the centre of the mass system), which favours the generation of PKA's with lower initial energies T and, hence, the generation of smaller displacement cascades. However, this effect is much smaller than in the case of radiation damage initiated by thermal neutrons (Table 1).

Radiation damage generated by <u>inelastic</u> neutron scattering can be computed from a relation equivalent to eq. 6. Due to the temporal separation between absorption and emission of a neutron, to a very good approximation the inelastic scattering is isotropic in the centre of the mass system. Therefore, by neglecting the energy dependence of the correction factors c_1 and c_2, the inner integral of eq. 6 can be calculated analytically. By assuming that there exists only one excitation state Q for the nucleus interacting with the neutrons this calculation yields:

$$dpa_{in} = c_1 \cdot c_2 \int_{E_n^{min}}^{\infty} \emptyset_f(E_n) \cdot t \frac{\Lambda E_n}{4E_d} \sigma_{in}(E_n) \left(1 - \frac{1+A}{2A} \cdot \frac{Q}{E_n}\right) dE_n \qquad (7)$$

$$\text{with } E_n^{min} = \frac{1+A}{A} \cdot Q \qquad (8)$$

E_n^{min} is the minimum energy a fast neutron must have in order to transfer the nucleus into the excited state with energy Q. The target nuclei possess, however, a number of discrete excitation levels. For higher excitation energies the excitation levels even form a continuum. For an accurate damage calculation the integrant of eq. 7 therefore has to be replaced by a sum over all excited states. However, the available experimental data are not sufficient to allow us a precise calculation. We therefore approximate the contribution of the inelastic scattering to the damage as follows: The average damage is computed from the number of displacements which would result if always the maximal possible excitation energy were transfered and those displacements which would result if always only the lowest level above the ground-state were excited. This approximation yields:

$$dpa_{in} = c_1 \cdot c_2 \int_{(1+A)/A \cdot Q_{min}}^{\infty} \emptyset_f(E_n) \cdot t \cdot \frac{\Lambda E_n}{4E_d} \cdot \frac{3}{4} \sigma_{in}(E_n) dE_n \qquad (9)$$

Calculating the displacements per atom from eqs. 6 and 9 yields $5 \cdot 10^{-3}$ dpa after irradiation to $6.3 \cdot 10^{18}$ n_f/cm^2 (corresponding to $1 \cdot 10^{19}$ n_{th}/cm^2 in the chosen reactor position). Because inelastic scattering takes places only for neutron energies beyond about 1 MeV, only about 25 % of all displacements are generated by inelastic neutron scattering. Due to the small total cross section of Cadmium for fast neutrons, shielding with 1 mm Cd does not lower the flux of fast neutrons. Therefore, the dpa_f-values are not influenced by shielding with Cd (Table 2).

Tabelle 2: Displacements per atom generated by thermal (dpa_{th}) and by fast neutrons(dpa_f) after irradiation to a total dose of $1 \cdot 10^{19}$ n_{th}/cm^2 ($E_n < 0.1$ eV) and $0.63 \cdot 10^{19}$ n_f/cm^2 ($E_n > 0.1$ MeV). (Cd) refers to the irradiation behind a 1 mm Cd-shielding of the particular alloy.

	dpa_{th}	dpa_f	$dpa = dpa_{th}+dpa_f$	dpa_{th}/dpa_f
$Fe_{40}Ni_{40}{}^{10}B_{20}$	2.8	0.005	2.805	560
$Fe_{40}Ni_{40}{}^{10}B_{20}$ (Cd)	0.07	0.005	0.075	14
$Fe_{40}Ni_{40}{}^{11}B_{20}$	0.026	0.005	0.031	5.2
$Fe_{40}Ni_{40}{}^{11}B_{20}$ (Cd)	0.00065	0.005	0.00565	0.13

RESULTS AND DISCUSSION

As shown in Table 2, n-irradiation of the amorphous alloys $Fe_{40}Ni_{40}{}^{10}B_{20}$ and $Fe_{40}Ni_{40}{}^{11}B_{20}$ either with or without Cd-shielding allows us on one hand to vary the damage rate and on the other hand to study the damage generated by different particles. Both irradiation conditions reveal a homogeneous damage throughout the whole specimen. In the ^{10}B-alloy, the damage is essentially generated by the capture of thermal neutrons by ^{10}B, regardless of whether the specimens were Cd-shielded or not. For shielded specimens, the damage rate is a factor 37 smaller. Upon n-irradiation of the ^{11}B-alloy without Cd-shielding the damage caused by α- and Li-particles is about 5.2 times as large as that generated by fast neutrons. Using a Cd-shielding, this ratio decreases to about 0.13. The damage rate for unshielded $Fe_{40}Ni_{40}{}^{11}B_{20}$ is only a factor 5.5 higher than for irradiation behind a Cd-shielding.

In Fig. 2 the variation of the relative strain at fracture, ε_f, with the displacement level is shown for the ^{10}B-alloy. Since the specimens were thermally embrittled, ε_f is about $1.5 \cdot 10^{-2}$ prior to the irradiation. With increasing dpa-level the specimens recover and, irrespective of the irradiation conditions (Cd-shielded or not) the ductility is completely restored. However, the damage level at which complete restoration has occurred, is found to be sligthly lower for the Cd-shielded specimens.

Upon irradiation of $Fe_{40}Ni_{40}{}^{11}B_{20}$ the ductility of Cd-shielded specimens also recovers at a lower damage level than in the case of unshielded irradiation (Fig. 3). By comparison to the ^{10}B-alloy, the influence of the shielding on the dpa-level at which the ductility of this alloy is completely restored is even more pronounced. These results clearly demonstrate that it is not the displacement level alone which controls exclusively the regain of the ductility of these alloys. In order to check, whether there exists a conceivable influence of the damage rate or the kind of particles generating the damage, it is necessary to discuss the reasons for the thermal embrittlement and the irradiation-induced restoration of the ductility.

As has been reported in a previous paper (5) the embrittlement of amourphous $Fe_{40}Ni_{40}B_{20}$ alloys at low temperatures is always associated with a slight increase of the density ρ, due to the loss of microscopic free volume. Indeed, theoretical considerations have emphasized the importance of free volume sites for ductile behaviour of amorphous alloys (6, 7). The subsequent n-irradiation of the thermally embrittled $Fe_{40}Ni_{40}B_{20}$ alloys revealed an irradiation induced decrease of ρ, i.e. a swelling. The ^{11}B-alloy for example, revealed a swelling of roughly 0.1 % at $2 \cdot 10^{-4}$ dpa for both irradiation conditions. This radiation-induced increase of the microscopic free volume is considered to be reponsible for the recovery of the ductility upon irradiation.

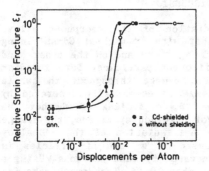
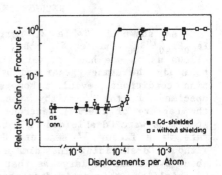

Fig. 2: The variation of the relative strain at fracture with
the dpa-level in n-irradiated
$Fe_{40}Ni_{40}{}^{10}B_{20}$ under a Cd-shielding and without shielding.

Fig. 3: The variation of the relative strain at fracture with
the dpa-level in n-irradiated
$Fe_{40}Ni_{40}{}^{11}B_{20}$ under a Cd-shielding and without shielding.

The conceivable influence of the damage rate on the restoration
of the ductility ought to be most obvious for the ^{10}B-alloy. The
damage rate affects the defect structure if the radiation damage can
partly recover or rearrange during the irradiation. A reduced damage
rate would then require a higher dpa-level to reveal the same property changes as irradiation with a high damage rate. However, since
irradiation of the ^{10}B-alloy at a smaller damage rate reveals the
ductility to be restored even at a lower dpa-level, a rate effect can
be ruled in. Therefore, the observed recovery of the ductility at
different dpa-levels rather seems to result from the different pro-
jectiles causing the damage. For the ^{11}B-alloy the difference of the
contributions from the two types of irradiation is more pronounced
with respect to the ^{10}B-alloy. Indeed, the differences in the damage
levels for complete recovery of ε_f are also more pronounced for
$Fe_{40}Ni_{40}{}^{11}B_{20}$. As for the ^{10}B-alloy, the ductility of this alloy is
found to recover at a lower dpa-level if irradiated behind a Cd-
shielding. Therefore, one may conclude that the displacements generated by fast neutrons restore the ductility more efficiently than
those created by α- and Li-particles. As shown in Table 1, upon thermal neutron irradiation, 98 % of the cascades involve only 1 to 10
displaced atoms, while 38 % of the cascades induced by fast neutrons
involve more than 100 displaced atoms. As the beneficial effect of
the irradiation is thought to be the generation of additional free
volume, displacements within a large cascade seem to yield a higher
amount of free volume per displaced atom than in a smaller one.

The ratio of the efficiency of the displacements due to thermal
and fast neutrons can be inferred from the different dpa-levels of
complete restoration of the ductility. According to Fig. 3 these

damage levels are $2.4 \cdot 10^{-4}$ and $5.5 \cdot 10^{-5}$ dpa for unshielded and Cd-shielded [11]B-specimens, respectively. Then the displacements by fast neutrons are about a factor 20 more efficient with respect to the recovery of ε_f than those displacements generated by thermal neutrons via the production of α- and Li-particles.

The Cd-shielding of the [10]B-alloy also alters the ratio dpa_{th}/dpa_f. However, under each of the two irradiation conditions, dpa_{th} contributes far more to the total damage than does dpa_f. Nevertheless, even from this experiment the different efficiencies can be deduced, and the displacements by fast neutrons also appear to be a factor 20 more efficient with respect to those dpa's due to thermal neutrons. The effect of the dpa-dependent recovery of the ductility can be described coherently for both alloys by the same efficiency ratio between the displacements resulting from thermal and from fast neutrons. However, the range of the dpa-levels required for complete recovery is quite different for the two alloys. At present the reason for this different behaviour of the two alloys remains unclear. However, it must be pointed out that the two alloys have also a different susceptibility to thermal embrittlement: the [10]B-alloy becomes brittle after aging at 280 °C whereas the [11]B-alloy embrittles only at 304 °C. Conceivably, the reasons for the higher susceptibility of $Fe_{40}Ni_{40}{}^{10}B_{20}$ to thermal embrittlement are the same which impede the irradiation-induced restoration of the ductility of this alloy.

REFERENCES

1. R. Wagner, R. Gerling and F.P. Schimansky,
   Journ. Non-Cryst. Solids 61 + 62 (1984) 1015.

2. F.P. Schimansky and R. Gerling, GKSS 83/E/32.

3. G.H. Kinchin and R.S. Pease, Rep. Progr. Phys. 18 (1955) 1.

4. M.J. Poole, Journ. Nucl. Energy 5 (1957) 325.

5. R. Gerling and R. Wagner, Scripta Met. 17 (1983) 1129.

6. A.S. Argon, Acta Met. 27 (1979) 47.

7. J.C.M. Li, Rapidly Quenched Metals IV, Vol. 2
   (The Japan Institute of Metals, Sendai 1982) p. 1335.

DEFECT MICROSTRUCTURE AND IRRADIATION STRENGTHENING IN Fe/Cu ALLOYS AND Cu BEARING PRESSURE VESSEL STEELS

R. Wagner, F. Frisius, R. Kampmann, and P.A. Beaven

GKSS Research Centre Geesthacht, Institut für Physik

Box 1160, D-2054 Geesthacht, FRG

ABSTRACT

The nature of the radiation damage and its dependence on Cu concentration have been studied in a series of Fe-Cu alloys and Cu containing pressure vessel steels, irradiated to 7×10^{19} n/cm^2 (E > 0.1 MeV) at 290 °C, using a combination of TEM, SANS, and compression testing. The magnitude of the irradiation hardening in the Fe-Cu-alloys is found to increase with the concentration of Cu in solution prior to irradiation, and is shown to be due to the formation of (i) Cu precipitates, and (ii) a finer dislocation loop distribution than in Cu-free Fe. Preliminary SANS measurements on a series of pressure vessel steels show that the defect microstructures formed during irradiation are more complex than in Fe-Cu alloys.

INTRODUCTION

During the past decade, several investigations have been concerned with the influence of alloying elements, or impurities, on the nature and distribution of the damage produced under neutron irradiation, and its implications to the embrittlement of reactor pressure vessel steels [1 - 3]. Empirically, the residual elements Cu and P have been found to be most detrimental, although it has not been possible to obtain a complete understanding of the changes in the radiation-induced microstructure which, in the presence of these elements, lead to the observed embrittlement. This is chiefly due to problems associated with TEM investigations of very small defects (frequently with radii < 2 nm) in ferromagnetic materials.

Smidt and Sprague [1] found, using TEM, that the number densities of dislocation loops and voids in a binary Fe - 0.3 at % Cu alloy, irradiated to 4.5×10^{20} n/cm² (> 1 MeV) at 280 °C, were a factor of ~ 4 higher than in pure Fe irradiated under identical conditions. After irradiation to 2.5×10^{19} n/cm², the increase in the yield strength in the Cu containing alloy was found to be much more pronounced than in other binary alloys e.g. Fe - 0.3 at % P. However, the radiation-induced strengthening after 4.5×10^{20} n/cm², was found to be about the same in pure Fe and in various binary alloys with 0.3 % Cu, Ni, V, or P.

Recently we have shown, by means of small angle neutron scattering (SANS), that precipitation of Cu-rich particles occurs during n-irradiation at 290 °C in Fe-Cu alloys [4], a feature not observed by Smidt and Sprague [1]. In the light of these results from SANS, the question arises as to what extent the observed irradiation hardening, or embrittlement, of Cu bearing ferritic alloys is caused by the precipitation hardening effect of copper particles, or, by the enhanced formation of radiation defects in the presence of Cu atoms in solid solution. We have therefore investigated the defect microstructure in a range of irradiated Fe-Cu alloys using a combination of TEM, SANS, and mechanical compression testing, in order to identify the irradiation hardening mechanism, and complement the experiments on pressure vessel steels.

EXPERIMENTAL DETAILS

Materials: Specimens for TEM, SANS, and compression testing were prepared from four different alloys containing 0, 0.29, 0.64 and 0.86 at % Cu; the interstitial impurity levels were determined by chemical analysis to be (in wt. ppm): C - 120 ppm, N - 20 ppm, P - 36 ppm, and 0 - 415 ppm. After homogenising at 840 °C and brine quenching, some specimens from each alloy were aged at 500 °C or 600 °C. (A full list of specimens is given in Table I.) Both homogenised, and pre-aged specimens were then irradiated to 7×10^{19} n/cm² (E > 0.1 MeV) at 290 °C in a temperature-controlled capsule in the FRG-2 reactor, as were the steel specimens (see Table II for compositions). Selected specimens were then subjected to post-irradiation annealing for various times at 400 °C.

Mechanical Testing: The 0.2 % offset yield strength (σ_0) of all specimens was determined from compression tests at a strain rate of 3×10^{-4} sec⁻¹. Fig. 1a shows the variation of σ with Cu content (c_0), for both the homogenised, and the pre-aged (200 h/ 500 °C) specimens, before (σ_0), and after, (σ_{0i}), irradiation. The increase in σ after irradiation is considerable even in the absence of Cu ($c_0 = 0$). For the Cu containing specimens the irradiation strengthening is greater the higher the Cu content, although the increase is less accentuated in pre-aged specimens (curve D), compared to homogenised specimens (curve C).

TEM: TEM observations were made on all specimens using standard imaging conditions for defect identification and quantitative measurement of defect radius and number density. (Foil thickness was determined using thickness fringes and/or the convergent beam diffraction technique.) The Cu free specimens, proved to be relatively straightforward in that dislocation loops and voids could be imaged separately; in Cu containing specimens, however, attempts to separate small voids, Cu precipitates, and loops, in order to make quantitative measurements of each has proved to be extremely difficult, particularly as the voids and Cu precipitates occur in size ranges near the limits of their visibility.

SANS: For a specimen containing N_0 identical particles with radius R the nuclear scattering length densities (n_{nuc}) and the magnetizations are assumed to be homogeneous in both the ferromagnetic α-matrix and the precipitates. By applying a strong homogeneous magnetic field to the specimen, all spins can be aligned in one direction perpendicular to the incident neutron beam. Then in the static approximation for a dilute system of randomly distributed spherical particles (or voids) the coherent scattering cross-section per unit volume is given by:

$$\frac{d\Sigma(\kappa,\alpha)}{d\Omega} = N_0 \left(\frac{4\pi}{3} R^3\right)^2 \left\{\Delta n_{nuc}^2 + \Delta n_{mag}^2 \cdot \sin^2\alpha\right\} \times$$
$$\times \left[3 \frac{\sin(\kappa R) - \kappa R \cos(\kappa R)}{(\kappa R)^3}\right]^2$$

where Δn_{nuc} and Δn_{mag} denote the differences in the nuclear and magnetic scattering length densities between matrix and particle or void; α is the angle between the magnetization of the specimens and the direction of the scattering vector κ.

The scattering intensity is measured as a function of scattering angle θ, by means of position-sensitive detectors parallel ($\parallel$) and perpendicular ($\perp$) to the applied magnetic field [4]. The ratio of the corresponding scattering cross-sections is then:

$$A \equiv \frac{(d\Sigma/d\Omega)\perp}{(d\Sigma/d\Omega)\parallel} = \frac{\Delta n_{nuc}^2 + \Delta n_{mag}^2}{\Delta n_{nuc}^2}$$

which gives A = 11.5 for Cu-particles and A = 1.4 for voids.

In the present study it was possible to measure the rather strong magnetic SANS from the Cu particles over an extended range of κ. Therefore, it was possible to determine not only a Guinier-radius but also a precipitate size distribution N(R). From the absolute scattering intensity, the precipitated volume fraction (f) could also be determined quite accurately.

551

(a) Fe-Cu alloys:

The SANS data are shown in Table I and are described here using three irradiated specimens as examples (specimens 1'-Fe, 2'-Fe - 0.29 % Cu (homog.), and 9'-Fe - 0.86 % Cu (20 h 600 °C) for which the measured scattering cross-sections $d\Sigma/D\Omega$ $(\kappa)/\|$ (nuclear) and $d\Sigma/d\Omega(\kappa)/\underline{\perp}$ (magnetic + nuclear) are shown in Fig. 2a, b, c. The theoretical cross-sections $d\Sigma/d\Omega/\underline{\perp}$ (upper curves), which have been calculated assuming a log-normal size distribution of spherical heterogeneities, fit the experimental curves extremely well.

For the irradiated iron $d\Sigma/d\Omega/\|$ (lower curve in Fig. 2a) is a factor of A = 1.39 smaller than $d\Sigma/d\Omega/\underline{\perp}$. This indicates the presence of defects (mean radius 0.85 nm) which may be voids (5 × 10^{16} cm^{-3}) or loops (3 × 10^{18} cm^{-3}). That the defects are voids with radii $\leq$ 1 nm and a number density 5 × 10^{16} cm^{-3} is clearly demonstrated by TEM (Fig. 3) with the interstitial component of the damage being in the form of loops of radius ~ 7.5 nm, having a number density ~ 2 × 10^{15} cm^{-3}.

Fig. 2b illustrates the corresponding scattering cross-sections for an irradiated Fe - 0.29 % Cu specimen (2'). For $\kappa \leq$ 9 × 10^{-2} Å^{-1}, the ratio A is found to be 11.5, clearly indicating the formation of Cu precipitates during irradiation. For larger values of κ, A decreases as a result of the increasing contribution to the scattering intensity from voids. Data analysis yields a mean radius R_2 = 2.04 nm, and a number density, N_2 = 5 × 10^{16} cm^{-3}, for the Cu particles, with the volume fraction found (0.2 %) indicating that decomposition is close to completion. The radius and number density of the voids, however, could only be estimated, due to the superimposed scattering from Cu particles in the range κ > 0.09 Å^{-1}, and gives similar values to those found in Fe. At this stage, data from TEM is limited to the interstitial loop distribution, which is found to have mean radius ~ 3-4 nm and a number density > 10^{16} cm^{-3} - significantly finer than that in the irradiated iron.

Pre-aging of the Fe - 0.86 % Cu alloy for 20 h/600 °C (specimen 9) gives rise to a precipitate structure (R_1 = 8.78 nm, N_1 = 1.5 × 10^{15} cm^{-3}), the size distribution of which is shown in Fig. 4; the volume fraction, f_1 = 0.56 %, implies a residual Cu content of ~ 0.26 % in the matrix, prior to irradiation. Consideration of the scattering curves in Fig. 2c (9, 9') reveals that this primary precipitate structure (dashed line) remains largely unaffected by the irradiation (see Table 1 and Fig. 4), whereas the kink in the scattering curve at κ ~ 6 × 10^{-2} Å^{-1} indicates that a secondary precipitate structure with R_2 = 2.67 nm and N = 2.2 × 10^{16} cm^{-3} is formed during irradiation. The additional scattering from voids is also evident for κ > 0.15 Å^{-1}, although a quantitative evaluation of R and N of the voids is not feasible.

The size distribution of the resulting <u>bimodal</u> precipitate structure is shown in Fig. 4, and reference to Table 1 reveals that such
bimodal precipitate distributions are a feature of all the specimens (6',7',9') that contained Cu precipitates prior to irradiation. Inspection of the volume fractions of the secondary precipitates in Table 1 demonstrates, furthermore, that under the chosen
irradiation conditions, <u>complete</u> precipitation occurs in all Fe-Cu
alloys irrespective of the initial state of decomposition.

(b) Pressure vessel steels:
    SANS data for the three pressure vessel steels investigated
are shown in Table II. A typical SANS curve is shown in Fig. 5 from
which it can be deduced that the ratio of the scattering cross-sections A, neither approaches the value of 11.5 expected for Cu precipitates nor the value of 1.4 associated with voids. Inspection of
Table II shows that the average A values (since A varies with scattering angle) lie in the range 2.6 - 4.2, indicating that the defect structures in the steels are more complex than in the binary
Fe-Cu alloys. Assuming that the nuclear scattering is chiefly due
to the presence of voids, allows the given values of void size,
number density and volume fraction to be calculated. These are
similar to those obtained in the Fe-Cu alloys. With increasing Cu
content there appears to be a tendency for the void distribution to
become somewhat coarser. The data derived from the nuclear + magnetic scattering curves refer to the distributions of all scattering
centres (precipitates, voids) in the specimen with sizes in the
range of scattering angles measured.

    In all three cases the measured volume fractions are greater
than the nominal Cu contents of the steels, indicating that the
curves cannot be interpreted as arising simply from distributions
of Cu precipitates and voids as is the case for the binary Fe-Cu
alloys. The tendency for A to become larger with increasing Cu
content in the steel, however, suggests that a scattering contribution from Cu precipitates may indeed be present. Further investigations of the nature of the scattering centres are clearly necessary
in order to interpret the curves more fully.

    Preliminary results showing the response of the steels to
post-irradiation annealing treatments are also shown in Table II.
These data (obtained from the nuclear + magnetic scattering curves)
show that during annealing the scattering centres increase slightly
in radius, and that their number densities and volume fractions
decrease. These changes appear to be largely associated with the
precipitate microstructure. Analysis of the nuclear scattering from
voids indicates that changes in void volume fraction are by comparison relatively small. The irradiated Fe control specimen (1')
has been more fully investigated, and changes in void volume fraction and radius during annealing at 400 °C are shown in Fig. 6. The
reduction in void volume fraction is clearly associated with a
reduction in number density with the void radius remaining roughly
constant (data derived from magnetic scattering).

DISCUSSION

The defect structure in irradiated iron with the given interstitial impurity level, consists of dislocation loops (L) and voids
(V) with mean number densities and radii of $N_L \sim 2 \times 10^{15}$ cm^{-3},
$R_L \sim 7.5$ nm and $N_V = 5 \times 10^{16}$ cm^{-3}, $R_V = 0.85$ nm. The number density of the voids and their volume fraction ($f_v = 1.5 \times 10^{-2}$ %) are
factors of ~ 120 and ~ 2, respectively, larger than the values for
iron irradiated to the higher fluence 4.5×10^{20} n/cm^2 reported in
ref. 1. The data are, however, in good agreement with recent TEM
results by Horton et al. [5]. Where feasible, analysis of SANS data
from Fe-Cu specimens yielded void structures similar to those in
pure iron. According to the TEM results, in the Cu bearing alloys
R_L of the dislocation loops is considerably smaller (~ 4 nm). Qualitatively, it appears that N_L increases with the amount (ΔC) of
copper dissolved in the matrix, although due to the 'black spot
contrast' from both the small loops and the Cu-particles at certain
diffraction conditions, a quantitative determination of N_L (ΔC) has
not been feasible. From SANS and Table 1 it is evident that the
chosen irradiation conditions lead to <u>complete</u> decomposition of all
the Fe-Cu alloys examined. During irradiation of the pre-aged specimens further decomposition proceeds by nucleation and growth of a
second precipitation structure rather than by growth of the already
pre-existing one. This suggests that heterogeneous nucleation of
Cu-particles occurs at radiation-induced defects.

Based on the given microstructural data, it is concluded that
Cu contributes to the observed radiation-strengthening (Fig. 1) in
two different ways: i) radiation-enhanced precipitation results
directly in precipitation hardening; ii) Cu atoms in solid solution
influence the nucleation of radiation defects, thus indirectly
increasing the obstacle density for dislocation motion. Since pre-
cipitation-hardening in Fe-Cu alloys is most effective for a large
number density of small Cu-particles (R $\leq$ 3 nm), it will contribute
to the total radiation-induced strength particularly at lower fluences before (radiation-enhanced) overaging occurs. Since the enhanced nucleation of (presumably) loops correlates only with the Cu
concentration in solid solution, the influence of Cu on radiation-
strengthening is particularly pronounced at smaller fluences before
precipitation is completed. Thus, both effects acting together
should be responsible for the rapid initial radiation strengthening
observed in Cu containing ferrites. Once precipitation is completed, a perceptible decrease in the radiation-strengthening rate is
expected; this has, in fact, been observed in Fe $-$ 0.3 % Cu at
fluences above 3×10^{19} n/cm^2 [1].

As far as the pressure vessel steels are concerned, it is not
possible at this stage to enter into a detailed discussion. Clearly, further work is necessary in order to determine the nature of
the scattering centres produced during irradiation, and their behaviour during post-irradiation annealing.

Table I. Mean radii (R), number densities (N), and volume fractions (f) of Cu-precipitates. Primed numbers denote irradiated specimens; indices 1, 2 refer to precipitates resulting from pre-aging and from irradiation. p : standard deviation of the logarithmic normal distribution.

Specimen	c_0 at % Cu	pre-aging	R_1 nm	$N_1 \times 10^{-16}$ cm^{-3}	f_1 %	p_1 %	R_2 nm	$N_2 \times 10^{-16}$ cm^{-3}	f_2 %	p_2 %
1, 1'	0	homog., aged	no precipitation				2.04	5.0	0.20	21
2, 2'	0.29	homog.					2.48	7.8	0.57	21
3, 3'	0.64	homog.					2.57	9.5	0.78	22
4, 4'	0.86	homog.								
5'	0.29	200 h/500 °C	5.63	4.0	0.38	28	2.14	4.3	0.21	23
6'	0.64	200 h/500 °C	4.39	1.0	0.51	34	1.53	4.9	0.09	26
7	0.86	200 h/500 °C	5.8	0.62	0.54	27				
7'			6,21	0.53	0.71	31	2.01	1.88	0.08	25
9	0.86	20 h/600 °C	8.78	0.15	0.56	30				
9'			9.06	0.14	0.58	31	2.67	2.2	0.21	19

Table II. Mean radii (R), number densities (N), and volume fractions of defects in irradiated pressure vessel steels.

Pressure Vessel Steel	Composition (wt %)					Condition	Radius R, nm	Number density N_v, cm^{-3}	Vol. fraction f, %	'A'
	Mn	Cr	Mo	Ni	Cu					
A. IAEO (French)	1.47	0.04	0.52	0.65	0.03	As Irradiated				
						Nuclear	0.9	1.7×10^{17}	0.05	
						Magnetic	0.82	$1.. \times 10^{18}$	0.25	2.6
						24 hrs.400 °C	0.92	4.7×10^{17}	0.16	
B. KS02 (plate)	0.95	0.5	0.57	1.3	0.09	As Irradiated				
						Nuclear	0.77	3×10^{17}	0.057	
						Magnetic	0.97	6.6×10^{17}	0.24	2.9
						24 hrs.400 °C	0.83	1×10^{18}	0.24	
						48 hrs.400 °C	1.05	3.7×10^{17}	0.18	
						96 hrs.400 °C	1.07	3.1×10^{17}	0.16	
C. KS07B (plate)	0.62	0.46	0.99	0.73	0.26	As Irradiated				
						Nuclear	1.2	5×10^{16}	0.037	
						Magnetic	1.15	4.7×10^{17}	0.3	4.2
						24 hrs.400 °C	1.17	3.7×10^{17}	0.25	
						48 hrs.400 °C	1.21	2.7×10^{17}	0.21	

REFERENCES

1. F.A. Smidt, and J.A. Sprague, Effects of Irradiation on Sub-
    structure and Mechanical Properties of Metals and Alloys,
    ASTM-STP-529, 1973, p. 78.
2. S.S. Brenner, R. Wagner and J. Spitznagel,
    Met. Trans. A, 1978, 9, 1761.
3. J. Ahlf, and F.J. Schmitt, J. Nucl. Mat., 1982, 105, 48.
4. F. Frisius, and D. Bünemann, Proc. Irradiation Behaviour of
    Metallic Materials for Fast Reactor Core Components,
    Corsica, French Atomic Energy Com., 1979, p. 247.
5. L.L. Horton, J. Bentley and K. Farrell,
    J. Nucl. Mat., 1982, 108 & 109, 222.
6. K.C. Russell, and L.M. Brown, Acta Met., 1972, 20, 969.

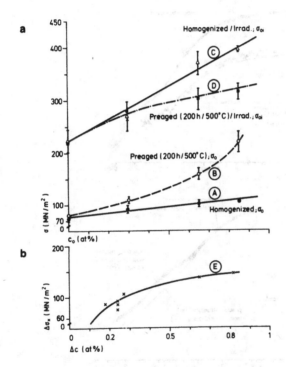

Fig.1.(a). Yield strengths of Fe-Cu alloys before and after
           irradiation.

      (b). Dependence of irradiation hardening on Cu super-
           saturation.

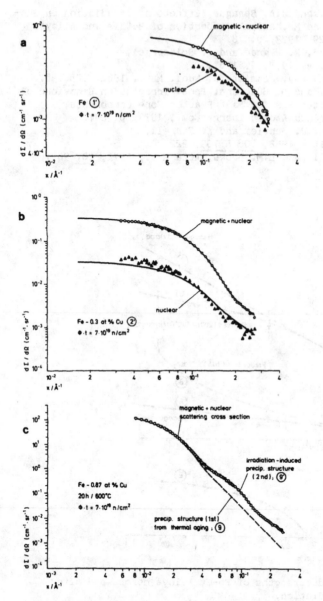

Fig.2. SANS cross-sections of irradiated specimens. (a) Fe, (b) Fe-0.29%Cu, and (c) Fe-0.86%Cu, pre-aged 20hrs. at 600°C.

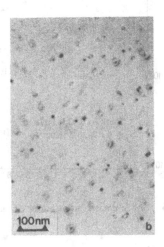

Fig.3. TEM micrographs of irradiated Fe, (a) voids, (b) loops.

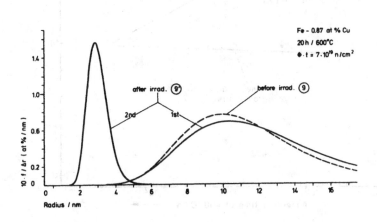

Fig.4. Precipitate size distributions for Fe-0.86%Cu pre-aged
for 20hrs. at 600°C (9), and after irradiation (9').

559

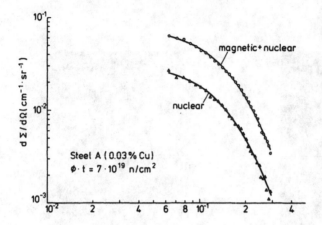

Fig.5. SANS cross-sections of an irradiated steel specimen.

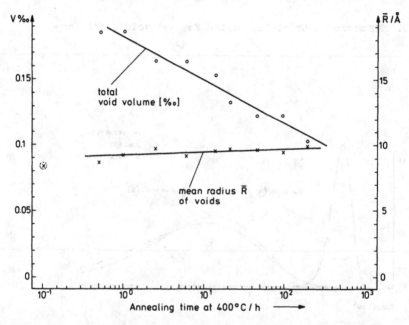

Fig.6. Changes in void distribution in Fe as a function of post irradiation annealing time at 400°C.

THE USE OF SILICON TRANSISTORS

AS DAMAGE MONITORS IN REACTOR NEUTRON METROLOGY

J. G. Williams and C. F. Hsun

University of Arkansas

Fayetteville, AR 72701, USA

ABSTRACT

Measurements of neutron induced damage to silicon bipolar transistors using Cf-252 sources have been performed
in order to establish a technique for use in fast neutron
metrology. A general purpose n-p-n bipolar transistor
(PN2222A) was chosen as the radiation damage monitor, and
the change in inverse d.c. current gain before and after
irradiation was chosen as the damage parameter for the
measurements. A controlled annealing cycle after retrieval
of transistors from the exposure and prior to gain measurements was carried out to overcome the problem of partial
annealing of damage during exposure.

The main finding of the investigation was that the
change in inverse d.c. current gain for transistors was a
linear function of the fast neutron fluence up to 10^{14}
n(1 MeV)/cm^2. The damage coefficient, defined as the ratio
of the measured damage parameter to the 1 MeV equivalent
fluence, was found to be $4.95 \times 10^{-16} \pm 1.41\%$ and $7.01
\times 10^{-16} \pm 2.6\%$ cm^2/n(1 MeV) for bare and D_2O-moderated Cf-252
spectra, respectively. High temperature annealing has also
been tried at temperatures up to 180°C. This treatment has
no effect on unirradiated transistors but removes about
66.7% of the damage from irradiated transistors.

Research sponsored by the Arkansas Power & Light Co.

INTRODUCTION

Solid-state devices such as diodes, transistors and integrated circuits are widely used in electronic systems. They consist of semiconductor materials that are quite sensitive to nuclear radiation, often to such an extent that the changes are of great engineering significance. The degradation of transistor current gain, measured by the change in the inverse current gain, is known to follow a linear relation with the fast neutron fluence. This degradation has application in monitoring potential damage to installed nuclear plant instrumentation and in other areas of fast neutron metrology. The objective of this study is to establish a method of fast neutron metrology by means of damage and annealing meaurement in silicon transistors. Measurements are relevant to pressue vessel and personnel dosimetry, as well as to the degradation of electronic equipment in the nuclear power industry.

METHODS OF ANALYSIS

Irradiations were performed at the SEFOR Calibration Center with intense californium sources encapsulated at Savannah River. The Cf-252 source used to obtain the results reported here had an effective mass of 39 mg ($9.07 \times 10^{10} \pm 3\%$ n/sec). In all of these irradiations, the transistors were exposed to either a bare or D_2O-moderated Cf-252 source at a distance of 10 cm and 25 cm. The D_2O sphere used in this study is a 30-cm diameter stainless steel shell (0.8 mm thick walls) containing the D_2O, designed by National Bureau of Standards (2). This sphere is covered with a 0.5 mm cadmium wall, which is used to absorb thermal neutrons. The change in inverse d.c. current gain before and after irradiation was selected as the damage parameter for irradiated transistors. The linear relation between the damage parameter and neutron fluence is described by the following equation:

$$\Delta(1/h_{FE}) = (1/h_o - 1/h_D) = K \, \Phi_{1 \text{ MeV}} \qquad [1]$$

where h_{FE} = transistor d.c. current gain
$\quad h_o$ = pre-irradiation current gain
$\quad h_D$ = post-irradiation current gain
$\quad K$ = damage coefficient, $cm^2/n(1 \text{ MeV})$
$\quad \Phi_{1 \text{ MeV}}$ = 1 MeV equivalent fluence, $n(1 \text{ MeV})/cm^2$, calculated according to the ASTM Standard Practice E722. (1)

Transistor current gain measurements were made at an ambient temperature of 22°C. Corrections for small deviations from this temperature were made by using the temperature coefficient of gain derived from a group of transistors, including some irradiated

562

transistors. The percentage change in gain per $^{\circ}$C was found to be approximately 0.4% per $^{\circ}$C.

After the exposure of the transistors to the fast neutron source, all the transistors were annealed at a temperature of 80°C for a period of two hours. This controlled annealing cycle after retrieval of transistors from the exposure and prior to gain measurements was carried out in order to remove the possible effects of variations in temperature during the irradiation period. This is expected to be significant in measurements to be made in power plant reactor buildings.

Some irradiated transistors were also annealed at higher temperatures. The annealing temperature ranged from 40°C to 180°C in steps of 20°C, annealing at each step for one hour. The percent damage remaining in the transistor after the annealing was calculated from the measured gains.

$$\Delta(1/h)_T/\Delta(1/h)_i = (1/h_{D,T} - 1/h_o) / (1/h_D - 1/h_o) \qquad [2]$$

where $\Delta(1/h)_T$ = Damage parameter after annealing at temp. T,

$\Delta(1/h)_i$ = Damage parameter before annealing, and

$h_{D,T}$ = Current gain after irradiation and after annealing at temperature T.

The annealing measurements were made in order to establish the temperature range at which the device can be used, in order to select procedures designed to minimize errors caused by ambient temperature annealing, and in order to investigate annealing as a diagnostic technique for electronic components after subjection to neutron exposure.

RESULTS

A linear dependence of transistor damage on equivalent fluence for both the bare and D_2O-moderated Cf-252 spectra was observed in the fluence range used. These results are plotted in Figure 1 and Figure 2 as graphs of transistor damage versus 1 MeV equivalent fluence. The damage coefficients, equal to the slope of the lines, for both the bare and D_2O-moderated irradiation experiments are shown in Table 1. These damage coefficients were calculated by performing a least squares linear regression of the data obtained after the 80°C annealing treatment.

The percent of damage removed by the 80°C annealing for both the bare and D_2O-moderated experiments was approximately 5% in addition to that which was removed by annealing at room temperature.

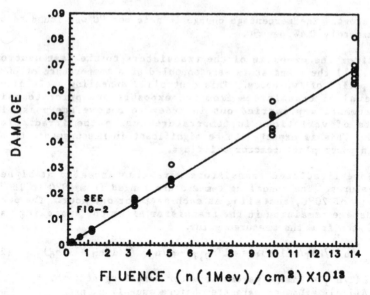

Figure 1 Transistor Damage Versus 1 MeV Equivalent Fluence
from Cf-252 Neutrons

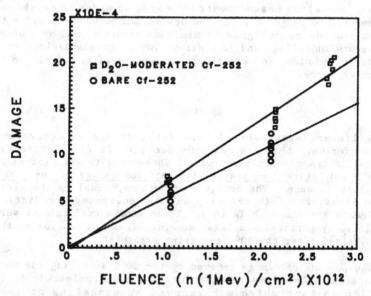

Figure 2 Transistor Damage Versus 1 MeV Equivalent Fluence
for Bare and D_2O-moderated Cf-252 Neutrons

TABLE 1

Damage Coefficient of PN2222A Bipolar Transistors
after 80°C Annealing

Experiment	No. of Transistors	Fluence Range	Damage Coefficient
Bare	56	up to 14 x 10^{13}	4.95 x $10^{-16} \pm 1.41\%$
D_2O-moderated	21	up to 0.28 x 10^{13}	7.01 x $10^{-16} \pm 2.6\%$

The annealing of transistors at elevated temperatures up to 180°C demonstrated that it is possible to remove a large portion of the neutron induced damage. The results are plotted in Figure 3 as the percent damage remaining in the transistor versus the annealing temperature. The annealing treatment showed an increase in damage recovery for the irradiated transistors as the annealing temperature increases. Despite the different radiation dosage in each transistor, all irradiated transistors have shown a very similar recovery percentage. Table 2 gives the average of percent damage recovery for each temperature.

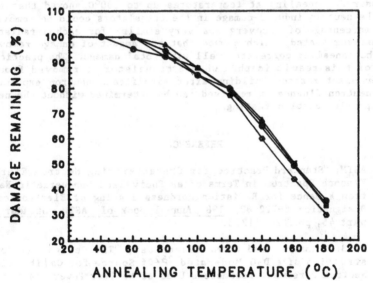

Figure 3 Effects of Isochronal Annealing on
Irradiated Transistors

TABLE 2

Percent of Damage Recovered at Each Annealing Temperature

Annealing Temperature (C^o)	40	60	80	100	120	140	160	180
Average Percent Recovery	0	1.2	5	13.6	20.9	35.6	51.9	66.7

CONCLUSIONS

The main finding in this investigation shows that the PN2222A bipolar transistors' neutron damage is a linear function of neutron fluence for 1 MeV equivalent fluences up to 10^{14} n/cm^2. The damage coefficient for D_2O-moderated Cf-252 neutrons was found approximately 40% higher than for bare Cf-252 neutrons. The reason for this large discrepancy needs to be resolved. The discrepancy may be due either to inadequate Kerma factor data in ASTM standard E-722 or an incorrect energy spectrum for the D_2O-moderated Cf-252 source.

The 80°C annealing treatment applied for all the calibrated transistors indicated that this treatment is a practical approach to overcome the problem of partial annealing of damage during the exposure. Annealing at temperatures up to 180°C showed that 67% of the neutron induced damage in the transistors could be removed. The percentage of recovery was very similar for each irradiated transistor tested, which showed that the amount of damage removed by the annealing correlated well with total damage. The practical use of this result is that when a transistor is removed from a power plant reactor building after significant neutron exposure, the neutron fluence it received can be determined without the need for pre-irradiation testing.

REFERENCES

(1) ASTM, "Standard Practice for Characterizing Neutron Energy Fluence Spectrum in Terms of an Equivalent Mononergetic Neutron Fluence for Radiation Hardness Testing of Electronics", Designation E-722-80, 1981 Annual Book of ASTM Standards, part 45, pp. 1267-1273.

(2) R. B. Schwartz and C. M. Eisenhauer, "The Design and Construction of a D_2O-Moderated ^{252}Cf Source for Calibrating Neutron Personnel Dosimeters Used at Nuclear Power Reactors", NUREG/CF-1204, U.S. Nuclear Regulatory Commission, Washington (1980).

(3) Krohn, J. L., "Neutron Damage to Bipolar Transistors in a Moderated Flux Field", Thesis presented at U. of Ark., Fayetteville, AR, December 1983.

PART VI
NEUTRON AND GAMMA
SPECTRUM DETERMINATION

CALCULATION OF THE ^{252}Cf NEUTRON SLOWING DOWN

SPECTRUM AND REACTION RATE RATIOS IN MODERATING MEDIA

Dariush Azimi-Garakani, Manijeh Rahbar and N. Fouladi-Oskoui

Department of Physics, University of Tehran

P.O. Box 11365-7693, Tehran, Iran

ABSTRACT

The slowing down spectrum and reaction-rate ratios of the Cf-252 spontaneous source in moderating media of the University of Tehran Irradiation Facility (UTIF) are calculated using the ANISN one-dimensional transport code. In these calculations the UKNDL 37 group structure cross section is used with isotropic scattering (P_o). The effect of the number of mesh intervals, degrees of approximation (S_N), irradiation facility geometry (spherical or cylindrical) and moderating materials on the shape of the neutron spectrum are studied in detail. The reaction cross sections set are derived from the ENDF-B/IV Nuclear Data Library. Twelve reactions are calculated and their ratios are found to the most fundamental reaction, U-235 (n, f).

INTRODUCTION

The University of Tehran Irradiation Facility (UTIF) has been designed and constructed with the aim of providing a neutron standard field for intercomparison measurements and computations. In this facility a Cf-252 spontaneous fission source is placed at the centre of a moderating medium to provide a standard and well-known spectrum. The experimental results have not yet been accomplished but the theoretical calculations have been performed using a one-dimensional discrete ordinates code.

J. P. Genthon and H. Röttger (eds.), Reactor Dosimetry, 571–577.
© *1985 ECSC, EEC, EAEC, Brussels and Luxembourg*

This paper describes the calculation of the slowing down spectrum and reaction-rate ratios in the UTIF system for two moderating materials.

METHODS OF CALCULATION

The Cf-252 neutron slowing down spectrum and reaction-rate ratios are the major concern of the standard neutron field measurements. The UTIF system is designed in such a way as to allow the optimal neutron spectrum charcterization by measurements but also by computation. The choice of square cylindrical geometry enabled a simple and more economic one-dimensional code to be used. It is assumed that the neutron slowing down flux in the UTIF system is nearly spherically symetrical. The departures from the symetry are caused by the variations of the source position along the tank vertical axis.

The neutron slowing down spectrum and reaction-rate ratios of the Cf-252 spontaneous fission source in moderating media are calculated using the ANISN one-dimensional transport code (1). In these calculations the ANISN-C version (2) was used with a 37 group data set derived from the UKAEA Nuclear Data Library by the GALAXY processing code (3).

RESULTS AND DISCUSSION

The effects of the number of mesh intervals, degrees of approximation, irradiation facility geometry and moderating materials are studied in detail to choose the optimal neutron spectrum characterization by computation.

The distributed source for the ANISN input was calculated using the widely used expression for neutron energy distribution of the Cf-252 spontaneous fission source (4)

$$\Phi(E) = 0.6672 \ E^{\frac{1}{2}} \ \exp \ (- \ 1.5 \ E \ / \ 2.13)$$

where the neutron energy, E, is in MeV. The fission spectrum was calculated by writing the computer program CHI which calculates the probability distribution of the function $\Phi(E)$ in such a way that the area under the curve is unity.

The effect of the number of mesh intervals on the shape of the slowing down spectrum was studied by chooing different number of intervals in the moderating region and it was found that the choice of 90 mesh intervals is adequate with a reasonably good confidence in results.

In the present calculations the effects of the S_4, S_8 and S_{16} approximations were studied only with isotropic scattering, P_o^{16}, since the cross sections with higher order components were not available. The comparison based on the calculations made with the $S_{16}P_o$, S_8P_o and S_4P_o approximations shows that the variation of the ratio $S_{16}P_o/S_8P_o$ is about 2.5% over the 37-group energy range, while that of the ratio S_8P_o/S_4P_o is about 6% in the same energy range. The computer processing unit (CPU) time for the $S_{16}P_o$ approximation was abput 36% longer than that of the S_8P_o approximation. Although the higher degrees of approximation could not have been tried because of the limited allocation of the CPU time, it is thought that the use of the $S_{16}P_o$ approximation is adequate for the present work.

The effect of the geometry of the system on the shape of the slowing down spectrum was studied by considering the cylindrical and spherical geometries. The calculation showed that the discrepancy in the total flux between these two geimetries was sufficiently small to justify the use of the sherical geometry. The total flux in the moderating regiondiffers by less than 1% in the two geometries and the CPU time for the spherical calculations is about 5.5% shorter than that of the cylindrical calculations.

In the present work the calculations were made for distilled water and paraffin moderating materials. The results showed that the shape of the spectrum in both cases remains almost unchanged that is the ratio of the flux per unit lethargy in water to that of paraffin varies from 1.2 to 1.5. Fig. (1) shows the neutron slowing down spectra in water and paraffin moderating materials.

The fast neutron flux orginating from the Californium source is sharply attenuated when passing through the moderating materials so that the neutron flux is virtually negligible outside the boundries of the UTIF ystem. In the present calculations, the neutron flux has been calculated up to the radial distance of 200 mm beyond the outer boundry of the system. The results showed that at this distance the relative neutron flux compared to that of the centre of the UTIF system is at the order of 10^{-6} and 10^{-7} for water and paraffin moderating material, respectively.

In the present calculations, the UKNDL/GALAXY 37 group structure was used and 12 reaction-rate ratios were calculated. Table I showes the 12 data file used for reaction cross sections. The last reaction with unity cross section reflects the average total flux in the moderating medium where the detector is placed. All reaction cross sections were taken from the Evaluated Nuclear Data File (ENDF/B-IV) library (5). The original 620 group structure data set which dad been reduced to 100 group structure by using the RATIF computer program (6) was then condensed to 37

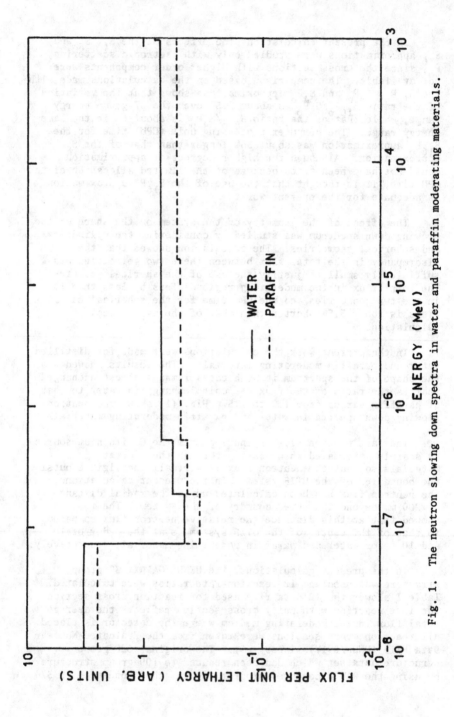

Fig. 1. The neutron slowing down spectra in water and paraffin moderating materials.

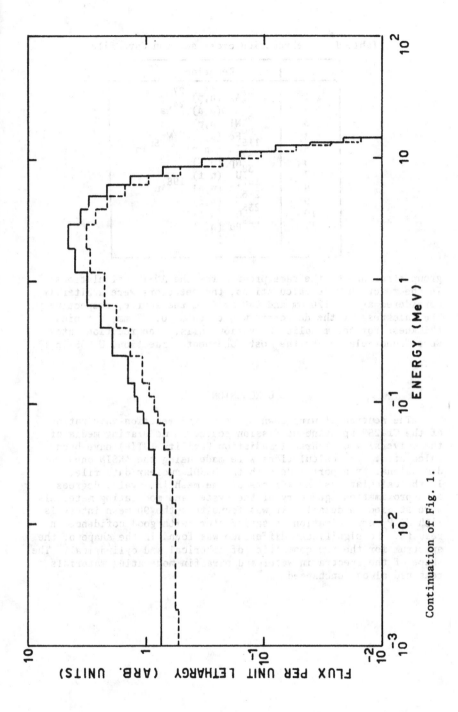

Continuation of Fig. 1.

575

Table I   The reaction cross section data file.

No	Reaction
1	^{27}Al (n,p) ^{27}Mg
2	^{27}Al (n,a) ^{24}Na
3	^{58}Ni (n,p) ^{58}Co
4	^{56}Fe (n,p) ^{56}Mn
5	^{115}In (n,n') ^{115m}In
6	^{237}Np (n,f)
7	^{238}U (n,f)
8	^{197}Au (n,g) ^{198}Au
9	^{236}U (n,f)
10	^{235}U (n,f)
11	^{239}Pu (n,f)
12	1.0

group structure by the same program for the ANSIN calculations.
In the reaction rate calculations, the detectors were arbitrarily
considered at the 120 mm and 240 mm from the source, respectively.
The thickness of the detectors was taken as 0.127 mm, a typical
thickness for the metalic activation foils. The reaction rates
were found relative to the most fundamental raection, U-235 (n,f).

CONCLUSIONS

The neutron slowing down spectrum and reaction-rate ratios
of the Cf-252 spontaneous fission source in moderating media of
the University of Tehran Irradiation Facility (UTIF) have been
calculated. The calculations were made using the ANSIN one-
dimensional transport code with the UKNDL nuclear data file.
In the calculations the effects of the mesh intervals, degrees
of approximation, geometry of the system and moderating materials
were studied in detail. It was found that the 90 mesh intervals
with $S_{16}P_0$ approximation is satisfactory with good cofidence in
results. No significant difference was found in the shape of the
spectrum for the two geometries of spherical and cylindrical. The
shape of the spectra in water and paraffin moderating materials
remained almost unchanged.

REFERENCES

1. W.W. Engle, Jr., A User's Manual for ANISN: A One-Dimensional Discrete Ordinates Transport Code with Unisotropic Scattering, U.S. AEC Report K-1693 (1967)

2. ANSIN-C, NEA Computer Programme Library (CPL), Ispra, Italy (1976)

3. W.M.M. Kerr, K. Parker and D.V.J. Williams, The Calculation of Group Averaged Neutron Cross Sections, UKAEA Report AWRE O-97/64 (1965)

4. W.L. ZijP, H.J. Nolthenius and J.H. Baard, "Ratio of Measured and Calculated Reaction Rates for some known Spectra", p. 265 in Neutron Cross Sections for Reactor Dosimetry, Proceedings of a Consultants' Meeting on Integral Cross Section Measurements in Standard Neutron Fields for Reactor Dosimetry, Vienna, 15-19 November, 1976, IAEA-208, Vol. I, International Atomic Energy Agency, Vienna (1978)

5. D. Garber, C. Dunford and S. Pearlstein, ENDF-102 Data Formats and Procedures for the Evaluated Nuclear Data File, ENDF, BNL-NCS-50496 (ENDF 102), (October 1975)

6. M. Rahbar, Testing of Inelastic Scattering Data by Integral Measurements in U-238 Spherical Shells, PhD Thesis, University of London, Imperial College (1978)

EXPERIMENTAL VALIDATION OF NEUTRON SPECTRA CALCULATED VIA TRANSPORT CODES

E. Borioli, G. Sandrelli

ENEL, Ente Nazionale per l'Energia Elettrica, Centro di Ricerca

Termica e Nucleare (CRTN), Milano, Italy

and

A. Cesana, V. Sangiust, M. Terrani

Politecnico di Milano, Istituto di Ingegneria Nucleare, Centro

Studi Nucleari E. Fermi (CESNEF), Milano, Italy

ABSTRACT

The results of discrete ordinate DOT calculations are compared with Multiple Foil Activation measurements performed at RB-2/TV research facility (Montecuccolino Laboratories, Bologna) and at Caorso power plant (Caorso, Piacenza).

Comparisons indicate that integral neutron fluxes can be predicted within about 10%, whereas experimental flux shapes in some cases may deviate significantly from the computed ones, especially at neutron energies higher than 2 MeV.

INTRODUCTION

The object of this work was the experimental check of the quality of the results obtainable with the two-dimension, discrete ordinate DOT code, when it is used in criticality or shielding computations to produce neutron guess spectra for adjustment procedures in reactor dosimetry.

J. P. Genthon and H. Röttger (eds.), Reactor Dosimetry, 579–587.
© *1985 ECSC, EEC, EAEC, Brussels and Luxembourg*

Two cases were investigated. The first test was performed on the relatively simple structure of the research facility RB-2/TV at the Montecuccolino Laboratories (Bologna), a 40 W thermal-fast reactor, where the computed results were checked in two positions inside the core.

For the second test the much more complex structure of the primary containment of Caorso power plant, a 2600 MWth BWR, was considered, with particular attention to the neutron environments in the cavity around the pressure vessel and at some penetrations piercing the sacrificial shield.

The Multiple Foil Activation (MFA) technique was adopted as a measuring technique. The dosimeters were chosen according to their sensitivity to the expected neutron fluxes as well as to the expected irradiation and cooling times.

This way, all reactions producing radionuclides with half-lives shorter than two days were discarded in the measurements at Caorso power plant.

The reaction rates per atom were determined by high-resolution gamma-ray counting. Isotopic abundances, half-lives and gamma-ray emission probabilities were drawn from Ref.1.

The computed fluxes were adjusted by means of the MINIS (2) and STAY'SL (3) codes in a 40-group energy structure (4) ranging from 10^{-10} to 18 MeV.

The cross-section library in the same group structure was derived from 620-group libraries based on either ENDF B/IV or ENDF B/V file, using the standard spectrum of the Sigma-Sigma structure as weight function.

RB-2/TV RESEARCH REACTOR EXERCISE

The facility, described in detail in Ref.5, mainly consists of a 29-cm-diam and 30-cm-high central fast zone, surrounded by a "neutron B_4C filter" 1.5-cm-thick and 120-cm-high, a "buffer zone" (32 cm i.d., 60 cm o.d., 90 cm in height), an annular graphite reflector and axial graphite "flux flatteners". The overall dimensions are 150 cm in diameter and 130 in height.

Two positions inside the core on the reactor mid-plane were selected for the exercise, the former at the center of the core (Pos.A) and the latter 9.0 cm apart (Pos.B). Hard fluxes were expected, with integral values of about $5 \cdot 10^8$ n/cm^{2}s at maximum reactor rating (6).

Reactor modelling and neutron flux computation

For the calculation of the neutron flux energy and space distribution DOT-3.5 E code (7) was employed in a criticality computation.

The facility was modelled in a cylindrical structure composed of 31 material zones. The spatial discretization was accomplished by using 55 radial and 52 axial meshes thickened next to the interfaces and the measure points, where intervals down to 1 mm were considered.

The P_1-S_4 approximation was adopted. The standard energy range from 10^{-10} to 14.9 MeV was divided into 39 groups (38 fast and 1 thermal), into which the cross sections were collapsed starting from the 200-group cross-section library of GGTC-E code. Calculated spectra were extrapolated up to 18 MeV by joining their high energy region with Watt's ^{235}U fission spectrum.

In both GGTC and DOT calculation all the downscattering terms were considered, whereas the upscattering terms were neglected, since only one thermal group was provided for the energy structure. The neutron spectrum calculated in Pos.A is shown in Fig.1. It turned out that neither the integral value nor the shape of the neutron fluxes change significantly inside a few-centimeter interval around the reactor mid-plane, and that the integral neutron flux in the Pos.B is slightly higher than that calculated in Pos.A . The perturbation induced by the measuring stations was checked and found negligible.

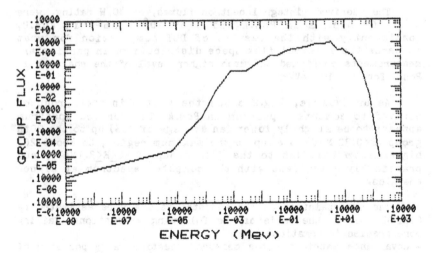

Fig.1 - RB-2/TV Reactor. Computed spectrum in Pos.A .

Experimental

The check of the calculated results was carried out using the following set of reactions :

^{23}Na (n,gamma) ^{27}Al (n,p) ^{63}Cu (n,gamma) ^{115}In (n,gamma)

^{24}Mg (n,p) ^{55}Mn (n,gamma) ^{103}Rh (n,n') ^{235}U (n,f)

^{27}Al (n,alfa) ^{58}Ni (n,p) ^{115}In (n,n')

The dosimeters, made of high purity materials, were 8.0-mm-diam and from 0.0025-mm-(Au) to 1.0-mm-thick (Mg,Al).

In each position, measurements were performed simultaneously about 1 cm above and below the reactor mid-plane. Ni foils were used as flux monitors.

Results and conclusions

The prediction of energy and space constant neutron flux around the reactor mid-plane was experimentally confirmed.

With regard to the adjustment procedure, the unfolding performed by MINIS code was stopped when a 4% mean deviation of measured from calculated reaction rates per atom was achieved, allowing at the same time all measured values to be fitted within the experimental uncertainties.

The derived integral neutron fluxes, at 40 W rating, were $5 \cdot 15 \ 10^8$ n/cm^{2}s in Pos.A and $5 \cdot 65 \ 10^8$ n/cm^{2}s in Pos.B, consistently with the results of DOT computation for what concerns the integral flux space distribution. In particular, measurements confirmed the much higher level of the spectrum in Pos.B for $E \leq 100$ eV.

As an example, Fig.2 shows the ratios in each group of derived to computed spectrum in Pos.A. The derived spectrum appears to be slightly lower (an average of 10%) up to the 20th group ($E \leq 0.32$ MeV), i.e. up to the maximum region, to about 20% higher from the 21st to the 28th group ($0.32 \leq E \leq 3.1$ MeV), and practically coincident with the computed spectrum at higher energies.

The uncertainties on the adjusted values were estimated by STAY'SL code on the basis of the following assumptions (failing more precise information):
- Covariance matrix for the measured reaction rates per atom of diagonal type, with uncertainties on the experimental values

ranging from 4% [^{23}Na (n,gamma), ^{58}Ni(n,p)] to 20% [^{235}U(n,f)].
- Covariance matrix for the computed spectrum groups, with 15% uncertainties on the diagonal for all groups, the first six excepted, for which higher uncertainties were assumed. In addition, a connection of gaussian type was assumed among groups, ranging from 80% correlation between contiguous groups to no correlation for j and j+5 groups.
- Covariance matrix for the cross-section groups of diagonal type, with an 8% uncertainty on each group.

The most important results were:
- Measured and calculated reaction rates per atom agree to a normalized χ^2 value very near the unity.
- Measurements add the most significant information to the computed results, for both Pos.A and Pos.B, in the range from the 7th to the 12th group ($0.30 \leq E \leq 6.00$ keV) and from the 25th to the 35th group ($1.2 \leq E \leq 10.0$ MeV), where the group uncertainties decrease from the assumed 15% to 9% on the average.
- The uncertainty on the derived integral flux turns out to be lower than 6% in both measure points.

  In conclusion, calculated and derived spectra appear to be in good agreement. In addition, both calculations and measurements show the same spatial trend of the neutron integral flux , both inside each measuring station and from one to the other measure point.

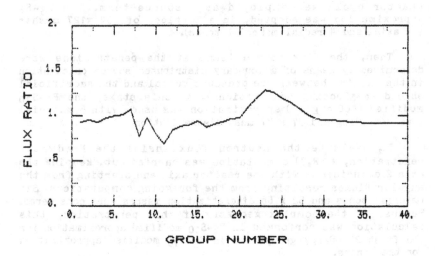

Fig.2 - RB-2/TV Reactor. Ratio of derived to computed spectrum in Pos.A.

CAORSO POWER PLANT EXERCISE

Measurements were planned just outside the sacrificial shield, at three penetrations corresponding to Feedwater, Water Level and Recirculation Inlet piping, and at four positions situated at different elevations in the cavity between the pressure vessel and the sacrificial shield. The outer positions were selected as near as possible the inner ones.

Outside the sacrificial shield relatively hard neutron fluxes with integral values ranging from $3 \cdot 10^5$ to $1 \cdot 10^8$ n/cm^{2}s were expected, whereas in the cavity the integral values of the calculated hard neutron fluxes ranged from 10^8 to 10^9 n/cm^{2}s.

Reactor modelling and neutron flux computation

The basic calculation of the neutron flux energy and space distribution was carried out by DOT 3.5-E code using the subdivision technique. The 39-group cross sections were collapsed from the 100-group DLC-37 library. Also in this calculation all the downscattering terms were considered and the upscattering was neglected.

Firstly, a (R,Z) volume distributed source calculation was performed for the reactor core, from which the flux distribution out of the core was obtained. The average power of the second reactor cycle was employed as a source term. The P_3-S_8 approximation was adopted, in a lattice of 54×127 meshes (14 axial and 4 radial material zones).

Then, the flux source terms at the penetrations were determined by means of a boundary distributed source calculation in the cavity between the pressure vessel and the sacrificial shield (neglecting penetrations). At this stage, the P_5-S_{10} modified (100 angles) approximation was used, with 5 material zone modelling and 37×77 mesh discretization.

To evaluate the neutron flux inside the Feedwater penetration, a (R,Z) computation was carried out, keeping the axis Z coincident with the reactor axis and starting from the angular fluxes resulting from the foregoing computations. Six material zones and 52×66 discretization meshes were considered. Because of the complex geometry of this penetration, this calculation was performed in P_5-S_{10} modified approximation for the first 20 energy groups and in P_3-S_{10} modified approximation for the others.

At the Water Level and Recirculation Inlet penetrations, the (R,Z) calculations were carried out with the axis Z coincident

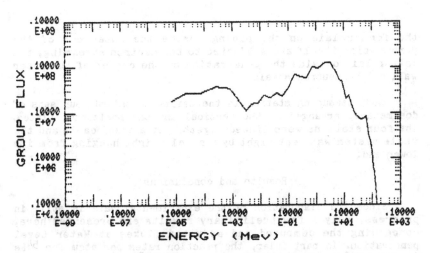

Fig.3 - Caorso Power Plant . Computed neutron flux at Water Level
penetration outside the sacrificial shield .

with the penetration axis. In both cases the P_5-S_{10} modified
approximation was adopted. Both structures were divided into 6
material zones, discretized into 35 x 30 and 39 x 20 meshes,
respectively.

As an example of the results of computations, the neutron
flux calculated at Water Level   penetration outside the
sacrificial shield is shown in Fig.3 for E $\geq$ 0.69 eV.

Experimental

Since one year's irradiation time was expected, the
following reactions were selected for neutron flux measurements:

^{54}Fe (n,p) ^{58}Ni (n,p) ^{109}Ag (n,gamma)

^{54}Fe (n,alfa) ^{63}Cu (n,alfa) ^{197}Au (n,gamma)

^{58}Fe (n,gamma) ^{94}Zr (n,gamma) ^{238}U (n,gamma)

^{59}Co (n,gamma) ^{93}Nb (n,n')

The dosimeters were from 8.0-mm- to 20.0-mm-diam and
0.015-mm- (Nb) to 3.0-mm-thick (Fe,Cu) at the penetrations, 8.0-
mm-diam and from 0.025-mm- (Au) to 1.0-mm-thick (Cu) in the
cavity. Some of them were enveloped in a Cd shield, 0.5-mm-thick.

Each penetration was equipped with two measuring stations,

the former laid on the piping inside the outer end of the penetration itself and subjected to the neutron streaming, the latter laid outside the penetration on the center of the outer wall of the neutron shield.

Each measuring station in the cavity contained four sets of dosimeters, arranged on two vertical and one horizontal plates. The four stations were linked together by a steel cable and the whole system was kept tight by a steel weight hanging from its lower end.

Results and conclusions

Since the elaboration of the experimental data is still in progress, only a few preliminary results are presented here, concerning the computed and adjusted fluxes at Water Level penetration. In particular, the reaction rates per atom for ^{54}Fe (n,alfa), ^{93}Nb(n,n'), ^{109}Ag(n,gamma) and ^{238}U(n,gamma) have not yet been worked out.

Any way, the remaining reaction rates per atom, determined for the two vertical and the two horizontal sets of dosimeters, allowed the isotropy of the neutron field to be verified in the cavity.

The neutron flux adjustment has been performed till now only by MINIS code. The unfolding procedure was stopped after one to

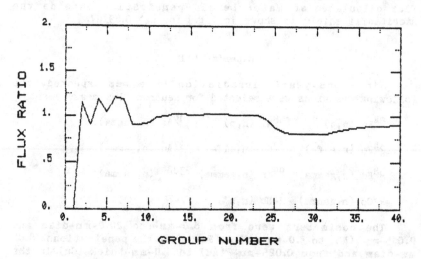

Fig.4 - Caorso Power Plant. Ratio of derived to computed spectrum at Water Level penetration .

three iterations, when an agreement within 5% was reached between measured and calculated reaction rates per atom. The results obtained at this stage can be summarized as follows:
- The derived integral flux in the cavity turns out to be $6.1 \cdot 10^8$ n/cm^{2}s, i.e. only 16% lower than the predicted one. Outside the sacrificial shield the derived integral flux is $1.1 \cdot 10^8$ n/cm^{2}s, i.e. only 10% lower than the predicted one.
- The adjusted spectrum in the cavity deviates significantly from the computed one below 1 keV and above 2 MeV : at high energy the derived groups are up to twice as high as the expected ones. On the contrary, outside the sacrificial shield adjusted and computed spectra agree within 10% on the average. Fig.4 shows the ratio of derived to computed spectrum.

Therefore it is possible to conclude that the preliminary results already show that computed and experimental neutron fluxes agree very well as far as the integral value is concerned. Also the flux shape is well fitted at the outer end of the Water Level penetration, whereas inside the cavity computations seem to require some improvements for reproducing the correct shape of the neutron flux.

AKNOWLEDGEMENT

The authors want to thank G.Alloggio and E.Brega of ENEL-Thermal and Nuclear Research Center for the RB-2/TV calculations and P.Barbucci and G.Mariotti of the same Center for the Caorso calculations.

REFERENCES

1) W.L.Zijp, J.H.Baard, Nuclear Data Guide for Reactor Neutron Metrology, Report EUR-7164 EN, ed. 1979 (1981).
2) G.Sandrelli; A.Cesana, V.Sangiust, M.Terrani, Impiego di Rivelatori Fissili per la Misura di Spettri di Neutroni. Parte II, Internal Report ENEL-DSR/CRTN N1/6 (1980).
3) F.G.Perey, Least-Squares Dosimetry Unfolding: The Program STAY'SL, Report ORNL/TM-6062, ENDF-254 (1977).
4) A.Cesana, G.Sandrelli, V.Sangiust, M.Terrani, Nucl.Sci.Eng., 82, 102 (1982).
5) P.Azzoni et.al., Nucl.Sci.Eng., 76, 70 (1980).
6) G.Sandrelli; A.Cesana, V.Sangiust, M.Terrani, Determinazione Sperimentale di Spettri Neutronici presso i Reattori TAPIRO e RB-2/TV, Internal Report ENEL-DSR/CRTN N1/83/01 (1983).
7) P.Barbucci, F.Di Pasquantonio, Implementation of Exponential Supplementary Equations on DOT-III and DOT 3.5 codes, NEA Newsletter,22, 15, Nuclear Energy Agency, France (1977).

NEUTRON AND GAMMA RAY FLUX CALCULATIONS FOR THE

VENUS PWR ENGINEERING MOCKUP

ARNOLD H. FERO

WESTINGHOUSE NUCLEAR TECHNOLOGY DIVISION

PITTSBURGH, PENNSYLVANIA U.S.A.

ABSTRACT

This paper describes the analysis of neutron and gamma ray fluxes in the VENUS PWR engineering mockup. This mockup experiment is unique in two ways. It is the first mockup to correctly represent the heterogeneities which exist in the PWR core peripheral fuel assemblies, core baffle, core barrel, and neutron pad. This is accomplished by using a representative PWR fuel assembly geometry (15 x 15) and full thickness Type 304 stainless steel reactor internals structures located with representative water gap spacing. The VENUS mockup also represents locally the stairstep geometry of the core periphery. Second, the VENUS mockup is extremely well characterized in terms of "as-built" dimensions, material compositions, and pin-by-pin core power distributions.

The analysis of VENUS was performed using the methods and procedures used to analyze commercial PWR's. Specifically, two-dimensional discrete ordinates transport theory calculations were performed using the DOT IIIW code. Calculations were run in both X-Y and R-theta geometries for a 90 degree sector of the mockup. Fixed distributed source calculations were run S_8-P_3 using the SAILOR (coupled 47 neutron – 20 gamma ray groups) cross-section library which was derived from the VITAMIN-C (coupled 171 neutron – 36 gamma ray groups) cross-section library. The finite height of the VENUS mockup was accounted for in the two-dimensional calculations by the use of group- and zone-dependent DB^2 terms derived from axial leakages calculated in a DOT R-Z geometry approximation of the mockup.

J. P. Genthon and H. Röttger (eds.), Reactor Dosimetry, 589–610.
© *1985 ECSC, EEC, EAEC, Brussels and Luxembourg*

Calculated results are presented for eight fast-neutron reactions and for gamma ray energy-deposition rates. The calculated effects of photo reactions on neutron dosimetry are also presented. Comparisons to measured neutron reaction rates that were available at the time of the analysis show agreement in the fuel and the steel regions (baffle and barrel) that is within $\pm$ 15 percent. In general, the agreement in the water zones is also within $\pm$ 15 percent. However, a couple of points are underpredicted by as much as 40 percent.

INTRODUCTION AND PURPOSE

The VENUS PWR engineering mockup benchmark (Mol, Belgium) was conceived to address several concerns. Prior experimental benchmark work in the USNRC sponsored Light Water Reactor Pressure Vessel Surveillance Dosimetry Improvement Program (LWR-PV-SDIP) has included a dedicated dosimetry-metallurgy irradiation at the Oak Ridge Research Reactor (ORR) Poolside Facility (PSF) at the Oak Ridge National Laboratory in Tennessee, as well as investigations of neutron penetration through the pressure vessel wall (ORNL poolside critical assembly benchmarks) and field perturbation by surveillance capsules (ORR surveillance dosimetry measurement facility benchmarks at the PSF). These benchmark experiments as well as their interrelationship with pressure vessel steel embrittlement surveillance issues are well documented (1, 2, 3, 4).

All of the above experiments are driven by MTR-type reactor cores and involve relatively simple, finite-slab geometry. Actual PWR cores, on the other hand, consist of arrays of low-enrichment (3-4 w/o U-235) fuel pins (on the order of 10 mm in diameter x 13 mm pitch). The fuel pins are grouped into assemblies of typically 15 x 15 pins (less control rods and instrumentation thimble). The overall azimuthal reactor geometry is complex: the core periphery and adjacent stainless steel baffle plates have a stairstep shape and are, in turn, surrounded by a cylindrical stainless steel core barrel and thermal shield (or neutron pads). Metallurgical specimens and neutron dosimetry are typically placed at surveillance positions located between the thermal shield and the pressure vessel.

The analysis and evaluation of the metallurgical specimens and neutron dosimetry and the subsequent prediction of pressure vessel property changes depends, in part, upon the radiation analyst's ability to model this complex geometry, including the detailed pin-by-pin variation of power in the core peripheral fuel assemblies. The VENUS PWR engineering mockup is a well characterized benchmark which can be used to assess the accuracy of present and future analysis methods in a realistic PWR geometry.

590

PHYSICAL PARAMETERS

The physical description of the VENUS Critical Facility, the "as-built" dimensions and chemical composition of all core and structural materials is provided in Section 1.0 of Reference 5. Figure 1 is a dimensioned illustration of the VENUS Model geometry with the experimental data locations identified. The VENUS grid coordinates referred to are on a grid pitch of 1.26 cm. The model is located in the (-X, -Y) quadrant of the VENUS grid and the model center is located at grid location (+2.5, +2.5). The 3/0 Fuel Region consists of 0.95 cm O.D. ZIRC-4 clad fuel pins with a 3.3 w/o U-235 enrichment. The 4/0 Fuel Region consists of 0.98 cm O.D. SS-304 clad fuel pins with a 4.0 w/o U-235 enrichment. Both fuel types are located on a 1.26 cm pitch.

The reference data for the absolute core power normalization was obtained from Section 5.2 of Reference 5. The recommended measured fission density is given for the midplane of the fuel pin located at the center of the 3/0 Fuel Region, VENUS grid position (-5, -5). This value is $2.81 \pm 06 \times 10^8$ fissions per centimeter of fuel per second at the reference nominal 100 percent power level. The power at this pin location relative to a core (2552 fuel pins) average of 1.0 is 1.361 (from the experimental power distribution map given in Section 5.1 of Reference 5). The full power normalization used for 90 degree X-Y and R-theta dots was

$$S01 = \frac{2.81 \times 10^8}{1.361} \frac{\text{fissions}}{\text{cm-sec}} \times \frac{2552}{4} \text{ cm of fuel} \times 2.47 \frac{n}{\text{fission}}$$

$S01 = 3.254 \times 10^{11}$ neutrons/second.

The experimental pin-by-pin core power distribution given in Section 5.1 of reference 5 was processed by the Westinghouse SORCERY Code into the X-Y and R-theta DOT mesh. SORCERY performs a fine mesh area weighting of pin-by-pin input.

TRANSPORT CROSS-SECTIONS

Microscopic cross-sections for Cr, Mn, Fe, Ni, Zr, U-235, and U-238 were obtained from the SAILOR (47-N, 20-G) cross-section library (6). Microscopic cross-sections for H, B, O, Si, Mo, Co and Sn were obtained from the BUGLE-80 (47-N, 20-G) cross-section library (7). Both libraries are derived from the VITAMIN-C (171-N, 36-G) cross-section library (8). In preparing the SAILOR cross-sections for fuel and steel constituents, the VITAMIN-C cross-sections were self-shielded in PWR fuel cell (0.95 cm Pin O.D.) and PWR steel/water downcomer geometries. The fuel cell pellet, clad, and moderator temperatures were 650°C, 400°C, and 310°C, respectively.

FIGURE 1
VENUS PWR ENGINEERING MOCKUP
KEY DIMENSIONS AND
LOCATION OF EXPERIMENTAL DATA POINTS

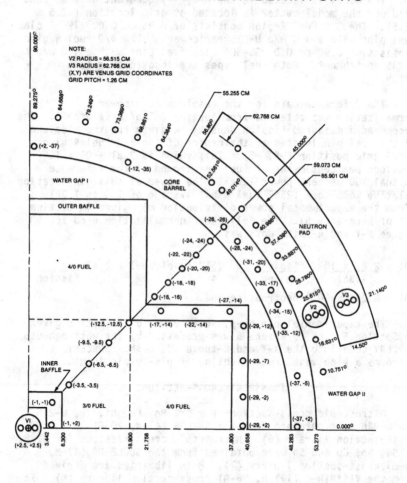

NOTE:
V2 RADIUS = 56.515 CM
V3 RADIUS = 62.768 CM
(X,Y) ARE VENUS GRID COORDINATES
GRID PITCH = 1.26 CM

The temperature for the steel/water regions was 320°C. These temperatures, while resulting in correct resonance region processing for the analysis of PWRs at power, are too high for the room temperature operating conditions of the VENUS mockup. However, as we do not yet have the ability to process the AMPX format VITAMIN-C library on our CRAY-1S computers, we must (temporarily) live with the small error that this introduces. For the fuel materials this is expected to be less than a 10 percent effect in the resonance region cross-sections. The effect is expected to be smaller for steel cross-sections. Note that VITAMIN-C, and therefore both SAILOR and BUGLE-80, is processed from ENDF-B/IV using a room temperature maxwellian weighting function linked to a 1/E at 0.125 eV. The two lowest energy groups in the VITAMIN-C and SAILOR structure (E < 0.414 eV) are thus best referred to as "room temperature thermal". This distinction is important when interpreting reactor condition "thermal fluxes" calculated using the SAILOR library. Note also that the VITAMIN-C and SAILOR coupled libraries do not include delayed fission product gamma rays and that this component must be treated separately in power reactor calculations.

NEUTRON AND GAMMA RAY RESPONSE FUNCTIONS

The neutron reaction cross-sections used are listed in Tables 1 and 2. The gamma ray reaction cross-sections used are listed in Table 3. The data for all of the neutron reactions except the Iron dpa response are taken from the SAILOR dosimetry file. This data, in turn, is a processing of ENDF/B-IV dosimetry cross-sections in a SAND-II Library collapsed over the spectrum at a 1/4 T position in a PWR pressure vessel. The Iron dpa response is from Reference 9 flat weighted into the 47 group structure.

The gamma ray energy deposition response was developed from the kerma factor data (eV-barn/atom) on the BUGLE-80 library, which in turn was obtained from Reference 10. The photoreaction data were taken from Reference 11.

The conversion of reaction rates to equivalent fission fluxes used the fission spectrum averaged cross sections listed below (12).

Np-237 (n,f) F.P.	1322	mb
U-238 (n,f) F.P.	294.7	mb
Al-27 (n,α) Na-24	0.693	mb
Fe-54 (n,p) Mn-54	77.8	mb
Ni-58 (n,p) Co-58	101	mb
Cu-63 (n,α) Co-60	0.540	mb
In-115 (n,n') In-115m	173	mb

TABLE 1

NEUTRON REACTION CROSS-SECTIONS (BARNS/ATOM)

Group	Upper Energy (Mev)	Np-237 (n,f)	U-238 (n,f)	Iron dpa
1	1.733 + 1	2.505	1.259 - 1	2.928 + 3
2	1.419 + 1	2.311	1.059 - 1	2.683 + 3
3	1.221 + 1	2.340	9.819 - 1	2.455 + 3
4	1.000 + 1	2.327	9.876 - 1	2.245 + 3
5	8.607	2.244	9.880 - 1	2.092 + 3
6	7.408	1.930	8.410 - 1	1.971 + 3
7	6.065	1.517	5.818 - 1	1.792 + 3
8	4.966	1.543	5.622 - 1	1.600 + 3
9	3.679	1.636	5.482 - 1	1.369 + 3
10	3.012	1.680	5.463 - 1	1.265 + 3
11	2.725	1.698	5.527 - 1	1.198 + 3
12	2.466	1.695	5.526 - 1	1.246 + 3
13	2.365	1.694	5.512 - 1	1.166 + 3
14	2.346	1.692	5.499 - 1	1.096 + 3
15	2.231	1.677	5.391 - 1	1.036 + 3
16	1.920	1.645	4.677 - 1	8.190 + 2
17	1.653	1.604	2.736 - 1	9.150 + 2
18	1.353	1.539	4.227 - 2	6.320 + 2
19	1.003	1.380	1.051 - 2	2.670 + 2
20	8.208 - 1	1.206	2.900 - 3	5.160 + 2
21	7.427 - 1	9.868 - 1	1.402 - 3	4.000 + 2
22	6.081 - 1	6.470 - 1	5.461 - 4	2.690 + 2
23	4.979 - 1	2.823 - 1	1.584 - 4	3.080 + 2
24	3.688 - 1	9.733 - 2	8.699 - 5	2.240 + 2
25	2.972 - 1	3.259 - 2	6.068 - 5	1.780 + 2
26	1.832 - 1	2.052 - 2	4.680 - 5	1.920 + 2
27	1.111 - 1	1.578 - 2	4.019 - 5	1.340 + 2
28	6.738 - 2	1.231 - 2	4.000 - 5	7.540 + 2
29	4.087 - 2	1.097 - 2	5.816 - 5	7.140 + 2
30	3.183 - 2	1.027 - 2	8.561 - 5	2.940 + 2
31	2.606 - 2	1.002 - 2	8.700 - 5	6.080 + 1
32	2.418 - 2	9.892 - 3	8.700 - 5	3.610
33	2.188 - 2	9.706 - 3	8.700 - 5	7.730
34	1.503 - 2	1.004 - 2	5.679 - 5	1.190 + 1
35	7.102 - 3	7.236 - 3	2.962 -11	7.360
36	3.355 - 3	1.429 - 3	1.132 - 9	3.820
37	1.585 - 3	2.302 - 2	4.207 - 4	1.170
38	4.540 - 4	3.698 - 2	1.461 - 8	9.360 - 2
39	2.144 - 4	6.123 - 2	1.044 - 8	1.360 - 1
40	1.013 - 4	9.101 - 2	1.244 - 8	2.060 - 1
41	3.727 - 5	2.248 - 2	1.958 - 8	1.100 - 1
42	1.068 - 5	1.016 - 2	3.087 - 8	6.090 - 1
43	5.043 - 6	3.994 - 3	4.771 - 8	9.250 - 1
44	1.855 - 6	9.345 - 3	7.137 - 8	1.470
45	8.764 - 7	1.381 - 2	5.002 - 8	2.120
46	4.140 - 7	0	0	3.380
47	1.000 - 7	0	0	7.650

TABLE 2
NEUTRON REACTION CROSS-SECTIONS (BARNS/ATOM)

Group	Upper Energy (MeV)	Al-27 (n,α) Na-24	Fe-54 (n,p) Mn-54	Ni-58 (n,p) Co-58	Cu-63 (n,α) Co-60	In-115 (n,n') In-115m
1	1.733 + 1	1.086 − 1	2.828 − 1	3.091 − 1	3.111 − 2	5.939 − 2
2	1.419 + 1	1.267 − 1	4.361 − 1	4.705 − 1	3.871 − 2	9.564 − 2
3	1.221 + 1	1.047 − 1	5.358 − 1	6.211 − 1	4.321 − 2	2.069 − 1
4	1.000 + 1	7.490 − 2	5.792 − 1	6.593 − 1	3.254 − 2	2.696 − 1
5	8.607	3.990 − 2	5.886 − 1	6.547 − 1	1.662 − 2	2.881 − 1
6	7.408	9.690 − 3	5.559 − 1	6.271 − 1	7.371 − 3	3.059 − 1
7	6.065	3.491 − 4	4.655 − 1	5.311 − 1	4.917 − 4	3.162 − 1
8	4.966	3.439 − 7	3.070 − 1	3.839 − 1	0	3.196 − 1
9	3.679	0	1.816 − 1	2.315 − 1	Balance Zero	3.139 − 1
10	3.012	Balance Zero	1.153 − 1	1.653 − 1		3.071 − 1
11	2.725		7.827 − 2	1.137 − 1		2.967 − 1
12	2.466		5.297 − 2	9.347 − 2		2.877 − 1
13	2.635		4.756 − 2	9.232 − 2		2.848 − 1
14	2.346		4.325 − 2	8.254 − 2		2.819 − 1
15	2.231		2.000 − 2	4.641 − 2		2.655 − 1
16	1.920		4.712 − 3	2.647 − 2		2.255 − 1
17	1.653		6.457 − 4	1.346 − 2		1.643 − 1
18	1.353		1.126 − 5	4.156 − 3		9.596 − 2
19	1.003		0	4.769 − 4		3.931 − 2
20	8.208 − 1		Balance Zero	1.741 − 4		2.198 − 2
21	7.427			4.983 − 5		1.277 − 2
22	6.081 − 1			7.802 − 6		6.040 − 3
23	4.979 − 1			1.041 − 6		2.019 − 3
24	3.688 − 1			4.897 − 8		1.388 − 4
25	2.972 − 1			0		0
				Balance Zero		Balance Zero

TABLE 3

GAMMA RAY REACTION CROSS-SECTIONS

Group	Energy (MeV)	w/gm-SS 304 γ/cm²-sec	Photoreaction Cross-Section (mb/atom)		
			Np-237 (γ,f)	U-238 (γ,f)	In-115 (γ,γ')
1	14.0 – 10.0	4.248 – 14	0	0	0
2	10.0 – 8.0	3.301 – 14	81.38	30.50	1.729
3	8.0 – 7.0	2.675 – 14	35.75	12.81	0.898
4	7.0 – 6.0	2.166 – 14	26.75	8.250	0.291
5	6.0 – 5.0	1.864 – 14	5.375	1.875	0.192
6	5.0 – 4.0	1.525 – 14	0	0	0.150
7	4.0 – 3.0	1.192 – 14	Balance	Balance	0.105
8	3.0 – 2.0	8.381 – 15	Zero	Zero	0.076
9	2.0 – 1.5	6.499 – 15			0.048
10	1.5 – 1.0	4.980 – 15			0.020
11	1.0 – 0.8	3.874 – 15			0
12	0.8 – 0.7	3.331 – 15			Balance
13	0.7 – 0.6	2.950 – 15			Zero
14	0.6 – 0.4	2.332 – 15			
15	0.4 – 0.2	1.659 – 15			
16	0.2 – 0.1	2.224 – 15			
17	0.1 – 0.06	5.458 – 15			
18	0.06 – 0.03	1.342 – 14			
19	0.03 – 0.02	5.374 – 14			
20	0.02 – 0.01	1.452 – 13			

TRANSPORT THEORY CALCULATIONS

Two dimensional discrete ordinates transport theory calculations were performed using the DOT IIIW code. Calculations were run in both X-Y and R-theta geometries for a 90 degree sector of the mockup. The DOT geometries are shown in Figures 2 and 3. The calculations were run $S_8 - P_3$ using the fixed distributed sources and cross-sections described above.

The finite height (50 cm) of the VENUS mockup was accounted for in the X-Y and R-theta DOT calculations by the use of group- and zone-dependent DB^2 terms derived from axial leakages calculated in a DOT R-Z geometry approximation of the mockup. Cylindrical ANISN calculations using the zone dependent experimental buckling values (5) show good fast neutron agreement with ANISN calculations using the DB^2 values from the R-Z DOT.

RESULTS OF CALCULATIONS

Neutron and photon spectra at detector locations V1, V2, and V3 are listed in Tables 4 and 5, respectively. Tables 6 through 8 present comparisons of calculated and experimental equivalent fission fluxes. The limited experimental data is from the 13th LWR-PV-SDIP meeting minutes (5). More data will be presented at this conference by others (5).

Within the steel regions the agreement with Np-237 (n,f) and U-238 (n,f) data is within ±10 percent, Ni-58 (n,p) is within ±15 percent, and In-115 (n,n') is generally underpredicted by 5-10 percent. In the water zones Np-237 (n,f) is underpredicted by 15-40 percent, and U-238 is within ±5 percent.

Tables 9 and 10 present the calculated data for the Al-27 (n,α), Cu-63 (n,α), Fe-54 (n,p) and Iron dpa reactions.

Tables 11 and 12 present gamma ray energy deposition rates and equivalent fission fluxes resulting from gamma ray induced reactions. The magnitude of the photoreaction contribution ranges up to about 3 percent for Np-237 (n,f), 6 percent for U-238 (n,f) and less than 1 percent for In-115 (n,n').

The only available experimental gamma ray data was that presented by Fabry, et al. (13). Based upon this data it would appear that the calculated gamma ray data is 30-50 percent low. More experimental gamma ray data are expected to be available at this meeting and at the 14th LWR-PV-SDIP meeting to be held in London, England on October 1-5, 1984.

597

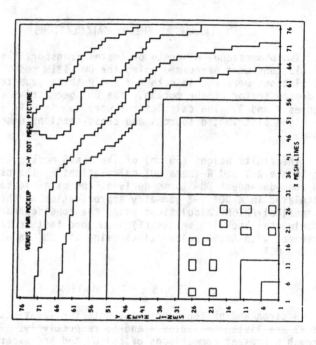

Figure 2
X-Y DOT Geometry

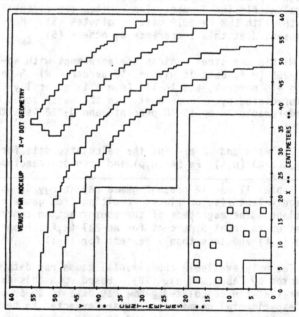

Figure 3
R-Theta DOT Geometry

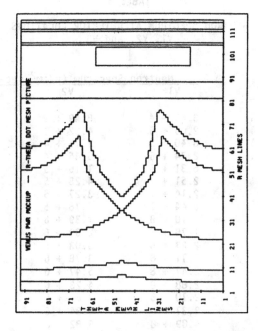

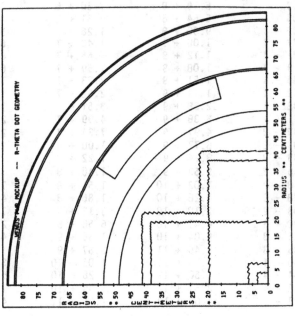

TABLE 4

NEUTRON SPECTRUM AT DETECTOR LOCATIONS
V1, V2, and V3

Group	Upper Energy (MeV)	Neutron Spectrum (n/cm^2-sec-MeV)		
		V1	V2	V3
1	1.733 + 1	3.65 + 4	8.49 + 2	3.11 + 2
2	1.419 + 1	2.09 + 5	4.92 + 3	1.82 + 3
3	1.221 + 1	8.40 + 5	1.75 + 4	6.13 + 3
4	1.000 + 1	2.74 + 6	5.45 + 4	1.87 + 4
5	8.607	6.31 + 6	1.16 + 5	3.84 + 4
6	7.408	2.51 + 7	4.28 + 5	1.38 + 5
7	6.065	2.14 + 7	3.27 + 5	1.01 + 5
8	4.966	6.44 + 7	8.78 + 5	2.71 + 5
9	3.679	1.10 + 8	1.39 + 6	4.36 + 5
10	3.012	1.79 + 8	2.39 + 6	8.13 + 5
11	2.725	2.27 + 8	3.03 + 6	1.06 + 6
12	2.466	2.71 + 8	3.78 + 6	1.35 + 6
13	2.365	3.47 + 8	5.47 + 6	2.00 + 6
14	2.346	2.88 + 8	4.23 + 6	1.64 + 6
15	2.231	2.70 + 8	3.81 + 6	1.59 + 6
16	1.920	3.37 + 8	4.84 + 6	2.08 + 6
17	1.653	4.09 + 8	5.82 + 6	2.70 + 6
18	1.353	5.19 + 8	7.60 + 6	3.98 + 6
19	1.003	6.64 + 8	9.61 + 6	4.77 + 6
20	8.208 − 1	9.00 + 8	1.26 + 7	5.75 + 6
21	7.427 − 1	1.00 + 9	1.48 + 7	8.80 + 6
22	6.081 − 1	1.12 + 9	1.65 + 7	8.59 + 6
23	4.979 − 1	1.08 + 9	1.59 + 7	8.81 + 6
24	3.688 − 1	1.51 + 9	2.23 + 7	1.27 + 7
25	2.972 − 1	1.79 + 9	2.62 + 7	1.09 + 7
26	1.832 − 1	2.38 + 9	3.51 + 7	1.66 + 7
27	1.111 − 1	3.38 + 9	4.99 + 7	2.15 + 7
28	6.738 − 2	4.97 + 9	7.34 + 7	2.65 + 7
29	4.087 − 2	6.77 + 9	1.00 + 8	2.36 + 7
30	3.183 − 2	8.24 + 9	1.22 + 8	2.10 + 7
31	2.606 − 2	9.53 + 9	1.40 + 8	1.18 + 7
32	2.418 − 2	1.02 + 10	1.51 + 8	7.49 + 7
33	2.188 − 2	1.26 + 10	1.86 + 8	4.16 + 7
34	1.503 − 2	2.10 + 10	3.12 + 8	7.48 + 7
35	7.102 − 3	4.41 + 10	6.56 + 8	2.32 + 8
36	3.355 − 3	9.04 + 10	1.35 + 9	4.94 + 8
37	1.585 − 3	2.46 + 11	3.67 + 9	1.09 + 9
38	4.540 − 4	7.03 + 11	1.05 + 10	3.03 + 9
39	2.144 − 4	1.50 + 12	2.26 + 10	5.87 + 9
40	1.013 − 4	3.57 + 12	5.39 + 10	1.50 + 10
41	3.727 − 5	1.08 + 13	1.65 + 11	4.69 + 10

TABLE 4 (Cont.)

NEUTRON SPECTRUM AT DETECTOR LOCATIONS
V1, V2, and V3

Group	Upper Energy (MeV)	Neutron Spectrum (n/cm^2-sec-MeV)		
		V1	V2	V3
42	1.068 – 5	3.07 + 13	4.72 + 11	1.27 + 11
43	5.043 – 6	7.24 + 13	1.13 + 12	2.67 + 11
44	1.855 – 6	1.76 + 14	2.76 + 12	5.56 + 11
45	8.764 – 7	3.69 + 14	5.85 + 12	9.45 + 11
46	4.140 – 7	2.02 + 15	3.30 + 13	2.28 + 12
47	1.000 – 7	4.79 + 16	1.14 + 15	1.61 + 13

TABLE 5

PHOTON SPECTRUM AT DETECTOR LOCATIONS
V1, V2, and V3

Group	Energy (MeV)	Photon Spectrum (γ/cm^2-sec-MeV)		
		V1	V2	V3
1	14.0 – 10.0	3.82 + 3	1.34 + 2	5.15 + 1
2	10.0 – 8.0	4.95 + 7	2.62 + 6	8.33 + 5
3	8.0 – 7.0	2.17 + 8	1.02 + 7	3.27 + 6
4	7.0 – 6.0	6.65 + 7	3.46 + 6	1.19 + 6
5	6.0 – 5.0	8.18 + 7	3.91 + 6	1.36 + 6
6	5.0 – 4.0	9.73 + 7	4.11 + 6	1.46 + 6
7	4.0 – 3.0	1.70 + 8	5.76 + 6	2.03 + 6
8	3.0 – 2.0	7.94 + 8	2.94 + 7	5.38 + 6
9	2.0 – 1.5	6.86 + 8	2.10 + 7	7.18 + 6
10	1.5 – 1.0	9.85 + 8	2.72 + 7	9.60 + 6
11	1.0 – 0.8	1.83 + 9	4.63 + 7	1.66 + 7
12	0.8 – 0.7	1.78 + 9	5.01 + 7	1.74 + 7
13	0.7 – 0.6	2.07 + 9	6.04 + 7	2.06 + 7
14	0.6 – 0.4	3.72 + 9	1.32 + 8	4.96 + 7
15	0.4 – 0.2	6.31 + 9	2.39 + 8	7.55 + 7
16	0.2 – 0.1	9.84 + 9	4.26 + 8	7.99 + 7
17	0.1 – 0.06	7.21 + 9	4.20 + 8	1.47 + 7
18	0.06 – 0.03	1.80 + 9	1.48 + 8	4.12 + 5
19	0.03 – 0.02	3.98 + 7	3.08 + 6	3.81 + 4
20	0.02 – 0.01	4.00 + 5	3.09 + 4	7.69 + 3

TABLE 6

COMPARISON OF CALCULATED AND MEASURED EQUIVALENT FISSION FLUXES IN THE STEEL REGIONS

POINT	Nb-237 (n,f) F.P.			U-238 (n,f) F.P.			Ni-58 (n,p) Co-58			In-115 (n,n') In-115m		
	CAL	EXP	C/E	CAL	EXP	C/E	CAL	EXP	C/E	CAL	EXP	C/E
INNER BAFFLE												
(-1, +2)	2.48+9	2.31+9	1.07	1.80+9	1.72+9	1.04	1.55+9	1.36+9	1.14	1.83+9	1.90+9	0.96
(-1, -1)	2.95+9	2.71+9	1.09	2.16+9	2.06+9	1.05	1.88+9	1.63+9	1.15	2.20+9	2.27+9	0.97
OUTER BAFFLE												
(-29, +2)	9.86+8	1.01+9	0.98	7.30+8	7.16+8	1.02	6.38+8	5.74+8	1.11	7.41+8	7.83+8	0.95
(-29, -2)	9.38+8			6.94+8	6.64+8	1.05	6.06+8	5.46+8	1.11	7.04+8	7.37+8	0.96
(-29, -7)	7.58+8	7.54+8	1.00	5.60+8	5.51+8	1.02	4.88+8	4.47+8	1.09	5.68+8	5.98+8	0.95
(-29, -12)	4.55+8	4.51+8	1.01	3.27+8			2.82+8	2.64+8	1.07	3.34+8	3.52+8	0.95
(-27, -14)	4.73+8	4.66+8	1.02	3.42+8	3.28+8	1.04	2.95+8	2.76+8	1.07	3.48+8	3.64+8	0.96
(-22, -14)	9.08+8	8.88+8	1.02	6.65+8	6.53+8	1.02	5.78+8	5.15+8	1.12	6.76+8	7.16+8	0.94
(-17, -14)	1.49+9	1.47+9	1.01	1.08+9	1.04+9	1.04	9.30+8	8.38+8	1.11	1.10+9	1.16+9	0.95
CORE BARREL												
(-37, +2)	1.12+8	1.11+8	1.01	8.75+7	9.02+7	0.97	7.94+7	8.11+7	0.98	8.71+7	9.72+7	0.90
(-37, -5)	1.02+8	1.03+8	0.99	7.51+7	7.85+7	0.96	6.61+7			7.58+7		
(-35, -12)	1.03+8	1.13+8	0.91	7.27+7	7.78+7	0.93	6.23+7	6.72+7	0.93	7.41+7	8.28+7	0.89
(-34, -15)	9.00+7	9.95+7	0.90	6.11+7	6.51+7	0.94	5.12+7			6.30+7		
(-33, -17)	8.11+7	8.80+7	0.92	5.51+7	6.08+7	0.91	4.62+7			5.67+7		
(-31, -20)	6.83+7	6.98+7	0.98	4.91+7	5.41+7	0.91	4.31+7	4.92+7	0.88	4.98+7	5.60+7	0.89
(-28, -24)	5.35+7	5.75+7	0.93	4.14+7	4.25+7	0.97	3.84+7			4.12+7		
(-26, -26)	5.23+7	5.62+7	0.93	4.12+7	4.28+7	0.96	3.85+7	4.36+7	0.88	4.08+7	4.50+7	0.91
NEUTRON PAD												
(21.14°)	9.55+6			6.56+6			5.67+6			6.70+6		
(45.0°)	6.80+6			4.80+6			4.34+6			4.86+6		

NOTE: Not Adjusted for Photoreaction

TABLE 7

COMPARISON OF CALCULATED AND MEASURED EQUIVALENT FISSION FLUXES IN THE WATER REGIONS

POINT	Np-237 (n,f) F.P.			U-238 (n,f) F.P.			N1-58 (n,p) Co-58			In-115 (n,n') In-115m		
	CAL	EXP	C/E	CAL	EXP	C/E	CAL	EXP	C/E	CAL	EXP	C/E
CENTER HOLE (+2.5, +2.5)	1.44+9	1.69+9	0.85	1.19+9	1.22+9	0.98	1.14+9	1.02+9	1.12	1.17+9	1.24+9	0.94
WATER GAP I												
(-16, -16)	9.70+8	1.02+9	0.95	7.49+8	7.37+8	1.02	6.86+8			7.49+8		
(-18, -18)	4.58+8	5.59+8	0.82	3.88+8			3.79+8			3.78+8		
(-20, -20)	2.29+8	3.68+8	0.62	2.08+8	2.07+8	1.00	2.13+8			1.99+8		
(-22, -22)	1.19+8	1.65+8	0.72	1.14+8			1.21+8			1.08+8		
(-24, -24)	7.05+7	1.02+8	0.69	6.61+7	6.61+7	1.00	6.98+7			6.27+7		
WATER GAP II												
(10.75°)	3.00+7	3.35+7	0.90	2.40+7	2.72+7	0.88	2.32+7			2.36+7		
(16.63°)	3.04+7	3.50+7	0.87	2.38+7	2.61+7	0.91	2.27+7			2.35+7		
(21.14°)	2.97+7			2.28+7	2.68+7	0.85	2.16+7			2.26+7		
(25.62°)	2.75+7	3.06+7	0.90	2.07+7	2.26+7	0.92	1.93+7			2.06+7		
(28.78°)	2.54+7	2.94+7	0.86	1.92+7	2.14+7	0.90	1.80+7			1.91+7		
(33.89°)	2.18+7	2.38+7	0.91	1.71+7	1.87+7	0.91	1.65+7			1.68+7		
(37.44°)	1.96+7	2.24+7	0.88	1.59+7	1.78+7	0.90	1.58+7			1.55+7		
(40.99°)	1.83+7	2.32+7	0.79	1.52+7			1.53+7			1.47+7		
(45.00°)	1.77+7	2.04+7	0.87	1.49+7	1.59+7	0.94	1.51+7			1.44+7		

NOTE: Not Adjusted for Photoreaction

TABLE 8

COMPARISON OF CALCULATED AND MEASURED EQUIVALENT FISSION FLUXES ALONG A 45° TRAVERSE

POINT	Np-237 (n,f) F.P.			U-238 (n,f) F.P.			Ni-58 (n,p) Co-58			In-115 (n,n') In-115m		
	CAL	EXP	C/E	CAL	EXP	C/E	CAL	EXP	C/E	CAL	EXP	C/E
CENTER HOLE (+2.5, + 2.5)	1.44+9	1.69+9	0.85	1.19+9	1.22+9	0.98	1.14+9	1.02+9	1.12	1.17+9	1.24+9	0.94
INNER BAFFLE (-1, -1)	2.95+9	2.71+9	1.09	2.16+9	2.06+9	1.05	1.88+9	1.63+9	1.15	2.20+9	2.27+9	0.97
3/0 FUEL												
(-3.5, -3.5)	4.15+9	3.87+9	1.07	3.60+9			3.52+9			3.52+9		
(-6.5, -6.5)	4.42+9	4.22+9	1.05	3.89+9			3.84+9			3.80+9		
(-9.5, -9.5)	4.24+9	3.99+9	1.06	3.72+9			3.66+9			3.63+9		
(-12.5, -12.5)	3.14+9	3.22+9	0.98	2.64+9			2.52+9			2.60+9		
WATER GAP I												
(-16, -16)	9.70+8	1.02+9	0.95	7.49+8	7.37+8	1.02	6.86+8			7.49+8		
(-18, -18)	4.58+8	5.59+8	0.82	3.88+8			3.79+8			3.78+8		
(-20, -20)	2.29+8	3.68+8	0.62	2.08+8	2.07+8	1.00	2.13+8			1.99+8		
(-22, -22)	1.19+8	1.65+8	0.72	1.14+8			1.21+8			1.08+8		
(-24, -24)	7.05+7	1.02+8	0.69	6.61+7	6.61+7	1.00	6.98+7			6.27+7		
BARREL (-26, -26)	5.23+7	5.62+7	0.93	4.12+7	4.28+7	0.96	3.85+7	4.36+7	0.88	4.08+7	4.50+7	0.91
WATER GAP II (55.255)	1.77+7	2.04+7	0.87	1.49+7	1.59+7	0.94	1.51+7			1.44+7		
NEUTRON PAD (62.768)	6.80+6			4.80+6			4.34+6			4.86+6		

NOTE: Not Adjusted for Photoreaction

TABLE 9
OTHER CALCULATED NEUTRON REACTIONS IN THE STEEL REGIONS

EQUIVALENT FISSION FLUXES

POINT	Al-27 (n,α) Na-24	Cu-63 (n,α) Co-60	Fe-54 (n,p) Mn-54	Iron dpa
INNER BAFFLE				
(-1, +2)	1.78 + 9	1.12 + 9	1.50 + 9	2.04 - 12
(-1, -1)	2.12 + 9	1.34 + 9	1.81 + 9	2.43 - 12
OUTER BAFFLE				
(-29, +2)	7.29 + 8	4.60 + 8	6.17 + 8	8.16 - 13
(-29, -2)	6.92 + 8	4.37 + 8	5.86 + 8	7.76 - 13
(-29, -7)	5.60 + 8	3.53 + 8	4.72 + 8	6.26 - 13
(-29, -12)	3.29 + 8	2.07 + 8	2.71 + 8	3.74 - 13
(-27, -14)	3.48 + 8	2.19 + 8	2.85 + 8	3.90 - 13
(-22, -14)	6.64 + 8	4.19 + 8	5.58 + 8	7.50 - 13
(-17, -14)	1.05 + 9	6.60 + 8	8.97 + 8	1.23 - 12
CORE BARREL				
(-37, +2)	1.14 + 8	7.08 + 7	7.77 + 7	9.45 - 14
(-37, -5)	9.30 + 7	5.77 + 7	6.42 + 7	8.40 - 14
(-35, -12)	8.50 + 7	5.28 + 7	6.02 + 7	8.44 - 14
(-34, -15)	6.94 + 7	4.30 + 7	4.92 + 7	7.32 - 14
(-33, -17)	6.34 + 7	3.93 + 7	4.44 + 7	6.59 - 14
(-31, -20)	6.37 + 7	3.94 + 7	4.18 + 7	5.62 - 14
(-28, -24)	6.22 + 7	3.83 + 7	3.78 + 7	4.47 - 14
(-26, -26)	6.33 + 7	3.90 + 7	3.80 + 7	4.39 - 14
NEUTRON PAD				
(21.14°) V3	9.95 + 6	6.04 + 6	5.51 + 6	7.78 - 15
(45.00°)	8.50 + 6	5.13 + 6	4.26 + 6	5.59 - 15

TABLE 10
OTHER CALCULATED NEUTRON REACTIONS IN
THE WATER AND THE FUEL

EQUIVALENT FISSION FLUXES

POINT	Al-27 (n,α) Na-24	Cu-63 (n,α) Co-60	Fe-54 (n,p) Mn-54	Iron dpa
CENTRAL HOLE				
(+2.5, +2.5)	1.51 + 9	9.44 + 8	1.13 + 9	1.30 - 12
FUEL				
(-3.5, -3.5)	4.20 + 9	2.66 + 9	3.48 + 9	3.60 - 12
(-6.5, -6.5)	4.60 + 9	2.91 + 9	3.80 + 9	3.85 - 12
(-9.5, -9.5)	4.36 + 9	2.77 + 9	3.63 + 9	3.69 - 12
(-12.5, -12.5)	2.97 + 9	1.88 + 9	2.49 + 9	2.70 - 12
WATER GAP I				
(-16, -16)	8.41 + 8	5.29 + 8	6.71 + 8	8.36 - 13
(-18, -18)	5.17 + 8	3.24 + 8	3.75 + 8	4.20 - 13
(-20, -20)	3.18 + 8	1.98 + 8	2.13 + 8	2.19 - 13
(-22, -22)	1.97 + 8	1.22 + 8	1.22 + 8	1.17 - 13
(-24, -24)	1.19 + 8	7.34 + 7	7.04 + 7	6.62 - 14
WATER GAP II				
(10.75°)	3.89 + 7	2.39 + 7	2.31 + 7	2.67 - 14
(16.63°)	3.74 + 7	2.30 + 7	2.25 + 7	2.68 - 14
(21.14°)	3.53 + 7	2.17 + 7	2.13 + 7	2.62 - 14
(25.62°)	3.14 + 7	1.93 + 7	1.90 + 7	2.41 - 14
(28.78°)	2.98 + 7	1.83 + 7	1.78 + 7	2.23 - 14
(33.89°)	2.87 + 7	1.75 + 7	1.63 + 7	1.92 - 14
(37.44°)	2.86 + 7	1.75 + 7	1.57 + 7	1.75 - 14
(40.99°)	2.86 + 7	1.75 + 7	1.53 + 7	1.63 - 14
(45.00°)	2.86 + 7	1.74 + 7	1.52 + 7	1.58 - 14

TABLE 11

GAMMA RAY DATA IN THE STEEL REGIONS

POINT	ENERGY DEPOSITION RATE (W/gm - SS 304)	PHOTOREACTION EQUIVALENT FISSION FLUXES								
		Np-237 (γ,f) F.P.			U-238 (γ,f) F.P.			In-115 (γ,γ') In-115m		
		CAL	C/E	(n+γ)/E	CAL	C/E	(n+γ)/E	CAL	C/E	(n+γ)/E
INNER BAFFLE										
(-1, +2)	3.63 - 5	1.78+7	0.008	1.08	2.88+7	0.017	1.06	3.62+6	0.002	0.96
(-1, -1)	3.20 - 5	1.33+7	0.005	1.09	2.14+7	0.010	1.06	2.83+6	0.001	0.97
OUTER BAFFLE										
(-29, +2)	1.23 - 5	5.64+6	0.006	0.98	9.13+6	0.013	1.03	1.17+6	0.001	0.95
(-29, -2)	1.18 - 5	5.29+6			8.56+6	0.013	1.06	1.10+6	0.001	0.96
(-29, -7)	9.66 - 6	4.35+6	0.006	1.01	7.04+6	0.013	1.03	8.98+5	0.001	0.95
(-29, -12)	6.76 - 6	3.40+6	0.008	1.02	5.51+6			6.83+5	0.002	0.95
(-27, -14)	7.69 - 6	3.68+6	0.008	1.02	5.95+6	0.018	1.06	7.52+5	0.002	0.96
(-22, -14)	1.29 - 5	5.93+6	0.007	1.03	9.59+6	0.015	1.03	1.23+6	0.002	0.94
(-17, -14)	1.82 - 5	8.57+6	0.006	1.02	1.39+7	0.013	1.05	1.78+6	0.002	0.95
CORE BARREL										
(-37, +2)	4.96 - 6	2.61+6	0.023	1.04	4.23+6	0.047	1.02	5.20+5	0.005	0.90
(-37, -5)	3.69 - 6	1.98+6	0.019	1.01	3.20+6	0.041	1.00	3.90+5		
(-35, -12)	3.17 - 6	1.73+6	0.015	0.93	2.80+6	0.036	0.97	3.40+5	0.004	0.90
(-34, -15)	2.67 - 6	1.44+6	0.014	0.92	2.33+6	0.036	0.98	2.83+5		
(-33, -17)	2.63 - 6	1.40+6	0.016	0.94	2.26+6	0.037	0.94	2.76+5		
(-31, -20)	2.82 - 6	1.41+6	0.020	1.00	2.29+6	0.042	0.95	2.84+5	0.005	0.89
(-28, -24)	2.75 - 6	1.25+6	0.022	0.95	2.02+6	0.048	1.02	2.59+5		
(-26, -26)	2.84 - 6	1.28+6	0.023	0.95	2.08+6	0.049	1.01	2.66+5	0.006	0.91
NEUTRON PAD										
(21.14°)	6.21 - 7	3.29+5			5.32+6			6.48+4		
(45.0°)	3.48 - 7	1.63+5			2.64+6			3.30+4		

TABLE 12

GAMMA RAY DATA IN THE WATER REGIONS

POINT	ENERGY DEPOSITION RATE (w/gm – SS 304)	Np-237 (γ,f) F.P.			U-238 (γ,f) F.P.			In-115 (γ,γ') In-115m		
		CAL	C/E	(n+γ)/E	CAL	C/E	(n+γ)/E	CAL	C/E	(n+γ)/E
CENTER HOLE (+2.5, +2.5)	3.75 – 5	1.36+7	0.008	0.86	2.21+7	0.018	0.99	3.01+6	0.002	0.94
WATER GAP I										
(-16, -16)	1.75 – 5	6.11+6	0.006	0.96	9.89+6	0.013	1.03	1.38+6		
(-18, -18)	1.56 – 5	4.50+6	0.008	0.83	7.27+6			1.09+6		
(-20, -20)	1.31 – 5	3.45+6	0.009	0.63	5.58+6	0.027	1.03	8.60+5		
(-22, -22)	1.06 – 5	2.69+6	0.016	0.74	4.35+6			6.60+5		
(-24, -24)	7.87 – 6	2.49+6	0.025	0.72	4.02+6	0.061	1.06	5.61+5		
WATER GAP II										
(10.75°)	1.85 – 6	7.62+5	0.023	0.92	1.23+6	0.045	0.93	1.58+5		
(16.63°)	1.76 – 6	7.66+5	0.022	0.89	1.24+6	0.047	0.96	1.57+5		
(21.14°)	1.66 – 6	7.39+5			1.20+6	0.045	0.90	1.51+5		
(25.62°)	1.58 – 6	7.12+5	0.023	0.92	1.15+6	0.051	0.97	1.44+5		
(28.78°)	1.54 – 6	6.91+5	0.023	0.89	1.12+6	0.052	0.95	1.40+5		
(33.89°)	1.48 – 6	6.37+5	0.027	0.94	1.03+6	0.055	0.97	1.31+5		
(37.44°)	1.45 – 6	5.98+5	0.027	0.91	9.67+5	0.054	0.95	1.24+5		
(40.99°)	1.43 – 6	5.78+5	0.025	0.81	9.34+5			1.20+5		
(45.00°)	1.43 – 6	5.71+5	0.028	0.89	9.24+5	0.058	0.99	1.19+5		

REFERENCES

1. E. P. Lippincott and W. N. McElroy, <u>LWR Pressure Vessel</u> <u>Surveillance Dosimetry Improvement Program</u>, Progress Reports, NUREG/CR-3391, (1983), NUREG/CR-2805 (1982), NUREG/CR-2345 (1981), NUREG/CR-1241 (1980), NUREG/CR-1240 (1979).

2. W. M. McElroy, <u>LWR Pressure Vessel Surveillance Dosimetry</u> <u>Improvement Program: PCA Experiments and Blind Test</u>, NUREG/CR-1861, July 1981.

3. F. B. K. Kam, <u>Proceedings of the Fourth ASTM-EURATOM</u> <u>Symposium on Reactor Dosimetry</u>, NUREG/CP-0029, July 1982.

4. H. Rottger, <u>Proceedings of the Third ASTM-EURATOM Symposium</u> <u>on Reactor Dosimetry</u>, EUR-6813, 1980.

5. <u>VENUS PWR Core Source and Aximuthal Lead Factor Experiments</u> <u>and Calculational Tests</u>, to be issued as NUREG/CR-3323. See also <u>Minutes of the 13th Light Water Reactor Pressure</u> <u>Vessel Surveillance Dosimetry Improvement Program</u> <u>(LWR-PV-SDIP)</u>, W. M. McElroy, et al., April 13, 1984.

   See also <u>VENUS PWR Engineering Mock-Up: Core Qualification</u> <u>Neutron and Gamma Field Characterization</u>, A. Fabry, et al., in this proceedings.

6. G. L. Simmons and R. W. Roussin, <u>SAILOR-Coupled,</u> <u>Self-Shielded, 47 Neutron, 20 Gamma Ray, P3, Cross Section</u> <u>Library for Light Water Reactors</u>, RSIC-DLC-76, March 8, 1983, ORNL Radiation Shielding Information Center.

7. R. W. Roussin, <u>BUGLE-80 - Coupled, 47 Neutron, 20 Gamma</u> <u>Ray, P3 Cross-Section Library for LWR Shielding</u> <u>Calculations, RSIC-DLC-75, June 1980, ORNL Radiation</u> <u>Shielding Information Center.</u>

8. R. W. Roussin, et al., <u>VITAMIN-C - 171 Neutron, 36 Gamma</u> <u>Ray Group Cross-Sections in AMPX and CCCC Interface Format</u> <u>for Fusion and LMFBR Neutronics</u>, RSIC-DLC-41, September 23, 1977, ORNL Radiation Shielding Information Center.

9. ASTM E693-79, <u>Practice for Characterizing Neutron Exposures</u> <u>in Ferritic Steels in Terms of Displacements Per Atom</u> <u>(dpa)</u>, 1983 Annual Book of ASTM Standards, American Society for Testing and Materials, Philadelphia, PA, 1983.

10. Y. Gohar and M. A. Abdou, <u>MACKLIB-IV-82 - 171 Neutron, 36 Gamma Ray Group Nuclear Response Function Library Calculated with MACK-IV From Cross-Section Data in ENDF/B-IV,</u> RSIC-DLC-60, January 15, 1982, ORNL Radiation Shielding Information Center.

11. V. V. Verbinski, et al., "Photo Interference Corrections in Neutron Dosimetry for Reactor Pressure Vessel Lifetime Studies," <u>NUCL. SCI. & ENG.</u> (75), Pg 159, 1980.

12. ASTM E706-(IIE)-81a, <u>Guide for Benchmark Testing of Reactor Vessel Dosimetry,</u> 1983 Annual Book of ASTM Standards, American Society for Testing and Materials, Philadelphia, PA, 1983.

13. A. Fabry, et al., "The Belgium BR3 Characterization Program and the VENUS Program for Core Source to PV Wall Fluence Verification," Pg. 197-211 of <u>Proceedings of the USNRC Eleventh Water Reactor Safety Research Information Meeting,</u> NUREG/CP-0048, Vol. 4 October, 1983.

CHARACTERIZATION OF FUEL DISTRIBUTION IN

THE THREE MILE ISLAND UNIT 2 (TMI-2) REACTOR

SYSTEM BY NEUTRON AND GAMMA-RAY DOSIMETRY

Raymond Gold, James H. Roberts,
Frank H. Ruddy, Christopher C. Preston,
James P. McNeece, Bruce J. Kaiser
and William N. McElroy

Westinghouse Hanford Company
Hanford Engineering Development Laboratory
Richland, Washington 99352

ABSTRACT

Neutron and gamma-ray dosimetry are being used for nondestructive assessment of the fuel distribution throughout the Three Mile Island Unit 2 (TMI-2) reactor core region and primary cooling system. The fuel content of TMI-2 makeup and purification Demineralizer A has been quantified with Si(Li) continuous gamma-ray spectrometry and solid-state track recorder (SSTR) neutron dosimetry. Results obtained from these gamma-ray and neutron dosimetry experiments were 1.3 ± 0.6 kg and 1.7 ± 0.6 kg, respectively, for the fuel content of TMI-2 Demineralizer A.

For fuel distribution characterization in the core region, results from SSTR neutron dosimetry exposures in the TMI-2 reactor cavity (i.e., the annular gap between the pressure vessel and the biological shield) will be presented. These SSTR results are consistent with the presence of a significant amount of fuel debris, equivalent to several fuel assemblies or more, lying at the bottom of the reactor vessel.

J. P. Genthon and H. Röttger (eds.), Reactor Dosimetry, 611–619.
© *1985 ECSC, EEC, EAEC, Brussels and Luxembourg*

INTRODUCTION

The resolution of technical issues generated by the accident at Three Mile Island Unit 2 (TMI-2) will inevitably be of long-range benefit. Determination of the fuel debris dispersal in the TMI-2 reactor system represents a major technical issue. In-reactor recovery operations, such as for the safe handling and final disposal of TMI-2 waste, quantitative fuel assessments are being conducted throughout the reactor core and primary coolant system. Recent data at TMI-2 indicate that the void in the core geometry constitutes about 26% of the volume, and much of the displaced fuel appears to have been reduced to rubble. It is possible that significant amounts of this fuel debris have been transported out of the core into off-normal locations. For planning recovery operations, it is important to know the location of the dispersed fuel debris.

TMI-2 fuel distribution assessments can be carried out non-destructively by neutron and gamma-ray dosimetry. In gamma-ray dosimetry, gamma-rays associated with specific fission products are measured. In neutron dosimetry, one measures neutrons generated from a combination of spontaneous fission, (α,n) reactions, and subcritical multiplication.

Existing constraints preclude the application of many routine dosimetry methods for TMI-2 fuel distribution characterization. These constraints arise from many origins, ranging from sensitivity and background considerations to practical day-to-day restrictions of TMI-2 recovery operations. Two highly specialized methods have been applied to overcome these constraints, namely solid-state track recorder (SSTR) neutron dosimetry and continuous gamma-ray spectrometry with a Si(Li) Compton spectrometer.

A general exposition on the applicability of SSTR neutron dosimetry for TMI-2 applications has already been published (1). Efforts to characterize the fuel distribution in the TMI-2 makeup and purification demineralizers with SSTR neutron dosimetry and Si(Li) gamma-ray spectometry have been successfully completed (2-5). This report summarizes results from these TMI-2 demineralizer experiments, as well as more recent SSTR dosimetry efforts in the TMI-2 reactor cavity (i.e., the annular gap between the pressure vessel and the biological shield) for characterization of the fuel distribution in the core region.

CHARACTERIZATION OF TMI-2 DEMINERALIZER A

Earlier efforts were directed toward fuel debris characterization of the makeup and purification demineralizers (A and B) that maintained reactor coolant water purity at TMI-2. Since both

Demineralizer A and B were on-line during the accident and high gamma-ray intensities were observed in the location of these demineralizers, significant amounts of fuel could have been trapped in the demineralizers' resin beds. Intense gamma-ray fields (>2000 R/h) in these demineralizer cubicles precluded personnel entry for neutron and gamma-ray dosimetry measurements.

Si(Li) Continuous Gamma-Ray Spectrometry

Access to Demineralizer A was obtained through a 6-in. x 9-in. penetration in the wall of Cubicle A, ∿8 ft above the tank. A special boom and winch assembly was fabricated to remotely position the collimated gamma-ray spectrometer inside the cubicle. The boom provided for horizontal movement of the spectrometer over the complete width of the cell. A swing arm and winch provided for lowering the spectrometer down both the north and south sides of the tank. Tape measures attached to the boom allowed for accurate positioning of the spectrometer inside the cubicle.

To reduce the intense background radiation from cesium-137, the Si(Li) detector was surrounded by a 5.5-in. x 8-in. lead shield. Two shields were used to provide different levels of background attenuation. Small diameter collimator holes in the shields' sides permitted accurate mapping of the geometrical source distribution within the demineralizer tanks. The shield weighed 78 lb and permitted operation in gamma fields up to 2000 R/h. Horizontal traverses were made across the top of the tank at the 321 ft-9 in. elevation, and vertical traverses were made down both the north and south sides of the tank.

Two sets of data were taken, one set with the collimator opening plugged (background) and the other set with the collimater opening toward the tank (foreground). Geometrical source distributions are obtained by subtracting the background data from the foreground data. This difference is the response due only to the uncollided gamma-rays that enter through the collimator opening. The collimator limits the detector's field of view to small diameter regions, thereby allowing the relative source intensity distribution to be geometrically mapped. On the other hand, flux distributions are obtained from the background data only. The flux at any location is then a function of the total source within the cell, not just emission from a small region.

Figures 1 and 2 display the relative source intensity distributions of cesium-137 and cerium-144 from the horizontal and vertical traverses carried out in Cubicle A. These data reveal non-uniform cesium-137 and cerium-144 source distributions.

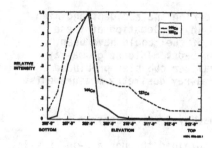 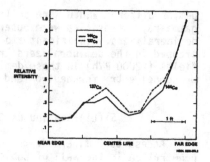

FIGURE 1. Relative Cerium-144
and Cesium-137 Source
Intensities from
Vertical Scan on
South Side of Demin-
eralizer A.

FIGURE 2. Relative Cerium-144
and Cesium-137 Inten-
sities from Horizontal
Scan ∿8 ft Over Demin-
eralizer A at 321 ft-
9 in. Elevation.

These horizontal source distributions are dramatically skewed with
the higher intensity toward the north side of the tank. The vertical distributions show cesium-137 and cerium-144 to be limited
to a region below the 309-ft elevation. There is virtually no
cerium-144 source above 309 ft, but some cesium-137 source is
present. The cesium-137 source above 309 ft may be due to residual contamination on the tank wall as the resin bed subsided
because of radiation damage and thermal degradation.

Background data obtained from the horizontal traverse were
used to determine the absolute content of cerium-144 in the tank.
Before the cerium-144 flux data could be used to determine the
amount of fuel in the demineralizer tank, it was necessary to
establish the amount of attenuating medium inside the tank. The
difference between a tank full of water and a dry tank results in
a significant difference, in fact up to two orders of magnitude,
in the calculation of the fuel content.

The shape of the Compton recoil electron spectrum from the
0.662-MeV cesium-137 gamma-ray was used to determine the water
equivalent attenuator in the tank. The foreground minus background spectrum at the 321 ft-9 in. elevation was used. This
spectrum was compared to spectra obtained from laboratory calibration experiments in which foreground minus background spectra
from a cesium-137 source were measured for various thicknesses of
water attenuator. In this way, it was demonstrated that the
cerium-144 source could be defined simply as a distributed source
in a water equivalent medium in the 2-ft region below the 309-ft
elevation. The vertical intensity distribution of the source has

already been shown in Figure 2. There is no additional attenuating medium above the 309-ft elevation.

Assuming the cerium-144 fission product does not migrate out of the fuel, the absolute activity of cerium-144 is directly related to the quantity of fuel present. Based on the observed source geometry and the measured absolute flux of the cerium-144 2.18-MeV gamma-rays, the fuel content of Demineralizer A was calculated to be 1.3 ± 0.6 kg. The principal factors contributing to the experimental error of this result are uncertainties in the attenuation coefficient of the cerium-144 2.18-MeV gamma-ray in both lead and water as well as the uncertainty in the cerium-144 fission product yield, which is based on fission product inventory calculations for TMI-2 fuel.

SSTR Neutron Dosimetry

Because of the intense gamma-ray fields present near the demineralizers, SSTR neutron dosimeters were attached to stringers that could be remotely positioned from outside the cubicle. Only Cubicle A was accessible. After a 29-day exposure, the SSTRs were transported to the National Reactor Dosimetry Center at the Hanford Engineering Development Laboratory (HEDL), where they were processed by etching with 49% HF at room temperature for 90 minutes. The developed tracks from selected dosimeters were then manually counted with the aid of a microscope. Figure 3 shows the location of the SSTR dosimeters, as well as the track density results obtained.

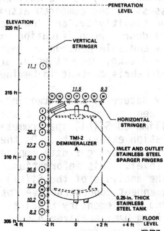

FIGURE 3. Location of SSTR Neutron Dosimeters on Horizontal and Vertical Stringers Remotely Positioned in Cubicle A. The underlined numbers in italics are the observed track densities in tracks/cm² at selected dosimeter locations.

Background measurements gave a track density of about 5 tracks/cm^2, whereas baseline measurements in Cubicle A were about 10 tracks/cm^2 (see Figure 3). This difference, ~5 tracks/cm^2, is due to room return neutrons. The room return response of the SSTR dosimeters was evaluated by calibration experiments in a concrete cubicle mockup at HEDL, using a californium-252 spontaneous fission source.

Based on this room return response, the SSTR neutron dosimetry result was 1.7 ± 0.6 kg of fuel in Demineralizer A. The dominant contributing factors to experimental error are the statistical uncertainties of the track density data, the uncertainty in the room return calibration constant, and the uncertainty in the neutron emission rate, which is based on actinide inventory calculations for the TMI-2 fuel.

In summary, the fuel debris content of Demineralizer A has been determined nondestructively by Si(Li) continuous gamma-ray spectrometry and SSTR neutron dosimetry. To our knowledge, gamma-ray spectrometry has never before been carried out under such adverse conditions, where the general radiation field intensity exceeded 2000 R/h. The track densities observed in the SSTR neutron dosimeters correspond to extremely low neutron flux intensities. In fact, the total neutron emission rate in Cubicle A, about 500 n/s, corresponds to an observed flux intensity on the order of 10^{-3} neutrons/cm^2·s, which is generally comparable with the intensity level of the cosmic-ray neutron flux at sea level. As a consequence, SSTR dosimetry is the only known method of neutron metrology possessing the combined attributes of passive applicability, extreme sensitivity, and low background response required for such fuel debris quantification experiments. Si(Li) gamma-ray spectrometry and SSTR neutron dosimetry results, namely 1.3 ± 0.6 kg and 1.7 ± 0.6 kg, respectively, are in excellent agreement for the fuel debris content in Demineralizer A.

The complementary nature of these two independent methods must be stressed. Indeed, source spatial distribution data obtained with the collimated Si(Li) spectrometer were used to guide the SSTR calibration experiments. On the other hand, SSTR evidence from the vertical stringer data implied that the tank was dry above the 309-ft level. This information, in turn, provided useful guidance for the analysis of the Si(Li) spectrometer data. Hence, these two independent nondestructive dosimetry methods provided concordant and complementary results. Some six months later, samples taken from Demineralizer A substantiated our conclusions that this tank was dry.

SSTR NEUTRON DOSIMETRY IN THE TMI-2 ANNULAR GAP

In more recent efforts, SSTR neutron dosimetry has been carried out in the TMI-2 reactor cavity (i.e., the annular gap between the pressure vessel and the biological shield) for characterization of the fuel distribution in the core region. The SSTR dosimeters used in the TMI-2 annular gap consisted of 1.91-cm diameter mica track recorders and asympototically thick (~0.0127 cm) uranium-235 foil. Two axial stringers were deployed in the annular gap, with 17 SSTR dosimeters located on each stringer. Of the 17 SSTR dosimeters, 14 were bare (i.e., aluminum covered) and 3 were cadmium covered. Axial locations (elevations) of these SSTR extended from the nozzles well above the core to the flow distributor plate well below the core. The region around the nozzles was of interest, since it has been speculated that some fuel debris might be lodged in the inlet or outlet nozzles.

Asimuthal locations of these two stringers, the east (E-SSTR) stringer and the west (W-SSTR) stringer, were chosen near the source range monitors (SRM). The count rates of the SRMs are roughly an order of magnitude higher than normal. The location of the SSTR stringers was chosen so that some insight into the origin of this high count rate might be provided by the SSTR dosimetry data.

These SSTR stringers were exposed for ~3 weeks in the TMI-2 annular gap, from August 19, 1983, until September 9, 1983. After retrieval, they were shipped to HEDL, processed, and scanned manually by optical microscopy (6).

The absolute thermal neutron flux obtained from these SSTR data are plotted in Figure 4 in comparison with the thermal flux anticipated for the TMI-2 reactor cavity. The curves, labeled M=2 and M=4, correspond to a core multiplication of 2 and 4, respectively. These curves were obtained by scaling of radiometric dosimetry conducted in the ANO-1 reactor cavity (7-8), a Babcock and Wilcox (B&W) plant of similar design to TMI-2.* Normal shutdown neutron multiplication for such a B&W plant is M=12. However, the high concentration of borated water and the redistribution of the core lower the multiplication at TMI-2 down to the approximate range: $2 < M < 4$.

These TMI-2 annular gap results differ significantly in shape and magnitude from the anticipated thermal flux. The TMI-2 thermal flux intensity exceeds the anticipated intensity by roughly an

---

*References 7 and 8, to be presented at this symposium, provide a more detailed description of these radiometric dosimetry experiments.

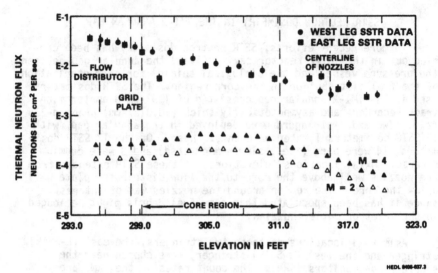

FIGURE 4. Thermal Neutron Fluxes in the TMI-2 Annular Gap.

order of magnitude at high elevations, and this difference grows with decreasing elevation to more than two orders of magnitude at the flow distributor elevation. In contrast with the axial symmetry expected about midplane, as is observed in the ANO-1 radiometric dosimetry data, the SSTR data for TMI-2 is clearly asymmetric.

Consequently, the SSTR data reveal the existence of a neutron transport phenomenon that dominates the neutron intensity and is not predicted by scaling measurements or calculations for a normal core configuration. The SSTR vertical profile of the neutron intensity is consistent with the presence of a significant amount of fuel debris, equivalent to several fuel assemblies or more, lying at the bottom of the reactor vessel. Neutrons from this quantity of fuel can pass essentially unmoderated out of the reactor vessel into the concrete cavity beneath the vessel, where they are moderated within the concrete and stream upward through the annular space between the vessel and the biological shield.

On the basis of this neutron streaming explanation of SSTR results in the TMI-2 annular gap, one would expect the count rate of the SRMs to be considerably higher than normal. The very poor signal-to-background ratio that exists at the SRM because of this neutron streaming seriously limits their applicability for in-core defueling operations.

618

REFERENCES

1. R. Gold, F. H. Ruddy, J. H. Roberts, C. C. Preston, J. A. Ulseth, W. N. McElroy, F. J. Leitz, B. R. Hayward and F. A. Schmittroth, "Application of Solid-State Track Recorder Neutron Dosimetry for Three Mile Island Unit 2 Reactor Recovery," Nucl. Tracks 7, 13-30 (1983).

2. J. P. McNeece, B. J. Kaiser, R. Gold and W. W. Jenkins, Fuel Content of the Three Mile Island Unit 2 Makeup Demineralizers, HEDL-7285, Hanford Engineering Development Laboratory, Richland, WA (February 1983).

3. F. H. Ruddy, J. H. Roberts, R. Gold, C. C. Preston and J. A. Ulseth, Solid-State Track Recorder Neutron Dosimetry Measurements for Fuel Debris Assessment of TMI-2 Makeup and Purification Demineralizer A, HEDL-TC-2492, Hanford Engineering Development Laboratory, Richland, WA (October 1983).

4. F. H. Ruddy, J. H. Roberts, R. Gold and C. C. Preston, "Applications of Solid-State Track Recorder Neutron Dosimetry for Fuel Debris Location in the Three Mile Island Unit 2 Makeup and Purification Demineralizers," 12th International Conference on Solid State Nuclear Track Detectors, Acapulco, Mexico, September 4-10, 1983.

5. R. Gold, J. H. Roberts, J. P. McNeece, B. J. Kaiser, F. H. Ruddy, C. C. Preston, J. A. Ulseth, and W. M. McElroy, "Fuel Debris Assessment for Three Mile Island Unit 2 (TMI-2) Reactor Recovery by Gamma-Ray and Neutron Dosimetry," 6th International Conference on Nondestructive Evaluation in the Nuclear Industry, Zurich, Federal Republic of Germany, November 27-December 2, 1983.

6. ASTM Standard E854-81, "Application and Analysis of Solid State Track Recorder (SSTR) Monitors for Reactor Vessel Surveillance," 1981 Annual Book of ASTM Standards, Part 45, "Nuclear Standards," American Society for Testing and Materials, Philadelphia, PA, 1981.

7. C. O. Cogburn, J. B. Williams and N. Tsoulfanidis, "Pressure Vessel Dosimetry at U.S. PWR Plants," 5th International ASTM-EURATOM Symposium on Reactor Dosimetry, Geesthacht, Federal Republic of Germany, September 24-28, 1984.

8. T. H. Newton Jr, C. O. Cogburn and J. G. Williams, "Use of Stainless Steel Flux Monitors in Pressure Vessel Surveillance," 5th International ASTM-EURATOM Symposium on Reactor Dosimetry, Geesthacht, Federal Republic of Germany, September 24-28, 1984.

A STUDY OF THE EMBRITTLEMENT OF REACTOR VESSEL

STEEL SUPPORTS

W. C. Hopkins

Bechtel Power Corporation

Gaithersburg, MD, USA

W. L. Grove

Science Applications
International Corporation
McLean, VA, USA

ABSTRACT

Neutron transport analyses have been done on three standard USA design pressurized water reactors to determine the fluence levels at the vessel supports. The shift in the nil-ductility transition temperature (ΔNDTT) at the supports has been estimated directly with the calculated fluence levels in accordance with ASTM Standard E706. Values of the displacement per atom unit (dpa) are also provided as an alternate estimate of the ΔNDTT.

The results indicate that supports with relatively high copper contents experience a large transition temperature shift. Further, the use of dpa values with E>1.0 MeV predict a higher transition temperature shift than using fluence with E>1.0 Mev.

1.0   CALCULATED NDTT SHIFT AT THE PWR SUPPORTS

The calculated NDTT shifts at the PWR supports are presented in Figures 1 through 3 as contours of iso-ΔNDTT as a function of distance from the core centerline for each model. These plots were obtained with the results of the DOT 4 (1) calculations and the following equation from ASTM Standard E706-IIF (2):

$$\Delta C_v 30_M = \left(\frac{\phi}{10^{19}} \right)^K (a+b\ Cu) \qquad \text{Eq. (1.1)}$$

J. P. Genthon and H. Röttger (eds.), Reactor Dosimetry, 621–628.
© 1985 ECSC, EEC, EAEC, Brussels and Luxembourg

where,

$\Delta C_v 30_M$ = the mean shift in degrees Farenheit of the NDTT as measured at the C_v 30 ft-lbf level

ϕ = neutron fluence (E>1 MeV) (neutrons/cm^2)

K = 0.31

a = 4.0

b = 540.0

Cu = copper content in weight percent (0.05 for Cu<.05, 0.4 for Cu>.4, value for .5<Cu<.4).

2.0   NSSS MODEL 1

As can be seen in Figure 1, the ΔNDTT at the PWR supports (cavity) ranged from 110°F at the core midplane to 28°F at the top of the pressure vessel. When this shift is combined with the support steel's initial NDTT of 50°F and copper content of .4%, the expected NDTT at the support after 32 years of operation varies from 160°F to 78°F.

3.0   NSSS MODEL 2

Figure 2 presents the iso-ΔNDTT contours at the PWR supports (cavity) for this model. In this case the shift ranged from 34°F at the core midplane to 10°F at the top of the pressure vessel. Since the initial NDTT is 0°F with .1% copper content for this steel, this temperature range is the expected NDTT after 32 years of operation. This is not as significant an increase as for NSSS Model 1.

4.0   NSSS MODEL 3

The iso-ΔNDTT contours at the PWR supports (cavity) for this model are shown in Figure 3. For this model the ΔNDTT ranged from 14°F at the core midplane to 6°F at the top of the pressure vessel. The NDTT for this steel, which has .05% copper and an initial NDTT of 10°F is 24°F to 16°F after 32 years of service. Of the three PWR models, this one has the smallest NDTT increase.

5.0 ΔNDTT AT PWR SUPPORTS BASED ON DPA (E>1.0 MeV)

The results printed in this section and shown in Table 2 are from calculations done using dpa (E>1.0 Mev) to predict the ΔNDTT at the supports of the PWRs. This was accomplished by utilizing the data in Table 1 (2). Although the data in this table is for the PCA 12/13 configuration, it is the only data of its type available at this time.

6.0 DISCUSSION OF RESULTS

Based on the DOT-4.2 calculated fluence (E>1.0 MeV), the estimated ΔNDTT for the NSSS-1 model's supports ranges from 110°F at the core midplane to 28°F at the top of the pressure vessel. For the NSSS-2 model the shift ranges from 23°F to 10°F; from 11°F to 6°F for the NSSS-3. Based on the calculated dpa (E>1.0 MeV) and the data from Table 1, the estimated ΔNDTT at the supports of the NSSS-1 ranges from 127°F at the core midplane to 39°F at the top of the pressure vessel. For the NSSS-2 model the shift ranges from 26°F to 15°F; from 13°F to 7°F for the NSSS-3 model. The difference in the NDTT shifts calculated with fluence (E>1.0 MeV) and dpa (E>1.0 MeV) indicates that dpa might be a better neutron exposure parameter (3).

7.0 REFERENCES

1. W. A. Rhodes and R. L. Childs, An Updated Version of DOT 4 One-and Two-Dimensional Neutron/Photon Transport Code, Oak Ridge National Laborary, Report No. ORNL-5851, 1982.

2. ASTM, "Standard Master Matrix for Light-Water Reactor Pressure Vessel Surveillance Standards," ASTM Designation E706-81a, ASTM Standards, Part 45, 1982.

3. J. D. Varsick, S. M. Schloss, and J. J. Koziol, "Evaluation of Irradiation Response of Reactor Pressure Vessel Materials," EPRI-NP-2720, November 1982.

TABLE 1

Change in dpa Exposure for a Fluence of 1.0×10^{19} a/cm^2 (E>1.0 MeV)

(In going from PCA/PSF[a] surveillance capsule location through the pressure vessel wall into the ex-vessel void box cavity)

Position	Fluence (E>1.0 MeV) n/cm^2	dpa[c]	$\left[\dfrac{dpa(<1.0)}{dpa(Total)}\right]$[c]	$\left[\dfrac{\phi(>0.1)}{\phi(>1.0)}\right]$
Surveillance	1.0×10^{19}	0.0151(1.00)[b]	0.23	2.2
Incident PV	1.0×10^{19}	0.0158(1.05)	0.32	2.8
1/4T	1.0×10^{19}	0.0178(1.18)	0.42	3.6
1/2T	1.0×10^{19}	0.0205(1.36)	0.54	5.0
3/4T	1.0×10^{19}	0.0243(1.61)	0.63	6.5
In Cavity	1.0×10^{19}	0.0252(1.67)	0.65	6.9
PWR 1/4T	1.0×10^{19}	0.0166(1.10)	0.42	3.1
Test reactor core	1.0×10^{19}	0.0141(0.93)	0.17	1.9

[a]For the 12/13 configuration.

[b]Normalization point.

[c]For these calculations, the DPA cross section below 0.1 MeV has been set equal to zero because of uncertainties associated with the computation of the low energy flux. The actual DPA exposure, therefore, in the PV positions and in the cavity would be somewhat, but not significantly, higher.

(Reference 2)

TABLE 2

ΔNDTT at PWR Supports Based on φ(E>1.0 MeV) and dpa (E>1.0 MeV)

PWR	A DOT-4.2 Fluence (E>1 MeV) (neutrons/cm²)*	C dpa	ΔNDTT(A)	B Table 1 Fluence** (neutrons/cm²)	ΔNDTT(B)
NSSS Model 1					
Core Midplane	1.08(18)	9.03(−4)	110°F	1.71(18)	127°F
Top of Pressure Vessel	1.24(16)	2.06(−5)	28°F	3.90(16)	39°F
NSSS Model 2					
Core Midplane	4.86(17)	3.96(−4)	23°F	7.50(17)	26°F
Top of Pressure Vessel	3.70(16)	6.81(−5)	10°F	1.29(17)	15°F
NSSS Model 3					
Core Midplane	3.08(17)	2.92(−4)	11°F	5.53(17)	13°F
Top of Pressure Vessel	4.89(17)	5.02(−5)	6°F	9.51(16)	7°F

*Assumed 40-year plant life and 80% capacity factor
**B = 1.89(21)*C

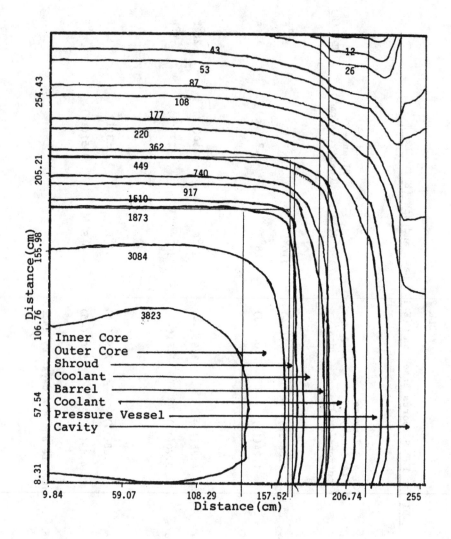

FIGURE 1

NSSS Model 1 Iso–ΔNDTT Contours (°F)

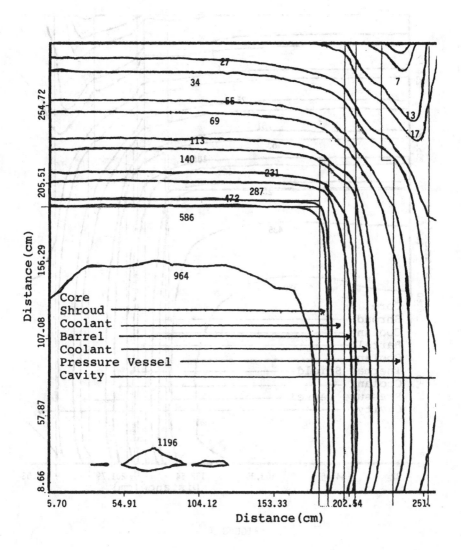

FIGURE 2

NSSS Model 2 Iso-ΔNDTT Contours ($^\circ$F)

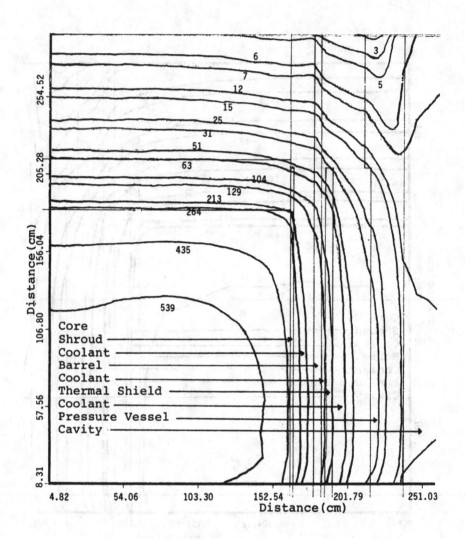

FIGURE 3

NSSS Model 3 Iso-ΔNDTT Contours (°F)

EVALUATION OF NEUTRON EXPOSURE

CONDITIONS FOR THE BUFFALO REACTOR

E. P. Lippincott, L. S. Kellogg and
W. N. McElroy

Westinghouse Hanford Company
Hanford Engineering Development Laboratory
Richland, WA USA

C. A. Baldwin

Oak Ridge National Laboratory
Oak Ridge, TN USA

ABSTRACT

A dosimetry irradiation was carried out at the
Buffalo Reactor in order to provide exposure data
under carefully controlled conditions. The data will
provide a basis for correlation of calculated neutron
exposure values with the experimental measurements.
The dosimetry irradiation was performed in a mockup
of an actual metallurgical assembly, which was carefully measured to ensure correct dimensions would be
used in the calculation. The calculation was performed using transport theory and detailed estimates
of fuel burnup throughout the core. Comparisons of
the calculation and experimental measurements indicate
good agreement and that exposure parameter values can
be determined to within 10 to 15% under typical
conditions.

J. P. Genthon and H. Röttger (eds.), Reactor Dosimetry, 629–637.
© *1985 ECSC, EEC, EAEC, Brussels and Luxembourg*

INTRODUCTION

The light water test reactor at the Nuclear Science and Technology Facility of the State University of New York at Buffalo is currently being used to irradiate specimens in in-core positions for NRC-sponsored metallurgical tests. It is important that the neutron exposures for these Buffalo tests be consistent with those determined for related irradiations in the Bulk Shielding Reactor (BSR) and Oak Ridge Reactor (ORR) reactors at Oak Ridge National Laboratory (ORNL). Therefore, a dosimetry irradiation, using Hanford Engineering Development Laboratory (HEDL) National Reactor Dosimetry Center dosimetry procedures and a neutronics calculation using ORNL calculational procedures were combined for an evaluation of a typical test condition.

DESCRIPTION OF EXPERIMENT

The Buffalo Reactor consists of a 6 x 6 array of elements (Figure 1). Of these elements, 30 are fuel, each containing 25 pins of 6% enriched uranium. The burnup in MWd/tonneU of each fuel element is indicated in Figure 1. The remaining six locations contain experiments, startup source, etc.; four of these locations are shown as blank in Figure 1 and consisted mostly of water at the time of the dosimetry irradiation.

The reactor is controlled by six control blades of 80% silver, 15% indium, and 5% cadmium that move in shroud tubes between Rows 4 and 5 and 2 and 3 as shown in Figure 1. These blades were all at least 90% withdrawn for the dosimetry irradiation and very little movement occurs during a typical irradiation because of the slow burnup of the core.

Two locations in the reactor are used for metallurgical irradiations, namely B-4 and C-2. The dosimetry irradiation was carried out in B-4 and lasted 48 hours. It was immediately adjacent in time to a metallurgical irradiation in the same location for which accurate fluence data was desired.

The metallurgical experiment was irradiated in an "ABC block," which consists of three separate steel sections as shown in Figure 2. Note that the top two sections (A and B) are smaller and have a water gap on one side to allow instrument tubes to carry thermocouple leads from the experiment.

The dosimetry pin was irradiated in a dummy ABC block with a hole down the center to contain the pin. The pin contained three cadmium-covered dosimetry spectral sets (S) positioned at the center of each of the three sections (see Figure 3). Bare gradient sets (G) were placed near the top and bottom of each section. The

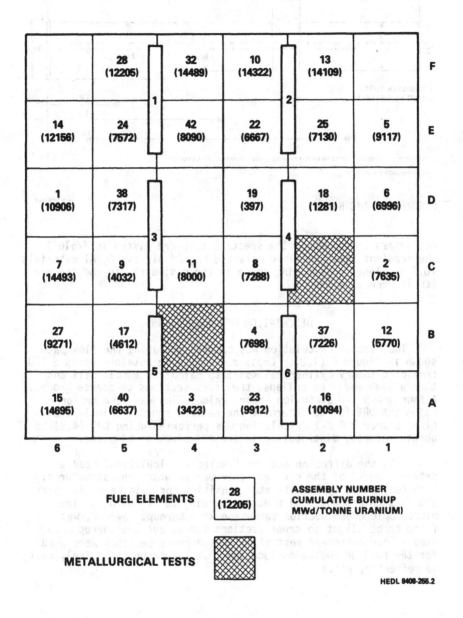

FUEL ELEMENTS

28
(12205)

ASSEMBLY NUMBER
CUMULATIVE BURNUP
MWd/TONNE URANIUM)

METALLURGICAL TESTS

HEDL 8408-255.2

FIGURE 1. Buffalo Reactor Core Layout.

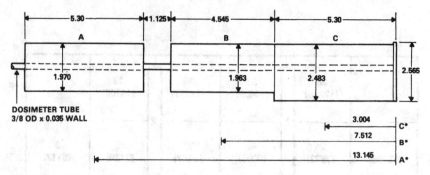

DOSIMETER TUBE
3/8 OD x 0.035 WALL

A
1.970

B
1.963

C
2.483

5.30
1.125
4.545
5.30

2.565

3.004   C*
7.512   B*
13.145   A*

*DISTANCE TO CENTER OF SPECIMEN GROUP (NOMINAL) AND CENTER OF DOSIMETRY SET

ALL DIMENSIONS IN INCHES

FIGURE 2. ABC Block.   HEDL 8408-256.3

dosimeters contained in the spectral sets are listed in Table 1;
the gradient sets contained only the Fe, Co/Al, and Ag/Al materials
(0.5 in. long). Remaining space in the dosimetry pin and the pin
itself were aluminum.

DESCRIPTION OF CALCULATION

The reactor calculation was carried out using the flow path
shown in Figure 4 (1,2). Cross sections were obtained using a 1-D
transport theory cylindrical model to calculate a fuel cell and
then a slab model to collapse the cross sections to coarse groups.
A four-group 3-D diffusion theory calculation was then performed
using VENTURE (3) to determine the neutron source. Finally, a
fixed source 2-D X-Y calculation was performed using DOT (4,5) to
obtain the flux distribution in the ABC block at midplane.

Both the diffusion and the transport calculations used a
detailed model of the core and surrounding water and structures.
Each fuel assembly had its atom densities for both heavy elements
and fission products calculated according to its burnup. The
microscopic cross section variation with burnup, however, was
found to be slight so cross sections for an average burnup were
used. Four different sets of collapsed cross sections were used
for the fuel depending on location (i.e., surrounded by fuel, next
to reflector, etc.).

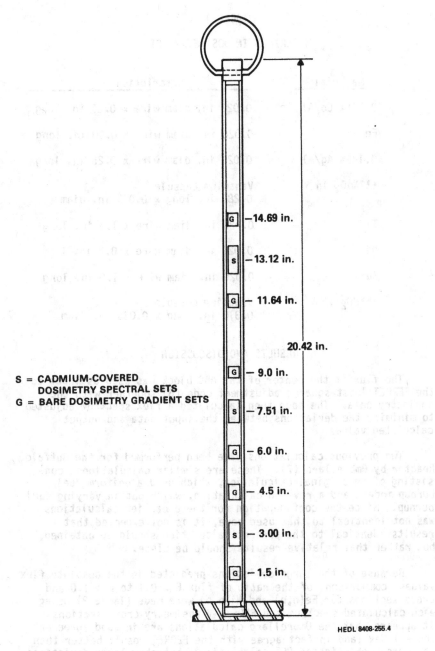

S = CADMIUM-COVERED
    DOSIMETRY SPECTRAL SETS
G = BARE DOSIMETRY GRADIENT SETS

— 14.69 in.

— 13.12 in.

— 11.64 in.

20.42 in.

— 9.0 in.

— 7.51 in.

— 6.0 in.

— 4.5 in.

— 3.00 in.

— 1.5 in.

HEDL 8408-255.4

FIGURE 3. Dosimetry Pin.

TABLE 1

MATERIALS IN DOSIMETRY SETS

Material	Description
0.116% Co/Al	0.020 in. diam wire x 0.25 in. long
Fe	0.020 in. diam wire x 0.50 in. long
0.145% Ag/Al	0.020 in. diam wire x 0.25 in. long
$^{237}NpO_2$ in V	Vanadium capsule 0.280 in. long x 0.035 in. diam
Ti	0.020 in. diam wire x 1.5 in. long
Ni	0.020 in. diam wire x 0.2 in. long
Cu	0.020 in. diam wire x 1.0 in. long
$^{238}UO_2$ in V	Vanadium capsule 0.310 in. long x 0.035 in. diam

RESULTS AND DISCUSSION

The flux at the center of the ABC block was used as input to the FERRET least-squares adjustment code (6) together with the dosimetry data. The code then calculated a flux spectrum adjusted to minimize the deviations between the input data and output calculated values.

Two previous calculations have been performed for the Buffalo Reactor by Ombrellaro (7). These are similar calculations, consisting of an original calculation, which used a uniform fuel burnup model, and a revised calculation, which put in varying fuel burnup. Since the configuration for these earlier calculations was not identical to that used here, it is not expected that results identical to the present calculation should be obtained, but rather that relative results should be close.

Because of the large deviations predicted in the absolute flux values, comparisons of the ratio of flux E > 0.1 to E > 1.0 and cross sections for Fe(n,p) and Ni(n,p) were made (Table 2) using each calculated spectrum and ENDF/B-V dosimetry cross sections. It appears that the Ombrellaro calculations are in good agreement above 1 MeV (and in fact agree with the FERRET result better than the more sophisticated DOT calculation), but show large deviations

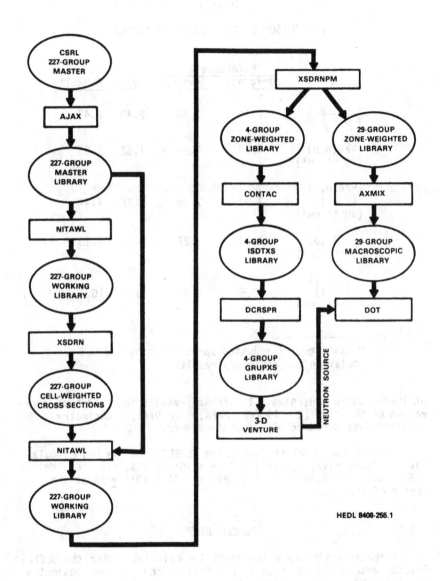

FIGURE 4. Cross-Section Processing and Calculational Path.

HEDL 8408-255.1

TABLE 2

COMPARISON OF BUFFALO CALCULATIONS

	Reference 7 Original	Revised	DOT	FERRET*
$\dfrac{\phi\ E > 0.1}{\phi\ E > 1.0}$	2.76	3.56	2.40	2.47
$\sigma Fe(n,p)$ $(10^{-26}\ cm^2)$	1.99	0.94	1.82	1.63
$\sigma Fe(n,p)$ $(E > 1)$ $(10^{-26}\ cm^2)$	7.93	7.06	8.37	7.47
$\sigma Ni(n,p)$ $(10^{-26}\ cm^2)$	2.64	1.27	2.43	2.20
$\sigma Ni(n,p)$ $(E > 1)$ $(10^{-26}\ cm^2)$	10.5	9.57	11.1	10.1

*Least-squares adjusted spectrum using DOT cal-
culation and dosimetry results.

at lower neutron energies. The FERRET results indicate that derivation of the flux (E > 1) from Fe(n,p) or Ni(n,p) dosimetry
results gives a reasonably accurate answer.

Results for absolute flux using FERRET with the DOT calcula-
tion as input gives ϕ(total) = 4.76 x 10^{13} + 15%, ϕ(E > 0.1 MeV) =
2.57 x 10^{13} + 11%, and ϕ(E > 1 MeV) = 1.04 x 10^{13} + 5% at the
center of the block.

CONCLUSIONS

A recent dosimetry measurement and calculation for the Buffalo
Reactor determined the flux (E > 1 MeV) at the midplane dosimetry
point to ∼5%. This indicates that accuracies of fast flux in
metallurgical capsules with dosimetry can typically be determined
to 10% accuracy when spatial variations are calculated. Previously used, but untested, calculations of the Buffalo Reactor are
indicated to be adequate for determining the fast flux from Fe or
Ni dosimetry measurements.

ACKNOWLEDGMENTS

The authors would like to thank Russ Hawthorne (MEA) and Martin Haas (ENSA) for helping with the experiment and for providing details of the experiment design and dimensions, the staff of the Buffalo Reactor for providing details on the reactor configuration and material composition, Brian Worley (ORNL) for calculating the isotopic changes with burnup, and W. Y. Matsumoto (HEDL) and J. M. Ruggles (HEDL) for assisting with the radiometric analysis. Helpful discussions with Mark Williams (LSU) and F. B. K. Kam (ORNL) are also acknowledged.

REFERENCES

1. W. E. Ford III, B. R. Diggs, L. M. Petrie, C. C. Webster, R. M. Westfall, CSRL-V: Processed ENDF/B-V 227-Group and Pointwise Cross-Section Libraries for Criticality Safety, Reactor, and Shielding Studies, NUREG/CR-2306, ORNL/CSD/ TM-160, Oak Ridge National Laboratory, Oak Ridge, TN (June 1982).

2. N. M. Greene, W. E. Ford III, J. L. Lucius, J. E. White, L. M. Petrie and R. Q. Wright, AMPX-II: A Modular Code System for Generating Coupled Multigroup Neutron-Gamma-Ray Cross-Section Libraries from Data in ENDF Format, ORNL/TM-3706, Oak Ridge National Laboratory, Oak Ridge, TN (December 1978).

3. D. R. Vondy, T. B. Fowler and G. W. Cunningham IV, The BOLD-VENTURE Computation System for Nuclear Reactor Core Analysis, Version III, ORNL-5711, Oak Ridge National Laboratory, Oak Ridge, TN (June 1976).

4. G. C. Haynes, The AXMIX Program for Cross-Section Mixing and Library Arrangement, RSIC-PSR-75, Oak Ridge National Laboratory, Oak Ridge, TN (June 1976).

5. W. A. Rhoads and R. L. Childs, An Updated Version of the DOT-IV One- and Two-Dimension Neutron/Photon Transport Code, ORNL-5851, Oak Ridge National Laboratory, Oak Ridge, TN (July 1982).

6. F. A. Schmittroth, FERRET Data Analysis Code, HEDL-TME 79-40, Hanford Engineering Development Laboratory, Richland, WA (September 1979).

7. P. A. Ombrellaro, R. A. Bennett, E. P. Lippincott and C. L. Long, "Buffalo Light Water Reactor Calculations," Dosimetry Methods for Fuels Cladding and Structural Materials, Proceedings of the Third ASTM-EURATOM Symposium on Reactor Dosimetry, EUR 6813, 589 (1980).

RECENT PROGRESS AND DEVELOPMENTS IN LWR-PV CALCULATIONAL
METHODOLOGY*

R. E. Maerker, B. L. Broadhead and M. L. Williams[†]

Oak Ridge National Laboratory

Oak Ridge, TN 37831

ABSTRACT

New and improved techniques for calculating beltline surveillance activities and pressure vessel fluences
with reduced uncertainties have recently been developed.
These techniques involve the combining of monitored in-
core power data with diffusion theory calculated pin-by-
pin data to yield absolute source distributions in R-θ
and R-Z geometries suitable for discrete ordinate transport calculations. Effects of finite core height, whenever necessary, can be considered by the use of a three-
dimensional fluence rate synthesis procedure. The
effects of a time-dependent spatial source distribution
may be readily evaluated by applying the concept of the
adjoint function, and simplifying the procedure to such
a degree that only one forward and one adjoint calculation are required to yield all the dosimeter activities
for all beltline surveillance locations at once. The
addition of several more adjoint calculations using
various group fluence rates as responses is all that is
needed to determine all the pressure vessel group
fluences for all beltline locations for any arbitrary
source distribution.

*Work performed for the Electric Power Research Institute under
research project 1399-1 with the Oak Ridge National Laboratory
and under Interagency Agreements 40-551-75 and 40-552-75, NRC FIN
No. B0415, and with the Martin Marietta Energy Systems, Inc.
under contract DE-AC05-840R21400 with the U.S. Department of Energy.
†Louisiana State University, Baton Rouge, LA 70803.

J. P. Genthon and H. Röttger (eds.), Reactor Dosimetry, 639–648.
© *1985 ECSC, EEC, EAEC, Brussels and Luxembourg*

INTRODUCTION

The original concept of the LEPRICON adjustment procedure (1,2,3) has been enlarged to include methods for obtaining á priori values of calculated surveillance dosimeter activities and pressure vessel fluence rates in PWR systems (4,5). New and improved techniques have been developed to handle various aspects of the general transport process, and have been used in the analysis of experiments performed in test reactor environments (6,7,8) whenever applicable as well as to power systems.

ABSOLUTE NEUTRON SOURCE DISTRIBUTIONS

For analysis of PWR surveillance dosimetry, it has been found adequate in the analysis of the Arkansas Nuclear One-Unit I reactor (ANO-1) to use in-core assembly power data determined from self-powered neutron detectors (SPND) in conjunction with diffusion theory calculations that provide reasonable intra-assembly power distributions for the peripheral assemblies. Thus,

$$P(x,y,z,m) = P(z,m)[p(x,y,m)/ \int_{V_m} p(x,y,m)dV] , \qquad (1)$$

where $P(x,y,z,m)$ is the power spatial distribution for peripheral assembly m, $P(z,m)$ is the axial power distribution as measured by SPND's in assembly m, and $p(x,y,m)$ is the diffusion theory calculated horizontal midplane power distribution for assembly m. By defining Eq. (1) to hold over some coarse time interval Δt_j and multiplying it by $F_i(\Delta t_i)$, the "instantaneous" fraction of the SPND power of duration $\Delta t_i, \Delta t_i \epsilon \Delta t_j$, which is extracted from the power history and reflects power level swings during operation, a suitable approximation to the peripheral power distribution as a function of time may be written as

$$P(x,y,z,m,\Delta t_i) = F_i(\Delta t_i)P_j(z,m,\Delta t_j)[p_j(x,y,m)/ \int_{V_m} p_j(x,y,m)dV]. \quad (2)$$

Throughout cycles 4 and 5 of this reactor, which have been the only ones analyzed so far, the virtually time-independent pin-by-pin peripheral assembly distributions were such that the core leakage averaged about 15-20% less than that from a flat distribution normalized to the same assembly powers. Thus, neglecting the fall-off in pin-by-pin power with increasing radius for the peripheral assemblies would result in typical overestimates of the order of 20% in calculated pressure vessel fluences. The conversion factor relating neutron production rate to the power expressed in Eq. (2) is burn-up dependent because of different values for ^{239}Pu and ^{235}U. The fission spectrum is also somewhat different for these two isotopes. Both of these effects tend to increase the high-energy fluence with time over that obtained if they were neglected, but their combined effect is probably significant (>5%) only for low leakage cores in which partially burned fuel is used in the peripheral assemblies.

EFFECTS OF THE FINITE CORE HEIGHT

These effects and their treatment have been discussed elsewhere (6,7,8,9) for near-midplane locations, but only corrections for PWR reactor cavity positions probably need to be considered. For the PCA and PSF experiments, however, even calculations at surveillance and pressure vessel midplane locations need to be corrected, since the core height is relatively small compared to detector distances from the core. The procedure is based on combining results from three transport calculations. For PWR applications,

$$\phi_g(R,\theta,Z) = \phi_g(R,\theta)\phi_g(R,Z)/\phi_g(R) \quad , \tag{3}$$

where

$$\phi_g(R,\theta) \text{ is the solution using } S(R,\theta) = \int_{-\infty}^{\infty} S(R,\theta,Z)dZ \quad , \tag{4}$$

$$\phi_g(R,Z) \text{ is the solution using } S(R,Z) = \int_{0}^{2\pi} S(R,\theta,Z)Rd\theta \quad , \tag{5}$$

and

$$\phi_g(R) \text{ is the solution using } S(R) = \int_{0}^{2\pi} Rd\theta \int_{-\infty}^{\infty} S(R,\theta,Z)dZ \quad . \tag{6}$$

In Eqs. (4-6), $S(R,\theta,Z)$ is the three-dimensional neutron source distribution expressed in cylindrical coördinates.

In applications to ANO-1, and presumably many other reactors, the ratio $\phi_g(R,Z)/\phi_g(R)$ was assumed to be independent of the source distribution because all axial loadings were virtually identical.

EFFECTS OF TIME-DEPENDENT SPATIAL SOURCE DISTRIBUTIONS

Source spatial distributions in general are functions of time, and both intracycle and intercycle variations may be important as they can affect pressure vessel fluence. A method based on the concept of the adjoint function has been developed and simplified which allows one to obtain not only surveillance activities at the end of irradiation and pressure vessel fluences for any arbitrary source distribution without the time and expense of performing new transport calculations each time, but also to investigate effects on pressure vessel fluence alone from different core loadings and fuel shuffling schemes.

Turning our attention first to calculating the dosimeter saturated activities $RR^d(t)$,

$$RR^d(t) = \int_{detector} d\bar{r} \int_0^\infty \phi(E,\bar{r},t)\sigma^d(E)dE = \int_{core} d\bar{r} \int_0^\infty \phi_d^*(E,\bar{r})S(E,\bar{r},t)dE, \quad (7)$$

where ϕ is the "forward" calculated fluence rate, ϕ_d^* is the adjoint fluence rate using σ^d, the reaction cross section for dosimeter d, as the adjoint source, and S is the core source.

$$\text{Since } S(E,\bar{r},t) = \chi(E)S(\bar{r},t) \quad , \quad (8)$$

where χ is a normalized fission spectrum,

$$RR^d(t) = \int_{core} S(\bar{r},t)d\bar{r} \int_0^\infty \chi(E)\phi_d^*(E,\bar{r})dE \quad . \quad (9)$$

We may rewrite Eq. (9) as

$$RR^d(t) = \sum_{k=1}^8 \int_0^\infty \chi(E)dE \int_{octant} \phi_d^*(k,\bar{r},E)S(k,\bar{r},t)d\bar{r} \quad , \quad (10)$$

where k denotes a given octant of the core.

Most fuel loading patterns exhibit the octal symmetry property $S(k,\bar{r},t) = S(\bar{r},t)$ for all k, leading to

$$RR^d(t) = \sum_{k=1}^8 \int_0^\infty \chi(E)dE \int_{octant} \phi_d^*(k,\bar{r},E)S(\bar{r},t)d\bar{r} \quad . \quad (11)$$

Equation (11), requiring a model of the complete core, may be greatly simplified if the adjoint function is calculated modeling only the nearest octant of the core together with imposing a perfectly reflecting boundary condition on the two limiting rays of the octant (see Fig. 1). Since, by symmetry, Fig. 1 is equivalent to no reflective boundary condition and eight symmetrically placed sources about the core (broken crosses in Fig. 1), one has, in terms of the reflected adjoint function $\phi_{d,R}^*$,

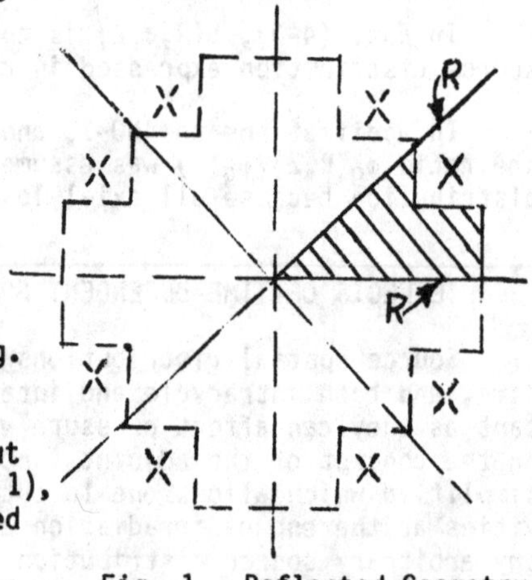

Fig. 1. Reflected Geometry Using One Octant of the Core

$$\int_0^\infty \chi(E)dE \int_{core} \phi_{d,R}^*(\bar{r},E)S(\bar{r},t)d\bar{r} = 8RR^d(t) , \tag{12}$$

and since all octants are identical,

$$\int_0^\infty \chi(E)dE \int_{octant} \phi_{d,R}^*(\bar{r},E)S(\bar{r},t)d\bar{r} = RR^d(t) . \tag{13}$$

Thus, subtracting Eq. (13) from Eq. (11) and interchanging the order of octant summing and integration,

$$\int_0^\infty \chi(E)dE \int_{octant} \{ \sum_{k=1}^8 \phi_d^*(k,\bar{r},E) - \phi_{d,R}^*(\bar{r},E)\}S(\bar{r},t)d\bar{r} = 0 , \tag{14}$$

and since $S(\bar{r},t)$ is arbitrary,

$$\sum_{k=1}^8 \phi_d^*(k,\bar{r},E) \equiv \phi_{d,R}^*(\bar{r},E) . \tag{15}$$

Thus, although the dosimeter location is in general asymmetric, the adjoint function readily calculated with reflective boundary conditions in octal geometry is the correct solution if the core loading is octally symmetric. The quantity $\chi(E)\phi_{d,R}^*(\bar{r},E)$ is the Green's function for the response function $\sigma^d(E)$ and a particular location of the detector.

Figure 2 illustrates the behavior of the ratio of the Green's functions of $^{237}Np(n,f)$ relative to $^{63}Cu(n,\alpha)$ for surveillance dosimeters located near the region of peak pressure vessel fluence in ANO-1. The five shaded quarter assemblies represent regions of greatest contribution to the dosimeter activities, and the near constancy of this ratio, which involves two completely different energy intervals, is to be observed. Similar conclusions can be inferred from the distributions in Fig. 3, where the dosimeters are located in the ANO-1 reactor cavity. One may conclude from Figs. 2 and 3 that the similar spatial dependence of the two Green's functions over the contributing part of the core for two such disparate response functions implies, by Eq. (13), the independence of the relative reaction rates with source distribution. Hence the relative fluence rates at a given surveillance or pressure vessel location also tend to be independent of spatial source distribution. Only the intensity of the total fluence rate is affected by the source distribution; the spectrum at a given location remains virtually independent of the source.

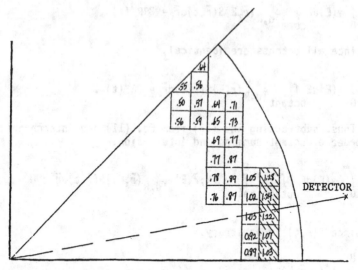

Fig. 2. Ratio of Weighted Source Importances of Quarter
Peripheral Core Assemblies Summed Over All Octants, ^{237}Np(n,f)/
^{63}Cu(n,α), in Units of 10^3. Detector at Surveillance Location in
Downcomer.

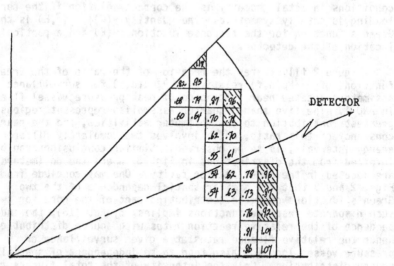

Fig. 3. Ratio of Weighted Source Importances of Quarter
Peripheral Core Assemblies Summed Over All Octants, ^{237}Np(n,f)/
^{63}Cu(n,α), in Units of 10^4. Detector in Reactor Cavity.

The constancy of the spectrum allows one to calculate all the dosimeter activities for a new source distribution by using only the one adjoint function and scaling the original dosimeter activities by the ratio of the new to the old activities obtained for the particular dosimeter whose response cross section was used in calculating the adjoint function. Thus the saturated activities $RR^d(t)$ and $RR^d(t_0)$ calculated from Eq. (13) using the sources $S(\bar{r},t)$ and $S(\bar{r},t_0)$ respectively may be used in conjunction with the activities $RR^d(t_0)$ obtained from a forward calculation performed earlier using the reference source $S(\bar{r},t_0)$ to provide new saturated activities $RR^d(t)$ for all the dosimeters:

$$RR^{\vec{d}}(t) = \{[RR^d(t)]_{adjoint}/[RR^d(t_0)]_{adjoint}\}[RR^{\vec{d}}(t_0)]_{forward} \quad (16)$$

As an illustration, we have calculated in the forward mode the activities in the ANO-1 cavity for each of two source distributions - one flat and the other modified by a pin-by-pin diffusion theory calculation. Taking the flat distribution as the new distribution and the pin-by-pin distribution as the original one, the new activities were calculated as well by scaling the original ones using each of two adjoint functions in Eq. (16). The results appear in Table 1.

Table 1. Comparison of Forward with Adjoint Scaled Saturated Activities in the ANO-1 Cavity for a Flat Intra-Assembly Source Distribution

Dosimeter	Using $^{63}Cu(n,\alpha)$ Adjoint	Using $^{237}Np(n,f)$ Adjoint	Forward	Discrepancy	
$^{63}Cu(n,\alpha)^{60}Co$	6.43-17†	6.74-17	6.43-17	0.0%	4.8%
$^{46}Ti(n,p)^{46}Sc$	8.85-16	9.29-16	8.92-16	1.0%	4.1%
$^{54}Fe(n,p)^{54}Mn$	4.86-15	5.10-15	4.97-15	2.3%	2.6%
$^{58}Ni(n,p)^{58}Co$	6.74-15	7.07-15	6.90-15	2.4%	2.5%
$^{238}U(n,f)F.P.$	2.83-14	2.98-14	2.93-14	3.5%	1.7%
$^{237}Np(n,f)F.P.$	4.92-13	5.17-13	5.17-13	5.1%	0.0%

†Read 6.43×10^{-17} reactions per second per atom, etc.

Table 1 indicates about the same error is incurred using either extreme adjoint function, but that using an intermediate one calculated from either the $^{54}Fe(n,p)$ or the $^{58}Ni(n,p)$ cross section should result in no worse than about a 2.5% error.

Even though the spectrum changes with detector location throughout the pressure vessel for a given source, no changes in these relative spectra are induced by changes in the source; i.e., these spectra are determined strictly by the neutron transport properties of the downcomer and pressure vessel. Hence, one can use Eq. (16) as well to use scaling factors at one location to obtain activities at another, as is illustrated in Table 2.

Table 2. Comparison of Forward with Downcomer Adjoint Scaled Saturated Activities in the ANO-1 Cavity for a Flat Intra-Assembly Source Distribution

Dosimeter	Using $^{63}Cu(n,\alpha)$ Adjoint*	Using $^{237}Np(n,f)$ Adjoint†	Forward	Discrepancy	
$^{63}Cu(n,\alpha)^{60}Co$	6.18-17‡	6.46-17	6.11-17	1.1%	5.8%
$^{46}Ti(n,p)^{46}Sc$	8.48-16	8.87-16	8.45-16	0.4%	4.9%
$^{54}Fe(n,p)^{54}Mn$	4.67-15	4.88-15	4.73-15	1.2%	3.3%
$^{58}Ni(n,p)^{58}Co$	6.47-15	6.76-15	6.57-15	1.6%	2.9%
$^{238}U(n,f)F.P.$	2.71-14	2.84-14	2.80-14	3.1%	1.5%
$^{237}Np(n,f)F.P.$	4.77-13	4.98-13	4.96-13	4.1%	0.4%

$$*(RR^d)_{FLAT,CAVITY} = \frac{\{(RR^{Cu})_{FLAT,SURV}\}_{ADJOINT}}{\{(RR^{Cu})_{PIN-BY-PIN,SURV}\}_{ADJOINT}} \cdot$$
$$\cdot \{(RR^d)_{PIN-BY-PIN,CAVITY}\}_{FORWARD}$$

†Similar to above with RR^{Np} replacing RR^{Cu}.
‡Read 6.18×10^{-17} reactions per second per atom, etc.

Combining Eqs. (2-6) with Eqs. (13) and (16) and decaying the resultant saturated activity to the end of irradiation yields the following expression for the activity when the power spatial distribution is a function of time:

$$A^{d'}(T) = e^{-\lambda_{d'}T}\left(\frac{RR^{d'}}{RR^d}\right) \cdot$$
$$\cdot \sum_{j=1}^{J} C_j \sum_{g=1}^{G} X_{g,j} \int_{octant} P_j(\bar{r})\phi^*_{g,d,R}(\bar{r})d\bar{r} \sum_{i=1}^{I(j)} F_i[1-e^{-\lambda_{d'}(t_{i+1}-t_i)}]e^{\lambda_{d'}t_{i+1}}, \tag{17}$$

where $A^{d'}(T)$ = Activity of dosimeter d' in dps/atom at end of irradiation,

$\lambda_{d'}$ = Decay constant of dosimeter d' in reciprocal days
T = Irradiation period in days,
$RR^{d'}$ = Saturated activity of dosimeter d' calculated in the forward mode using the fluence rate synthesis procedure expressed by Eq. (3) with some reference source,
RR^d = Saturated activity of dosimeter d calculated in the adjoint mode in R-θ geometry using the same reference source as above,

$\qquad$ J = Number of distinct relative spatial distributions used over T,

$\qquad$ C_j = Conversion factor relating neutron source and assembly power in units of neuts/sec per MW,

$\qquad$ $x_{g,j}$ = Relative number of neutrons in group g produced per fission event, normalized to unity,

$\qquad$ $\phi^*_{g,d,R}$ = Adjoint flux in group g (i.e., $g=1$ corresponds to lowest energy group) calculated in an adjoint octally reflected R-θ run using the response of dosimeter d, in disintegrations per atom per neutron per cm^3,

$\qquad$ P_j = The x,y power density distribution in the peripheral core integrated over z based on SPND data and supplemented by diffusion theory pin-by-pin calculations, in MW per cm^3,

$\qquad$ $I(j)$ = Number of subintervals describing "instantaneous" total power history for each SPND interval j,

$\qquad$ F_i = Fraction of total SPND power in fine time interval i due to "instantaneous" power level changes,

$\qquad$ t_i, t_{i+1} = Beginning and end of time subinterval i relative to beginning of irradiation, in days.

Although they will not be detailed here, similar considerations based on adjoint functions for the various group fluence rates in the pressure vessel can lead to expressions for the group fluences arising from time dependent spatial source distributions. In agreement with conclusions deduced from the behavior of the various reaction rates, the group fluence rates were found to scale with predictable accuracies ranging from two to five percent when the scaling takes into account, as it should, the dependence of the group fluence rates on the source fission spectrum as well (10).

REFERENCES

1. R. E. Maerker, J. J. Wagschal and B. L. Broadhead, Development and Demonstration of an Advanced Methodology for LWR Dosimetry Applications, EPRI NP-2188, Interim Report (1981).

2. R. E. Maerker, M. L. Williams, B. L. Broadhead, J. J. Wagschal and C. Y. Fu, Revision and Extension of the Database in the LEPRICON Dosimetry Methodology, Second Interim Report, to be published as an EPRI report (1984).

3. J. J. Wagschal, R. E. Maerker and B. L. Broadhead, "Surveillance Dosimetry: Achievements and Disappointments," Proc. Fourth ASTM-EURATOM Symposium on Reactor Dosimetry, Vol. 1, NUREG/CP-0029, CONF-82032/V1, NBS (1982).

4. R. E. Maerker, B. L. Broadhead, B. A. Worley, M. L. Williams and J. J. Wagschal, Application of the LEPRICON System to the ANO-1 Reactor, Third Interim Report, to be published as an EPRI Report (1985).

5. J. J. Wagschal, R. E. Maerker, B. L. Broadhead and M. L. Williams, "Unfolded ANO-1 Fluxes Using the LEPRICON Methodology," Proceedings this Conference.

6. W. N. McElroy, (Ed.), LWR Pressure Vessel Surveillance Dosimetry Improvement Program: PCA Experiments and Blind Test, NUREG/CR-1861, HEDL-TME 80-87, R5, Hanford Engineering Development Laboratory (1981).

7. R. E. Maerker and B. A. Worley, "Calculated Spectral Fluences and Dosimetry Activities for the Metallurgical Blind Test Irradiations at the ORR-PSF," Proceedings this Conference.

8. R. E. Maerker and B. A. Worley, Activity and Fluence Calculations for the Startup and Two-Year Irradiation Experiments Performed at the Poolside Facility, NUREG/CR-3886, ORNL/TM-9265 (1984).

9. R. E. Maerker and M. L. Williams, "Calculations of the Westinghouse Perturbation Experiment at the Poolside Facility," Proc. Fourth ASTM-EURATOM Symposium on Reactor Dosimetry, Vol. 1, NUREG/CP-0029, CONF-820321/V1, NBS (1982).

10. B. L. Broadhead and R. E. Maerker, "Sensitivities of the Flux Spectrum in the Cavity of a PWR to Variations in the Core Source Distribution," Trans. Am. Nucl. Soc. 46, 659 (1984).

A NEW ADJUSTMENT CODE BASED ON THE BAYES' THEORY

COMBINED WITH THE MONTE-CARLO TECHNIQUE

M.Nakazawa, N.Ueda, T.Taniguchi[*] and A.Sekiguchi

Department of Nuclear Engineering, University of Tokyo

[*]The Institute of Applied Energy, Tokyo, JAPAN

ABSTRACT

A new adjustment code using the Monte-Carlo technique, based on the Bayes' theory, has been developed to estimate the non-negative neutron flux and spectra from multiple activation detectors.

The Monte-Carlo technique is used to generate the initial multi-group neutron spectrum as the random number series from a priori probability density function (p.d.f.). And final neutron spectra and its variance-covariance matrix are obtained through statistical estimations of these random spectrum considering each conditional probability.

An emphasis is placed on the ability to estimate a posteriori p.d.f. rigorously for every differential flux and integral quantities, which are thought to be a little different from the standard normal distribution.

Using a lognormal distribution with J-log type unfolding method[1], this new adjustment code has been satisfactorily applied to a typical test problem to which ordinary algorithms are sometimes to give unphysical negative answers due to its large a priori uncertainty.

INTRODUCTION

General least-squares methods have been well developed for and applied to recent data evaluation methods, such as neutron unfolding problems.

J. P. Genthon and H. Röttger (eds.), Reactor Dosimetry, 649–656.

Some types of unfolding methods are based on Bayes' theory for the use of a priori information including uncertainties and correlations. And in these methods, it is a convenient way to use lognormal distribution as a priori p.d.f. for the problems in which several statistical variables have very large uncertainty and are needed to consider the physical constraint of non-negativeness [2].

In these non-linear least-squares methods, however, for the use of an asymmetric distribution, it is very difficult to appreciate the states of estimated values with only two parameters such as their mean values and variances (one sigma). The essential point of this paper is a development of a formal basis for the evaluation of a posteriori p.d.f. of neutron flux and integral quantities with the Monte-Carlo technique.

This evaluation of a posteriori p.d.f. can be obtained through statistical Bayesian estimation procedures considering a conditional probability weight for each initial neutron spectra sampled as random number series from a priori p.d.f..

At first, basic principles are introduced, then, the algorithm of this non-linear neutron adjustment method combined with the Monte-Carlo technique is described, in which some test calculations of sample problem is also presented.

BASIC PRINCIPLES

Generally, neutron unfolding problem is given in the following relation between the experimental reaction-rate data $R_i \pm \Delta R_i$ $(i=1,N)$ and the unknown spectra ϕ_g $(g=1,G)$,

$$R_i \pm \Delta R_i = \sum_{g=1}^{G} (\sigma_{ig} \pm \Delta \sigma_{ig})(\phi_g \pm \Delta \phi_g) , \tag{1}$$

where G is the maximum number of energy group and σ_{ig} means the g-th group reaction cross-section value of the i-th experimental reaction-rate R_i.

In condition that ϕ_g satisfying Eq. (1) exists, present problem is to estimate the linear integral quantity I and its uncertainty ΔI represented with a window function W_g $(g=1,G)$,

$$I = \sum_{g=1}^{G} W_g \phi_g . \tag{2}$$

In general least-squares methods, the Bayesian estimation theory gives one useful way to combine a priori information with the experimental relation of eq. (1) to obtain a posteriori values,

$$\Pi(\vec{\phi}|\vec{R},\vec{\phi}_0) = P(\vec{R}|\vec{\phi}) \cdot \Pi_0(\vec{\phi}|\vec{\phi}_0) , \tag{3}$$

where $\Pi_0(\vec{\phi}|\vec{\phi}_0)$ is a priori p.d.f. of initial spectra around $\vec{\phi}_0$, $P(\vec{R}|\phi)$ is a conditional p.d.f. which gives the weight of each initial spectra considering the agreement with the experimental reaction-rates $\vec{R}$, and $\Pi(\vec{\phi}|\vec{R},\vec{\phi}_0)$ is a posteriori p.d.f. which determines the probability of final spectra $\vec{\phi}$.

Monte-Carlo technique is used to obtain the trial neutron spectra as the random number series from a priori p.d.f. such as the truncated gaussian distribution, the poisson distribution and the lognormal distribution.

If neutron flux has no correlations between different energy groups, random numbers can be independently sampled using the simple Monte-Carlo technique. On the other hand, in case that variance-covariance matrix M_z of statistical values $\vec{z}$ includes non-zero value covariances, a symmetric matrix M_z must be transformed into diagonal matrix T with some unitary matrix U for the generation of correlated random number series,

$$M_z = UTU^t . \tag{4}$$

With these preparations, correlated random number series $\vec{z}$ can be generated by performing orthogonal transformation of uncorrelated random number series $\vec{y}$, which has a distribution function of the diagonal variance matrix T around $\vec{y}_0$,

$$\vec{z} = U\vec{y} , \tag{5}$$

$$\vec{y}_0 = U^t\vec{z}_0 . \tag{6}$$

The two types of a priori p.d.f. are tested for a new adjustment technique, which are the truncated gaussian distribution and the lognormal distribution.

The sampling method of random number series $\vec{z}$ from the truncated gaussian distribution consists of three steps;
  (1) generation of random number series $\vec{y}$ from the normal distribution,
  (2) orthogonal transformation from $\vec{y}$ into $\vec{z}$,
  (3) rejection of the random number series $\vec{z}$ including negative value components and adoption of $\vec{z}$ which satisfies non-negative condition.

The sampling efficiency of this method may be rather worth, because many random numbers series are finally rejected due to their including negative value components.

On the other hand, for a non-negativeness of random numbers can be always guaranteed, in the following formulations, the lognormal distribution is adopted as a priori distribution, which means efficient use of selected random numbers without any rejection.

MONTE-CARLO ADJUSTMENT TECHNIQUE

The first step of neutron adjustment method with Monte-Carlo technique begins with the generation of the initial random spectrum $\vec{\phi}^{(m)}$ as a random number series from a priori p.d.f., where (m) means the m-th trial.

As mentioned before, when a priori p.d.f. is given as the lognormal distribution, correlated logarithmic spectrum $z_g^{(m)}$ (g=1,G) should be generated by orthogonal transformations of uncorrelated random number series $y_g^{(m)}$ basing on eqs. (4)-(6), where mean value y_{0g} and variance T_{gg}.

With the crude Monte-Carlo technique, an enormous amount of trial spectrum may be needed. Practically, however, from the limitation of computer system, some additional method is requested to improve the efficiency of Monte-Carlo technique.

In this study, J-log type unfolding method in NEUPAC-83[1], based on a non-linear functional representation, is applied to a initial selected random spectra $\vec{\phi}^{(m)}$ for the modification,

$$\vec{\phi}'^{(m)} = J\text{-log}[\vec{\phi}^{(m)}] . \tag{7}$$

Then, the conditional probability $P(\vec{R}|\vec{\phi}'^{(m)})$ corresponding to $\vec{\phi}'^{(m)}$ can be given below, in which the Taylor's series expansion method is applied for the expression of f_i and $f_i^{(m)}$ assuming the lognormal distribution for the other statistical variables such as experimental reaction-rates and reaction cross-section values,

$$P(\vec{R}|\vec{\phi}'^{(m)}) = \eta_f \cdot \exp[-\tfrac{1}{2}(\vec{f}-\vec{f}^{(m)})^t M_f^{-1}(\vec{f}-\vec{f}^{(m)}) , \tag{8}$$

$$f_i = \ln R_i - \ln \sum_{g=1}^{G}\sigma_{ig}\phi_g^* , \tag{9}$$

$$f_i^{(m)} = \sum_{g=1}^{G}u_{ig}^*(z_g'^{(m)} - z_g^*) , \tag{10}$$

$$u_{ig}^* = \sigma_{ig}\phi_g^* / \sum_{g=1}^{G}\sigma_{ig}\phi_g^* , \tag{11}$$

where η_f is the normalization constant, M_f is a variance-covariance matrix of f, and ϕ_g^* is the Taylor's series expansion point.

In this new adjustment code, an emphasis is placed on the ability to estimate a posteriori p.d.f. for each group flux and for every integral quantities that are thought to be asymmetric distributions under physical constraint of non-negativeness.

Now, the expectation operator for the variable x can be defined by

$$\langle x \rangle = \sum_{m=1}^{M} P(\vec{R}|\vec{\phi}'^{,(m)}) \cdot x^{(m)} / \sum_{m=1}^{M} P(\vec{R}|\vec{\phi}'^{,(m)}) , \tag{12}$$

where M is the sum of trials.

Then, neutron group flux values and their variance-covariance matrix can be given as follows,

$$\phi_g = \langle \phi'_g \rangle , \tag{13}$$

$$\text{cov.}(\phi_g, \phi_k) = \langle \phi'_g \phi'_k \rangle - \langle \phi'_g \rangle \langle \phi'_k \rangle . \tag{14}$$

In a similar way, integral quantities and their correlations can be formulated.

Finally, a posteriori probability distribution function $P(I^{(s)})_{\Delta I^{(s)}}$ of integral quantity I, integrated its p.d.f. over the s-th separated interval $[I^{(s)}, \Delta I^{(s)}]$, can be represented using above expectation operator $\langle x \rangle$,

$$P(I^{(s)})_{\Delta I^{(s)}} = \langle \delta^{(s)}(I) \cdot I \rangle , \tag{15}$$

where $\delta^{(s)}(I)$ means a weighting-function, which gives a value of 1 or 0, whether $I^{(m)}$ is contained between the s-th interval of $[I^{(s)}, \Delta I^{(s)}]$ or not, respectively.

PRACTICAL APPLICATION

A sample problem is a dosimetry data analysis in the Japanese Fast Experimental Reactor "JOYO". Detail of the JOYO dosimetry experiments are described elsewhere[3]. The adjustment calculations are carried out using five kinds of reaction-rate data such as $^{45}Sc(n,\gamma)$, $^{46}Ti(n,p)$, $^{59}Co(n,\gamma)$, $^{63}Cu(n,\alpha)$ and $^{237}Np(n,f)$, where input neutron spectra calculated by DOT-3.5 is given in 20 energy group numbers which are expanded into 103 energy group numbers for the cross-section library in NEUPAC-83 code[1]. And an initial 103 x 103 variance-covariance matrix is defined with following relation[4],

$$\text{cov.}(\phi_{0g}, \phi_{0k}) = V_g V_k \cdot P_{gk} \tag{16}$$

$$P_{gk} = (1-\theta) \cdot \delta_{gk} + \theta \cdot \exp[-(g-k)^2/2\gamma^2] \tag{17}$$

where V_g is the variance of the g-th group flux, θ is the variance-covariance ratio and γ is the group correlation factor. In this test calculation $\theta = 0.8$ and, $\gamma = 2.0$ are used.

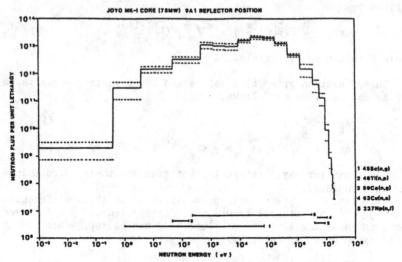

NEUTRON SPECTRUM

JOYO MK-I CORE (75MW) 9A1 REFLECTOR POSITION

Fig. 1   Ajusted Neutron Spectra of Japanese Fast
Experimental Reactor "JOYO"

Adjusted   neutron   spectra is shown in Fig. 1, where 90 %
confidence level of each reaction cross section is also presented.
The present J-log Monte-Carlo result does not   show remarkable
difference from the J1 type adjustment method in NEUPAC-83 based
on linear functional representation[1].

The   estimated uncertainty of iron dpa are shown   in Fig. 2,
where the NEUPAC J-1 method can be found to give a great
overestimated answer, in comparison with the present J-log Monte-
Carlo method.   The   more   efficient corrections for   the
overestimation are found in the case of the large a priori
uncertainty.

A posteriori probability distribution function of iron 'dpa
values are typically presented in Fig. 3, in which the result of
J-log Monte-Carlo is obtained after 100,000 trials of random
selected spectrum for the initial uncertainty of 200 % in each
group flux. The present J-log Monte-Carlo method is found to give
a meaningful probability for the non-negative quantities, while
the NEUPAC J-1 method shows a large possibility of negative value
answers.

The above sample problem is computed using a system of Super
Computer HITAC S-810 model 20. CPU time of this calculation is
about 9 seconds for 10,000 trials, and about 70 seconds for
100,000 trials.

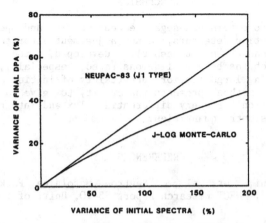

Fig. 2 Dependency of Estimated Iron DPA's Uncertainty
        on Uncertainty of Initial Neutron Spectra

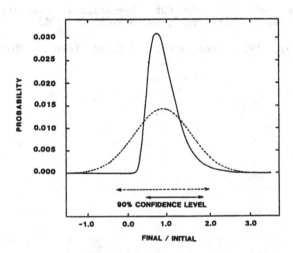

Fig. 3 A Posteriori Probability Distribution Function
        of Iron DPA
        —— estimated by J-log Monte-Carlo
        - - - estimated by NEUPAC-83 J-1 type
        Interval inference of 90 % confidence level
        is also presented, where the linear method
        gives the negative answer of lower end of
        5 % level.

CONCLUSION

In order to obtain non-negative neutron flux and spectra from multiple activation detectors, a new adjustment algorithm "J-log Monte-Carlo" has been successfully developed, the effective feature of which has been well demonstrated, especially, for the problem including large a priori uncertainty of initial spectra.

And this method presents the ability to give a rigorous p.d.f. estimation for every differential flux and integral values relating to dosimetry quantities.

REFERENCES

1. T.Taniguchi, et. al., "Neutron Unfolding Package Code NEUPAC-83", NEUT Research Report 83-10, Univ. of Tokyo, Sep. 1983

2. F.Schmittoroth, "A Method for Data Evaluation with Lognormal Distributions", N.S.E., 72, 19-34, 1979

3. Proceeding of 2nd PNC/DOE Specialists' Meeting on Collaborative Dosimetry Test, SAO13 FWG81-01, 1981

4. W.N.McElroy, "PCA Experiments and Blind Test", NUREG/CR-1861, 1981

EVALUATION OF UNCERTAINTIES OF ^{235}U FISSION SPECTRUM

M. PETILLI

ENEA, Casaccia, C.P. 2400, 00100 ROMA A.D. (Italy)

and

D. M. GILLIAM

National Bureau of Standards, Washington, D.C. 20234

ABSTRACT

This paper presents a first attempt of uncertainty evaluation for the ^{235}U fission spectrum. The covariance matrix, in 15 energy groups, has been obtained from fission rate measurements performed in Mol cavity in collaboration between NBS and SCK/CEN.

The evaluation has been done by mean of multiple unfolding, using the code EOLO. A flux transfer from ^{252}Cf spectrum has been applied to obtain the fluence rate in Mol cavity. This resulted to be equal to: $1.675 \times 10^8 \pm 3.9$ % (n · cm^{-2} · sec^{-1}) at a power level of 600 kW.

I. INTRODUCTION

The importance of uncertainties of nuclear data, such as cross sections and spectra, is undoubted.

Actually there is a big effort, between evaluators of cross section data, in order to generate or, if already existing, to improve their covariance matrices. Instead, at the moment, there are only few informations about the uncertainties of fundamental neutron fields for dosimetry,[1] and the data files containing these benchmark spectra do not include any error.[2]

In this paper the covariance matrix of ^{235}U fission spectrum induced by thermal neutron is evaluated.

This has been done by the analysis of fission rates measured in

J. P. Genthon and H. Röttger (eds.), Reactor Dosimetry, 657–665.
© 1985 ECSC, EEC, EAEC, Brussels and Luxembourg

the experiment MARK IIA performed in Mol cavity. The experimental data are updated at June 1984, because their evaluation was not still completed at the time of data delivering.[3]

The method adopted for the analysis is the multiple unfolding, which is a contemporary adjustment of different spectra and different cross sections.[4,5] The multiple unfolding is particulary useful in this case because as object of the experiment there was also the evaluation of fluence rate in Mol cavity by mean of a flux transfer from ^{252}Cf spectrum.

The covariance matrix of the ^{235}U fission spectrum is coming from the balance of solved equations containing the covariance matrix of the analized experiment and all the fluxes and cross sections involved in the experiment, in multigroup form.

II. EXPERIMENTAL DATA

The experiment Mark II-A consists in some measurements of fission rates, using a fission chamber shielded by Cd, with ^{239}Pu, ^{235}U, ^{238}U and ^{237}Np as fissile isotope for the deposit. The measured values are in Table 1.

Table 1. Measured fission rates (fiss/g · monitor count)

^{239}Pu	^{235}U	^{238}U	^{237}Np
55.02 ±2.5%	36.91 ±1.8%	9.511 ±1.9%	41.69 ±1.7%

To evaluate the covariance matrix of the experiment a detailed description of all error components is required. These are contained in Table 2. Then the covariance matrix of the experiment is obtained by an iteration, for each contribute to the error, of the expression giving the covariance matrix of derived data:

$$N' = S \cdot N \cdot S^t \tag{1}$$

Table 2. Error components of fission rate measurements. (%)

	^{239}Pu	^{235}U	^{238}U	^{237}Np
1. Reproducibility	.3	.22	.2	.2
2. Reactor background	2.	1.	.07	.03
3. Fissions in other isotopes	.1	.1	-	-
4. Wall return	.5[a]	.36[a]	.05	.1
5. Difference in wall return	.01	.01	.01	.01
6. Cd + Al scattering	.4[b]	.4[b]	.4[b]	.4[b]
7. Fission chamber scattering	-	-	.4	.2
8. Deposit mass	.5	.6	1.3	1.1
9. Extrapolation to zero pulse	.25 (.15)[b]	.5 (.2)[b]	.6 (.3)[b]	.25 (.15)[b]
10. Self absorption	.35 (.15)[b]	.35(.2)[b]	.4 (.2)[b]	.35 (.15)[b]
11. Epithermal absorption in wall return corrections	.3[b]	.3[b]	.3[b]	.3[b]
12. Source contribution	.1[b]	.1[b]	.1	.1~
13. Source absorption	1.	1.	1.	1.

[a] correlation factor equal to .5 for indicated isotopes
[b] "   "   "   " 1.0   "   "
N.B. The values in the brackets are for ratio measurements.

In the Eq. 1, N' is the covariance matrix of parameters derived by some others having N as covariance matrix, and S is the sensitivity matrix of the transformation. The matrix N' coming from one iteration becomes the input matrix N in the next iteration.[6]

The correlation matrix of the experiment is in Table 3.

Table 3. Correlation matrix of measured fission rates. (%)

^{239}Pu	100			
^{235}U	14	100		
^{238}U	10	18	100	
^{237}Np	8	14	14	100

III. CALCULATION PROCEDURE

The calculation is a combination of multiple unfolding and flux transfer.

First a flux transfer from ^{252}Cf spectrum is done to obtain the neutron intensity in Mol cavity. The applied relation is:

$$\Phi = R \cdot \frac{1}{\bar{\sigma}^{exp} (\chi_{Cf})} \cdot \frac{\bar{\sigma}^{calc} (\chi_{Cf})}{\bar{\sigma}^{calc} (\chi_{25})} \qquad (2)$$

The data used are in Table 1 and Table 4.[1,8]

Table 4. Average fission cross sections for E > .4 eV . (barn)

	^{239}Pu	^{235}U	^{238}U	^{237}Np
$\bar{\sigma}^{exp} (\chi_{Cf})$	1.824±1.9%	1.216±1.6%	.326 ±2.0%	1.365±2.0%
$\bar{\sigma}^{calc} (\chi_{Cf})$	1.792±2.8%	1.236±2.0%	.3136±2.0%	1.352±9.2%
$\bar{\sigma}^{exp} (\chi_{25})$	1.832±3.0%	1.216±1.6%	.309 ±2.6%	1.344±4.0%
$\bar{\sigma}^{calc} (\chi_{25}$ from NBS)	1.785±2.8%	1.236±2.0%	.2946±1.1%	1.322±9.3%
$\bar{\sigma}^{calc} (\chi_{25}$ from ENDF)	1.791±2.8%	1.236±2.0%	.3052±1.1%	1.347±9.3%

The difference , between Table 1 and Ref. 1, in the errors of calculated cross sections, is due to the fact that TASHI propagates the cross section uncertainties of data files, instead the errors given in Ref. 1 are essentially based on the uncertainties of ^{235}U and ^{252}Cf fission spectra.

For the calculation, the ratio, with respect to ^{235}U, are evaluated for the fission rates of different isotopes expressed as fissions/count,[†] and for the results of flux transfer. These ratios and their covariance matrix, together with the experimental data of Table 4 supposed to be not correlated, are used as input for the multiple unfolding done by mean of code EOLO which solves the equations:

$$P' - P = N_P \cdot S^\dagger \cdot (N_A + N_A\circ)^{-1} \cdot (A - A\circ) \qquad (3)$$

$$N_{P'} - N_P = -N_P \cdot S^\dagger (N_A + N_A\circ)^{-1} \cdot S \cdot N_P^\dagger \qquad (4)$$

The vector P contains the fission spectrum of ^{235}U and ^{252}Cf, and the fission cross sections of ^{239}Pu, ^{235}U, ^{238}U and ^{237}Np, in 15 groups, obtained by TASHI from IRDF multigroup form. For the fission spectrum of ^{235}U, the NBS evaluation has been used.[2]

From the multiple unfolding calculaion the requested covariance matrix of the ^{235}U fission spectrum is now obtained. The results are in Table 5 and Table 6.

The EOLO gives also the adjusted average fission cross sections in the ^{235}U and ^{252}Cf spectra. They are given in Table 7.

After the multiple unfolding calculation, the flux transfer is again applied to obtain, with the adjusted values of calculated fission cross sections, the neutron intensity in Mol cavity, coming from the different measured fission rates.

The obtained values, given in Table 8, are then combined by code BOLIK[9] and one obtaines, for the neutron intensity of Mol cavity:

$$\Phi = 11963 \pm 2.5\% \qquad n/(cm^2 \cdot count)$$

which, at a power level of 600 kW, corresponds to a fluence rate:

[†] obtained by data of Table 1 multiplied by $A/\mathcal{N}$, where: A = atomic number, and $\mathcal{N}$ = 6.022x10^{22} = Avogadro's number.

Table 5. ^{235}U fission spectrum in 15 groups. ($\Phi \cdot \Delta E$)

Group	Energy	NBS	ENDF/B-5	EOLO adjustment
1	0.4 eV	3.2×10^{-10}	3.3×10^{-10}	3.3×10^{-10}
2	1.0	1.3×10^{-8}	1.4×10^{-8}	1.4×10^{-8}
3	10.	.00040	.00044	.00045
4	1.0 keV	.0131	.0130	.0134
5	0.1 MeV	.166	.149	.136
6	0.6	.273	.268	.266
7	1.4	.206	.209	.219
8	2.2	.136	.142	.146
9	3.0	.101	.106	.108
10	4.0	.0542	.0565	.0575
11	5.0	.0283	.0288	.0299
12	6.0	.0208	.0212	.0222
13	8.0	.00528	.00550	.00554
14	11.0	.00046	.00047	.00048
15	13.0	.00019	.00012	.00012
–	18.0	–	–	–

Table 6. Uncertainties of ^{235}U fission spectrum. (%)

	Error	Correlation matrix													
1	20.2	100													
2	20.2	2	100												
3	19.1	2	2	100											
4	20.7	2	2	2	100										
5	17.4	2	2	3	-5	100									
6	9.3	4	4	5	1	-60	100								
7	9.0	5	5	5	5	6	7	100							
8	9.5	4	4	5	4	12	12	-13	100						
9	9.8	4	4	4	4	10	11	-7	-4	100					
10	10.2	4	4	4	4	8	10	1	2	4	100				
11	10.3	4	4	4	4	7	10	5	5	6	6	100			
12	10.3	4	4	4	4	8	11	5	5	6	6	7	100		
13	9.8	4	4	4	4	5	9	10	9	9	8	8	8	100	
14	16.8	2	2	3	2	3	5	6	5	5	5	5	5	5	100
15	12.0	3	3	4	3	4	8	8	7	7	7	7	7	7	4	100

$$\phi = 1.675 \times 10^8 \pm 3.9\% \qquad n \cdot cm^{-2} \cdot sec^{-1}$$

with a calibration value of 14000 ± 200 counts/sec at the nominal power of 600 kW.

Table 7. Adjusted valued of average fission cross sections.
(E > .4 eV)   (barn)

	^{239}Pu	^{235}U	^{238}U	^{237}Np
$\bar{\sigma}$ (χ_{25})	1.809±1.6%	1.216±1.1%	.3111±1.7%	1.368±2.0%
$\bar{\sigma}$ (χ_{Cf})	1.815±1.6%	1.224±1.2%	.3239±4.3%	1.366±9.6%

Table 8. Neutron density in Mol cavity. ($n \cdot cm^{-2} \cdot count^{-1}$)

Isotope	Value	Correlation matrix (%)			
^{239}Pu	12015 ± 4.6%	100			
^{235}U	11924 ± 3.2%	10	100		
^{238}U	12007 ± 5.2%	8	6	100	
^{237}Np	11996 ±10.1%	4	4	19	100

IV. CONCLUSIONS

This has been the first attempt of the evaluation of covariance matrix for the ^{235}U fission spectrum. It has been done by mean of multiple unfolding, that is an adjustment procedure regarding, in this case, the covariance matrix.

Because the code EOLO uses the Least Squares method, the final result is depending also on the informations of covariance matrix before the analized experiment was performed, which, in this case, were very poors. Therefore, all the informations for the covariance matrix of the ^{235}U fission spectrum are coming from the analized experiment, and everyone knows that only one experiment cannot be exhaustive when new data have to be evaluated.

Nevertheless, the obtained results give some degree of knowledge about the covariance matrix of the ^{235}U fission spectrum, which can be used as input data for next adjustments with other experiments.

New experiments with threshold detectors seem to be useful for more information of covariances at low and high energy.

Moreover, specially in view of flux transfer applications, ratio measurements between cross sections in ^{235}U and ^{252}Cf fission spectra, could be very interesting, also for information of covariances between the two spectra.

ACKNOWLEDGMENTS

The authors are pleased to thank the NBS team and A. FABRY for all useful discussions. Moreover, this collaboration between ENEA and NBS has been thanks to the suggestions, sincerely appreciated, of W. MANNHART.

REFERENCES

1. E.P.Lippincott, W.N.McElroy, "LWR Pressure Vessell Surveillance Dosimetry improvement Program.",Quarterly Progress Report, Oct. 1983 – Dec. 1983, NUREG/CR-3391, Vol. 4, R5(1984)

2. D.E.Cullen,N.Kocherov, P.K.McLaughlin, "The International Reactor Dosimetry File (IRDF-82)", IAEA-NDS-41(1982) and IAEA-NDS-48(1982)

3. D.M.Gilliam, J.A.Grundl, G.P.Lamaze, E.D.McGarry and A.Fabry, "Proceedings of Fifth ASTM-Euratom Symposium on Reactor Dosimetry", to be published

4. M.Petilli, "A new analysis of the experiment for measurement of $\Phi > 1$ MeV in the pressure vessel cavity of U.S. light water power reactor Arkansas.", Proceedings of the Fourth ASTM-Euratom Symposium on Reactor Dosimetry, NUREG/CP-0029(1982), Vol. 1, p. 545, and ENEA-RT/FI(82)10, 1982

5. M.Petilli, Code EOLO, to be published as ENEA-RT/FI

6. M.Petilli, "How to evaluate the variance-covariance matrix of an integral experiment", CNEN-RT/FI(80)18, 1980

7. M.Petilli, Code TASHI, to be published as ENEA-RT/FI

8) J.A.Grundl, and D.M.Gilliam, "Fission Cross Section Measurements in Reactor Physics", American Nuclear Society, Anneal Meeting, Dedroit, June 1983

9. M.Petilli, "BOLIK: A Code to Evaluate Correlated Data Starting by Multiple or Ratio Values.", ENEA-RT/FI(83)5, 1983

EVALUATION OF GAMMA-HEATING RATES IN THE JMTR CORE

(Benchmark Calculation)

K. Sakurai[†] and N. Yamano[††]

Japan Atomic Energy Research Institute

Ibaraki-ken 319-11, Japan

Gamma-heating rates and its dependence on core configurations were calculated by using RADHEAT-V4 code system. DOT-3.5 code was used for neutron and gamma transport calculation. The calculated gamma-heating rates were compared with the TLD measured gamma-heating rates, and the calculational accuracy was evaluated. The compiled group constants were 120 energy group structure with P-5 and S-48.

The calculated gamma-heating rates were in good agreement with the measured gamma-heating rates. The C/E's obtained at locations G-6, -10, K-6, -10 and H-7, -9, J-7, -9 irradiation holes in fuel region were 1.00 and 1.06, respectively. The average C/E's at locations Be-1 (11 row F ~ M), Be-2 (12 row F ~ M) and Al-1 (13 row F ~ M) irradiation holes were 1.12, 1.25 and 1.19, respectively.

INTRODUCTION

The Japan Materials Testing Reactor (JMTR) is a high flux reactor with a rated power of 50 MW. The maximum thermal neutron flux (< 0.6826 eV) is about 4×10^{14} n/cm^2 sec, and the maximum fast neutron (> 1.0 MeV) is about 4×10^{14} n/cm^2 sec. About 60 capsules have been

---

† Institute of Nuclear Safety Japan, Mita 1-4-28, Tokyo
†† Nuclear Fuel Industries Ltd., Tosabori 2-2-7, Osaka

J. P. Genthon and H. Röttger (eds.), Reactor Dosimetry, 667–680.
© 1985 ECSC, EEC, EAEC, Brussels and Luxembourg

irradiated in each operation cycle, and the capsule arrangement was different from cycle to cycle.

Neutron spectrum and gamma-ray spectrum in the irradiation field are very important information for a multi-purpose reactor such as the JMTR. The former is for evaluation of neutron fluence, while the latter is for evaluation of gamma-heating rate (GHR).

The measurement and the evaluation of neutron spectra in the JMTR have been performed in the last 10 years (1,2). The GHR's have been measured by using calorimeters (3) and thermoluminescence dosimeters (TLD) (4,5).

GHR's and its dependence on core configurations were calculated by using RADHEAT-V4 code system (6,7,8). DOT-3.5 code (9) was used for neutron and gamma transport calculation. The calculated GHR's were compared with the TLD measured GHR's, and the calculational accuracy was evaluated.

In this paper, we briefly present the calculation method, accuracy evaluation method and dependence on capsule arrangements of GHR.

CALCULATIONAL METHOD

The core configuration of the JMTR is shown in Fig. 1. The samples have been irradiated for more than 1 cycle (24 days) in the fuel region, the first beryllium reflector region (Be-1), the second beryllium reflector region (Be-2) , the first aluminum reflector region (Al-1), the Oarai Water Loop nos. 1 and 2 (OWL-1 and -2), the Oarai Gas Loop no. 1 (OGL-1) and the Neutron Control Facility (NCF). Irradiations of less than 1 cycle have been performed in the Hydraulic Rabbit nos. 1 and 2 (HR-1 and -2).

The aluminum element in the fuel region is 77.2 mm x 77.2 mm in width and 750 mm in length and has irradiation hole with 42 mm in diameter at locations G-6 and -10, K-6 and -10. The irradiation capsule of 40 mm in diameter is set in the irradiation hole. The aluminum element at the locations H-7 and -9, J-7 and -9 in the fuel region has four irradiation holes, and the irradiation capsule of 30 mm in diameter is set in each irradiation hole. The beryllium frame at locations E-6 to E-13 and N-6 to N-13 has irradiation holes of 32 mm in diameter, and the irradiation capsule of 30 mm in diameter is set in the irradiation holes. The beryllium frame at the location from G-5 to L-5 has irradiation holes of 38 mm in diameter, and the irradiation capsule of 36 mm in diameter is set in the irradiation

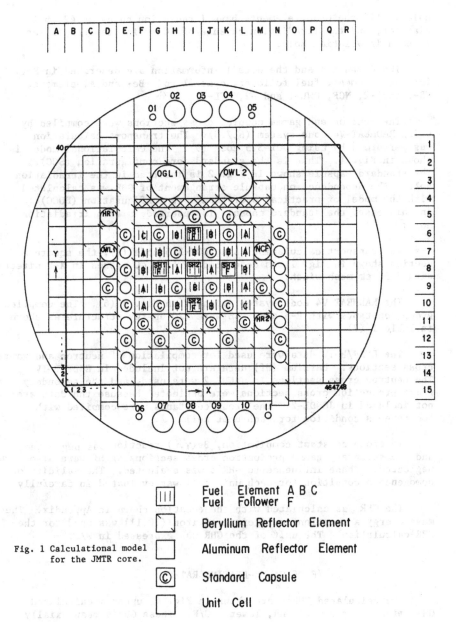

A B C D E F G H I J K L M N O P Q R

01 02 03 04 05

OGL 1   OWL 2

HR 1

OWL 1

NCF

HR 2

SR1 TF   SH1 TF   SH2 TF   SH3 F   SR2 TF

40

Y

X

1 2 3

3 2

6 1 2 3   46 47 48

1
2
3
4
5
6
7
8
9
10
11
12
13
14
15

06 07 08 09 10 11

Fig. 1 Calculational model
for the JMTR core.

||| Fuel Element A B C
     Fuel Follower F

     Beryllium Reflector Element

     Aluminum Reflector Element

ⓒ   Standard Capsule

     Unit Cell

holes. All the other elements have irradiation holes of 42 mm in diameter, and the irradiation capsule of 40 mm in diameter is set in each irradiation hole.

The dimension and the detail information are described in Ref. 5 for fuel element, fuel follower, control rod, Be- and Al-element, HR-1 and -2, NCF, OWL-1 and -2, OGL-1.

The neutron and gamma coupled cross sections were compiled by using RADHEAT-V4 code system (6,7,8). The transport calculation was performed by using DOT-3.5 code (9). The calculational model is shown in Fig. 1. This is the standard core configuration (SCC). The standard capsule shown in Fig. 2 is inserted in the irradiation hole. The dependence on capsule arrangement of GHR was calculated with the model of practical operation core configuration (POCC). In this case, the standard capsule is inserted in each irradiation hole.

In transport calculation using DOT-3.5 code (9), the square section shown in Fig. 1 was described by 48 x 40 meshes in X-Y directions with 2.57 cm mesh width.

The RADHEAT-V4 code system is shown in Fig. 3 (6). The compiled group constants with P-5 and S-48 were 120 energy structure as shown in Table 1 (7).

The ENDF/B-IV data were used for compilation of neutron and gamma cross sections. Hafnium (Hf) data are not included in ENDF/B-IV. The neutron cross sections in JENDL-2 were used, and the secondary gamma production cross sections were neglected because the data are not included in JENDL-2. The group constants were compiled with homogeneous condition for each unit cell.

In group constant compilation, Be(γ,n) reaction was neglected, and the secondary gamma production cross sections of Hf were also neglected. These influence to GHR's was evaluated. The validity of homogeneous condition for each unit cell was evaluated in carefully.

The GHR was calculated with the equation shown in Appendix. The mass energy absorption coefficient of iron (10,11) was used for the GHR calculation. The unit of the GHR was expressed in W/g.

C/E OF GAMMA-HEATING RATES

The calculated GHR's are shown in Fig. 4, upper : calculated GHR, middle : measured GHR, lower : C/E. These GHR's mean axially peak values.

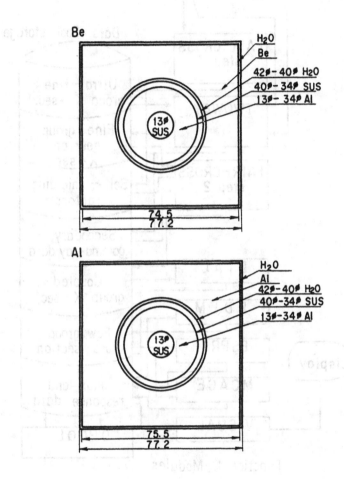

Fig. 2 Unit cell with standard capsule.

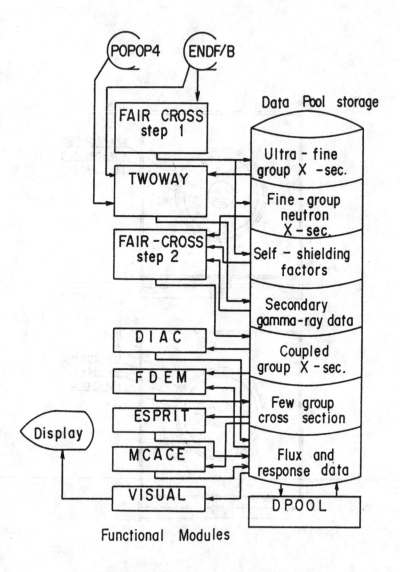

POPOP4 ENDF/B

Data Pool storage

FAIR CROSS step 1

TWOWAY

FAIR-CROSS step 2

Ultra-fine group X-sec.

Fine-group neutron X-sec.

Self-shielding factors

Secondary gamma-ray data

DIAC

Coupled group X-sec.

FDEM

ESPRIT

Few group cross section

Display

MCACE

Flux and response data

VISUAL

DPOOL

Functional Modules

Fig. 3 Hierarchy of the RADHEAT-V4 code system (6).

Table 1 Energy group structure (8).

(a) NEUTRON ENERGY GROUP (EV)

NO.	LOWER	UPPER	NO.	LOWER	UPPER
1	1.4550E+07	1.6487E+07	51	2.8088E+04	3.1828E+04
2	1.2840E+07	1.4550E+07	52	2.4788E+04	2.8088E+04
3	1.1331E+07	1.2840E+07	53	2.1875E+04	2.4788E+04
4	1.0000E+07	1.1331E+07	54	1.9305E+04	2.1875E+04
5	8.8250E+06	1.0000E+07	55	1.7036E+04	1.9305E+04
6	7.7880E+06	8.8250E+06	56	1.5034E+04	1.7036E+04
7	6.8729E+06	7.7880E+06	57	1.1709E+04	1.5034E+04
8	6.0653E+06	6.8729E+06	58	9.1188E+03	1.1709E+04
9	5.3526E+06	6.0653E+06	59	7.1017E+03	9.1188E+03
10	4.7237E+06	5.3526E+06	60	5.5308E+03	7.1017E+03
11	4.1686E+06	4.7237E+06	61	4.3074E+03	5.5308E+03
12	3.6788E+06	4.1686E+06	62	3.3546E+03	4.3074E+03
13	3.2465E+06	3.6788E+06	63	2.6126E+03	3.3546E+03
14	2.8650E+06	3.2465E+06	64	2.0347E+03	2.6126E+03
15	2.5284E+06	2.8650E+06	65	1.5846E+03	2.0347E+03
16	2.2313E+06	2.5284E+06	66	1.2341E+03	1.5846E+03
17	1.9691E+06	2.2313E+06	67	9.6112E+02	1.2341E+03
18	1.7377E+06	1.9691E+06	68	7.4852E+02	9.6112E+02
19	1.5335E+06	1.7377E+06	69	5.8295E+02	7.4852E+02
20	1.3534E+06	1.5335E+06	70	4.5400E+02	5.8295E+02
21	1.1943E+06	1.3534E+06	71	3.5357E+02	4.5400E+02
22	1.0540E+06	1.1943E+06	72	2.7536E+02	3.5357E+02
23	9.3014E+05	1.0540E+06	73	2.1445E+02	2.7536E+02
24	8.2085E+05	9.3014E+05	74	1.6702E+02	2.1445E+02
25	7.2440E+05	8.2085E+05	75	1.3007E+02	1.6702E+02
26	6.3928E+05	7.2440E+05	76	1.0130E+02	1.3007E+02
27	5.6416E+05	6.3928E+05	77	7.8893E+01	1.0130E+02
28	4.9787E+05	5.6416E+05	78	6.1442E+01	7.8893E+01
29	4.3937E+05	4.9787E+05	79	4.7851E+01	6.1442E+01
30	3.8774E+05	4.3937E+05	80	3.7267E+01	4.7851E+01
31	3.4218E+05	3.8774E+05	81	2.9023E+01	3.7267E+01
32	3.0197E+05	3.4218E+05	82	2.2603E+01	2.9023E+01
33	2.6649E+05	3.0197E+05	83	1.7603E+01	2.2603E+01
34	2.3518E+05	2.6649E+05	84	1.3710E+01	1.7603E+01
35	2.0754E+05	2.3518E+05	85	1.0677E+01	1.3710E+01
36	1.8316E+05	2.0754E+05	86	8.3153E+00	1.0677E+01
37	1.6163E+05	1.8316E+05	87	6.4760E+00	8.3153E+00
38	1.4264E+05	1.6163E+05	88	5.0435E+00	6.4760E+00
39	1.2588E+05	1.4264E+05	89	3.9279E+00	5.0435E+00
40	1.1109E+05	1.2588E+05	90	3.0590E+00	3.9279E+00
41	9.8037E+04	1.1109E+05	91	2.3824E+00	3.0590E+00
42	8.6517E+04	9.8037E+04	92	1.8554E+00	2.3824E+00
43	7.6351E+04	8.6517E+04	93	1.4450E+00	1.8554E+00
44	6.7379E+04	7.6351E+04	94	1.1254E+00	1.4450E+00
45	5.9462E+04	6.7379E+04	95	8.7642E-01	1.1254E+00
46	5.2475E+04	5.9462E+04	96	6.8256E-01	8.7642E-01
47	4.6309E+04	5.2475E+04	97	5.3158E-01	6.8256E-01
48	4.0868E+04	4.6309E+04	98	4.1399E-01	5.3158E-01
49	3.6066E+04	4.0868E+04	99	1.5183E-01	4.1399E-01
50	3.1828E+04	3.6066E+04	100	3.5238E-04	1.5183E-01

(b) PHOTON ENERGY GROUP (EV)

NO.	LOWER	UPPER
1	1.2000E+07	1.4000E+07
2	1.0000E+07	1.2000E+07
3	8.0000E+06	1.0000E+07
4	6.5000E+06	8.0000E+06
5	5.0000E+06	6.5000E+06
6	4.0000E+06	5.0000E+06
7	3.0000E+06	4.0000E+06
8	2.5000E+06	3.0000E+06
9	2.0000E+06	2.5000E+06
10	1.6600E+06	2.0000E+06
11	1.3300E+06	1.6600E+06
12	1.0000E+06	1.3300E+06
13	8.0000E+05	1.0000E+06
14	6.0000E+05	8.0000E+05
15	4.0000E+05	6.0000E+05
16	3.0000E+05	4.0000E+05
17	2.0000E+05	3.0000E+05
18	1.0000E+05	2.0000E+05
19	5.0000E+04	1.0000E+05
20	2.0000E+04	5.0000E+04

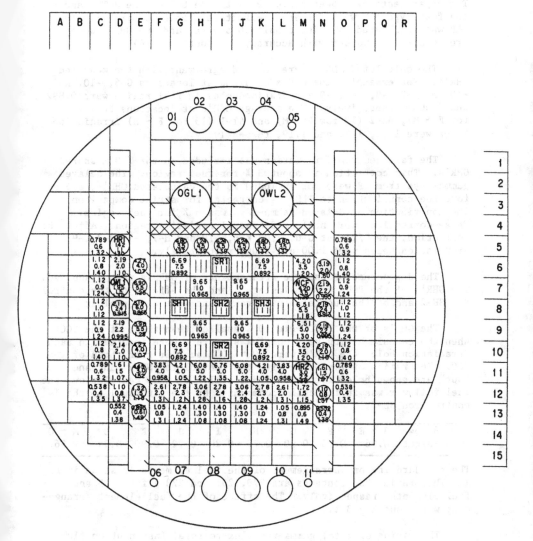

Fig. 4 Gamma-heating rates for the JMTR core in W/g.
upper : calculated GHR, middle : measured GHR,
lower : C/E (iron equivalent).

The calculations were performed for the SCC shown in Fig. 1.
The experiments have been performed with the SCC of the JMTRC and
the four type TLD with different effective atomic number (4). The
TLD was calibrated by using cobalt-60 source, and the GHR's were
iron equivalent values with accuracy of about 15 % (4).

The calculated GHR's were in good agreement with the measured
GHR's. For exsample, the C/E's obtained at locations G-6, -10, K-6,
-10 and H-7, -9, J-7, -9 irradiation holes in fuel region were 0.892
and 0.965, respectively. The average C/E's at locations Be-1 (11
row F ~ M), Be-2 (12 row F ~ M) and Al-1 (13 row F ~ M) irradiation
holes were 1.12, 1.25 and 1.19, respectively.

The fast neutron irradiation was included in thr TLD measured
GHR's. This correction is about 3 % for fuel region. The delayed
gamma-rays from FP were not included in the calculated GHR's, therefore the contribution to TLD was corrected for strict comparison.
The contribution of delayed gamma-rays was 6 % for fuel region.
These corrections were neglected other irradiation region. After the
correction, the C/E's at G-6, -10, K-6, -10 and H-7, -9, J-7, -9
irradiation holes were 1.00 and 1.06, respectively.

The dependence on capsule arrangements of GHR is next discussed.
The GHR's of the POCC were calculated, and the effects of the control
rod SH-2 and of the fuel element arrangement were also evaluated.

The GHR's of the POCC were 10 % smaller than those of the SCC.
When the capsule having very complex structure were inserted in each
irradiation hole, the GHR's were 10 ~ 30 % smaller than those of the
SCC. The ratios of the GHR's with the control rod SH-2 "in" and
"out" are shown below for 8 row B ~ Q cell. In this calculation, the
fuel follower was in the cell accompanied by the withrawal of the
control rod SH-2.

8 row	B ~ E	F	G	H	I	J	K	L	M ~ Q
in/out	0.96	0.95	0.90	0.72	0.20	0.72	0.90	0.95	0.96

The standard in-core arrangement of the fuel elements is shown in Fig.
1. The uranium-235 contents are 279, 251 and 223 g for A, B and C
fuel elements, respectively. The effect of the fuel element arrangement was about 2 ~ 3 %.

The ratios of total gamma-ray flux to total fast neutron flux
(>0.6826 eV) and to total thermal neutron flux (<0.6826 eV) were
calculated for the SCC. The obtained former ratios were 2, 3, 6 ~ 9
and 8 ~ 9 for fuel region, Be-1, Be-2, Al-1 irradiation holes,
respectively. On cne other hand, the obtained latter ratios were
7, 4, 3, 2 ~ 3 and 3 for H-7, -9, J-7, -9 irradiation holes, G-6,
-10, K-6, -10 irradiation holes, Be-1, Be-2 and Al-1 irradiation holes,
respectively.

DISCUSSION AND CONCLUSION

(γ,n) reaction was not considered in the radiation transport calculation. The cross sections of the Be(γ,n) reaction is very small (12), therefore the influence to the GHR's is negligible small. The up-scattering of neutrons from thermal energy group was not considered in compiling of the group constants for RADHEAT-V4 code system. The influence of up-scattering was about 7 % for thermal neutron flux, but it is difficult to estimate its effect to GHR.

The group constants were compiled from the base data of ENDF/B-IV. The Hf data were not included in ENDF/B-IV and therefore the neutron cross sections of Hf in JENDL-2 were used in compiling the group constants, but the secondary gamma production cross sections were neglected because the data were again included in JENDL-2. The control rod of the JMTR is square tube type containing Hf, and its volume in the cell is about 18.6 %. When the secondary gamma production cross sections were neglected, the influence to the GHR in fuel region was estimated to be about 3 ~ 7 %.

When the calculation was performed for the full core, the each cell must be homogenized in practice. The GHR's at HR-1, -2, NCF, OWL-1, -2 and OGL-1 were 40 ~ 80 % higher than the TLD measured GHR's and the calorimeter measured GHR's, and for these cells, heterogeneous calculation becomes necessary. The mesh division of these cell was 10 ~ 20 for X-direction and 10 ~ 20 for Y-direction. In the heterogeneous calculation, the C/E's at these cells were about 1.2 ~ 1.3. The GHR's of HR-1, -2, NCF and OWL-1 in Fig. 4 result from the heterogeneous calculation. The measured GHR's and the C/E's of OWL-2 and OGL-1 are not included in Fig. 4, because the measured GHR's were only measured by the calorimeters (5).

The mass energy absorption coefficient of iron (10,11) was used for the GHR calculation. As shown in Appendix, μ-k/ρ or μ-en/ρ is used for the GHR calculation, but the μ-s/ρ should be used for more strict evaluation (13). When the KX-rays are considered and the bremsstrahlung is neglected, μ-k/ρ is used. When the KX-rays are neglected, μ-s/ρ is used. The KX-rays and the bremsstrahlung are neglected in the calculation with ANISN, DOT and MORSE codes, μ-k/ρ is usually used. In this case, the μ-s/ρ should be used. When the KX-rays and the bremsstrahlung are considered, μ-en/ρ is used. In this work of GHR calculation, μ-en/ρ was used, because the influence of the bremsstrahlung was considered to be small. When μ-s/ρ is used, the GHR's become 1 ~ 2 % larger than the values shown in Fig. 4.

As shown in Appendix, the uncertainty of the evaluated GHR depend on that of the transport calculation of gamma-ray spectrum and of the mass energy absorption coefficient. The uncertainty of the latter is about 2 % (11). The dominant uncertainty factor comes from the transport

677

calculation. In the calculation by DOT-3.5, the uncertainty of gamma-ray spectrum depends on that of the neutron cross sections and the secondary gamma production cross sections, but the dominant uncertainty is that of the secondary gamma production cross sections. In the GHR calculation, there are many uncertainty factor, but the accuracy is about 0 ~ 30 % for the JMTR core. The calculated GHR's were slightly larger than the TLD measured GHR's.

The calculations were performed by using a high speed computer M-380 at the Japan Atomic Energy Research Institute (JAERI). The used computer memory was about 8000 kbits for 18 mixtures with 120 energy groups and $P_5 -S_{48}$, JMTR full core modelling and DOT-3.5 two dimensional calculation with X-48 and Y-40 meshes. The inner iteration number, the CPU time and the convergence accuracy are as follows.

inner iteration number	CPU time (min.)	convergence accuracy (%)
2	18	n ; 30~40, γ; 6~80
10	33	n ; 2~ 8, γ; 2~ 8
20	34	n ; 2~ 8, γ; 2~ 8

ACKNOWLEDGMENT

The authors would like to express their grateful acknowledgment to Dr. S. Tanaka at JAERI for his helpful discussion on the evaluation of the gamma-heating rate.

Scincere thanks are also expressed to Prof. A. Sekiguchi and Prof. M. Nakazawa at the University of Tokyo, Prof. I. Kimura at Kyoto University for their helpful discussion at the Research Committee on Data Evaluation of Neutron Irradiation in Atomic Energy Society of Japan.

REFERENCES

1. I. Kondo and K. Sakurai, "Experimental Evaluation of Reactor Neutron Spectrum in Irradiation Field", J. Nucl. Sci. Technol. 18, 461(1981).

2. K. Sakurai, "Measurement and Evaluation of Neutron Spectra above 0.1 MeV in the JMTR", Nucl. Instr. and Methods 213, 359(1983).

3. T. Hayashi, et al., "Measurements of Gamma-Heat in the JMTR", J. Nucl. Sci. Technol. 9, 133(1972).

4. K. Takeda, et al., "Measurement of Gamma-Heating Rates with TLD in the JMTR", in preparation (1984).

5. Dep. of JMTR Project, JMTR Irradiation Handbook, JAERI-M 83-053 (February 1983).

6. N. Yamano, K. Koyama and K. Mimami, "Development of Integrated Shielding Analysis Code System RADHEAT-V4", Proc. 6th. Conf. on Radiation Shielding, May 16-20, 1883, Vol. 1, 331(1983).

7. N. Yamano, JSD1000 ; Multi-Group Cross Section Sets for Shielding Materials, JAERI-M 84-038 (March 1984).

8. N. Yamano, Graphs of Neutron Cross Sections in JSD1000 for Radiation Shielding Safety Analysis, JAERI-M 84-053 (March 1984).

9. W. A. Rhoades, et al., DOT-3.5 Two Dimensional Discretes Radiation Transport Code, CCC-276 (1977).

10. E. Storm and H. I. Israel, Photon Cross Sections from 0.001 to 100 MeV for Elements 1 through 100, LA-3753 (november 1967).

11. J. H. Hubbell, "Photon Mass Attenuation and Energy-absorption Coefficients from 1 keV to 20 MeV", Int. J. Appl. Radiat. Isot. 33, 1269(1982).

12. Prepared by ANL, Reactor Physics Constants, ANL-5800 (2nd ed.), p. 647(1963).

13. S. Tanaka and N. Sasamoto, "Gamma-Ray Absorbed Dose Measurements Media with Plural Thermoluminescent Dosimeters Different Atomic Number", J. Nucl. Sci. Technol. (to be published)

$$D_M (\text{Ir}) = \int_0^{E_{\gamma,max}} \emptyset (E_\gamma, \text{Ir}) \cdot E_\gamma \frac{\mu_x(E_\gamma)}{\rho} dE_\gamma$$

$\emptyset (E_\gamma, \text{Ir})$: photon flux having energy E_γ (photons/cm^2·sec·MeV)

$\mu_x(E_\gamma)/\rho$: energy absorption coefficient (cm^2/g)

(1) μ_a/ρ : mass absorption coefficient

$$\frac{\mu_a}{\rho} = \frac{\tau}{\rho} + \frac{\sigma_c}{\rho} \frac{E_e}{E_\gamma} + \frac{k}{\rho}$$

τ/ρ : photoelectric mass attenuation coefficient

σ_c/ρ : Compton

k/ρ : mass attenuation coefficient for pair production

E_e : average energy of the Compton electron per Compton scattering

(2) μ_κ/ρ : mass energy transfer coefficient

$$\frac{\mu_\kappa}{\rho} = \frac{\tau}{\rho} \left(1 - \frac{\overline{E_f}}{E_\gamma} \right) + \frac{\sigma_c}{\rho} \frac{E_e}{E_\gamma} + \frac{k}{\rho} \left(1 - \frac{2mc^2}{E_\gamma} \right)$$

$\overline{E_f}$: average energy emitted as fluorescent radiation per photon absorbed

$$\frac{\mu_s}{\rho} = \frac{\tau}{\rho} + \frac{\sigma_c}{\rho} \cdot \frac{E_e}{E} + \frac{k}{\rho} \left(1 - \frac{2mc^2}{E_\gamma} \right)$$

(3) μ_{en}/ρ : mass energy absorption coefficient

$$\frac{\mu_{en}}{\rho} = \frac{\mu_\kappa}{\rho} (1 - g)$$

g : bremsstrahlung factor

LSL-M1 AND LSL-M2: TWO EXTENSIONS OF THE LSL ADJUSTMENT

PROCEDURE FOR INCLUDING MULTIPLE SPECTRUM LOCATIONS

F.W. Stallmann

Oak Ridge National Laboratory

Oak Ridge, Tennessee, USA

ABSTRACT

Most current adjustment procedures, including LSL
(1), can adjust only one spectrum with dosimetry located
at the point of the input spectrum. Many radiation
experiments have dosimetry at more than one location, and
fluence or damage exposure values are desired for locations other than those covered by dosimetry. Thus, the
use of single-spectrum dosimetry to these experiments
causes considerable loss of information and introduces
large uncertainties. Two extensions of the LSL code to
cover multiple-spectra adjustment are discussed. Each
extension has different restrictions and covers a different range of applications.

INTRODUCTION

For an accurate determination of neutron fluences in test and
power reactors, both transport calculations and dosimetry measurements are needed. Adjustment methods provide the analytical tool
for combining the two. As a result, best estimates for fluences
and damage parameter values with improved uncertainties can be
obtained. The output shows also whether the input data are consistent within the bounds of the input uncertainties. The majority of the currently available adjustment methods (2) allow
adjustment of only one spectrum at a time, which is restricted to
the location of a dosimeter set. However, the fluence spectra of
interest, for instance in metallurgical test specimen or inside

the pressure vessel wall of power reactors, are at locations without dosimeters and some form of extrapolation is necessary. Also, sensors with different response functions are often placed at different locations, so that different parts of the spectrum are adjusted in different adjustment runs depending on the monitor set used. These problems can be solved by processing simultaneously several spectra making use of the high correlation between the calculated spectra within the same transport calculation. Since a larger amount of input information is used, uncertainties are decreased at the same time.

The benefits of multiple spectrum adjustment come at a price. Each calculated group fluence and each dosimetry measurement adds an entry to the covariance matrix, with the consequence that the storage requirement goes up roughly with the square of the entries and the computing time with the third power of the dosimetry data. Thus, some restrictions are necessary to keep the computing requirement reasonable. For the LSL-M2 version, the number of spectra is restricted to $N_s \times N_g \leq 400$, where N_s is the number of spectra and N_g is the number of energy groups, that is roughly 10-20 spectra with 20-40 energy groups. The number of dosimeters is restricted to 100. The energy group structure must be the same for each spectrum and the largest number of energy groups, N_g, should not exceed $N_g = 50$. These are reasonable restrictions which will suffice for most applications.

A larger number of spectra can be processed without substantial increase in computations if the correlations between any two spectra at different locations are essentially the same, not dependent on the space coordinates. This is not an unreasonable assumption, since the errors in transport calculations will affect equally all fluence spectra, which are not too far apart in space, say, all spectra within a metallurgical irradiation capsule or a pressure vessel surveillance capsule. A "super matrix" is then created having as elements the covariance matrices from each individual spectrum. This super matrix can be inverted explicitly provided the assumption about spectrum correlations holds. Only the covariance matrices for the individual spectra need to be inverted; their size is the number of different dosimetry sensors at a given location which may differ from one location to the other. This method is named LSL-M1, since it was the first extension of the original LSL method (1). The mathematical details are discussed in Ref. 3. In both extensions of the LSL method calculated fluxes may be introduced either by absolute magnitude or as relative spectra, allowing for a free scale factor, which fits the calculated spectrum to the dosimetry measurements in the least squares sense. There is also a choice of reaction rates or equivalent fission fluxes or any mixture thereof as input for dosimetry measurements.

Output consists of integral parameters, such as fluence > 1.0 MeV or dpa with uncertainties in the form of relative standard deviation and correlation. These can be obtained at any location, whether there are dosimeters at this location or not. Unlike the original, single-spectrum, LSL method, no iterations are performed in the multiple spectra LSL methods in order to keep the computation effort down. This means that the adjusted values are only approximately consistent if adjustments are large. This is of little consequence, however, since large adjustments are always connected with large uncertainties, which exceed potential inconsistencies.

PROGRAMMING CONSIDERATIONS

The two multiple spectrum adjustment methods, LSL-M1 and LSL-M2, are intended primarily for application to test and research reactors. That means that the program can accept a large variety of configurations with few restrictions to the input and output data. However, the user has to provide not only the input spectra and input dosimetry measurements but must also supply variances and covariances for these data. While this does not exclude application to pressure vessel surveillance and extrapolation procedures, the much more comprehensive LEPRICON (7) methodology is preferable in this case.

The programs are written in standard FORTRAN for interactive use on the DEC-10 system. Separate files are needed for the reaction rate, spectrum, and cross section data and for the corresponding variance-covariance information. An additional file is needed to contain the response functions for the damage parameter values. Data sets are identified by names for reactions and locations, so that the information need not be given in any specific order, and the same cross-section files can be used for several adjustment problems. Auxiliary programs have been written to create LSL files from more basic information. The primary ones are:

- FPROC - a general purpose program which interpolates, extrapolates, and converts group fluences in a given energy group structure to any other structure using a 620-group auxiliary spectrum for interpolation and/or extrapolation.

- PUFF (4) - which has been developed earlier to read ENDF cross-section and covariance files (5) and convert them to a specified energy group structure.

- ACT - converts count rates to reaction rates for any given irradiation history including possible spectrum changes during irradiation.

Both LSL-M1 and LSL-M2 have been thoroughly tested and successfully applied (3,6). Documentation and release is expected to be
completed by the end of 1985.

REFERENCES

1. F.W. Stallmann, "LSL - A Logarithmic Least Squares Adjustment
   Method," Proc. of the Fourth ASTM-EURATOM Symposium on Reactor
   Dosimetry, NUREG/CR-0029, Nuclear Regulatory Commission,
   Washington, DC, 1983.

2. "ASTM E944-83, Standard Practice for Application of Neutron
   Spectrum Adjustment Methods in Reactor Surveillance, 1983
   Annual Book of Standards, Vol. 12.02, American Society for
   Testing and Materials, Philadelphia, PA, 1983.

3. F.W. Stallmann, C.A. Baldwin, and F.B.K. Kam, Neutron Spectral
   Characterization of the 4th Nuclear Regulatory Commission
   Heavy Section Steel Technology 1T-CT Irradiation Experiment:
   Dosimetry and Uncertainty Analysis, NUREG/CR-3333,
   ORNL/TM-8789, Oak Ridge National Laboratory, Oak Ridge, TN,
   July 1983.

4. J.D. Smith, III and B.L. Broadhead, PUFF-2, Determination of
   Multigroup Covariance Matrices from ENDF/B-V Uncertainty File,
   RSIC-PSF-157, Radiation Shielding Information Center, Oak
   Ridge, TN.

5. F.H. Perey, The Data Covariance Files for ENDF/B-V,
   ORNL/TM-5938, Oak Ridge National Laboratory, Oak Ridge, TN,
   1977.

6. F.W. Stallmann, Determination of Damage Exposure Parameter
   Values in the PSF Metallurgical Irradiation Experiment,
   NUREG/CR-3814, ORNL/TM-9166, Nuclear Regulatory Commission,
   Washington, DC, 1984.

7. R.E. Maerker, J.J. Wagschal, and B.L. Broadhead, Development
   and Demonstration of an Advanced Methodology for LWR Dosimetry
   Applications, EPRI NP-2188, (Project 1399-1, Interim Report),
   Electric Power Research Institute, Palo Alto, CA, 1981.

SYSTEMATIC STUDY ON SPECTRAL EFFECTS IN THE

ADJUSTMENT CALCULATIONS USING THE NEUPAC-83 CODE

T.Taniguchi[1] , N.Ueda[2] , M.Nakazawa[2] and A.Sekiguchi[2]

1. The Institute of Applied Energy

2. Dept. of Nucl. Eng., University of Tokyo, Japan

ABSTRACT

The improvements of the neutron unfolding package code NEUPAC-83 were made sucessfully on several points such as an introduction of non-negative estimation algorithm (J-log method) and a preparation of the dosimetry cross section with covariance library.

In addition to these improvements, systematic study has been made by the NEUPAC-83, focussing our attention on the spectral covariance effects in the adjustment calculations. The results may be summarized as follows:
(1) adjusted integral values are influenced considerably by the prior probability density function of initial neutron spectrum, especially by the correlation factor, and (2) the determination method of statistically permitted area of the solutions should be applied more effectively in the adjustment calculation.

INTRODUCTION

The current research activities in dosimetry studies have been mainly concentrated to make clear and improve the accuracy of the current dosimetry techniques, which are basing on the reaction-rate measurements, nuclear cross section evaluations and adjustments, and unfolding procedures. Especially present status of the uncertainty evaluation in dosimetry analysis goes toward the introduction of covariances in the adjustment calculation.

J. P. Genthon and H. Röttger (eds.), Reactor Dosimetry, 685–691.
© 1985 ECSC, EEC, EAEC, Brussels and Luxembourg

As for the reaction-rate measurements, the uncertainty evaluation has been investigated in detail for a long time and has been made clear. On the other hand, as for the nuclear cross sections, it is very actively being investigated. However, an important problem has been left unresolved in the adjustment calculation. The problem lie in the uncertainty estimates of neutron spectrum by the neutronics calculations.

As is generally known, the multiple foil activation technique relies on some estimate of the neutron spectrum prior to adjustment. Uncertainties of input neutron spectrum are the most troublesome and typically are the largest of error in the adjustment calculations, they tend to be rather subjective, based on well-arranged experiments. From these reasons, one generally has to make an educated guess concerning spectral uncertainties.

From the above mentioned point of view, we have developed the neutron unfolding package code NEUPAC-83 and have investigated systematically the spectral covariance effects on the adjusted solutions, and have listed the necessary care to be taken and problems encountered on applying it to the adjustment calculations in this study.

NEUTRON UNFOLDING PACKAGE CODE "NEUPAC-83"

The neutron unfolding package code NEUPAC-83 is based on the J-1 type unfolding method (1,2), which is formulated by a variational representation of the integral quantity.

Main features of this code are an usage of more complete covariance data of the measured reaction-rate values, the dosimetry cross sections and the initial neutron spectrum, and an estimation of spectral parameter together with its uncertainty. And it was mathematically proven that the solutions of the NEUPAC-83 were in agreement with that of the standard linear least-squares method such as the STAY'SL code (2).

In this study, the improvements of this code were made sucessfully on several points. One is a development of non-negative estimation method, named as J-log, which is based on the non-linear functional representation coupled with a lognormal distribution as prior probability density function (2). Detail explanation of J-log algorithm is omitted here due to an insufficient space. As was expected, J-log method can improve considerably the uncertainties of initial group flux due to adding the physical constraint. Further, from comparisons of the linear estimation method (J-1) with the non-linear estimation method (J-log) through a sample problem, it can be seen that the J-1 method gives an overestimation for output uncertainty, especially for the problem with larger uncertainties of initial neutron spectrum than about 50%, and that these two solutions are almost equivalent when the initial uncertainties are less than about 50%.

Another is a preparation of the dosimetry cross section with covariance library. It has been generated from the ENDF/B-V dosimetry file by the NJOY system (3) and was represented with 103 energy group structure. And it was expressed by the Boxer format developed by the LASL for the reduction of data, because it is difficult to handle easily as input data since the covariance matrix of cross section needs enormous amount of memory area even for one reaction.

SYSTEMATIC STUDY ON SPECTRAL COVARIANCE EFFECTS

Problems Relating to The Uncertainties of Calculational Spectrum

As stated before, uncertainties of the neutron spectrum are not directly available from the present neutronics calculations. Consequently, they become the most troublesome and the largest of error origins in the adjustment calculation.

In general, the origins of uncertainty of calculated neutron spectra may be roughly summarized as follows:
(1) microscopic cross sections and the production of group constants
(2) calculational model
(3) numerical method (discrete ordinate transport calculation)
(4) fission spectrum, and so on.
These are basically caused by the discrete approximation of data. In fact, there are nothing at all about the systematic procedures to estimate an accuracy of neutron spectrum calculation. The current sensitivity studies are insufficient due to the lack of nuclear data covariance file.

We cannot generate the prior probability density function of initial neutron spectrum because the uncertainties of calculational spectrum are not known. In the NEUPAC-83, practically, the prior probability density function is assumed artifically by using the normal distribution function. However, in such a case one should pay an attention to use the aritificial spectral covariance.

As a simple expample, Figure 1 shows a spectral covariance effect. This is a dependency of chi-square value in the statistical check of initial neutron spectrum on the correlation width defined later. As is shown in this figure, it is considered that there is the effect of the correlation sensitively.

Procedures of Systematic Study

A sample is a dosimetry analysis of the JOYO MK-I Core (4). The adjustments were based on five reaction-rates ($^{45}Sc(n,\gamma)$, $^{46}Ti(n,p)$, $^{59}Co(n,\gamma)$, $^{63}Cu(n,\alpha)$ and $^{237}Np(n,f)$) measurements. As an initial neutron spectrum, the DOT-3.5 calculated results in the 20 group energy structure were used.

Next, let us explain the procedures of systematic study briefly in order.

(1) assume the covariance matrix of initial neutron spectrum as follows:

$$\text{cov.}(\phi_{0g}, \phi_{0k}) = V^2\left[(1-\theta)\cdot\delta_{gk} + \theta\cdot\exp(-(g-k)^2/2\gamma^2)\right]$$

where V gives uncertainty for each group flux. The first term in large parenthesis gives purely random uncertainties and second gives short-range correlation over a range γ (called as the correlation width in this study.). θ means the amplitude of correlation and was chosen to be 0.8 in this study.

(2) obtain an allowable region of uncertainty parameter set (γ, V) from the viewpoint of the statistical consistency by the mean of chi-square test.

$$\chi^2 = \sum_{i,j}(R_i^m - R_i^c)(\overline{\Delta R_i^m \Delta R_j^m} + \overline{\Delta R_i^c \Delta R_j^c})^{-1}(R_j^m - R_j^c)$$

where $\overline{\Delta R_i^c \Delta R_j^c} = \iint \overline{\Delta\sigma_i(E_1)\Delta\sigma_j(E_2)}\phi_0(E_1)\phi_0(E_2)dE_1 dE_2 +$
$$\iint \sigma_i(E_1)\sigma_j(E_2)\overline{\Delta\phi_0(E_1)\Delta\phi_0(E_2)}dE_1 dE_2$$

(3) make adjustment calculations for each selected parameter set by the NEUPAC-83 code.

(4) determine the statistically permitted area of solutions for neutron exposure parameter. These results are expressed by using the bias factor α defined as follows:

$$\text{Bias Factor } \alpha = \frac{\text{Initial value}}{\text{Adjusted value}}$$

Results

Figure 2 shows an allowable region of uncertainty parameter set (γ,V) of initial neutron spectrum, which pass the confidence level (upper level is 90% and lower is 10%.) of chi-square test. The statistically permitted area for the adjusted total neutron flux and neutron flux greater than 1MeV is shown in Figures 3 and 4 respectively. There is found that a great difference of the spread of adjusted solution between the two figures, depending on each objective values such as total flux and flux greater than 1 MeV. In the case of total neutron flux, the distribution of the ratio of adjusted value to initial value is spread widely throughout a range of 1.1 to 2.3. On the other hand, in the case of neutron flux greater than 1MeV, the spread is narrow throughout 0.85 to 1.1. This fact means that spectral adjustments are made mainly depending on the detector coverages, which are set in the energy region higher than 1 MeV in the present problem. And it can be found the interesting minimums at near 10-20 of the correlation width γ in Figures 1 and 2. Here, let us define the degree of adjustment freedom to the initial neutron spectrum as follows:

$$\text{degree of adjustment freedom} = \frac{\text{energy group number } (=103)}{\text{correlation width } \gamma}$$

It is considered that these minimums occurs at point where the degree of freedom of adjustment calculation becomes nearly equal to the number of activation detectors used in this study.

CONCLUSIONS

It can be concluded from simple investigations that adjusted values are influenced considerably by the prior probability density function of initial neutron spectrum, and that the determination method of statistically permitted area of the solutions is very effective in the adjustment calculations, although we cannot definetly say about the spectral covariance effects to the adjusted solutions.

At present stage, there are many points which are uncertain. As a more detailed investigation, the J-log monte-carlo method(5) will be applied to this study. Also a theoretical investigation to explain such complicated phenomena is important.

REFERENCES

1. M.Nakazawa and A.Sekiguchi, "Several Applications of J1-Unfolding Method of Multiple-Foil Data to Reactor Neutron Dosimetry", The 3rd ASTM-EURATOM Symp. on Reactor Dosimetry, Ispra 1979.

2. T.Taniguchi et al., " Neutron Unfolding Package Code NEUPAC-83 ", NEUT Research Report 83-10, 1983.

3. R.E.MacFarlane et al., "The NJOY Nuclear Data Processing System", Volume 1 and 2, LA-9303-M, ENDF-32, May 1982.

4. Proceeding of 2nd PNC/DOE Specialists' Meeting on Collaborative Dosimetry Test, SA013 FWG81-01, 1981.

5. M.Nakazawa et al., "A New Adjustment Code Based on The Bayes' Theory Combined with The Monte-Carlo Technique", The 5th ASTM -EURATOM Symp. on Reactor Dosimetry, Geesthacht, Sept. 1984

FIGURE 1

Dependency of $\mathcal{X}^2$-value on Correlative Width γ of Initial Covariance Matrix

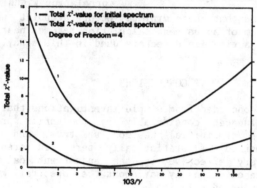

1 — Total $\mathcal{X}^2$-value for initial spectrum
2 — Total $\mathcal{X}^2$-value for adjusted spectrum
Degree of Freedom = 4

FIGURE 2

Parameter (γ, V) Allowable Through $\mathcal{X}^2$-Test

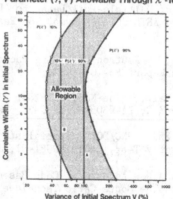

FIGURE 3

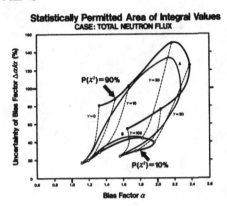

Statistically Permitted Area of Integral Values
CASE: TOTAL NEUTRON FLUX

FIGURE 4

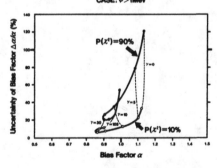

Statistically Permitted Area of Integral Values
CASE: $\phi > 1\text{MeV}$

NEUTRON ENERGY SPECTRUM CALCULATIONS IN THREE PWR

Nicholas Tsoulfanidis, D. Ray Edwards, Charles Abou-Ghantous, Keith Hock and Frank Yin

Nuclear Engineering Department, University of Missouri-Rolla
Rolla, Missouri 65401

ABSTRACT

Transport calculations of neutron energy spectra were performed for three PWR: ANO-1, ANO-2 and McGuire-1. The fluxes were calculated in R-θ geometry utilizing the eighth core symmetry. Taking the R-θ flux as a basis, a 3-D flux was synthesized by multiplying the R-θ flux with a "leakage correction" factor obtained from a 2-D R-Z flux divided by a 1-D flux. Using cross sections derived from VITAMIN-C, the neutron spectrum was obtained in terms of 26 energy groups. The calculated neutron spectra were used to compute reaction rates for several threshold reactions which were in turn compared with measured ones obtained by irradiating foils placed at several in-vessel and ex-vessel locations. Integral fluxes for neutron energy above 100 keV and above 1 Mev, as well as corresponding fluences for the lifetime of the plant, have been obtained for the location in front of the PV.

INTRODUCTION

Transport calculations of neutron energy spectra for three Pressurized Water Reactors (PWR) are presented at several in-vessel and ex-vessel locations. Specifically, data and results of this work refer to the following PWR: 1) Arkansas Nuclear One Unit 1 (ANO-1), a B&W PWR owned and operated by the Arkansas Power and Light Co (APL); 2) Arkansas Nuclear One Unit 2 (ANO-2), a CE PWR, also owned and operated by APL; 3) McGuire-1, a Westinghouse PWR owned and operated by Duke Power Co.

J. P. Genthon and H. Röttger (eds.), Reactor Dosimetry, 693–701.

The objective of this work is to develop a methodology for the most accurate calculation of the neutron flux (and fluence) hitting the Pressure Vessel (PV) of a PWR. Comparison of the calculation with experimental results is provided when such results are available. The outcomes of measurements come from two positions: a) The Surveillance Specimen Tube (SST), located in front of the PV, and b) The cavity surrounding the PV. For ANO-1, the main experimental results come from fast reaction threshold foils irradiated in the cavity. For ANO-2, experimental results exist from both the SST (in-vessel) and cavity (ex-vessel) measurements. For McGuire-1, only preliminary experimental results exist (as of August 1984).

METHOD OF CALCULATIONS

The neutron energy spectrum has been calculated using 1-D and 2-D SN transport codes. The 1-D calculations were performed using the codes ANISN and XSDRNPM. The 2-D calculations, both in R-θ and R-Z geometry were executed using the code DOT, version IV.2. The basic cross section data for this work came from ENDF/B-IV in the form of the VITAMIN-C library (1). After introducing a 1-D model of the reactor under study into XSDRNPM, the 171 neutron group VITAMIN-C cross sections were collapsed into a 47 group library. Another 47-group library, also based on VITAMIN-C, is the SAILWR library developed by SAI (2). For our work, we picked the U-238 and plutonium isotope cross sections from the SAILWR library and the cross sections for all the other isotopes from VITAMIN-C. The composite 47-neutron group set that resulted from this mixture, was collapsed into a 26-neutron group library which was used for the transport calculation. All computations were performed with a P_3 expansion of the scattering cross section and a S_8 angular quadrature.

For the 2-D R-θ geometry, one eighth of the core (a 45 degree slice) was modeled, divided into 52 angular intervals (ANO-1 and McGuire-1) or 50 angular intervals (ANO-2). About 100 mesh intervals were used in the radial direction (3). Dimensions of various core components of ANO-1 and ANO-2 were taken from drawings C-188 and M-2877, respectively. Dimensions for McGuire were provided by the Duke power Co. Reflective boundary conditions were used along the 0 and 45 degree lines, void boundary conditions were used at the edge of the R direction. The source option of DOT was employed with the neutron source calculated from direct power measurements at the plant.

A correction of the R-θ flux for the leakage along the Z-axis was implemented by calculating "leakage factors" by position and energy group and subsequently multiplying the R-θ flux by these factors according to the equation

$$\text{Leakage Factor} = \Phi_g(R,Z)/\Phi_g(R)$$

$$\Phi_g(R,\theta,Z) = \Phi_g(R,\theta)*(\text{Leakage Factor})$$

RESULTS

Tables I to III give the calculated flux at four different locations for ANO-1, ANO-2 and McGuire-1. The energy group structure is also shown in these tables. At any single location, the shape of the spectrum is essentially the same for all reactors.

The variation of the neutron flux as a function of the angle O is shown in Fig. 1. The angular variation of the flux shows the following characteristics: (a) It depends on the particular fuel loading and specifically on the power produced in the outermost row of fuel assemblies. (b) It changes as the neutrons move through the pressure vessel. (c) It becomes smoother in the cavity, i.e., the angular variation is much less pronounced in the cavity. (d) It is not generic for all plants and it changes even for the same reactor from cycle to cycle (This is a consequence of item (a) above).

Taking the maximum of the flux in front of the PV and assuming continuous operation of 40 years with a capacity factor of 80%, the neutron fluence above 0.1 MeV and 1.0 MeV hitting the PV is shown in Table IV for the three reactors ANO-1, ANO-2 and McGUIRE-1. Reaction rates, calculated and measured ones, are given in Table V. The agreement is within 20% or less for the fast neutron threshold reactions except for the SST position of ANO-2 for which the difference is close to 30%. For ANO-2, independent measurements and calculations have been performed by Battelle and are presented at this conference (4).

ACKNOWLEDGEMENTS

The authors gratefully acknowledge the financial aid provided by EPRI and, in particular, wish to thank Thomas Passell for his continuous support and many helpful discussions and suggestions. They are also grateful to the Arkansas Power & Light Co and the Duke Power Co for providing all the plant data necessary for the calculations.

695

REFERENCES

1. R.W.Roussin, C.R.Weisbin, J.E.White, N.M.Greene, R.Q.Wright and J.B.Wright, "VITAMIN-C: The CTR Proceed Multigroup Cross Section Library for Neutronic Studies", ORNL/RSIC-37 (1980).
2. G.L.Simmons, SAI Company, La Jolla CA. (private communication)
3. N.Tsoulfanidis et al, "Radiation Metrology Techniques, Data Bases, and Standardization", Proceedings of the 4th ASTM-EURATOM Symposium on Reactor Dosimatry, NBS, Gaithersburg MD, March 1982, pp. 519-532.
4. M.P.Manahan, A.R.Drosenfield, C.W.Marschall and M.P.Landow, Battelle's Columbus Laboratories Reactor Vessel Surveillance Service Activities, 5th ASTM-EURATOM Symposium on Reactor Dosimetry, GKSS, Geesthacht, FRG, September 1984.

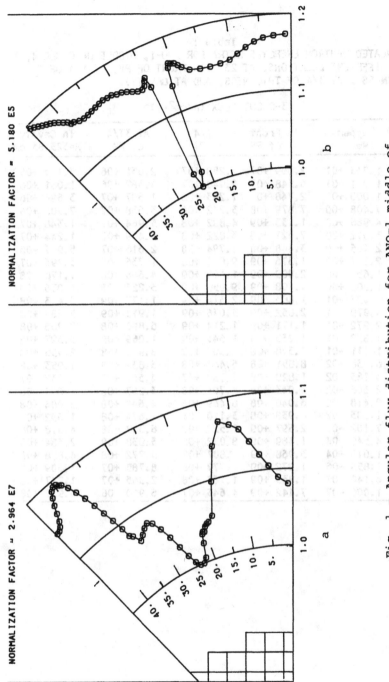

NORMALIZATION FACTOR = 2.964 E7

NORMALIZATION FACTOR = 5.180 E5

a

b

Fig. 1 - Angular flux distribution for ANO-1 middle of cycle 4, in R-θ geometry; a) in front of P.V.; b) in cavity at foil position.

Table I

CALCULATED NEUTRON ENERGY SPECTRA FOR ANO-1, MIDDLE OF CYCLE 4, AT FOUR DIFFERENT LOCATIONS, AT 23° (IN FRONT OF PV, AT 1/4 OF THICKNESS, AT 3/4 OF THICKNESS, AND AT CAVITY).

3-D DOT FLUX (n/cm$^2 \cdot$s)

GROUP	E$_{upper}$ (MeV)	In Front of PV	At T/4 of PV	At 3T/4 of PV	In Cavity R=323.83 cm
1	1.733 +01	2.699 +07	1.258 +07	2.051 +06	5.175 +05
2	1.221 +01	6.548 +07	2.940 +07	4.382 +06	1.061 +06
3	1.000 +01	2.748 +07	1.205 +08	1.607 +07	3.694 +06
4	7.408 +00	7.879 +08	3.242 +08	3.638 +07	7.908 +06
5	4.966 +00	1.133 +09	4.882 +08	5.964 +07	1.348 +07
6	3.012 +00	7.621 +08	3.822 +08	5.409 +07	1.244 +07
7	2.466 +00	3.448 +08	1.794 +08	2.616 +07	5.967 +06
8	2.307 +00	1.376 +09	9.010 +08	1.736 +08	3.893 +07
9	1.653 +00	2.000 +09	1.688 +09	4.666 +08	1.176 +08
10	1.003 +00	1.108 +09	9.956 +08	3.227 +08	1.036 +08
11	7.427 −01	1.863 +09	2.314 +09	1.177 +09	3.413 +08
12	4.979 −01	2.052 +09	3.036 +09	1.915 +09	5.289 +08
13	2.972 −01	1.171 +09	1.214 +09	6.890 +08	2.908 +08
14	1.832 −01	1.235 +09	1.643 +09	1.069 +09	3.327 +08
15	1.111 −01	8.378 +08	1.303 +09	8.928 +08	2.556 +08
16	6.738 −02	8.851 +08	6.444 +08	3.832 +08	1.953 +08
17	3.183 −02	1.820 +08	7.730 +07	4.531 +07	2.832 +07
18	2.606 −02	3.043 +08	4.703 +08	3.287 +08	1.094 +08
19	2.418 −02	6.040 +08	6.469 +08	4.641 +08	2.004 +08
20	1.503 −02	6.938 +08	3.110 +08	1.674 +08	1.652 +08
21	7.102 −03	2.697 +09	1.528 +09	6.367 +08	4.918 +08
22	4.540 −04	1.489 +09	8.032 +08	3.038 +08	2.364 +08
23	1.013 −04	3.968 +09	1.807 +09	5.272 +08	4.828 +08
24	1.855 −06	1.272 +09	3.777 +08	8.785 +07	1.309 +08
25	4.140 −07	1.835 +09	1.767 +08	2.365 +07	1.247 +08
26	1.000 −07	7.442 +09	4.645 +07	6.920 +06	3.937 +08

Table II

CALCULATED NEUTRON ENERGY SPECTRA FOR ANO-2, MIDDLE OF CYCLE 4, AT FOUR DIFFERENT LOCATIONS, AT 20° (IN FRONT OF PV, AT 1/4 OF THICKNESS, AT 3/4 OF THICKNESS, AND IN CAVITY).

3-D DOT FLUX $(N/CM^2 \cdot S)$

GROUP	E_{upper} (MeV)	In Front of PV	At T/4 of PV	At 3T/4 of PV	At Foil Position
1	1.733 +01	7.373 +07	3.189 +07	5.924 +06	1.622 +06
2	1.221 +01	2.330 +08	9.728 +07	1.659 +07	4.384 +06
3	1.000 +01	1.265 +09	5.148 +08	7.927 +07	1.990 +07
4	7.408 +00	4.565 +09	1.724 +09	4.029 +08	5.359 +07
5	4.966 +00	7.486 +09	2.910 +09	4.029 +08	9.773 +07
6	3.012 +00	5.313 +09	2.383 +09	3.752 +08	9.108 +07
7	2.466 +00	2.438 +09	1.129 +09	1.831 +08	4.406 +08
8	2.307 +00	9.928 +09	5.725 +09	1.195 +09	2.754 +08
9	1.653 +00	1.492 +10	1.104 +10	3.189 +09	8.080 +08
10	1.003 +00	8.237 +09	6.600 +09	2.302 +09	7.108 +08
11	7.427 −01	1.460 +10	1.602 +10	8.005 +09	2.258 +09
12	4.979 −01	1.567 +10	2.041 +10	1.204 +10	3.328 +09
13	2.972 −01	8.603 +09	8.400 +09	4.581 +09	1.932 +09
14	1.832 −01	9.282 +09	1.138 +10	6.923 +09	2.164 +09
15	1.111 −01	6.773 +09	8.073 +09	5.007 +09	1.585 +09
16	6.738 −02	6.212 +09	4.538 +09	2.509 +09	1.320 +09
17	3.183 −02	9.359 +08	4.206 +08	2.304 +08	2.006 +08
18	2.606 −02	2.570 +09	3.218 +09	2.150 +09	7.005 +08
19	2.418 −02	4.319 +09	4.496 +09	3.083 +09	1.360 +09
20	1.503 −02	4.580 +09	2.194 +09	1.196 +09	1.119 +09
21	7.102 −03	1.829 +10	9.711 +09	4.194 +09	3.355 +09
22	4.540 −04	9.975 +09	4.705 +09	1.853 +09	1.626 +09
23	1.013 −04	2.731 +10	1.172 +10	3.715 +09	3.458 +09
24	1.855 −06	8.374 +09	2.297 +09	6.348 +08	9.823 +08
25	4.140 −07	1.073 +10	8.887 +08	1.851 +08	1.025 +09
26	1.000 −07	4.889 +10	8.640 +08	1.291 +08	3.623 +09

Table III
CALCULATED NEUTRON ENERGY SPECTRA FOR MCGUIRE-1 CYCLE 1, AT FOUR
DIFFERENT LOCATIONS AT 20°(IN FRONT OF PV, AT 1/4 OF THICKNESS,
AND AT 3/4 OF THICKNESS, AND AT CAVITY).

3-D DOT FLUX

GROUP	E_{upper} (MeV)	In Front of PV	At T/4 of PV	At 3T/4 of PV	At Foil Position
1	1.733 +01	5.062E+07	2.105E+07	3.374E+06	3.020E+05
2	1.221 +01	1.567E+08	6.249E+07	9.117E+06	7.764E+05
3	1.000 +01	8.316E+08	3.236E+08	4.212E+07	3.293E+06
4	7.408 +00	2.893E+09	1.048E+09	1.139E+08	8.135E+06
5	4.966 +00	4.424E+09	1.662E+09	1.940E+08	2.014E+07
6	3.012 +00	3.001E+09	1.331E+09	1.793E+08	2.193E+07
7	2.466 +00	1.360E+09	6.255E+08	8.702E+07	1.135E+07
8	2.307 +00	5.295E+09	3.126E+09	5.775E+08	9.912E+07
9	1.653 +00	7.603E+09	5.901E+09	1.572E+09	4.448E+08
10	1.003 +00	4.142E+09	3.448E+09	1.125E+09	4.510E+08
11	7.427 -01	6.947E+09	8.317E+10	4.178E+09	2.546E+09
12	4.979 -01	7.404E+09	1.087E+10	6.870E+09	4.565E+09
13	2.972 -01	4.298E+09	4.316E+09	2.495E+09	2.449E+09
14	1.832 -01	4.468E+09	6.069 +09	4.085E+09	3.419E+09
15	1.111 -01	3.380E+09	4.375E+09	3.079E+09	2.604E+09
16	6.738 -02	3.362E+09	2.360E+09	1.480E+09	2.024E+09
17	3.183 -02	6.466E+08	2.369E+08	1.473E+08	3.793E+08
18	2.606 -02	1.153E+09	1.697E+09	1.260E+09	1.490E+09
19	2.418 -02	2.209E+09	2.410E+09	1.866E+09	2.812E+09
20	1.503 -02	2.657E+09	1.170E+09	7.405E+08	1.946E+09
21	7.102 -03	1.016E+10	5.142E+09	2.643E+09	5.942E+09
22	4.540 -04	5.600E+09	2.547E+09	1.197E+09	2.867E+09
23	1.013 -04	1.473E+10	5.865E+09	2.142E+09	5.993E+09
24	1.855 -06	4.670E+09	1.151E+09	3.590E+08	1.680E+09
25	4.140 -07	6.696E+09	4.592E+08	9.626E+07	1.772E+09
26	1.000 -07	2.712E+10	3.376E+08	4.336E+07	5.392E+09

TABLE IV

CALCULATED NEUTRON FLUENCE AT THE CORE MIDPLANE AND IN FRONT OF
PV FOR THE REACTORS LIFETIME (40 YEARS, 80% CAPACITY FACTOR).

neutron fluence (n/cm^2)

	ANO-1 at 23°	ANO-2 at 10°	McGUIRE-1 at 20°
E > 1.0 MeV	6.833E+18	5.535E+19	2.585E+19
E > 0.1 MeV	1.452E+19	1.266E+20	5.422E+19

TABLE V
REACTION RATES AT THE REACTOR MID-PLANE, IN REACTIONS PER ATOM PER SECOND.

ANO-1 At The Foil Position In Cavity (23°)

reaction	UMR	Measured[@]	C/E[*]
Cu-63(n,a)	2.012E-19	1.70E-19	1.18
Ti-46(n,p)	2.649E-18	2.83E-18	0.93
Ni-58(n,p)	1.797E-17	1.95E-17	0.92
Fe-54(n,p)[$]	1.295E-17	1.34E-17	0.96

ANO-2 At The SST Middle Compartiment

reaction	UMR	Measured[#]	C/E
Cu-63(n,a)	1.185E-16	8.82E-17	1.34
Ti-46(n,p)	2.045E-15	1.56E-15	1.31
Ni-58(n,p)	1.419E-14	1.06E-14	1.34
Fe-54(n,p)[$]	1.102E-14	8.23E-15	1.34

ANO-2 At Foil Position In The Cavity (21°)

reaction	UMR	Measured[@]	C/E
Cu-63(n,a)	1.036E-16	8.71E-19	1.19
Ti-46(n,p)	1.560E-17	1.34E-17	1.16
Ni-58(n,p)	1.310E-16	9.94E-17	1.22
Fe-54(n,p)[$]	8.563E-17	7.03E-17	1.22

McGUIRE At Foil Position (3-1-1) In the Cavity

reaction	UMR	Mesurment[@]	C/E
Cu-63(n,a)	4.871E-19	4.31E-19	1.13
Ti-46(n,p)	6.886E-18	6.47E-18	1.06
Ni-58(n,p)	4.995E-17	4.31E-17	1.16
Fe-54(n,p)[$]	3.551E-17	2.88E-17	1.23

[@] University of Arkansas.
[*] Calculated to experimental values ratios.
[$] Cadmium covered foil.
[#] Battelle, Columbus Laboratories.

ON THE CONVERSION OF COARSE GROUP SPECTRA TO FINE GROUP SPECTRA
(USING A CONTINUITY PRINCIPLE)

J. Végh*

Netherlands Energy Research Foundation ECN, Petten, The Netherlands.

*Guest from Central Research Institute for Physics, Budapest, Hungary.

ABSTRACT

An approach is described for converting neutron spectra given in coarse group representations to fine group
structures. The method is based on a spline interpolation
imposed on the cumulative integral function of the coarse
group spectrum. During the conversion the fluence rates
between the original energy boundaries remain unchanged,
and the resulting fluence rate distribution has a continuous first derivative. Various types of neutron spectra
(CFRMF, ORR, YAYOI), given in different coarse group
structures, were converted to the SAND-II 640 group representation by the procedure. The obtained spectra were
used to condense fine group cross-section data to the
original group structures. The spectrum averaged cross-
sections calculated from the different coarse group representations of the same neutron spectrum show a good
agreement with the reference values. The proposed method
therefore can be used to construct an appropriate weighting function, if reaction rates are to be calculated in
coarse group representations.

INTRODUCTION

Fluence rate conversion methods, recently used in connection
with spectrum adjustment and other neutronics calculations, are far
from optimal, if applied to spectra given in coarse (broad) group
structures. The applied procedures (e.g. log-log or cubic spline

J. P. Genthon and H. Röttger (eds.), Reactor Dosimetry, 703–710.
© 1985 ECSC, EEC, EAEC, Brussels and Luxembourg

interpolation imposed on point data representing the spectrum) often fail to conserve the original physical information, i.e. the broad group fluence rates and the spectrum shape. Other procedures, which are 'conservative' in the above mentioned sense, often introduce physically unjustified structure in the resulting spectrum. Furthermore, if reaction rates (e.g. displacement or gas-production rates) are to be calculated in a coarse group structure, a suitable weighting spectrum has to be applied to condense the fine group cross-section data. If the applied weighting function is not a proper representation of the given neutron spectrum, then the condensed cross-section values may deviate considerably from their real values. This effect was shown for various spectra, see e.g. (1) for the ORR and YAYOI, (2) for the CFRMF, STEK-4000 and HFR, (3) for the CFRMF. The interpolation method given in the present paper was elaborated to avoid these difficulties: the intention was to produce a physically acceptable smooth spectrum shape, while conserving the original physical information, as well.

SPLINE INTERPOLATION IMPOSED ON THE CUMULATIVE INTEGRAL FUNCTION OF THE COARSE GROUP SPECTRUM

The coarse group spectrum is given by the group energy boundaries (E_i, i=1,2,...,n+1 in ascending sequence) and the corresponding fluence rate per unit lethargy values (ϕ_u^i, i=1,2,...n). The cumulative integral function of the spectrum (i.e. the fluence rate as a function of lethargy) is defined by

$$\phi(u_i) = \int_{u_i}^{u_1} \phi_u(u)du = \sum_{k=1}^{i-1} \phi_u^k (u_k - u_{k+1}) \quad (i=2,3,...,n+1) \quad (1)$$

where $u_i = \ln(E_{n+1}/E_i)$ for i=1,2,...,n+1; and $u_1 = \ln(E_{n+1}/E_1) = u_{max}$.

An appropriate F(u) interpolation function, which conserves the coarse group fluence rates, must fulfil the $F(u_i) = \phi(u_i)$ condition for all i. If a piecewise polynomial interpolation scheme is to be used, then at least cubic interpolation must be applied to $\phi(u)$, in order to have a continuous first derivative for $\phi_u(u)$. In (4) a piecewise cubic F(u) was proposed to solve the conversion problem, the method was mainly based on the theory outlined in (5). However, the cubic spline interpolation occasionally produces extraneous inflection points in the resulting curve. In our case these extra inflection points lead to unacceptable negative fluence rates. Different procedures are available to eliminate these irregularities (see e.g. (5)), in the present approach the 'spline in tension' concept is used, which contains also the cubic spline as special case (6). The F(u) interpolation function has the following form in each coarse group (i=1,2,...,n):

$$F_i(u) = C_{1i} + C_{2i}(u-u_i) + C_{3i} e^{p(u-u_i)} + C_{4i} e^{-p(u-u_i)} \quad (2)$$

where $p \geq 0$ is the 'tension parameter', which determines the allowed undulation of the resulting function.

The $F(u)$ exponential spline is constructed to fulfill the following conditions:

$$F_i(u_{i+1}) = \phi(u_{i+1}) \ (i=1,2,\ldots,n); \quad F_i'(u_{i+1}) = F_{i+1}'(u_{i+1}) \ (i=1,2,\ldots,n-1) \tag{3}$$

$$F_i(u_{i+1}) = F_{i+1}(u_{i+1}) \ (i=1,2,\ldots,n-1); \quad F_i''(u_{i+1}) = F_{i+1}''(u_{i+1}) \ (i=1,2,\ldots,n-1)$$

The fluence rate conservation and the continuity conditions give $4n-3$ equations for the $4n$ unknown C_{ij} coefficients. Three additional equations can be established as follows:

$$F_1(u_1) = \phi(E<E_1) \tag{4}$$

$$F_1'(u_1) = s_1 \quad \text{and} \quad F_n'(u_{n+1}) = s_n \tag{5}$$

where $\phi(E<E_1)$ is the fluence rate below the first energy boundary. The s_1 and s_n slopes are estimated values from the available boundary condition (i.e. spectrum shape) information.

After $F(u)$ is constructed, the fluence rate of the group determined by the arbitrary $x_1 > x_2$ lethargies is simply given by $F(x_2) - F(x_1)$. Around the two endpoints the interpolation procedure is extremely sensitive to the proper choice of the endslopes. Improper estimations for s_1 and/or s_n may lead to negative fluence rates, as in the case of a cubic spline interpolation. However, using appropriate s_1 and s_n slopes, these difficulties can be avoided.

The resulting linear system of the C_{ij}-s is strictly diagonally dominant and can be solved by straightforward elimination without pivoting (see (6)).

CALCULATIONS FOR TESTING THE MERITS OF THE CONVERSION METHOD

A computer program, STELLA, was written to realize the described interpolation function for an arbitrary neutron spectrum. For the construction and evaluation of the spline two 'tailor-made' routines CURV1 and CURV2 of (6) were used, the applied tension factor was 1,0. The test examples were the ORR, YAYOI and CFRMF spectra. The ORR and YAYOI were available in 100, 50 and 20 group representations (see (1)). The CFRMF spectrum was taken from (7) in 620 groups, containing non-zero fluence rate values above 10^{-7} MeV. The spectrum part above 10^{-6} MeV was used in our calculations as reference, the high end was extrapolated to 20 MeV by a Watt fission spectrum. The CFRMF spectrum is particularly suitable for data testing purposes, since real fine details are present in the spectrum between 10^{-6} MeV and $2,7 \cdot 10^{-3}$ MeV (see figs. 1 and 4). The 640 group CFRMF spectrum was condensed to the 68 and 26 coarse group structure of (3), by the program FITOCO (2).

All the coarse group spectra were converted to 640 groups by the
spline method. The s_n endslope was estimated in each case assuming
a Watt fission spectrum shape. For the YAYOI and CFRMF s_1 was estimated assuming a $\phi_E(E) \sim E^{\frac{1}{2}}$ shape at the low energy end, while the
thermal part of the ORR was covered by a Maxwellian- joined to a
1/E-distribution. Figure 2 gives the CFRMF spectrum obtained from
68 groups, figs. 6 and 5 show the fine group representation of ORR
obtained from 20 and 100 groups, respectively.

The obtained fine group spectra were used to calculate spectrum
averaged cross-section values, in order to investigate their reproducibility from the different group structures. For the ORR and YAYOI
the reactions were those of the REAL-80 exercise (1); for the CFRMF
the reactions tested in (3) were applied. The fine group cross-
section data were taken from (8), which is mainly based on ENDF/B-V
(version 2) data. The response regions of the applied reactions are
given in figure 4 and 5 for the CFRMF and ORR, respectively.

For each reaction the vertical bar indicates the median response
energy, the horizontal bars correspond to the 90%, 60% and 30%
response regions. The intercomparison of the spectrum averaged cross-
section data, calculated from the different coarse group representations of the test spectra, is given in table 1.

DISCUSSION

The intercomparison given in table 1 shows, that for the great
majority of the reactions the calculated average cross-sections are
identical in each representation of the spectrum. The spline interpolation therefore can be used to construct a suitable weighting
function even in those cases, when the number of groups is limited
(i.e. 26 or 20). However, for certain reactions, of which the response
originates from a very narrow energy region (e.g. $^{59}Co(n,\gamma)$-CFRMF
and $^{197}Au(n,\gamma)$-ORR), the average cross-section calculated from a
limited number of groups may deviate significantly from its real
value. These local deviations originate from the fact, that no
interpolation method can recover the fine spectral information,
which was lost by the averaging procedure during the fluence rate
condensation. This loss of information is illustrated in figure 3
for the CFRMF. Obviously, the detailed information which is present
in the CFRMF 640 group spectrum cannot be reconstructed from a much
coarser group structure by any interpolation method. Only some
integral quantities (coarse group fluence rates) can be recovered
reliably. Despite the deficiencies indicated by an asterisk in table 1, the spline method generally gives much better agreement between the average cross-sections calculated from the different representations, than the recently applied other conversion procedures.
As it was shown in (1) and (4), these methods give much higher deviations in general, especially for reactions with high threshold
energies.

REFERENCES

1. W.L. Zijp et al.: "Final report on the REAL-80 exercise",
   Report ECN-128 (February 1983).

2. H.Ch. Rieffe: "FITOCO. A program for the conversion of fine
   group flux density and cross-section data to coarse group values",
   Report ECN-92 (April 1981).

3. R.A. Anderl et al.: "INEL integral data-testing report for
   ENDF/B-V dosimeter cross-sections",
   Report EGG-PHYS-5608 (1981).

4. J. Végh: "On the use of a spline interpolation method for converting coarse group spectra to fine group representations",
   Report ECN-84-026 (February 1984).

5. Carl de Boor: "A Practical Guide to Splines",
   Applied Mathematical Sciences, Vol. 27 (Springer Verlag, New
   York, 1978).

6. A.K. Cline: "Scalar- and planar-valued curve fitting using
   splines under tension",
   Communications of the ACM, Vol. 17, No. 4 (1974).

7. R.A. Anderl et al.: "Addendum to integral data-testing report
   for ENDF/B-V dosimeter cross-sections",
   Report EGG-PHYS-5668 (1982).

8. W.L. Zijp, H.J. Nolthenius, G.C.H.M. Verhaag: "Cross-section
   library DOSCROS84 (in a 640 groups structure of the SAND-II
   type)",
   To be published as ECN-report (1984).

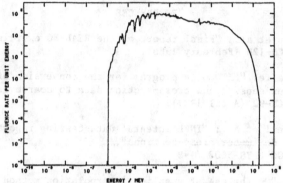

Fig.1. CFRMF spectrum in 640 groups (7).

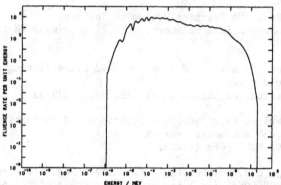

Fig.2. CFRMF spectrum converted from 68 groups.

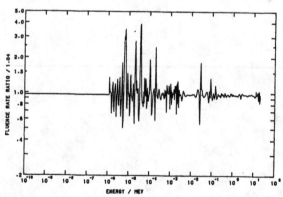

Fig.3. Ratio of the spectra plotted in figs. 1 and 2.

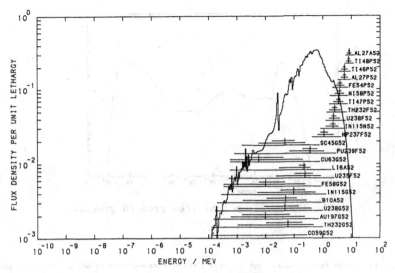

Fig.4. Response regions for the CFRMF spectrum.

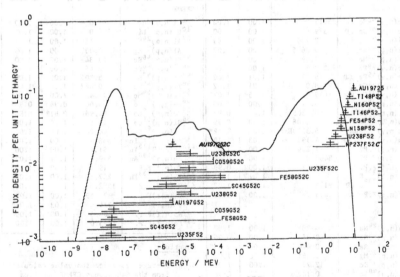

Fig. 5. Response regions for the ORR spectrum.

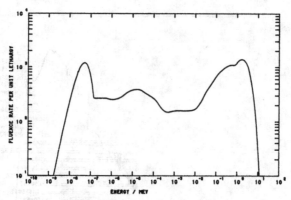

Fig.6. ORR spectrum converted from 20 groups.

Table 1. Intercomparison of calculated spectrum averaged cross-section values for the test spectra.

reaction name	σ_{100} (in mb)	$\dfrac{\sigma_{50}}{\sigma_{100}}$	$\dfrac{\sigma_{20}}{\sigma_{100}}$	reaction name	σ_{640} (in mb)	$\dfrac{\sigma_{68}}{\sigma_{640}}$	$\dfrac{\sigma_{26}}{\sigma_{640}}$
ORR				**CFRMF**			
$^{45}Sc(n,\gamma)^{46}Sc$	5913.0	1.00	1.00	$^{6}Li(n,\alpha)^{3}H$	891.0	1.00	1.00
$^{45}Sc(n,\gamma)^{46}Sc$ CD	329.6	1.00	1.00	$^{10}B\ (n,\alpha)^{7}Li$	1598.0	1.00	1.01
$^{46}Ti(n,p)^{46}Sc$	2.706	1.00	0.99	$^{27}Al(n,\alpha)^{24}Na$	0.1583	1.00	1.01
$^{48}Ti(n,p)^{48}Sc$	0.0695	1.00	1.02	$^{27}Al(n,p)^{27}Mg$	0.9391	1.00	1.00
$^{54}Fe(n,p)^{54}Mn$	19.99	1.00	1.00	$^{45}Sc(n,\gamma)^{46}Sc$	23.34	1.00	1.00
$^{58}Fe(n,\gamma)^{59}Fe$	263.5	1.00	1.00	$^{46}Ti(n,p)^{46}Sc$	2.455	1.00	1.00
$^{58}Fe(n,\gamma)^{59}Fe$ CD	26.37	1.01	1.00	$^{47}Ti(n,p)^{47}Sc$	5.305	1.00	1.00
$^{59}Co(n,\gamma)^{60}Co$	9398.0	1.01	1.02	$^{48}Ti(n,p)^{48}Sc$	0.062	1.00	1.00
$^{59}Co(n,\gamma)^{60}Co$ CD	1659.0	1.05*	1.07*	$^{54}Fe(n,p)^{54}Mn$	18.42	1.00	1.01
$^{58}Ni(n,p)^{58}Co$	26.21	1.00	1.00	$^{58}Fe(n,\gamma)^{59}Fe$	6.607	0.96	0.95*
$^{60}Ni(n,p)^{60}Co$	0.6305	1.00	0.99	$^{59}Co(n,\gamma)^{60}Co$	84.96	0.81*	0.86*
$^{197}Au(n,2n)^{196}Au$	0.8408	1.00	0.99	$^{58}Ni(n,p)^{58}Co$	24.54	1.00	1.00
$^{197}Au(n,\gamma)^{198}Au$	6.3530E+4	1.02	1.12*	$^{63}Cu(n,\gamma)^{64}Cu$	44.20	1.00	1.01
$^{197}Au(n,\gamma)^{198}Au$ CD	4.1950E+4	1.03	1.18*	$^{115}In(n,\gamma)^{116}In$	338.40	1.00	1.01
$^{235}U\ (n,f)F.P.$	1.2540E+5	1.00	1.00	$^{115}In(n,n')^{115}In$	51.43	1.00	1.00
$^{235}U\ (n,f)F.P.$	8460.0	1.00	0.99	$^{197}Au(n,\gamma)^{198}Au$	379.0	1.01	1.02
$^{238}U\ (n,f)F.P.$	82.42	1.00	0.99	$^{232}Th(n,f)F.P.$	19.14	1.00	1.00
$^{238}U\ (n,\gamma)^{239}U$	1.003E+4	1.01	1.02	$^{232}Th(n,\gamma)^{233}Th$	257.8	1.00	1.01
$^{238}U\ (n,\gamma)^{239}U$ CD	9195.0	1.01	1.02	$^{235}U\ (n,f)F.P.$	1552.0	1.00	1.00
$^{237}Np(n,f)F.P.$ CD	414.2	1.00	1.00	$^{238}U\ (n,f)F.P.$	79.59	1.00	1.00
				$^{238}U\ (n,\gamma)^{239}U$	216.6	1.03	1.06*
YAYOI				$^{237}Np(n,f)F.P.$	606.5	1.00	1.00
$^{23}Na(n,\gamma)^{24}Na$	0.385	1.00	1.00	$^{239}Pu(n,f)F.P.$	1773.0	1.00	1.00
$^{24}Mg(n,p)^{24}Na$	0.955	1.00	0.96				
$^{27}Al(n,\alpha)^{24}Na$	0.474	1.00	0.95*				
$^{27}Al(n,p)^{27}Mg$	2.361	1.00	1.00				
$^{47}Ti(n,p)^{47}Sc$	12.77	1.00	1.00				
$^{48}Ti(n,p)^{48}Sc$	0.182	1.00	0.96				
$^{55}Mn(n,\gamma)^{56}Mn$	4.254	1.00	1.00				
$^{56}Fe(n,p)^{56}Mn$	0.623	1.00	0.98				
$^{59}Co(n,\alpha)^{56}Mn$	0.096	1.00	0.96				
$^{58}Ni(n,p)^{58}Co$	58.77	1.00	1.00				
$^{115}In(n,n')^{115}In$	113.20	1.00	1.00				
$^{186}W\ (n,\gamma)^{187}W$	48.95	1.00	1.00				
$^{197}Au(n,\gamma)^{198}Au$	120.40	1.00	1.00				

Notation:

σ_{640}: average cross-section value calculated from the 640 groups CFRMF spectrum given in (7).

σ_{100} (68, 50, 26 or 20): average cross-section calculated from the 640 groups spectrum obtained from the 100 (68, 50, 26 or 20) coarse group representation by the spline method.

* : outstanding ($\geq$ 5%) deviation compared to the reference value.

CALCULATION OF THE NEUTRON SOURCE DISTRIBUTION

IN THE VENUS PWR MOCKUP EXPERIMENT

M.L. Williams, P. Morakinyo; F.B.K. Kam; L. Leenders, G. Minsart, and A. Fabry

Louisiana State Univ.; Oak Ridge National Lab.; and CEN/SCK

Baton Rouge, LA, USA; Oak Ridge, TN, USA; and Mol, Belgium

ABSTRACT

The VENUS PWR Mockup Experiment is an important component of the Nuclear Regulatory Commission's program goal of benchmarking reactor pressure vessel (RPV) fluence calculations in order to determine the accuracy to which RPV fluence can be computed. Of particular concern in this experiment is the accuracy of the source calculation near the core-baffle interface, which is the important region for contributing to RPV fluence.

Results indicate that the calculated neutron source distribution within the VENUS core agrees with the experimental measured values with an average error of less that 3%, except at the baffle corner, where the error is about 6%. Better agreement with the measured fission distribution was obtained with a detailed space-dependent cross-section weighting procedure for thermal cross sections near the core-baffle interface region. The maximum error introduced into the predicted RPV fluence due to source errors should be on the order of 5%.

INTRODUCTION

Radiation embrittlement of reactor pressure vessels (RPV) has recently been a concern of the nuclear industry because of the possibility that rapid cooling could lead to the failure of a brittle vessel. The U.S. Nuclear Regulatory Commission (USNRC) and the U.S. and European nuclear industries are currently conducting studies to determine the ability of PWR vessels to withstand severe thermal shocks without compromising their integrity. One of the major components of this research consists of benchmarking RPV fluence determination methods, since the RPV fluence is a determining factor in the degree of radiation embrittlement. An important part of the on-going RPV benchmark studies is called the "VENUS PWR Engineering Mockup Experiment."

Earlier benchmark experiments have focused on validating the accuracy of ex-core transport calculations to predict neutron fluence; however, they did not address the problem of determining the core fission source distribution which drives the RPV fluence calculation. Of particular concern is the accuracy of the source calculation near the core-baffle interface which is the important region for contributing to RPV fluence. The PWR Engineering Mockup Experiment was designed primarily to address this problem. The experimental work is being performed by CEN/SCK (Centre d'Etudie de l'Energie Nucleaire/Studie Centrum voor Kernenergie) at the VENUS Critical Facility in Mol, Belgium, while the calculational study is being done by both Mol and by the Louisiana State University Nuclear Science Center under subcontract to the Oak Ridge National Laboratory (ORNL).

The primary objective of this study is to compute the VENUS core neutron source distribution and compare with measured values to contribute to USNRC's program goal of validating RPV fluence calculations. The calculated fission source is then used as a fixed source for dosimeter calculations in order to ascertain the expected accuracy to which fluence can be predicted from a core eigenvalue calculation coupled with ex-core transport calculations.

EXPERIMENTAL CONFIGURATIONS

The PWR Benchmark Configuration in the VENUS Critical Facility is shown in Fig. 1. The central portion of the geometry is water, surrounded by a 2.858-cm thick inner steel baffle. The inner core zone in the immediate vicinity of the inner baffle contains 752 zircaloy-clad 3.3% enriched fuel cells, with 48 pyrex rods interspersed among them. The outer core zone contains 1800 steel-clad, 4.0%-enriched fuel cells. The core itself is surrounded by a 2.858-cm thick outer steel baffle, a water reflector, a 4.972-cm thick steel core barrel, a water gap, a neutron pad, and the reactor pool. (The neutron pad is not shown in Fig. 1.)

The configuration shown in Fig. 1 was selected by Mol as the core loading best suited for the realization of the required measurements in the fuel zones, reflector, barrel, and up to the

1/4 CORE OF THE VENUS MODEL
• Dosimeter Locations (Coordinates)

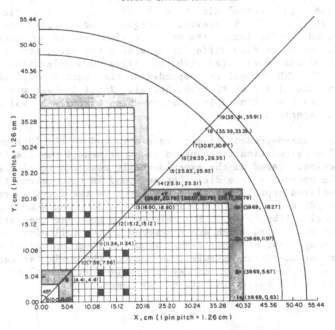

Fig. 1. One-fourth core of the VENUS model.

neutron pad. The distribution of pyrex rods in the inner zone
of the core permits criticality without boron in water, and it
shifts the power peak towards the core edges, thereby improving
the core power distribution for the ex-core measurement. Details
of the VENUS experiment configuration can be found in Ref. 3.

CALCULATIONAL METHODOLOGY

A 218-group cross section set based on ENDF/B-IV data (4) was
used as the fine-group library. The AMPX (5) modular system was
used for all the cross-section processing calculations. Resonance
self-shielding was done with the Nordheim integral method as in
the NITAWL (6) code, and the cell homogenization was performed
with fluxes obtained from one-dimensional discrete ordinates calculations (XSDRN-PM) (7) of the 3 and 4% fuel cells and the pyrex-
rod cell. The resulting 218-group resonance-shielded, cell-
averaged cross sections were collapsed to 10 broad groups using
zone-averaged fluxes computed for a one-dimensional model of the
VENUS core.

The DOT IV (8) discrete ordinates transport theory code was used to perform a two-dimensional X-Y calculation of the critical eigenvalue. This calculation used the 10-group cross-section library discussed in the previous paragraph and was performed with a P_3 Legendre expansion of the cross sections and an S_8 quadrature set. The weighted flux differencing scheme was used, and the calculation was accelerated with the diffusion acceleration option. The VENUS model corresponded to the reactor midplane, extending radially past the core barrel. Axial leakage was treated with a buckling approximation using a single B^2 value obtained experimentally.

The space-dependent fission rate obtained in the 10-group eigenvalue calculation is compared to experimental results in the next section. Based on the experimentally measured total fission rate, the neutron source distribution found in this calculation was normalized appropriately and was input as the fixed source in a second DOT X-Y calculation which used the 56-group ELXSIR cross-section library (9). The calculation provided the various dosimeter activities which are discussed in the following section. The 56-group cross sections rather than the 10-group values were used for the dosimeter calculations because the dosimeter reactions have relatively high energy thresholds, and the 10-group structure is too coarse in this energy range.

RESULTS

While performing the 218-group one-dimensional transport calculations, it was observed that the thermal neutron spectrum significantly hardens near the interface of the core and the outer baffle (10). The rapid change in the thermal spectrum within the last 2 cm of the core was found to have a noticeable effect on the collapsed thermal cross-section values used in the two-dimensional calculations. In order to account for the spatial variation in the thermal cross sections, it was necessary to use a separately weighted set of collapsed cross sections at approximately every one-quarter centimeter within the last 2.52 cm (i.e., 2 cell widths) of the core. The spatial variation in the collapsed U-235 thermal fission cross section is shown in Table 1. The value for ^{235}U σ_f varies about 7% over this distance.

The effective multiplication factor for the two-dimensional X-Y calculations was determined to be $k_{eff} = 0.996$. This value was underpredicted by about one-half of one percent. The low value of k_{eff} is consistent with other LWR lattice studies which show that the ENDF/B-IV cross sections tend to underestimate the eigenvalue due to excessive U-238 capture estimates.

714

A comparison of the calculated and measured relative power distribution of the VENUS model is shown in Fig. 2 (10). The average agreement between calculation and experiment is within 3% error, with an uncertainty of about 1.5% in the measurements. The worst agreement has an error of 6.5%, and it occurs in a cell near the baffle corner. Disagreements of up to 3% can also be found at locations near the pyrex rods. The error introduced into the computed RPV fluence by these source discrepancies should be on the order of 5% or less.

A comparison of the calculated and measured dosimeter values is shown in Table 2. At most locations, the agreement between calculation and experiment is very good - nearly always within 10% and much better at many locations. The ^{237}Np results are an exception; the computed values are 10-17% higher than measured values in the core, and up to 30% lower in the water gap between the outer baffle and the barrel. There is also an odd value for ^{235}U in the inner baffle for which the measured value is about 35% higher than our calculation. We suspect an experimental problem here.

CONCLUSIONS

The space-dependent neutron fission rate in the VENUS core, particularly at the core periphery, can be accurately calculated with discrete ordinates transport theory. A high degree of accuracy can be obtained by using space-dependent cross-section weighting for ^{235}U thermal fission in the important outer baffle-core interface region, although this procedure may not be practical for power reactor analysis.

Comparison of calculation with measured relative power distribution indicates that the spatial neutron source can be computed to an accuracy of within 6% error near the important core-baffle region and an average agreement of about 3% error for the whole in-core area.

When the computed neutron source distribution is used in transport calculations of dosimeter activation, the results are found to agree very well with experimental measurements, except at a few locations at which the experimental values are suspect. The agreement for most dosimeters is usually better than 7%. The ^{237}Np values only agree to 10-17% within the core, however.

Overall, it is concluded that determination of the core source with a calculational procedure of this type is an adequate approach for pressure vessel fluence calculations.

Table 1. Variation of U-235 thermal fission cross section*

Distance from outer baffle (cm)	U-235 group 10/10 fission cross section (barns)
0.252	250.35
0.504	255.79
0.756	259.64
1.008	262.40
1.260	264.55
1.512	266.28
2.520	269.41
2.52-18.90	278.80

*These are collapsed values and are applicable to the 4.0% fuel region only.

Table 2. C/E values for dosimeter results

Dosimeter location (see Fig. 1)	^{115}In(n,n')	^{58}Ni(n,p)	^{238}U(n,f)	^{237}Np(n,f)	^{235}U(n,f)
1	0.99	1.08	1.06	1.07	-----
2	0.99	1.08	1.08	-----	-----
3	0.98	1.05	1.05	1.09	0.97
4	0.98	1.03	-----	1.10	0.98
5	0.98	1.02	1.07	1.10	-----
6	0.96	1.07	1.04	1.10	-----
7	0.96	1.05	1.06	1.08	0.95
8	0.98	1.07	1.01	0.97	-----
9	0.99	1.10	1.08	1.17	0.62
10	-----	-----	-----	1.14	-----
11	-----	-----	-----	1.12	-----
12	-----	-----	-----	1.14	-----
13	-----	-----	-----	1.13	-----
14	-----	-----	-----	1.01	-----
15	-----	-----	-----	0.90	-----
16	-----	-----	-----	0.70	-----
17	-----	-----	-----	0.82	-----
18	-----	-----	-----	0.76	-----
19	0.93	0.83	0.98	1.01	-----

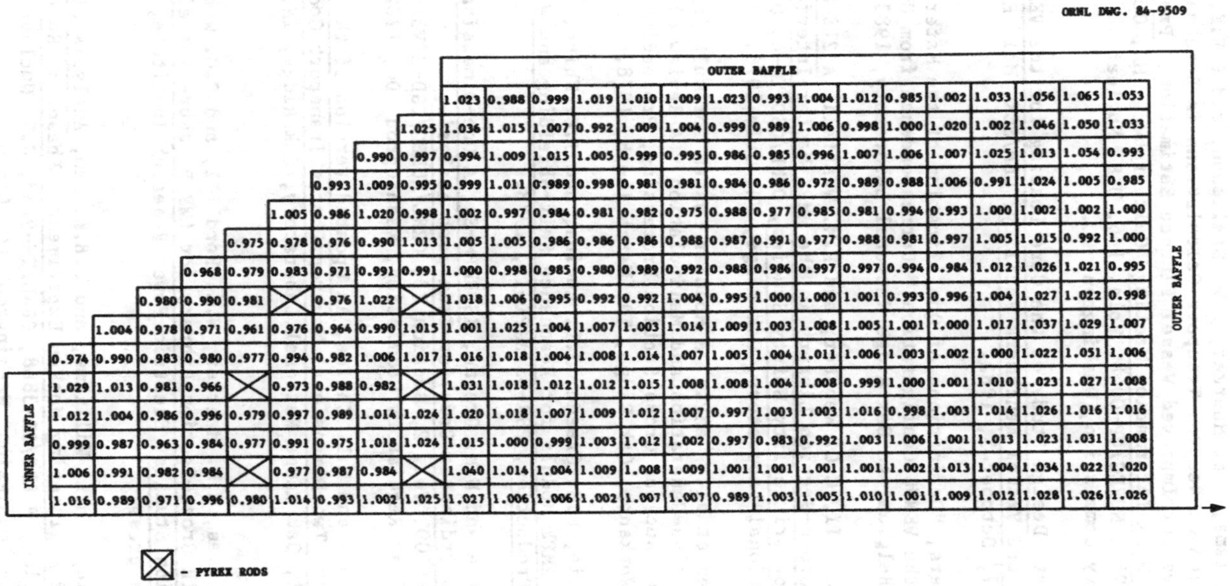

OUTER BAFFLE

INNER BAFFLE

⊠ - PYREX RODS

Fig. 2. Comparison of calculated and measured (C/E) relative power distribution.

REFERENCES

1. M.L. Williams, R.E. Maerker, F.W. Stallmann, and F.B.K. Kam, "Validation of Neutron Tranport Calculations in Benchmark Facilities for Improved Vessel Fluence Estimation," Proc. of the 11th WRSR Information Meeting, Gaithersburg, MD, October 24-28, 1983, NUREG/CP-0048, Vols. 1-6, U.S. Nuclear Regulatory Commission, Washington, DC.

2. G. Minsart, Design Study of the Core Loading for the VENUS PWR Pressure Vessel Benchmark Facility, CEN/SCK, Mol, Report 380/82-27, October 5, 1982.

3. L. Leenders, "Definitions of Qualification of the Materials Used in the VENUS Configuration," Correspondence from CEN/SCK in Mol, Belgium, to Oak Ridge National Laboratory, 1983.

4. W.E. Ford, III, C.C. Webster, and R.M. Westfall, A 218-Group Neutron Cross-Section Library in the AMPX Master Interface Format for Criticality Safety Studies, ORNL/CSD/TM-4, Oak Ridge National Laboratory, Oak Ridge, TN, July 1976.

5. N.M. Green et al., "AMPX: A Modular System for Multigroup Cross-Section Generation and Manipulation," A Review of Multigroup Nuclear Cross-Section Processing, Proceedings of a Seminar-Workshop, Oak Ridge, TN, March 14-16, 1978.

6. L.M. Petrie, N.M. Greene, J.L. Lucius, and J.E. White, NITAWL: AMPX Module for Resonance Self-Shielding and Working Library Production, PSR-63/AMPX-II, 1978.

7. L.M. Petrie and N.M. Greene, XSDRNPM-S: A One-Dimensional Discretes Ordinates Code for Transport Analysis, NUREG/CR-0200, Vol. 2, Section F3, ORNL/NUREG/CSD-2/V3/R1, U.S. Nuclear Regulatory Commission, Washington, DC, 1982.

8. W.A. Rhoades and R.L. Childs, An Updated Version of the DOT IV One- and Two-Dimensional Neutron/Photon Transport Code, ORNL-5851, Oak Ridge National Laboratory, Oak Ridge, TN, 1982.

9. M.L. Williams, R.E. Maerker, W.E. Ford III, and C.C. Webster, The ELXSIR Cross-Section Library for LWR Pressure Vessel Irradiation Studies, Electric Power Research Institute, EPRI report in press.

10. P. Morakinyo, M.L. Williams, and F.B.K. Kam, Analysis of the VENUS PWR Engineering Mockup Experiment - Phase I: Source Distribution, NUREG/CR-3888, ORNL/TM-9238, U.S. Nuclear Regulatory Commission, Washington, DC, (in press).

NEW ASPECTS IN LEAST SQUARES ADJUSTMENT METHODS

E.M. Zsolnay[x], H.J. Nolthenius[+], E.J. Szondi[x], W.L. Zijp[+]

[x]Nuclear Reactor of the Technical University, BME,
H-1521 Budapest, Hungary
[+]Netherlands Energy Research Foundation, ECN,
NL-1755 ZG Petten, The Netherlands

ABSTRACT

Radiation damage estimates are of limited use unless they are accompanied by a realistic estimate of their inaccuracy. The results of the REAL-80 international interlaboratory exercise have shown that discrepancies in prediction of radiation damage parameters and unrealistic values in the uncertainties of these data originate partly from using incomplete and/or unrealistic input data in the calculations, partly from incorrect (or not optimal) processing of the available data by the relevant neutron spectrum adjustment codes.

In this paper new developments to improve the least squares neutron spectrum adjustment and obtain more realistic uncertainty information will be presented. Experiences with linear and nonlinear least squares adjustment will be described in combination with the effect of different input covariance information.

INTRODUCTION

In the frame of the REAL-80 international interlaboratory exercise aimed to investigate the performance of different laboratories and their adjustment codes in predicting radiation damage parameters several problems of this field have been

J. P. Genthon and H. Röttger (eds.), Reactor Dosimetry, 719–728.
© 1985 ECSC, EEC, EAEC, Brussels and Luxembourg

detected (1, 2). It became clear that the uncertainty information supplied to the neutron spectrum and related integral data was not realistic, while one has to realize that radiation damage estimates are of limited economic value unless they are accompanied by an estimate of their accuracy (3). One important source of the difficulties was the rather scarce availability of realistic, physics based covariance information for the input data set of a neutron spectrum adjustment. On the other side, no detailed analysis of uncertainty data (on their derivation, reliability and the role of the input covariance information) has been performed until now.

The participants of the IAEA consultans' meeting on the assessment of the results of the REAL-80 project (3) strongly supported a recommandation that spectral adjustments must be performed with computer codes which properly account for and propagate uncertainties.

To come nearer to the solution of all these problems new aspects of the generalized least squares neutron spectrum adjustment and analysis of the related uncertainty data are described in this paper.

NEW DEVELOPMENTS IN THE LEAST SQUARES ADJUSTMENT

Mathematical approaches

The common definition of the generalized least squares approach of the neutron spectrum adjustment can be given by the following expression:

$$\chi^2(\underline{P}) \longrightarrow \min \qquad (\underline{P} \geqslant 0) \qquad (1)$$

where $\quad \underline{P} \quad$ parameter vector $\quad \underline{P} = \begin{bmatrix} \underline{\phi} \\ \underline{S} \end{bmatrix}$

$\underline{\phi}$ vector of group fluence rate values
$\underline{S}$ vector of cross section values $\mathfrak{S}_{ij}$

$$\chi^2 = \begin{bmatrix} \underline{P}_o - \underline{P} \\ \underline{A}_m - \underline{A}_c \end{bmatrix}^T \begin{bmatrix} \underline{V}(\underline{P}_o) & 0 \\ 0 & \underline{V}(\underline{A}_m) \end{bmatrix} \begin{bmatrix} \underline{P}_o - \underline{P} \\ \underline{A}_m - \underline{A}_c \end{bmatrix} \qquad (2)$$

$\underline{P}_o$ best available information on $\underline{P}$
$\underline{A}_m$ vector of the measured reaction rates
$\underline{A}_c$ vector of the calculated reaction rates based on $\underline{P}$
$\underline{V}$ covariance matrix

The commonly used, least squares neutron spectrum adjustment codes (4-7, etc.) apply a linear approximation for the above problem by neglecting the cross product term in the covariance matrix of the calculated reaction rates (1, 4, 8):

$$\underline{\underline{V}}(\underline{A}_c) = \underline{\underline{V}}_{SP} + \underline{\underline{V}}_{XS} + \underline{\underline{V}}_{NL} \tag{3}$$

where $\underline{\underline{V}}_{SP}$ spectrum contribution

            $\underline{\underline{V}}_{XS}$ cross section contribution

            $\underline{\underline{V}}_{NL}$ cross product term

$$(\underline{\underline{V}}_{NL})_{ik} = \sum_j \sum_l \text{cov}(\phi_j, \phi_l)\, \text{cov}(\sigma_{ij}, \sigma_{kl}) \tag{4}$$

The same approximation is usually present also in the uncertainty calculations of spectrum characteristic integral data (eg radiation damage parameters) as well. By the way, it is not guaranteed at all that the neglection of the cross product terms is allowed also in cases if there are present many covariance terms with a large value and also with negative sings.

In earlier studies the importance of the normalization procedure preceeding the energy dependent neutron spectrum adjustments has been shown (9) and a modified version of the STAY'SL program (4) was elaborated and used in the present analysis. The calculations by this program are done in two steps: first the optimal normalization factor is derived by the generalized least squares method, then the neutron spectrum adjustment is performed.

A better solution might be to determine the best value of the normalization constant together with the other parameters in one generalized least-squares procedure where also the contribution of the cross product terms defined by Eq (4) are taken into account. For this purpose, a new program EAGLE[*] based on the generalized least-squares theory and giving the maximum likelihood estimate on the parameter $\underline{P}$ in Eq (1) has been developed and is under testing.

Calculations-results

For better understanding the differences in mathematical modelling of the least squares neutron spectrum adjustment

---

[*]EAGLE = Elaborated Adjustment by Generalized Least-Squares Estimate

procedure and the numerical treatment of uncertainty estimates, calculations with different versions of the STASY'SL code were performed.

Refering to Eq (3) it was a special interest to investigate the validity of the linear approximation as a function of the applied covariance data. Therefore both a "linear" and "nonlinear" version of the STAY'SL code were established (called STAYFO and STAYNL, resp.) and several runs were made with a large variety of covariance information for the input neutron spectra and reaction cross sections. In all cases the optimum normalization factor determined by the generalized least squares procedure was used. The uncertainty propagation and the effect of the covariance information on the results of the neutron spectrum adjustment with special emphases on the uncertainty estimates were analysed.

Data of the REAL-80 project were applied as input (1, 2) together with the CFRMF input data set (10, 11).

In frame of the REAL-80 exercise two different neutron spectra - the one of the thermal ORR reactor and the YAYOI fast reactor - were considered, and artificially generated correlation information in form of Gaussian band matrices was used both for the input neutron spectra and the reaction cross sections (see Figure 1). During the calculations in some cases this information was substituted by arbitrarily selected matrices showing stronger correlations and much structure for the input spectra (Figure 2), while covariance data derived from the ENDF/B-V Dosimetry File Version 2 were applied in case of the cross sections. In rather arbitrary way a value of 1 % was assumed for the cross correlations between the various cross section sets. Similar approximation was used in case of the reaction rate covariance matrices both for REAL-80 and CFRMF spectra.

Some typical results are presented in Table 1. Different spectrum modifications (defined as the ratio of the output and input spectrum) are shown by Figure 3. Table 2 presents the standard deviation of input reaction rates for some characteristic cases.

CONCLUSIONS

Analysing the results of calculations outlined above we arrived at the following conclusions:

1) The input covariance information of the neutron spectrum adjustment has appreciable influance on the normalization and energy dependent spectrum modifications (see Table 1 and Figure 3). Nevertheless, the integral parameters (like $\phi_{>1 \text{ MeV}}$, R_{Ni}, R_{dpa}) - which are of interest from radiation damage point of view - do

not reflect this effect. They are only very slightly altered in comparison with their uncertainties.

2) In the reactor vessel surveillance positions usually only a small number of dosimetry detectors with a rather limited response region is present. Then the neutron spectrum adjustment is mainly restricted to the normalization. As the spectrum normalization is strongly dependent on the input covariance data, special attention to the quality of this information has to be paid.

3) The uncertainty values of integral (damage) parameters determined here are sensitive to the input covariance data (see Table 1 and 2). Nevertheless, the change of the cross section covariances from REAL-80 to ENDF/B-V Version 2 had only a small effect.

4) The uncertainty values of spectrum characteristic integral parameters like ϕ_{tot}, $\phi_{> 1MeV}$, $\phi_{> 1MeV}$ reflect the uncertainties in the solution spectrum. At the same time, the main contribution to the uncertainty of reaction rate in nickel and displacement rate in iron originates from their cross sections (see Table 1).

5) No realistic covariance information for the dpa cross sections is presently available. For realistic uncertainty estimates these data have to be evaluated.

6) The consistency of the system - expressed by the χ^2 value - is highly dependent on the quality of the input covariance information. Care has to be taken of deriving these data. In our case for example the χ^2-value became more credible by decreasing the uncertainty of the input spectrum by a factor of two (see data in Table 1). This is an indication that the estimate of input fluence rate covariances for the spectra of the REAL-80 exercise was rather pessimistic.

7) The contribution of the cross product terms in the co-variance matrix of the calculated reaction rates in the cases considered was not important (see Table 2).

REFERENCES

1. W.L.Zijp et al.: Final Report on the REAL-80 Exercise. Report ECN-128, INDC(NED)-7, BME-TR-RES-6/82. Netherlands Energy Research Foundation. (February 1983)

2. W.L.Zijp, E.M. Zsolnay, H.J. Nolthenius, E.J, Szondi, G.C.H.M. Verhaag: Intercomparison of Predicted Displacement Rates Based on Neutron Spectrum Adjustments (REAL-80 Exercise). Nuclear Technology, to be published (Sept.1984)

3. D.E.Cullen (editor): Proceedings of the IAEA Consultants' Meeting on the Assessment of the Results of the REAL-80 Project on Neutron Spectrum Unfolding Codes and Planning for Continuation of This Project, 13-15 June 1983, Vienna, Austria.

4. F.G. Perey: Least Squares Dosimetry Unfolding. the Program STAY'SL. Report ORNL/TM-6062, ENDF-254. ORNL, Oak Ridge. (October 1977)

5. M.Najzer, B.Glumac: "Unfolding of REAL-80 Sample problems by ITER-3 and STAY'SL codes." p. 1171 in the Proceedings of the 4th ASTM-EURATOM Symposium on Reactor Dosimetry, 22-26 March 1982. Report NUREG/CP-0029. Washington, BC, 1982.

6. M.Sasaki, M. Nakazawa: NEUPAC Unfolding Code for Use with Neutron Spectra from Activation data of Dosimeter Foils.Report PSR-177. ORNL RSIC, Oak Ridge (January 1981).

7. F.W. Stallmann: "LSL. A Logarithmic Least-Squares Adjustment Method". p. 1123 in the Proceedings of the 4th ASTM-EURATOM Symposium on Reactor Dosimetry, 22-26 March 1982. Report NUREG/CP-0029 Washington, DC, 1982.

8. M.Kovács, E.M. Zsolnay, E.J. Szondi: "Analysis of a Mathematical Method Used in Multifoil Activation Neutron Spectrometry" p. 248 in Mathematical Models in Physics and Chemistry and Numerical Methods of Their Realization., Editors: A.A. Samarskij, I. Kátai, Teubner Verlag, Leipzig, 1984.

9. E.M. Zsolnay et al: Effect of Normalization on the Neutron Spectrum Adjustment Procedure. Report ECN-139, BME-TR-RES-7/83. Netherlands Energy Research Foundation. (October 1983).

10. J.M. Ryskamp et al: Sensitivity and a Priori Uncertainty Analysis of CFRMF Central Flux Spectrum. Report EGG-PHYS-5243. Idaho National Engineering Laboratory. (September 1980)

11. H.J. Nolthenius, W.L.Zijp: Consistency between Data from the ENDF/B-V Dosimetry File and Corresponding Experimental Data for Some Fast Neutron Reference Spectra. Report ECN-103. Netherlands Energy Research Foundation. (November 1981).

Table 1

Input characteristics and uncertainties in integral data obtained from the solution neutron spectra

Identification		Input covariances			χ^2	Normalization factor		std Φ_{tot} in %		output uncertainties in %						Corresp.Figure No.
neutron spectrum	adjustment code	spectrum variance	correlation	cross sections		f_0	std %	input	output	$\Phi_{>.1MeV}$	$\Phi_{>1MeV}$	R_{Ni58P} φ-contr.	R_{Ni58P} 6-contr.	R_{dpa} φ-contr.	R_{dpa} 6-contr.	
ORR	STAYFO	REAL-80	REAL-80	REAL-80	3.85	1.06	12.0	12.7	5.6	9.3	6.6	5.2	8.7	4.9	11.9	3a
ORR	STAYFO	REAL-80/2	REAL-80	REAL-80	8.71	1.05	6.6	6.4	3.2	5.2	5.0	4.4	8.5	3.6	11.9	3b
ORR	STAYFO	REAL-80	045	REAL-80	6.72	1.00	5.2	18.1	5.8	8.3	5.0	4.3	8.8	4.9	11.9	3c
ORR	STAYFO	REAL-80	045	ENDF/B-V	10.4	1.00	4.8	18.1	3.5	7.1	4.9	4.2	7.0	4.6	12.1	3d
ORR	STAYNL	REAL-80	REAL-80	ENDF/B-V	5.28	1.05	12.0	12.7	5.6	9.3	6.8	5.2	6.8	5.0	12.0	–
ORR	STAYNL	REAL-80	045	ENDF/B-V	9.78	1.00	5.0	18.1	3.5	7.3	5.0	4.1	7.0	4.6	12.1	–
YAYOI	STAYFO	REAL-80	REAL-80	REAL-80	7.16	1.23	12.0	13.7	7.0	7.8	12.7	8.2	8.4	7.6	12.2	3a
YAYOI	STAYFO	REAL-80/2	REAL-80	REAL-80	11.5	1.18	7.2	6.9	4.5	4.8	7.5	7.4	9.6	5.1	12.3	3b
YAYOI	STAYFO	REAL-80	Y47	REAL-80	7.80	1.27	8.3	9.2	5.7	6.0	10.5	6.9	8.4	7.0	12.1	3c
YAYOI	STAYFO	REAL-80	Y47	ENDF/B-V	7.48	1.31	8.0	9.2	5.3	5.6	10.3	6.7	6.4	6.6	12.1	3d
YAYOI	STAYNL	REAL-80	REAL-80	ENDF/B-V	6.96	1.23	12.0	13.7	7.0	7.8	12.7	8.4	8.4	7.7	12.3	–
YAYOI	STAYNL	REAL-80	Y47	ENDF/B-V	7.21	1.31	8.1	9.2	5.3	5.6	10.4	6.9	6.4	6.7	12.1	–
CFRMF	STAYNL	CFRMF	CFRMF	ENDF/B-V	34.2	1.02	1.9	.1	.1	.5	2.1	2.3	6.5	.9	11.5	–

Table 2

The role of the cross product terms

Neutron spectrum	$\underline{V}(\phi_o)$	$\underline{V}(\underline{6})$	Réaction	Partial standard deviation in %			Total std %
				flux contr.	cross sect. contr.	cross prod. contr.	
ORR	REAL-80	ENDF/B-V	FE58G+Cd	27.3	11.2	6.1	30.1
			TI46P	24.1	12.6	3.1	27.4
			TI48P	27.2	10.4	3.0	29.3
			NI60P	26.5	7.8	2.4	27.7
			FE58G	32.1	6.0	2.0	32.7
			NP237F+Cd	15.1	9.4	1.6	17.9
			AU197G+Cd	48.2	3.0	1.5	48.3
			others	...	...	<1.0	...
YAYOI	REAL-80	ENDF/B-V	NA23G	18.5	12.0	4.2	22.4
			TI48P	36.1	10.3	3.9	37.7
			TI47P	25.5	11.3	2.9	28.0
			IN115N	20.3	12.1	2.6	23.8
			AL27P	32.1	5.8	2.5	32.7
			AL27A	38.6	5.4	2.3	39.0
			CO59A	37.0	4.3	1.8	37.3
			FE56P	36.4	4.4	1.8	36.7
			AU197G	16.8	4.8	1.2	17.5
CFRMF	CFRMF	ENDF/B-V	TI46P	8.2	12.6	1.04	15.1
			TI48P	8.5	10.3	0.91	13.4
			TI47P	7.9	11.3	0.89	13.8
			IN115N	6.7	12.0	0.86	13.8
			NI58P	7.9	6.5	0.53	10.8
			others	...	...	<0.50	...

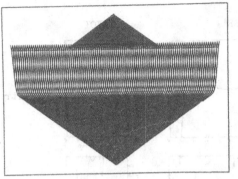

thermal end   fast end

Figure 1.
Gaussian band matrix used in the calculations

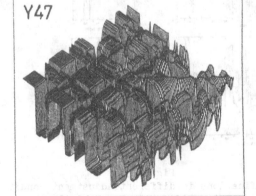

Figure 2.
Correlation matrix derived in course of the REAL-80 exercise
for the ORR and YAYOI neutron spectrum, coded by O45 and Y47,
respectively

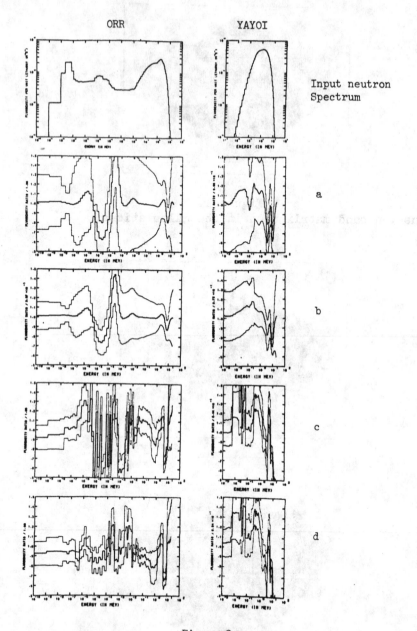

ORR   YAYOI

Input neutron
Spectrum

a

b

c

d

Figure 3.
Spectrum modifications in different adjustment conditions (for
data identification see Table 1).

PLANS FOR A REAL-84 EXERCISE

W.L. Zijp[1], E.M. Zsolnay[2], E.J. Szondi[2],

H.J. Nolthenius[1] and D.E. Cullen[3]

[1]Netherlands Energy Research Foundation ECN, Petten.
[2]Nuclear Training Reactor, Technical University (BME), Budapest.
[3]International Atomic Energy Agency, Nuclear Data Section, Vienna.

ABSTRACT

After the final report of the REAL-80 exercise
plans are being developed to start an other international
calculational exercise, originally called REAL-84, for
improving the accuracy of radiation damage predictions in
neutron environment.

This document reports on the outcome of an IAEA consultants meeting on REAL-80 results and REAL-84 plans,
and gives indications about the present situation and
prospects.

INTRODUCTION

After publication of the final report on the REAL-80 exercise
((1), summarized in (2)), an IAEA Consultants' Meeting was convened
(Vienna, 13-15 June 1983) in order to assess the results of the
REAL-80 project on spectrum adjustment codes, and to plan a
continuation of that project (under the project name REAL-84).

This paper summarizes the outcome of that meeting (3) and indicates the problems encountered in the preparatory phase, and the
prospects in the vision of the evaluators of the REAL-80 intercomparison.

LESSONS FROM THE REAL-80 EXERCISE

1. Spectra adjusted by means of different approaches can have
   remarkable differences in shape, and still can predict correct

J. P. Genthon and H. Röttger (eds.), Reactor Dosimetry, 729–736.

integral parameters (i.e. displacement rates in steel, and activation rates in nickel).

2. The adjustment procedures did not show deficiencies in predicting these reaction rates. However, some adjustment codes are not capable to take into account the input covariance matrices.
3. Also some codes cannot make estimates of output uncertainties.
4. The correlation matrices for the output spectra for ORR and YAYOI obtained by a generalized least squares adjustment procedure (STAY'SL type) were clearly different from those obtained by a Monte Carlo procedure (modified SAND-II type). Up till now this Monte Carlo technique did not take into account the input correlations.
5. Realistic (i.e. physics based) covariance information both for input spectra and cross-section data are hardly available.
6. Many available codes to derive multigroup cross-section data from the ENDF/B files introduced large inaccuracies in the multigroup cross-section data.
7. Not always correct procedures were used to convert cross-section data and spectrum data from one group structure to another. Procedures for collapsing multigroup data (spectra, cross-sections and covariances) should be distributed.
8. The spectrum adjustment codes based on a generalized least squares principle have the future, since they can take into account covariance matrix information.
9. There is a need for (input) spectrum covariance matrices, based on neutron physics calculations.
10. The input data for a neutron spectrum adjustment should show consistency.
11. The definition of the spectrum normalization factor plays an important role in the adjustment procedure.

DEVELOPMENT OF GENERALIZED LEAST SQUARES CODES

The evaluators of the REAL-80 exercise think that existing adjustment codes based on the generalized least squares principle can be improved with respect to several aspects.
1. The codes should test the consistency of the input data, and give messages where input data have unlikely values.
2. The codes should use a well chosen and well defined spectrum normalization factor (4). Preferably the normalization factor should be an adjustable parameter.
3. For thermal reactor spectra maybe a preadjustment is necessary to adjust independently the fast and thermal parts of the neutron spectrum, so that the prenormalization factor for the thermal part can differ from the prenormalization factor for the fast part. Maybe two normalization factors, with a smooth joining function for the intermediate region, could give some improvement in particular cases.
4. It seems easy to introduce an exact formula for the propagation

of variances and covariances (as outlined in appendix 4 of the final REAL-80 report (1)).

5. The adjustment code should have an easy coupling to various processing codes.
6. Since all non-linear least squares procedures need first estimates for the adjustable parameters, one could think of an automatic improvement procedure for the first estimates, when the input uncertainties are quite large (e.g. a generalized SIMPLEX procedure (5)).
7. Improvement seems also possible with respect to the sensitivity to the number of digits in the input data.

Various of these aspects will be incorporated in a new computer code EAGLE (Elaborated Adjustment by Generalized Least-Squares Estimates), which is being developed at BME (Budapest) and will be tested at ECN (Petten). It turned out that this development work took much more time than expected.

AIMS FOR REAL-84

In view of the conclusions of the REAL-80 exercise it does not seem necessary to repeat the exercise with the same aims for establishing the state of the art of the performance of the various neutron spectrum adjustment codes.

The IAEA Consultants Meeting recommended (in agreement with the recommendations of the INDC meeting, 16-20 May 1983, Rio de Janeiro) that a new exercise should be performed with the aim of improving the assessment of accuracies in radiation damage predictions by various laboratories. The emphasis should be on radiation damage to pressure vessels and related nuclear technology. One should realize that a 1% change in damage prediction can lead to a 1% change in pressure vessel life-time. The neutron energy range of interest is therefore primarily below 20 MeV. The long term aims of REAL-84 will be to strive towards establishment of standardized procedures and data for use in spectrum adjustment calculations. The short term aims will be improvement in the available data, particularly spectra and cross-section covariance information. In addition, the REAL-84 exercise will allow participants to assess and validate the accuracy of the methods and codes that they are presently using.

The joint effort of the participants could contribute in solving some basic mathematical and physical problems recently encountered in neutron spectrum adjustments for selected neutron spectra.

MEANS AND METHODS

The REAL-84 exercise will have the form of an international comparison of the outcomes of various generalized least squares adjustment procedures, as submitted by interested laboratories for a few well defined spectrum cases. It is expected, that the IAEA Nuclear Data Section will be responsible for the organization, while the evaluators of the REAL-80 exercise are in principle again willing to participate in the advisory work at the start and in the evaluation work later on.

In the exercise the organizers will provide the participating laboratories also with available computer codes for processing cross-section data, for conversion of group structures and for collapsing covariance matrices. The exercise will imply the choice of test cases for neutron spectrum adjustment, the definition of the numerical data, and the preparation of the most relevant (i.e. recent and reliable) data on damage cross-sections.

REQUIREMENTS FOR REAL-84

There are a few basic requirements for the definition of the test cases:
- Representative spectra with good covariance information;
- Damage cross-sections of good quality, accompanied by covariances;
- Experimental reaction rates for a set of activation detectors, accompanied by their covariance information.

PREPARATORY ACTIONS

The preparatory phase exists in collecting, developing and improving the necessary input data, thus making available improved data, particularly covariance information.
Four spectra are considered for inclusion in REAL-84; these include the following representative spectra:
- a prompt fission spectrum;
- a surveillance spectrum;
- a fusion spectrum, incident on first wall;
- an accelerator spectrum.

Also other spectra (CFRMF, ORR, YAYOI, spallation source) have been discussed. The main problem is to have reliable, physics based information on the spectrum covariance matrix.
With respect to the damage in materials the Consultants Meeting identified the following materials of interest:
Fe, Ni, Cr, Zr, Ti, V, Al (Al_2O_3), Nb, Mo, Cu, C (graphite), W and stainless steel. In addition also the H and He production are of importance.

With respect to neutron metrology reactions the ENDF/B-V cross-section data and covariance data should be used. It is recognized that these covariance data are not necessarily ideal, but they are the best data currently available.

Inclusion of these covariance data within the REAL-84 exercise will provide useful data testing information which can be used to improve future versions of ENDF covariance data.
The preparatory phase includes also the following actions, definitions of numerical input data, the consistency check of input data, and the preparation of various processing codes already mentioned. For the preparatory work the help of other experts and laboratories was, and still is, essential. Moreover volunteers to assist in the preparatory actions and in the future evaluation work are welcomed.

EXPECTED ACTIONS FROM THE PARTICIPANTS

The candidate participants will be asked
- to perform neutron spectrum adjustments with a generalized least squares code for one or more of the spectrum data sets provided; and derive output spectrum covariance information;
- to calculate spectrum characteristics (like ϕ_{tot}, ϕ_{th}, $\phi(E>0.1MeV)$, $\phi(E>1MeV)$, ϕ_{Ni}, etc.), accompanied by corresponding uncertainties;
- to calculate some integral reaction rate parameters, like displacement rate in steel and specified metals and compounds, H and He production rate in steel and specified metals and compounds, accompanied by corresponding uncertainties.
- to evaluate, where necessary, missing or incomplete covariance information.

The participants have to apply numerical input data, consistent with the ENDF/B-V dosimetry and gas production files, either by directly converting the files or by making use of processed data in multigroup form. The participants can make use of their own group conversion procedures and interpolation routines.

PROGRESS

Spectra. The following spectrum information, including covariance information has been reported:
- the ORR spectrum, with a covariance matrix in 100 groups (1);
- the YAYOI spectrum, with a covariance matrix in 100 groups (1);
- the ^{235}U fission neutron spectrum, with a covariance matrix in 24 groups (6);
- the ^{252}Cf fission neutron spectrum, with a covariance matrix in 24 groups (6);
- the CFRMF neutron spectrum, with a covariance matrix in 26 groups (7).

It is expected that very soon spectrum information on the ORR-PSF pressure vessel simulation experiment (established in the LWR-PU surveillance program) will be available. Spectrum information for other spectra (CTR first wall; accelerator spectrum) has been solicited.

Cross-sections. Several laboratories are now preparing multi-group covariance matrix information through processing of the ENDF/B-V data. Special effort is involved in the treatment of the resonance parameter data as present in file 32.
Some mutual checks on the results should be performed for a few spectra (fast and thermal), and for a chosen group structure. There is agreement on the performance of such checks between ORNL (Stallman) and ECN (Zijp).

Adjustment codes. As mentioned above, the adjustment code EAGLE is under development at the Technical University Budapest.

Damage models. At ANL (Greenwood) a software package (8) has been established, which can deal with damage calculations using specified input spectra and a damage model as described by Odette (9).

Problems encountered. The spectrum correlation matrices for ^{235}U and CFRMF mentioned above have very high correlation coefficients which are published in 2 or 3 significant digits. In this situation the uncertainties in the integral parameters (such as ϕ_{tot}) could not be determined due to the accumulated rounding errors (e.g. negative variance values and correlation coefficients larger than 1 were obtained).

It turned out that the establishment of good quality output data for the considered input data sets is far from an easy and smooth procedure.
With the reported data at present available to the authors it seems possible to distribute to candidate participants the following input data sets.
- ^{235}U fission neutron spectrum in 24 groups with spectrum covariance matrix data;
- CFRMF neutron spectrum in 26 groups with spectrum covariance matrix data;
- and maybe the ORR-PSF spectrum for the pressure vessel steel irradiation experiment.
In view of the problems encountered, it is thought that looking at some problems by the participants is part of the exercise, because the solution of this type of problems is part of the procedure for arriving at neutron spectrum adjustment. It is therefore proposed that the participants have some freedom to make justified modifications in the input data, in case they can give good arguments to improve the quality of the data sets (e.g. to arrive at a better

consistency or to introduce better cross correlations between measured reaction rates or between cross-sections. If this is acceptable, the REAL-84 exercise could deal with the spectra mentioned above. The success of such an exercise depends largely on the participants actions to modify and improve (or not) the input data sets, before making a routine adjustment of the data with their own computer code.

SUGGESTED TIME SCHEDULE

Sept. '84 Presentation of plans at 5th ASTM-EURATOM symposium.
Collection of names of candidate participants.
Collection of data sources.
If possible, no corrections for neutron self-shielding should be required.
ECN can supply ENDF/B-V cross-section and uncertainty data in special group structures.

Febr. '85 Distribution of input data sets to interested laboratories.

June '85 Receipt of participants contributions by evaluators.

Spring '86 Overview of results presented by evaluators.

ACKNOWLEDGEMENT

Part of the work described in this document has been carried out under contract to the European Commission and has been financed by the JRC budget.

REFERENCES

1. W.L. Zijp, É.M. Zsolnay, H.J. Nolthenius, E.J. Szondi, G.C.H.M. Verhaag, D.E. Cullen, C. Ertek, "Final Report on the REAL-80 Exercise", Report ECN-128. Also INDC(NED)-7 and BME-TR-RES-6/82. Netherlands Energy Research Foundation ECN, Petten, February 1983).

2. W.L. Zijp, É.M. Zsolnay, H.J. Nolthenius, E.J. Szondi, G.C.H.M. Verhaag, "Intercomparison of Predicted Displacement Rates Based on Neutron Spectrum Adjustments (REAL-80 Exercise)", Nuclear Technology, to be published (Sept. 1984?).

3. D.E. Cullen (editor), "Proceedings of the IAEA Consultants' Meeting on the assessment of the results of the REAL-80 project on cross-section unfolding codes and planning for continuation of this project", Vienna, 13-15 June 1983.

4. É.M. Zsolnay, W.L. Zijp, H.J. Nolthenius, "Effect of Normalization on the Neutron Spectrum Adjustment Procedure", Report ECN-139. Also BME-TR-RES-7/83. (Netherlands Energy Research Foundation ECN, Petten, October 1983).

5. S.N. Deming, L.R. Parker Jr., "A review of SIMPLEX optimization in analytical chemistry", CRC Critical Reviews in Anal. Chem. 7 (1978), p. 187-202.

6. R.E. Maerker, J.J. Wagschal, B.L. Broadhead, "Development and Demonstration of an Advanced Methodology for LWR Dosimetry Applications", Report EPRI-NP-2188 (Electric Power Research Institute, Palo Alto, December 1981).

7. J.M. Ryskamp, R.A. Anderl, et al., "Sensitivity and a priori uncertainty analysis of the CFRMF central flux spectrum", Report EGG-PHYS-5243 (EG&G, Idaho, September 1980).

8. L.R. Greenwood, "Neutron source characterization and radiation damage calculations for materials studies", Journal Nucl. Materials 108 & 109 (1982), p. 21-27.

9. G.R. Odette, D.R. Doiron, "Neutron-energy dependent defect production cross-sections for fission and fusion applications", Nuclear Technology 29 (1976) p. 346-368.

PART VII
BENCHMARKS, REFERENCE AND
STANDARD SPECTRA

A COMPARATIVE ANALYSIS OF THE OAK RIDGE PCA AND NESDIP PCA 'REPLICA' EXPERIMENTS USING THE LONDON ADJUSTMENT TECHNIQUE

M Austin, A Dolan, A F Thomas

Rolls-Royce & Associates Ltd

P O Box 31, Raynesway, Derby

The Oak Ridge Pool Critical Assembly (PCA) experiment comprises a benchmark for the evaluation of physics-dosimetry techniques used in the assessment of fast neutron damage parameters in operating Light Water Reactor pressure vessels. To extend the range of application of the PCA experiment, the NESTOR Shielding and Dosimetry Improvement Programme (NESDIP) is underway at Winfrith, in the UK. Phase 1 of NESDIP consisted of measurements in a 'Replica' of the PCA array in the 12/13 configuration. Comparative results of the analysis of both experiments yield an opportunity to study the effect of a fission plate neutron source, as used in NESDIP, as opposed to a distributed source.

A Monte-Carlo transport code was used for the analyses and least squares adjustments were made using the code suite LONDON. Results are prese nted for the calculations and several data adiustments. These indicate possible inadequacies in differential data used for the calculations.

J. P. Genthon and H. Röttger (eds.), Reactor Dosimetry, 739–750.
© *1985 ECSC, EEC, EAEC, Brussels and Luxembourg*

INTRODUCTION

This paper compares the results of similar analyses carried out on the Oak Ridge Pool Critical Assembly (PCA) (1) array and the NESTOR Shielding and Dosimetry Improvement Programme (NESDIP) PCA Replica Array (2). Phase 1 of the NESDIP programme was carried out with a view to establishing the neutronic equivalence of the two arrays. The results of the analyses presented here indicate the extent to which the two are indistinguishable. Monte Carlo transport calculations were used to calculate the neutron spectra throughout both arrays in the 12/13 configuration. As a consequence of the difficulty in accurately assessing neutron fluxes at deep penetrations in steel/water arrays, and a tendency to under rather than overestimate high energy neutron flux levels, a linear least squares technique has been developed to allow for the statistical adjustment of calculated reaction rates through reactor pressure vessels using measured reaction rates in the cavity between a reactor pressure vessel and primary shield tank. Both benchmark arrays have been used to validate the sequence of calculations which it is intended to apply to the estimation of reactor pressure vessel damage rate. In cases where measured and calculated responses agree well, the method produces reduced uncertainties in estimates of responses over those produced by calculation only, and underpredictions of damage rate are generally avoided.

As a by-product the method produces maximum liklihood data adjustments to the integral data involved in the calculation. Such adjustments depend, in detail, on the variance-covariance information available for the differential data included in the Monte Carlo calculation. As such, adjustments depending on small amounts of measured data are shown to be limited in reliability. The reliability of data adjustments is shown to improve as more measured data is involved.

DETAILS OF THE CALCULATIONAL SEQUENCE USED

The Monte-Carlo Code McBEND (3) was used to calculate neutron fluxes and hence detector reaction rates at all on-axis locations at which detector activation measurements were made. McBEND uses energy group data in a fine group scheme derived from the UKNDL database. Calculations on the AEEW NESDIP PCA (12/13) array used both 300 and 8000 group data. The results were statistically indistinguishable. The calculation performed on the Oak Ridge PCA (12/13) array used the finer Group Structure. The calculations were run for energies down to 0.11 Mev, since this represents the threshold of the measurement with the lowest energy response. In order to accelerate particle scoring it is necessary to supply an importance distribution in both space and energy. In order to score particles of all energies equally well an inverse of the neutron flux spectrum at the void box location was used as the source to an adjoint diffusion calculation using the ADC model (4).

The sources were represented using a Watt fission spectrum. Responses for ^{115}In(n,n') ^{115m}In, ^{103}Rh(n,n') ^{103m}Rh, ^{58}Ni(n,n) ^{58}Co and ^{27}Al(n,α) ^{24}Na were scored in the Oak Ridge PCA (12/13) calculation while the calculation of the AEEW NESDIP PCA (12/13) array contained data to score the first two and ^{32}S(n,p) ^{32}P. The choice was limited due to the lower flux levels in this latter array. Both calculations were used to give predictions for atom displacement rate (DPA) and flux > 1 Mev. The calculation of the Oak Ridge PCA also gave values of flux > 0.1 Mev. The detector responses were taken from the IRDF 82 library derived largely from ENDFB V. They were included into McBEND in 620 energy groups for the calculation of the Oak Ridge PCA but were collapsed into the scoring energy group scheme of 25 groups using a slowing down spectrum in water as a weighting spectrum for the calculations of the PCA Replica. It is possible to calculate other responses by folding the scored group fluxes with the cross sections but the values obtained are less accurate. Values of the calculated to measured responses are given for the Oak Ridge PCA (12/13) array in Table 1 and for the AEEW NESDIP PCA (12/13) replica array in Table 2.

THE ADJUSTMENT TECHNIQUE

A computer code LONDON was used to adjust the differential nuclear data used in the Monte-Carlo calculation in order to bring measured and calculated responses into statistical agreement. The methodology used in LONDON is similar to that described in (5) but, as it is intended for use as a design tool, responses are adjusted rather than individual group fluxes. Four classes of differential data are adjusted each of which is considered to be independent from the others,

740

namely the source distribution, transmission and detector group cross-sections and Monte-Carlo stochastic uncertainties. In cases where the measured and calculated data are not compatible, as indicated by a value of χ^2 greater than expected, the uncertainties assumed to exist on chosen classes of differential nuclear data are scaled up to give the expected value. Values of χ^2 smaller than expected do not result in a reduction of the uncertainties assumed. The sensitivities of integral responses to the transmission data were obtained using a perturbation package included in McBEND (6). Variance covariance data were obtained from the published literature (7, 8, 9, 10, 11) and largely originate from END/F-BV.

The transmission cross sections adjusted were Fe(N,N), Fe(N,X), O(N,N), O(N,α), O(N,N'), H(N,T), Cr(N,N), Ni(N,N) and Ni(N,X), for both adjustments and C(N,N) and C(N,X) also for the AEEW NESDIP PCA Replica adjustment. A 15 energy group scheme was used for the adjustment of detector cross sections (5) while a 9 group scheme was used for the transmission data (10). Since adequate uncertainty data was not found for uranium and aluminium transmission cross sections these were not included. The exposure parameters are treated as standards so no uncertainties in responses were included for these. Only seven responses could be included into McBEND, a restriction which no longer exists. For this reason some possibilities were excluded. For example the sensitivities of the transmission data to flux $>$ 0.1 Mev was not determined for the AEEW NESDIP PCA replica adjustment.

Due to the highly correlated nature of the Watt fission spectrum (5), a coarse 5 energy group scheme was used for all the adjustments described here. Due to lack of data only 3 spacial regions were considered for the Oak Ridge PCA core source adjustment. This was allowed to vary independently. Care was taken to characterise the AEEW NESDIP PCA Replica source in great detail, the thin plate source which is only 0.6 cms thick was modelled in McBEND using 220 special regions. The main source of inaccuracy was the overall source level so a solid variance-covariance matrix was used to describe this source causing any adjustment to effect the level but not the shape of the source distribution. The source was described using a 9 spacial regions for the adjustment.

Three data adjustments are described briefly here and the results are compared in terms of predicted damage rates through the reactor pressure vessel simulators in the two arrays. The first two adjustments use measured data in the cavity simulator (void box) to predict damage parameters throughout the RPV simulator while the third uses considerably more measured data at a range of locations to predict the same parameters. The results of this, more reliable, adjustment are compared with those arising in the first two cases, both in terms of predicted damage parameters and individual differential nuclear data adjustments.

THE RESULTS OF THE ADJUSTMENTS

The three adjustments described are:

(i) An adjustment of the ^{115}In(n,n') ^{115m}In, ^{58}Ni(n,p)^{58}Co ^{27}Al(n,) ^{24}Na, ^{103}Rh(n,n') ^{103m}Rh, flux $>$ 1 Mev, flux $>$ 0.1 Mev and DPA responses at the RPV A4(T/4) location using the measured ^{27}Al(n,) ^{24}Na and ^{115}In(n,n') ^{115m}In responses in the A7 (void box) for the Oak Ridge PCA (12/13) array.

(ii) An adjustment of the ^{115}In(n,n') ^{115m}In, ^{103}Rh(n,n') ^{103m}Rh, ^{54}Fe(n,p) ^{54}Mn, ^{56}Fe(n,p) ^{56}Mn, ^{32}S(n,p) ^{32}P, flux $>$ 1 Mev and DPA responses at the RPV A4(T/4) location using the measured ^{115}In(n,n') ^{115m}In, ^{103}Rh(n,n') ^{103m}Rh and ^{32}S(n,p) ^{32}P responses in the A7 (void box) for the AEEW NESDIP PCA (12/13) array.

(iii) A combined adjustment using the A3 to A7 locations for the Oak Ridge array with some of that available for the AEEW Replica (A6, A7 locations) for all measured responses to predict damage parameters throughout the RPV simulator.

In the first two cases attention is concentrated on those responses involved in both adjustments. Variance scaling was only required in the second adjustment in which all classes of data were scaled, except the stochastic uncertainties.

Tables 3 and 4 summarise the main results from the first adjustment using the data for the Oak Ridge PCA (12/13) array. Table 3 shows the ratio of calculated to measured and adjusted to measured responses at the void box and A4 locations for this array, resulting from adjustment 1. Table 4 shows the contribution of the variances of the different classes of differential data to the variances on the responses at the void box and A4 locations.

741

Tables 5 and 6 show the same results for the AEEW NESDIP PCA Replica (12/13) array. In order to compare the two arrays the change in the ratio of measured results for the ^{115}In(n,n')^{115m}In responses and in the calculated responses as a function of position appears in Table 7, the ratio being normalised to unity at the A4 location.

Table 8 shows a selection of spectral indices for both arrays. Only the ratio of the ^{103}Rh(n,n')^{103m}Rh/^{115}In(n,n)'^{115}In responses allow a comparison of the arrays and this is of limited interest since it compares responses integrated over similar energy ranges. Table 9 shows the ratio of adjusted to measured responses for all the included measurements. Comparison with Table 1 shows the extent to which the calculated values have been brought to agreement with measurement. Only the adjusted value for ^{27}Al(n,α)^{24}Na lies further than one standard deviation from the measured value. Tables 10 and 11 may be compared with tables 3, 4, 5 and 6 which give the same results for the first two adjustments. Figures 1 to 4 show histograms indicating the largest individual data adjustments.

DISCUSSION OF RESULTS

Tables 1 and 2 indicate the accuracy which can be achieved in calculating fluxes throughout a steel water array using a Monte-Carlo neutron transport code. As indicated by a value of ^{2} the agreement between experiment and calculation is good for both arrays, if all data is included, and would indicate overestimation rather than underestimation of the uncertainties on differential nuclear data. Analyses using an in-situ adjustment code (12) indicate that the various responses measured at such locations throughout the RPV simulator are consistent for both arrays. Measured responses in the void box are less consistent particularly in the Oak Ridge PCA (12/13) array. The tables further indicate a tendency towards overall progressive underestimation with penetration through the RPV simulator blocks.

Comparison of the results of the first two adjustments tell similar stories in that underestimation of responses in the void boxes leads to upward adjustments of responses at the A4 location. The agreement of calculated and measured responses at the A4 location is good particularly in the case of the AEEW NESDIP PCA Replica and the upward adjustments clearly represent overestimates of the damage parameters which are to be estimated. Tables 3 and 5 indicate, however, that if a 2 confidence level is to be used most of the pessimism is removed from the calculation due to reduced variances on the damage estimates after adjustment. A comparison of Tables 4 and 6 indicate the uncertainties to transmission data dominate the overall uncertainties in both calculations.

In both cases the uncertainty in the iron inelastic cross section is the dominant contributor to this component. The larger contribution of the stochastic uncertainties in the analysis of the Oak Ridge PCA array is due to fewer particles being scored at a given location for this Monte-Carlo run. The decision as to the contribution these should be allowed to make is largely economic. The other main difference relates to the larger contribution of detector response uncertainties of the AEEW NESDIP PCA array, largely contributed by the ^{103}Rh(n,n') ^{103m}Rh and ^{32}S(n,p) ^{32}P detectors.

Table 7 highlights the interesting fact that the arrays appear to be indistinguishable as judged by the fall off of measured ^{115}In(n,n') ^{115m}In response throughout the array while calculation predicts a slower fall in response for the Oak Ridge PCA array. Some effect of this nature could be expected due to a harder spectrum emerging from the core source than the plate one resulting from the enhanced contributions of high energy neutrons produced deeper into the core. High energy background present in the AEEW NESDIP array could explain the non observation of this difference, but a 4% background was removed from all responses to account for this background. A more detailed calculation could be performed to remove the background, but effects of ~4% could not explain the differences observed in Table 7.

The final adjustment was performed to give the 'best possible' estimates of damage parameters through the RPV simulator and to produce more reliable data adjustments. Even when using the larger range of measured data included here, small adjustments to differential data have been treated with circumspection. This is particularly true since it was not possible to include correlations between the stochastic uncertainties on any of the calculated responses.

Table 7 indicates that the final adjustment reduces the discrepancy between the fall off of ^{115}In(n,n')^{115m}In response in the two arrays through the RPV simulator and into the void box. The change must be due largely to adjustments to the spacial source distribution although changes to transmission data could contribute (the composition of the thermal shield for example is not identical in the two arrays).

Figure 4 shows the adjustment to the front quarter of the PCA core and it can be seen that the Watt fission spectrum has been slightly softened. Relative to this adjustment the next quarter depth is adjusted similarly, and the back half of the core has a relative reduction to level of ~2%. It would have been more realistic to impose symmetry on the core adjustment by defining adjustment regions in a symmetrical manner. In this case the overall reduction in level at the back of the core would probably be smaller. A relative increase in source level of about 3% was observed for the AEEW NESDIP replica array.

The changes to the dosimetry responses are shown in figures 1 and 2 where it can be seen that the ^{115}In(n,n')^{115m}In and ^{103}Rh(n,n') ^{103}Rh responses are increased while the three higher energy responses have been reduced. The ^{32}S(n,p) ^{32}P/^{115} In(n,n') ^{115m}In spectral index would be changed by 15% by these adjustments bringing calculation and experiment into agreement. The largest change to the transmission data was a reduction to the iron inelastic cross section, particularly in the high energy range. The adjustments to the iron transmission data appear in figure 3.

CONCLUSIONS

Monte-Carlo calculations of on axis responses through the RPV simulator and into the void box for both Oak Ridge PCA (12/13) and AEEW NESDIP PCA replica (12/13) arrays give reasonable agreement with experiment. Some tendency to progressive underestimation with penetration of the RPV simulator is observed in both cases and measurements are less self-consistent within the void box than through the steel. A comparison of the ^{115}In(n,n') ^{115m}In response through both arrays suggests some differences between the two arrays.

Data adjustments based on measured data in the void box location only for each array separately lead to pessimistic predictions of damage parameters through the RPV but when a 2 σ confidence interval is added to the predicted values most of the pessimism due to excessive upward adjustments is removed.

A single data adjustment based on 23 measurements from both arrays shows that it is possible to move all calculated values to within one standard deviation of measurement except for the ^{27}Al(n,α) ^{24}Na measurement in the void box which remains a little low.

Correlations between stochastic uncertainties were not taken into account. These certainly exist and would effect the detailed data adjustments so only overall trends in the adjustment should be deemed relevant.

The iron inelastic scattering cross section has been significantly decreased, especially at high energy. The ^{115}In(n,n')In and ^{103}Rh(n,n')^{103m}Rh responses were significantly increased while those for ^{58}Ni(n,p) ^{58}Co, ^{27}Al(n,α) ^{24}Na and ^{32}S(n,p) ^{32}P were reduced. The fission spectrum was softened a little while the source level of the AEEW NESDIP PCA replica fission plate was increased by ~2% relative to that of the Oak Ridge PCA.

TABLE 1

COMPARISON OF REACTION RATES CALCULATED USING THE MONTE CARLO CODE McBEND AND MEASURED FOR THE OAK RIDGE PCA (12/13) ARRAY

REACTION	LOCATION IN ARRAY					
	A2	A3M	A4	A5	A6	A7
^{115}In(n,n') ^{115m}In	0.99 ±0.07*	1.07 ±0.07	0.96 ±0.07	0.95 ±0.06	0.92 ±0.07	0.96 ±0.10
^{103}Rh(n,n') ^{103m}Rh	-	-	0.97 ±0.06	0.92 ±0.07	0.91 ±0.07	-
^{58}Ni(n,n') ^{58}Co	1.03 ±0.06	1.04 ±0.06	0.93 ±0.06	0.91 ±0.06	0.94 ±0.07	-
^{27}Al(n,α) ^{24}Na	0.98 ±0.06	0.99 ±0.07	0.97 ±0.07	0.93 ±0.06	0.87 ±0.07	0.67 ±0.09
^{238}U(n,f)FP	-	1.10 ±0.07	1.01 ±0.12	1.00 ±0.11	0.94 ±0.11	1.13 ±0.07
^{237}Np(n,f)FP	1.27 ±0.07	0.99 ±0.07	0.92 ±0.11	0.94 ±0.11	0.93 ±0.11	1.01 ±0.09

* In calculating the uncertainty in this ratio the uncertainties in measurements are of equivalent fission flux only. The uncertainties in fission average cross sections are small. The uncertainties on calculation are stochastic only ie uncertainties in differential nuclear data are not considered. Measurement and calculation are considered to be independent.

TABLE 2

COMPARISON OF REACTION RATES CALCULATED USING THE MONTE CARLO CODE McBEND AND MEASURED FOR THE AEEW NESDIP PCA REPLICA (12/13) ARRAY (using results calculated at AEEW) (CALCULATED/MEASURED)

REACTION	LOCATION IN ARRAY			
	A4	A5	A6	A7
^{115}In(n,n') ^{115m}In	1.01 ±0.10	-	0.88 ±0.09	0.83 ±0.09
^{103}Rh(n,n') ^{103m}Rh	1.01 ±0.09	-	0.97 ±0.08	0.85 ±0.08
^{32}S(n,p) ^{32}P	1.03 ±0.11	-	1.02 ±0.09	1.00 ±0.08

TABLE 3

COMPARISON OF DAMAGE PARAMETERS BEFORE AND AFTER ADJUSTMENT WITH LONDON AND MEASURED FOR THE OAK RIDGE PCA (12/13) ARRAY AT THE A4 LOCATION NORMALISED TO CALCULATED VALUE

	Calculated	Calculated +2σ	Adjusted	Adjusted +2σ
Flux > 1 Mev	1.0 ±12%	1.22	1.07 ±11%	1.30
Flux > 0.1 Mev	1.0 ±13%	1.26	1.03 ±10%	1.24
DPA	1.0 ±10%	1.24	1.04 ±10%	1.25

TABLE 4

RELATIVE CONTRIBUTION TO TOTAL VARIANCE OF RESPONSES
AT THE A4 AND A7 LOCATIONS OF VARIANCE IN
NUCLEAR AND EXPERIMENTAL DATA
FOR THE OAK RIDGE PCA (12/13) ARRAY

LOCATION	DETECTOR CROSS SECTIONS	TRANSMISSION CROSS SECTIONS	SOURCE	M/C STOCHASTICS	MEASUREMENT
A7	14%	68%	11%	6%	-
A4 (before adjustment)	26%	65%	5%	4%	-
A4 (after adjustment)	21%	42%	8%	20%	9%

χ^2 at void box before adjustment = 0.6

TABLE 5

COMPARISON OF DAMAGE PARAMETERS BEFORE AND AFTER ADJUSTMENT
WITH LONDON FOR THE WINFRITH PCA REPLICA (12/13) ARRAY
AT THE A4 LOCATION (x10^{-24} PER WATT OF FISSION PLATE POWER)

Response	Calculated	Calculated +2σ	Adjusted	Adjusted +2σ
Flux > 1 Mev	1.0 ±16.3%	1.33	1.07 ± 9.2%	1.27
DPA	1.0 ±15.2%	1.30	1.06 ± 10.2%	1.27

TABLE 6

RELATIVE CONTRIBUTION TO TOTAL VARIANCE OF RESPONSES IN
THE AEEW PCA REPLICA (12/13) CONFIGURATION AT THE
A4 AND A7 LOCATIONS FROM VARIANCE IN NUCLEAR AND
EXPERIMENTAL DATA

	DETECTOR	TRANSMISSION	SOURCE	M/C STOCHASTIC	MEASUREMENT
A7	10.1%	77.9%	11.5%	0.5%	-
A4 (before adjustment)	25.9%	56.4%	17.0%	0.8%	-
A4 (after adjustment)	50.6%	23.2%	12.3%	6.2%	7.7%

χ^2 at void box before adjustment was 1.48.

A variance scale factor of 1.98 was applied to detector transmission and source uncertainties.

TABLE 7

COMPARISON OF THE ^{115}In(n,n') ^{115m}In REACTION RATE AS MEASURED AND CALCULATED FOR THE OAK RIDGE PCA (12/13) AND WINFRITH REPLICA PCA (12/13) ARRAYS

Ratio	A4	A5	A6	A7
$\left(\dfrac{M_{ORR}/M_{REP}}{M_{ORR}/M_{REP}}\right)_{A4}$	1.00		0.98	1.01
$\left(\dfrac{C_{ORR}/C_{REP}}{C_{ORR}/C_{REP}}\right)_{A4}$	1.00	1.02	1.05	1.18
$\left(\dfrac{A_{ORR3}/A_{REP3}}{A_{ORR3}/A_{REP3}}\right)_{A4}$	1.00		0.95	1.05

Glossary of Terms:
M – measurement
C – calculation
A – adjustment

subscript
ORR – for the Oak Ridge PCA (12/13) array
REP – for the AEEW NESDIP PCA (12/13) array

TABLE 8

COMPARISON OF REACTOR RATE RATIOS FOR OAK RIDGE PCA (12/13) AND AEEW NESDIP PCA (12/13) REPLICA

^{27}Al(n,α)^{24}Na/^{58}Ni(n,p)^{58}Co

Location	Oak Ridge C	E	C/E	AEEW C	E	C/E
A3M	1.20-2	1.26-2	0.95	1.11-2	--	--
A4	1.30-2	1.25-2	0.95	1.09-2	--	--
A6	1.27-2	1.35-2	1.03	1.11-2	--	--
A7	1.31-2	--	--	1.21-2	--	--

^{103}Rh(n,n')^{103m}Rh/^{58}Ni(n,p)^{58}Co

Location	Oak Ridge C	E	C/E	AEEW C	E	C/E
A3M	6.3	-	-	7.3	--	--
A4	10.2	9.9	1.04	12.3	--	--
A6	19.1	1.98	0.96	22.6	--	--
A7	20.5	-	-	25.57	--	--

^{103}Rh(n,n')^{103m}Rh/^{115}In(n,n')^{115m}In

Location	Oak Ridge C	E	C/E	AEEW C	E	C/E
A3M	4.17	4.41	0.95	4.19	--	--
A4	5.15	5.06	1.02	5.28	5.26	1.00
A6	7.05	7.16	0.98	7.34	6.69	1.10
A7	6.95	6.63	1.05	8.07	7.79	1.04

^{32}S(n,p)^{32}P/^{115}In(n,n')^{115m}In

Location	Oak Ridge C	E	C/E	AEEW C	E	C/E
A3M	0.433	-	-	0.377	-	-
A4	0.320	-	-	0.316	0.275	1.15
A6	0.229	-	-	0.205	0.179	1.15
A7	0.210	-	-	0.195	0.161	1.21

TABLE 9

THE RATIO OF ADJUSTED TO MEASURED RESPONSES RESULTING FROM
A COMBINED ADJUSTMENT OF THE OAK RIDGE PCA (12/13) AND
AEEW NESDIP PCA (12/13) MEASURED RESPONSES

Response	Location Oak Ridge PCA (12/13)					AEEW NESDIP PCA REPLICA (12/13)	
	A3M	A4	A5	A6	A7	A6	A7
^{115}In(n,n')^{115m}In	1.02	0.98	0.99	0.98	0.99	1.00	0.99
^{103}Rh(n,n')^{103m}Rh	-	0.99	0.97	0.97	-	1.01	0.99
^{58}Ni(n,p)^{58}Co	0.99	0.97	0.98	1.01	-	-	-
^{27}Al(n,α)^{24}Na	0.97	1.00	1.01	1.01	0.91	-	-
^{32}S(n,p)^{32}P	-	-	-	-	1.00	1.00	1.00

TABLE 10

% CONTRIBUTION TO TOTAL VARIANCE TO THE RESPONSES FOR ALL
LOCATIONS DUE TO THE VARIANCE IN NUCLEAR AND EXPERIMENTAL DATA
AS INDICATED IN ADJUSTMENT TO BOTH ARRAYS

	Detector	Transmission	Source	M/C Stochastics	Measurement
before adjustment +	48%	41%	6%	5%	-
before adjustment *	-	75%	19%	6%	-
after adjustment *	16%	28%	9%	22%	25%

+ This gives contributions to the calculated detector
responses for all 7 locations involved (Oak Ridge
PCA A3M-A7, AEEW REPLICA A6, A7)

* This gives the contributions to the calculated
(adjusted) damage parameters for the Oak Ridge PCA
locations (A3M-A7) before (after) adjustment

TABLE 11

COMPARISON OF DAMAGE PARAMETERS BEFORE AND AFTER ADJUSTMENT
WITH LONDON FOR OAK RIDGE PCA (12/13) ARRAY THROUGHOUT
RPV SIMULATOR ADJUSTING DATA FROM BOTH ARRAYS

	Calculated	Calculated $+ 2\sigma$	Adjusted	Adjusted $+ 2\sigma$
Flux > 1 Mev				
A3M	1.00 ± 0.10	1.20	1.04 ± 0.07	1.18
A4	1.00 ± 0.11	1.22	1.04 ± 0.08	1.20
A5	1.00 ± 0.12	1.24	1.03 ± 0.08	1.19
A6	1.00 ± 0.12	1.24	1.03 ± 0.09	1.21
A7	1.000 ± 0.13	1.26	1.04 ± 0.09	1.22
Flux > .1				
A3M	1.00 ± 0.10	1.20	0.97 ± 0.06	1.09
A4	1.00 ± 0.13	1.26	0.98 ± 0.07	1.12
A5	1.00 ± 0.14	1.28	0.99 ± 0.07	1.13
A6	1.00 ± 0.16	1.32	1.03 ± 0.08	1.15
A7	1.00 ± 0.16	1.32	1.03 ± 0.08	1.15
DPA				
A3M	1.00 ± 0.09	1.18	0.97 ± 0.06	1.09
A4	1.00 ± 0.12	1.24	0.98 ± 0.07	1.12
A5	1.00 ± 0.13	1.26	0.99 ± 0.08	1.15
A6	1.00 ± 0.13	1.26	1.00 ± 0.08	1.16
A7	1.00 ± 0.14	1.28	1.00 ± 0.08	1.16

REFERENCES

1. W N McElroy LWR Pressure Vessel Surveillance Dosimetry Improvement Program: PCA Experiments and Blind Test HEDL NUREG/CR-1861 July 1981

2. A K McCracken Phase 1 Winfrith Measurements and Calculations NUREG/CR 3324 AEE Winfrith UK 1983

3. D E Bendall and R J Brissenden McBEND Program User Guide WRS Modular Code Trans Am Nucl Soc 41 pp 332-333 AEEW UK 1982

4. Modified form of Diffusion Theory for Use in Calculating Neutron and Gamma-Ray Penetration in Practical Shields. Proceedings of the Fourth International Conference on Reactor Shielding Paper B2-7, Paris France October 1972

5. R E Maerker Development and Demonstrations of an Advanced Methodology for LWR Dosimetry Applications EPRI WP 2188 December 1981

6. M G C Hall DUCKPOND-A Perlurbation Monte-Carlo Package and its Applications Proc NEA Specialists Meeting on Nuclear Data and Benchmarks for reactor Shielding Paris October 1980

7. OECD Compilation of Threshold Reaction Neutron Cross Sections for Neutron Dosimetry and Other Applications EA WDC 93 U February 1974

8. W L Zijp and J H Baard. Nuclear Data Guide for Reactor Neutron Metrology Part I and II. EUR 7167 Netherlands August 1979

9. J D Dorchler and C R Weisbin Computation of Multigroup Cross Section Covariance Matrices for Several Important Reactor Materials ORNL 5318 (ENDF-235), Oak Ridge, TN 1977

10. D W Muir and R J LaBauve CIVFILS A30 Group Covariance Library Based on ENDF/B V LA-87-33-MS Los Alamos National Laboratory, Los Alamos, NH, March 1981

11. Strohmaier, Stagesen and V Voniach Evaluation of the Cross Sections for the Reaction $^{19}F(n,2n)^{18}F$, $^{32}P(n,p)^{31}Si$, $^{93}Nb(n,n')^{93m}Nb$ and $^{103}Rh(n,n')^{103m}Rh$. Energie Physik Mathematik Gmbff Karlsruhe 1980

Fractional adjustments to detector cross sections resulting from combined adjustment of the Oak Ridge PCA (12/13) array and the AEEW NESDIP PCA (12/13) replica array

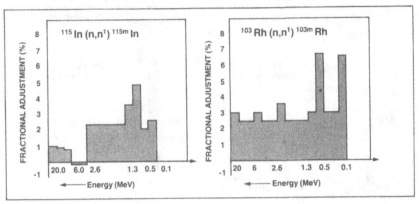

FIG. 1

Fractional adjustments to detector cross sections resulting from combined adjustment of the Oak Ridge PCA (12/13) array and the AEEW NESDIP PCA (12/13) replica array

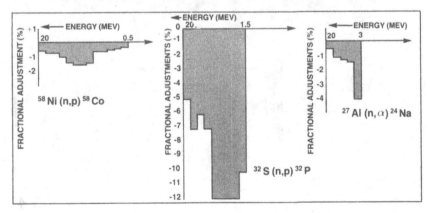

FIG.2

Fractional adjustments to iron transmission cross sections resulting from combined adjustment of the Oak Ridge PCA (12/13) array and the AEEW NESDIP PCA (12/13)

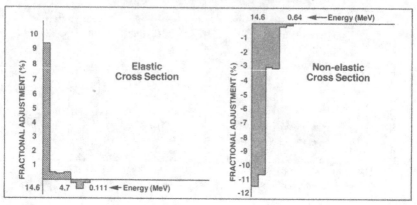

FIG.3

Fractional adjustment of front quarter depth of PCA core source resulting from combined adjustment of the Oak Ridge PCA (12/13) array and the AEEW NESDIP PCA(12/13) array

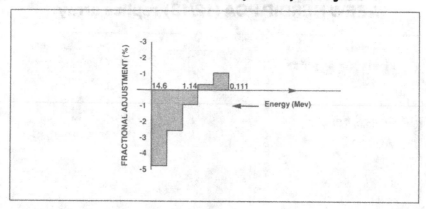

FIG.4

THE NESSUS REFERENCE FIELD

IN THE NESTOR REACTOR AT WINFRITH

M D Carter, I J Curl, M F Murphy and A Packwood

UKAEA WINFRITH, UK

1   INTRODUCTION

Reference radiation fields play an important role in the evaluation and monitoring of measurement techniques used by nuclear metrologists in research and industry.   These well-defined, stable environments are either located close to long-lived neutron sources such as Cf-252 or inside critical assemblies.   A key role of these fields is the intercomparison of the measurement techniques used in low flux benchmark experiments and those employed for power reactor monitoring; because of the wide range of detector sensitivities it is rarely possible to perform the comparison in a single field.

In the new UK reference field, NESSUS, which has been commissioned in the NESTOR reactor at Winfrith, the fluxes can be varied over a dynamic range of 3×10^4 and it is possible to perform precision measurements with nearly all neutron and gamma-ray integral measurement techniques – with the exception of the high dose damage monitors.

This paper describes the facility and outlines the procedures adopted to define the radiation environment.

2   DESCRIPTION OF THE NESSUS FACILITY AT NESTOR

The NESSUS facility provides access to the centre of the inner graphite reflector of the NESTOR reactor which has an annular core and an external graphite reflector as shown in Figure 1.   The NESSUS cavity is located in a sample holder made from graphite which is loaded into the inner reflector via an aluminium thimble of internal diameter 67 mm.   When in position the cavity is located at the core mid-height.   The size of the cavity is limited only by the external dimensions of the cylindrical sample holder; for special applications it can be manufactured up to a practical diameter of 50 mm.   For the reference measurements a cavity with a height of 87 mm and a diameter of 29 mm was used.

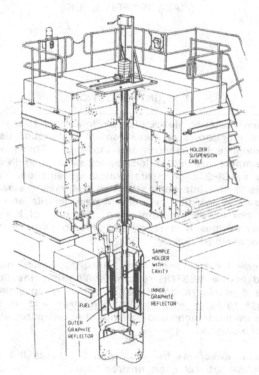

FIG 1 THE NESSUS FACILITY

3 DEFINITION OF THE NEUTRON FIELD

The neutron spectrum above 0.1 MeV was first calculated using the Monte Carlo code McBEND(1). The geometric model is shown in Figures 2 and 3. The annular core of the reactor has fuel loaded in two rings. The model includes the 24 MTR-type fuel-elements in the inner ring but omits two elements which reside in the outer ring; these are replaced by graphite blocks. Adjoint Monte Carlo calculations enabled the importance of this simplification to be estimated and small corrections were applied. The spatial source distributions were taken from the original NESTOR commissioning measurements utilising the fission spectrum defined by the Watt distribution:

$$\chi(E) = 0.453 \ e^{-1.0123E} \ \sinh \sqrt{2.1893E}$$

The calculation was performed in the point cross-section mode using an 8000-group data set processed from the UKNDL nuclear data library. The spectrum was scored in 25 energy groups.

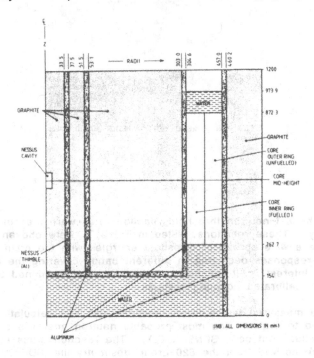

FIG 2 ELEVATION OF McBEND MODEL FOR NESTOR (NOT TO SCALE)

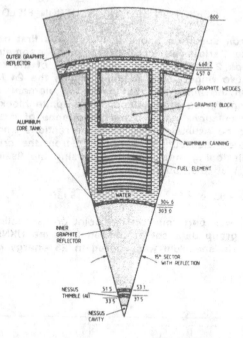

OUTER GRAPHITE
REFLECTOR

800

460 2
457 0

GRAPHITE WEDGES

GRAPHITE BLOCK

ALUMINIUM
CORE TANK

ALUMINIUM CANNING

FUEL ELEMENT

WATER

304 6
303 0

INNER
GRAPHITE
REFLECTOR

15° SECTOR
WITH REFLECTION

NESSUS
THIMBLE (AI)

51 5   53 1

33 5   37 5

NESSUS
CAVITY

(NB ALL DIMENSIONS IN mm)

FIG. 3 PLAN OF McBEND MODEL FOR NESTOR (NOT TO SCALE)

Eight fast-neutron threshold reaction-rates were measured in the cavity. These reactions, listed in Table 1, were chosen as they have a wide spread of threshold energies with the main detector responses occurring in different bands covering the energy range of interest. All reaction-rates have been determined using absolutely-calibrated counting systems.

The measured fast-neutron reaction-rates and calculations were used to evaluate the most probable neutron energy spectrum via the adjustment code SENSAK (2). The reaction cross-sections used were derived from the 620-group dosimetry file IRDF-82 and group-averaged using the unadjusted NESSUS spectrum.

SENSAK seeks maximum likelihood consistency between
measurement and calculation by adjustment of all statistically
quantifiable parameters. To achieve C/E ratios of unity for all the
reactions in this case, SENSAK exercised only small spectral and
cross-sectional adjustments. The unadjusted cross-sections were
used with the adjusted spectrum to determine the reaction-rates
shown in Table 1, and the deviation of the C/E ratios from unity
essentially demonstrates the degree of cross-section adjustment
exerted by SENSAK. The adjusted NESSUS spectrum is presented
in Table 2. In Table 3 is a list of NESSUS spectrum-averaged
cross-sections for a selection of the threshold detectors in the
IRDF-82 data set.

TABLE 1

COMPARISON OF CALCULATED TO MEASURED REACTION
RATES IN NESSUS

Reaction	C/E
Al27(n,α)Na24	1.03±0.06
Ti46(n,p)Sc46	0.96±0.10
Fe54(n,p)Mn54	1.04±0.07
S32(n,p)P32	1.04±0.09
Ni58(n,p)Co58	1.02±0.09
U238(n,f)	1.01±0.06
In115(n,n')In115^m	0.93±0.08
Rh103(n,n')Rh103^m	0.94±0.05

TABLE 2

THE NESSUS HIGH ENERGY NEUTRON SPECTRUM

Group	Upper Bound (MeV)	Flux*	Group	Upper Bound (MeV)	Flux*
1	1.4918E+01	3.535E-05	14	2.7250E+00	1.593E-02
2	1.2214E+01	2.732E-04	15	2.4660E+00	6.089E-03
3	1.0000E+01	1.015E-03	16	2.3460E+00	2.718E-02
4	8.1873E+00	1.469E-03	17	2.0190E+00	2.836E-02
5	7.0469E+00	3.233E-03	18	1.8270E+00	3.173E-02
6	6.3763E+00	5.756E-03	19	1.6530E+00	2.648E-02
7	5.4881E+00	6.599E-03	20	1.4950E+00	3.022E-02
8	4.9659E+00	7.455E-03	21	1.3530E+00	1.035E-01
9	4.4930E+00	1.169E-02	22	1.0026E+00	8.990E-02
10	4.0660E+00	8.170E-03	23	7.4274E-01	9.916E-02
11	3.6790E+00	8.885E-03	24	5.5023E-01	2.432E-01
12	3.3290E+00	1.139E-02	25	2.4724E-01	2.219E-01
13	3.0110E+00	1.037E-02		1.1109E-01	

*Normalised to unit spectrum above 0.1 MeV.
Multiply by 1.78E+09 to obtain 1kW levels

TABLE 3

NESSUS SPECTRUM AVERAGED CROSS-SECTIONS FROM
IRDF-82

Reaction	Spectrum Averaged Cross-Section (Barns)
103Rh(n,n')103Rh^m	4.00,-1
93Nb(n,n')93Nb^m	7.44,-2
115In(n,n')115In^m	7.70,-2
232Th(n,f)FP	2.90,-2
238U(n,f)FP	1.22,-1
47Ti(n,p)47Sc	7.51,-3
46Ti(n,p)46Sc	3.51,-3
31P(n,p)31Si	9.03,-3
58Ni(n,p)58Co	3.40,-2
64Zn(n,p)64Cu	1.12,-2
32S(n,p)32P	2.21,-2
54Fe(n,p)54Mn	2.55,-2
27Al(n,p)27Mg	1.31,-3
56Fe(n,p)56Mn	2.91,-4
59Co(n,α)56Mn	3.92,-5
63Cu(n,α)60Co	1.59,-4
24Mg(n,p)24Na	3.88,-4
27Al(n,α)24Na	1.82,-4
48Ti(n,p)48Sc	7.42,-5

Below 0.1 MeV the spectrum in this graphite environment is
well thermalised and a measurement of the Mn55(n,γ)Mn56
reaction-rate has shown that the Westcott flux at full power is
3×10^{11} n.cm^{-2}s^{-1} when the cavity is empty.   When samples are
loaded into the cavity some thermal flux depression can be
expected.   Foil measurements can, however, be performed easily
in NESSUS and measurements of the low energy flux are made with
each irradiation if required.

4   DEFINITION OF THE GAMMA-RAY FIELD

The gamma-ray field was defined by the normalisation of a
calculated spectrum so that it reproduced the measured exposure-
rate.   This calculation included gamma-ray sources from prompt
fission, fission-product decay and thermal neutron capture.   The
calculated spectrum is given in Table 4.

There are considerable problems in making precision gamma-
ray dosimetry measurements in reactor environments due to the
high fluxes of thermal neutrons which can induce secondary
gamma-ray sources in the measurement device.   Fast neutron
interactions can further enhance the measured signal.   The CEGB
at their Berkeley Nuclear Laboratories have undertaken the
development of mixed field gamma-ray dosimetry techniques as part
of their work on AGR graphite energy-deposition (3).   This work

has led to the development of custom-built graphite walled
ionisation-chambers free from significant thermal neutron capture
gamma-ray sources (4) and to the development of beryllium oxide
as a passive dosimeter (5). At an earlier stage of the NESSUS
development programme graphite calorimetry measurements were
performed (6) in which the combined neutron and gamma-ray
energy-deposition rates were recorded. To permit a direct
comparison of the three measurement techniques an ionisation
chamber which closely models the calorimeter has been produced
and a demountable calorimeter has been provided with access for
BeO thermoluminescent dosimeters. The neutron component of the
energy-deposition rate has been estimated from the neutron
spectrum. The gamma-ray exposure-rate in NESSUS contains a
component from fission product decay which is dependent on
reactor history. A measurement procedure has been adopted in
NESSUS whereby the exposure-rates are deduced from a 20 minute
power irradiation, timed from 37% of full power, followed by a 5
minutes decay after shut-down. During the rise to power a
constant doubling time of 25 seconds is maintained. A close
check is kept on the core gamma-ray background from long-lived
fission products and this contribution is subtracted from the
measured exposure-rates. Corrections are applied for thermal and
fast neutron responses and, in the case of BeO, for gamma-ray
energy response of the thermoluminescent light output. The
exposure-rates per NESTOR kW deduced from the various
instruments are given in Table 5 and give an average exposure-rate
of $1.0 \times 10^4 Rh^{-1}$ per kW. The calorimetry measurements, after
adjustment for the neutron energy-deposition component and for
positional differences in the measurement location, are in good
agreement with this figure. Further development of the absolute
calibration of the calorimeter is required before final comparisons
can be made.

TABLE 4

THE NESSUS GAMMA-RAY SPECTRUM ESTIMATED FROM
DOT CALCULATIONS

Group	Upper Bound (MeV)	Flux*	Group	Upper Bound (MeV)	Flux*
1	14.00	6.06E+02	11	1.66	2.36E+08
2	12.00	4.27E+03	12	1.33	3.46E+08
3	10.00	5.40E+04	13	1.00	2.84E+08
4	8.00	5.86E+07	14	0.80	4.02E+08
5	6.50	4.67E+07	15	0.60	7.39E+08
6	5.00	1.46E+08	16	0.40	6.07E+09
7	4.00	1.74E+08	17	0.30	1.15E+09
8	3.00	1.33E+08	18	0.20	4.15E+09
9	2.50	2.27E+08	19	0.10	6.27E+09
10	2.00	1.76E+08	20	0.05	7.05E+08
				0.02	

*Normalised to NESTOR power of 0.8kW

TABLE 5

GAMMA-RAY EXPOSURE-RATE MEASUREMENTS IN NESSUS

Measurement Technique	Measured Exposure-Rate (Rh⁻¹ at 100W)
BeO in graphite	970±24
6cc BNL graphite walled ionisation chamber	1052±35
Calorimeter ionisation chamber	1014±35

5   OPERATIONAL PROCEDURES

NESSUS can be used to irradiate both passive and active detectors including calorimeters such as those described in Reference 6 which measure down to the limit of about $10\mu W/g$. The variety of materials which may be irradiated is restricted only by factors such as: the total absorption worth, the fissile content, and the usual problems associated with the handling of hazardous materials. The maximum permissible absorption worth corresponds to a total dilute absorption cross-section worth of 12 cm^{-1}. For black absorbers, such as cadmium, a maximum surface area of 25 cm^2 is permitted. Fissile samples are permitted with an upper limit that corresponds to 2 g of dilute U235. At present no liquid samples can be irradiated and the operating temperature of the sample must remain below 70°C. It is expected that the restrictions on the use of liquids and on the sample temperatures will be relaxed when more operating experience has been gained. The level of irradiation is limited by the radiation hazard associated with activated species in the sample on its removal from NESSUS.

6   POWER RANGE AND STABILITY

The useable power range of NESTOR extends from a few watts to 30 kW. At low power the long-lived fission-product component becomes important for gamma dosimetry measurements. The core power has been demonstrated to be linear with the indicated power using S32(n,p)P32 activation monitors and fission chambers in the cavity.

To monitor run-to-run reproducibility, sulphur detectors are irradiated simultaneously in reference positions at the top and base of the cavity. For multiple irradiations at high powers nickel monitors are also used. Ten irradiations covering powers from 1 to 30 kW over a period of 1 month resulted in a standard deviation on the monitor measurements of 1%. The reactor is dismantled every two years for inspection and maintenance after which it is necessary to re-establish the calibration of the reactor instrumentation using foil measurements both in NESSUS and in positions external to the core.

7   SUMMARY

There is a continuing programme of irradiations in NESSUS and to date the facility has been used for the investigations listed below.

1   Quality control of integral fast neutron detectors.   A NESSUS Metrology Quality Assurance Certificate (MQAC) has been devised to standardise the documentation of the irradiations, subsequent counting and analysis.

2   Inter-laboratory comparison of neutron metrology techniques as part of LWR and AGR shielding and dosimetry programmes.

3   The study of the $^{93}Nb(n,n')^{93m}Nb$ reaction which is used for power reactor monitoring. The preliminary results of Gayther et al (7), who are measuring the differential cross-section, give rather better agreement with the reaction-rate measured in NESSUS than the theoretical curve published by Strohmaier (8).

4   Development of micro-calorimetry techniques.

5   Methods testing of the calculational route for the determination of energy-deposition in graphite moderator reactors.

6   Activation analysis to determine ppm levels of trace elements in samples of materials.

7   Radioisotope production.

8   The precalibration of the reactor instruments has allowed sample absorption cross-sections to be determined from reactivity measurements.

The scope and variety of these applications demonstrates the versatility of NESSUS; the excellent reproducibility observed justifies the extensive programme of work already undertaken to define the radiation environment.   An important characteristic of the facility is the wide dynamic range which enables power reactor monitors such as niobium with low sensitivities to be compared directly with the high-sensitivity reactions utilised for benchmark measurements in low power facilities. The next step in the programme is to pursue the intercomparison with reference fields outside the UK, as an important link in the exchange of reference information.

8   REFERENCES

1   D E Bendall and R J Brissenden.   Program Users Guide –
     McBEND. COSMOS Ref Set Doc/147.

2   A K McCracken.   "Few Channel Unfolding in Shielding – The
     SENSAK Code".   Proceedings of the Third ASTM–Euratom
     Symposium on Reactor Dosimetry, ISPRA (1979).

3   S J Kitching, T A Lewis and T S Playle.   "In-Core Dosimetry
     in CAGR – Measurements on Power Reactors and Laboratory
     Facilities".   Gas–Cooled Reactors Today, BNES, LONDON
     1982.

4   S J Kitching and T A Lewis.   "An Ionisation Chamber for
     Measuring Gamma Exposure Rates in Mixed Radiation Fields".
     TPRD/B/0348/N83.

5   P J Heffer and T A Lewis.   "The Use of Beryllium Oxide
     Thermoluminescence Dosimeters for Measuring Gamma
     Exposure Rates".   To be presented at 5th ASTM–EURATOM
     Symposium on Reactor Dosimetry.

6   I J Curl.   Energy Deposition in the "NESTOR" Reactor. Ph. D
     Thesis, University of London, 1983.

7   D B Gayther et al.   "Activation Measurements of the Cross-
     Section for 93Nb(n,n')93m Nb Reaction".   Neutron
     Interlaboratory Seminar, PTB Braunschweig, 6–8 June 1984.

8   B Strohmaier et al.   "Evaluation of the Cross–Section for the
     Reactions of 19F(n,2n)18F, 31P(n,p)31Si,
     93Nb(n,n')93m Nb, and 103Rh(n,n')103m Rh".
     Physik Daten, No 13-2 (1980).

ACKNOWLEDGEMENTS

    The authors wish to acknowledge the assistance of
Mr T A Lewis of the CEGB, Berkeley Nuclear Laboratories who
carried out the gamma-ray exposure measurements.

COMPARISONS OF THEORETICAL AND EXPERIMENTAL NEUTRON SPECTRA, 115In(n,n') AND FISSION RATES, IN THE CENTRE OF THREE SPHERICAL NATURAL URANIUM AND IRON SHELL CONFIGURATIONS, LOCATED AT BR1

G. De Leeuw-Gierts, S. De Leeuw (SCK/CEN)

D.M. Gilliam (NBS)

ABSTRACT

Three spherical configurations of iron and uranium shells have been studied. The configurations were a 1-cm thick natural uranium shell, a 1-cm thick natural uranium shell with an inner 7-cm thick iron shell and a 1-cm thick natural uranium shell with an inner iron shell of 14-cm thickness. For the measurements, the shells were located at the centre of a hollow cavity, 100-cm in diameter, in the vertical graphite thermal column of the BR1 reactor.

The central neutron spectra were calculated by means of the DTF-IV code, using the 208-group KEDAK-3 library, and by means of the ANISN code, using the 171-group VITAMIN-C library. Central neutron spectra, measured by the proton-recoil and $^6Li(n,\alpha)t$ spectrometry techniques, are compared to the theory between $\sim$ 100 keV and 5 MeV.

Mean fission cross-sections of ^{240}Pu, ^{237}Np, ^{234}U, ^{235}U, ^{236}U and ^{238}U were deduced from the calculations. Their ratios with respect to ^{238}U are compared to measurements made with NBS dual fission chambers.

INTRODUCTION

Measurements, performed in the centre of three spherical

U_{nat}-Fe configurations to test the Fe-group cross-sections used in reactor calculations, were previously described in ref. 1.

Fission and $^{115}In(n,n')$ reaction rates have been obtained since. The spectral ^6Li measurements and corresponding spectral calculations have been pursued and improved.

DESCRIPTION OF THE EXPERIMENT

Fission and $^{115}In(n,n')$ Reaction Rate Measurements

^{240}Pu, ^{237}Np, ^{234}U, ^{235}U, ^{236}U and ^{238}U fission measurements, with NBS dual fission chambers [2], and $^{115}In(n,n')$ measurements were performed in the centre of a 1 cm thick U_{nat} shell (outer diameter : 34.5 cm), a 1 cm U_{nat} with 7 cm inner Fe shell and a 1 cm U_{nat} with 14 cm inner iron shell, placed in the centre of the 1 m diameter cavity hollowed in the thermal neutron graphite column of BR1.
Reaction rate ratios and spectral indices were deduced. In foils were also irradiated in the Ortec ^6Li neutron spectrometer to estimate its perturbation.

Differential Neutron Spectrometry Measurements

Previous, (n,p) measurements with SP2 proportional counters and $^6Li(n,\alpha)t$ measurements with a "home" made spectrometer in the 1 cm U_{nat} + 7 cm Fe and 1 cm U_{nat} + 14 cm Fe configurations were achieved [1]. $^6Li(n,\alpha)t$ spectrometry measurements up to $\sim$ 6.5 MeV, neutron energy, were pursued in the 1 cm U_{nat} and 1 cm U_{nat} + 7 cm Fe configuration, with a Ortec ^6Li neutron spectrometer. The ^6Li measurements were improved, as described in [3], by the use of a ND 6600 analysing system (DAS). The improvement resulted principally from the reduction of the γ-background and the simultaneous recording on a single system of the E_t and $E_\alpha + E_t$ distributions.

CALCULATIONS

The 40-gr DTF-IV code, with the BR2 library was used to determine the fission source, taken for the calculation of the central neutron spectra with the 208-gr DTF-IV, KEDAK-3 and KEDAK-3 Intermediate (i.e. $\sigma_{in,i,j}$ of KEDAK-2 for ^{238}U) libraries, and with the 171-gr ANISN - VITAMIN-C codes. An error was found in the earlier published ANISN calculations that were therefore repeated by G. Minsart.

For comparison with the ^{6}Li results, complementary DTF-IV calculations have been done, taking into account the bulkiness of the Ortec counter.

The mean cross-sections for comparison with the reaction rate measurements were calculated with the ENDF/B-5 dosimetry file.

RESULTS AND DISCUSSION

The theoretical ANISN and DTF-IV central spectra for the 1 cm U_{nat} and 1 cm U_{nat} + 14 cm Fe configurations are drawn in fig. 1, together with the Fe inelastic cross-sections of KEDAK-3 and ENDF/B-4 (DLC-41B). The 1 cm U_{nat} driver is well predicted by the 171-gr ANISN - VITAMIN-C code.

Because inaccuracies in the inelastic ^{238}U cross-sections and/or in $\sigma_{in\ i,j}$ have a reduced importance in the shape of the central neutron spectrum for a shell of only 1 cm thickness, the KEDAK-3 calculated spectrum is as good in agreement with the experimental results as the ANISN spectrum [2].

In fig. 2 the ratio of the ^{6}Li spectra, 1 cm U_{nat} over 1 cm U_{nat} + 7 cm Fe, is compared to the corresponding DTF-IV - KEDAK-3 and ANISN - VITAMIN-C values.

In table 1 measured and calculated spectral indices σ_x/σ_{238_U} are compared.

In table 2 experimental and theoretical reaction rate ratios are intercompared. Two ratios are given for In (irradiations with and without the Ortec spectrometer) to be able to estimate experimentally the spectrometer perturbation factor on the ^{6}Li spectral ratios. The results observed confirm the theoretical DTF-IV calculation with perturbation of table 1.

Fig. 3 illustrates the bulkiness of the Ortec spectrometer and shows also the drawing of a new designed spectrometer, as used in the Venus measurements; box and detector supporting discs are minimized and adaptable to the nature of the environmental material, metallic or hydrogeneous.

The ratios of the calculated spectral indices to the measured ones are reported in table 3.

In the neutron energy range ~ 100 keV - 6 MeV, where the applied techniques are sensitive, there is a good agreement between the experimental spectral indices and those calculated with the ENDF/B-5 dosimetry file and respectively the DTF-IV - KEDAK-3 central spectra and the ^{6}Li neutron spectra. A discrepancy with the theoretical data deduced from the ANISN - VITAMIN-C results, increasing with the Fe thickness, is observed and is consistent with the conclusions drawn from the analogous PCA (ORNL) experiment (table 4).

The ^{6}Li measurements up to 10 MeV, performed to confirm the present indication of a too high σ_{in} Fe cross-section above ~ 5 MeV, of importance too in the frame of fusion technology, are almost achieved.

REFERENCES

[1] G. and S. De Leeuw, pp. 971-979, EUR 6813 EN-FR V2.

[2] G. and S. De Leeuw, pp. 755-767, CONF 820321 V2.

[3] J.A. Grundl et al., Nucl. Technol. 25, p. 237, 1975.

ACKNOWLEDGMENT

The authors wish to thank F. Cops for his assistance in the measurements and D. Langela for their interpretation and his theoretical calculations.

Table 1. Comparison of measured and calculated spectral indices $\bar{\sigma}_x/\sigma_{238U}$ for the 1 cm U_{nat}, 1 cm U_{nat} + 7 cm Fe and 1 cm U_{nat} + 14 cm Fe configuration

The last four columns are grouped under the heading: **Spectral indices calculated with the ENDF/B-5 dosimetry file and the** — with "208 gr DTF-IV spectrum" spanning the KEDAK-3 and KEDAK-3_I (2) columns, and "171 gr ANISN spectrum" over the VITAMIN-C column.

Configuration	Isotope	NBS double fission chamber results	^{6}Li spectral results (1)	208 gr DTF-IV spectrum — KEDAK-3	208 gr DTF-IV spectrum — KEDAK-3_I (2)	171 gr ANISN spectrum — VITAMIN-C
1 cm U_{nat}	^{237}Np	5.11 ± 5.9 %	5.63	5.28	5.33 / 5.42 (3)	5.38
	^{234}U	4.49 ± 5.8 %	4.88	4.56	4.62 (3) / 4.70 (3)	4.66
	^{240}Pu	5.39 ± 7.0 %	6.01	5.61	5.71 (3) / 5.79 (3)	5.64
1 cm U_{nat} + 7 cm Fe	^{237}Np	7.58 ± 1.8 %	8.04	7.68	7.92 (3)	8.22
	^{234}U	6.76 ± 3.4 %	7.13	6.76	6.98 (3)	7.29
	^{236}U	2.44 ± 3.5 %	2.33	2.32	2.34 (3)	2.38
	^{240}Pu	7.31 ± 5.8 %	8.37	8.04	8.29 (3)	8.54
1 cm U_{nat} + 14 cm Fe	^{237}Np	10.84 ± 2.2 %	11.03	10.93	11.0	12.26
	^{234}U	9.88 ± 3.4 %	9.95	9.84	9.90	11.38
	^{236}U	2.85 ± 3.4 %	2.64	2.64	2.63	2.77
	^{240}Pu	11.08 ± 2.2 %	11.61	11.47	11.5	12.82

(1) Neutron spectrum measured by means of the ^{6}Li neutron spectrometry technique between 100 keV and 6 MeV and extrapolated by means of the DTF-IV KEDAK-3 spectrum.

(2) The KEDAK-3 Intermediate library is used for ^{238}U i.e. same σ_{in} as in KEDAK-3 but matrix $\sigma_{in\,i,j}$ of KEDAK-2.

(3) The perturbation due to the bulkiness of the Ortec spectrometer is taken into account in the calculation.

Table 2. Experimental and theoretical reaction rate ratios

Reaction ratios	Experimental reaction ratio		Ratios deduced with the ENDF/B5 dosimetry file from the neutron spectra obtained with								
			the ^6Li neutron spectrometry technique (3)			208 gr DTF-IV - KEDAK-3			171 gr ANISN - VITAMIN-C		
	$^{237}Np(n,f)$	$^{238}U(n,f)$ $^{115}In(n,n')$	^{237}Np	^{238}U	^{115}In	^{237}Np	^{238}U	^{115}In	^{237}Np	^{238}U	^{115}In
1 cm U_{nat}	1.61	2.39 2.16									
1 cm U_{nat} + 7 cm Fe		2.28 (1)									
1 cm U_{nat} + 7cm Fe	1.32	1.89 2.09 (2)	1.56	2.22	2.08	1.57	2.28	2.12	1.57	2.40	2.31
1 cm U_{nat} +14cm Fe		1.66 (2) 1.71	1.34	1.80	1.66	1.29	1.83	1.68	1.32	1.96	1.91

Table 3. Calculated/measured spectral indices

	CONFIGURATIONS								
	1 cm U_{nat}			1 cm U_{nat} + 7 cm Fe			1 cm U_{nat} + 14 cm Fe		
ISOTOPE	DTF IV (a) KEDAK-3 NBS results	ANISN (b) VITAMIN-C NBS results	DTF IV (c) KEDAK-3 ^6Li results (3)	(a)	(b)	(c)	(a)	(b)	(c)
^{237}Np	1.03	1.05	0.98	1.01	1.08	0.98_5	1.01	1.13	0.99
^{234}U	1.02	1.04	0.98	1.00	1.08	1.00	1.00	1.15	0.99
^{236}U				0.95	0.97_5	0.98	0.93	0.97	1.00
^{240}Pu	1.04	1.05	0.99	1.10	1.17	0.99	1.03_5	1.16	0.99

(1) Reaction rate used in the 1 cm U_{nat} configuration takes into account the maximum ^{235}U contamination amount.

(2) The In reaction rate in the 1 cm U_{nat} configuration is deduced from the In sample irradiated in the Ortec Li-6 spectrometer.

(3) The accuracy in flux level may reach 5 %, because of dead time corrections.

Table 4. Comparison of the PCA and the U_{nat} + Fe experiments

$^{237}Np/^{238}U$ in PCA	Geometry	Exp. A. FABRY et al.	ANISN - VITAMIN-C	ANISN/EXP.
Results deduced from the equivalent fission fluxes published in (1)	1/4 T (5 cm Fe)	1.69	1.81	1.07
	1/2 T (10 cm Fe)	2.17	2.40	1.11
	3/4 T (15 cm Fe)	2.74	3.22	1.17

$^{237}Np/^{238}U$ in the U_{nat} + Fe experiment	Geometry	Exp. D. GILLIAM	ANISN - VITAMIN-C	ANISN/EXP.
This report	1 cm U_{nat}	5.11	5.37	1.05
	1 cm U_{nat} + 7 cm Fe	7.58	8.19	1.08
	1 cm U_{nat} + 14 cm Fe	10.8	12.19	1.14

(1) A. Fabry et al.
NUREG/CR-1861, HEDL-TME 80-87.

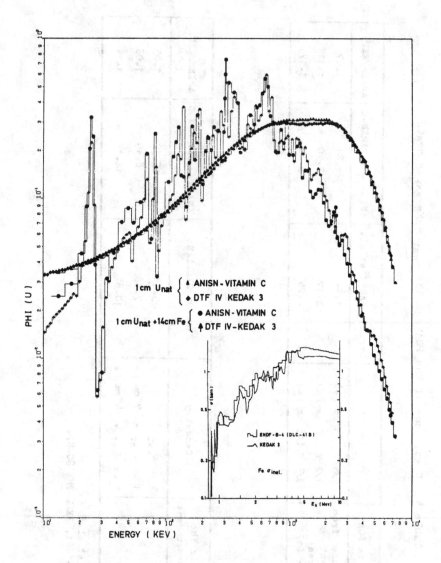

Fig. 1. Central 1 cm U_{nat} and 1 cm U_{nat} + 14 cm Fe neutron
spectra calculated by the 208 gr DTF-IV – KEDAK-3
and 171-gr ANISN – VITAMIN-C codes. The insert
compares the σ_{in} of Fe in the ENDF/B-4 and KEDAK-3
libraries.

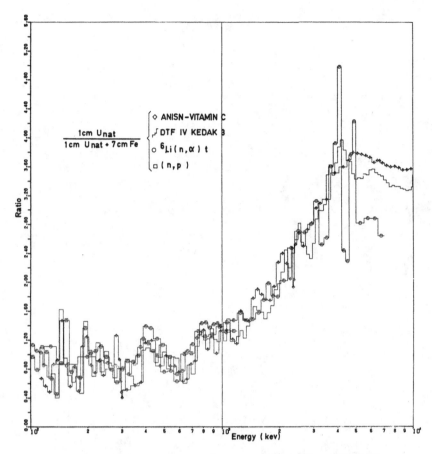

Fig. 2. 1 cm U_{nat}/1 cm U_{nat} + 7 cm Fe   DTF-IV – KEDAK-3,
ANISN – VITAMIN-C, ^{6}Li(n,α)t and (n,p) spectral ratios.

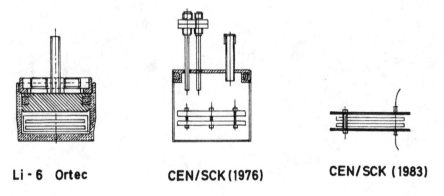

Li - 6 Ortec   **CEN/SCK (1976)**   **CEN/SCK (1983)**

Fig. 3.

VENUS PWR ENGINEERING MOCK-UP :

CORE QUALIFICATION, NEUTRON AND GAMMA FIELD CHARACTERIZATION

A. Fabry, L. Leenders, R. Menil, G. Minsart, H. Tourwé
S. De Leeuw, P. De Regge, J. Debrue (SCK/CEN)

E.D. McGarry (NBS)

T.A. Lewis (CEGB)

C. Barr-Wells, M. Austin (RR&A)

ABSTRACT

The VENUS PWR Engineering Mock-up experiment is part
of the "Belgian PWR-Pressure Vessel Surveillance
Program", sponsored by the Belgian Utilities. Through
a cooperation agreement it contributes also to the
"LWR-Pressure Vessel Surveillance Dosimetry Improvement Program" supported by the U.S. Nuclear Regulatory
Commission. The VENUS pressure vessel mock-up simulates the reflector geometry and the core boundary
shape of a generic 3-loop power plant. The VENUS core
itself is made of a regular pattern of fuel pins typical of the modern PWR power reactors. The outgoing
neutron spectrum is thus close to realistic situations. The paper includes typical experimental results concerning the pin-to-pin power distribution,
the fast neutron propagation out of the core and the
gamma energy deposition in selected positions of the
core baffle. The theoretical analysis scheme is presented together with a comparison between calculated
and measured data.

FRAME OF THE WORK, OBJECTIVES

The VENUS PWR Engineering Mock-up experiment is part of the "Belgian PWR-Pressure Vessel Surveillance Program", sponsored by the Belgian Utilities. Through a cooperation agreement it contributes also to the "LWR-Pressure Vessel Surveillance Dosimetry Improvement Program" supported by the U.S. Nuclear Regulatory Commission.

An early phase of the program at the ORNL Pool Critical Assembly addressed the neutronic validation of LWR-PVS lead factors in a slab arrangement of a thermal shield plus pressure vessel simulator. Actual LWR-PVS lead factors also involve significant azimuthal variations whose calculational accuracy depends upon a correct estimate of the source distribution near the core boundary and a correct modelling of that boundary.

The VENUS pressure vessel mock-up simulates the reflector geometry and the core boundary shape of a generic 3-loop power plant. The VENUS core itself is made of a regular pattern of fuel pins typical of the modern PWR power reactors. The outgoing neutron spectrum is thus close to realistic situations.

Most measurements were made on an interlaboratory basis with the participation of U.S. (HEDL, NBS) and of U.K. (CEGB, RR&A) laboratories. All results are normalized to an absolute calibration of the core power distribution and benchmark-field referencing was applied to equivalent fission flux determinations (link to the MOL Cavity Uranium-235 Fission Spectrum Standard Neutron Field and the NBS 252Cf Fission Spectrum Standard Neutron Field) and to gamma energy deposition measurements (MOL Gamma Benchmark Field, CEGB and RR&A reference fields). A particular attention was paid to verify sensor calibrations and to check material quality assurance.

The present paper includes the following items :

- core description and quality assurance;
- experimental and calculational pin-to-pin core power distributions;
- experimental and calculated equivalent fission neutron flux values for a number of locations in the core, in the baffles, in the barrel and in the neutron pad;
- experimental and calculated gamma doses in selected locations in the core baffles, in the barrel and in the neutron pad.

Other measurements, namely neutron and gamma-ray spectrometry, are still being analysed.

The different steps of the study, still under progress, are shown on figure 1 in a blockdiagram, where the connexions between experiments and calculations are also indicated. Most of the items are commented in the following paragraphs and illustrated in the figures, where a few typical results are displayed.

CORE DESCRIPTION

The 2-zone cruciform core loading has been designed in order to simulate the outer part of actual PWR reactors, while providing clean conditions for the measurements; it is illustrated on fig. 2. The central zone is made of 3.3 w/o ^{235}U enriched Zircaloy-clad oxide fuel rods, with PYREX tubes disseminated in the lattice in a similar way as the guide tubes in modern PWR assemblies. The outer zone is made of 4.0 w/o ^{235}U enriched rods with SS-304 cladding. The fuel pitch is uniform. Safety aspects and scram are based on the fast water dump system, so that no local perturbations are introduced in the lattice, except for the insertion of a few fine regulating rods at locations diametrically opposed to the measurement sector.

All the dimensions were carefully checked, as well as the relative positions of the baffle, of the barrel and of the neutron pad. Specimens were taken from all the components and analysed, in order to provide detailed compositions, including most of the impurities. Detailed specifications of the VENUS PWR mock-up have been issued.

The experimental program covered a period of six months. Redundant monitoring systems were installed in VENUS and in the BR1 thermal column where the reference neutron and gamma fields are located, in order to guarantee a good consistency of the results finally normalized at the VENUS nominal power level.

CORE POWER DISTRIBUTION

A special run in the experimental program has been executed to provide a reference power map in one eight of the core, in front of the barrel. A number of fresh fuel pins, distributed in this sector, were irradiated at the maximum flux level during one day, and then scanned in an automatic counting device. At the end of the experiments, a large number of fuel pins irradiated during the whole program, were also scanned, and both maps were used to define a pin-to-pin fission density distribution across the considered core sector.

In parallel, theoretical studies of the configuration were performed. The detailed analysis of the fission density distribution was made with the 2-Dimension transport code DOT 3.5 in XY geometry and S8 approximation; the 10-group cross-sections were obtained by space-dependent collapsing of the SCK/CEN 40-group library (spectra yielded by 1-Dimension 40-group problems). Accurately measured vertical bucklings ($\pm$ 0.5 %) were used in these calculation.

Both theoretical and experimental maps were normalized to a same average and compared. Although significant gradients are observed in the 4 w/o enriched zone, as shown on figure 3, the comparison of the full maps indicates a very good agreement, except for a few pins close to the outer baffle and also around the PYREX rods (fig. 4).

The absolute core average fission density at the mid-plane level was determined by four independent methods, for the reference nominal power level.

Method	Fission density ($cm^{-1} s^{-1}$)
Gamma scanning(counting of the ^{140}Ba-^{140}La 1.6 MeV gamma peak) calibrated on the basis of absolute neutron density measurements (Dy, Au foils activated between fuel pellets) in a reference VENUS rod sample irradiated in the BR1 thermal column.	$(2.13 \pm 0.04) \cdot 10^8$
Gamma scanning calibrated on the basis of absolute 1.6 MeV gamma activity measurements on VENUS fuel pellets irradiated in the VENUS core and in the BR1 thermal column.	$(2.06 \pm 0.07) \cdot 10^8$
Gamma scanning calibrated on the basis of radiochemical analysis of VENUS fuel pellets irradiated in the VENUS core (several fission products).	$(2.08 \pm 0.04) \cdot 10^8$
Calibrated miniature fission chambers measurements between fuel pins (corrected with computed disadvantage factors).	$(2.03 \pm 0.06) \cdot 10^8$
Weighted average value and observed dispersion (90 % confidence interval).	$(2.10 \pm 0.04) \cdot 10^8$

This value has been used to scale the computed fission density (and hence all the theoretical results).

FAST FLUX DISTRIBUTIONS

Equivalent fission neutron flux measurements have been performed throughout the core, the baffles, the barrel, the thermal shield and numerous other reflector locations of the VENUS "core source" to "pressure vessel wall fluence" benchmark. Results were obtained for the threshold reactions ^{237}Np(n,f), ^{103}Rh(n,n'), ^{115}In(n,n'), ^{238}U(n,f) and ^{58}Ni(n,p). The activation technique has been used, except for the fissionable isotopes which involve miniature fission chambers and solid state track recorders. All measurements are benchmark-field referenced to the MOL Cavity Uranium-235 Fission Spectrum Standard Neutron Field and are normalized to the absolute core power. The ^{235}U benchmark-field flux has, furthermore, been re-verified by flux-transfer measurements from the NBS ^{252}Cf Fission Spectrum Standard Neutron Field. All the available data are compared to DOT (R,θ) transport theory analyses accomplished by SCK/CEN (S8 P3 computation with 17-group cross-sections derived from the VITAMIN-C 171-group library by space-dependent collapsing based on spectra yielded by 1-Dimension 171-group ANISN calculations). The fixed fission source for this run was the computed distribution obtained in the previous step (XY model) normalized on the absolute fission density as determined experimentally and transposed into the Rθ grid by a home-made Fortran program. The space-dependent fast neutron axial bucklings were also measured (to ± 1 %) and used in this calculation.

Typical experimental results are illustrated on figs 5 and 6 and the comparison between theory and experiment, shown on fig. 7, indicates that accuracy goals for azimuthal lead factor predictions are reasonably met.

GAMMA FIELD CHARACTERIZATION

The motivations for this part of the VENUS program are :

1) The interest in quantifying heating sources in LWR surveillance capsule environments.

2) The general lack of qualified experimental data for energy-deposition rates in PWR steel internals which is needed to understand the behaviour of structural components in PWR power plants.

3) The requirement to validate calculated corrections for the gamma-ray response of some LWR neutron dosimeters.

Interlaboratory integral gamma measurements have been performed in the VENUS core baffles, in the barrel and in the thermal shield, under near-equilibrium conditions with respect to
delayed fission-product emission. This was achieved by automatic insertion of the gamma sensors after three hours of operation. Various kinds of thermoluminescent dosimeters were used
by experimenters; i.e. BeO by CEGB, ^{7}LiF by RR&A and Al_2O_3 by
SCK/CEN.

Reference exposures in Cobalt-60 fields as in one of the
MOL Black-Absorber Capture-Gamma Standard Fields were made to
verify sensor calibrations and check material quality assurance;
also thermal neutron response corrections were experimentally
assessed. Fast neutron response corrections and cavity corrections are not yet included; they are expected to not exceed
10 %.

A first gamma transport theory calculation has been performed (DOT XY run, S8 P5) using the gamma source derived from
capture and fission rates corresponding to the DOT XY 10-group
neutron run, and the EURLIB-4 gamma cross-section library.
Gamma doses at the experimental locations were then obtained
by conversion of the gamma fluxes.

After scaling to the absolute power level, the computed
and measured dose rates were compared. Some of these results
are illustrated on fig. 8. Although this part of the work is
still in a preliminary stage of the evaluation, the agreement
is most often better than 10 % in the outer baffle and also at
larger distances from the core (barrel, pad).

CONCLUSIONS

This overview of VENUS PWR Mock-up results shows that the
main features governing the power distribution, the neutron and
the gamma fields at the periphery of a power reactor are well
represented with this mock-up. The quality assurance on the core
geometry and on the compositions, and the interlaboratory effort
for the measurements lead to valuable reference data for computer
code validation in support to vessel surveillance programmes.
Moreover the calculational procedures adopted by SCK/CEN for the
analysis of this mock-up give encouraging results, in view of the
PWR-Pressure vessel characterization. Along this line, papers
from ORNL and WESTINGHOUSE at this Symposium (Poster Session II)
are to be mentioned. The complete results and the final analysis
will be reported in a joint NRC - SCK/CEN publication (NUREG/
CR 3323, BLG 568).

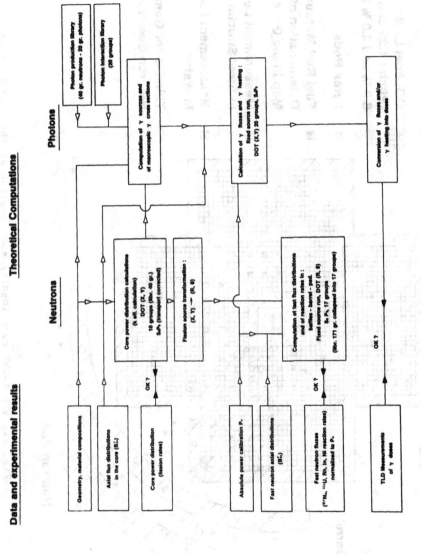

Fig. 1. Block-diagram of the work for the VENUS-LWR Benchmark Experiment.

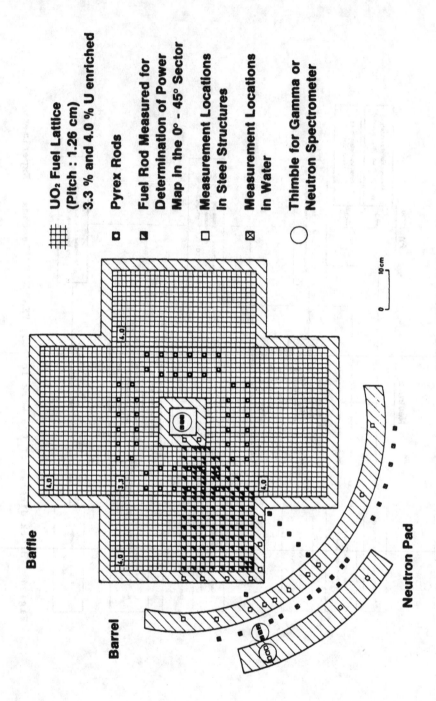

Baffle

Barrel

Neutron Pad

UO₂ Fuel Lattice
(Pitch : 1.26 cm)
3.3 % and 4.0 % U enriched

Pyrex Rods

Fuel Rod Measured for
Determination of Power
Map in the 0° - 45° Sector

Measurement Locations
in Steel Structures

Measurement Locations
in Water

Thimble for Gamma or
Neutron Spectrometer

Fig. 2. VENUS-LWR core loading and location of the measurements.

Fission rate distribution in the corner assembly.
(core average = 1.0)

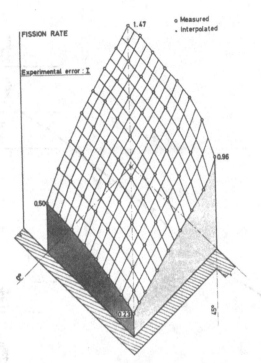

Fig. 3.

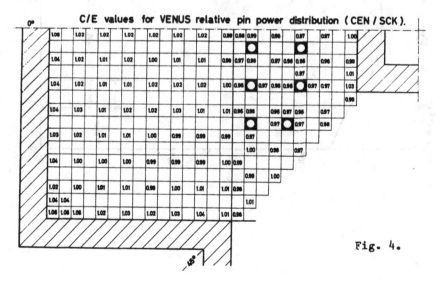

Fig. 4.

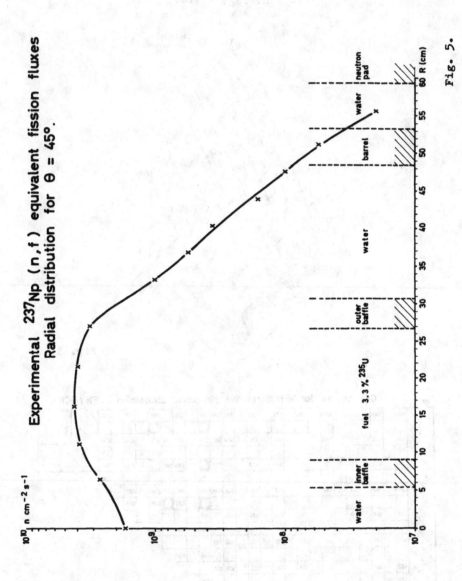

Experimental ^{237}Np (n,f) equivalent fission fluxes
Radial distribution for $\theta = 45°$.

Fig. 5.

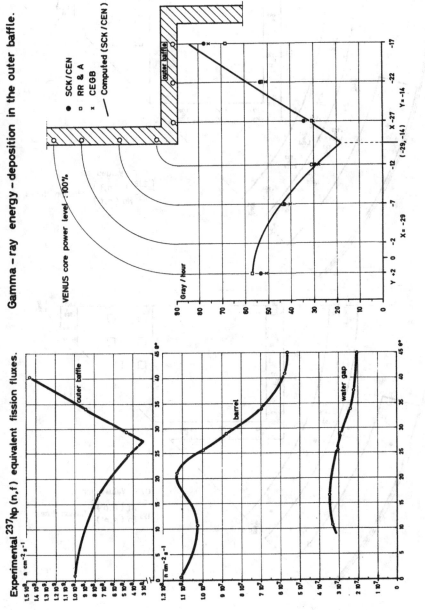

Gamma – ray energy – deposition in the outer baffle.

Experimental ^{237}Np (n,f) equivalent fission fluxes.

Fig. 8.

Fig. 6.

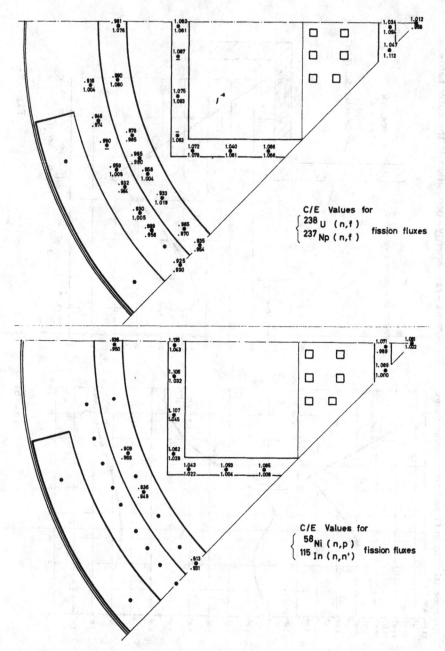

Fig. 7. Comparison of computed and measured
equivalent fission fluxes.

EVALUATION OF NEUTRON FLUX IN THE POOL CRITICAL ASSEMBLY

E. P. Lippincott, F. H. Ruddy, R. Gold,
L. S. Kellogg and J. H. Roberts

Westinghouse Hanford Company
Hanford Engineering Development Laboratory
Richland, WA USA

ABSTRACT

A recently completed series of experiments in
the Pool Critical Assembly (PCA) at Oak Ridge National
Laboratory (ORNL) provided extensive neutron flux
characterization of a mockup pressure vessel configuration. Considerable effort has been made to understand the uncertainties of the various measurements
made in the PCA and to resolve discrepancies in the
data. Additional measurements are available for
similar configurations in the Oak Ridge Reactor-
Poolside Facility (ORR-PSF) at ORNL and in the NESDIP
facility in the U.K. Comparisons of these results,
together with associated neutron field calculations,
enable a better evaluation of the actual uncertainties and realistic limits of accuracy to be assessed.
Such assessments are especially valuable when the
accuracy improvements of benchmark referencing are to
be included and extrapolations to new configurations
are made.

INTRODUCTION

The Pool Critical Assembly (PCA) is a major test facility
that has been established for validation of analytical tools
needed for Light Water Reactor (LWR) pressure vessel wall

J. P. Genthon and H. Röttger (eds.), Reactor Dosimetry, 783–790.
© 1985 ECSC, EEC, EAEC, Brussels and Luxembourg

neutronic projections. The facility consists of a critical assembly reactor core and a simulated thermal shield/pressure vessel wall/cavity arrangement designed to mockup actual reactor pressure vessel/surveillance configurations.

The PCA facility has served three purposes:

1) to assess the accuracy of reactor physics computations for representative configurations

2) to assess and improve the accuracy of dosimetry techniques and

3) to provide measurements to determine the preferred final configuration for a pressure vessel mockup irradiation of metallurgical specimens in the ORR-PSF.

A detailed physical description of the PCA is provided in Reference 1. Three configurations have been studied. The first (12/13 configuration) consisted of a nominal 12-cm water gap next to the reactor core face, a 5.9-cm thick thermal shield, a nominal 13-cm watergap, and a 22.5-cm thick pressure vessel mockup. The second (8/7 configuration) was the same except that the watergaps were changed to a nominal 8 and 7 cm, respectively. The third (4/12 SSC configuration) added a simulated surveillance capsule behind the thermal shield and changed the water gaps to 4 and 12 cm, respectively. This final configuration was close to that used in the ORR irradiation.

In the previous PCA report (1) least-squares analyses of the 12/13 and 8/7 configurations were reported. These least-squares analyses utilized the FERRET (2) code to derive a "best fit" neutron flux spectrum at each location using as input a calculated neutron flux spectrum; integral solid state track recorder (SSTR), fission chamber (FC), radiometric monitor (RM), nuclear research emulsion (NRE), and neutron spectrum measurements, together with reaction cross sections and uncertainty estimates including covariances. In this first study, discrepancies were found in the data and, in general, data discrepancies larger than the quoted experimental uncertainty were prevalent.

In this paper, an updated least-squares analysis is discussed that focuses on more recent results. Several modifications to the least-squares procedure have been made to simplify the analysis. These include treating each spectrum separately (and not referencing the data to the fission spectrum) and collapsing cross-section correction factors for each spectrum to improve the broad group cross-section accuracy. Differential data were not included in the present HEDL study.

Calculated neutron spectra for the PCA 12/13 and 4/12 SSC configurations were obtained from ORNL. The 12/13 calculation was a 47-group SSC calculation covering the entire neutron spectrum extending to thermal energies. However, the 4/12 SSC calculation used 102 groups to cover only the energy range from 0.098 to 16.5 MeV (similar to the calculation for the PSF). This is an adequate procedure since no reactions used in the least-squares analysis had any significant response below ∿0.4 MeV.

Although the absolute calculated flux has little impact on the least-squares result, it lends confidence to the result if the calculation and adjusted results are close. Comparisons of the calculated and adjusted flux >1.0 MeV are shown in Table 1. It is noted that the calculation is quite close to the measurements and relative variations within the block are less than 12% for both configurations.

Radiometric data (measured at HEDL and Mol) were converted to reactions per second per core neutron for input to FERRET. Uncertainties were increased from the quoted measurement uncertainties to cover relative normalization and systematic effects. The radiometric uncertainties used were all 6% to 7%, and the data were found to be consistent with the a priori input calculational results.

TABLE 1

COMPARISON OF CALCULATED AND MEASURED FLUX >1.0 MeV

Midplane Location	Calculated Flux >1.0 MeV (n/cm² per Core Neutron)	FERRET Flux >1.0 MeV (n/cm² per Core Neutron)	FERRET/ Calculated Flux
PCA 12/13 Configuration			
1/4 T	4.13 E-8	4.58 E-8	1.11
1/2 T	1.90 E-8	2.21 E-8	1.16
3/4 T	8.00 E-9	9.82 E-9	1.23
PCA 4/12 SSC Configuration			
1/4 T	2.18 E-7	2.34 E-7	1.07
1/2 T	1.00 E-7	1.15 E-7	1.14
3/4 T	4.33 E-8	4.87 E-8	1.12

Cross sections for the radiometric and fission reactions were obtained from ENDF/B-V. The rhodium reaction was not used. Covariances were also obtained from ENDF/B-V or other sources and are the same as those being used for the PSF analyses. Cross-section uncertainties do not have a significant impact on the PCA results.

Fission rates were taken as averages of the HEDL SSTR measurements and NBS fission chamber results (supplied by E. D. McGarry). The SSTR and FC results were found to have a systematic bias of ~10% (which is larger than the individual experimental uncertainties), and the uncertainty in the average rates was, therefore, taken to be 11%. These values were also converted to fissions per second per core neutron for input to FERRET. If only SSTR results were available, as with ^{232}Th, the SSTR fission rate value was used but the uncertainty was assumed to still be 11%.

The fission rates for ^{237}Np and ^{238}U were found to be consistent with the radiometric data, but the ^{232}Th fission rate was consistently low (14% to 16% compared with the average fission rate data, and 6% compared with SSTR data). This could indicate a slight problem with the ^{232}Th cross section or may be in part due to an overestimation of the photofission contribution.

The input data are not adequate to unambiguously choose between the SSTR and FC results. A comparison of results using SSTR data and averaged SSTR and FC data is shown in Table 2. Use of the SSTR data results in a decrease in exposure parameter values of 4% to 7%. A corresponding effect in the opposite direction occurs if the FC data are used. If the uncertainties in the fission rate data are lowered to 5% to 7% (as could be justified using each set of data alone), the uncertainty on flux (E > 1.0) is lowered from 7% to 5%, on flux (E > 0.1) from 14% to 13%, and for dpa from 8% to 7% for the typical case.

The NRE data were measured in the form of integral data. These data were converted to reactions per core neutron for input to FERRET. Group cross sections were constructed for FERRET for the I- and J-integral values.* For energy groups above the threshold, the group-averaged cross section for the I-integral is given by

$$\bar{\sigma}_i = \int_{E_1}^{E_2} \frac{\sigma_{np}}{E} \phi(E)dE \bigg/ \int_{E_1}^{E_2} \phi(E)dE \ . \tag{1}$$

---

*See Reference 1 for a derivation of the I- and J-integrals.

TABLE 2

COMPARISON OF FERRET-DERIVED EXPOSURE PARAMETER VALUES FOR AVERAGE* AND SSTR FISSION RATES FOR PCA 12/13 CONFIGURATION

Midplane Location	Value Based on Average Fission Rate	Value Based on SSTR Fission Rate	SSTR-Average Difference (%)
ϕ(E > 1.0 MeV) n/cm² per Core Neutron			
1/4 T	4.58 E-8	4.36 E-8	-5
1/2 T	2.21 E-8	2.08 E-8	-6
3/4 T	9.82 E-9	9.39 E-9	-5
ϕ(E > 0.1 MeV) n/cm² per Core Neutron			
1/4 T	1.44 E-7	1.39 E-7	-4
1/2 T	9.73 E-8	9.12 E-8	-7
3/4 T	5.93 E-8	5.63 E-8	-5
dpa per Core Neutron			
1/4 T	7.45 E-29	7.16 E-29	-4
1/2 T	4.22 E-29	3.96 E-29	-7
3/4 T	2.25 E-29	2.14 E-29	-5

*Linear average of SSTR and FC results.

where E_1 and E_2 are the energy boundaries of group i and σ_{np} is the proton scattering cross section. The I-integral is then calculated by summing $\sigma_i \phi_i$ over all groups (similar to the other reaction rates). The corresponding J-integral group averaged cross section is

$$\bar{\sigma}_j = \int_{E_1}^{E_2} \frac{\sigma_{np}}{E} (E - E_{th}) \, \phi(E) dE \bigg/ \int_{E_1}^{E_2} \phi(E) dE \qquad (2)$$

where E_{th} is the energy threshold. For groups below the threshold, $\sigma_j = 0$ and, for the group including the threshold, a suitable interpolation is used.

The NRE data for the three 12/13 positions were found to be consistently biased with respect to the other data. Both the I- and J-integrals appeared self-consistent but high by about 50%. No explanation for this bias is currently available. In contrast, previously reported (1) preliminary measured-perturbed* I- and J-integral results for the 1/4-T position for the 8/7 configuration were ~20% low. At the time this earlier analysis was performed, the 20% difference was concluded to be in reasonable agreement since this was about a 2σ error.

Because this result is not understood, the NRE data were not used in the derivation of the recommended exposure parameter values. However, an investigation was carried out to determine the effect of the NRE data on the exposure parameters and uncertainties. If the data were renormalized and the quoted experimental uncertainties are used, the NRE data increase the flux (E > 1.0) by about 3% and decrease the flux (E > 0.1) by about the same margin. The dpa is not affected significantly. The uncertainties on flux (E > 1.0) and dpa are lowered significantly to ~3%, and the uncertainty on flux (E > 0.1) is lowered to 6%. Thus, the NRE data have the potential to make a very real contribution to measurement accuracy. It should be noted, however, that the improvement observed here may be overestimated since the nine emulsion integral results (for the approximate thresholds of 0.40, 0.45, 0.50, 0.60, 0.65 and 0.70) were treated as a set of independent measurements.

The HEDL exposure parameter results for the three PCA configurations derived using the FERRET least-squares analysis are given in Table 3. The following uncertainty assumptions were made for this analysis:

---

*No perturbation correction was made for the effect of the proton-recoil chamber (~2.5-cm diameter cylinder) in which the first set of NREs were exposed. The British (3) have studied the perturbation effect of rather large voids (4.7- and 6.6-cm diameter holes) for the "PCA 12/13 Replica" experiment at the 1/4-T position and find ratios of solid/hole for Rh of 0.92 + 3% and 0.85 + 3%, for In of 0.83 + 3% and 0.80 + 3%, and for S of 0.77 + 3% and 0.76 + 3%, respectively. The perturbation correction for the NRE measurements in the ~2.5-cm diameter cylindrical proton recoil chamber would be expected to be smaller. For the present series of NRE exposures discussed in this report, the irradiations were accomplished in small Cd boxes to further minimize perturbation effects.

TABLE 3

RECOMMENDED EXPOSURE PARAMETER VALUES FOR PCA BENCHMARK CONFIGURATIONS BASED ON FERRET SAND II ANALYSIS

Location	Flux >1.0 MeV (n/cm² per Core Neutron)	Flux >0.1 MeV (n/cm² per Core Neutron)	dpa per Core Neutron
	8/7 Configuration		
1/4 T	2.56 E-7 (±6%)	8.56 E-7 (±14%)	4.26 E-28 (±8%)
1/2 T	1.23 E-7 (±7%)	5.66 E-7 (±15%)	2.42 E-28 (±10%)
3/4 T	5.65 E-7 (±8%)	3.19 E-7 (±16%)	1.25 E-28 (±11%)
	12/13 Configuration		
1/4 T	4.58 E-8 (±7%)	1.44 E-7 (±14%)	7.45 E-29 (±8%)
1/2 T	2.21 E-8 (±7%)	9.73 E-8 (±15%)	4.22 E-29 (±10%)
3/4 T	9.82 E-9 (±7%)	5.93 E-8 (±15%)	2.25 E-29 (±11%)
	4/12 SSC Configuration		
1/4 T	2.34 E-7 (±7%)	9.34 E-7 (±14%)	4.08 E-28 (±10%)
1/2 T	1.15 E-7 (±7%)	6.17 E-7 (±14%)	2.37 E-28 (±11%)
3/4 T	4.87 E-8 (±9%)	3.57 E-7 (±14%)	1.23 E-28 (±11%)

1) Conservative estimates of data uncertainties were used but normalization error was only partly included since this uncertainty is extensively correlated in the independent data measurements. The normalization uncertainty should, therefore, be combined with the uncertainty values in Table 3 if a total uncertainty is desired. Uncertainties in fission rates were not taken from estimated uncertainties in the measurements but rather from the difference between the SSTR and FC results. When only one set of measurements was available (e.g., ^{232}Th fission rates), the uncertainty was increased to be consistent with the other fission rate uncertainties.

2) The input flux-spectra were assumed to have a large normalization uncertainty (100%), and the input group fluxes were assigned an uncertainty of 25% with a short-range correlation of 0.8 extending over a width of six groups (details of this formulation are discussed in Section 4.2 of Ref. 1). Using these assumptions, the flux magnitude was typically adjusted 5% to 15% and group fluxes were relatively shifted up to about 15% by FERRET. These results provide confidence that the uncertainties in Table 3 are realistic.

The present status of the data allows for PCA exposure parameters to be determined to 1σ accuracies of 7% to 16%. Data inconsistencies are present that currently limit the improvement of this accuracy. If these inconsistencies can be removed, improvements in accuracy by a factor of 2 could be attained. The PCA data then could be applied to further improve the presently quoted 8% to 12% uncertainties for the PSF physics-dosimetry results as well as contributing to an improved overall accuracy for derived exposure parameter values for LWR surveillance programs.

REFERENCES

1. W. N. McElroy, Ed., LWR-PV-SDIP: PCA Experiments and Blind Test, NUREG/CR-1861, HEDL-TME 80-87, NRC, Washington, DC (July 1981).

2. F. A. Schmittroth, FERRET Data Analysis Code, HEDL-TME 79-40, Hanford Engineering Development Laboratory, Richland, WA (September 1979).

3. J. Butler, M. D. Carter, I. J. Curl, M. R. March, A. K. McCracken, M. F. Murphy and A. Packwood, The PCA Replica Experiment Part I: Winfrith Measurements and Calibrations, NUREG/CR-324, AEEW-R1736, Part I, NRC, Washington, DC (January 1984).

THE U.S. U-235 FISSION SPECTRUM STANDARD NEUTRON FIELD REVISITED

E. D. McGarry, C. M. Eisenhauer, D. M. Gilliam, J. A. Grundl
and G. P. Lamaze

U.S. National Bureau of Standards, Gaithersburg, Maryland, USA

ABSTRACT

As use is made of a standard neutron field, formerly unidentified needs and ways to improve its performance became apparent. This paper presents improvements in calibration techniques, results of Monte Carlo calculations which better define scattering corrections, and new handling procedures which improve reproducibility and decrease radiation exposure to personnel. Also, an application of the NBS ^{235}U fission spectrum to test consistency among laboratories who analyze surveillance dosimetry for reactor pressure vessel exposures is summarized.

NBS ^{235}U CAVITY FISSION SOURCE IRRADIATION FACILITY

The U.S. standard ^{235}U fission neutron spectrum at the National Bureau of Standards in Gaithersburg, Maryland, operates in the center of a 30-cm diameter spherical cavity located in the graphite thermal column of the NBS Research Reactor. The facility is frequently called the Cavity Fission Source (CFS). The upper view in Figure 1 shows, in detail, the CFS assembly[1] containing neutron sensors and the lower view shows its location within the thermal column cavity. Two disks of ^{235}U metal (16-mm dia x 0.13-mm thick) are placed above and below a cylindrical cadmium pill box 0.076-cm thick, which encloses approximately six passive neutron sensors (nomimally 1.27-cm diameter x 0.025-cm thick). The neutron sensors to be irradiated are held in the center of the assembly by light-weight aluminum pieces.

J. P. Genthon and H. Röttger (eds.), Reactor Dosimetry, 791–800.
© 1985 ECSC, EEC, EAEC, Brussels and Luxembourg

Fission fluence rates of ~2 x 10^{10} n/cm^2·s are obtained between the ^{235}U source disks at a separation distance of one centimeter. Applications include detector calibrations for reactor dosimetry and fission cross section measurements.

The absolute source strength, and therefore the fluence rate, of the CFS is dependent upon reactor power level. Consequently, a fluence monitor that is independently calibrated is required for each irradiation. Both the 70.8-day ^{58}Ni(n,p)^{58}Co reaction and the 4.5-hr ^{115}In(n,n')^{115m}In reaction together with a fission-chamber or a power-level monitor have been used. The CFS fluence is tied through these monitors back to known ^{252}Cf fission neutron fields.

^{252}Cf Standard Neutron Fields at NBS and Their Calibrations

The source for the NBS ^{252}Cf spontaneous fission neutron field is a bead of ^{252}CfO$_2$ in a light-weight aluminum and stainless steel capsule. The source is suspended 1.6 m from the nearest reflecting surface in a low-scatter environment. The free-field fluence rate for such a ^{252}Cf source is then a function of only source strength, source-to-detector distance, time, and appropriate geometry-dependent scattering corrections.

All neutron fluence scales in NBS standard neutron fields are traceable to NBS-I, the national standard, radium-beryllium, photoneutron source. The most recent calibrations of its source strength were made in 1961 and 1978[2]. The calibrations agree to within 0.25%. The 4π neutron emission rate of ^{252}Cf fission sources is determined in a 1.2 m diameter, manganous-sulfate bath by comparing induced ^{56}Mn activity with that of NBS-I. This technique is capable of yielding a ^{252}Cf source strength with an uncertainty of ±0.9%.

As mentioned, the ^{252}Cf field is used as an intermediate step in maintaining traceability of other NBS neutron fields to NBS-I. The procedure to accomplish this is known as neutron "fluence transfer."

Fluence Transfer

In the fluence transfer technique, a neutron sensor response is measured first in a standard neutron field and then in the field to be calibrated. For example, transfer from a ^{252}Cf standard neutron field to the NBS ^{235}U CFS has been accomplished by the ^{115}In(n,n')^{115m}In reaction. Foils of known purity are irradiated to a certified fluence in the californium field. The indium foils are then analyzed with a radioactivity detector having a reproducible geometry. The factor Φ_{Cf}/R_{Cf} (n·cm^{-2} per counts·s^{-1}·gm^{-1}) is thereby established for the neutron sensor.

Here R_{Cf} could refer to any reproducible response such as counting rate, reaction rate, or fission rate in the field of interest.

The same indium neutron sensor, or a different indium sensor with a known mass, is then exposed to a certified fluence in the ^{235}U field and the desired fluence , Φ_{235_U} is given by:

$$\Phi_{235_U} = \frac{\Phi_{Cf}}{R_{Cf}} \cdot R_{235_U} \cdot \frac{\bar{\sigma}_{Cf}(In)}{\bar{\sigma}_{235_U}(In)} \tag{1}$$

The spectrum-averaged, cross-section-ratio term on the extreme right of Eq. (1) is more accurately known than the cross sections themselves because errors in the scalar magnitudes are eliminated by the division. For the $^{115}In(n,n')^{115m}In$ reaction, the cross section ratio has been calculated to be 1.048 ±1.6%, using ENDF/B-V differential cross section data and the NBS evaluations of the ^{252}Cf and ^{235}U fission spectra. However, different analytical forms of the ^{235}U spectrum give calculated ratios which are as small as 1.015. The experimental value for the ratio is 1.031 ±2.1%.[3] To circumvent this uncertainty in the fluence transfer process, the $^{239}Pu(n,f)$ reaction has replaced $In(n,n')$ for fluence transfer based upon new experiments performed at a ^{235}U cavity fission source in Mol, Belgium.[3] For ^{239}Pu, the spectrum-averaged cross section ratio is 1.003 ±0.2%, with only a 0.3% uncertainty resulting from different descriptions of the spectra.

Fluence Transfer Through the Belgian ^{235}U Fission Spectrum

Fig. 2 shows the ^{235}U cavity fission source assembly at the SCK/CEN Laboratory, Mol, Belgium.[4] This assembly incorporates a 100-cm dia. cavity as opposed to the 30-cm dia. cavity for the NBS ^{235}U field. The significance of the larger cavity is that there are substantially fewer low-energy (wall returned) neutrons in the Belgian field. Also, the larger source permits exposure of high quality, light-weight ^{239}Pu deposits in an NBS fission chamber. This chamber cannot be used in the smaller volume NBS Cavity Fission Source.

Fig. 3 depicts the present and former (indium "only") fluence transfer procedures from the NBS ^{252}Cf fission spectrum. With the exception of differences in the scattering corrections in the Belgian and NBS ^{235}U fields, their energy spectra are identical. Therefore, details of the cross section of the reaction chosen as the transfer instrument do not matter (see Eq. 3 at bottom of Fig. 3). Scattering differences in the two fields require a net adjustment of almost 4%; however, this correction has been calculated to ±1.4%.

Tables I and II summarize the uncertainties in the fluence transfer process. The total uncertainties for In(n,n') and for ^{58}Ni(n,p) are the uncertainties of Tables I and II taken in quadrature. Respectively, these are 2.0% and 2.3%.

Neutron Scattering and Removal Effects

Corrections must be made for effects of neutron scattering and removal because of the variety of materials in close proximity to each neutron sensor undergoing irradiation in a cavity fission source. For example, there are cadmium, aluminum in support structures, uranium in the fission disks, and various elements in other dosimeters. The corrections are a complex function of the cross sections (or thresholds) of the dosimetry reactions as well as the position of a particular sensor relative to all other materials. The present corrections for the NBS CFS are based upon rather extensive Monte Carlo calculations* of indium, nickel and aluminum dosimeters in a detailed model of the cavity fission source. The results are summarized in Fig. 4.

INTRALABORATORY CONSISTENCY OF RADIOMETRIC SENSOR ANALYSES

Between July 1978 and March 1979, twenty-seven nickel foils were irradiated in the NBS ^{235}U CFS, activating the ^{58}Ni(n,p)^{58}Co reaction. Subsequently, the relative foil activities were measured at NBS and then most were distributed internationally to test consistency among laboratories that frequently analyze radiometric dosimeters.

Results were submitted by eighteen laboratories. A number of factors made a direct comparison of fluence measurements difficult:

(1) Although a standard reporting form was distributed, some participants reported only ^{58}Co activity. Also, several laboratories submitted more than one result, from either measurements on multiple gamma counters or from an exchange of foils with some other participant.

(2) To ease the "burden" of having to calculate a spectrum-averaged cross section, the value of 102 mb was also distributed. This value was obtained by integrating ENDF/B-V data over the NBS evaluation of the ^{235}U fission spectrum. Many participants preferred to calculate and use their own cross section getting values up to 111 mb. The current experimental value is 110 ±2.5%.[2]

(3) Complex scattering corrections (up to ~2% in magnitude) could not be made until lengthy Monte Carlo calculations were completed.

*The calculations were performed by Dr. P. Sorenson at the Los Alamos Scientific Laboratory.

794

(4) The results from the indium fluence-rate monitors were discovered to be uncertain to ~2% because of flux variations due to reactor shim-arm movements which effect gradients in the thermal column but were not evident from the reactor power history charts.

To directly intercompare activity measurement capability, the ratio of the reported activity to the NBS-measured relative activity was determined. The results are shown in Fig. 5. The error bars are those reported by the laboratory, if available, or an arbitrarily assigned ±2%, if not. The mean value of the measurements is shown with an associated ±2% inner band. The outer lines show the actual standard deviation of ±4.7%. The largest difference from the mean is 12%. The associated laboratory requested a second nickel foil which when analyzed agreed with the NBS measurement to ~3%. The large discrepancy was never resolved. A final report on the NBS fluence values for the individual nickel foils must still be sent to the participants. A less accurate comparison of fluence results was published in Ref. 5 in 1982.

REFERENCES

1. Compendium of Benchmark and Test Region Neutron Fields for Pressure Vessel Irradiation Surveillance; Standard Neutron Spectra. NRC NUREG/CR-0551, Progress Report--July-Sept. 1978 (Dec. 1978).

2. "Neutron Fluence and Cross Section Measurements for Fast Neutron Dosimetry," G. P. Lamaze, D. M. Gilliam, E. D. McGarry, NBS, Gaithersburg, Maryland; A. Fabry, SCK/CEN Laboratory, Mol, Belgium; Proc. of Int'l. Conf. Nucl. Meth. and Environ. and Energy Res., San Juan, PR (April 1984) (Proc. to be published by Department of Energy).

3. "Cross Section Measurements in the ^{235}U Fission Spectrum Neutron Field," D. M. Gilliam, J. A. Grundl, G. P. Lamaze, C. M. Eisenhauer, NBS, Gaithersburg, Maryland; A. Fabry, SCK/CEN Laboratory, Mol, Belgium (Sept. 1984). (This ASTM/EURATOM Symposium.)

4. "The Mol Cavity Fission Spectrum Standard Neutron Field and Its Applications," A. Fabry, G. Minsart, F. Cops, and S. DeLeeuw; Proc. Fourth ASTM/EURATOM Symposium on Reactor Dosimetry, Washington, DC (March 1982).

5. E. D. McGarry, "Requirements for Referencing Reactor Pressure Vessel Surveillance Dosimetry to Benchmark Neutron Field" in Nuclear Data for Radiation Damage Assessment and Related Safety Aspects, Proc. of the Advisory Group Meeting, Vienna, Austria, Oct. 12-16, 1981; IAEA-TECDOC-263 (1982).

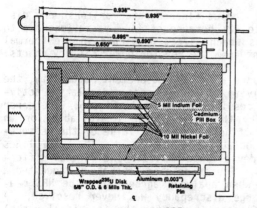

NBS ^{235}U CAVITY FISSION SOURCE

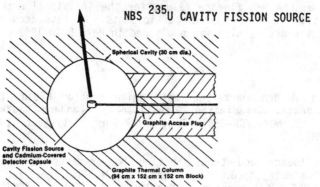

Figure 1.

Belgian (SCK/CEN) ^{235}U Fission Spectrum Field in BR1 Reactor Thermal Column

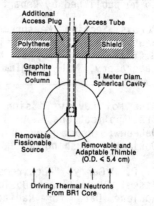

Figure 2.

TABLE I.

Summary Of Uncertainties In Fluence Transfer Measurements

● For Transfer From NBS ^{252}Cf Field to Belgian ^{235}U Field

Source of Uncertainty	Contribution (%)
Source Strength of ^{252}Cf	0.9
^{252}Cf-Source to ^{239}Pu-Deposit Distance	0.6
Statistics of Fission Counting	0.1
Precision of 2π Fission Counting Corrections	0.4
^{239}Pu Cross-Section Ratio in ^{252}Cf Field and ^{235}U Field	0.1
Scattering Corrections for ^{239}Pu in ^{252}Cf Field	0.7
Wall Return Corrections in Belgian ^{235}U Field	0.6
Scattering Corrections for ^{239}Pu In Belgian ^{235}U Field	0.1
Total Uncertainty in Fluence Transfer = (Contributions Taken in Quadrature)	1.5%

TABLE II.

Summary Of Uncertainties In Fluence Transfer Measurements

● For Transfer From Belgian to NBS ^{235}U Field

Source of Uncertainty	Contribution (%)	
	For ^{115}In (n, n′)	For ^{58}Ni (n, p)
Statistics of Gamma Counting	0.2	0.4
Scattering + Wall Return in Belgian ^{235}U Field	0.9	1.4
Scattering + Wall Return in NBS ^{235}U Field	0.8	0.8
Fluence-Rate Gradient in NBS ^{235}U Field	0.4	0.4
Total Uncertainty in Fluence Transfer = (Contributions Taken in Quadrature)	1.3%	1.7%

797

Present and Former Procedures for Calibration Of The Neutron Fleunce Rate In The NBS ²³⁵U Cavity Fission Source

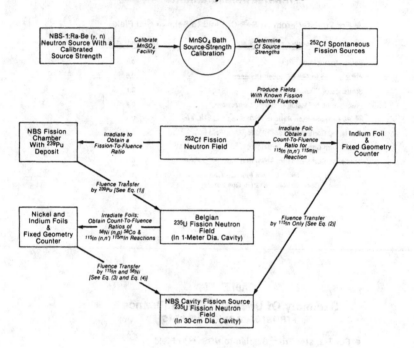

$$\varnothing_{\text{U-235}}^{\text{SCK/CEN}} = \left[\frac{\varnothing_{cf}}{F_{cf}^{\text{Pu-239}}}\right] \bullet F_{\text{U-235}}^{\text{Pu-239}} \bullet \left[\frac{\overline{\sigma}_{cf}^{\text{Pu-239}}}{\overline{\sigma}_{\text{U-235}}^{\text{Pu-239}}}\right] \tag{1}$$

$$^{\text{Old}}\varnothing_{\text{U-235}}^{\text{NBS}} = \left[\frac{\varnothing_{cf}}{R_{cf}^{\text{In-115}}}\right] \bullet R_{\text{U-235}}^{\text{In-115}} \bullet \left[\frac{\overline{\sigma}_{cf}^{\text{In-115}}}{\overline{\sigma}_{\text{U-235}}^{\text{In-115}}}\right] \tag{2}$$

$$^{\text{New}}\varnothing_{\text{U-235}}^{\text{NBS}} = \left[\frac{\varnothing_{\text{U-235}}^{\text{SCK/CEN}}}{R_{\text{U-235 (SCK)}}^{\text{Ni-58}}}\right] \bullet R_{\text{U-235 (NBS)}}^{\text{Ni-58}} \tag{3}$$

$$^{\text{New}}\varnothing_{\text{U-235}}^{\text{NBS}} = \left[\frac{\varnothing_{\text{U-235}}^{\text{SCK/CEN}}}{R_{\text{U-235 (SCK)}}^{\text{In-115}}}\right] \bullet R_{\text{U-235 (NBS)}}^{\text{In-115}} \tag{4}$$

Figure 3.

NBS Cavity Fission Source

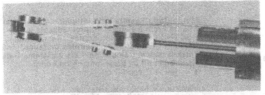

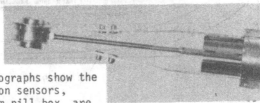

The above two photographs show the way in which neutron sensors, inside of a cadmium pill box, are reproducibly inserted between two highly radioactive fission disks.

The photograph to the right shows "cold" fission disks being mounted. The cylindrical assembly is a lead shield which will be used for their removal after the irradiation.

The photograph to the left shows the fission disks and cadmium-covered neutron sensors mounted inside of a graphite cavity, which will be inserted into the thermal column of the NBS Research Reactor.

Improved Handling Mechanism for the NBS Cavity Fission Source

Figure 6.

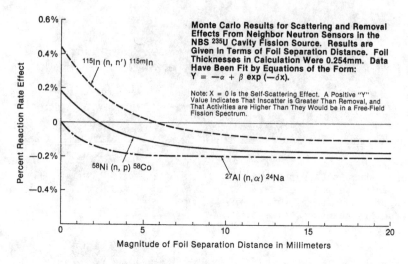

Figure 4.

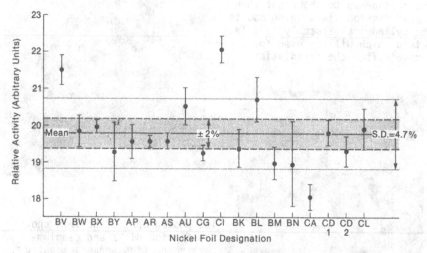

Results of an intercomparison of activity measurements of the 70.8-day activity from the $^{58}Ni(n,p)^{58}Co$ reaction induced in a certified fluence irradiation in the NBS Cavity Fission Source.

Figure 5.

RECENT EXPERIMENTS ON CF-252 SPECTRUM-AVERAGED NEUTRON
CROSS SECTIONS

W. Mannhart

Physikalisch-Technische Bundesanstalt

Bundesallee 100, D-3300 Braunschweig, Fed. Rep. of Germany

ABSTRACT

Spectrum-averaged cross sections of the reactions
V-51(n,p), V-51(n,α), Cu-65(n,2n), Zn-64(n,p), I-127(n,2n)
and Hg-199(n,n') were measured relative to Ni-58(n,p) or
In-115(n,n'). The data were compared with other experiments
and with calculations. A complete uncertainty covariance
matrix was derived for the data. The correlation of the
present data with previous experiments is also shown. This
information and the data of a few recent other experiments
were used to update the evaluation of a complete set of
experimental spectrum-averaged cross sections determined in
the standard neutron field of Cf-252.

INTRODUCTION

In reactor metrology, neutron monitor reactions are used for
measuring neutron fluence and fluence rate. A combination of various
neutron reactions, each of which is sensitive to a different neutron
energy range, gives additional information on the shape of neutron
spectra. These techniques are used, for example, in reactor pressure
vessel surveillance metrology with the aim of obtaining estimates of
the radiation damage limiting the lifetime of structural components.

The reliability of conclusions drawn with the above described
procedures strongly depends on the validity of the nuclear data
involved in the method. For a specific neutron monitor, the relation
between the experimentally determined reaction rate and the derived
fluence rate is given by the spectrum-averaged neutron cross section

which for its part depends on σ(E), the corresponding energy-dependent cross section.

Certain well-defined standard neutron fields allow the direct measurement of spectrum-averaged cross sections and these results can be used for the validation of σ(E) data of neutron monitors before applying them in other neutron fields. In this context, the standard neutron field of spontaneous fission of Cf-252 is of special importance. It can be realized in a very clean form, i.e. almost free of spectrum distortions requiring extensive corrections which would limit the obtainable precision of experiments in such fields.

The present work continues the nuclear data standardization program (1,2) performed at the Physikalisch-Technische Bundes-anstalt. The selection of the reactions is based on various aspects. For V-51(n,p) a σ(E) experiment recently became available which filled a gap in the data for this reaction below 13 MeV neutron energy (3). A similar experiment for V-51(n,α) is in progress (4). Vanadium is used as structure material in fast-neutron fission reactors and is also a candidate for similar purposes in future fusion devices. Under high neutron fluences the hydrogen and helium production of both reactions causes radiation damage due to the embrittlement of the material. The reaction Hg-199(n,n') belongs to the group of low-threshold neutron reactions. Compared with other reactions such as Nb-93(n,n') and Rh-103(n,n'), this reaction produces gamma-rays of 158 keV and 374 keV instead of only low-energy X-rays, which makes it preferable to the others as the radioactivity measurement is possible with the usual detectors and does not require special detectors and techniques. The reactions Cu-65(n,2n) and I-127(n,2n) extend earlier systematics of high-threshold reactions (1). No experiments with either reaction have as yet been performed in the Cf-252 field. Finally, the reaction Zn-64(n,p) has been re-measured to validate the experiments carried out to date.

EXPERIMENTAL PROCEDURE

The experimental technique was almost the same as that used in earlier experiments (1,2). Irradiations were carried out at an outdoor irradiation facility at a position 12 m above ground. The source strength of the Cf-252 neutron source was 4.7×10^9 s^{-1} on April 1, 1982. The neutron source was encapsulated in a double zircaloy cylinder with outer dimensions of 10 mm in diameter and 10 mm in height. The wall thickness of the cylinder was 1.5 mm. Irradiations were performed in a sandwich arrangement. Each sample investigated was placed between two foils of nickel or indium acting as neutron flux density monitors. The sandwich packets were almost touching the convex surface of the source cylinder. In most cases

Table 1: Sample and decay parameters of the various neutron reactions

Reaction	Chemical form	Purity	Thickness (mm)	Isotopic abundance %	Decay parameters			
					$T_{1/2}$	E_γ (keV)	I_γ	Ref.
$^{51}V(n,p)^{51}Ti$	V metal	99.8 + %	0.500	99.75	5.76(1) m	320.1	0.930(4)	(5)
$^{51}V(n,\alpha)^{48}Sc$	V metal	99.8 + %	0.500	99.75	43.7(1) h	983.5 1037.5 1312.1	1.000(5) 0.975(5) 1.000(5)	(6)
$^{58}Ni(n,p)^{58}Co$	Ni metal	99.99+ %	0.250	68.3	70.78(10)d	810.8	0.9945(1)	(7)
$^{65}Cu(n,2n)^{64}Cu$	Cu O powder	enriched	-	65: 99.70 (2) 63: 0.30 (2)	12.701(2)h	511.0	0.3574(36)	(6)
$^{64}Zn(n,p)^{64}Cu$	Zn metal	99.99+ %	0.250	48.6	12.701(2)h	511.0	0.3574(36)	(6)
$^{115}In(n,n')^{115}In^m$	In metal	99.999 %	0.125	95.7	4.486(4)h	336.2	0.459(1)	(7)
$^{127}I(n,2n)^{126}I$	$I_2 O_5$ powder	99.9 %	-	100	13.02(7)d	388.6 666.3	0.341(27) 0.331(25)	(8)
$^{199}Hg(n,n')^{199}Hg^m$	Hg O powder	99.9 %	-	16.85	$42.6(5)^{a)}$ m	158.4 374.1	$0.523(5)^{b)}$ 0.123(5)	a: (9) b: (10)

the samples were high-purity metallic disks, 10 mm in diameter and between 0.125 mm and 0.5 mm thick. The chemical and physical size of the various materials is summarized in columns 2 to 4 of table 1. Powders were pressed in cylindrical lucite containers with inner dimensions of 0.8 mm in length and 10 mm in diameter. The wall thickness of the lucite cylinders was 0.2 mm at the front and rear plane and 1 mm at the envelope. The monitor foils were directly attached to the cylinder planes on both sides.

The induced radioactivity was measured with a Ge(Li) detector of a volume of 130 cm^3 and a nominal efficiency of 28 %. With radioactive standard sources photopeak efficiency curves were determined at source distances of 16 mm, 53 mm and 110 mm from the detector face. Total efficiencies were also determined to allow summing corrections of radionuclides emitting coincident photons. All measurements in this experiment were conducted at the 16 mm position. This position was calibrated mainly with monoenergetic standard sources. In addition, various reference sources emitting coincident photons were calibrated at more distant positions (with small summing corrections) and transferred to the 16 mm position to verify the correctness of the summing corrections applied. All calibrations were performed with nearly massless point sources. By slightly varying the distance of these sources from the detector and by placing them in positions excentric from the detector symmetry axis, corrections were determined to take into account the extended volume of the measured samples. Self-absorption within the samples was also corrected for.

A special calibration was carried out to measure the positron emitter Cu-64 produced via Cu-65(n,2n) and Zn-64(n,p). To ensure complete annihilation the samples were placed inside closed copper containers with a wall thickness of 1.5 mm. This special configuration was calibrated with a Na-22 source, the decay rate of which was determined via the 1275.5 keV gamma radiation. The 511 keV photons from positron annihilation contribute as a relatively strong component to the detector background due to natural radioactivity. The correction for this was 0.4 % for the Zn-64(n,p) samples and 1.5 % for the Cu-65(n,2n) samples. For the Zn-64(n,p) samples an additional 0.2 % correction of the positrons from the Zn-65 decay resulting from the concurring reactions Zn-64(n,γ) and Zn-66(n,2n) was necessary.

In converting the measured count rates into reaction rates the decay parameters listed in table 1 were used. These data were taken from a variety of different sources (5 - 10) to ensure a most complete representation of the best data available.

CORRECTIONS AND UNCERTAINTIES

The reaction product Cu-64 is produced with the Cu-65(n,2n) reaction as well as with the Cu-63(n,γ) reaction. For natural copper the (n,γ) process dominates over the (n,2n) process due to a ratio of about 2.25 : 1 of the isotopic abundances of Cu-63 to Cu-65 and is further enhanced by a Cf-252 spectrum-averaged cross section of (10.39 ± 0.30) mb (2) for Cu-63(n,γ) compared with about 0.66 mb for Cu-65(n,2n). To allow the measurement of Cu-65(n,2n), a material highly enriched with Cu-65 was chosen (see also table 1). Even with the present enrichment of 99.7 % of Cu-65, the contribution of Cu-63(n,γ) was not negligible. The contribution of this reaction, i.e. the spectrum-averaged cross section of Cu-63(n,γ) weighted with the ratio of the isotopic abundances, was 0.0313 mb. This correction of 4.7 % of the final result was subtracted from the measured data. The relative uncertainty contribution due to this correction was 0.35 %.

For the Hg-199(n,n') too, reactions were competing, namely Hg-198(n,γ) and Hg-200(n,2n), leading to the same reaction product. For a U-235 neutron spectrum the contributions of both reactions to the Hg-199(n,n') cross section were estimated by Kobayashi et al. (11). The contributions were 0.8 mb for the (n,γ) process and 1.12 mb for the (n,2n) process. For the Cf-252 neutron spectrum the effect of the (n,γ) reaction was estimated to be in first order the same as for the U-235 spectrum. From systematics of the calculated Cf-252/U-235 spectrum-averaged cross section of (n,2n) reactions plotted as a function of the Q-values, a factor of 1.70 of the ratio was derived for the Hg-200(n,2n) reaction with a Q-value of -8.029 MeV. Based on this, the total correction of the (n,γ) and

804

(n,2n) process to the spectrum-averaged cross section of
Hg-199(n,n') was 2.7 mb corresponding to 0.93 % of the final result.

Besides the counting statistics, the main uncertainty
contributions of the experiment were due to the scattering
correction and the photopeak efficiency calibration. All
measurements of cross sections given here were relative
measurements. The experimental determination of ratios has some
advantage over absolute measurements, due to a compensating effect
in the corrections and their uncertainty contribution.

The irradiation geometry was optimized as regards the
scattering effects. The results of (n,γ) reactions measured with
this geometry (1,12) showed, for example, that a thermal
perturbation component due to backscattered neutrons was of a
negligible order of magnitude. Nevertheless, there are small
perturbations of the neutron spectrum due to the source
encapsulation, the samples and their support structures (thin
aluminum cylinders of only 0.3 mm wall thickness). The perturbations
were of the order of 1 % depending on the neutron energy. Details
are given elsewhere (2). The scattering correction is a function of
the energy response range and was considered individually for each
reaction. The uncertainty attributed to this correction was with 1 %
of the same order of magnitude as the correction itself. The
influence of this correction on a cross section ratio strongly
depends on the overlap of the energy response ranges of the
individual reactions. A correlation pattern of the scattering
corrections was established based on the mean overlap of the energy
response ranges. For example, for the reactions Hg-199(n,n') and
In-115(n,n') this overlap is 0.95, i.e., for the measured ratio the
contribution due to the scattering corrections is strongly reduced
with this large correlation. The net effect of the total correction
was thus only 0.32 %.

The relative uncertainty of the photopeak efficiency
calibration was 1.5 %. Correlations between the efficiencies at
different photon energies were derived from the experimental
calibration procedure. Different correlation length were used,
depending on the photon energy. At low energies the corrrelations
were of a "short-range" type and at higher energies they were more
of a "long-range" type. This procedure is a good approximation of
the experimental reality. A compensating effect similar to that for
the scattering correction also occurs for the photopeak efficiency
in a ratio experiment. The effect depends on the difference of the
photon energies of the reaction products. For the ratio
V-51(n,p)/In-115(n,n'), for example, the difference is between
320.1 keV of Ti-51 and 336.2 keV of In-115m. Instead of the maximum
possible uncertainty contribution due to the efficiency calibration
of 2.12 %, the contribution to this ratio was only 0.60 %.

RESULTS AND DISCUSSION

The measured cross section ratios of this experiment are listed
in the last six rows of table 2. The covariance matrix of the data
is given in the form of relative standard deviations and a correlation matrix. Due to the similarity of the present experiment with
earlier measurements (1,2) the covariances of these data are also
shown in the table. This information has not been quoted before. The
correlations between the experiments are remarkable, indicating the
analogy of the experimental procedure.

The present data were normalized with cross sections of
(118 ± 3) mb and (195 ± 5) mb (12) of the monitor reactions
Ni-58(n,p) and In-115(n,n'), respectively, to allow a comparison
with other experiments and with calculations, and are given in
table 3. For the calculated data the segment-adjusted spectrum
representation of Grundl and Eisenhauer (13) was used. In each case
it is indicated from where the $\sigma(E)$ data used in the calculation
were taken.

For the reactions Cu-65(n,2n) and I-127(n,2n) no other
experiments were available. The present data can therefore only be
compared with calculations based on $\sigma(E)$ data from ENDF/B-V. The
agreement between the present measurement of Cu-65(n,2n) and the
calculation is within 1.4 % and is surprisingly good.

For the reaction I-127(n,2n) the calculation is about 12 %
higher than the experiment. The ENDF/B-V data of this reaction are

Table 2: Measured ratios and uncertainties (experiment (1,2) and present data)

Reaction ratio	<σ> - ratio	Rel. Std. Dev. %	Correlation matrix (x 100)
^{19}F(n,2n)/^{58}Ni(n,p)	1.381E-4	3.64	100
^{24}Mg(n,p)/^{27}Al(n,α)	1.998E+0	2.40	-1 100
^{27}Al(n,p)/^{115}In(n,n')	2.459E-2	2.45	-23 0 100
^{55}Mn(n,2n)/^{58}Ni(n,p)	3.458E-3	2.00	27 -1 4 100
^{59}Co(n,p)/^{58}Ni(n,p)	1.424E-2	2.49	21 0 -12 13 100
^{59}Co(n,α)/^{58}Ni(n,p)	1.877E-3	1.52	18 -1 21 42 27 100
^{59}Co(n,2n)/^{58}Ni(n,p)	3.441E-3	2.19	23 -1 0 45 5 27 100
^{58}Ni(n,2n)/^{58}Ni(n,p)	7.572E-5	3.64	31 0 -23 25 28 16 21 100
^{63}Cu(n,γ)/^{115}In(n,n')	5.328E-2	3.18	23 0 29 6 0 7 5 3 100
^{63}Cu(n,α)/^{58}Ni(n,p)	5.686E-3	3.00	23 0 -15 15 51 31 8 33 1 100
^{63}Cu(n,2n)/^{115}In(n,n')	9.390E-4	3.98	27 -1 28 20 3 14 19 11 38 4 100
^{90}Zr(n,2n)/^{58}Ni(n,p)	1.873E-3	2.68	25 0 -2 39 21 27 29 26 4 20 15 100
^{51}V(n,p)/^{115}In(n,n')	3.266E-3	1.54	3 1 32 5 15 39 4 3 4 18 13 4 100
^{51}V(n,α)/^{58}Ni(n,p)	3.285E-4	2.17	35 -1 -10 47 58 53 35 38 5 55 16 44 20 100
^{65}Cu(n,2n)/^{58}Ni(n,p)	5.636E-3	2.62	66 -1 -32 40 29 30 43 60 31 34 36 31 5 53 100
^{64}Zn(n,p)/^{58}Ni(n,p)	3.486E-1	2.13	58 0 -40 4 31 6 0 29 33 32 27 10 0 28 81 100
^{127}I(n,2n)/^{58}Ni(n,p)	1.757E-2	2.58	25 -1 -39 34 8 30 30 23 -16 18 -3 21 5 31 39 15 100
^{199}Hg(n,n')/^{115}In(n,n')	1.511E+0	1.22	-1 0 19 -2 0 4 -2 -1 12 1 10 -2 6 -2 -2 0 12 100

Table 3: Cf-252 spectrum-averaged neutron cross sections (in mb)

| Reaction | Experiment | | | Calculation | σ (E) from: |
	Present	Other	Ref.		
^{51}V(n,p)	0.637 ± 0.018	0.93 ± 0.10	(14)	0.5454	ENDF/B-V
		0.713± 0.059	(15)	0.6796	(3)
^{51}V(n,α)	(3.88 ± 0.12)x10^{-2}	(4.30 ± 0.20)x10^{-2}	(14)	3.739 x 10^{-2}	ENDF/B-V
^{65}Cu(n,2n)	0.665 ± 0.023	-		0.6560	ENDF/B-V
^{64}Zn(n,p)	41.1 ± 1.3	39.4 ± 1.0	(16)	39.23	(20)
		46.4 ± 2.3	(14)		
		41.8 ± 1.7	(15)		
		36.2 ± 1.5	(17)		
		41.3 ±(2.9)	(18)		
^{127}I(n,2n)	2.07 ± 0.07	-		2.316	ENDF/B-V
^{199}Hg(n,n')	295 ± 8	168 ± 6	(19)	245.1	(21)

based on an evaluation from 1972 (22) and are identical with
ENDF/B-IV. The evaluator mentions (22) that some of the data used in
the evaluation are based on gamma emission probabilities other than
the values given in table 1. To take this into account would lower
the σ(E) data by a factor of 0.925 (22). If this normalization is to
be valid for the whole evaluation, the calculated spectrum-averaged
cross section becomes 2.14 mb, which is relatively close to the
experimental result of 2.07 mb. It should also be mentioned that the
photon emission probabilities of the 388.6 keV and the 666.3 keV
photons used in the present analysis (table 1) have relatively large
uncertainties. In the present experiment the results obtained with
both photon lines agreed within 2.5 %. In the determination of the
I-127(n,2n) cross section the mean value was formed and an
additional uncertainty component of 1.7 % was attributed to the
final results instead of the uncertainties shown in table 1.

In the present experiment on the Zn-64(n,p) reaction the
radioactivity of Cu-64 was measured via the 511 keV positron
annihilation radiation. In two other experiments (16,17) the same
method was used. There is some confusion due to the different values
used for the positron emission probability. The values accepted at
present for this is 17.87 % (see table 1). When the other
experiments are renormalised to this value, one experiment (16)
changes from 39.4 mb to 40.6 mb and the other (17) from 36.2 mb to
39.1 mb. Both renormalised data agree fairly well with the present
result of (41.1 ± 1.3) mb. In the remaining experiments (14,15,18)
the 1345.9 keV gamma-ray line of Cu-64 emitted with a fraction of
only 0.49 % per decay was analyzed. Two of these experiments (15,18)
show a good agreement with the statistically better experiments in
measuring positron annihilation. The uncertainty quotation of one
experiment (18), given in table 3 in brackets, comprises only the

statistical component. The result of the earlier experiment of Deszö and Csikai (14) is, in comparison with their recent data (18), substantially higher. This is probably due to spectrum perturbations which seem to be eliminated with an improved scattering-free irradiation geometry (18). The evaluation of the σ(E) data used in the calculated spectrum-averaged cross section is based on a value of 19.3 % for the positron emission (20). The renormalisation increases the result by a factor of 1.08 to a calculated value of 42.4 mb which is only 3 % higher than in the present experiment.

The present result obtained with the V-51(n,p) reaction shows no agreement with other experiments (14,15). The value of 0.93 mb (14) is probably too high for the same reasons as discussed in the case of the Zn-64(n,p) reaction. However, it is puzzling that within the uncertainty the other experiment (15) shows no overlap with the present data. Neither do the calculated results pour any ligth on the situation. The calculated value based on ENDF/B-V is about 14 % lower than that in the present experiment. The ENDF/B-V evaluation (23) represents the (n,p) cross section of elemental vanadium. With the low isotopic abundance of 0.30 % of V-50, the influence of the V-50(n,p) reaction remains small and the elemental cross section should be essentially identical with the V-51(n,p) cross section. Due to a lack of data below 13 MeV neutron energy, the ENDF/B-V evaluation (23) is based mainly on theoretical model calculations. The response of the Cf-252 spectrum-averaged cross section above 13 MeV is only 2 %, i.e. the calculated value depends strongly on the adequacy of the model of the σ(E) data. This situation has recently been improved with experimental σ(E) data of V-51(n,p) measured between reaction threshold and 9.3 MeV (3). However, the Cf-252 spectrum-averaged cross section calculated with this data is also higher by a factor of 1.067 than the value of the present experiment of (0.637 ± 0.018) mb. An analogy to this is given by comparing similar results obtained in a U-235 fission neutron field. Here, the author recently measured a spectrum-averaged value of 0.503 mb (24). The value calculated with the recent σ(E) data (3) and a Watt representation of the U-235 spectrum is 0.5366 mb (3). The bias factor between calculation and experiment is 1.067, exactly the same as for the Cf-252 data. The reason for this common bias between the spectrum-averaged data and the σ(E) data (3) is not at present understood.

In the case of the reaction V-51(n,α) the result of the only other experiment available (14) is about 11 % higher than the present data of 0.0388 mb. The ENDF/B-V data used in the calculation were also for elemental vanadium and also based mainly on theoretical models (23) similar to V-51(n,p). However, in contrast to V-51(n,p), here the calculated value is only 3.6 % lower than in the present experiment.

In the measurement of the reaction Hg-199(n,n'), the decay of the excited state was followed over the photon lines of 158.4 keV and 374.1 keV. In the experiment the self-absorption correction within the sample was of the order of 50 % for the 158.4 keV radiation. As a correct determination of such a large correction would require an exact calculation of the sample-to-detector geometry and because the detector's position within its housing was not sufficiently known, the activity measurement was confined to the analysis of the 374.1 keV photon line with a self-absorption correction of only 6.5 %. While the photon emission probability of the 158.4 keV photon line is relatively well established in the literature, the same is not true of the 374.1 keV line. The values of this line scatter between 12.3 % and 15.4 % (9). Here, a value of 12.3 % (see table 1) was accepted to allow a comparison with other data based on this value. The present data did not take into account the uncertainty of this value. The experimentally obtained cross section of Hg-199(n,n') is (295 ± 8) mb. Another experiment (19) gives 168 mb for Hg-198(n,γ) plus Hg-199(n,n'). No details of this experiment are given which would allow an interpretation. Of more interest is the comparison with the calculated result based on σ(E) data measured by Sakurai et al. (21). The calculated value of 245.1 mb is substantially lower than in the present experiment. The same is valid for similar results in the U-235 neutron field where Kobayashi et al. (11) measured a value of (277 ± 15.4) mb and the calculation gave a value of 238.3 mb (21). It remains an open question whether the experimental-to-calculated ratio of 1.204 in the Cf-252 neutron field and of 1.162 in the U-235 neutron field can be traced back to problems with the self-absorption correction in the σ(E) experiment (21) where the loss factor was 80.0% for the 158.4 keV photon line and 32.4 % for the 374.1 keV photon line.

EVALUATION OF SPECTRUM-AVERAGED DATA

Since 1979, the experimental basis of Cf-252 spectrum-averaged cross section data has been evaluated with generalized least-squares techniques (25). From time to time the evaluation has been updated by the inclusion of recent data (2). The present evaluation comprises the data of table 2. In addition, a recent experiment (26) performed at the National Bureau of Standards (NBS) has been included. The earlier NBS experiment in fission cross section measurements, already contained in the evaluation (2), has recently been revised (27,28). The results were modified due to improvements in the neutron source strength determination, the fission deposit mass determination and the scattering correction. The earlier data were dropped from the evaluation and replaced by the revised values.

The result of the evaluation with a complete covariance matrix is shown in table 4. The present evaluation supersedes earlier versions. Based on 70 data points, the data of 35 different neutron

Table 4 : Result of the Cf-252 evaluation

Reaction	$\langle\sigma\rangle$ (mb)	Std. Dev. %	Correlation matrix (× 100)
^{19}F(n,2n)	1.613E-2	3.40	100
^{24}Mg(n,p)	1.998E+0	2.42	17 100
^{27}Al(n,p)	4.885E+0	2.14	15 29 100
^{27}Al(n,α)	1.017E+0	1.47	30 56 46 100
^{32}S(n,p)	7.262E-1	3.50	12 24 22 34 100
^{46}Ti(n,p)	1.409E-1	1.76	24 34 35 58 23 100
^{47}Ti(n,p)	1.929E-1	1.66	26 37 38 62 25 60 100
^{48}Ti(n,p)	4.251E-1	1.89	22 31 35 52 21 66 54 100
^{51}V(n,α)	6.493E-1	1.95	22 30 55 46 20 37 39 38 100
^{51}V(n,α)	3.904E-2	2.22	39 24 28 41 18 34 37 31 38 100
^{56}Mn(n,2n)	4.079E-1	2.34	40 24 30 40 18 33 36 30 27 62 100
^{54}Fe(n,p)	8.692E-1	1.34	31 42 44 69 30 57 61 52 45 48 33 100
^{56}Fe(n,p)	1.466E+0	1.77	33 44 49 73 32 61 66 56 47 51 34 88 100
^{58}Ni(n,2n)	1.766E-2	1.30	33 19 15 09 25 11 21 23 20 19 29 21 32 100
^{58}Ni(n,p)	8.941E-3	3.59	21 22 37 16 30 33 28 32 60 34 36 19 27 32 100
^{59}Co(n,α)	1.692E+0	2.49	34 29 45 50 22 41 44 41 58 29 31 45 60 41 29 100
^{59}Co(n,2n)	2.220E-1	1.86	37 25 37 16 31 23 33 28 27 18 22 37 54 66 30 43 100
^{59}Co(n,p)	4.055E-1	2.52	21 20 48 11 33 18 19 27 25 11 10 22 31 49 50 40 31 100
^{63}Cu(n,γ)	1.045E-1	3.24	34 27 30 44 19 36 39 34 41 28 28 48 51 31 28 52 42 28 100
^{63}Cu(n,α)	6.893E-1	1.98	24 27 45 19 36 31 19 29 40 20 22 32 37 41 42 37 33 41 39 100
^{65}Cu(n,2n)	1.845E-1	3.98	59 26 23 45 11 21 21 20 23 18 16 20 27 15 20 31 31 20 36 26 100
^{65}Cu(n,2n)	6.587E-1	2.24	48 36 29 60 25 49 53 45 49 40 40 62 66 36 25 49 37 23 52 20 42 100
^{64}Zn(n,p)	4.063E-1	1.64	33 19 23 19 24 27 29 24 27 26 32 42 46 35 16 36 31 16 36 16 21 41 100
^{67}Zn(n,2n)	2.212E-1	2.90	28 42 57 66 27 53 56 67 41 38 31 55 51 26 24 44 32 21 46 21 09 55 16 100
^{115}In(n,n')	1.622E-2	1.57	18 27 33 43 17 34 37 30 32 23 23 23 30 20 28 28 26 16 24 08 08 36 14 29 100
^{115}In(n,γ)	1.257E-2	2.23	31 28 43 17 53 28 34 28 34 27 22 32 29 34 36 29 37 28 32 22 11 44 21 28 16 100
^{127}I(n,2n)	1.976E-2	1.37	26 37 47 61 24 52 47 44 36 32 41 42 39 26 24 41 32 16 42 12 16 52 21 41 26 37 100
^{197}Au(n,γ)	2.071E+0	2.75	21 33 42 54 21 42 41 46 32 31 26 32 37 22 35 32 36 22 28 20 09 36 16 35 24 29 36 100
^{197}Au(n,2n)	7.686E-1	1.59	21 32 32 51 20 51 46 42 27 29 20 29 36 16 20 29 25 12 29 10 08 29 13 34 23 23 31 28 100
^{199}Hg(n,n')	5.511E+0	1.83	26 28 35 44 26 35 36 35 28 28 23 32 41 24 26 36 29 18 37 15 06 36 15 27 23 26 37 27 31 100
^{238}U(n,γ)	2.966E-2	1.20	21 32 41 52 21 44 43 42 25 31 20 29 37 22 22 36 25 16 36 13 07 31 11 32 24 26 32 23 21 20 100
^{235}U(n,f)	1.210E+3	1.58	10 17 23 15 14 23 24 21 16 19 16 21 27 10 16 25 16 09 22 11 06 26 09 27 10 17 22 10 17 15 23 100
^{237}Np(n,f)	1.361E+3	1.63	09 12 16 19 08 24 19 18 13 17 14 23 29 12 15 23 19 11 22 08 07 26 08 29 17 19 23 17 19 17 19 71 100
^{238}U(n,f)	3.257E+2	1.37	08 10 13 17 07 15 14 14 12 16 12 16 21 09 07 17 08 06 20 09 05 26 06 26 15 15 32 09 15 20 20 67 79 100
^{239}Pu(n,f)	1.812E+3	1.37	10 13 16 22 10 20 20 17 15 18 15 20 26 14 10 20 10 08 19 14 08 27 10 26 20 20 26 12 20 15 20 69 67 100

reactions were evaluated. The evaluation resulted in a chi-square of 35.3 with 35 degrees of freedom. For 20 of the 35 reactions, the relative uncertainty of the Cf-252 spectrum-averaged cross section data is less than 2 % and for 28 reactions less than 2.5 %.

REFERENCES

1.   W. Mannhart, p. 429 in Nuclear Data for Science and Technology, K.H. Böckhoff, Ed., D. Reidel Publ. Comp., Dordrecht, Holland, (1983).
2.   W. Mannhart, Proc. 4th ASTM-EURATOM Symp. Reactor Dosimetry, NUREG/CP-0029, Vol. 2, p. 637 (1982).
3.   D.L. Smith, J.W. Meadows, I. Kanno, "Measurements of the V-51(n,p) reaction cross section from threshold to 9.3 MeV by the activation method", Report ANL/NDM-85 (June 1984).
4.   D.L. Smith, priv. communication (1984).
5.   Nuclear Data Sheets, 23, 163 (1978).
6.   W.L. Zijp, J.H. Baard, "Nuclear data guide for reactor neutron metrology", Report EUR 7164 (1979).
7.   U. Schötzig, H. Schrader, "Recommended decay parameters", Report PTB-Ra-16 (June 1984).
8.   Nuclear Data Sheets, 32, 227 (1982).
9.   Nuclear Data Sheets, 24, 57 (1978).
10. Table of Isotopes, 7th edition, C.M. Lederer and V.S. Shirley, Eds. John Wiley and Sons, New York, (1978).
11. K. Kobayashi, S. Yamamoto, I. Kimura, "The U-235 fission spectrum averaged cross section for the Hg-199(n,n') reaction", Ann. Rep. Res. Reactor Inst. Kyoto Univ. 14, 1 (1981).
12. W. Mannhart, W.G. Alberts, Nucl. Sci. Eng. 69, 333 (1979).
13. J. Grundl, C. Eisenhauer, Techn. Doc. IAEA-208, Vol. I, p. 53 (1978).
14. Z. Deszö, J. Csikai, Proc. 4th All Union Conf. Neutron Physics, Atomizdat Moscow, Vol. 3, p. 32 (1977).
15. K. Kobayashi, I. Kimura, W. Mannhart, J. Nucl. Sci. Techn. 19, 341 (1982).
16. W.G. Alberts, E. Günther, M. Matzke, G. Rassl, Proc. 1st ASTM-EURATOM Symp. Reactor Dosimetry, EUR 5667, Part I, p. 131 (1977).
17. H. Benabdallah, G. Paic, J. Csikai, p. 421 in Nuclear Data for Science and Technology, K.H. Böckhoff, Ed., D. Reidel Publ. Comp., Dordrecht, Holland (1983).
18. Z. Deszö, J. Csikai, p. 418 in Nuclear Data for Science and Technology, ibid
19. M. Buczko, Z.T. Böde, J. Csikai, Z. Deszö, S. Juhasz, H.M. Al-Mundheri, G. Petö, M. Varnagy, Proc. Int. Symp. Cf-252 Utilization, CONF-760436, Vol. II, p. IV - 19 (1979).
20. S. Tagesen, H. Vonach. B. Strohmaier, Phys. Data, 13-1, 1 (1979).

21. K. Sakurai, H. Gotoh, S. Yamamoto, K. Kobayashi, I. Kimura, J. Nucl. Sci. Techn. 19, 775 (1982).
22. R. Sher, p. 167 in ENDF/B-V Dosimetry File, B.A. Magurno, Ed., BNL-NCS-50446 (April 1975).
23. P. Guenther, D. Havel, R. Howerton, F. Mann, D. Smith, A. Smith, J. Whalen, "Fast neutron cross sections of vanadium and an evaluated neutronic file", Report ANL/NDM-24 (May 1977).
24. W. Mannhart, "Spectrum-averaged neutron cross sections measured in the U-235 fission-neutron field at Mol", contribution to this conference.
25. W. Mannhart, F.G. Perey, Prod. 3rd ASTM-EURATOM Symp. Reactor Dosimetry, EUR 6813, Vol. 2, p. 1016 (1980).
26. G.P. Lamaze, E.D. McGarry, F.J. Schima, p. 425 in Nuclear Data for Science and Technology, K.H. Böckhoff, Ed., D. Reidel Publ. Comp., Dordrecht, Holland (1983).
27. J. Grundl, D. Gilliam, D. McGarry, C. Eisenhauer, P. Soran, Report INDC (NDS) - 146, p. 237 (1983).
28. J.A. Grundl, D.M. Gilliam, Report INDC (NDS) - 146, p. 241 (1983).

SPECTRUM-AVERAGED NEUTRON CROSS SECTIONS MEASURED IN THE U-235 FISSION-NEUTRON FIELD IN MOL

W. Mannhart

Physikalisch-Technische Bundesanstalt (PTB)

Bundesallee 100, D-3300 Braunschweig, Fed. Rep. of Germany

ABSTRACT

The cross sections of 17 neutron reactions of importance in reactor metrology were measured in a U-235 neutron field. The measurements were carried out relative to Ni-58(n,p) and In-115(n,n'), which acted as neutron fluence monitors. The experimental procedure, the corrections applied and the uncertainties attributed to the data are described. Due to the fact that the technique used to measure the induced radioactivity has much in common with that used in similar experiments in the Cf-252 neutron field, correlations between both standard neutron fields were established. These data and a complete covariance matrix of the present experiment are given.

INTRODUCTION

In reactor pressure vessel surveillance dosimetry, the prediction of spectral neutron fluences is based on the measurement of neutron monitors exposed in appropriate locations. The responses of such monitors determined in well-known neutron standard fields can be used to reduce the uncertainty in the driving fission spectrum and the monitor cross sections (1,2). This fuller information can then be applied in the analysis of the more complex and application-oriented fields for spectrum adjustment purposes. As the neutron spectrum of U-235 is the driving source of such fields, the disadvantage that no complete uncertainty information on data measured in the U-235 neutron field is available (1,2) has been emphasized. The present experiment is a first attempt to close this gap. Neutron reaction response ratios were determined in the U-235 neutron field and besides a complete covariance, a cross-correlation with similar Cf-252 data (3-5) was also derived.

Ratio measurements were performed which have the advantage of partially compensating many of the uncertainty sources and of reducing large corrections to relatively small correction ratios. The cross sections of Al-27(n,p) and V-51(n,p) were measured relative to In-115(n,n') and the cross sections of Mg-24(n,p), Al-27(n,α), Ti-46(n,p), Ti-47(n,p), Ti-48(n,p), V-51(n,α), Fe-54(n,p), Fe-56(n,p), Co-59(n,p), Co-59(n,α), Co-59(n,2n), Ni-58(n,2n), Zr-90(n,2n), In-115(n,n') and Au-197(n,2n) were measured relative to Ni-58(n,p).

EXPERIMENTAL PROCEDURE

The features of the U-235 neutron field of the Belgian reactor BR1 at Mol have already been described elsewhere (6). The field is generated in the center of a spherical cavity 1 m in diameter inside the vertical thermal column of the reactor. Access to the position of the field is given by a 2 m long cadmium tube, 33 mm in outer diameter and 1 mm thick, which can be loaded with the irradiation samples from the top of the reactor. The source is a cylinder of U-235 wrapped in an aluminum cladding and directly mounted to the outside of the cadmium tube. In the present experiment the source configuration "Mark II A", was used. Here the material was enriched to 89.66 % U-235. The height of the uranium cylinder was 79.6 mm, the wall thickness was 0.18 mm. The thickness of the aluminum cladding was 0.7 mm at the inner side of the cylinder and 0.75 mm outside. The sample holder was a cylindrical aluminum tube, 200 mm long and closed at the bottom. The outer diameter was 12 mm and the wall thickness was 1 mm. The samples were disk-shaped, high purity metallic foils 10 mm in diameter and between 0.125 mm and 0.5 mm thick. They were fixed into position at half height of the sample holder cylinder by two aluminum cylinders, 10 mm in outer diameter and 0.5 mm thick, which were inserted into the main cylinder below and above the samples. The sample holder was mounted to a long teflon rod, fixed concentrically in the cadmium tube, and was removable during reactor operation. The sample was positioned exactly at half height of the U-235 source cylinder. A layout of the irradiation geometry is shown in fig. 1.

At a reactor power of 600 kW, eleven irradiations of periods of between one and twelve hours were performed. One irradiation run of 19 hours, interrupted by two intervals each of 16 hours, was conducted at a power of 3.2 MW. The fluence rate of thermal neutrons driving the fission source was monitored with a U-235 double fission

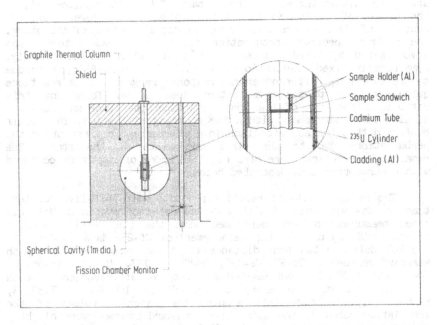

Graphite Thermal Column

Shield

Sample Holder (Al)

Sample Sandwich

Cadmium Tube

^{235}U Cylinder

Cladding (Al)

Spherical Cavity (1m dia.)

Fission Chamber Monitor

Fig. 1 Layout of the irradiation geometry

chamber (7) containing two deposits of 358.9 microgram and 7.00 microgram. In the experiment only the fission events of the smaller deposit were counted. The monitor was located in the thermal column (see fig. 1) at a position sufficiently distant from the cavity center to avoid perturbations and was checked to ensure a direct proportionality to the fluence rate at the sample position (8). The monitor rates showed a high constancy of the fluence level. The maximum variation within one irradiation period was 3 %, mainly due to a slow increase of the fluence rate with the core temperature. The monitor rates averaged over the irradiation periods on different days were constant within 1.1 %. The fission monitor rates referring to the irradiation history were used in the calculation of the activity build-up factors.

Samples were irradiated by means of the sandwich technique. Each sample was placed between two foils (of the same diameter) of nickel or indium acting as neutron fluence monitors. With this technique the small radial gradient of the flux density at the cavity center (6) was automatically compensated. Stacks of sandwiches up to a maximum thickness of 4 mm were irradiated simultaneously. The axial gradient over a stack of 4 mm was 0.26 %, based on purely geometrical considerations. Experimentally an additional symmetrical gradient was found due to scattering and absorption within the stack. For a stack of 4 mm the monitor foil in the middle of the stack showed a response 0.8 % lower than the response of the monitors located at the top and bottom of the stack. The neutron spectrum perturbation due to the stack material was investigated by forming a stack of gold and nickel foils only, symetrically mixed. The ratio of Au-197(n,2n) to Ni-58(n,p) is due to the extremely different energy response ranges of both reactions quite sensitive to spectrum perturbations. It was found that this ratio as a function of the position within the stack was constant within 0.34 % which was within the counting statistics. The result confirmed that the sandwich technique eliminates in first order the perturbations due to the other components of the stack. The remaining corrections were of a neglible order of magnitude compared with other corrections described below.

The radioactivity of reaction products with half-lives of less than one day was measured with a Ge(Li) detector system in Mol. All other measurements were performed with the PTB Ge(Li) detector system routinely used in the measurement of Cf-252 data (3 - 5). At the Mol detector two near distance positions were calibrated with standard sources of Co-57, Ce-139, Sr-85, Cs-137, Mn-54, Zn-65, Co-60 and Y-88. The corresponding photopeak efficiencies of both positions at a gamma-ray energy of 600 keV were 1.45E-2 and 7.30E-3, respectively. To take into account the extended volume of the irradiation samples, the point-like standard sources were slightly varied in their position, and radial and axial correction factors were determined. The technique was quite similar to that described in ref. (5) for Cf-252 measurements.

Decay parameters were taken mainly from ref. (9). For the reactions V-51(n,p), Ni-58(n,p) and In-115(n,n') the data given in table 1 of ref. (5) were used.

CORRECTIONS AND UNCERTAINTIES

The reaction rates measured in the cavity center cannot directly be referred to the U-235 neutron spectrum. Due to the structure materials (see fig. 1) perturbations are caused which result in a neutron spectrum deviating from the U-235 spectrum. This effect was determined with one-dimensional transport calculations (S8P3). For the structure materials ENDF/B-IV cross sections in a 171 group structure (VITAMIN-C) were used in the calculation. Fig. 2 shows the resulting spectral fluence rate relative to the unperturbed case. Based on this data, correction factors were determined to relate the measured reaction rates to a pure U-235 neutron spectrum. The reaction cross-sections used in this calculation were taken mainly from the ENDF/B-V dosimetry file and in a few cases from other sources (identical with the references given in table 5). In the present experiment only threshold

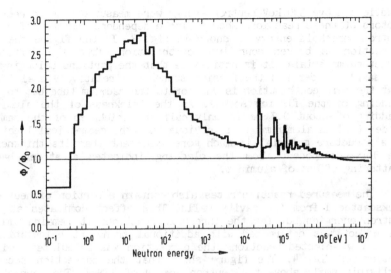

Fig. 2 Ratio of the perturbed-to-unperturbed spectral fluence rate

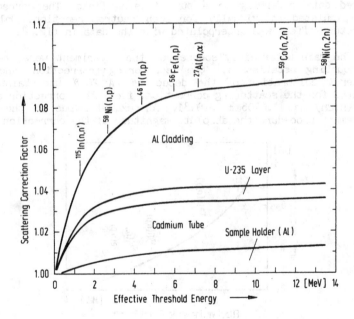

Fig. 3 Contributions to the scattering correction from various
structure materials

reactions above 330 keV neutron energy were measured. The correction factors of these reactions show a simple dependence on the effective neutron threshold energy as shown in fig. 3. In the figure the net correction is broken down into contributions from the different structure materials. It is noticeable that the portions belonging to the sample holder and the fission source are relatively small, and that the main contribution is due not to the cadmium tube but to the cladding of the fission source. As the thickness of the aluminum cladding of about 0.7 mm is only half the thickness of the sample holder (1.5 mm aluminum) it is obvious that the geometrical location of a structure material is much more important than its thickness. The large component from the cladding indicates a strong back-scattering effect of aluminum.

The measured reaction rates also contain a portion of neutrons backscattered from the cavity walls. This effect dominates at low neutron energies but for the threshold reactions it remains small. The effect was determined with additional transport calculations. For a few selected reactions the correction was calculated and is plotted in fig. 4. The figure shows that the correction becomes negligibly small above the neutron energy of 4 MeV. The correction factors of the various neutron reactions are summarized in table 1. Multiplication of the measured reaction rates with these factors produced data referring to a pure fission field. The corrections were calculated individually for each neutron reaction. Only the wall return effect was interpolated with the data in fig. 4.

The main uncertainty sources of the experiment were, besides the counting statistics, the scattering correction and the efficiency calibration of the detectors. A 20 % uncertainty was estimated for the scattering correction, i.e., the correction of Au-197(n,2n) was 1.0906 ± 0.0181, for example. Due to the ratio measurement procedure the absolute magnitude of the corrections is

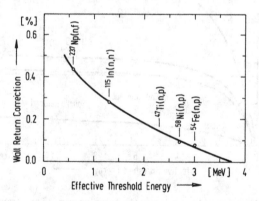

Fig. 4 Magnitude of the wall return correction

Table 1: Correction factors for scattering due to the structure
materials and for wall return

Reaction	Scattering	Wall return	Reaction	Scattering	Wall return
$^{115}In(n,n')$	1.0431	0.9972	$^{48}Ti(n,p)$	1.0873	1
$^{47}Ti(n,p)$	1.0615	0.9985	$^{59}Co(n,\alpha)$	1.0873	1
$^{58}Ni(n,p)$	1.0654	0.9991	$^{24}Mg(n,p)$	1.0877	1
$^{54}Fe(n,p)$	1.0673	0.9992	$^{27}Al(n,\alpha)$	1.0881	1
$^{59}Co(n,n)$	1.0785	1	$^{51}V(n,\alpha)$	1.0896	1
$^{27}Al(n,p)$	1.0789	1	$^{197}Au(n,2n)$	1.0906	1
$^{46}Ti(n,p)$	1.0794	1	$^{59}Co(n,2n)$	1.0919	1
$^{51}V(n,p)$	1.0815	1	$^{90}Zr(n,2n)$	1.0922	1
$^{56}Fe(n,p)$	1.0857	1	$^{58}Ni(n,2n)$	1.0926	1

diminished. Correlations between the corrections of various neutron
reactions allowed for this.

The relative uncertainty of the photopeak efficiency
calibration was 2.0 % for the Mol detector and 1.5 % for the PTB
detector. Based on the experimental calibration method, correlations
between the efficiencies at different gamma-ray energies were
established for each detector as described in ref. (5). Due to
principal differences in the calibration there were no correlations
between the two detectors.

RESULTS

The cross section ratios measured in this experiment are given
in table 2 with a complete uncertainty covariance matrix. The low
uncertainties quoted for the reaction ratios Fe-54(n,p)/Ni-58(n,p)
and Co-59(n,2n)/Ni-58(n,p) can easily be understood with regard to
the correlations. The reaction Co-59(n,2n) leads to the same
reaction product as the reaction Ni-58(n,p). In a ratio measurement
the photopeak efficiency and its uncertainty contribution is fully
elimimated. For the ratio Fe-54/Ni-58 the measured gamma-ray
energies of 843.8 keV of Mn-54 and of 810.8 keV of Co-58 are
relatively similar and result in a partially compensated uncertainty
contribution of the efficiency. This ratio also shows a strong
compensation of the scattering correction due to the similar energy
responses of both reactions (see table 1).

The common use of the PTB Ge(Li) detector for the present U-235
cross section ratio measurements as well as for similar Cf-252 data
(5) established correlations between both experiments. These
correlations are shown in table 3, they are not negligible. The
correlation pattern shows some peculiarities which will be briefly
reviewed. In the ratio Mg-24(n,p)/Al-27(n,α) measured in the Cf-252

Table 2: Measured cross section ratios in the ^{235}U neutron field and uncertainties

Reaction ratio	Cross section ratio	Rel. Std. Dev. %	Correlation matrix (x 100)
^{24}Mg(n,p)/^{58}Ni(n,p)	1.376E-2	2.87	100
^{27}Al(n,p)/^{115}In(n,n')	2.074E-2	3.41	8 100
^{27}Al(n,α)/^{58}Ni(n,p)	6.481E-3	2.98	89 8 100
^{46}Ti(n,p)/^{58}Ni(n,p)	1.064E-1	1.88	35 12 34 100
^{47}Ti(n,p)/^{58}Ni(n,p)	1.626E-1	2.19	36 -3 35 27 100
^{48}Ti(n,p)/^{58}Ni(n,p)	2.766E-3	2.14	43 11 41 75 27 100
^{51}V(n,p)/^{115}In(n,n')	2.641E-3	2.81	12 22 12 15 -3 16 100
^{51}V(n,α)/^{58}Ni(n,p)	2.213E-4	2.52	36 9 36 64 23 72 13 100
^{54}Fe(n,p)/^{58}Ni(n,p)	7.381E-1	0.75	13 4 13 36 11 30 6 26 100
^{56}Fe(n,p)/^{58}Ni(n,p)	1.004E-2	3.04	40 46 38 34 34 39 11 33 13 100
^{59}Co(n,p)/^{58}Ni(n,p)	1.297E-2	2.04	41 11 39 77 36 76 13 65 31 39 100
^{59}Co(n,α)/^{58}Ni(n,p)	1.477E-3	3.57	35 39 34 29 29 34 9 29 11 69 33 100
^{59}Co(n,2n)/^{58}Ni(n,p)	1.052E-3	1.51	31 15 31 29 0 42 23 39 9 27 24 25 100
^{58}Ni(n,2n)/^{58}Ni(n,p)	3.844E-5	4.39	29 5 28 25 23 31 8 28 9 26 34 23 25 100
^{90}Zr(n,2n)/^{58}Ni(n,p)	9.453E-4	2.22	30 10 30 47 12 52 15 47 29 27 41 24 50 23 100
^{115}In(n,n')/^{58}Ni(n,p)	1.746E+0	2.73	29 -10 28 21 41 22 -13 19 9 27 29 23 0 19 9 100
^{197}Au(n,2n)/^{58}Ni(n,p)	3.212E-2	2.50	50 9 49 41 41 49 13 44 15 46 47 40 42 35 38 33 100

field and in the ratio Co-59(n,2n)/Ni-58(n,p) measured in the Cf-252 and the U-235 field, the efficiency is eliminated due to the same reaction products of the ratio components. Thus these ratios should show a zero correlation between U-235 and Cf-252. The reaction rates of Al-27(n,p), V-51(n,p) and In-115(n,n') in the U-235 experiment were measured with the Mol detector, due to their short-lived reaction products. The Al-27(n,p)/In-115(n,n') and V-51(n,p)/In-115(n,n') ratios of U-235 therefore show no correlation with any Cf-252 data. The zero correlation of the Cf-252 ratios of Cu-63(n,γ)/In-115(n,n'), Cu-63(n,2n)/In-115(n,n'), V-51(n,p)/In-115(n,n') and Hg-199(n,n')/In-115(n,n') with all U-235 ratios with the exception of Au-197(n,2n)/Ni-58(n,p) is due to similar reasons. The correlation of the Cf-252 ratios with the last row of table 3 is due to the similar gamma-ray energies of 320.1 keV, 336.2 keV, 355.7 keV and 374.1 keV of the reaction products of V-51(n,p), In-115(n,n'), Au-197(n,2n) and Hg-199(n,n'), respectively.

COMPARISON WITH OTHER DATA

The ratios Fe-54(n,p)/Ni-58(n,p) and In-115(n,n')/Ni-58(n,p) of table 2 can be directly compared with results from the literature. This comparison is given in table 4. In a recent NBS experiment (10), the U-235 spectrum-averaged cross sections of Ni-58(n,p) and Fe-54(n,p) were determined. The data are based on a fluence transfer from the Cf-252 neutron field to the U-235 neutron field via the reaction In-115(n,n'). After elimination of the uncertainty contributions belonging to this transfer and taking into account the correlations between the reaction rates (see also table 6) a reaction rate ratio of 1.380 ± 0.023 of Ni-58(n,p) to Fe-54(n,p) was

Table 3: Correlation matrix (x 100) between U-235 and Cf-252 spectrum-averaged cross section ratios. The horizontal Cf-252 ratios refer to table 2 of ref. (5) and the vertical U-235 ratios to table 2 of this work.

U-235 \ Cf-252	19F(2n)/58Ni(p)	24Mg(p)/27Al(α)	27Al(p)/115In(n')	55Mn(p)/58Ni(p)	59Co(p)/58Ni(p)	59Co(α)/58Ni(p)	59Co(2n)/58Ni(p)	58Ni(2n)/58Ni(p)	63Cu(γ)/115In(n')	63Cu(α)/58Ni(p)	63Co(2n)/115In(n')	90Zr(α)/58Ni(p)	51V(2n)/115In(n')	51V(α)/58Ni(p)	65Co(2n)/58Ni(p)	64Zn(p)/58Ni(p)	127I(2n)/58Ni(p)	199Hg(n')/115In(n')
24Mg(p)/58Ni(p)	22	-30	0	3	23	5	0	22	0	24	0	7	0	21	30	37	11	0
27Al(p)/115In(n')	0	0	0	0	0	0	0	0	0	0	0	0	0	0	0	0	0	0
27Al(α)/58Ni(p)	21	-28	0	3	22	4	0	21	0	23	0	7	0	20	29	36	11	0
46Ti(p)/58Ni(p)	16	-16	0	10	47	14	0	18	0	39	0	22	0	54	22	28	-0	0
47Ti(p)/58Ni(p)	28	-39	0	4	30	6	0	28	0	31	0	10	0	27	39	48	14	0
48Ti(p)/58Ni(p)	16	-17	0	9	48	12	0	21	0	40	0	20	0	55	23	28	-0	0
51V(p)/115In(n')	0	0	0	0	0	0	0	0	0	0	0	0	0	0	0	0	0	0
51V(α)/58Ni(p)	14	-15	0	8	41	11	0	18	0	34	0	17	0	47	19	24	-0	0
54Fe(p)/58Ni(p)	7	10	0	24	19	32	0	7	0	17	0	19	0	24	9	11	0	0
56Fe(p)/58Ni(p)	20	-28	0	3	21	4	0	20	0	22	0	7	0	19	28	35	10	0
59Co(p)/58Ni(p)	22	-25	0	9	64	13	0	31	0	53	0	20	0	58	30	37	3	0
59Co(α)/58Ni(p)	17	-24	0	3	18	4	0	17	0	19	0	6	0	17	24	30	9	0
59Co(2n)/58Ni(p)	0	0	0	0	0	0	0	0	0	0	0	0	0	0	0	0	0	0
58Ni(2n)/58Ni(p)	14	-19	0	2	21	3	0	28	0	24	0	5	0	17	20	24	7	0
90Zn(2n)/58Ni(p)	7	-3	0	9	20	12	0	7	0	17	0	19	0	23	10	12	-0	0
115In(n')/58Ni(p)	23	-31	0	3	24	5	0	23	0	25	0	8	0	22	31	39	11	0
197Au(2n)/58Ni(p)	25	-67	0	4	26	5	0	25	-25	27	-20	8	-5	24	34	42	42	1

Table 4: Comparison of U-235 spectrum-averaged cross section ratios

Reaction ratio	Cross section ratio	Ref.
^{58}Ni(n,p)/^{54}Fe(n,p)	1.380 ± 0.023	(10)
	1.346 ± 0.030	(11)
	1.355 ± 0.010	Present
^{58}Ni(n,p)/^{115}In(n,n')	0.574 ± 0.029	(12)
	0.573 ± 0.016	Present

obtained. This value is fairly consistent with the ratio experiment of Fleming and Spiegel (11) and with the present data. The reaction ratio of Ni-58(n,p) to In-115(n,n') has been carefully determined in a previous experiment by Fabry and Czock (12). The result is almost identical with the present data.

The experimentally determined ratios of table 2 were normalized to allow a direct comparison with other data. In the normalization the cross section value of (109 ± 3) mb for the Ni-58(n,p) reaction of the recent NBS experiment (10) was used. The ratios measured relative to In-115(n,n') were not normalized with a cross section value taken from the literature but with the value of (190.3 ± 7.3) mb obtained from the In-115(n,n')/Ni-58(n,p) ratio measured in this experiment. The reason for this was an uncorrected contribution of photoactivation in the measured reaction rates of In-115(n,n'). The effect due to this is probably small but nevertheless the In-115(n,n') cross section in table 5 and the ratios relative to In-115(n,n') in table 2 may change slightly after applying the correction. The chosen procedure eliminated the influence of this neglected correction in the Al-27(n,p) and V-51(n,p) cross sections given in table 5.

In table 5, the agreement of the present data with Fabry's results (13) is in most cases remarkable. For the reaction Fe-54(n,p) the present value of (80.5 ± 2.3) mb should also be compared with the NBS result of (79 ± 3) mb (10). It should be noted that the present data of Fe-56(n,p) and Co-59(n,α) both leading to Mn-56 are systematically higher than Fabry's data (13). Large discrepancies are given only for Ni-58(n,2n) and Zr-90(n,2n). A comparison with similar data measured in the Cf-252 neutron field (4) gives more support to the lower values of the present experiment than to the data given by Fabry (13). The data of Kobayashi and Kimura (14) show a general tendency to be lower than those of the present experiment and also lower than Fabry's data (13). An explanation for this is probably the systematically low values of the cross sections of the monitor reactions Al-27(n,α), Ni-58(n,p) and In-115(n,n') used in this experiment (20). Some of the data in ref. (14) were measured in the core of the YAYOI reactor and correspond only approximately to a pure U-235 neutron field.

<u>Table 5:</u> U-235 spectrum-averaged neutron cross sections (in mb)

| Reaction | Experiment | | | Calculation* $\sigma(\bar{E})$ from: | | |
	Present work	Fabry et.al.(13)	Kobayashi et.al.(14)	ENDF/B-V	Other	Ref.
^{24}Mg(n,p)	1.50 ± 0.06	1.48 ± 0.082	1.38 ± 0.07		1.51	(15)
^{27}Al(n,p)	3.95 ± 0.20	3.86 ± 0.25	3.65 ± 0.20	4.26		
^{27}Al(n,α)	0.706± 0.028	0.705± 0.040		0.719	0.686	(16)
^{46}Ti(n,p)	11.6 ± 0.4	11.8 ± 0.75		11.2		
^{47}Ti(n,p)	17.7 ± 0.6	19.0 ± 1.4		22.5		
^{48}Ti(n,p)	0.302± 0.010	0.300± 0.018		0.281		
^{51}V(n,p)	0.503± 0.024		0.456± 0.023*	0.430	0.537	(17)
^{51}V(n,α)	(2.41 ± 0.09)x10^{-2}			2.23 x 10^{-2}		
^{54}Fe(n,p)	80.5 ± 2.3	79.7 ± 4.9	78.1 ± 3.7	81.0		
^{56}Fe(n,p)	1.09 ± 0.04	1.035± 0.075	1.02 ± 0.05	1.035		
^{59}Co(n,p)	1.41 ± 0.05				1.44	(18)
^{59}Co(n,α)	0.161± 0.007	0.143± 0.010	0.131±0.006*	0.150		
^{59}Co(n,2n)	0.202± 0.006			0.183		
^{58}Ni(n,2n)	(4.19 ± 0.22)x10^{-3}	(5.77 ± 0.31)x10^{-3}		2.80 x 10^{-3}	3.06 x 10^{-3}	(19)
^{90}Zr(n,2n)	0.103± 0.004	0.247± 0.017			0.0788	(15)
^{115}In(n,n')	190.3 ± 7.3	189 ± 8		179.2		
^{197}Au(n,2n)	3.50 ± 0.13		3.00 ± 0.16	3.22		

* Based on the ENDF/B Watt spectrum representation

* Measured in the YAYOI core

The calculated spectrum-averaged cross section values shown in table 5 are based on the ENDF/B representation of the U-235 neutron spectrum by a Watt distribution. Besides the ENDF/B-V cross section data, $\sigma(E)$ data from other sources were also used. For the reactions V-51(n,p) and V-51(n,α) the ENDF/B-V data are based on elemental vanadium. The calculated-to-experimental (C/E) data of both reactions in the U-235 neutron field show the same behaviour as similar data in the Cf-252 neutron field (5) did. For V-51(n,p) the bias factor of the present experiment with the calculated value, based on the $\sigma(E)$ data of ref. (17), is the same as in the Cf-252 neutron field (5). The better agreement of the present experiment with Al-27(n,α) with the calculated value based on ENDF/B-V than with the value based on ref. (16) is probably a fictitious one. The data of this reaction measured in the Cf-252 field (4) are more biased in favour of the $\sigma(E)$ data of ref. (16) than the ENDF/B-V

<u>Table 6:</u> Covariances of the NBS experiment (10)

Neutron Field	Reaction	<σ> (mb)	Rel. Std. Dev. %	Correlation matrix (x 100)				
Cf-252	^{54}Fe(n,p)	89	1.83	100				
	^{58}Ni(n,p)	121	1.82	91	100			
	^{115}In(n,n')	196	1.94	47	47	100		
U-235	^{54}Fe(n,p)	79	3.13	40	40	38	100	
	^{58}Ni(n,p)	109	2.69	46	47	45	85	100

data. This inversion between both fields seems to be caused by the Watt distribution which underestimates the high energy tail of the U-235 neutron spectrum. This conclusion is further substantiated by comparing the C/E values of the present experiment of other high threshold reactions such as Co-59(n,2n), Ni-58(n,2n), Zr-90(n,2n) and Au-197(n,2n) with similar values in the Cf-252 neutron spectrum (4). For example, the C/E value of Ni-58(n,2n) with regard to the $\sigma(E)$ data set of Winkler et al. (19) is 0.947 in the Cf-252 neutron field compared with the value of 0.730 of the present data in the U-235 neutron field.

CONCLUSIONS

In the recent past, methodologies have been developed (1,2) which promised an essential improvement of the reactor metrology basis. However, due to a lack of or incomplete data their application was limited. It is the aim of the present work to fill one of the existing gaps. The data obtained in this experiment represent one of the first data sets measured in the U-235 neutron field with a complete uncertainty description of form of a covariance matrix. This information should ease the transfer between neutron standard fields such as U-235 and Cf-252 and the neutron fields generated in reactor pressure vessels. Another recent experiment answered the same purpose: due to its very detailed uncertainty listing, the NBS experiment with U-235 and Cf-252 data (10) contains information similar to that given here. For completeness, the corresponding covariance matrix of this experiment, derived from the given data, is shown in table 6. The combination of these data with the data in tables 2 and 3 of this work, with corresponding information given in table 2 of ref. (5) and with the data shown in ref. (21), should form a sound basis for future progress in reactor pressure vessel metrology.

ACKNOWLEDGEMENTS

The author is deeply indebted to the members of the Reactor Physics Group of the S.C.K./C.E.N. (Mol) for their strong support and kind hospitality during the course of this experiment. The help of Dr. A. Fabry in performing the neutron transport calculations and the support of Dr. H. Tourwé in the radioactivity measurements is greatly appreciated.

REFERENCES

1. J.J. Wagschal, R.E. Maerker, B.L. Broadhead, Proc. 4th ASTM-EURATOM Symp. Reactor Dosimetry, NUREG/CP-0029, Vol I, p. 79 (1982).

2. R.E. Maerker, J.J. Wagschal, B.L. Broadhead, "Development and demonstration of an advances methodology for LWR dosimetry applications", Report EPRI NP-2188 (December 1981).

3. W. Mannhart, Proc. 4th ASTM-EURATOM Symp. Reactor Dosimetry, NUREG/CP-0029, Vol. 2, p. 637 (1982).

4. W. Mannhart, p. 429 in Nuclear Data for Science and Techn., K.H. Böckhoff, Ed., D. Reidel Publ. Comp., Dordrecht, Holland (1983).

5. W. Mannhart, "Recent experiments of Cf-252 spectrum-averaged neutron cross sections", contribution to this conference.

6. A. Fabry, G. Minsart, F. Cops, S. De Leeuw, Proc. 4th ASTM-EURATOM Symp. Reactor Dosimetry, NUREG/CP-0029 Vol. 2, p. 665 (1982).

7. J.A. Grundl, D.M. Gillian, N.D. Dudey, R.J. Popek, Nucl. Techn. 25, 237 (1975).

8. A. Fabry, R. Mevil, F. Cops, Internal memo, Mol (April 1983).

9. W.L. Zijp, J.H. Baard, "Nuclear data guide for reactor neutron metrology", Report EUR 7164 (1975).

10. G.P. Lamaze, E.D. McGarry, F.J. Schima, p. 425 in Nuclear Data for Science and Technology, K.H. Böckhoff, Ed., D. Reidel Publ. Comp., Dordrecht, Holland (1983).

11. R. Fleming, V. Spiegel, Proc. 2nd ASTM-EURATOM Symp. Reactor Dosimetry, NUREG/CP-0004, Vol. 2, p. 953 (1977).

12. A. Fabry, K.H. Czock, "Measurement of integral cross section ratios in two dosimetry benchmark neutron fields", Report INDC (IAE)-005/G (December 1974).

13. A. Fabry, W.N. McElroy, L.S. Kellogg, E.P. Lippincott, J.A. Grundl, D.M. Gilliam, G.E. Hansen, IAEA-TEDOC-208, Vol. 1, p. 233 (1978).

14. K. Kobayashi, I. Kimura, Proc. 3rd ASTM-EURATOM Symp. Reactor Dosimetry, EUR 6813, Vol. 2, p. 1004 (1980).

15. S. Tagesen, H. Vonach, B. Strohmaier, Physics Data 13-1 (1979).

16. S. Tagesen, H. Vonach, Physics Data 13-3 (1981).

17. D.L. Smith, J.W. Meadows, I. Kanno, "Measurement of the V-51 (n,p) reaction cross section from threshold to 9.3 MeV by the activation method", Report ANL/NDM-85 (June 1984).

18. D.L. Smith, J.W. Meadows, Nucl. Sci. Eng. 60, 187 (1976).

19. G. Winkler, A. Pawlik, H. Vonach, A. Paulsen, H. Liskien, p. 400 in Nuclear Data for Science and Technology, K.H. Böckhoff, Ed., D. Reidel Publ. Comp., Dordrecht, Holland (1983).
    A. Pawlik, G. Winkler, Report INDC (AUS)-9/L (1983).

20. I. Kimura, K. Kobayashi, S.A. Hayashi, S. Yamamoto, H. Gotoh, H. Yagi, IAEA-TECDOC-208, Vol. 2, p. 265 (1978).

21. W. Mannhart, F.G. Perey, Proc. 3rd ASTM-EURATOM Symp. Reactor Dosimetry, EUR 6813, Vol. 2, p. 1016 (1980).

UNFOLDED ANO-1 FLUXES USING

THE LEPRICON METHODOLOGY*

J.J. Wagschal**, R.E. Maerker,[†]B.L. Broadhead,[†] & M.L. Williams[†]

**Racah Institute of Physics Hebrew University
91904, Jerusalem Israel
[†]Engineering Physics & Mathematics, Oak Ridge National Laboratory
P.O.B. X, Oak Ridge TN 37831, U.S.A.

ABSTRACT

Calculated fluxes at a one-quarter depth midplane
location in the pressure vessel of the ANO-1 reactor were
adjusted using the LEPRICON methodology. The adjustment
was based on the LEPRICON benchmark database in combination with the ANO-1 cavity activation measurements.
The LEPRICON database has been expanded and now also
includes the PSF field, of which the analysis has been
recently completed. The measured, calculated and adjusted
ANO-1 activations are presented together with their
corresponding uncertainties. However, the main purpose
of this work is to obtain more credible flux values in
the pressure vessel. The pressure-vessel flux adjustments
and the reduction in the pressure vessel flux uncertainties
are elaborated on in the concluding section.

INTRODUCTION

The LEPRICON (Least-squares EPRI CONsolidation) methodology
has been developed over a period of a few years (1,2,3). The system
predicts the absolute fluence levels as a function of energy in the
pressure vessel of an LWR from the analysis of measurements performed
at some other readily accessible surveillance locations. LEPRICON is

*Research sponsored by the Electric Power Research Institute under
project RP 1399

unique in the field of few-group spectral unfolding in that it provides, in addition to a more reliable estimate of the extrapolated spectral fluences, justifiable reduced uncertainties of these fluences as well.

Since the overall better prediction of the spectrum and exposure levels in the pressure vessel, i.e. the prediction with reduced uncertainties, means that the uncertainty in the damage from steel embrittlement is also reduced, the application of this methodology should result in more reliable predictions of reactor-vessel lifetimes, and consequently in the possible extension of these lifetimes as well.

Among the unique features of this methodology is the capability to combine the LWR surveillance data with data derived from analysis of dosimetry benchmark measurements. The adjustment of the calculated pressure-vessel group fluxes is thus based on a consistent analysis of all relevant data. Hence, the development of a sound dosimetry-benchmark database is of utmost importance.

The development of the LEPRICON database is documented in two EPRI reports. The first (1) describes the technique, introduces along the way 20 benchmark experiments, and describes these experiments in detail. These experiments, however, fall short of satisfying the requirements that such experiments should: they are not particularly sensitive to the iron inelastic cross section, and they do not lead to meaningful adjustments in dosimeter cross sections which are at present of most common usage by the dosimetry community.

The second report (3) is a follow-up of the initial benchmark analysis. It includes the analysis of 17 additional benchmarks, such that the complete set of 37 experiments now spans a relatively broad range of dosimeter reactions, spectra, and steel thicknesses. These 37 experiments also provide measurements at several locations in the simulated pressure vessel together with measurements at a simulated surveillance location, which combination adds another dimension to the analysis.

The complete set of benchmark experiments includes 10 measurements in ^{252}Cf fields, 4 fission-ratio measurements in the National Bureau of Standards Intermediate-energy Standard Neutron Field (ISNF), 9 measurements at two locations in the Oak Ridge National Laboratory (ORNL) Pool Critical Assembly (PCA), and 14 measurements at three locations in the ORNL Pool Side Facility (PSF). The latter two facilities produce prototypic fields in the neighborhood of pressure-vessel mock-ups.

A major deficiency of the earlier LEPRICON database, as well as of the original ENDF/B-V dosimetry cross-section library, was an almost complete absence of cross-correlation information, which information is very significant in the adjustment procedure. A new library of dosimetry cross sections which has cross-correlations included in its "global" covariance matrix has been developed for LEPRICON (4).

A consistency analysis of the 37 benchmark measurements in-cluded in the LEPRICON database indicates that three of these should probably be omitted when the benchmarks are used in conjunction with the revised differential data (cross sections, covariances etc.). Thus 34 measurements comprise the benchmark database which is to be simultaneously adjusted with a set of power reactor sur-veillance measurements.

Up to now the new extended database was only applied to the flux unfolding at two locations in the pressure-vessel mock-up of the PSF (3). The unique PSF dosimetry measurements at these two locations in the pressure vessel, in addition to measurements at a simulated surveillance location (a situation unrealizable in power reactors), allowed the unfolding of the group fluxes at the pressure vessel locations with in situ measurements. The results of the standard LEPRICON unfolding, namely adjustment with surveillance data only plus "extrapolation" to the pressure vessel, are in a rather good agreement with the results obtained by the above, pre-ferable but generally impractical, method.

The recent dosimetry experiments in the pressure-vessel cavity of the Arkansas Power and Light reactor ANO-1 (Arkansas Nuclear One-1) (5) have now made the application of the advanced LEPRICON methodology to the flux unfolding of a real operating power reactor possible. Such an application is presented here for the first time.

In the following sections the results of the analysis of the ANO-1 fluxes by the LEPRICON system will be presented and discussed. However, such an analysis necessitates a rather extensive and varied set of input data. We shall therefore precede the presentation of our results by a brief description of these data with a particular emphasis on the critical evaluation of their uncertainties.

ANALYSIS OF ANO-1

The analysis of any reactor begins with a careful calculation of the energy-space-time distribution of the flux. Recent progress in, and new techniques for these calculations are described by Maerker et. al. (6). Such calculations should be obviously based on the best nuclear input data available. The adjusted data generated by LEPRICON from the selected 34 benchmark experiments could already serve as such input, even without additional measurements. However, the inclusion of additional surveillance dosimetry measurements can only further improve the flux and fluence estimates, and depending on the quality of these measurements and the reliability of the calculations of the corresponding quantities, might significantly reduce their uncertainties.

The complete analysis of a given set of surveillance dosimetry experiments necessitates the measured values of the responses under discussion and their uncertainty matrix, the corresponding calculated values and their uncertainties, and the sensitivities of the calculated responses to their respective input parameters (cross sections, bias factors, etc.).

In order to modify the flux estimate at any point, and in particular at the point of highest anticipated radiation damage in the pressure vessel, i.e. the T/4 location, sensitivities of the flux at that point to all relevant parameters are needed as well.

The data employed in the current ANO-1 analysis will be discussed next.

ANO-1 Dosimetry Experiments

Four ANO-1 dosimetry experiments are analyzed in this work with the aim of improving fast-neutron fluence estimates at the T/4 position in the pressure vessel horizontal midplane. These experiments are denoted by I-5, I-7, I-8 and I-9 in Ref. 5, where details on their irradiation history can be found.

The experiments were carried out at different times in the pressure-vessel cavity, and cover different fuel cycles. The foil packages, each containing ^{63}Cu, ^{46}Ti, ^{58}Ni, ^{54}Fe and ^{237}Np foils, and in experiment 5 also a ^{238}U foil, were located at the level of the horizontal midplane in two different vertical detector wells. Each detector well was located at a distance of 324 cm from the core center. Experiments 5 and 7 were performed in the same vertical well located at azimuth angle 157° from the North. Experiments 8 and 9 were performed in a well located diametrically opposite the former well, i.e. at azimuth angle 337°. The 21 measured activation values are listed in Table I (E values).

The calculated values of ANO-1 fluxes, and those of the activations in particular, evolve from a series of sophisticated calculations. These involve source distributions in peripheral core assemblies, one and two dimensional forward and adjoint transport calculations, effects of different detector wells and more. The progress in the calculational methodology is discussed by Maerker et. al. (6), and the actual calculations will be detailed in a forthcoming report (7).

Table I. Comparison of Measured (E) and Calculated (C)
Activations for Four Midplane Experiments in ANO-1.

Reactions are given per target nucleus per source neutron.

Exp		^{63}Cu(n,α)	^{46}Ti(n,p)	^{58}Ni(n,p)	^{54}Fe(n,p)	^{238}U(n,f)	^{237}Np(n,f)
	E	1.39-41	3.46-39	2.92-38	5.61-39	1.84-38	2.70-37
5	C	1.11-41	2.92-39	2.45-38	4.63-39	1.66-38	2.63-37
	C/E	.80	.84	.84	.83	.90	.97
	E	9.30-42	2.85-39	2.33-38	4.28-39		2.75-37
7	C	8.94-42	2.43-39	2.05-38	3.76-39		2.50-37
	C/E	.96	.85	.88	.88		.91
	E	1.21-41	2.79-39	2.58-38	5.62-39		2.74-37
8	C	1.03-41	2.12-39	1.99-38	4.37-39		2.02-37
	C/E	.85	.76	.77	.78		.74
	E	4.63-41	5.10-39	4.23-38	1.62-38		2.65-37
9	C	3.67-41	3.67-39	3.18-38	1.25-38		2.16-37
	C/E	.79	.72	.75	.77		.81

The calculated values corresponding to the measured activations
are also listed in Table I (C values). A comparison of the measured
and calculated values shows that the calculated values are always
smaller than the corresponding measured values and that the discrepancies in experiments 8 and 9 are bigger than the others.

ANO-1 Measurement Uncertainties

The measurements' covariance matrix elements for the four experiments in ANO-1 arise from several different uncertainty sources,
some of which are correlated. The uncertainties reported by J.G.
Williams (8) will be documented in Ref. 7. The major uncertainties
are:

(a) 3.6% in the power-to-neutron-source conversion factor and in
     the total power of the core. This is needed since LEPRICON
     expresses the measured and calculated activations per source
     neutron. This uncertainty is common to all measurements, i.e.

it contributes fully correlated uncertainties to all measurements.

(b) 5% in the gamma-ray detector efficiency. This introduces partial correlations among the different foils in the same experiment ranging from zero for most of these foils to 0.9 between ^{58}Ni and ^{54}Fe.

(c) 1%-2% in counting. These uncertainties are for each measurement, of course, completely uncorrelated.

(d) 5%-10% in the reaction-and-experiment-dependent bias factors which are meant to correct results for photofissions.

(e) <5% in other reaction-and-experiment-dependent bias factors.

The relative uncertainties expressed as Percent Standard Deviations (PSD) and the lower triangle of the correlation matrix are presented in Table II. The detailed information on all correlations will be documented in Ref. 7. However, the evaluation of two variances and the corresponding correlation is illustrated here.

The percent standard deviation of the $^{237}Np(n,f)$ measurements in experiments 8 or 9 is calculated as follows:

$$11.79\% = \left[(3.6\%)^2 \text{ from a} +(5\%)^2 \text{ from b} +(1\%)^2 \text{ from c} +(10\%)^2 \text{ from d}\right]^{\frac{1}{2}},$$

and α, the correlation between the $^{237}Np(n,f)$ measurements in experiments 8 and 9, assuming a 0.5 correlation for gamma-ray detector efficiency for similar foils, is obtained in the following way:

$$\alpha = \frac{3.6\% \times 3.6\% \text{ from a} + 5\% \times 5\% \times 0.5 \text{ from b} + 10\% \times 10\% \times 1 \text{ from d}}{11.79\% \times 11.79\%} = 0.90.$$

A small correlation exists between the LEPRICON database benchmark measurements and ANO-1 measurements, due to the coversion factor between power and neutron source, but it can be safely neglected.

ANO-1 Calculational Uncertainties

The introduction of correction factors, more commonly known as bias factors, to modify measured or calculated values by taking into account some additional effects (photofission, for instance, which was mentioned in the preceding section) is a common practice. Obviously, there is also an uncertainty in these bias factors which has to be taken into account. Sometimes there are additional uncertainties in the calculations (due to modeling or geometrical sources) which have to be taken into account, although no correction factors

Table II. Correlation Matrix for ANO-1 Measurements.

	PSD	Experiment 5						Experiment 7					Experiment 8					Experiment 9				
		Cu	Ti	Ni	Fe	U238/8	Np	Cu	Ti	Ni	Fe	Np	Cu	Ti	Ni	Fe	Np	Cu	Ti	Ni	Fe	Np
Cu	6.95	1.0																				
Ti	8.19	.23	1.0																			
Ni (05)	8.00	.23	.39	1.0																		
Fe	8.00	.23	.39	.56	1.0																	
U238/8	11.79	.16	.13	.14	.14	1.0																
Np	8.00	.23	.20	.20	.20	.54	1.0															
Cu	6.95	.66	.23	.23	.23	.16	.23	1.0														
Ti	8.19	.20	.57	.20	.20	.13	.20	.46	1.0													
Ni (07)	8.00	.23	.25	.79	.29	.14	.20	.23	.20	1.0												
Fe	8.00	.23	.25	.29	.79	.14	.20	.23	.20	.56	1.0											
Np	8.00	.23	.20	.20	.20	.44	.79	.23	.20	.20	.20	1.0										
Cu	6.95	.66	.23	.23	.23	.16	.23	.66	.20	.23	.23	.23	1.0									
Ti	6.46	.29	.48	.31	.31	.17	.25	.29	.24	.81	.31	.25	.29	1.0								
Ni (08)	6.24	.30	.32	.61	.37	.18	.26	.30	.25	.51	.37	.26	.30	.63	1.0							
Fe	6.24	.30	.32	.37	.61	.18	.26	.30	.25	.37	.61	.26	.30	.63	.91	1.0						
Np	11.79	.16	.13	.14	.14	.30	.44	.16	.13	.14	.14	.44	.16	.17	.18	.18	1.0					
Cu	6.95	.66	.23	.23	.23	.16	.23	.66	.20	.23	.23	.23	.66	.29	.30	.30	.16	1.0				
Ti	6.46	.29	.48	.31	.31	.17	.25	.29	.24	.31	.31	.25	.29	.61	.40	.40	.17	.29	1.0			
Ni (09)	6.24	.30	.32	.61	.37	.18	.26	.30	.25	.61	.37	.26	.30	.40	.65	.48	.18	.30	.63	1.0		
Fe	6.24	.30	.32	.37	.61	.18	.29	.30	.25	.37	.61	.26	.30	.40	.48	.65	.18	.30	.63	.91	1.0	
Np	11.79	.16	.13	.14	.14	.30	.54	.16	.13	.14	.14	.54	.16	.17	.18	.18	.90	.16	.17	.18	.18	1.0

are introduced. In the adjustment module of LEPRICON all these uncertainty sources are handled as additional data, most frequently
as unit bias factors with energy- and space-dependent covariance
matrices.

The introduction of the ANO-1 activation measurements as
additional input data to LEPRICON involved a careful uncertainty
analysis of both measurements and calculations. The uncertainty
analysis of the measured values resulted in the covariance matrix
given in Table II, and that of the calculated values in turn resulted in the introduction of several flux bias factors and their
covariance information. Some of these bias factors will be discussed
here. All of them, however, are listed in Table III.

The 3D flux synthesis, which is part of the LEPRICON calculational methodology, necessitates the introduction of a bias factor
to handle the finite core height. This flux-unit-bias-factor applies
to ANO-1, as well as to the PCA and PSF which are part of the LEPRI-
CON benchmark database. The PSD of each energy-group flux at the
ANO-1 pressure-vessel cavity is 3.5%, and the correlation between
any two group fluxes   (known as "auto-correlation", in contra-
distinction to "cross-correlations" with other bias factor) is 0.75.
The cross-correlation between the ANO-1 cavity finite-core-height

bias factors and the PCA and PSF finite-core-height bias factors at all locations range from 0.525 to 0.675.

The finite-core-height bias factors are the only bias factors in the ANO-1 analysis with a correlation to bias factors in the 34-experiment LEPRICON database.

Experiments 8 and 9 were irradiated in a vertical monitor well containing lead. The approximate method adopted in simulating the cylindrical well geometry introduced the lead-well-transmission unit bias factors for the group fluxes with a 5% PSD and auto-correlations of 0.7 between any two group fluxes.

The bias factors discussed so far have the same PSD for all group fluxes. However, the 1% uncertainty in the steel density, which is also represented by flux unit bias factors, turns out to induce group dependent PSDs ranging from 1.9 to 3.6 at the T/4 position and from 2.8 to 6.6 in the pressure-vessel cavity.

Deserving special mention is the uncertainty in the relative source distribution in the peripheral fuel assemblies, which in turn influences the group-fluxes at T/4 and in the cavity.

RESULTS AND DISCUSSION

The LEPRICON system with its adjunct database, comprising 34 benchmark measurements, was applied to the ANO-1 T/4 flux analysis. We mention in passing that the point of peak flux at the one quarter depth horizontal midplane in the ANO-1 pressure vessel is at an azimuth of 169° from the North, which is different from the azimuth of the surveillance measurement, due to the proximity of peripheral fuel assemblies.

The relevant ANO-1 experimental data discussed in the preceding sections, were added to the LEPRICON database, and the combined data were input to its adjustment module.

The adjustment resulted in modified parameters, $\alpha'=\alpha+\delta\alpha$, and in corresponding modified covariance matrices, $C_{\alpha'\alpha'}$. The scalar product of the T/4 flux sensitivities, S_ϕ, and the parameter modifiers, $\delta\alpha$, gives the flux modification, $\delta\phi=S_\phi.\delta\alpha$, and the reduced uncertainty matrix of the modified flux, $\phi'=\phi+\delta\phi$, $C_{\phi'\phi'}$, is given by the so called "sandwich rule", $C_{\phi'\phi'} = S_\phi C_{\alpha'\alpha'} S_\phi^\dagger$.

The T/4 high energy flux adjustments (>0.1 MeV), as functions of the experiments taken into account in the adjustment process, the "participating" experiments, and their corresponding PSDs are given in Table IV. The top 15 energy groups were considered. The PSDs in the third column are the a-priori uncertainties in the group fluxes

Table III. The Flux Bias Factors Used in the Calculation of the ANO-1 Group Fluxes. (A variation range indicates an energy dependence)

Bias Factor Definition	PSD		Correlations	
	T/4	Cavity	Auto	Cross
Finite Core Height (3-D Synthesis)[†]	0	3.5	0.75	-
Lead Well Transmission (#8 and #9 only)	0	5.0	0.7	-
Cavity Wall Composition (#5 and #7 only)	0	0-15.6	0.7	1.0 between #5 or #7 with
Cavity Wall Composition (#8 and #9 only)	0	0- 8.9	0.7	#8 or #9
Steel Density (1%)	1.9-3.6	2.8-6.6	0.9	0.27
Water Density (2%)	3.7-7.2	3.7-7.1	0.9	0.81
PV Out of Roundness (as-built, 1cm H_2O)	6.8-10.9	6.7-10.7	0.9	0.81
Relative Source Distribution[††]	4.1-4.6	2.9- 3.1	1.0	∿0.8 in same experiment

[†]This bias factor has a cross-correlation with its counter parts in PCA and PSF ranging from 0.525 to 0.675

[††]For all peripheral assemblies combined. Results of adjustment not sensitive to possible correlations between experiments.

due to propagation of parameter uncertainties, including modeling and methods uncertainties. These uncertainties are always higher than 15% and get as high as 20%. Going from left to right each column shows the flux modifications and the uncertainties in the modified flux, as more and more experiments participate in the adjustment.

It is interesting to note that the adjustments are pretty much established when only the 34-adjunct LEPRICON benchmark experiments are considered. The participation of the first ANO-1 surveillance dosimetry experimental data further reduces the uncertainties, but hardly changes the flux modifications. The additional ANO-1 activation measurements have only a marginal effect on the flux modifiers and even on the flux uncertainties.

The importance of the benchmark measurements and of the various

Table IV. Relative Percent Flux Adjustments and Flux Uncertainties for ANO-1 Exp. 9 as Function of "Participating" Experiments.

Group	E_L MeV	None	34 Bench- marks	ANO-1 +Exp.5	ANO-1 +Exp.7	ANO-1 +Exp.8	ANO-1 +Exp.9
1	11.05	0.0±20.2	-3.1±12.4	1.3±10.0	-2.4± 9.8	-1.6± 9.8	-0.3± 9.6
2	8.187	0.0±18.9	-6.8±12.3	-1.4± 9.6	-5.6± 9.3	-4.8± 9.3	-3.5± 9.1
3	6.065	0.0±17.5	20.6±12.4	21.0± 9.3	18.8± 9.1	19.5± 9.0	20.5± 8.8
4	4.066	0.0±18.0	22.7±13.2	22.4± 9.9	20.4± 9.7	21.2± 9.7	22.1± 9.4
5	3.012	0.0±17.8	20.7±13.8	20.6±10.4	18.5±10.2	19.3±10.1	20.2± 9.9
6	2.592	0.0±16.2	8.7±14.1	9.2±10.6	7.3±10.4	8.3±10.3	9.1±10.1
7	2.123	0.0±16.4	5.9±14.5	6.4±10.9	4.6±10.8	5.7±10.7	6.4±10.4
8	1.827	0.0±16.2	6.5±14.4	7.1±10.8	5.3±10.7	6.4±10.6	7.1±10.3
9	1.496	0.0±16.2	6.4±14.4	6.9±10.9	5.2±10.7	6.3±10.6	7.1±10.4
10	1.225	0.0±16.1	6.4±14.4	7.1±10.9	5.3±10.7	6.4±10.6	7.2±10.4
11	.9072	0.0±16.5	6.4±14.6	7.1±10.9	5.3±10.7	6.4±10.6	7.2±10.4
12	.6081	0.0±15.3	8.0±13.9	8.3±10.6	7.0±10.5	8.0±10.5	8.7±10.2
13	.3688	0.0±15.3	8.6±14.0	8.8±11.0	7.5±10.9	8.6±10.8	9.3±10.6
14	.2128	0.0±15.2	8.9±14.0	9.0±11.1	7.7±11.0	8.7±11.0	9.5±10.7
15	.1111	0.0±15.3	8.8±14.2	8.9±11.3	7.6±11.2	8.6±11.1	9.3±10.9

ANO-1 activation experiments in the T/4 flux adjustments and in the uncertainty reduction was discussed above. Since in ANO-1, as in all other power reactors, there are no activation measurements in the pressure vessel itself it is hard to comprehend the results. However, since the same procedure also adjusts the flux at the ANO-1 cavity, for which we have measured values, we can see the effect of the various experiments participating in the adjustment on the calculated activations in the cavity. In Table V we can notice the narrowing of the gap between the measured (E) and the recalculated, after adjustment, (A) activations and in particular the dramatic reduction in the uncertainty of the adjusted activations for all measurements in experiment 9.

CONCLUDING REMARKS

The first attempt to modify by the LEPRICON system the one quarter depth flux in the pressure vessel of a working power reactor is described in this work. The analysis necessitated the preparation of additional input data to the LEPRICON adjustment module. These data were prepared for some ANO-1 activation measurements and included new bias factors pertinent to ANO-1, their correlations and sensitivities of ANO-1 T/4 and cavity fluxes to all parameters. The sensitivities were calculated with the aid of the FORSS system (9) and some were also derived internally in LEPRICON using parametrization techniques. Comparison indicated the success of the parametrization (based on PCA), thus reducing the user's interfacing

3. R.E. Maerker, M.L. Williams, B.L. Broadhead, J.J. Wagschal and C.Y. Fu, Revision and Expansion of the Database in the LEPRICON Dosimetry Methodology, EPRI Report EPRI NP- (1984).

4. C.Y. Fu and D.M. Hetrick, "Experience in Using the Covariances of Some ENDF/B-V Dosimetry Cross Sections: Proposed Improvements and Addition of Cross-Reaction Covariances," p 877 in Radiation Metrology Techniques, Data Bases, and Standardization, F.B.K. Kam, Editor NUREG/CP-0029 (July 1982).

5. C.O. Cogburn, J.G. Williams, "Pressure Vessel Dosimetry at the Arkansas Nuclear Plants," Trans. Am. Nucl. Soc., 45 p.589 (1983).

6. R.E. Maerker, B.L. Broadhead, M.L. Williams, "Recent Progress and Developments in LWR-PV Calculational Methodology," this conference (1984).

7. R.E. Maerker, B.L. Broadhead, B.A. Worley, M.L. Williams and J.J. Wagschal, Application of the LEPRICON System to the ANO-1 Reactor, EPRI NP- (1985).

8. J.G. Williams private communication to REM.

9. J.L. Lucius, C.R. Weisbin, J.H. Marable, J.D. Drischler, R.Q. Wright and J.E. White, A Users Manual for the FORSS Sensitivity and Uncertainty Analysis Code System, ORNL-5316 (ENDF-291) (1981).

Table V. Relative Percent Discrepancies Between Measured (E), Calculated (C) and Adjusted (A) Exp. 9 PV-Cavity Activations as Function of "Participating" Experiments. The PSDs are of the Original Discrepancy and of the Adjusted Activations Respectively.

Reaction	$\frac{E-C}{C}$		$\frac{E-A}{C}$			
		34 Bench-marks	+ANO-1 Exp.5	+ANO-1 Exp.7	+ANO-1 Exp.8	+ANO-1 Exp.9
$^{63}Cu(n,\alpha)$	4.0±25.2	-1.5±14.1	- 8.3±7.1	-0.3±5.6	-0.3±5.5	-0.3±5.5
$^{46}Ti(n,p)$	17.3±26.2	-3.5±14.3	- 8.0±6.9	-0.9±5.5	-1.0±5.4	-1.0±5.4
$^{58}Ni(n,p)$	13.7±25.1	-8.3±14.8	-10.6±7.1	-4.7±5.7	-4.8±5.6	-4.4±5.6
$^{54}Fe(n,p)$	13.8±25.6	-8.3±14.8	-10.7±7.2	-4.6±5.8	-4.4±5.7	-4.0±5.6
$^{237}Np(n,f)$	10.0±21.7	6.7±14.7	6.6±7.5	9.0±6.5	9.9±6.4	10.9±6.4

with LEPRICON.

The flux adjustments based on the 34-adjunct LEPRICON benchmark database and on a varying number of ANO-1 activation experiments all had χ^2/I (chi squared per degree of freedom) values in the agreeable range of 1.0 to 1.08.

The flux modifications were essentially obtained even without the use of ANO-1 activations. The inclusion of ANO-1 experiments in the adjustments also practically did not change the adjustments in the LEPRICON database benchmarks. However, the employment of the ANO-1 activation measurements substantially reduced the flux uncertainties.

ACKNOWLEDGEMENT

It is a pleasure to thank Prof. Y. Yeivin for his many helpful comments in reviewing the manuscript.

REFERENCES

1. R.E. Maerker, J.J. Wagschal, and B.L. Broadhead, Development and Demonstration of an Advanced Methodology for LWR Dosimetry Applications, EPRI Report EPRI NP-2188 (December 1981)

2. J.J. Wagschal, R.E. Maerker and B.L. Broadhead, "Surveillance Dosimetry: Achievements and Disappointments," p. 79 in Radiation Metrology Techniques, Data Bases, and Standardization, F.B.K. Kam, Editor, NUREG/CP-0029 (July 1982).

PART VIII
NUCLEAR DATA

CONSISTENCY CHECK OF IRON AND SODIUM CROSS-SECTIONS WITH INTEGRAL BENCHMARK EXPERIMENTS USING A LARGE AMOUNT OF EXPERIMENTAL INFORMATION

R.-D. Bächle, G. Hehn, G. Pfister;
Institut für Kernenergetik und Energiesysteme
University of Stuttgart, Germany
G. Perlini, W. Matthes; JRC, EURATOM, Ispra, Italy

ABSTRACT

Single material benchmark experiments are designed to check neutron and gamma cross-sections of importance for deep penetration problems. At various penetration depths a large number of activation detectors and spectrometers are placed to measure the radiation field as completely as possible. The large amount of measured data in benchmark experiments can be evaluated best by the global detector concept applied to nuclear data adjustment. But the trial to adjust for the first time a large number of multigroup cross-sections suffered by slow convergence of the method. So the effort had been concentrated to improve the convergence, resulting in a new iteration procedure, which has been implemented now in the modular adjustment code ADJUST-EUR. A theoretical test problem has been deviced to check the total program system with high precision. The method and code are going to be applied for validating the new European Data Files (JEF and EFF) in progress.

---

1. VALIDATION OF NEW DATA COMPILATIONS

For the compilations of evaluated nuclear cross-sections in progress called JEF (Joint European File) and EFF (European Fusion File) the validation work has been initiated. On a recent specialists' meeting of NEA/OECD at Paris a series of benchmark experiments have been selected and compiled for data testing /1/. For the first time an extensive use of single material deep penetration experiments is planned, which offer high sensitivities to differential cross-sections. Compared to the classical procedure of cross-section measurements by single collisions in thin material foils the new integral experiments for data testing are designed as deep penetration problems comprising for example neutron transport through slabs of 1.5 m of iron or 4 m of sodium. In these cases multiple scattering dominates resulting in an increased sensitivity of the measured integral quantities to the cross-sections of the penetrated material.

For the most important materials more than one dozen integral experiments of medium and deep penetration depths are now available. Most of them have been completed or well documented very recently /1/. For validating the reactor file JEF in the epithermal intermediate and fast neutron energy region all experiments with neutron converter plates, Californium-252 sources and well defined reactor spectra are suited. For the fusion file EFF additionally some experiments with 14 MeV neutron sources are available. Of course, with single material deep penetration experiments the neutron cross-sections are checked in a form, which is best suited for radiation shielding. For shielding more experiments would be welcome, which allow the validation of gamma production data, being the most uncertain data presently. The evaluation of deep penetration experiments can be performed in multigroup form using the results of one- and two-dimensional S_N-codes combined with a program for perturbation calculations. A direct check of point cross-sections is possible with linear pertubation methods applying Monte Carlo codes. So both deterministic and stochastic methods are available now and will be applied for the program mentioned. Recent advances and results of the deterministic method are presented here.

An essential part for all calculations is, that the error information as well as the correlations are provided completely. Compared with older methods of data validation essential improvements have been achieved, which can be used with advantage for the validation work planned.

2. IMPROVEMENTS OF THE CODE ADJUST-EUR

In single material deep penetration experiments the neutron field is measured as completely as possible resulting in a large amount of experimental information, which can be evaluated best by the global detector concept applied to nuclear data adjustment /2/. But the trial to adjust for the first time a large number of multigroup cross-sections suffered by slow convergence of the method. So the effort had been concentrated to improve the conver-

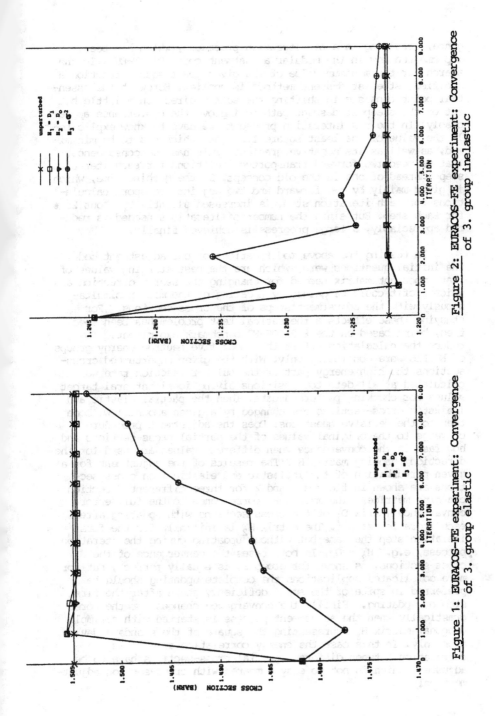

Figure 1: EURACOS-FE experiment: Convergence of 3. group elastic

Figure 2: EURACOS-FE experiment: Convergence of 3. group inelastic

gence, resulting in a new iteration procedure, which has been implemented now in the modular adjustment code ADJUST-EUR. In the search for the minimum value of the given least square function a modified "steepest descent method" is applied. Herewith the essential point consists in shifting the search direction a little bit out of the steepest descent path to improve the convergence appreciably. In the new interation procedure we have to know explicitely the value of the least square function, which has to be minimized, as well as its negative gradient. This has the consequence, that we need two forward transport calculations for each iteration step instead of one in the old concept. So the machine time, which is given mainly by two forward and two adjoint transport calculations for each iteration step, is increased slightly by about 25 % for each step. But since the number of iterations needed is reduced appreciably, a large progress is achieved finally.

In testing the above modifications of our adjustment code, the initial questions were, which are the best starting values of that important matrix needed for changing the search direction and which profit could be gained by updating of the matrix simultaneously with the adjustment steps of the cross-sections. For studying these effects a theoretical test problem has been specified, being based on the EURACOS-FE benchmark experiment. To reduce the calculation effort the upper five neutron energy groups of EURLIB were considered only. With the given unperturbed cross-sections the high energy part of the sulphur reaction rate was calculated at all detector positions giving ideal integral target values for checking perturbed data. Then the partial elastic and inelastic cross-sections are changed by a given amount. We come then to the decisive questions: Does the adjustment procedure converge to the original values of the partial cross-sections and how fast is the convergence when different values are used for the direction changing matrix H ? The results of the adjustment for a given perturbation of the inelastic and elastic iron cross-sections are shown in figures 1 and 2 for three different direction changing matrices. The matrix H_1 corresponds to the full error convariance matrix D_i of the cross-sections with updating after each iteration step i. The matrix H_2 is initially for the first iteration step the same but without updating during the iteration process, e.g. $H_2 = D_0$. In both cases the convergence of the cross-sections, as shown for group 3, is equally perfect, but for more complicated applications the complete updating should be preferred in spite of the small deficiency shown after the first step of updating. Finally the convergence changes for the worse drastically when the adjustment process is started with a simple diagonal matrix H_3 representing the square of the standard deviation only. In this case the energy correlation as well as the correlation between different partial cross-sections have to be adjusted, which is not so easy compared with cross-section adjustment only.

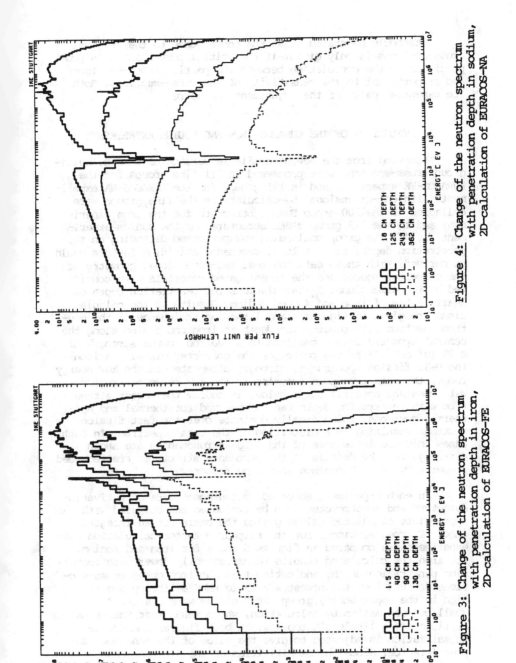

Figure 3: Change of the neutron spectrum with penetration depth in iron, 2D-calculation of EURACOS-FE

Figure 4: Change of the neutron spectrum with penetration depth in sodium, 2D-calculation of EURACOS-NA

845

It should be stated, that the convergence to the original
cross-sections is only given, if the initial perturbation is proportional to the correlation between the partial cross-sections
and proportional to the sensitivity of the cross-sections. Both
are essential parts of the adjustment procedure.

3. EVALUATION OF THE EURACOS IRON AND SODIUM EXPERIMENTS

Starting from the newest available version of ENDF/B, multigroup cross-sections were processed in 271 fine groups for the
EURACOS-FE experiment and in 171 groups for the EURACOS-NA experiment. With one-dimensional S_N-calculations the fine groups were
collapsed to the 100 group EURLIB structure for the iron experiment and to the 113 group BABEL structure for the sodium experiment /3, 4/. The group collapsing was performed depending on the
penetration depth, so that local resonance shielding factors could
be derived. With these data the real two-dimensional geometry of
the converter source and the investigated materials were considered by two-dimensional S_N-calculations for penetration depth of
nearly 1.5 m of iron and 4 m of sodium. Results of the calculations are given in figures 3 and 4 showing the change of the neutron spectrum with penetration depth in iron and sodium along the
central symmetry axis normalized to a neutron source strength of
$8.20 \cdot 10^7$ $cm^{-3}s^{-1}$ at the center of the converter plate. In iron
the U-35 fission spectrum is strongly attenuated in the MeV energy
range, the main transport occuring in the keV resonance region
with constant spectral shape below. In sodium the lowest attenuation with penetration depth can be observed for thermal and epithermal neutrons, which finally dominate over the fast fission
neutrons resulting in a distortion of the total spectrum. In both
cases only the influences of the largest resonances are shown
directly, but the details of the resonance structures are regarded
properly by local resonance shielding factors.

In each experiment about 400 measurement data of activation
detectors and spectrometers can be evaluated and compared with the
appropriate calculated values giving the basis for an integral
check of cross-sections. For the sulphur detector calculation and
experiment are compared in figures 5 and 6 for iron and sodium.
For iron the calculated results become steadily lower with increasing penetration depth, indicating that our iron group cross-sections derived from the newest available ENDF/B-library are too
high in the upper energy groups. For sodium we see a trend to a
small overestimation by calculation, which means that the cross-
sections are a little bit too low. In both figures the source
normalization is adjusted to give the value of the innermost sulphur detector. Since the sensitivities of the detector results to
partial cross-sections are high in deep penetration experiments,
the shown values of relative deviation between calculation and

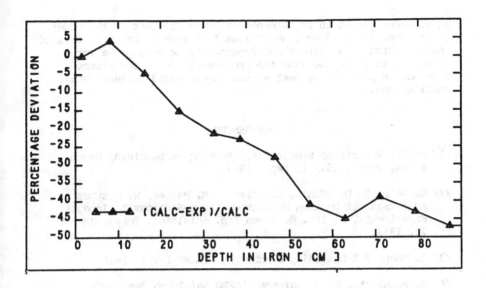

Figure 5: Deviation of sulfur rates in EURACOS-FE

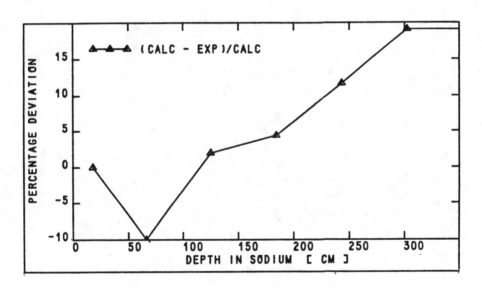

Figure 6: Deviation of sulphur rates in EURACOS-NA

measurement result in adjustments of cross-sections in the order ot one percent and lower, which lies full within the error bars of the evaluated data. With these experiments we have got a very sensitive tool for checking the cross-section and covariance information proposed as best values in evaluated nuclear data compilations.

REFERENCES

/1/ NEACRP Restricted Specialists' Meeting on Shielding Benchmarks, Paris, 10./11. Sept. 1984.

/2/ G. Hehn, R.-D. Bächle, G. Pfister, M. Mattes, W. Matthes, Adjustment of Neutron Multigroup Cross-Sections to Integral Experiments, 6. Int. Conf. on Rad. Shielding, Tokyo, May 16-20, 1983.

/3/ G. Hehn, PWR Shielding Benchmark, NEACRP-L-264, 1983.

/4/ G. Palmiotti, M. Salvatores, LMFBR Shielding Benchmark, NEACRP-L-263, 1983

BENCHMARK EXPERIMENT FOR NEUTRON TRANSPORT IN NICKEL

J. Burian, B. Janský, M. Marek, J. Rataj

Nuclear Research Institute

Řež near Prague, Czechoslovakia

ABSTRACT

A description of an experiment providing the verification of the accuracy of the available neutron cross section for use in the transport calculations of the deep penetration of neutrons is presented. The neutron leakage spectrum from the nickel sphere of diameter 50 cm with ^{252}Cf neutron source in the sphere centre was measured. The set of spectrometers consisted of a stilben scintilator, a hydrogen proportional counter and Bonner balls. The calculation of neutron transport in the nickel sphere was performed with the modified version of ANISN-JR code using the multigroup cross section libraries EURLIB-4 and VITAMIN-C. Comparison of the calculated and experimental results have been utilized to determine the accuracy of the transport calculations using these two different cross section data sets for practical purposes.

INTRODUCTION

The stainless steel constitutes the major portion of the construction materials used in nuclear reactor shielding. Therefore it is essential that accurate results should be obtained to verify transport of neutron not only through the stainless steel but also through

849

J. P. Genthon and H. Röttger (eds.), Reactor Dosimetry, 849–855.
© 1985 ECSC, EEC, EAEC, Brussels and Luxembourg

all important ingredients which form stainless steel.
To test such calculations, benchmark experiments for
nickel were carried out in the NRI, Řež. The comparison
of calculational and experimental solution have been
utilized to determine the accuracy of the obtained
results for transport calculations.

DESCRIPTION OF THE EXPERIMENT

A very precise information can be obtained from
the experiments in the spherical geometry with a neutron
point source in the centre of sphere.

High purity nickel sphere of diameter 50 cm was
investigated. A ^{252}Cf neutron source was used because
of its relatively smooth and well-known spectrum of
neutrons. The source with stainless steel cover of the
cylindrical shape 10 mm high and 4 mm in diameter has
emission about 10^9 neutrons per second. The sphere was
placed to the centre of the measuring room, the room
scattered background correction was made by means of
a shadow cone.

The following methods for neutron spectra measurement were used:
1. a proton recoil spectrometer with a stilben
   scintilator for the energy range 0.5 MeV –
   15 MeV
2. a proton recoil spectrometer with a cylindrical
   proportional hydrogen counter SNM-38 for the
   energy range 20 keV – 700 keV
3. a moderating spheres spectrometer (Bonner balls)
   for the energy range from thermal energy to
   10 MeV.

CALCULATIONS

The calculations in spherical geometry were performed using the modified Japanese version of the discrete
ordinate transport code ANISN-JR /1/. To determine
a suitable order of angular quadrature, an order of the
scattering cross section expansion and the spatial mesh
sizes, some tests were carried out which established
a satisfactory mode of operation. The $S_{12}P_5$ approximation was found quite sufficient.

The ^{252}Cf neutron source was considered as the
holow sphere of diameter 0.8 cm. The neutron spectrum

was approximated by the Maxwell distribution with
T = 1.427 MeV and corrected by a factors recommended in
Ref. /2/.

The two following multigroup cross section data
libraries were used:
1. 100-group library EURLIB-4 /3/
2. 171-group library VITAMIN-C /4/.

RESULTS

Fig. 1 shows the intercomparison of both calculations using the different data libraries. VITAMIN-C
group structure enables to determine the energy structure of neutron spectrum more precisely, especially in
the resonance regions. The greatest difference of results
is in the energy region between 20 keV and 100 keV and
in the energy region between 4 MeV and 7 MeV, where no
resonances are in the calculation using the EURLIB-4
data library.

In Fig. 1 the total cross section of both libraries
are also given to show the correlation between the resonances of the cross section and the energy structure
of neutron flux. The differences in the cross sections
are evident, especially in the energy range round 0.1MeV.

Fig. 2 compares results of measurement and calculation using the data library VITAMIN-C. The two maxima
between 4 MeV and 7 MeV are strongly above the measurement, in the energy region between 160 keV and 300 keV
there is a double underestimation in the calculation.
The shape of the cross section in the energy region
between 40 keV and 100 keV corresponds much better experimental results than those obtained using EURLIB-4
data.

Fig. 3 compares results of measurement and calculations using the data library EURLIB-4. In the energy
region above 4 MeV there is also overestimation in the
calculation but no resonances. In the energy region
between 100 keV and 300 keV there is a double underestimation, too.

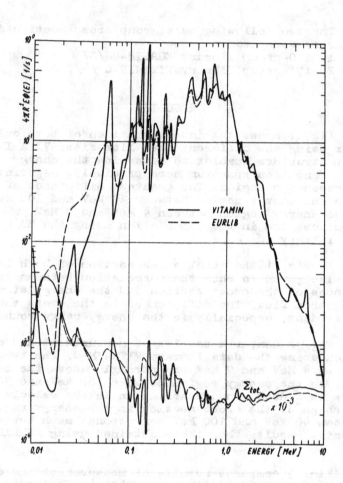

Fig. 1
The comparison of calculated neutron spectra in
pure nickel with data libraries VITAMIN-C and
EURLIB-4.

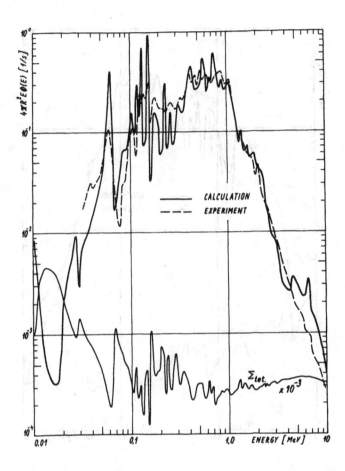

Fig. 2
The comparison of results of measurement and
calculation using the data library VITAMIN-C
for pure nickel.

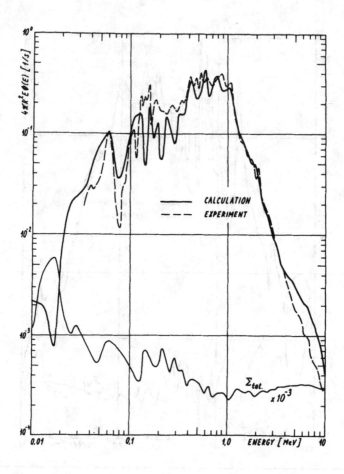

Fig. 3
The comparison of results of measurement and
calculation using the data library EURLIB-4
for pure nickel.

REFERENCES

1. J. Rataj, <u>The modifications of program ANISN</u>,
   Report ÚJV-6137, 1982 (in Czech)

2. Starostov at al., <u>Voprosy atomnoj nauki i techniki
   - jadernyje konstanty</u>, Vol. 2/37, Obninsk 1980

3. EURLIB-4 - The 120 Group Coupled Neutron and Gamma
   Data Library, NEA Data Bank, 1978

4. VITAMIN-C, The CTR Processed Multigroup Cross
   Sections Library for Neutronic Studies, ORNL/RSIC-37,
   July 1980

CONSISTENCY OF CROSS-SECTION DATA IN INTEGRAL EXPERIMENTS

A. Cesana, V. Sangiust and M. Terrani

Politecnico di Milano, Istituto di Ingegneria Nucleare,
Centro Studi Nucleari E. Fermi (CESNEF), Milano, Italy

and

G. Sandrelli

ENEL, Ente Nazionale per l'Energia Elettrica, Centro di
Ricerca Termica e Nucleare (CRTN), Milano, Italy

ABSTRACT

A neutron spectrum adjustment with the multiple
foil activation technique has been performed on
two fast neutron facilities.
The results are analyzed and discussed with the
aim of deriving some conclusions on the reliabi
lity of the employed cross sections

INTRODUCTION

In the present paper we report and discuss the results of neutron spectrum adjustment performed with the
multiple foil activation technique in two irradiation fa
cilities. The first one was a B_4C filter (1) placed near
the core of the L54 reactor, Politecnico di Milano, and
the other was the TAPIRO reactor (2), ENEA-Casaccia (Roma). In both cases the neutron flux shape is of fast
reactor type. The well known adjustment codes SAND II(3)
and STAY'SL (4) were used, both in a 40-group structure,
which has been shown (5) adequate for spectra of this
kind.

An analysis of the results is given which tries to
derive some conclusions on possible systematic errors of

J. P. Genthon and H. Röttger (eds.), Reactor Dosimetry, 857–866.
© *1985 ECSC, EEC, EAEC, Brussels and Luxembourg*

the experimental reaction rates per atom and on the cross-section reliability of the employed reactions.

MEASUREMENT AND INPUT DATA

A list of the reactions considered is given in Table I, while Figs 1a and 1b show the input spectra for the two experiments. Both these spectra were calculated by DOT 3.5 code (6) in 40 groups.

High-resolution γ-ray counting and the nuclear data reported in ref 7 were used to determine the absolute reaction rates per atom. The results are also shown in Table I.

The cross sections for the adjustment procedure were obtained from the ENDF/B-IV and ENDF/B-V files, collapsing the 620-group data into the 40-group structure and using the input spectra as weight functions.

ADJUSTMENT OF THE SPECTRA

The spectrum adjustment was performed with STAY'SL and SAND II codes, written for a PDP 11/23 computer, both in a 40-group version.

As well known, STAY'SL requires information on the uncertainties of the input flux, of the cross sections and of the reaction rates per atom. Only the last point was analyzed in detail, considering the different errors and their correlations (8). As for the input flux uncertainties, an artificial matrix of gaussian shape was used with errors on the diagonal of the order of 20-40%, depending on the energy zone. The cross-section uncertainties were again given in a somewhat artificial way, assuming no correlation among different reactions and a diagonal matrix with errors derived from ref 9 for each reaction. The use of SAND II code is much simpler and the results reported were usually obtained after 5-6 iterations, with a final mean deviation of 5-6% of experimental from calculated reaction rates per atom.

As an example, we report in Figs 2a and 2b the ratios of output to input fluxes, obtained with SAND II code for both experiments. All 25 reactions were employed in the B_4C spectrum, while in the TAPIRO spectrum ^{47}Ti (n,p), ^{103}Rh(n,n') and ^{63}Cu(n,γ) could not be used. The same adjustment performed with STAY'SL code showed a similar trend, but for a smoother shape clearly imposed by the intrinsic characteristics of the code.

In the following the use of SAND II code was preferred, owing to its simplicity and quick response, considering also that a rational use of STAY'SL requires a lot of information on input fluxes and cross sections not yet available to us.

ANALYSIS OF THE RESULTS

An attempt is made here to derive some conclusions on the role of the different reactions, considering the influence of the uncertainties on the experimental reaction rates per atom and of the adopted cross sections, obviously in the energy region where a significant response is present.

The analysis of the trend of well known parameters, such as the fractional reaction rate of a given detector in the different energy groups and the fractional response of the different detectors in a given energy group, was of a great utility in this procedure. Both these parameters are easily obtained from the cross-section set and from the adopted input fluxes: some examples of them are shown in Figs 3a, 3b and 4a, 4b, for the B_4C spectrum.

Many adjustment procedures were performed for both experiments, eliminating one or more reactions and trying to clarify their influence on the shape of the adjusted flux.

The intercalibration procedure was also adopted for some reactions: the experimental reaction rates of one or more reactions in the TAPIRO experiment were derived using as a standard the B_4C spectrum, adjusted neglecting the same reactions. This procedure is here justified since the two spectra are not very different in the energy shape.

For some reactions the adjustment procedure showed questionable results; the following situations may happen:

a) The calculated reaction rate per atom after some iterations fits very well to the experimental one, but strong local distorsions are produced in the adjusted flux. This behaviour is typical of a reaction whose fractional response is concentrated in one or few energy groups.
b) The calculated reaction rate per atom cannot fit satisfactorily to the experimental one, in spite of possible distortions of the adjusted spectrum.
c) Independently of the quality of the final results of the adjustment, when a reaction is neglected, its reaction rate per atom, calculated in the adjusted spec –

trum, is remarkably different from the experimental one?

These effects may or may not disappear with a particular choice of the cross sections and/or with the adoption of the intercalibration.

In some cases it was not possible to decide if the anomalous behaviour of a reaction was due to the cross section or to nuclear decay parameters influencing the experimental reaction rate.

CONCLUSIONS

For 16 reactions a good agreement (within 5%) was achieved in both spectra between calculated and experimental reaction rates per atom, and none of the effects listed above occurred. ENDF/B-IV and ENDF/B-V files, in spite of some differences, turned out to be equivalent for the adjustment procedure, at least within the sensitivity limits of the experiments.

For the reaction ^{47}Ti(n,p) a good agreement (<5%) is found using only ENDF/B-V file, while the opposite happens for ^{58}Ni(n,p).

In some cases, ^{27}Al(n,p), ^{115}In(n,n'), ^{63}Cu(n,γ) and ^{103}Rh(n,n'), a bad consistency (10 ÷ 15%) was found with both libraries. Because of uncertainties on the experimental reaction rates, no definite conclusions can be drawn about the reliability of the cross-section data.

However, the intercalibration procedure, possible only for the first two reactions, remarkably improved the agreement.

Finally for three reactions, ^{23}Na(n,γ), ^{58}Fe(n,γ) and ^{197}Au(n,γ), a very bad consistency (up to 50%) was found in both spectra, adopting the ENDF/B-IV file. For the first reaction the discrepancy completely disappeared adopting the ENDF/B-V file or the intercalibration procedure.
For ^{58}Fe(n,γ) the discrepancy was cancelled by the inter calibration procedure, but we could not yet test the ENDF/B-V file.
For the reaction ^{197}Au(n,γ) no significant improvement was achieved with ENDF/B-V file nor by the intercalibration procedure. Since it is difficult to question the nuclear decay data or the experimental determination of the activity, we conclude that, obviously in the energy range of significant response (from $\sim$ 0.3 keV up to $\sim$ 0.7 MeV), the cross section for this reaction cannot be considered reliable.

REFERENCES

1. P.Barbucci, A.Cesana, G.Sandrelli, V.Sangiust, M.Terrani and S.Terrani, " Generation of a Neutron Reference Field by B_4C filtering of a Reactor Spectrum", En.Nucl. 26, 542 (1979)

2. A.D'Angelo, M.Martini, M.Salvatores, " The TAPIRO Fast Source Reactor as a Benchmark for Nuclear Data Testing", En.Nucl. 20, 614 (1973)

3. W.N.Mc Elroy, S.Berg and G.Gigas, "Neutron-Flux Spectral Determination by Foil Activation", Nucl.Sci.Eng. 27, 533 (1967)

4. F.G.Perey, "Least-Squares Dosimetry Unfolding: The Program STAY'SL", ORNL/TM-6062, ENDF-254, Oak Ridge National Laboratory (1977)

5. A.Cesana, V.Sangiust, M.Terrani and G.Sandrelli,"The Group Structure of Cross-Section Libraries in Neutron Activation Spectrometry", Nucl.Sci.Eng.82, 102 (1982)

6. DOT-3.5, Two Dimensional Discrete Ordinates Radiation Transport Code, available from the Reactor Shielding Information Center, Oak-Ridge National Laboratory (1976)

7. W.L.Zijp, J.H.Baard, Nuclear Data Guide for Reactor Neutron Metrology, Report EUR-7164 EN, ed 1979 (1981)

8. M.Petilli, How to Evaluate the Variance-Covariance Matrix of an Integral Experiment , Italian Report CNEN-RT/FI(80) 18 (1980)

9. N.J.C.M. van der Borg, H.J.Nolthenius and W.L.Zijp, Covariances of the Data of the ENDF/B-V Dosimetry File, Nederland Report ECN-80-091 (1980)

TABLE I

List of employed reactions and experimental reactor rates per atom

N	Reaction	Experimental B$_4$C	TAPIRO	Cross Section sources
1	NA23G	2.83E-16	3.24E-16	IV,V
2	MG24P	4.07E-17	4.64E-17	IV,V
3	AL27A	1.89E-17	2.18E-17	IV,V
4	AL27P	1.01E-16	1.16E-16	IV,V
5	TI47P	6.31E-16	----	IV,V
6	TI48P	7.74E-18	9.21E-18	IV,V
7	FE54P	2.29E-15	2.83E-15	IV.V
8	FE58G	1.01E-15	1.58E-15	IV
9	MN55G	5.00E-15	6.43E-15	IV,V
10	NI58P	3.25E-15	3.71E-15	IV.V
11	CU63G	8.20E-15	----	IV,V
12	RH103N	3.00E-14	----	IV,V
13	IN115N	7.53E-15	1.08E-14	V
14	IN115G	4.11E-14	8.30E-14	IV.V
15	AU197G	6.23E-14	1.73E-13	IV,V
16	U233F	3.16E-13	6.04E-13	IV
17	U235F	2.11E-13	3.98E-13	IV,V
18	NP237F	6.75E-14	1.16E-13	IV,V
19	PU238F	1.67E-13	3.17E-13	IV
20	PU240F	7.80E-14	1.35E-13	IV
21	PU241F	2.80E-13	5.20E-13	IV
22	AM241F	6.87E-14	1.08E-13	IV,V
23	U238F	1.02E-14	1.35E-14	IV,V
24	TH232F	2.56E-15	3.26E-15	IV,V
25	PU239F	2.23E-13	4.40E-13	IV,V

SAND40: INPUT SPECTRUM

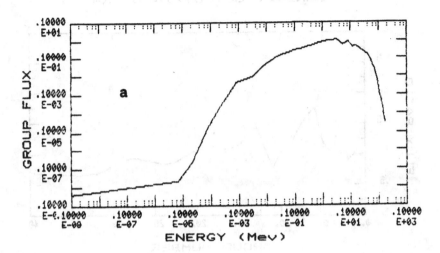

SAND40: INPUT SPECTRUM

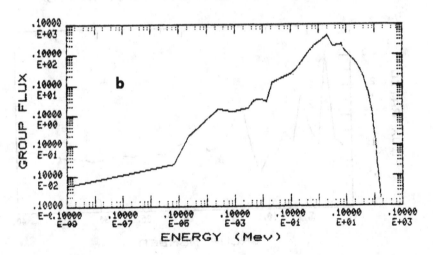

Fig.1   Input spectra: (a) B_4C, (b) TAPIRO

SAND40: OUTPUT/INPUT RATIO

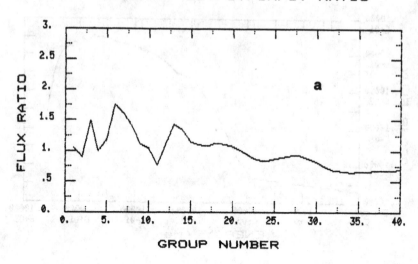

SAND40: OUTPUT/INPUT RATIO

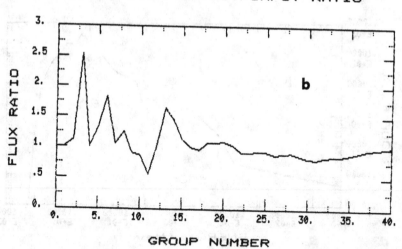

Fig.2   Output/input flux ratios: (a) B_4C, (b) TAPIRO

864

NA-23G5

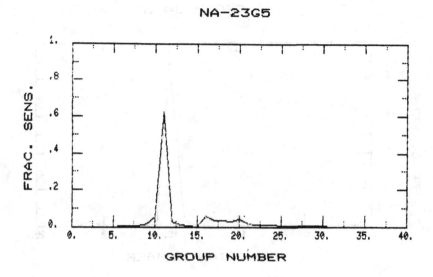

AU-197G5

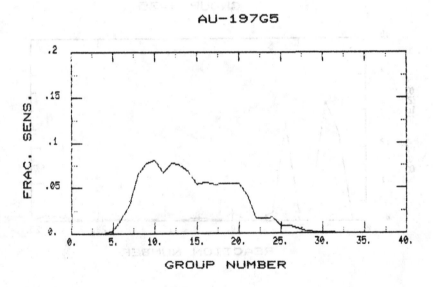

Fig.3   Two fractional reaction rates in the B_4C spectrum

GROUP J= 3

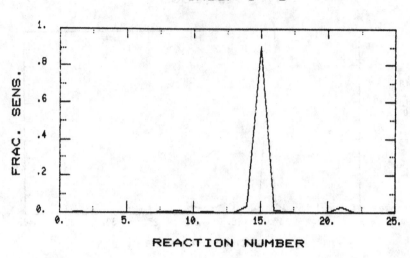

GROUP J=35

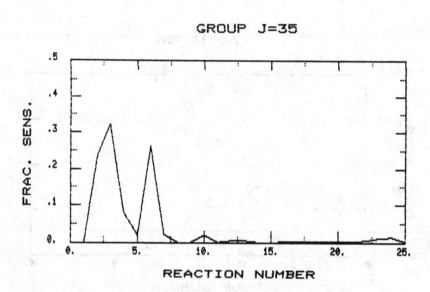

Fig.4   Two fractional group responses in the B_4C spectrum

CROSS SECTION MEASUREMENTS IN THE ^{235}U FISSION SPECTRUM
NEUTRON FIELD

D. M. Gilliam, J. A. Grundl, G. P. Lamaze, and E. D. McGarry

National Bureau of Standards, Gaithersburg, MD 20899 USA

A. Fabry

Centre D'Etude de L'Energie Nucleaire/Studiecentrum Voor Kernergie

Mol, Belgium

ABSTRACT

In the Cavity Fission Spectrum Standard Neutron Field
of the BR-1 Reactor at the Belgian CEN/SCK Laboratory, a
series of cross section measurements was made in which the
neutron fluence rate was determined by a transfer measurement from the known fluence rate in the vicinity of a
calibrated ^{252}Cf neutron source at the U.S. National
Bureau of Standards. The absolute calibration of the
^{252}Cf neutron source was established by means of a manganese bath intercomparison with the standard neutron source
NBS-I. Fission cross sections are reported for ^{238}U,
^{235}U, ^{233}U, ^{239}Pu, ^{240}Pu, ^{237}Np, and ^{232}Th. Also included
in the measurements were the ^{115}In(n,n') cross section and
the ^{58}Ni(n,p) cross section.

INTRODUCTION

The Cavity Fission Spectrum Standard Neutron Field at the
Belgium CEN/SCK BR-1 reactor is a unique facility that has been
made available for international cooperative efforts in improvement
of neutron metrology. The large size of the thermal column of BR-1
permits a 100-cm diameter cavity without sacrifice in adequacy of
thermalization; and the large diameter of the spherical cavity

J. P. Genthon and H. Röttger (eds.), Reactor Dosimetry, 867–875.
© *1985 ECSC, EEC, EAEC, Brussels and Luxembourg*

minimizes wall return corrections while permitting a large working volume for active instruments within the central fission neutron field.

The purposes for the present work were primarily (1) to provide fission-spectrum-averaged cross sections for use in dosimetry standardization for reactor pressure vessel surveillance programs, and (2) to provide integral tests for evaluated cross section tabulations such as ENDF/B-V.

Measurements were made in three configurations of of the BR-1 Cavity Fission Spectrum Field. These configurations differed in the thickness and diameter of the ^{235}U cylindrical source foil, so that corrections for cavity wall return, neutron scattering, and neutron absorption could be tested for consistency under a variety of conditions. Fission rates were observed by means of NBS fission ionization chambers with deposits from the NBS collection of fissionable isotope mass standards. Radiometric analysis of activated foils was performed by both the NBS and CEN/SCK Laboratories.

FACILITY AND INSTRUMENTATION DESCRIPTION

The Mol Cavity Fission Spectrum Standard Neutron Field facility has been previously described in detail (1). The arrangements employed in the present work are shown in Figure 1. In the configuration called Mark II, there was no aluminum cladding around the ^{235}U foil. In Mark IIA and Mark III, the aluminum clad thickness was 0.070 cm inside the uranium, and 0.075 cm outside.

All measurements, including both active fission rate measurements and the irradiation of ^{58}Ni and ^{115}In activation detectors, were made with deposits or foils mounted in an NBS fission chamber. The absolute calibrations of the NBS fission chamber and the fissionable isotope mass standards have been described in previous papers (2,3).

EXPERIMENTAL METHOD AND CORRECTIONS

The fluence rate in the Mol ^{235}U fission neutron field was established by means of the ^{239}Pu fission rate at the Mol facility relative to that in the NBS ^{252}Cf Spontaneous Fission Standard Neutron Field. Only a very minor correction factor (0.997) was required to account for the difference in the spectrum-averaged ^{239}Pu(n,f) cross section in the respective fission neutron spectra. (This transfer method could have been based equally well on the ^{235}U fission rate, without changing any of the present results by more than 1.3%.) The absolute fluence rate at the NBS ^{252}Cf facility is

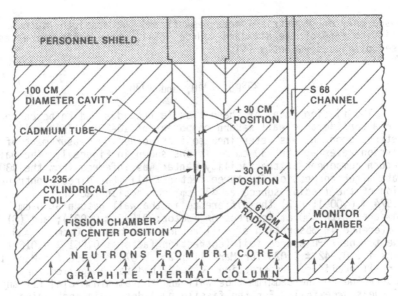

FIGURE 1A. BR1 CAVITY FISSION SPECTRUM STANDARD NEUTRON FIELD FACILITY

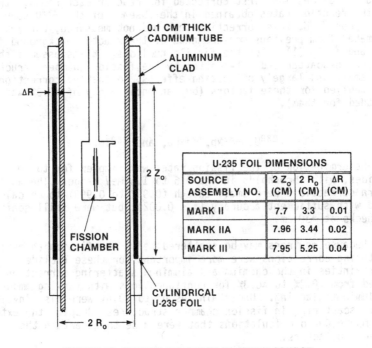

U-235 FOIL DIMENSIONS			
SOURCE ASSEMBLY NO.	$2 Z_o$ (CM)	$2 R_o$ (CM)	ΔR (CM)
MARK II	7.7	3.3	0.01
MARK IIA	7.96	3.44	0.02
MARK III	7.95	5.25	0.04

FIGURE 1B. FISSION SPECTRUM NEUTRON SOURCE AND DETECTOR

based only on geometrical measurements and the absolute neutron emission rate of the standard source NBS-I, whose calibration has been discussed recently (4).

^{239}Pu, ^{235}U, and ^{233}U

The fission rates of these fissile nuclides had to be corrected for a considerable wall return component of 10-20%. This wall return component was determined to first order by measurement of the fission rates at the ±30 cm positions shown in Figure 1. The wall return component at the facility center was inferred from the ±30 cm data after correction for three factors: (1) the direct contribution of the fission source at ±30 cm (correction factor: 0.974 ± 0.005); (2) the difference in the wall return at 0 cm compared with ±30 cm (correction factor: 0.995 ± 0.001); and (3) the difference in absorption of wall return neutrons at 0 cm or ±30 cm due to the ^{235}U source foil (correction factors: 0.961 ± 0.016 for Mark II; 0.942 ± 0.023 for Mark IIA; and 0.875 ± 0.050 for Mark III). For all nuclides, a correction ($\leq$ 4.5%) had to be applied for scattering in the cadmium tube wall and the aluminum clad (if any) of the source foil. For the fissile nuclides, this scattering correction introduced negligible uncertainty. Reaction rate data for all nuclides was first corrected for reactor background; that is, the reaction rates obtained in the absence of the ^{235}U source foil. For ^{233}U, this correction term was not measured, but was estimated from previous measurements of the reactor background for ^{235}U and ^{239}Pu (1). For the fissile nuclides, it was assumed that neutron in-scatter and out-scatter by the fission chamber structures were small and largely offsetting effects, so that no corrections were applied for these factors (but an uncertainty of ±0.3% was included for them).

^{238}U, ^{237}Np, ^{240}Pu, and ^{232}Th

Determination of the fission rates was simpler for these nuclides, with thresholds in the 0.6 to 1.5 MeV range. The wall return corrections were small enough (0.992 to 0.999) to be calculated with sufficient accuracy (± < 0.002), but were still confirmed by checks at ±30 cm.

Because neutrons may be scattered below the fission threshold, scattering corrections were more important for these nuclides. Uncertainties in the cadmium and aluminum scattering corrections ranged from ±0.2% to ±0.8% for configurations with varying amounts of aluminum cladding. Uncertainties up to ±0.4% were also involved due to scattering in fission chamber structures, despite the extensive Monte Carlo calculations that were made to ascertain the correction factors.

$^{58}\text{Ni}(n,p)$ and $^{115}\text{In}(n,n')$

These activation reactions have thresholds at about 2.1 and 1.2 MeV respectively, and therefore are similar to the higher threshold fission reactions in regard to wall return and reactor background perturbations. The reactor background for $^{115}\text{In}(n,n')$ was inferred from the ^{238}U background, in proportion to the respective (γ,γ') and (γ,f) cross sections. Both reactor background and wall return were estimated to be negligible for the $^{58}\text{Ni}(n,p)$ reaction. The combined scattering uncertainties were larger for these isotopes (± 0.014 for ^{58}Ni, and ± 0.009 for ^{115}In).

RESULTS AND CONCLUSIONS

Table 1 gives the experimental results in three forms: as reaction rates per nucleus in the first column, as cross sections relative to $^{239}\text{Pu}(n,f)$ in the second column, and as cross sections by the fluence rate transfer in the third column. If other cross section ratios are of interest to the reader, they may be obtained by taking the ratios from the respective reaction rates in the first column. Table 1 shows the results for three configurations of the Mol facility separately, while Table 2 combines these in a weighted average and gives comparisons with ENDF/B-V calculated values. Table 3 expresses the present results as ratios to cross sections previously measured in the ^{252}Cf neutron field (3) and compares these ratios with calculated values from ENDF/B-V.

A few particularly important discrepancies between the experimental results and calculated values deserve mention. The ^{115}In and ^{58}Ni cross sections which are very important to reactor dosimetry standardization seem to be in error in the ENDF/B-V tabulations by about $(5.9 \pm 2.1)\%$ and $(6.2 \pm 2.4)\%$ respectively, as indicated in the columns for the ENDF/B-V Watt or Madland-Nix (5) spectra in Table 2. These discrepancies show the importance of using standard neutron fields as benchmarks for dosimetry measurements, when accuracies of $\leq 5\%$ are required. (Note that Table 3 shows these discrepancies consistently in both ^{252}Cf and ^{235}U field measurements.) Another significant discrepancy is seen in the fission cross section ratio for $^{235}\text{U}(n,f)$ relative to $^{239}\text{Pu}(n,f)$ in Table 2. The experimental value of 0.660 has an estimated uncertainty of only $\pm 1.0\%$, while the value of this ratio obtained from calculation with ENDF/B-V data differs from the experimental value by about 4.5%. Since the $^{235}\text{U}(n,f)$ cross section is considered a standard, this 4.5% discrepancy should be examined for its implications regarding the evaluation of $^{235}\text{U}(n,f)$ data. Finally it may be noted from Table 3 that for the highest threshold reaction, $^{58}\text{Ni}(n,p)$, the ENDF/B-V Watt and Madland-Nix ^{235}U spectra seem to calculate the change in response between the ^{252}Cf and ^{235}U spectra more accurately than does the older NBS evaluation of the ^{235}U spectrum (6).

Table 1. Fully Corrected Experimental Results from Various
Experimental Configurations

Isotope and Reaction	$\left[\dfrac{\text{Nuclear Reactions}}{(\text{nuclei})(\text{monitor})}\right]$ (10^{-21})	$\left[\dfrac{\text{Cross Section}}{\sigma_f(^{239}\text{Pu})}\right]$	Cross Section by Flux Transfer (mb)
MARK IIA CENTER*			
^{239}Pu (n,f)	21.65 + 0.9%	1.0	(1818 ± 1.9%)[†]
^{235}U (n,f)	14.28 ± 1.0%	0.660 ± 1.1%	1199 ± 1.9%
^{238}U (n,f)	3.727 ± 1.8%	0.172 + 1.8%	313 ± 2.4%
^{237}Np (n,f)	16.26 ± 1.3%	0.751 + 1.4%	1366 ± 2.1%
^{240}Pu (n,f)	15.86 ± 1.3%	0.733 ± 1.4%	1332 ± 2.1%
^{233}U (n,f)	23.21 ± 2.5%	1.072 ± 2.6%	1949 ± 3.1%
^{232}Th (n,f)	0.988 ± 2.6%	0.0456 ± 2.6%	83 ± 3.1%
^{115}In (n,n')	2.258 ± 1.2%	0.1043 ± 1.4%	190 ± 2.1%
^{58}Ni (n,p)	1.325 ± 1.7%	0.0612 ± 1.8%	111 ± 2.4%
MARK III Center*			
^{239}Pu (n,f)	24.57 ± 1.4%	1.0	(1818 ± 1.9%)[†]
^{235}U (n,f)	16.04 ± 1.7%	0.653 ± 2.0%	1190 ± 2.6%
^{238}U (n,f)	4.201 ± 1.8%	0.171 ± 2.1%	312 ± 2.6%
^{237}Np (n,f)	18.26 ± 1.3%	0.743 ± 1.8%	1355 ± 2.4%
MARK II Center*			
^{239}Pu (n,f)	15.64 ± 0.8%	1.0	(1818 ± 1.9%)[†]
^{235}U (n,f)	10.39 ± 1.0%	0.664 ± 1.0%	1211 ± 1.9%
^{238}U (n,f)	2.675 ± 1.6%	0.171 ± 1.6%	312 ± 2.3%
^{237}Np (n,f)	11.67 ± 1.2%	0.746 ± 1.4%	1360 ± 2.1%

*These results apply to a 0.5-inch (1.27 cm) diameter foil
positioned as shown in Figure 1. These results have not been
corrected to correspond to a point detector at the facility
center.

[†]This value is taken from NBS measurements in the ^{252}Cf spectrum
and a calculated ratio, based on ENDF/B-V cross section data and
the NBS evaluation of the ^{252}Cf and ^{235}U neutron spectra.

Table 2. Combined Results and Comparisons with Values Calculated from ENDF/B-V Cross Section Data and Various Evaluations of the ^{235}U Fission Neutron Spectrum.

Isotope and Reaction	⎡Experimental Cross Section (10^{-27} cm^2)⎤ and ⎡Ratio Relative to $\sigma_f(^{239}$Pu)⎤	Ratio of Calculated Cross Section or Cross Section Ratio to the Corresponding Experimental Value		
		ENDF/B-V Watt Spectrum	Madland-Nix "Exact" Spectrum	NBS Evaluated ^{235}U Fission Spectrum
^{239}Pu (n,f)	(1818 ± 1.9%)	0.985	0.986	0.982
	1.0	1.	1.	1.
^{235}U (n,f)	1200 ± 1.9%	1.030	1.029	1.030
	0.660 ± 1.0%	1.046	1.044	1.049
^{238}U (n,f)	312 ± 2.3%	0.978	0.984	0.945
	0.1714± 1.7%	0.994	0.999	0.963
^{237}Np (n,f)	1359 ± 2.1%	0.991	0.997	0.973
	0.747 ± 1.4%	1.007	1.012	0.991
^{240}Pu (n,f)	1332 ± 2.1%	1.015	1.021	0.995
	0.733 ± 1.4%	1.030	1.035	1.013
^{233}U (n,f)	1949 ± 3.1%	0.978	0.978	0.981
	1.072 ± 2.6%	0.993	0.992	0.999
^{232}Th (n,f)	83 ± 3.1%	0.904	0.907	0.872
	0.0456± 2.6%	0.918	0.921	0.889
^{115}In (n,n')	190 ± 2.1%	0.942	0.947	0.913
	0.1043± 1.4%	0.958	0.963	0.931
^{58}Ni (n,p)	111 ± 2.4%	0.946	0.937	0.910
	0.0612± 1.8%	0.958	0.948	0.924

Table 3. Ratio of Measured Reaction Cross Sections in the ^{235}U and ^{252}Cf Fission Neutron Fields and Comparisons with Ratios Calculated from ENDF/B-V Cross Sections and Various ^{235}U Spectra.

Isotope and Reaction	$\left[\dfrac{\sigma \text{ in } ^{235}\text{U}}{\sigma \text{ in } ^{252}\text{Cf}}\right]$ Experimental Result	Calculated Ratio[†]/Experimental Result		
		ENDF/B-V Watt for ^{235}U	Madland-Nix "Exact" for ^{235}U	NBS Evaluation for ^{235}U
^{239}Pu (n,f)	(0.997)	1.002	1.004	1.[*]
^{235}U (n,f)	0.987 ± 1.9%	1.013	1.012	1.013
^{238}U (n,f)	0.957 ± 2.2%	1.017	1.023	0.982
^{237}Np (n,f)	0.995 ± 2.1%	1.001	1.007	0.983
^{240}Pu (n,f)	0.996 ± 2.7%	1.001	1.007	0.982
^{233}U (n,f)	1.030 ± 3.1%	0.972	0.972	0.975
^{232}Th (n,f)	0.928 ± 2.7%	1.035	1.039	0.999
^{115}In (n,n')	0.969 ± 2.1%	1.016	1.021	0.984
^{58}Ni (n,p)	0.917 ± 2.7%	1.006	0.997	0.968

[*]Exactly 1. due to flux transfer normalization.

[†]The NBS evaluation of the ^{252}Cf fission neutron spectrum is used for all the calculated values.

REFERENCES

1. A. Fabry, G. Minsart, F. Cops, and S. De Leeuw, "The Mol Cavity Fission Spectrum Standard Neutron Field and its Applications," Proceedings of the Fourth ASTM-EURATOM Symposium on Reactor Dosimetry, Washington, D.C. (March, 1982).

2. J.A. Grundl, D.M. Gilliam, N.D. Dudey, and R.J. Popek, "Measurement of Absolute Fission Rates," Nuclear Technology, 25, 237 (1975).

3. J.A. Grundl and D.M. Gilliam, "Fission Cross Section Measurements in Reactor Physics and Dosimetry Benchmarks," Transactions of the American Nuclear Society, 44, 533 (1983).

4. G.P. Lamaze, D.M. Gilliam, E.D. McGarry, and A. Fabry, "Neutron Fluence and Cross Section Measurements for Fast Neutron Dosimetry," Proceedings of the Fifth International Conference on Nuclear Methods in Environmental and Energy Research, Mayaguez, Puerto Rico, April 1984. To be published by the U.S. Department of Energy.

5. R.J. La Bauve and D.G. Madland, "Comparison of Measured and Calculated Integral Cross Sections for the Thermal Fission of ^{235}U," Transactions of the American Nuclear Society, 42 (June 1982). Tabulation by private communication.

6. J.A. Grundl and C.M. Eisenhauer, "Fission Spectrum Neutrons for Cross Section Validation and Neutron Flux Transfer," Proceedings of a Conference on Nuclear Cross Sections and Technology, NBS Special Publication 425, U.S. Department of Commerce, Washington, D.C. (March 1975).

SPECTRUM-INTEGRATED HELIUM GENERATION CROSS SECTIONS FOR ^{6}Li AND^{10}B IN THE INTERMEDIATE-ENERGY STANDARD NEUTRON FIELD

B. M. Oliver and Harry Farrar IV
Rockwell International Corporation
Canoga Park, California 91304, USA

and

D. M. Gilliam
National Bureau of Standards
Gaithersburg, Maryland 20234, USA

and

E. P. Lippincott
Westinghouse Hanford Company
Richland, Washington 99352, USA

ABSTRACT

The spectrum-integrated helium generation cross sections for ^{6}Li and ^{10}B have been determined for the Intermediate-Energy Standard Neutron Field (ISNF) at the U.S. National Bureau of Standards. Helium concentrations were measured by precise high-sensitivity gas-mass spectrometric analysis of vaporized small encapsulated and unencapsulated crystalline samples of natural boron, enriched ^{10}B, and enriched ^{6}LiF. The cross section results are consistent with previously obtained data for BIG-10, CFRMF, Sigma Sigma, and the Fission Cavity. A comparison with cross sections calculated using the ENDF/B-V file and a calculated neutron spectrum continue to indicate a need for revision to the ^{10}B, and perhaps ^{6}Li, cross sections above ~0.1 MeV.

J. P. Genthon and H. Röttger (eds.), Reactor Dosimetry, 877–885.
© 1985 ECSC, EEC, EAEC, Brussels and Luxembourg

INTRODUCTION

Previous measurements of the ^{6}Li(n,He) and ^{10}B(n,He) reactions in benchmark neutron fields have indicated consistent discrepancies when compared with calculated values (1,2). These earlier measurements were conducted in the Coupled Fast Reactivity Measurements Facility (CFRMF) at the Idaho National Engineering Laboratory (3), in the 10% enriched U-235 critical assembly (BIG-10) at the Los Alamos National Laboratory (4), and in the Sigma Sigma and Fission Cavity fields of the BR1 reactor at the CEN/SCK laboratories in Mol, Belgium (5). Extensive analysis of the BIG-10 and CFRMF data (6) indicated that the discrepancies noted for the ^{6}Li(n,He) reaction (C/E ~0.93) are a factor of 2 to 3 larger than the integral test uncertainties, whereas the discrepancies noted for the ^{10}B(n,He) reaction (C/E ~0.86) are a factor of ~4 larger than uncertainties. The conclusions reached from the analyses in Reference 6 were that revisions were required to the ^{6}Li cross section in the region of the 250 keV resonance, and to the ^{10}B cross section above ~10 keV.

In order to provide additional benchmark data for these two reaction cross sections, both of which are considered to be standards, a further irradiation of ^{6}LiF and ^{10}B helium accumulation fluence monitors (HAFMs) was conducted in ISNF in September 1982.

EXPERIMENTAL DETAILS

ISNF Facility. The ISNF facility at NBS is similar in concept to the Sigma Sigma facility at CEN/SCK (5). The facility operates in a 30-cm diameter spherical cavity in the graphite thermal column of the research reactor at NBS. The ISNF, which is driven by eight enriched ^{235}U disks 1 cm inside the cavity surface, has a fast-reactor-like neutron energy spectrum. This spectrum is regarded as a standard because it is accurately calculable by one-dimensional transport theory codes using only the ^{235}U fission spectrum and a small set of well-established cross section data (7,8). The working volume of the facility is a spherical 5 cm diameter inner region inside a spherical shell of ^{10}B (~95% ^{10}B), as shown in Figure 1. This shell, which is held in position in the center of the cavity by a lightweight aluminum structure, screens out the thermal and epithermal components of a graphite-moderated fission neutron spectrum.

HAFM Samples. Twenty-seven bare and encapsulated samples of natural boron, enriched ^{10}B, and enriched ^{6}LiF were irradiated for 45.8 hours in ISNF. The encapsulated ^{6}LiF and ^{10}B HAFMs were taken from the same stockpile originally fabricated earlier for Sigma Sigma (2). The natural boron and empty HAFM capsules

878

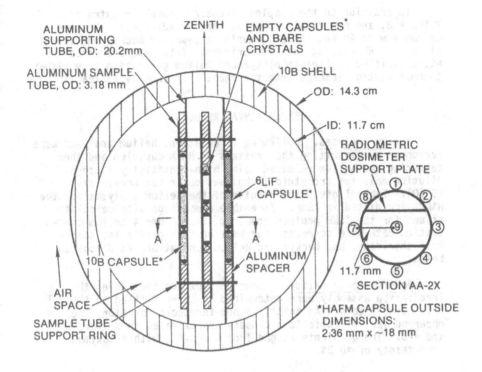

ALUMINUM SUPPORTING TUBE, OD: 20.2mm

ALUMINUM SAMPLE TUBE, OD: 3.18 mm

ZENITH

EMPTY CAPSULES* AND BARE CRYSTALS

10B SHELL

OD: 14.3 cm

ID: 11.7 cm

RADIOMETRIC DOSIMETER SUPPORT PLATE

6LiF CAPSULE*

A   A

10B CAPSULE*

ALUMINUM SPACER

AIR SPACE

SAMPLE TUBE SUPPORT RING

11.7 mm

SECTION AA-2X

*HAFM CAPSULE OUTSIDE DIMENSIONS: 2.36 mm x ~18 mm

FIGURE 1. ARRANGEMENT OF HAFMs IN ISNF

were newly fabricated from the same material lots as used for Sigma Sigma. The HAFM capsules were manufactured from thin-wall (0.076-mm), 2.36-mm-OD, Type 304 stainless steel tubing. The finished length after electron beam welding of each end was ~17.8 mm.

Loading Configuration. The loading arrangement of bare crystals and encapsulated HAFMs in ISNF is also shown in Figure 1. Nine thin-walled aluminum tubes holding the individual encapsulated and bare crystal HAFM samples were located in the central region of the ISNF irradiation volume. The assembly was designed to maximize the symmetry of the neutron field and to minimize the interaction between the HAFMs. The boron and LiF HAFM capsules were located symmetrically about midplane in Pins 1, 3, 5, 7, and 9. The empty capsules and bare crystals of ^{10}B and ^{6}LiF were located in the remaining four pins. Six pairs of Au and In radiometric foils for flux gradient and flux normalization measurements, respectively, were attached to the dosimeter support plates shown in Figure 1.

In addition to the samples discussed above, an extra pair of ^{6}LiF, ^{10}B, and empty HAFM capsules were irradiated in the same assembly with Au and In foils during a separate background irradiation in ISNF without the ^{235}U drivers. This separate irradiation verified negligible background helium generation from either thermal neutron leakage from the reactor or from (γ,α) reactions.

EXPERIMENTAL RESULTS

Helium Analyses. Following irradiation, helium analyses were performed by vaporizing the crystals or HAFM capsules and then measuring the helium released using high-sensitivity isotope-dilution mass spectrometric techniques described previously (9,10). Table 1 gives the results of the helium analyses for the HAFMs irradiated in ISNF. To enable comparison with calculated values for the ISNF neutron spectrum, the Column 4 helium concentration data were corrected for neutron flux gradients, sample self-shielding, and background helium generation, as discussed below.

Neutron Perturbations. Neutron flux gradients over the irradiation assembly were determined by curve fitting to the gold foil Au(n,γ) data. As this is a non-threshold reaction, the observed spatial variations closely approximated those for ^{6}Li and ^{10}B. The gradients ranged from 0 to -4%, with a maximum uncertainty of ±0.3%.

Neutron self-shielding corrections were computed using a calculated ISNF spectrum in combination with ENDF/B-V ^{6}Li(n,He) and ^{10}B(n,He) cross sections. The corrections were based on first-flight transport theory approximations (11). For the bare (unencapsulated) boron and ^{6}LiF crystals, a spherical sample of equal mass was assumed. For the ^{10}B and ^{6}LiF HAFM capsules these calculations used a reduced effective density and diameter for the cylindrical array of crystals (1,2). The self-shielding corrections also included an additional small correction to account for flux depression from nearest-neighbor interactions. Combined self-shielding/flux depression corrections ranged from ~1.3% for the bare ^{6}LiF crystals to ~7.9% for the ^{10}B HAFM capsules. Estimated uncertainty in the combined corrections is ±20% of the correction value.

Helium generation in the ^{6}LiF from ^{19}F(n,He) reactions, based on numerous HAFM measurements from FFTF (12), was calculated to be ~0.2%. Helium generation measured in the HAFM capsules irradiated in the background run was very low, averaging 1.7 ± 1.3 x 10^9 atoms, or <1% of the total helium generated by the HAFMs in the main irradiation.

TABLE 1

MEASURED HELIUM CONCENTRATIONS IN ISNF HAFMs

Sample Name	Sample Type and Isotope	Isotopic Enrichment (%)	^{4}He Concentration (appb)		
			Measured[a]	Corrected[b]	Mean
6Li-E1	^{6}LiF	99.10	0.1042	0.1068	
6Li-R1	Capsules	±0.10	0.1005	0.1002	
6Li-X1			0.1028	0.1044	0.1040
6Li-Z1			0.1059	0.1092	±0.0036
6Li-B2			0.0996	0.0999	
6Li-F2			0.1014	0.1036	
6Li-20	^{6}LiF Bare	99.10	0.1001	0.1051	
6Li-21	Crystals	±0.10	0.1014	0.1025	
6Li-22			0.1019	0.1060	0.1047
6Li-23			0.0994	0.1049	±0.0019
6Li-24			0.1012	0.1026	
6Li-25			0.1027	0.1073	
10B-R3	^{10}B	93.01	0.2035	0.2253	
10B-E3	Capsules	±0.03	0.2102	0.2272	
10B-B3			0.2048	0.2249	0.2270
10B-H3			0.2071	0.2293	±0.0017
10B-X3			0.2099	0.2269	
10B-C4			0.2077	0.2281	
NB-S5	Natural	20.02	0.2215	0.2348	0.2348
NB-K5	Boron	±0.06	0.2283	0.2354	±0.0006
NB-Q5	Capsules		0.2229	0.2343	
10B-20	^{10}B Bare	93.01	0.2057	0.2294	
10B-21	Crystals	±0.03	0.2117	0.2274	
10B-22			0.2068	0.2269	0.2285
10B-23			0.2092	0.2316	±0.0025
10B-24			0.2107	0.2250	
10B-25			0.2101	0.2307	

[a]Measured helium concentration in atomic parts per billion with respect to ^{6}Li or ^{10}B.

[b]Measured helium concentration corrected for flux gradients, self-shielding and background. The ^{6}LiF HAFMs have an additional 0.2% correction to account for helium generation from ^{19}F(n,He) reactions.

Corrected Helium Concentrations. Final corrected helium concentrations are listed in Column 5 of Table 1. Combined random and systematic uncertainties in the corrected helium data are estimated to be ~1.5% for ^{6}Li and ~2% for ^{10}B. Examination of the data indicates excellent consistency in the corrected helium concentrations between the different groups of samples. The mean standard deviation between the two groups of lithium and three groups of boron HAFMs is 0.5% and 1.8%, respectively, even though different amounts and configurations of material were used in each case. Most of the ~3% difference between the corrected helium

TABLE 2
MEASURED AND CALCULATED ^{6}Li AND ^{10}B
HELIUM GENERATION CROSS SECTIONS IN ISNF

Reaction	Experimental σ(mb)[a]	Calculated σ(mb)[b]	C/E
^{6}Li(n,He)	831 ± 25	789	0.95
^{10}B(n,He)	1831 ± 60	1667	0.91
^{235}U(n,f)	1605 ± 35	1602	1.00

[a]Measured cross section (and estimated 1σ total
quadrature uncertainty) from helium data assuming
a total neutron fluence of 1.257 ± 0.033 x 10^{14} n/cm^2.
[b]Calculated spectrum-averaged cross section (see text).

concentrations for the natural boron and the enriched ^{10}B samples
is attributed to uncertainties in the combined self-shielding/flux
depression corrections.

Measured Cross Sections. The measured spectrum-averaged
cross sections were calculated from the mean helium generation
data (Table 1) using a total neutron fluence of 1.257 ± 0.033 x
10^{14} n/cm^2. This neutron fluence was obtained by relating the
measured activation rates for the six indium foils to similar data
obtained during a separate calibration run in ISNF where the
absolute ^{235}U fission rate was determined with a fission chamber.
The neutron fluence in the calibration run, in turn, was obtained
by a flux transfer method which relates the absolute ^{235}U fission rate to corresponding fission rates observed for carefully
intercompared ^{235}U deposits at the ^{252}Cf facility (13). Fluence
rates at well-determined radii at the NBS ^{252}Cf facility are known
on an absolute basis by comparison of the ^{252}Cf source to the
standard neutron source NBS-I via the manganese bath method (13).

COMPARISONS WITH CALCULATED CROSS SECTIONS

The measured spectrum-averaged helium generation cross sections for ^{6}Li, ^{10}B, and ^{235}U are compared with calculated values
in Table 2. The calculated cross sections were determined using a
150-group spectrum for ISNF calculated at LANL using ONEDANT (14)
with ENDF/B-V boron and carbon cross sections, and an average
ENDF/B-V and NBS-evaluated fission spectrum driving source. The
ENDF/B-V dosimetry cross section file for ^{6}Li and ^{10}B was used for

TABLE 3

COMPARISON OF C/E VALUES FOR ^{6}Li AND ^{10}B IN FIVE BENCHMARK FIELDS

Reaction	C/E Ratios				
	BIG-10	CFRMF	Sigma Sigma	Fission Cavity	ISNF
^{6}Li(n,He)	0.90	0.95	0.93	1.00	0.95
^{10}B(n,He)	0.85	0.86	0.88	0.91	0.91
^{235}U(n,f)	1.00	1.01	-	-	1.00

the total helium generation calculations. Uncertainty in the calculated cross sections has not been rigorously assessed, but is thought to be between 3 and 4%. The final column of Table 2 shows that the ratios of the calculated-to-experimentally measured (C/E) cross sections in ISNF are 0.95 for ^{6}Li and 0.91 for ^{10}B.

Table 3 shows that the C/E ratios for ISNF are consistent with the C/E values from four other benchmark fields. The data in Table 3 also show a consistent discrepancy in the ^{10}B measurements relative to ^{235}U. The mean C/E for ^{10}B-to-^{235}U is 0.87 ± 0.03. This value is also consistent with numerous comparisons of ^{10}B and ^{235}U reactivity worths conducted at ANL in the Zero Power Plutonium Reactor (ZPPR) and other critical assemblies. Although the C/E's for ^{10}B and ^{235}U worths individually can vary significantly as a function of sample geometry, calculational methods, etc., the ratio of C/E's for ^{10}B and ^{235}U worths are consistently about 0.90 (see, for example, Ref. 15).

In summary, the present data for ISNF show discrepancies between measured and calculated ^{6}Li and ^{10}B cross sections consistent with those measured in four other benchmark fields. Further, comparisons of ^{10}B and ^{235}U reaction rates (and worths) indicate a consistent offset in the spectrum-integrated ^{10}B, and to a lesser extent, ^{6}Li, cross section relative to ^{235}U. These combined results continue to indicate a need for upward revision to the ^{10}B, and perhaps ^{6}Li, cross sections above ~0.1 MeV.

ACKNOWLEDGEMENTS

The authors wish to acknowledge technical contributions to this work by M. E. McKee and W. R. Marley of Rockwell International, and by C. M. Eisenhauer of the National Bureau of Standards, who provided helpful comments on the ISNF neutron spectrum. We also wish to express our thanks to P. B. Hemmig and J. W. Lewellen of the U.S. Department of Energy for their continued interest and support of this work. This work was supported under DOE Contracts DE-AT03-81SF11561 at Rockwell International and DE-AC14-76FF02170 at Westinghouse Hanford Company.

REFERENCES

1. H. Farrar IV, B. M. Oliver, and E. P. Lippincott, "Helium Generation Reaction Rates for ^{6}Li and ^{10}B in Benchmark Facilities," Proc. 3rd ASTM-EURATOM Symp. on Reactor Dosimetry, EUR 6813 EN-FR, Vol. 1, p 552 (1980).

2. B. M. Oliver, H. Farrar IV, E. P. Lippincott, and A. Fabry, "Spectrum-Integrated Helium Generation Cross Sections for ^{6}Li and ^{10}B in the Sigma Sigma and Fission Cavity Standard Neutron Fields," Proc. 4th ASTM-EURATOM Symp. on Reactor Dosimetry, NUREG/CP-0029, Vol. 2, p 889 (1982).

3. JW Rogers, D. A. Millsap, and Y. D. Harker, "CFRMF Neutron Field Spectral Characterization," Nucl. Technol., 25, 330 (1975).

4. E. J. Dowdy, E. J. Lozito, and E. A. Plassman, "The Central Neutron Spectrum of the Fast Critical Assembly BIG-TEN," Nucl. Technol., 25, 381 (1975).

5. A. Fabry, G. DeLeeuw, and S. DeLeeuw, "The Secondary Intermediate-Energy Standard Neutron Field at the Mol Facility," Nucl. Technol., 25, 349 (1975).

6. R. A. Anderl, D. A. Millsap, JW Rogers, and Y. D. Harker, INEL Integral Data-Testing Report for ENDF/B-V Dosimeter Cross Sections, Idaho National Engineering Laboratory Report EGG-PHYS-5608 (October 1981).

7. C. M. Eisenhauer and J. A. Grundl, "Neutron Transport Calculations for the Intermediate-Energy Standard Neutron Field (ISNF) at the National Bureau of Standards," Proc. Int. Symp. on Neutron Standards and Applications, NBS Special Publication No. 493, U.S. Department of Commerce, Washington, D.C. (1977).

8. B. L. Broadhead and J. Wagschal, "The NBS Intermediate-Energy Standard Neutron Field (ISNF) Revisited," Trans. Am. Nucl. Soc. 33, (November 1979).

9. H. Farrar IV, W. N. McElroy, and E. P. Lippincott, "Helium Production Cross Section of Boron for Fast-Reactor Neutron Spectra," Nucl. Technol. 25, 305 (1975).

10. B. M. Oliver, James G. Bradley, and Harry Farrar IV, "Helium Concentrations in the Earth's Lower Atmosphere," Geochim. et. Cosmochim. Acta (in press, 1984).

11. K. M. Case, F. de Hoffman, and G. Placzek, Introduction to the Theory of Neutron Diffusion, Los Alamos Scientific Laboratory, Los Alamos, New Mexico, Vol. 1 (June 1953).

12. B. M. Oliver, Harry Farrar IV, J. A. Rawlins, and D. W. Wootan, Threshold Helium Generation Reaction Rate Measurements in FFTF, Rockwell International Report ESG-DOE-13497 (September 1984).

13. V. Spiegel, C. M. Eisenhauer, D. M. Gilliam, J. A. Grundl, E. D. McGarry, I. G. Schroeder, W. E. Slater, and R. S. Schwartz, "^{235}U Cavity Fission Neutron Field Calibration via the ^{252}Cf Spontaneous Fission Neutron Field," Proc. IAEA Consultants Meeting on Neutron Source Properties, INDC(NDS)-114/GT, p 301 (1980).

14. R. D. O'Dell, F. W. Brinkley, Jr. and D. R. Marr, User's Manual for ONEDANT: A Code Package for One-Dimensional, Diffusion-Accelerated, Neutron-Particle Transport, Los Alamos National Laboratory Report LA-9184-M (February 1982).

15. R. W. Schaefer and R. B. Bucher, "Calculated and Measured Reactivities in the U9 Critical Assemblies," Proc. 1980 Conf. on Advances in Reactor Physics and Shielding, NUREG/CP-0034, Vol. 1, p 93 (1980).

PART IX
GENERAL INTEREST

BOUCLE 4 CRAYONS COMBUSTIBLES EAU LEGERE IRRADIEE EN PERIPHERIE DU REACTEUR OSIRIS TECHNIQUES DE DOSIMETRIE MISES EN OEUVRE

A. Alberman, C. Morin, G. Simonet

IRDI/DERPE/Services des Piles de Saclay - C.E.N./SACLAY

91191 GIF-Sur-Yvette Cedex   (France)

RESUME

La boucle ISABELLE 4 permet l'irradiation de 4 crayons combustibles aux conditions thermohydrauliques des réacteurs à eau, dans le réflecteur du réacteur OSIRIS (70 MW) de SACLAY. Les expériences actuelles couvrent notamment l'interaction pastille-gaine (sauts de puissance, ...). Chaque crayon peut être défourné et examiné (C.N.D.) dans la piscine d'OSIRIS. On présente la technique de dosimétrie combustible développée dans le réacteur ISIS (maquette d'OSIRIS) :
. calcul des enrichissements pour une puissance identique à l'avant et à l'arrière de la grappe ;
. irradiation simultanée de crayons combustibles démontables en maquette instrumentée en collectrons, et de poudre UO_2 en colonne thermique (référence) ;
. mesure de puissance par comptage relatif de sources liquides ;
. pilotage de la boucle par collectrons.

SUMMARY

The ISABELLE 4 loop is designed for experimental irradiation of 4 LWR fuel pencils under thermohydraulic power reactors conditions. The loop is operated in OSIRIS (70 MW) test reactor reflector at SACLAY. Present studies cover pellet-clad interaction (PCI). Each pencil can be removed, tested (NDT) and reloaded in OSIRIS pool.The fuel dosimetry technique has been developped in the ISIS reactor (OSIRIS neutronic model) :
. enrichments determination allowing equivalent power from front and back bundle ;
. simultaneous irradiation of removable fuel pencils in SPND equipped ISABELLE mock-up, and of UO_2 powder in a standard neutron field (thermal column) ;
. power measurement by means of relative liquid sources counting ;
. loop monitoring with Self Powered Neutron Detectors.

J. P. Genthon and H. Röttger (eds.), Reactor Dosimetry, 889–896.
© *1985 ECSC, EEC, EAEC, Brussels and Luxembourg*

I - INTRODUCTION

Le développement des programmes de qualification des combustibles, pour réacteurs de puissance à eau pressurisée, dans des conditions poussées de fonctionnement, a conduit à de nombreuses irradiations en réacteurs de recherche.

En périphérie du réacteur OSIRIS (70 MW), deux dispositifs recréant l'environnement des crayons combustibles en réacteurs de puissance (150 bars, 320°C) ont été mis en oeuvre :
. étude de dépôts sur gaine en fonction de la chimie de l'eau (boucle IRENE),
. interactions gaine-combustible en fonction du taux de combustion et pour différentes conceptions de crayons (boucle ISABELLE).

Dans chaque dispositif, une grappe "carrée" de 4 crayons (au pas de 12,6 mm) a été irradiée. Le problème posé par la dosimétrie a été de calculer les enrichissements des crayons de sorte à :
. obtenir le niveau de puissance requis (la boucle est positionnée en milieu de périphérie),
. compenser le gradient de flux neutronique (avant/arrière) afin d'obtenir une puissance identique sur les 4 crayons.

Les précisions requises sont de l'ordre de 5 %. L'ajustement du niveau de puissance est obtenu par avance ou recul de la boucle, pilotée par sondes neutroniques (collectrons). Les dosimétries combustibles réalisées sur maquettes des boucles dans le réacteur ISIS ont permis de tester la validité des calculs et d'étalonner les sondes en fonction de la puissance nucléaire.

Dans ce qui suit, on décrit les techniques mises en oeuvre pour qualifier la boucle ISABELLE dont la description est donnée ailleurs /1/.

II - PRINCIPE GENERAL DE LA METHODE DE DOSIMETRIE COMBUSTIBLE

Soit à mesurer le taux de réaction de l'isotope i dans un échantillon :

$$R_i = n_i \int_{(E, \vec{r})} \varphi (E, \vec{r}) \, \sigma_i (E, \vec{r}) \, d E \, d\vec{r} \qquad (1)$$

Dans le cas où la section efficace est en $1/v$, et en supposant que la répartition énergétique est la même en tout point du combustible, l'expression ci-dessus se ramène au produit d'une section efficace à 2 200 m/s par le flux thermique moyen. Expérimentalement, cette grandeur peut être mesurée à l'aide de détecteurs par activation placés dans le combustible.

Pratiquement, on apportera des corrections sur la section efficace par rapport à la loi en 1/v, on pourra aussi évaluer la contribution épithermique du flux en mesurant les rapports cadmium de détecteurs par activation. Dans le cas des matériaux fissiles ou fertiles possédant une importante intégrale de résonance, la méthode analytique que l'on vient d'évoquer est d'une application très délicate (compte tenu en particulier de l'autoprotection des résonances).

L'application d'une méthode globale, consistant à mesurer directement un taux de réaction sur le combustible lui-même, s'est avérée nécessaire pour l'étude de combustible faiblement enrichi et très fortement dilué (combustible RHT), les captures résonantes de l'uranium 238 pouvant alors contribuer jusqu'à 90 % au taux de capture total de cet isotope. La méthode a été systématisée et appliquée pour les isotopes n'exigeant pas de "correction" si importante (uranium 235 et plutonium 239).

Le principe de la méthode /2/ est le suivant : une colonne thermique dans laquelle on irradie un échantillon fertile ou fissile de très faible densité (section efficace en $1/v = \sigma_0$, pas d'autoprotection et flux conventionnel à 2 200 m/s = $\emptyset_0$) conduira au taux de réaction :

$$R_i 0 = n_i \int_v \frac{\sigma_0 v_0}{v} \, n \, (v) \, v \, d \, v = n_i \, \sigma_0 \, \emptyset_0 \qquad (2)$$

Le taux de réaction dans un échantillon contenant le même isotope mais irradié dans l'environnement de la boucle ISABELLE, sera décrit par l'expression (1) que l'on cherche à expliciter.

Si les deux irradiations ont lieu simultanément et si le spectre thermique et le spectre de pile sont supposés induire les mêmes produits de fission (ou de capture) suivant le même rendement, la comparaison des activités spécifiques, après irradiation, des deux échantillons, permettra de connaître le rapport entre le taux de réaction cherché, en pile d'essai, et le taux de réaction de référence, en colonne thermique, calculé à partir de la connaissance de la section efficace thermique et de la mesure du flux.

L'expérience comportera les trois étapes suivantes :
a) irradiation :
Simultanément, on dispose de l'emplacement à ISIS équipé d'une maquette de la boucle renfermant les échantillons de combustible en grandeur réelle, et d'une colonne thermique en graphite, du réacteur d'enseignement ULYSSE de Saclay, dans laquelle sont placés des échantillons fissiles (poudre d'UO_2) et des détecteurs de flux thermique (Au). Les deux irradiations simultanées s'opèrent à bas niveau (fluence thermique 5.10^{12} à 2.10^{13} n.cm^{-2}).

CALIBRATION OF LIQUID SOURCE
(URANYL NITRATE) IN SEALED CUPEL.

DOSIMETERS COUNTING
automatic multi-channel analyser
INTERTECHNIQUE IN 96.

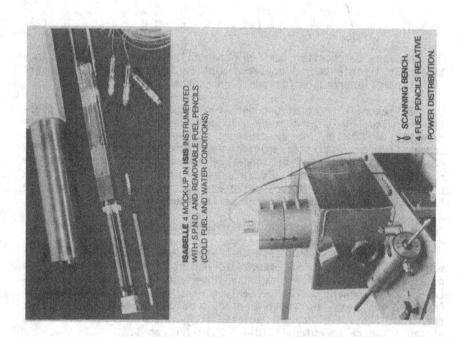

ISABELLE 4 MOCK-UP IN ISIS INSTRUMENTED
WITH S.P.N.D. AND REMOVABLE FUEL PENCILS
(COLD FUEL AND WATER CONDITIONS).

SCANNING BENCH.
4 FUEL PENCILS RELATIVE
POWER DISTRIBUTION.

b) gammamétrie :

Un banc de gammamétrie spécial permet juste après l'irradiation à ISIS d'obtenir la puissance relative entre les crayons de la maquette.

c) préparation des sources :

Les échantillons, provenant du réacteur d'essai et de la colonne thermique, sont transformés de telle sorte qu'ils conduisent à des solutions calibrées de nitrate d'uranyle disposées dans des conteneurs de mêmes caractéristiques géométriques.

d) comptage :

Le signal détecté est consitué par l'activité des produits de fission (ou de capture). On obtient le rapport R_j/R_{j0} des 2 taux de réaction explicités ci-dessus ; le flux et la section efficace du matériau en colonne thermique étant connus, on en déduit la quantité R_j cherchée.

Ce schéma, très général permet de constater le caractère global de la méthode proposée, les seules grandeurs absolues à introduire sont propres à la colonne thermique et à des matériaux fissiles ou fertiles très dilués donc ne comportant pas les corrections habituelles d'autoprotection et de spectre nécessaires dans le cas de mesures réalisées à l'aide de détecteurs par activation sur le combustible en pile lui-même.

Remarque :

Parallèlement à la détermination du taux de réaction, on mesure toujours un signal délivré par des collectrons, ou détecteurs auto-courants, sensibles essentiellement au flux thermique et placés en périphérie de la capsule. La puissance sera donc ramenée au courant délivré par les collectrons qui occuperont, dans l'expérience réelle, une place homologue à celle qu'ils occupent dans la maquette. Le signal collectron constitue donc le paramètre de normalisation entre la dosimétrie et l'irradiation réelle.

III - DOSIMETRIE DE LA BOUCLE ISABELLE

On applique la méthode à la détermination de puissance individuelle de chaque crayon de la boucle (fig.1). On obtient, de la sorte, les coefficients :

P_j (puissance nucléaire) / i_j (courant collectron)

$j = 1, 2, 3, 4$

Ces coefficients, déterminés "à froid" dans la maquette, sont corrigés par un calcul relatif de diffusion expérience / maquette pour tenir compte principalement de l'effet de température (la correction K_j est inférience à 5 % dans ce cas).

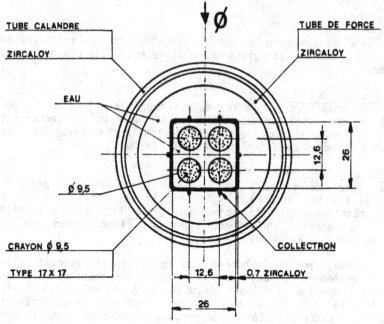

Fig. 1 - Coupe transversale ISABELLE

III.1 - Choix des enrichissements

Préalablement à la dosimétrie, on doit calculer le couple d'enrichissements avant/arrière afin de compenser le gradient de flux en périphérie du réacteur.

La méthode de calcul retenue /3/ consiste à déterminer les constantes macroscopiques de la "cellule UO_2 + Zr + H_2O" à l'aide du code de transport APOLLO /4/ développé au C.E.A.

On compare ensuite les taux de fission calculés dans chaque cellule par un code de diffusion 2D, décrivant le réacteur OSIRIS et la boucle ISABELLE en périphérie. Ceci permet de déterminer :
. l'effet d'ombre entre les 4 crayons et la dissymétrie de puissance,
. la valeur absolue de puissance ISABELLE normalisée à la puissance du réacteur.

Le couple d'enrichissements ^{235}U retenu est le suivant :
$$AV : 2,9 \% \qquad AR : 4,1 \%$$
Ces valeurs permettent d'obtenir jusqu'à 500 W/cm environ au plan de flux maximum.

III.2 - Résultats de la dosimétrie

Les crayons démontables, équipés de billettes UO_2 d'enrichissements indiqués ci-dessus, ont été irradiés à ISIS dans la maquette ISABELLE. Le tableau I résume tous les résultats :

	ISABELLE OSIRIS	Maquette ISIS		
	Calcul	Calcul	Dosimétrie combustible	Gammamétrie
Puissance $\frac{\text{Avant}}{\text{Arrière}}$	1	1,02	0,99	0,99

Tableau I - Dosimétrie ISABELLE

III.3 - Précision de la méthode

Au niveau de la détermination de puissance à bas flux, les causes d'erreur sont peu nombreuses et faciles à évaluer et ne concernent que la mesure de flux dans la colonne thermique, l'évaluation des masses des échantillons et la concentration des sources liquides, la perte d'activité des produits de fission au cours de la dissolution et enfin le comptage.

Globalement, nous estimons, jusqu'à ce stade, la précision à 3 % (cas des crayons d'UO_2), à laquelle il faut ajouter l'erreur sur l'énergie de fission déposée dans le combustible. Dans notre cas, un calcul d'atténuation gamma par la méthode de MONTE CARLO a permis de définir l'énergie déposée à 180 MeV/fission (tenant compte du rayonnement gamma du coeur du réacteur et de la présence des 4 crayons).

Si on indexe la puissance au courant délivré par les collectrons, on doit prendre en compte l'erreur sur la mesure de ce courant, environ 2 %. Compte tenu de l'incertitude sur le recalage maquette-expérience par le calcul, on estime que l'incertitude globale reste inférieure à 5 %.

Il est ensuite beaucoup moins facile de donner ne serait-ce qu'un ordre d'idée de la précision quand on transpose les mesures réalisées jusque là dans la maquette vers la pile OSIRIS. L'indexation puissance-courant peut être modifiée par l'environnement de l'expérience. Pour éviter ce type d'incertitude, les collectrons sont disposés à faible distance du crayon combustible, en sachant cependant que l'influence du crayon sur les collectrons peut ne pas être négligeable si la distance entre ces deux éléments n'est pas suffisante.

On tient compte de cela par un calcul d'évolution de la cellule permettant de corriger les coefficients P/i de la consommation relative de l'UO_2 et les collectrons Ag.

A titre indicatif, l'écart entre détermination absolue de puissance par cette méthode et bilan thermique global, est resté inférieur à 8 % pour la première irradiation (taux de combustion atteint : 3 000 MWj/tU).

REFERENCES

1. M. LUCOT "La boucle ISABELLE 4 : boucle d'essai PWR déchargeable et rechargeable dans la piscine du réacteur OSIRIS"
   CR Conférence Internationale sur la Technologie des Irradiations, Grenoble, Septembre 1982, EUR.8429, édité par P. VON DER HARDT, Reidel Publishing Co, 1983

2. R. BAUDRY et autres "Détermination des paramètres d'irradiation de combustibles en pile d'essai", CEA-R-4513, 1973

3. M. SARSAM "Neutronique des boucles d'irradiation (boucle IRENE)"
   Thèse de 3ème cycle, ORSAY, 1980

4. A. BOIVINEAU "Module APOLLO" Rapport SERMA/SPM/435/DR, 1980

QUALIFICATION DANS MELUSINE DE L'EVOLUTION NEUTRONIQUE

DE COMBUSTIBLES DE LA FILIERE A EAU LEGERE.

D.BERETZ - J.GARCIN - G.DUCROS - D.VANHUMBEECK
Service des Piles. Centre d'Etudes Nucléaires de Grenoble.
85 X - 38041 GRENOBLE Cedex - France.

P.CHAUCHEPRAT
Laboratoire de Physique des Réacteurs à eau. CEN CADARACHE.
B.P. 1 - 13115 SAINT PAUL LEZ DURANCE.   France

ABSTRACT

MELUSINE, a swimming pool type reactor, in Grenoble, for research and technological irradiations is well fitted to the neutronic evolution qualification of the LWR fuel. Several configurations have been projected for this purpose. In one of them, the experimental fuel assembly is inserted in the core itself. Thus, with an adjustment of the lattice pitch, representative neutron spectrum locations are available. The re-loading management and the regulation mode flexibility of MELUSINE lead to reproductible neutronic parameters configurations without restricting the reactor to this purpose only. Under these conditions, simple calculations can be carried out for interpretation, without taking into account the whole core. An instrumentation by Self Power Neutron Detectors (collectrons) gives on-line information on the fluxes at the periphery of the device. When required by the neutronicians, experimental pins can be unloaded during the irradiation process and scanned on a gammametry bench immersed in the reactor-pool itself, before their isotopic composition analysis. Thus, within the framework of neutronic evolution qualification, are studed fuel pins for advanced assemblies for the light water reactors or their derivatives, with large advantages over irradiations in power reactors.

I-INTRODUCTION.

La qualification de l'évolution neutronique de combustibles nouveaux, pour les réacteurs à eau sous pression (REP) demande l'irradiation prolongée d'expériences calculables, dans des conditions représentatives de la filière. De telles expériences dans les réacteurs de puissance manquent de souplesse. MELUSINE, réacteur de recherche et d'irradiation de type piscine (CEA - Centre d'Etudes Nucléaires de Grenoble) a des atouts qui lui permettent de mener à bien ce genre d'irradiations sans pour autant spécialiser le réacteur à ce seul usage. Ils sont basés sur l'adaptabilité du réacteur et sur l'instrumentation qui peut être associée à l'expérience. De ce fait, les calculs d'interprétation peuvent être relativement simples. Après une brève description du réacteur et de l'instrumentation associée, on illustrera, avec l'exemple de l'expérience GEDEON, les qualités du processus de qualification.

II-LE REACTEUR MELUSINE

II-1 Description.

MELUSINE est une pile de type piscine, à combustible MTR enrichi à 93% en ^{235}U. Irradiations technologiques et pour la recherche fondamentale font partie de ses missions : production de radioéléments, dopage de silicium, irradiation d'aciers, d'échantillons instrumentés, production de faisceaux sortis de neutrons (diffusion neutronique, neutronographie..). Les flux atteints sont de l'ordre de 5 x 10^{13} cm^{-2}s^{-1} et permettent d'atteindre les objectifs fixés. MELUSINE est complémentaire d'une pile voisine, de même type, SILOE, conçue pour atteindre des flux plus élevés et irradier, tant dans son coeur que dans son réflecteur (H$_2$O), des dispositifs complexes.

Le contrôle du coeur de MELUSINE est réalisé à l'aide de barres fourchettes en hafnium. La réfrigération forcée est de circulation descendante. Un surplus de débit primaire existant par rapport au strict nécessaire pour la réfrigération des éléments MTR, et un mode de fonctionnement souple, typiquement en cycles continus de 23 à 30 jours permettent l'irradiation de grappes de crayons combustibles occupant l'emplacement de un à neuf éléments (seize éléments moyennant quelques modifications, à faible coût, du circuit primaire), sans nécessiter de boucle de réfrigération spéciale.

II-2 Adaptation des configurations du coeur.

Le coeur de MELUSINE est adaptable aux besoins de ce type d'irradiation, tout en étant compatible avec les autres missions. L'étude de la configuration se fait selon le schéma suivant :
(i)choix, à priori, d'une géométrie répondant aux exigences de l'expérience et du réacteur (fonctionnement, sécurité,

autres irradiations).

(ii) calcul des constantes neutroniques des différents milieux par le code APOLLO /1/ : pour l'assemblage, calcul multicellule (1 cellule par crayon) ; les paramètres des éléments MTR sont tirés de calculs d'évolution jusqu'aux taux de combustion convenables.

(iii) les paramètres sont introduits, avec une description fine du coeur, dans le code NEPTUNE (code de diffusion) /2/ qui fournit : la puissance dégagée dans les crayons MTR, l'efficacité des barres de contrôle, les pics de puissance des éléments MTR, le flux en tous points du coeur et du réflecteur.

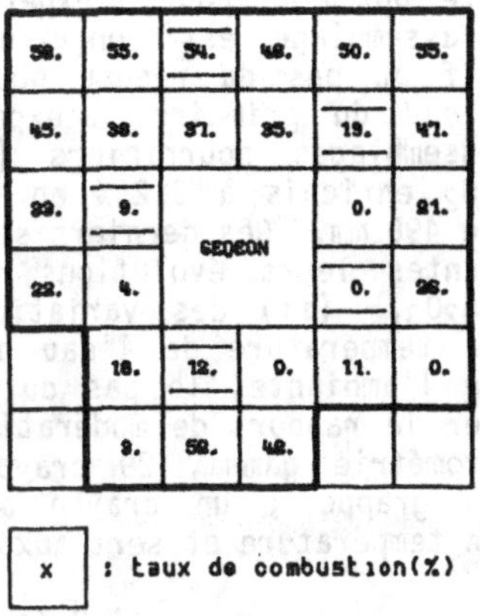

x : taux de combustion(%)

Figure 1 : configuration du coeur de MELUSINE avec l'expérience GEDEON

L'analyse de ces résultats, confrontés aux exigences de départ, permet éventuellement d'affiner la configuration : on itère alors le calcul. Pour l'expérience GEDEON /3/ était recherchée la meilleure uniformité possible de la nappe de flux : pour cela, le dispositif a été placé au centre même du coeur (figure 1). Les pics de puissance dans les éléments MTR ne sont pas majorés par rapport à un coeur compact. Les flux dans les canaux et dans le réflecteur sont peu diminués.

II-3 Stabilité et reproductibilité des flux.

Le coeur est calculé à l'équilibre : il est donc reproductible à chaque rechargement, et de ce fait la configuration neutronique également. Le niveau de puissance obtenu dans l'assemblage (figure 2) est proche de celui de la filière (170 Wcm^{-1} en moyenne dans un REP 900 MWe). Le pilotage, par cinq barres, a été adapté de façon à maintenir constante la répartition de la nappe de flux : trois d'entre elles servent à la compensation xenon et d'usure du combustible. Elles sont gérées en simili-rideau à partir de mesures de flux thermique en périphérie du dispositif (Cf § III-2).

III-EXEMPLE DE DISPOSITIF ET INSTRUMENTATION ASSOCIEE.

III-1 Dispositif

Les dispositifs d'irradiation de grappes combustibles peuvent être divers. Nous décrivons dans ce paragraphe celui qui a été optimisé pour l'expérience GEDEON. Le boîtier qui maintient l'assemblage est conçu pour adapter la géométrie du dispositif au pas du réseau nourricier. La puissance est évacuée par l'eau du primaire. L'expérience occupe l'emplacement de quatre assemblages nourriciers (figure 1). Il comporte 161 crayons UO_2 enrichis à 3,25% en ^{235}U et 4 UO_2-Gd_2O_3 /4/, d'une hauteur de 496 mm. Ces derniers sont disposés de manière à rendre indépendantes leurs évolutions vis-à-vis (i) des autres crayons UO_2-Gd_2O_3, (ii) des variations éventuelles du spectre nourricier. La température de l'eau dans le coeur de MELUSINE étant proche de l'ambiante, le pas du réseau a été choisi de façon à retrouver le rapport de modération des REP. En vue de mesures par spectrométrie gamma, 29 crayons sont extrayables sans démontage de la grappe ; un crayon amovible supplémentaire est instrumenté en température et sert aux essais de démarrage.

III-2 Instrumentation associée.

Sur le boîtier du dispositif sont disposés douze collectrons au rhodium /6/ en trois nappes horizontales. On en tire, en continu pendant l'irradiation les flux en périphérie de l'expérience ; ceux-ci permettent d'ajuster le mouvement des barres pour conserver la répartition de puissance dans l'assemblage.

Le programme expérimental, défini par les neutroniciens, prévoit l'examen de certains crayons, par scrutation gamma, au cours de l'irradiation. Pour cela, après arrêt du réacteur, les crayons concernés, extraits de l'assemblage, sont transportés sous la protection biologique de l'eau, jusque sur le banc de mesures immergé, où l'examen débute quelques heures après la sortie du coeur. Aussitôt après les mesures, l'irradiation peut reprendre. Un pré-dépouillement partiel, "en ligne", des résultats permet de déceler des anomalies éventuelles. Le banc de scrutation permet de déplacer le râtelier (conçu pour une parfaite reproductibilité de positionnement des crayons), de façon automatique devant un système de collimation traversant le mur de la piscine (verticalement, précision de 0,1 mm pour une course de 700 mm, horizontalement précision de 0,01 mm). Le système de collimation interchangeable

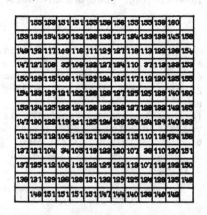

Figure 2 : puissance de fission (W.cm-1) dans GEDEON au plan médian (1 cellule par crayon)

et très reproductible est associé à une chaine de spectrométrie gamma utilisant un détecteur Ge(HP). Cet ensemble offre une dynamique de comptage de l'ordre de 10^5, autorisant aussi bien le contrôle de combustible frais très actifs que la mesure sur des crayons peu actifs (par exemple, combustible empoisonné ou peu irradié).

Cette méthode permet, par la mesure quantitative des produits de fission de période s'étalant de quelques heures à plusieurs dizaines d'années, d'accéder à des résultats variés : distribution axiales et radiales dans le combustible, analyse des migrations, analyse tomographique, mesure de la puissance dégagée et du taux de combustion.

IV-DES SITUATIONS CALCULABLES ET INTERPRETABLES.

IV-1 Calculs.

Les expériences sont conçues dès l'origine avec un double objectif : une représentativité neutronique la plus fidèle possible, une interprétation précise à l'aide de calcul relativement simples. Des mesures expérimentales, notamment en début de vie, peuvent permettre d'ajuster les paramètres d'un calcul neutronique performant adapté au problème. Une procédure d'évolution plus simple, compatible avec les moyens informatiques et d'un coût raisonnable, peut être ensuite définie si nécessaire. Les résultats au temps zéro sont comparés au calcul précédent.

Pour l'expérience GEDEON par exemple, les calculs sont menés avec le code APOLLO (théorie du transport). La configuration choisie permet un calcul en symétrie 1/8. Dans la zone nourricière du coeur MELUSINE, les hétérogénéités fines sont homogénéisées ; le dispositif est discrétisé en autant de cellules que de crayons. Le domaine énergétique est réparti en sept macrogroupes et la validité de ce découpage testée par rapport à une discrétisation fine /4/. Le calcul de référence utilise le module MARSYAS du code APOLLO qui effectue un calcul exact des probabilités de collision à deux dimensions. Les résultats montrent dans la zone centrale de

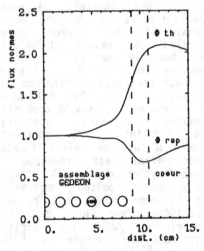

Figure 3 : traverse diagonale de flux dans GEDEON sans gadolinium (calcul APOLLO-MARSYAS)

l'assemblage (7 x 7 crayons), un flux plat et un spectre à l'équilibre (figure 3). Le spectre est proche de celui des REP. Les effets de températures sont analysés. L'interaction neutronique entre crayons gadoliniés est très faible au-delà de deux cellules UO_2 : on peut ainsi irradier et décharger plusieurs crayons expérimentaux indépendamment les uns des autres. Ce calcul est en bon accord avec celui mené pour définir le coeur (Cf § I-2). Il est enfin comparé aux mesures effectuées par spectrométrie gamma sur le dispositif après quelques jours d'irradiation (figure 4). L'accord est satisfaisant aux emplacements expérimentaux. Les écarts en périphérie sont interprétés par des calculs portant sur les effets azimutaux (dissymétrie du coeur, barres de contrôle).

Pour les calculs en évolution, plusieurs modules du code APOLLO ont été testés en optimisant la géométrie, et les résultats au temps zéro ont été comparés aux résultats précédents. L'évolution proprement dite est menée avec un découpage énergétique suffisamment fin (99 groupes) pour rendre compte des perturbations entraînées par les crayons expérimentaux.

IV-2 Interprétation.

La qualification des données neutroniques est rendue possible grâce à une définition correcte des expériences soustendue par la représentativité des calculs.

A des instants déterminés de l'irradiation GEDEON, où la sensibilité aux erreurs potentielles du calcul d'évolution est maximale, des crayons expérimentaux sont retirés de l'assemblage. L'analyse isotopique des corps lourds (U, Pu, Nd) permet de déterminer le taux de combustion atteint. L'analyse isotopique du gadolinium est confrontée aux résultats du calcul d'évolution : ainsi est réalisée la qualification du calcul d'évolution de ce poison consommable. De la même façon, ont pu être précisées, par une expérience antérieure /6/, les sections efficaces de capture de ^{239}Pu, ^{242}Pu, ^{243}Am, ^{244}Cm, dans un spectre de neutrons de réacteur à eau.

					0.000
				1.232 6.7	1.380
			1.014 2.7	1.151	1.321
		0.887 -1.4	0.850	1.125	1.289
	0.934	0.901 0.0	0.972	1.119	1.285
1.000 0.0	0.887	1.004 2.0	1.043	1.126 6.1	1.287

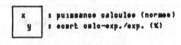

x : puissance calculée (normée)
y : écart calc-exp./exp. (%)

—— : crayon expérimental

Figure 4 : comparaison calculs (APOLLO-MARSYAS) - expérience (gammamétrie) en début de vie de GEDEON.

V-CONCLUSION.

L'expérience GEDEON est un exemple de ce qui peut être réalisé dans le réacteur MELUSINE, dans le cadre de la qualification neutronique d'évolution de combustibles pour la filière à eau légère. Le dispositif et le réacteur sont adaptés de façon à répondre aux exigences des neutroniciens concernant la représentativité à la filière et le caractère calculable; grâce à la souplesse du pilotage et à l'instrumentation qu'il est possible d'associer à l'expérience, la situation reste à tout moment calculable et interprétable. Ces caractéristiques, associées à une récupération aisée des crayons quand on le désire, sont des avantages importants par rapport à l'irradiation dans un réacteur de puissance.

REFERENCES :

/1/ APOLLO. Code multigroupe de résolution de l'équation du transport pour les neutrons thermiques et rapides.
A.HOFFMANN et Al. Note CEA N 1610 - 1973.

/2/ NEPTUNE, un système modulaire pour le calcul des réacteurs à eau légère.
A.KAVENOKY. - B.I.S.T. N° 212 p.7 - Mars 1976.

/3/ GEDEON. Une expérience d'irradiation pour la qualification des calculs d'évolution des poisons consommables au gadolinium.
P. CHAUCHEPRAT et Al.
ANS Winter Meeting. WASHINGTON. Novembre 1984. A paraître.

/4/ Contribution à la qualification du calcul du gadolinium dans les réacteurs à eau.
P. CHAUCHEPRAT. - Thèse 3ème cycle. Orsay. Juin 1982.

/5/ Collectrons au rhodium ; sensibilité aux neutrons et aux gamma; précision. H. PETITCOLAS et Al. - A paraître.

/6/ Détermination expérimentale des sections efficaces isotopes de Pu, Am et Cm dans un spectre de neutrons de réacteur à eau. M.DARROUZET et Al. - Int. Conf. on Nuclear Data for Science and Technology. ANTWERP. September 1982.

COMPARISON OF MEASURED AND CALCULATED REACTION RATES

FROM D$_2$O-MODERATED ^{252}Cf NEUTRONS

W. Brandon, C. Cogburn, R. Culp, R. Hamblen, J. Williams

University of Arkansas

Fayetteville, AR 72701

ABSTRACT

Activation detectors and an N.B.S. fission chamber have been used to measure reaction rates at several distances between 25 cm and 150 cm from the D$_2$O-moderated ^{252}Cf source at the University of Arkansas. This source is a duplicate of the 15 cm radius D$_2$O-moderated ^{252}Cf source used at the U.S. National Bureau of Standards. The reactions measured were ^{197}Au(n,γ), ^{115}In(n,n'), ^{32}S(n,p), ^{58}Ni(n,p), ^{235}U(n,f) and ^{237}Np(n,f). Corrections for room-scatter, for air-scattering and for non-1/r^2 effects near the source were applied to obtain the expected free-field reaction rates. Cross-sections from the ENDF/B V file and the D$_2$O-moderated source spectrum from the Monte Carlo calculation of H. Ing and W. G. Cross were used to calculate reaction rates which are compared with the measured results.

Except for ^{237}Np, the measured and calculated reaction rates agreed within the limits of experimental error. The ^{237}Np(n,f) reaction gave a C/E ratio of 0.89, suggesting the need for an upward adjustment of at least 20% in the fluence betwen 0.6 MeV and 1.6 MeV.

Research sponsored by the Arkansas Power & Light Co.

INTRODUCTION

The 15 cm radius D_2O-moderated ^{252}Cf source at the U.S. National Bureau of Standards was developed as a calibration and test source having a spectrum which resembles that found in the vicinity of light water reactors. This source has been adopted as the standard for calibration of personnel neutron dosimeters in the United States, and has been proposed as a calibration standard by the International Standards Organization. At the University of Arkansas, a duplicate of the N.B.S. source is used for dosimeter calibration and LWR research in an irradiation facility constructed in the refueling cell of the decommissioned SEFOR fast reactor.

In the summer of 1981, a series of reaction rate measurements were performed using the University of Arkansas moderated source in order to gain information about the D_2O-moderated spectrum and to study the wall-return characteristics of the SEFOR calibration facility. In 1982, H. Ing and W. G. Cross calculated the leakage spectrum for the D_2O-moderated ^{252}Cf source using the 05R Monte Carlo code (1), and in 1983 Schwartz, Eisenhauer and Grundl of the N.B.S. used fission chamber measurements to experimentally verify this calculated spectrum (2).

The aim of this paper is to compare measured reaction rates from the University of Arkansas with reaction rates calculated using the Ing and Cross spectrum, and to compare measured $^{235}U/^{237}Np$ fission ratios with values reported by Schwartz et al.

EXPERIMENTAL METHOD

A National Bureau of Standards double fission chamber containing ^{235}U and ^{237}Np deposits was used to measure fission rates at 25 cm, 50 cm and 100 cm from the University of Arkansas D_2O-moderated ^{252}Cf source. The construction and operation of the fission chamber is described in detail by Grundl et al (3) and will be outlined only briefly here. The chamber consists of two back-to-back ionization chambers each connected to an independent pulse counting circuit. The basic measured quantity is the count rate of fission fragment pulses above a discriminator set high enough to eliminate electronic noise and counts due to non-fission events. Several small corrections must be applied to the measured count rate in order to obtain the true fission count rate. This corrected count rate is then divided by the number of target nuclei in the deposit to obtain the fission rate per atom.

For the measurements with the University of Arkansas source, the true fission count rate was found by applying corrections to the raw data for: 1) extrapolation to zero fission pulse height,

2) absorption of fission fragments in the fission deposits, 3) fission in the ^{239}Pu impurity in the ^{237}Np deposit, and 4) neutron elastic scattering and absorption in the deposit backings and the chamber hardware. The last correction was made by taking the geometric mean of two counts with a 180-degree rotation of the chamber between counts so that each deposit alternately faced toward and away from the source.

In addition to the fission chamber measurements, activation detectors were used to measure ^{197}Au(n, γ), ^{32}S(n,p), ^{58}Ni(n,p) and ^{115}In(n,n') reaction rates at distances between 25 cm and 150 cm from the moderated source. The detectors were irradiated for six days using the same Savannah River Series 100 primary ^{252}Cf source capsule as for the fission chamber measurements. This source (#311Z) had an emission rate of 1.36×10^{10} n/s $\pm2\%$ and, assuming 11% absorption in the Cd covering of the D_2O moderator, produced a free-field fluence rate of 4.3×10^4 n/cm^{2}s at the 150 cm measurement position. Reaction rates were determined by gamma-ray counting of the gold, nickel and indium foils using a 15% Ge(Li) detector. Sulfur pellets were counted using an internal proportional counter.

CORRECTIONS TO DATA

Several corrections must be applied to the measured reaction rates in order to obtain free-field reaction rates which can be compared with calculated values derived from the Ing and Cross spectrum. The experimental free-field reaction rate is given by

$$R_e = R_m \cdot F_r / F_a \cdot F_n \cdot F_w \qquad [1]$$

where F_w is a correction for response to wall-scattered neutrons,
  F_a is a correction for response to air-scattered neutrons,
  F_n is a correction for departure from the $1/r^2$ relation near the source due to changes in the spectrum, and
  F_r is a correction for small deviations from the nominal source-to-detector distance.

The largest of these is the correction for response to wall-scattered neutrons. The wall-return correction factor, F_w, is the ratio of response to both free-field and wall-scattered fluence to the free-field response alone. This correction can be determined experimentally using a procedure developed by Eisenhauer and Schwartz (4) which is based on the assumption that the measured reaction rate R_m is the sum of a response to direct source neutrons which varies as $1/r^2$ and a response to wall-scattered neutrons which is a constant.

The air-scattering correction factor, F_a, is the ratio of the detector response in air at distance r from the source to the detector response at the same position in vacuum. Air-scattering correction factors were calculated by folding reaction cross sections with spectra for the D_2O-moderated source in air and in vacuum which were calculated using the ANISN code.

The non-$1/r^2$ correction factor, F_n, serves to correct for the increase in detector response near the surface of the D_2O-moderator sphere. Using ANISN calculated spectra for the source in vacuum and assuming that the source exhibits $1/r^2$ behavior at distances of 150 cm or greater, the non-$1/r^2$ correction factors were found for each reaction using

$$F_n = (R(r) \cdot (150)^2) / (R(150) \cdot (r)^2) \qquad [2]$$

where R(r) is the calculated free-field reaction rate at distance
r (cm) from the center of the source, and
R(150) is the calculated free-field reaction rate at 150 cm.

The nominal distance correction, F_r, was applied to the activation detector measurements to correct for small displacements of individual detectors from the reference measurement distance. Actual source-to-detector distances for the individual foils were measured and factors for correcting the measured reaction rates for deviations from the 'nominal' position were found using

$$F_r = d_x^2 / d_N^2 \qquad [3]$$

where d_x is the 'true' source-to-foil distance, and
d_N is the nominal measurement distance (i.e. 25 cm, 50 cm, etc.)

COMPARISON OF MEASURED AND CALCULATED VALUES

Spectrum averaged cross sections for each reaction were calculated using the code DETAN (5) to fold cross sections from the ENDF/B V dosimetry file with the Ing and Cross spectrum. The spectrum averaged cross sections were then multiplied by a free-field fluence rate to obtain a calculated reaction rate. The free-field fluence rate at each measurement distance was calculated using

$$\phi = 0.885 \, Q / 4\pi r^2 \qquad [4]$$

where Q is the source emission rate, n/s
r is the source-to-detector distance, cm, and

the factor 0.885 accounts for the absorption of thermalized neutrons in the cadmium shell which covers the D_2O-moderator sphere.

The observed calculated-to-experimental reaction rate ratios are listed in Table 1. The uncertainties for the average C/E ratios are total uncertainties found by propagating the experimental errors through the equations used to determine the C/E ratios. The uncertainties considered were those associated with: 1) the source emission rate, 2) counting statistics, 3) detector efficiency (for the activation foils), 4) foil mass, and 5) the wall return correction factor.

Indicated in Figure 1 for each reaction is the energy interval within which 90% of the response to the Ing and Cross spectrum occurs. The median response energy is also indicated within each interval. The $^{235}U(n,f)$ and $^{197}Au(n,\gamma)$ reactions respond primarily to neutrons in the low energy end of the spectrum and the C/E ratios in Table 1 suggest that the calculated spectrum slightly underestimates the fluence in this region. The $^{115}In(n,n')$ (threshold 1.0 MeV), $^{32}S(n,p)$ (threshold 2.9 MeV), and $^{58}Ni(n,p)$ (threshold 2.9 MeV) reactions all show close agreement with the calculated results. However, the $^{237}Np(n,f)$ reaction (threshold 600 keV) yields an average C/E ratio of only 0.89. If the measured results are correct, then the calculated spectrum significantly underestimates the fluence in the energies "seen" by the $^{237}Np(n,f)$ reaction.

Table 2 shows that the $^{235}U/^{237}Np$ fission ratio measured with bare ^{252}Cf at the University of Arkansas agrees well with the N.B.S. result (2). The calculated ratios for both the bare and the D_2O-moderated source also agree with the N.B.S. values. However, the $^{235}U/^{237}Np$ measurement with the moderated source gives a ratio about 3% smaller than the value reported by Schwartz et al.

The close agreement between calculated and measured results for $^{115}In(n,n')$, $^{32}S(n,p)$ and $^{58}Ni(n,p)$ indicates that any modification of the calculated spectrum to improve the ^{237}Np "C/E" ratio must occur between the 600 keV threshold for $^{237}Np(n,f)$ and the 1.0 MeV threshold of the $^{115}In(n,n')$ reaction. Since less than 24% of the $^{237}Np(n,f)$ response occurs in this region, the fluence would have to be increased significantly to produce an 11% change in the $^{237}Np(n,f)$ "C/E" ratio.

Table 1

Ratios of Calculated to Experimental Reaction Rates
for the University of Arkansas D_2O-Moderated ^{252}Cf Source

	25 cm	50 cm	75 cm	100 cm	150 cm	Average
$^{197}Au(n,\gamma)$	0.916	0.937	0.973	0.918	0.938	0.936 +5.7%
$^{235}U(n,f)$	0.966	0.956	-----	0.962	-----	0.961 +2.3%
$^{237}Np(n,f)$	0.876	0.910	-----	0.888	-----	0.891 +3.4%
$^{115}In(n,n')$	0.950	1.067	-----	-----	0.997	1.001 +5.6%
$^{32}S(n,p)$	0.999	1.014	1.015	1.013	1.007	1.010 +4.6%
$^{58}Ni(n,p)$	0.997	0.997	-----	-----	-----	0.997 +4.8%

Table 2

Comparison of $^{235}U/^{237}Np$ Fission Ratios

University of Arkansas Data

Source	Exp. Ratio	Calc. Ratio
D_2O-Mod	31.2	33.6
Bare	0.884	0.913

N.B.S. Data (2)

Source	Exp. Ratio	Calc. Ratio
D_2O-Mod	32.1	33.6
Bare	0.873	0.914

CONCLUSIONS

From the results of fission rate measurements with ^{235}U, ^{238}U, ^{237}Np, and boron-shielded ^{239}Pu, Schwartz et al proposed the following adjustment factors for the Ing and Cross spectrum covering four broad energy intervals:

1 eV	- 3.4 keV	0.98 ± 0.05
3.4 keV	- 600 keV	1.03 ± 0.08
600 keV	- 1.6 MeV	1.21 ± 0.14
1.6 MeV	- 10 MeV	0.89 ± 0.04

These energy intervals are shown in Figure 1 for comparison with the response intervals of the reactions measured at the University of Arkansas. Comparison of measured and calculated $^{237}Np(n,f)$ reaction rates at the University of Arkansas indicates that the fluence between 600 keV and 1.6 MeV in the Ing and Cross spectrum should be increased by at least 20%. However, calculated-to-experimental ratios for the high energy threshold reactions $^{115}In(n,n')$, $^{32}S(n,p)$, and $^{58}Ni(n,p)$ do not support the need for an 11% decrease in the fluence above 1.6 MeV as proposed by Schwartz et al.

REFERENCES

1. H. Ing and W. G. Cross, "Spectral and Dosimetric Characteristics of a D_2O-Moderated ^{252}Cf Calibration Facility", Health Physics 46, 97-106 (1984).

2. R. B. Schwartz, C. M. Eisenhauer, and J. A. Grundl, "Experimental Verification of the Neutron Spectrum from the NBS D_2O-Moderated ^{252}Cf Source", NUREG/CR-3399, National Bureau of Standards, Washington, D.C. (1983).

3. J. A. Grundl, D. M. Gilliam, N. D. Dudey, and R. J. Popek, "Measurements of Absolute Fission Rates", Nuclear Technology 25, 237 (1975).

4. C. M. Eisenhauer, R. B. Schwartz, and T. Johnson, "Measurement of Neutrons Reflected from the Surfaces of a Calibration Room", Health Physics 42, 489-495 (1982).

5. Private communication to W. W. Sallee, University of Arkansas, from C. M. Eisenhauer, N.B.S., September 5, 1981.

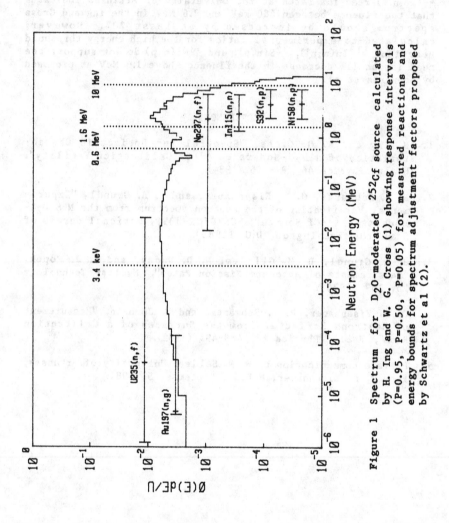

Figure 1 Spectrum for D₂O-moderated ²⁵²Cf source calculated by H. Ing and W. G. Cross (1) showing response intervals (P=0.95, P=0.50, P=0.05) for measured reactions and energy bounds for spectrum adjustment factors proposed by Schwartz et al (2).

DOSIMETRY WORK AND CALCULATIONS IN CONNECTION WITH THE
IRRADIATION OF LARGE DEVICES IN THE HIGH FLUX MATERIALS
TESTING REACTOR BR2 : FUEL BURN-UP ASPECTS IN CORRELATION
WITH THE OTHER DOSIMETRY DATA

J. Debrue, Ch. De Raedt, P. De Regge, L. Leenders,
H. Tourwé, A. Verwimp

SCK/CEN, Mol, Belgium

H. Farrar IV, B.M. Oliver

Rockwell International Corporation, California, USA

ABSTRACT

At the 1982 ASTM-EURATOM Symposium on Reactor Dosimetry
the application of various dosimetry techniques to determine the irradiation conditions of large devices in the
BR2 reactor was discussed (1) with, as example, a device
containing 19 sodium cooled fuel pins irradiated under
cadmium screen. The measurements have since been completed
by post-irradiation destructive radiochemical analyses and
non-destructive examinations of the fuel pins. These results are compared with the calculated data and with the
burn-up values deduced from heat balances. They are also
correlated with the other dosimetry data.

INTRODUCTION. THE MOL 7D LOOP

For about the last twenty years the high flux reactor BR2 has
been involved in the testing of fast reactor fuel pins. In (1), a
comparison was made between neutronic calculation results and some
post-irradiation data for MOL 7D. Data concerning the burn-up are
discussed in the present paper.

The MOL 7D loop (fig. 1) was irradiated in BR2 for 16 calen-
dar months (295 E.F.P.D.). It contained 19 sodium cooled UO_2-PuO_2
fuel pins (with Pu/(U+Pu) = 29.8 wt% and $^{235}U/U$ = 83.0 at%) surrounded by a hexagonal AISI-316 "wrapper tube" with vertical grooves.

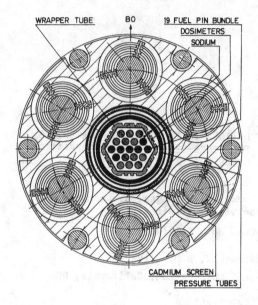

WRAPPER TUBE   BO   19 FUEL PIN BUNDLE
DOSIMETERS
SODIUM

CADMIUM SCREEN
PRESSURE TUBES

These contained thermo-couples as well as – in a long SS micro tube in one groove – a large number of neutron dosimeter wires and helium fluence monitors. The fuel pin bundle was further surrounded by pressure tubes, a Cd screen and six small "driver" fuel elements located in a large cy-lindrical Al plug, the whole occupying the central 200 mm Ø channel H1 of BR2.

Fig. 1. MOL 7D in the H1 channel of BR2.

DETERMINATION OF THE BURN-UP BY DESTRUCTIVE ANALYSIS

The fuel pins of interest in this section are the right four shaded black in fig. 1. Several samples were taken, located axially at various positions, the samples located at + 35 mm above the geo-metrical mid-plane of the reactor being those nearest to the maximum flux plane, located at about – 41 mm.

The heavy nuclide concentrations in the samples, as measured after the irradiation, are in reasonably good agreement with the concentrations calculated as explained below. This is illustrated in table I for the intermediate pin sample at + 35 mm. For ^{236}U the agreement is rather bad.

Table I. Heavy nuclide concentrations in the intermediate pin at start and end of irradiation, arbitrarily normalized to ^{238}U

Nuclide	^{234}U	^{235}U	^{236}U	^{238}U	^{239}Pu	^{240}Pu	^{241}Pu	^{242}Pu
BOL	0.0388	5.08	0	1.00	1.99	0.358	0.158	0.0306
EOL (meas.)	0.0335	4.61	0.176	1.00	1.81	0.380	0.139	0.0312
EOL (c/m)	1.003	0.994	1.104	1.000	1.007	0.998	0.935	1.036

The number of neodymium fission product atoms was measured for all the neodymium isotopes. The data analysis allowed a determination of the fission product contribution of ^{235}U, ^{239}Pu and ^{241}Pu (69 %, 26.5 % and 4.5 %, respectively), and hence the corresponding weighted fission yields : 5.315 (for mass 143), 4.831 (144), 3.540 (145), 2.778 (146), 1.677 (148) and 0.787 (150). The resulting burn-up values, in atom percent fissions per initial heavy atom (at % FIMA) indicate good agreement between the data obtained from the different neodymium isotopes : e.g. for the intermediate pin : $^{143+144}Nd$: 7.532, $^{145+146}Nd$: 7.403, ^{148}Nd : 7.426, ^{150}Nd : 7.465.

The burn-up was also calculated from ^{144}Ce : it showed a systematic deviation of - 6.5 % with respect to the Nd data, probably due to the in-pile decay correction factor as derived from the power history although an uncertainty of $\pm$ 3 % can be associated with the ^{144}Cs measurements themselves.

In fig. 2 the burn-up values of the four fuel pins at axial level + 35 mm, obtained from mass spectrometric measurements of Nd

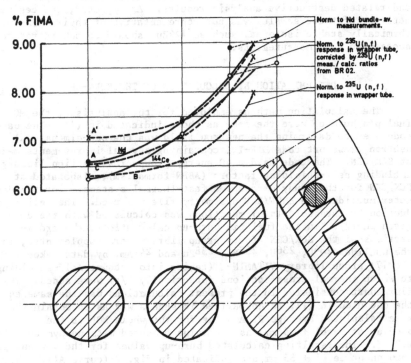

Fig. 2. Pin-to-pin burn-up distribution in the MOL 7D sodium loop : comparison of measured (——) and calculated (---) values.

(average value of the neodymium isotopes measured) and from radiochemical measurements of ^{144}Ce are drawn, together with the calculated values (see below) : the bundle-averaged value (at level
+ 35 mm), as measured with Nd, amounts to 8.38 at % FIMA. The maximum burn-up obtained in the MOL 7D experiment, viz in the outer
"corner" pin at axial level - 41 mm amounts to (on the basis of
the Nd results) 9.35 at % FIMA, which is very close to the value
deduced from the thermohydraulic heat balance made for the loop,
viz 9.42 at % FIMA (= 89.8 MWd/tM). The absolute errors, including
systematic errors, of the Nd measurements are of the order of 2 %.

NON-DESTRUCTIVE TECHNIQUES

The fuel pins of interest here are the seven shaded black in
fig. 1. Careful integration of the ^{137}Cs activity over the whole
length of the pins (1) gave pin-to-pin average burn-up values in
good agreement with the destructive measurements (neodymium), in
spite of cesium migration. Absolute burn-up measurements were tried,
using as monitors pin samples from previous irradiation experiments
and related destructive analysis results. As expected, the cesium
activity of these samples was not representative of their burn-up.
Chemically stable isotopes, such as ^{154}Eu, should be considered to
succeed in such measurements.

DETERMINATION BY CALCULATION OF THE BURN-UP

The calculation methods employed for the analysis of the MOL 7D
fuel pin burn-up were the same as those indicated in (1). The basic
code used to determine the neutron field was the one-dimensional
neutron transport code DTF-IV, containing several improvements made
at SCK/CEN. The code used a 40 energy group cross-section library,
including self-shielding factors (ABBN formalism), elaborated at
SCK/CEN for the study of coupled fast-thermal systems. For the dosimeter nuclides the ENDF/B V dosimetry file was used. The helium production by threshold (n,α) reactions was calculated with the data
given in (2). In the fuel pin burn-up calculations, the cross-
sections of the SCK/CEN forty group library were supplemented, for
the nuclides ^{234}U, ^{236}U, ^{237}Np, ^{238}Pu and ^{241}Am, by data taken from
the 171 group library VITAMIN-C (taking into account self-shielding),
to which thermal cross-sections (practically 1/v) were added. The
flux chart considered in the present calculations was the same as
the one used in (1), re-normalized in such a way as to obtain the
same calculated average FIMA value for the whole MOL 7D fuel pin
bundle (at level + 35 mm) as the value measured with neodymium, viz
8.38 %. The resulting calculated burn-up values for the various pin
samples at level + 35 mm are indicated in fig. 2 (curve A'). The
corresponding calculated nuclide concentrations are given in table I.

Concerning the shape of the radial burn-up distribution throughout the fuel pin bundle (fig. 2), a strong dip is measured in the central fuel pin, which does not appear in the calculated distribution. This was already observed earlier (1) for the Co (low energy) reaction rate in the fuel pin claddings : see fig. 3, taken from (1) and (3) (In the present fig. 3, the ^{235}U fissions are renormalized by the factor 1.023, in order to take account of the self-shielding of the ^{235}U dosimeter in the wrapper tube, neglected in (1) and (3)). In fig. 3 also the distribution of the ^{235}U fission rate measured in BRO2 (3) can be compared with the calculated one : the (rather small) dip in the central pin is again not rendered by the calculations. As mentioned above, the distribution of the burn-up throughout the

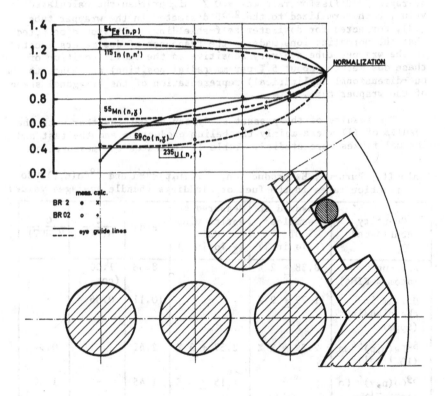

Fig. 3. Variation of reaction rates sensitive to low (^{235}U, ^{55}Mn and ^{59}Co) and high (^{115}In and ^{54}Fe) energy neutrons from the position allocated to the fluence monitors to different positions in the fuel pin bundle.

fuel pin bundle obtained with the destructive techniques discussed
in the present paper is in rather good agreement with the non-destructive relative measurement related in (1) (measured with ^{137}Cs
on whole fuel pins).

In (1) the calculated flux chart was normalized to the fission
rate in the ^{235}U dosimeter located in the wrapper tube. With this
normalization, and taking into account the correction for self-
shielding of this dosimeter mentioned above, the calculated burn-up
value, averaged over the fuel pin bundle at the level + 35 mm would
have amounted to 7.46 at % FIMA. The 8.38 at % FIMA value actually
measured is 12.4 % higher (because of the disappearance of fissile
atoms in the fuel pins, the flux chart has to be increased by 13.2 %
to obtain the 8.38 at % FIMA). These ratios can be compared with
what was observed for BR02 (see fig. 3), where the measured bundle-
averaged ^{235}U fission rate was 9.0 % higher than the calculated
value, both normalized to the ^{235}U dosimeter in the wrapper tube
((3), corrected for dosimeter self-shielding). It should be noted
that the normalizations made to the responses of dosimeters located
in the wrapper tube are very sensitive to the radial position of
these dosimeters (about 5 % per mm radial position) and hence to the
one-dimensional (cylindrical) representation of the hexagonal shape
of the wrapper tube.

The results of the present study can be put together with the
results of (1), where mainly the helium production and the fast and
thermal fluxes were studied. Table II indicates the measured and

Table II. Burn-up, He production, $^{54}Fe(n,p)^{54}Mn$ and $^{59}Co(n,\gamma)^{60}Co$
reaction rates in the fuel or claddings (bundle averaged values)

Quantity considered	Measured on samples DT ($\pm$ 1σ in %)	Measured on entire pins NDT ($\pm$ 1σ in %)	Calc.	$\dfrac{C}{M(DT)}$	$\dfrac{C}{M(NDT)}$
Burn-up (atom % FIMA)	8.38 $\pm$ 2 %	–	8.38	1.00 (norm.)	–
Helium in the claddings (appm)	11.79 $\pm$ 2 %	–	10.11	0.86	–
$^{54}Fe(n,p)^{54}Mn$ ($10^{-11}s^{-1}$)	2.39 $\pm$ 3 %	2.57 $\pm$ 3 %	2.42	1.01	0.94
$^{59}Co(n,\gamma)^{60}Co$ ($10^{-9}s^{-1}$)	–	1.15 $\pm$ 3 %	1.45	–	1.26

calculated burn-ups, helium productions, fast fluxes and thermal fluxes. The calculated values correspond to the normalization of the flux chart adopted in the present paper viz such that calculated and measured bundle-averaged burn-up values coincide. With this normalization (1.132 x 1.023 times higher than that adopted in (1)), the fast flux in the fuel pin claddings and the fuel pin burn-up are in excellent agreement. The calculated helium production is about 14 % too low, which could in part be due to uncertainties concerning the threshold reaction cross-sections and the N content in the claddings. The two-step nickel reaction only plays a small role (indeed, of the calculated 10.11 appm, 6.12 are due to threshold reactions in all constituents except N, 1.83 to threshold reactions in N, 1.91 to the ^{10}B reaction and 0.25 to the two-step nickel reaction). The calculated cobalt (thermal + epithermal) reaction rate is 26 % higher than the measured one.

GENERAL PROCEDURE FOR NEUTRON CHARACTERIZATION OF EXPERIMENTS

The distributions over the MOL 7D fuel pin bundle of the burn-up, the helium production and the damage (practically proportional to the fast fluence) are the three main neutron-related characteristics of the experiment to be determined.

The relative pin-to-pin distributions, as calculated, are in good agreement with the measured ones (destructive analysis), except for the strong thermal dip in the central pin, not rendered in the calculations. Non-destructive gamma scannings can also be performed and are in good agreement with the destructive results.

The absolute value of the three quantities considered can be determined by various methods :

A. by destructive measurements (most accurate, but also the most expensive method)

B. by calculations, with neutron flux charts normalized to the responses of adequate dosimeters located in the wrapper tube (cheapest method)

C. idem as B, with correction factors obtained from BRO2 measurements (this method obviously necessitates the irradiation in BRO2 of a mock-up of the loop considered).

Table III gives a summary of the results obtained with the three methods. For the burn-up determination - next to the thermo-hydraulic heat balance method, not further discussed in this paragraph - method C constitutes an important improvement with respect to method B (see also fig. 2). For the fast flux (damage) determination method B appears to be satisfactory. For the He determination method C does not bring an appreciable improvement : there remains a 16 % underestimation of the He production.

Table III. Neutron-related characteristics (bundle averaged values) of the MOL 7D irradiation as obtained with various methods

Quantity considered	Method used (see text)	Bundle-averaged flux correction factor from BR02	Value obtained absol. units	Value obtained rel. to A
Burn-up of the fuel pins (at % FIMA)	A.		8.38	1.00
	B. norm. to $^{235}U(n,f)$		7.46	0.89
	C.	$1.09 = ^{235}U(n,f)$	8.09	0.97
Helium production in the claddings (appmHe)	A.		11.79	1.00
	B. norm. to $^{54}Fe(n,p)$ for thresh. react. and to $^{235}U(n,f)$ for therm. react.		9.85	0.81
	C.	$0.99 = ^{115}In(n,n')$ $1.09 = ^{235}U(n,f)$	9.92	0.84
Fast flux ($10^{14}n/cm^2s$)	A.		3.00	1.00
	A_o. (non-destr. γ-scanning)		3.22	1.07
	B. norm. to $^{54}Fe(n,p)$		3.03	1.01
	C.	$0.99 = ^{115}In(n,n')$	3.00	1.00

This indicates the accuracy levels which can be reached according to the extent of the work (dosimetry, calculation, destructive examination). The accuracy increases clearly from method B, to C and to A, according to the effort devoted. Method B (and C) could be improved, as to the prediction of the helium formation, if fuel cladding samples are inserted as HAFM's (helium accumulation fluence monitors) in the dosimeter tube.

The present philosophy is applicable to loops other than MOL 7D also.

REFERENCES

1. Ch. De Raedt et al., "Dosimetry work and calculations in connection with the irradiation of large devices in BR2." Fourth ASTM-EURATOM symp. on reactor dosimetry, March 22-26, 1982. NUREG/CP-0029, Vol. 1, 229-244.

2. E.P. Lippincott et al., "He production in reactor materials," Conf. on Nucl.Cross-Sections & Techn.,Wash. DC, March 3-7,1975.

3. J. Debrue et al., "Dosimetry work in connection with irradiations in the high flux materials testing reactor BR2," First ASTM-EURATOM symp. on reactor dosimetry, JRC, Petten, September 22-26, 1975. EUR 5667 e/f, Part I, 700-737.

USE OF THRESHOLD ACTIVATION DETECTORS TO OBTAIN NEUTRON KERMA

FOR BIOLOGICAL IRRADIATIONS

C. Eisenhauer, J. Grundl   C. Cassapakis and V. Verbinski

National Bureau of Standards Science Applications, Inc.

Gaithersburg, MD 20899 USA   P.O. Box 2351, La Jolla, CA 92038 USA

ABSTRACT

Fission and non-fission activation foils have been irradiated in the experimental room at the Armed Forces Radiobiology Research Institute (AFRRI) in Bethesda, Maryland, in order to characterize the neutron field there. The field, which is generated by neutrons from a TRIGA MARK-F reactor adjacent to the room, is used for radiobiological experiments. The neutron exposure parameter, neutron tissue kerma rate per unit reactor power, is derived from individual detector measurements and an *a priori* spectrum obtained from neutron transport calculations. Beginning with conventional averaging of single-detector results, the analysis proceeds to a least-squares adjustment which avoids spectrum interpolation.

I. PHYSICAL DESCRIPTION OF FIELD

The neutron source for the radiation field considered in this paper is a TRIGA MARK-F pool-type research reactor with both pulsed and steady state capabilities[1] located at the Armed Forces Radio-biology Research Institute (AFRRI). It is licensed by the U.S. Nuclear Regulatory Commission to operate at steady-state power levels up to a maximum of 1.0 MW (thermal).

There are two exposure rooms available to users. The field described in this paper is in Exposure Room Number 1 (ER1) at 1 m from the center of the reactor core. This room is 6.1 m by 6.1 m by

J. P. Genthon and H. Röttger (eds.), Reactor Dosimetry, 921–928.
© *1985 ECSC, EEC, EAEC, Brussels and Luxembourg*

2.6 m high. A semi-cylindrical portion of the reactor tank projects
into the south wall. When the reactor core is adjacent to ER1 the
center of the core is approximately in the plane of the south wall.
The six interior concrete surfaces of the room are lined with 0.3 m
of wood to minimize neutron activation in the concrete. The inner
surface of the wood is painted with gadolinium oxide to reduce the
thermal neutron background in the room.

II. ACTIVATION DETECTOR MEASUREMENTS

Measurements in ER1 with threshold and non-threshold activation
detectors have been reported by Verbinski et al.[2] They were made
with thirteen different types of detectors, some of which were surrounded by boron to reduce response from low-energy neutrons. Eight
of these reactions were initially examined because the detector
response functions were available on the ENDF/BV Dosimetry File.[3]
The energy range over which each of eight detectors responds is
shown in Fig. 1. The tick marks show the 95, 50, and 5 percentile
response energies for the calculated spectrum in the AFRRI room.

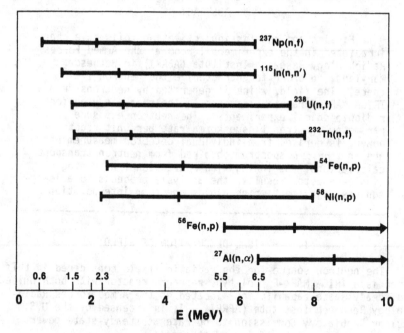

Fig. 1 Lines indicating the 95th, 50th, and 5th percentile energies
of response for different reactions in the neutron spectrum
in AFRRI exposure room 1.

Since the pairs ^{238}U-^{232}Th and ^{54}Fe-^{58}Ni are redundant in their spectrum coverage, ^{232}Th and ^{58}Ni were eliminated and the main analysis performed with the six remaining detectors.

III. AVERAGE KERMA RATE FROM INDIVIDUAL DETECTORS AND AN *A PRIORI* SPECTRUM

These measurements were analyzed in increasing degrees of analytic refinement in order to infer a value of the neutron kerma rate. These analyses can be grouped into two general methods. In the first method the kerma rate was obtained from the individual detector responses and from integral cross sections derived from an *a priori* neutron spectrum calculated by Verbinski et al.[4] The following expression was used to determine the kerma rate, K_r, for each reaction:

$$K_r = \overline{K}\phi = \frac{\overline{K} R_r}{\overline{\sigma}} = \frac{\overline{K} R_r}{\sigma(> E^r_{95})\psi(> E^r_{95})/0.95} \tag{1}$$

where $\overline{K}$ is the spectrum-averaged fluence-to-kerma conversion factor,

ϕ is the neutron fluence rate,

R_r is the measured average reaction rate for each detector,

$\overline{\sigma}$ is the spectrum-averaged reaction cross section,

E^r_{95} is the energy above which 95% of the detector response occurs,

$\overline{\sigma}(> E^r_{95})$ is spectrum-averaged cross section for energies greater than E^r_{95}, and

$\psi(> E^r_{95})$ is the fraction of the fluence rate above E^r_{95}.

Table I shows the results of this analysis.

Kerma rates K_r deduced for each detector are shown in the lower section of Table I. If all the detector responses, cross sections, and spectrum were correct, all kerma rate values would be the same. An unweighted average of these values is 1.69 ± 0.07 mGy s^{-1} kW^{-1}. A more refined estimate of K_r can be obtained if each estimate is weighted by the relative energy range spanned by that detector. For this weighting factor we have chosen the kerma coverage factor W_{K_r} defined for each detector by

$$W_{K_r} \equiv \int_{E^r_{95}}^{\infty} K(E)\phi(E)dE / \int_{0.4eV}^{\infty} K(E)\psi(E)dE \quad . \tag{2}$$

Table I. Detector Response Parameters and Kerma Rates Calculated from Eq. (1).

$$\overline{K} = 1.965 \times 10^{-8} \text{ mGy cm}^2$$

Detector	E^r_{95} (MeV)	$\sigma(> E^r_{95})$ (barns)	$\psi(> E^r_{95})$	R_r
^{237}Np	0.633	1.579	.531	8.51-17
^{115}In	1.213	0.276	.410	1.16-17
^{238}U	1.515	0.549	.358	1.66-17
^{54}Fe	2.424	0.257	.228	5.37-18
^{56}Fe	5.568	0.0272	.0371	8.31-20
^{27}Al(n, α)	6.513	0.0386	.0190	5.85-20

Detector	K_r (Tissue) (mGy s^{-1} kW^{-1})	W_{K_r}	K_r W_{K_r} (mGy s^{-1} kW^{-1})
^{237}Np	1.88	.884	1.67
^{115}In	1.91	.754	1.44
^{238}U	1.58	.683	1.08
^{54}Fe	1.71	.483	0.83
^{56}Fe	1.54	.092	0.14
^{27}Al(n, α)	1.49	.049	0.07
$\sum$	10.11	2.945	5.23
K	$\dfrac{10.11}{6} = 1.69$		$\dfrac{5.23}{2.945} = 1.78$

The use of such a weighting factor neglects any correlations due to the overlapping responses of the detectors (see Fig. 1). When estimates are weighted by the values of W_{K_r} listed in Table I, a weighted mean kerma rate of 1.78 ± 0.07 mGy s^{-1} kW^{-1} is obtained. The uncertainty assigned to the kerma rate is the standard deviation of the mean, and, as such, is not related to any evaluation of uncertainties in experimental reaction rates or calculated cross sections.

IV. AVERAGE KERMA RATE FROM DETECTOR RESPONSES AND AN ADJUSTED SPECTRUM

The second method of estimating kerma rate is to adjust the calculated spectrum on the basis of spectral indices obtained from the set of threshold detectors. Observed and calculated spectral indices are indicated in Table II. Nominal errors of 5% have been

assigned to the observed indices. A 5-group spectrum adjustment and
error propagation was made with the NBS SPAD code, based on the six
detector measurements. In this code, which involves no spectrum
interpolation, the square of the difference between measured and
calculated spectral indices is minimized, with the restriction that
the number of energy groups be equal to or less than the number of
detectors with distinguishable energy response. Uncertainties in
the 5-group cross sections due to shape were set equal to the
nominal 5% chosen for the observed indices.

Table II. Observed and Calculated Spectral Indices

Spectral Index	Observed	Obs/Calc
$^{238}U/^{237}Np$	0.195 ± 5%	0.832
$^{115}In/^{237}Np$	0.137 ± 5%	1.011
$^{54}Fe/^{237}Np$	0.063 ± 5%	0.904
$^{56}Fe/^{237}Np$	0.00098 ± 5%	0.811
$^{27}A\ell(n, \alpha)/^{237}Np$	0.00069 ± 5%	0.784

Boundaries of the energy group structure were tailored to the
threshold energies of the detectors. Thus, energy bounds of 0.6,
1.5, 2.3, 5.5, and 6.5 MeV correspond approximately to threshold
energies for ^{237}Np, ^{238}U or ^{115}In, ^{54}Fe, ^{56}Fe, and $^{27}A\ell(n, \alpha)$,
respectively.

The fraction of fluence rate below 0.6 MeV, not included in the
analysis because of uncertain low-energy detector response, was
fixed at its value in the *a priori* spectrum. Table III shows the
results of this analysis. The first two columns show the normalized
a priori spectrum ($\phi_{og}/\Sigma\phi_{og}$) and the normalized adjusted group spectrum ($\phi_{ag}/\Sigma\phi_{ag}$). The third column gives the adjusted absolute
fluence rate spectrum in units of 10^8 cm^{-2} s^{-1} kW^{-1}. The absolute
fluence rate was obtained by calculating a fluence rate $\phi_r = R_r/\bar{\sigma}$
from each reaction rate R_r and $\bar{\sigma}$ (averaged over the adjusted
spectrum), weighting each value of ϕ_r by W_ϕ obtained by setting
$K(E) \equiv 1$ in eq. (2), (weighting by spectrum coverage), and taking
the average for the six reaction rates. The fourth column gives the
group-averaged fluence-to-kerma conversion factor calculated from
the *a priori* spectrum. The last column gives the group kerma rates
obtained from the adjusted spectrum. The resulting spectrum-
averaged kerma rate is

$$K_{adj} = 1.91 \pm 0.05 \text{ mGy s}^{-1} \text{ kW}^{-1} \quad .$$

The uncertainty assigned to K_{adj} is obtained by propagation of the
nominal 5% uncertainties in the observed spectral indices (Table II)
and in the detector cross sections for each reaction, and combination of these uncertainties in quadrature.

Table III. Adjustment Parameters

E_B (MeV)	g	$\phi_{og}/\Sigma\phi_{og}$ (*a priori*)	$\phi_{ag}/\Sigma\phi_{ag}$ (adjusted)	$\phi\times10^{-8}$ (cm^{-2}s^{-1}kW^{-1})	$\overline{K}_g\times10^8$ (mGy cm^2)	Kerma Rate (mGy s^{-1}kW^{-1})
0						
	1	.4617	(.4617)[a]	.4665	0.468	.218
0.6						
	2	.1780	.2467±20%	.2491	2.304	.574
1.5						
	3	.1148	.0673±52%	.0680	3.000	.204
2.3						
	4	.2068	.1944±13%	.1964	3.876	.761
5.5						
	5	.0195	.0155±47%	.0157	4.672	.073
6.5						
	6	.0192	.0144± 9%	.0146	5.286	.077
18.0						
	Σ			1.010		1.907±2.8%

[a]The fluence rate below 0.6 MeV is fixed at its value in the *a priori* spectrum.

Table IV. Adjusted Fluence Rate Spectrum and Kerma Rate for Different Combinations of Input Detectors

E_B (MeV)	g	6 detectors (excl. Th, Ni)	5 detectors (excl. Th, Ni, U)	5 detectors (excl. Th, Ni, In)	8 detectors
0					
	1	(.4617)	(.4617)	(.4617)	(.4617)
0.6					
	2	.2467	.2621	.1530	.2527
1.5					
	3	.0673	.0535	.1638	.0614
2.3					
	4	.1944	.1912	.1886	.1945
5.5					
	5	.0155	.0170	.0178	.0155
6.5					
	6	.0144	.0145	.0150	.0143
18.0					
Kerma Rate (mGy s^{-1} kW^{-1})		1.91	1.90	1.97	1.90

The effect of removing or adding detectors in the analysis has been examined. Table IV shows the adjusted fluence rate spectrum obtained with four different combinations of detectors. The inconsistency in the observed spectral indices involving ^{238}U and ^{115}In, which have similar response thresholds (See Fig. 1) is reflected in large differences in the adjusted spectrum when one or the other is excluded.

In order to see how sensitive the adjusted spectrum is to the procedures contained in the SPAD code, the spectrum can be compared with those obtained from unfolding codes which use finer energy groups and hence impress spectrum interpolation schemes. Using the same six observed detector responses and the same *a priori* spectrum, P. Lippincott of HEDL and F. Stallman of ORNL used the FERRET and LSL codes, respectively, to obtain adjusted spectra. The adjusted spectra[5,6] and kerma rates are shown in Table V. The kerma rates from HEDL and ORNL are lower than the NBS kerma by 10%, 11%, and 6%.

Table V. Adjusted Fluence Rate Spectrum and Kerma Rate using Different Adjustment Codes

E_B (MeV)	SPAD(NBS)[a] 6 Groups	FERRET(HEDL)[5] 53 Groups	LSL(ORNL)[6] 37 Groups	6 groups
0	$(.462)$[b]	.457	.465	.462
0.6	$.247 \pm .049$	.187	.181	$.215 \pm .047$
1.5	$.067 \pm .034$	.118	.115	$.095 \pm .022$
2.3	$.194 \pm .024$	.203	.203	$.195 \pm .019$
5.5	$.016 \pm .008$	.017	.018	$.018 \pm .005$
6.5	$.014 \pm .001$	.018	.017	$.016 \pm .002$
18.0				
Kerma Rate (mGy s^{-1} kW^{-1})	1.91 ± 0.05	1.72	1.70	1.79

[a] The propagated errors correspond to nominal uncertainties of 5% in observed spectral indices, group cross sections, and cross section scales.

[b] The fluence rate below 0.6 MeV is kept fixed at its value in the *a priori* spectrum.

V. COMPARISON WITH OTHER MEASUREMENTS

The kerma deduced from threshold detector measurements can
be compared with direct measurements of kerma rate by Goodman[8] in
the same neutron field with tissue equivalent ionization chambers.
His measured values at two different times were 1.95 and
2.08 mGy s^{-1} kW^{-1}, with an estimated uncertainty of 5%. The
average of these values of kerma differs from that measured with
reaction rates by about 6%, well within the uncertainties quoted for
each measurement. Comparisons with kerma rates obtained from other
adjustment codes (Table V), however, show differences as large as
11%. Using the latter figure as an estimate of uncertainty, it can
be said that the kerma rate in the AFRRI exposure room ER1 at 1 m
from the center of the reactor is

$$K = 1.91 \pm 11\% \text{ mGy s}^{-1} \text{ kW}^{-1}.$$

VI. ACKNOWLEDGMENTS

We wish to thank P. Lippincott of HEDL and F. Stallman of ORNL
for calculating adjusted spectra with their unfolding codes. This
work was supported by the Defense Nuclear Agency, U.S. Department of
Defense.

REFERENCES

1. Sholtes, J. A., Jr. and Moore, M. L., Reactor Facility, Armed
   Forces Radiobiology Research Institute, AFRRI Technical Report
   AFRRI-TR81-2 May 1981.

2. Verbinski, V. V., Cassapakis, C. C., Hagan, W. K., Ferlic, K.,
   and Daxon, E., Radiation Field Characterization for the AFRRI
   TRIGA Reactor, Volume 1, Baseline Measurements and Evaluation
   of Calculational Data, 1 June 1981.

3. ENDF/BV Dosimetry File.

4. Verbinski, V. V. and Cassapakis, C. G., Calculation of the
   Neutron and Gamma-Ray Environment in and around the AFRRI TRIGA
   Reactor, Volume II, 1 June 1981.

5. P. Lippincott (private communication).

6. F. Stallman (private communication).

7. Compendium of Benchmark Neutron Fields for Reactor Dosimetry,
   Standard Neutron Field Entries, ^{252}Cf and ^{235}U Fission Neutron
   Sources, Intermediate-Energy Standard Neutron Field (ISNF),
   Evaluated and compiled at the National Bureau of Standards.

8. L. Goodman (private communication).

STATUS OF RECOVERY - THREE MILE ISLAND UNIT 2

REACTOR HEAD REMOVAL

R. L. Freemerman, W. C. Hopkins, and R. L. Rider

Bechtel North American Power Corporation

Gaithersburg, MD, USA

ABSTRACT

On July 24, 1984 the head of the reactor pressure vessel of Three Mile Island Unit 2 was successfully removed. This major milestone comes after the successful refurbishment and load tests of the polar crane in February 1984. The polar crane was the only apparatus available to remove the 159.5-ton reactor pressure vessel head to its shielded storage stand. Prior to head removal, valuable data on the damaged reactor core was obtained using sonar and remote video techniques and also solid state track recorders. Data was also taken both before and after the head removal to assess the impact on the radiological environment. Indications are that generally dose rates in the containment have dropped.

POLAR CRANE REFURBISHMENT

The large overhead polar crane in the containment building was subject to high temperatures following the accident. Visual inspections and electrical tests indicated the crane was structurally sound but sustained some corrosion to exposed surfaces and electrical wiring and components. A refurbishment plan was conducted which involved a complete inspection of all load-bearing components and a replacement of the brakes to ensure complete safety once the crane was put back into operation. Because of the difficulty and large number of worker hours to repair and replace electrical wiring and components to the

J. P. Genthon and H. Röttger (eds.), Reactor Dosimetry, 929–935.
© 1985 ECSC, EEC, EAEC, Brussels and Luxembourg

original design configuration, some modifications to the power and control circuits were designed and installed, and selected relays and other electrical parts were replaced. This was accomplished with a minimum of person-rem radiation exposure, despite radiation levels near 100 mr/hr in the work areas. Following these inspections and modifications, a load test of the crane was successfully completed and the crane was returned to service.

CORE DATA

Radiation measurements (1) were taken prior to head lift to assess the dose rate levels for the underside of the head, the control rod mechanisms, and the top of the plenum head. Analyses (2, 3) of that data indicated dose rates from 100 to 1,000 rem per hour with a corresponding estimated plateout of Cs-137 at 1,000 micro-curies per square centimeter. The cesium, normally soluble in water, appears to be chemically bound to the surface (4).

Sonar measurements (5) and new clear video pictures have provided a vast improvement in the knowledge of the degree of core damage. It is now estimated that about 5/12 of the original core has been totally displaced as a result of the accident, leaving a large void at the top of the core. A small amount of fuel has, in fact, been found in filters in the reactor coolant letdown and purification system (these systems were operating during the accident). This displaced fuel is now believed to be located in the bottom of the reactor pressure vessel and in other parts of the reactor coolant system such as the main reactor piping, pressurizer, coolant pumps and steam generators. The presence of fuel in the bottom of the vessel was confirmed by recent measurements (6, 7) with solid state track recorders (6). All these data were taken in preparation for activities leading up to and following head removal.

THE HEAD LIFT

The cleanup at TMI-2 reached a major milestone on July 24, 1984, when the reactor vessel head was removed. This is the first major step in the disassembly of the reactor leading to removal of the damaged core. The milestone, previously scheduled for August 1984, was achieved ahead of schedule.

Preparations for head removal had begun more than two years ago with the development of modifications to plant systems and tooling. The accident had rendered some of the plant systems inoperable and resulted in much higher radiation levels in the containment than would be seen during a normal plant outage, even though decontamination of the surfaces and installation of shielding around the top of the reactor vessel head had been performed to help reduce these fields. A small processing subsystem was designed and installed to clean the water in the vessel and a new, long-term seal was developed for the canal seal plate. Tooling was modified or new tools developed to perform head removal and later operations in the higher radiation fields.

The 60 large studs that hold the reactor pressure vessel head in place were then detensioned and removed. These studs had been somewhat corroded during the five years since the accident and were expected to be difficult to remove. While some of the studs required several attempts, the detensioning was completed without incident.

Due to the expected rise in the radiation levels within the containment as the head was being lifted and moved to its storage stand, many of the activities utilized remote techniques assisted by video monitoring. The first step involved aligning the crane over the head and attaching the adjustable lifting pendants. After the crane was raised to apply some load and to check the pendants, then lowered to make adjustments (this took five attempts), the head was ready to be removed. The crane operations were performed by an operator stationed behind shielding inside the containment who monitored the lift from several video cameras. The head was then lifted 33 inches, which allowed a clearance of 6 inches to clear the alignment keys. Radiation measurements and camera inspection under the head indicated conditions were lower than predicted to allow the removal to proceed.

A large contamination bag was remotely slipped under the head and secured. The head was then raised (see Figure 1) and moved to its storage stand on the 347'-6" Elevation in the containment. The stand had been shielded with sand bags and water columns to reduce radiation levels after the head was stored. Some minor difficulties were experienced during the transfer. The signals from two cameras on the head, used for remote alignment when being lowered onto the stand, were lost during the transfer, and the temporary shielding placed on the head for the transfer may have caused an uneven mass distribution on the head. These problems made the alignment of the crane difficult, and the transfer was temporarily halted with the head behind the shielding but suspended off the stand. A temporary scaffold was erected to allow workers to climb up to the top of the head service structure and to complete setting of the head onto the stand.

With the head removed from the reactor vessel, the top of the large plenum assembly was exposed. A ring called the internals indexing fixture (IIF), was then installed onto the reactor vessel flange, which was then filled with water to provide 5 feet of water shielding over the plenum. A work platform with additional integral shielding was then placed on the IIF. This work platform will be used to support later inspection and removal operations. The total personnel exposure used for the preparations for head lift was about 270 man-rem and about 10 man-rem for the actual head removal operation.

POST LIFT RADIATION CONTAINMENT LEVELS

Table 1 reprises the radiation levels on the operating deck of the TMI-2 containment both before and after the removal of the reactor pressure vessel head. As can be seen, the dose rates have dropped with the head removed. This is primarily due to two factors. One is the fact that the head and its contaminated control rod drive service structure has been stored inside a shielded reactor vessel head stand. Secondly, the installation of the internals indexing fixture provides additional shielding from the upper surface of the plenum where dose rates of about 1,000 rem per hour have been recorded. These lower dose rates will greatly contribute to the overall project ALARA man-rem goal of being near the lower end of the predicted range of 13,000 to 46,000 man-rem.

TABLE 1

TMI-2 Containment Dose Rates on El. 347'
Before and After Head Lift

Location	Dose Rate Before (millirem/hour)	Dose Rate After (millirem/hour)
Head Storage Stand	100	50*
South End of Refueling Pool	50	30
East Canal Walkway	40	30

*Sand Shields for Head in Place

THE NEXT STEP

The next major step will be the removal of the 55-ton plenum assembly from the reactor vessel. Underwater television pictures reveal that most of the fuel assembly upper end fittings and many stubs of partial fuel rods remain suspended in the plenum assembly in spite of the large void in the upper portion of the damaged core. A series of inspections are under way to determine how difficult it will be to remove the stuck end fittings. Attempts will be made to knock the fittings out of the plenum. The results of these activities will determine the contamination control provisions that will be required for the removal and storage of the plenum assembly.

The accident produced temperatures at the top of the core in excess of 2000°F. The possibility exists that some distortion has occurred to the plenum assembly which could interfere with removal operations. Another major factor in planning for the removal of the plenum is the outer row of fuel assemblies that are partially intact, but likely have their upper end fittings stuck in the plenum in the same manner as the fittings above the core void. The perimeter end fittings cannot be separated until the plenum is lifted. In order to control the initial lift of the plenum, a set of four hydraulic jacks have been designed. The jacks will be installed and the plenum will be carefully raised in increments. Potential areas of interference will be monitored as the initial lift is made. The lift of the plenum will continue until it clears the normal elevation of the fuel assembly end fittings. The lift will be stopped when the plenum is raised about nine inches. With the plenum supported by the jacks, the remaining end fittings will be knocked free, at which time the plenum will be ready for its final lift and transfer to a storage position in the containment building refueling canal.

Other preparations are being made and the design is proceeding for fuel canisters and the fuel removal tooling in anticipation of the start of defueling in mid-1985.

REFERENCES

1. GPUNC Technical Planning Department, Data Report on Underhead Data Acquisition Program, TPO/TMI-110, Revision 0, March 1984, Section 2.0.

2. GPUNC Technical Planning Department, Data Report on Dose Modeling of Underhead Source, TPO/TMI-042, Revision 1, December 1983.

3. N. L. Osgood, W. C. Hopkins, H. K. Peterson, "Analysis of Radiation Measurements Taken Inside the TMI-2 Reactor Vessel," Paper No. 340, 29th Annual Meeting of the Health Physics Society, June 3-8, 1984, New Orleans, LA.

4. TPO/TMI-110, op. cit., Section 4.0.

5. TPO/TMI-110, op. cit., Section 9.0.

6. TMI-2 TAAG Meeting of October 31 and November 1, 1983, "Reactor Cavity Neutron Flux Measurements--SSTR Results," Westinghouse Hanford Company (HEDL).

7. TPO-TMI-110, op. cit., Section 11.0.

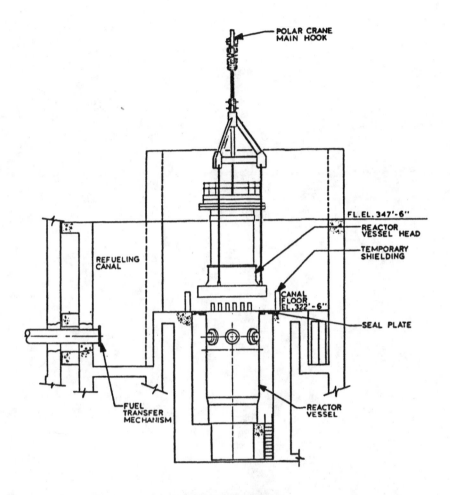

POLAR CRANE
MAIN HOOK

FL.EL.347'-6"

REACTOR
VESSEL HEAD

TEMPORARY
SHIELDING

REFUELING
CANAL

CANAL
FLOOR
EL.322'-6"

SEAL PLATE

FUEL
TRANSFER
MECHANISM

REACTOR
VESSEL

REACTOR VESSEL
HEAD REMOVAL
SECTION VIEW LOOKING EAST

FIGURE 1

ON THE UTILIZATION
OF ΣΣ-ITN BENCHMARK FIELD

I.Garlea, C.Miron, C.Roth, D.Dobrea, T.Musat

Central Institute of Physics

Bucharest, ROMANIA

ABSTRACT

There are described the ΣΣ-ITN system and its neutronic characteristics. There are presented the works for calibrating fission chambers, recoil proton counters and fissionable detectors associated with solid state track recorders (SSTR). In this spectrum have been also determined the fission yields. The ΣΣ-ITN spectrum has been used as reference one for neutron dosimetry in the channels of TRIGA ROMANIA reactor, at the 14 MeV neutron generator, at plasma focus installations and for measurements of fission densities in irradiated Romanian fuel. There was also performed in ΣΣ-ITN the analysis by radioactivation of some archeological objects and ores.

THE ΣΣ-ITN SYSTEM DESCRIPTION

The ΣΣ-ITN system is placed in the graphite thermal column of VVRS-INPE reactor, in a spherical cavity, 500 mm in diameter and the centre at 1530 mm from the reactor core. Within this cavity in graphite the neutron spectrum is thermal (1), having the following characteristics:
- neutronic temperature: 305 ± 7 K;
- Cd ratio for Gold: 400 ± 10;
- intensity: $10^3 - 5 \times 10^9$ n/cm^2 s.

The fission source of ΣΣ-ITN system is a multilayer sphere made out of metallic Uranium, having an inner spherical

J. P. Genthon and H. Röttger (eds.), Reactor Dosimetry, 937–942.
© 1985 ECSC, EEC, EAEC, Brussels and Luxembourg

experimental cavity, lined by a filter of boron carbide vibrocompacted in Aluminium. The ΣΣ ITN dimensions are identical to those of ΣΣ Mol (2), the both devices being made by the same materials. The experimental cavity could be acceded by two channels (vertical and horizontal) that pass through the plugs of the Uranium sphere. Some vacuum tight devices provided with locks allow handling and storage of Uranium sphere inside a lead cell.

The neutronic characterization of the system has been achieved by calculations, differential and integral measurements. Using the ANISN code (3) and DLC-2D/100G library (4) have been calculated the neutronic spectrum and have been appraised the influence of the various materials in its composition.

The neutron spectrum measurements by recoil proton spectrometers supplied a neutronic characterization over the range 5 keV - 1.5 MeV, performed the frame of scientific cooperation with CEN/SCK Mol Belgia (5) and KfK Rossendorf GDR(6). The absolute reaction rates and integral cross sections (7) averaged on the ΣΣ spectrum measured for some (n,f), (n,p), (n,α), (n,γ) and (n,n') reactions have been used to get the neutron spectrum by unfolding method, over the range 10^{-10} - 18 MeV (8). The program performed by means of fission chambers absolutely calibrated in ΣΣ Mol and ΣΣ -ITN spectra allowed the setting of an absolute fission scale for these spectra and has indicated their neutronic identity (9).

The calculations and measurements made for ΣΣ-ITN spectrum characterization showed that:
- the average energy of spectrum is 0.791 MeV;
- the integral flux value above 10 keV is 3.4×10^9 n/cm^2 s and above 1 MeV - 8.1×10^8 n/cm^2 s (for a full power of VVRS-INPE reactor of 2 MW).

CALIBRATION OF DETECTORS AND INSTRUMENTATION

The ΣΣ -ITN system generates a neutron spectrum in the intermediate energy range, similar to that in the fast reactors. It is thus appropriate for testing the integral cross sections for the reactors used in the neutron dosimetry. Nuclear data regarding 18 activation and 7 fission reactions have been tested in this spectrum and this has permitted the selection of detectors used in the neutronic measurements for characterization of the steady-state core of TRIGA ROMANIA reactor (10).

The ΣΣ and thermal standard spectra (1) have been used for calibration of instrumentation utilized in neutron dosimetry techniques, as follows:
- recoil proton counters (1,5,6). The stainless steel spherical

proportional counters having an inner diameter of 29.5 mm,
a wall thickness of 0.25 mm and filled by CH_4 or H_2 at the
2 or 4 atm.pressure, have been tested in $\Sigma\Sigma$ spectrum and
used for measurements in XC-1 channel (air-filled) in TRIGA.
- fission chambers. The miniature fission chambers Saclay type,
have been absolutely calibrated in $\Sigma\Sigma$ spectra. There was performed the quality control of fissionable deposits, determining the mass of main isotope and fissionable impurities.
The calibration errors for these deposits are: 1.6% for
$235U$, 2.4% for $238U$ and $237Np$, 1.8% for $239Pu$ and $233U$,
1.9% for $241Pu$ and 5% for $232Th$ (11).
- fissionable detectors. Pure metallic disks for Al alloys,
delivered by ORNL, have been calibrated in $\Sigma\Sigma$-ITN, in order
to use them associated with SSTR for fission rate measuring.
In the next table there are given the results of calibration, of number of nuclei that have left tracks on the
SSTR, per cm².

TABLE 1

Detectors		Nuclei/cm²	Errors (%)
$235U$	10% alloy in Al	3.02×10^{17}	2.4
$235U$	20% alloy in Al	6.57×10^{17}	2.5
$239Pu$	5% alloy in Al	1.48×10^{17}	2.6
$239Pu$	10% alloy in Al	3.01×10^{17}	2.7
$238U$	metallic	8.32×10^{18}	2.9

- nuclear fuel disks. These have been exposed in $\Sigma\Sigma$-ITN and
thermal standard spectra. There was determined the number
of fissionable nuclei in the layer that produces detectable
fissions on the SSTR. The disks are cut from Romanian fuel
pallets (natural Uranium, sintered UO_2). The average value
for the thin fuel disks is 6.10×10^{18} nuclei/cm² ($\pm 2.5\%$).

In order to measure the flux spectrum and fission densities
we have determined the fission yields for the (n,f) reactions of
$235U$, $238U$ and $239Pu$ for the following fission products: $95Zr$,
$103Ru$, $140Ba$, $131I$.

APPLICATIONS. The above described calibrations have allowed
us to use the detectors for some applications: characterization of
some exposure channels in TRIGA ROMANIA reactor, determination of
fission density in tested nuclear fuel, measuring of flux-spectrum
in plasma focus installations and flux monitoring of a 14 MeV neutron generator.

The TRIGA ROMANIA reactor, provided with exposure channels in the 14 thermal MW steady-state core, is intended for fuel and structural material testing. The XC-1 channel, located in the core centre, has a major importance for irradiation tests, having the highest neutronic flux. The determination of neutron flux-spectrum has been performed by multiple foil method, using a large number of absolutely measured fission and activation rates. The input spectral shapes using unfolding method, have been built from calculations and experimental results (measurements by means of recoil proton conters, in the range 0.3-1.9 MeV). The neutron flux-spectrum has been obtained by means of the computer codes SANDII, CRYSTAL BALL and STAY'SL by using the associated nuclear data libraries ENDF/B IV and V, Dosimetry files.

The monitoring system has used the data regarding the thermal power as well as the informations obtained from the fission chambers installed as monitors for these measurements. The irradiation have been performed by means of a hydropneumatic rabbit.

The results of measurements for the XC-1 channel are given in the following table, for two situations: "unpoissoned" and "poissoned" core, at full power of reactor (14 thermal MW). The energy ranges (in MeV) are: 10^{-10} - 18, 10^{-10} - 1.25×10^{-6} and 1.25×10^{-6}-18.

TABLE 2

Core	Channel	Flux-spectrum ($\times 10^{14} n/cm^2 s$)		
poissoned	water-filled	3.89 ± 7.3%	2.49	1.40
unpoissoned	water-filled	4.95 ± 6.8%	3.28	1.67
	air-filled	2.34 ± 6.5%	0.64	1.70

Three independent techniques have been used to obtain the fission density fuel elements made in Romania, irradiated in loops and capsules in TRIGA reactor (14 thermal MW):
- SSTR method - Both the disks of Romanian fuel and Makrofol foils have been irradiated in a dismountable fuel element. The fuel disks have been absolutely calibrated (in number of fissionable nuclei that have left tracks on solid state detector). The track density has been counted by means of image analyzer Quantimet 720.
- measurement of the neutron flux-spectrum in fuel by multiple foil method - The fission density has been determined from the absolute value of neutron flux and the cross sections averaged over the spectrum in fuel.
- high resolution gamma spectrometry for some fission products with well known yields - The measurements have been performed on the Romanian fuel disks, by means of 100 cm³ crystal Ge-Li, absolutely calibrated in efficiency.

The measurements have been made at low power as well as at full power, using a neutronic flux monitoring system with absolutely calibrated fission chambers.

The results at full power of reactor, for XC-2 channel, are presented in the following table:

TABLE 3

Power	Method	Power density (W/cm)
15.3 MW	Fission product	152.24 ± 5.2%
W	SSTR	148.30 ± 4.5%
W - kW	Flux spectrum	151.23 ± 6.5%

The flux of fast neutrons emitted by plasma focus installations of Central Institute of Physics Romania has been measured by means of solid state track recorders associated to the same absolutely fissionable detectors. The measure procedure is based on the analysis of damage fields created in a dielectric material by fission fragments emitted by fissionable deposits exposed to the neutronic bombardment. The used fissionable deposits have been calibrated in the reference $\Sigma\Sigma$ and standard thermal spectra to determine the number of fissionable nuclei within a thin layer from surface of detector that produces tracks.

The accurate determination the activation cross sections in the $\Sigma\Sigma$-ITN reference spectrum has allowed the analysis by radioactivation of some archeological objects and concentrates of ores.

REFERENCES

1. I.Garlea - Doctor Thesis, Bucharest, Romania (1979)

2. A.Fabry, G.De Leeuw, S.De Leeuw - The Secondary Intermediate Energy Standard Neutron Field at the Mol $\Sigma\Sigma$ Facility, in Nuclear Techn., 25, 349 (1975)

3. W.W.Engle - A One-dimensional Discrete Ordinates Transport Code with Anisotropic Scattering, Report K-1693, March (1973)

4. * * * - Documentation for DLC-2D/100 G Data Library. Radiation Shielding Inf. Centre, ORNL, USA (1973)

5. I.Garlea, A.Fabry, S.De Leeuw, G.De Leeuw - Application of the Methan Filled Spherical Recoil Proton Proportional Counters in the CEN/SCK Mol Secondary Standard Neutron Spectrum Facility. IAEA Spec.Meeting "Fast Reactor Spectrum

and their interpretation". A.N.L., USA (1974)

6. D.Albert, W.Hansen, W.Vogel, I.Garlea, C.Miron, C.Roth -
    "Investigation of the neutron Spectrum in the Reference Spec-
    trum ΣΣ-ITN by means of Spherical Proton Recoil Counters.
    Rep. RPP-16/80 (1980).

7. I.Garlea, C.Miron, M.Lupu, P.Ilie, A.Thurzo, N.Stanica,
    F.Popa - Measuring of a Few Integral Data in the Neutron
    Field. Rev.Roum.Physique, 4, 409 (1978)

8. I.Garlea, N.Stanica, M.Selariu, C.Miron, G.Fodor - Determi-
    narea spectrului neutronic absolut cu ajutorul codului SANDII
    din masuratori absolute de rate de reacţie in sistemul ΣΣ-ITN
    Studii si Cercet.de Fizica, No.1, 59 (1979)

9. I.Garlea, C.Miron, A.Fabry - Intercomparison of Fundamental
    Fission Rates and Ratios in the Intermediate Energy Stan-
    dard Neutron Fields at the ΣΣ-ITN and Mol ΣΣ Facilities. Rev.
    Roum.Physique, Tome 22, 6, 627 (1977)

10. I.Garlea, C.Miron, T.Muşat, C.Roth - Flux Measurements in
    XC-1 Channel, TRIGA Unperturbed Core, at Full Power (14 MW)
    External rep. IRNE, (1980)

11. A.Fabry, I.Garlea - Quality Control and Calibration of
    Miniature Fission Chambers by Exposure to Standard Neutron
    Fields. Application to the Measurements of Fundamental
    Integral Cross Section. Rep. IAEA-208, vol.II, p.291 (1978)

12. C.Miron, I.Garlea, C.Dumitrescu, V.Zoiţa - A Proposal to
    Determine Neutron Flux-Spectrum at Romanian Plasma Focus
    Installations, by SSTR method. Paper presented at XIII
    International Symposium on Nuclear Physics, Gaussig, GDR,
    Nov. (1983).

NEUTRON AND GAMMA RAY DOSE STUDIES

IN CAGR INSTRUMENTATION AND FUEL COMPONENTS

P J H Heffer

Central Electricity Generating Board (CEGB)

Berkeley Nuclear Laboratories, Berkeley, GL13 9PB, England

ABSTRACT

In all reactor types, it is important to know radiation
energy deposition or damage rates in various components so that
the safe and efficient operation of the plant can be ensured.
The size and geometric complexity of these components can vary
greatly and Monte Carlo methods have been adopted in the UK for
such dose assessments because of the flexibility they offer in
accurate modelling. However as the detail of the modelling is
increased it is important to ensure that the dosimetric
assumptions on which the calculations are based remain valid.
In particular, for small regions comprising materials of
different type secondary charged particle equilibrium is not
always achieved and the effect of secondary electron processes
must be included in any complete gamma transport analysis.

Within the core of Commercial Advanced Gas-Cooled Reactors
(CAGR) there are a number of examples of such situations and two
are illustrated. In the first, a dosimetric study of
calorimeters used to monitor moderator energy deposition is
described and the effect of secondary electron transport is
shown to be an important part of this analysis. In the second
example, electron transport and internal conversion processes
are considered in the estimate of dose rates in gadolinium oxide
burnable poison toroids which form part of some CAGR fuel. It
is shown that the neglect of these effects can introduce
systematic underpredictions of radiation doses in excess of 30%
in specific circumstances.

J. P. Genthon and H. Röttger (eds.), Reactor Dosimetry, 943–949.
© 1985 ECSC, EEC, EAEC, Brussels and Luxembourg

INTRODUCTION

Direct measurement of the detailed radiation distribution within the graphite moderator of Commercial Advanced Gas-cooled reactors is not technically possible and so assessments of energy deposition and radiation damage are carried out theoretically, using Monte Carlo calculational techniques. The methods are applied over large regions of homogeneous material and though the geometry is sometimes relatively complex, standard dosimetric assumptions are always valid. However in all CAGR cores there is at least one instrument thimble containing a calorimeter which provides a dose reference point in the plant to compare with the theoretical predictions. These calorimeters are physically small and the graphite heatsink is surrounded by complex structural steel and has thermocouples and electrical heater wires buried within it. Very detailed modelling is required to account accurately for the perturbations to the local radiation field caused by this extra material, both in terms of geometry and dosimetric assumptions. The extension of the standard dose analysis route to include the detailed representation of such instruments is described and the effect of secondary electron transport in redistributing the primary energy deposition is calculated.

In the second part of the paper, these electron transport techniques are applied to dose assessments of small toroids of Gadolinium Oxide burnable poison, used to hold down reactivity during the initial stages of burn-up of CAGR fuel. These components, which comprise only 0.32% of the fuel weight can account for up to 10% of the neutron absorption in the fuel element or nearly 1% of the total power. The analysis was required to demonstrate that the very localised heat production in the poison rings does not result in them attaining unacceptably high temperatures.

METHOD

In keeping with UK practice in shielding and heating methods Monte Carlo techniques have been used for all main particle transport calculations in this study. The neutron and gamma ray work has been carried out using the code SPARTAN (1) and the electron transport with BETA (2), both codes having three dimensional capability.

A Watt fission source spectrum which has been tested
against measured reaction rate data was used throughout this
work and gamma production data was devrived from UK recommended
datasets (3). Point particle interaction data has been taken
from the DICE (4) library for neutrons and from the GAMBLE (5)
library for gamma rays. All interaction data has been used
explicitly with no approximations to either the energy or
angular distributions. The spartial source distributions have
been generated either from the Monte Carlo simulations
themselves or taken from lattice calculations carried out using
the WIMS (6) suite.

The calculation of the general cell energy distributions
are carried out in a straight-forward manner using an unbiased
game though with energy cutoffs to exclude thermal neutron
tracking. The radial geometry used is shown in Figure 1a and
the axial in Figure 1b. The mean radial energy deposition
distributions are presented in Figure 2 and it is these types of
distribution which form the basis of core assessment models.

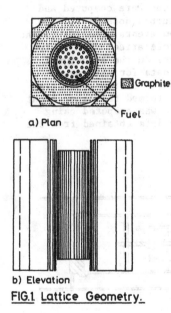

a) Plan

b) Elevation

FIG.1 Lattice Geometry.

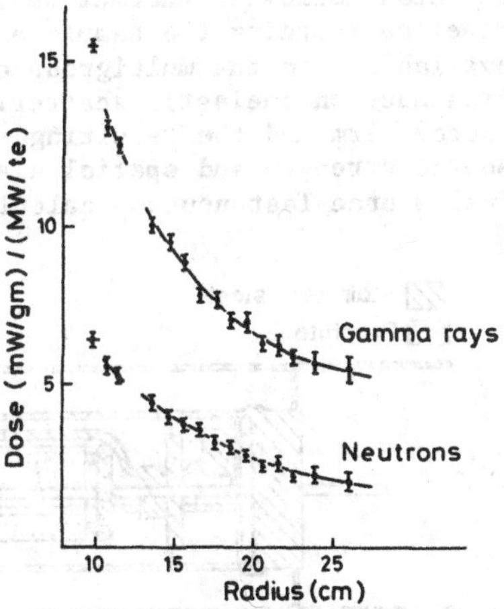

FIG.2. Lattice Energy Deposition.

A two part approach was adopted to extend the dose analysis to include the core calorimeters. An explicit geometric model of the instrument, shown in Figure 3 was generated and represented as the central feature of a four channel supercell model based on the lattice of Figure 1. Directionally biased calculations were used to obtain particle crossing information at a radial boundary 5cm from the instrument. This data was used to generate starting fluxes for a set of calculations describing the instrument and its local graphite only. As well as fast neutron and gamma ray fluxes, a multigroup thermal neutron calculation was carried out in the same way to obtain detailed information on the thermal neutron capture rate distribution within the instrument assembly. This approach was chosen since it proved impractical to bias the calculations sufficiently to generate accurate dose information directly and the detail of information required for the subsequent analysis precluded the use of either adjoint or more conventional 2-dimensional smeared calculations.

Using the restricted geometry representation, the dose to the calorimeter sample was calculated separately for each of the different source terms. The fast neutron and gamma ray doses from the fuel and moderator capture sources were computed and these were combined to define the unperturbed dose level. To this was added contributions arising from events occuring in the instrument itself. The dominant local dose arises from photon emission following thermal neutron capture in the structural steel surrounding the sample and source data for this was available from the multigroup calculations. Photon production from neutron inelastic scattering events formed the other local source term and the resulting sample dose was computed using source strength and spatial distribution data obtained from the fuel source fast neutron calculations.

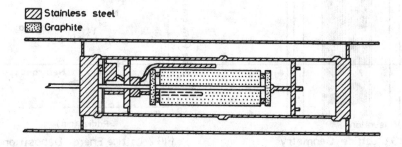

FIG. 3.   Calorimeter Geometry.

These dose assessments are based on the assumption that energy deposition occurs at the point of particle interaction as determined by the particle transport simulation. For neutron calculations, this is valid in all but the most specialist applications. For photon events, however, electron recoil ranges are significant compared with the typical dimensions of the calorimeter and the subsequent transport of these electrons redistributes the calculated primary photon energy deposition. The magnitude of this effect in the CAGR calorimeters was calculated by modelling the geometry of Figure 3 in the code BETA, though without any material outside the outer steel sheath. Electron transport calculations were then carried out based upon spectrum source terms generated during the SPARTAN photon transport simulations. Electron interaction data was taken from the compilations of Berger and Seltzer (7) or derived from the Bethe formulation (8). The condensed history transport model assumed 2% energy loss per path segment and delta ray production above 10keV was treated explicitly. Forty seven separate geometrical regions were treated in this model.

The results of these calculations show that the radiation dose generated in the calorimeter sample from reactor radiation sources is enhanced significantly by local sources generated within its own structure. The base level dose is increased by 38±2% by thermal neutron capture photon sources and by a further 3.2±.3% from the inelastic scattering photon production. These two sorts of event are the only ones normally considered in assessments of this type. The effect of secondary electron transport increases the base sample dose by a further 5.7±.3%, an effect of almost double that predicted from neutron inelastic scattering processes.

A further example of the need to include electron production and transport effects in predictions of component dose rates is provided by the calculations performed as part of the thermal assessment of the burnable poison rings placed in the support grids and braces of some CAGR fuel. The rings comprise a steel sheath of outer and inner radii of 0.178 and 0.142cm filled with gadolinium oxide on a circle diameter of 9.50cm. The locally generated energy deposition dominates the dose experienced by these rings and the assumptions made in the dose analysis give rise to markedly different energy deposition values. In the first part of the assessment, gamma ray transport calculations were carried out using SPARTAN representing one ring in isolation with the source defined in the outer 0.002cm of the poison volume, in keeping with the expected start-of-life thermal neutron capture distribution. This defined the standard gamma ray calculation base case dose. As well as gamma ray energy deposition information, electron recoil source spectra were generated in these runs and subsequent BETA transport calculations were performed using the same data and tracking assumptions as above. These results showed that there was a considerable redistribution of the energy deposition within the toroid which altered the ratio of poison to sheath heating rate but more importantly that due to electron leakage, only 75±3% of the primary deposited energy was retained within the ring.

The next stage of the assessment made allowance for the probability of internal conversion occuring as a competing reaction to gamma emission from Gadolinium. Data for this was derived from gamma emission line intensities taken from standard Nuclear Data Sheets combined with line conversion probabilities taken from the data of Segre (9). The evaluation showed that the overall conversion rate was 2.1%. This apparently low value is, however very significant since there is a high probability of this energy being retained within the poison ring, while the proportion of photon energy being absorbed is only of order 4%. This extra electron source term was combined with the recoil terms above and revised dose calculations were carried out. The primary gamma ray dose predictions for this case used a modified gamma source strength and spectrum which had been adjusted to maintain consistency with the increased electron energy source term. The results showed that the overall energy deposition in the poison ring was enhanced by 34±3% compared with the initial gamma only calculation. Neglect of the internal conversion effect in electron transport estimation would suggest a dose reduction of 25±3%.

To complete the exercise proper account was taken of the actual local geometry of the poison toroids, examples of which are shown in cross-section in Figure 4. Full gamma ray and electron transport modelling was carried out for these cases. The results show that for case a) the poison heating in each wire rises by 11.5±1.5% compared with the isolated case. For cases b) and c) the central toroid experiences the greatest dose enhancement, with increases of 25±2% and 46±3% respectively.

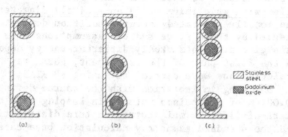

FIG.4.   Poison Toroid Locations in Brace.

CONCLUSIONS

Monte Carlo methods used within the United Kingdom for energy deposition assessments have sufficient flexibility to model a wide variety of reactor problems both in terms of geometry and nuclear data. However the modelling must also extend to the correct dosimetric representation of the system under examination.

The assumption of energy deposition at a point which is inherent in all transport simulations of this kind is shown not to be valid when considering core instrumentation and small components, when secondary particle ranges become comparable with their dimensions. To illustrate this effect, calculations including the effect of secondary electron transport processes for CAGR calorimeters and burnable poison fuel components have been compared with standard dose evaluations and important differences have been found.

ACKNOWLEDGEMENT

This paper is published by permission of the Central Electricity Generating Board.

REFERENCES

1. Bending R C and Heffer P J H. The Generalised 3D Monte Carlo Particle Transport Code SPARTAN. CEGB Report RD/B/N3178
2. Jordan T M. BETA II/B Time Dependent Monte Carlo Bremsstrahlung and Electron Analysis Code RSIC CCC 117, Oak Ridge National Laboratory
3. Sidebotham E W. Spectra of Energy Released from Thermal Neutron Capture British Report TRG 2189(R)
4. Parker J B. Preparation of Nuclear Data for Monte Carlo Calculations British Report AWRE O-77/66
5. Bendall D E and Parker J B. GAMBLE A Gamma Ray Nuclear Data Library British Report JNPC/SWP/N121
6. Askew J R, Fayers F J and Kemshall P B. A General Description of the Lattice Code WIMS Journal British Nuclear Energy Society Oct.1966
7. Berger M J and Seltzer S M. Tables of Energy Losses and Ranges of Electrons and Positrons U.S.Report NASA SP-3012 (1964)
8. Bethe H and Heitler W. Proc. Roy. Soc. A146 83 (1934)
9. Segre E. Experimental Nuclear Physics III, J.Wiley and Sons, N.Y. (1959).

DEVELOPMENTS IN HEALTH PHYSICS DOSIMETRY

J R A Lakey and R E Alexander

Royal Naval College and the Nuclear Regulatory Commission

Greenwich SE10 9NN, UK and Washington DC 20555, USA

ABSTRACT

The joint session on dosimetry organised by the Health
Physics Society and the American Nuclear Society in New Orleans
in June 1984 provided a demonstration of the relationship
between the health physics and reactor dosimetry. This paper
applies the papers given at the joint sessions to assess the
progress that has been made in the transfer of ideas and
techniques between these fields of study. The growing interest in
ALARA and the need for transferable standards provides some of the
incentives to seek improved dosimetry in health physics. In
addition the possibility of an increase in neutron quality
factor and the application of realistic anatomical phantoms in
Health Physics stimulate developments in dosimetry. The paper
reviews developments in dosimetry in the context of these areas
of interest to the health physicist.

INTRODUCTION

There is little direct interaction, between the dosimetrist
engaged in radiological protection and his counterpart in reactor
physics, although the detection techniques used in both fields of
study have similar roots. The reason for this gap is that the
two applications of dosimetry have been developed to meet
different needs. The Health Physicist requires robust, multi-
purpose, mass-usage detector devices and the Reactor Dosimetrist
designs an highly specific device, often with the necessity to
operate for a long period of time in a hostile environment. The

J. P. Genthon and H. Röttger (eds.), Reactor Dosimetry, 951–958.
© 1985 ECSC, EEC, EAEC, Brussels and Luxembourg

problems which arise in an operating reactor have led to the need for the reactor dosimetrist to examine carefully the basis of his measurement and analysis procedures. This has led to a re-definition of the target accuracies required of his measurement and this has led to the development, in the USA, of an inter-leaving set of ASTM standards which underpin the development work in reactor dosimetry. ASTM does not possess any regulatory significance in the UK, but similar efforts are now being made to bring together the project-based elements of reactor dosimetry in the UK, in an attempt to produce a more standardised approach.

Standardisation in radiological protection is well advanced in the UK although the emphasis lies in radiology rather than in engineering applications of physics. The recent IRPA Congress in Berlin was an opportunity for a review of the way ahead for Radiation Protection. Dunster and Webb (1) reviewed ICRP 26 which they described as - "a coherent self-consistent system of limiting exposure, the biological bases are clearly set out, the objectives are defined, and the method of achieving them is explained". Implementation of these recommendations varies between countries but ALARA, "as low as reasonable achievable" can have an impact on radiation protection standards demanding, in the USA at least the use of "state of the art" technology so that the best available dosimetry has to be used. The authors discussed the revision of T65D dose estimates of Hiroshima and Nagasaki nuclear weapons which has probably invalidated an important basis for the assessment of human risk due to neutron exposure. At the same time the linear dose effect relationship is being questioned, especially for low-LET radiation where the relationship could be linear/quadratic. Thus the interpretation between observed effects at high dose could exaggerate the risk at low dose. The present dose limits are therefore more conservative at low LET so a larger quality factor may be needed for high LET.

Personal dosimetry involves the measurement of energy deposition, although the interpretation of results is achieved by different means than those used in reactor dosimetry. If an attempt is made to transfer the technology of reactor dosimetry analysis to the personnel monitoring field, a number of pertinent questions need to be asked.

(1)   Is there an incentive for improvement in measurement accuracy?

(2)   Are the detectors currently used in health physics sufficiently stable and well-characterised to provide a basis for significant improvement?

(3)   Is the health physics measurement, which may be conducted
      in an extremely variable environment, such that measure-
      ment uncertainties would mask any improvements in
      efficiency?

REACTOR DOSIMETRY

Austin and Lakey (2) described the UK reactor dosimetry
programme of the past eight years which has lead to improvements
in dosimetry accuracy. Three main areas of development are:-

(i)   Careful standardisation and calibration of detector
        material.

(ii)   Evaluation of detector materials or assemblies by use
        of standard 'benchmark' fields.

(iii)   Careful analysis of results by the judicious use of
        both measurement and calculations, together with a
        thorough evaluation of the uncertainties associated
        with the analysis.

Gold expressed the reactor shielding viewpoint (3) and said
that the progress in reactor shield dosimetry contrasts with the
needs of Health Physics in which the specialists are implicitly
accountable to the public who hope for benefits as a result of the
practice. The attitude to accuracy is therefore different to
that of the shield experimentalist who has had to consider mixed
radiation fields intensively. USA studies of the light water
reactor pressure vessels are a good example of progress. In 1960
the aim was to achieve 5-15% accuracy but the standard deviation
was often as high as 30%. By 1980 the use of bench mark
facilities gave a dramatic improvement - a factor of 2 in standard
deviation was achieved by a ten fold increase in effort and with
recent studies for the fast breeder reactor programme in-core
measurements are reaching 5-10% accuracy. The parametric and
design-confirmation shielding measurements made at the Tower
Shielding Facility (TSF) in Oak Ridge (4) give an example of the
effort expended. The facility used a powerful reactor (up to
1 MW thermal) with spectral modifiers to provide neutron spectra
close to those associated with the various parts of a reactor. The
source allows measurements through shields of full reactor thickness.
The detectors include spectrometers and dosimeters for both neutrons
and gamma rays. A large exclusion area provided much flexibility.
    The extensive information generated by shielding research is
processed by the Reactor Shielding Information Centre (RSIC) at Oak
Ridge National Laboratory, Oak Ridge, Tennessee. The RSIC issues
a monthly newsletter which is the most important means of communication with the user community (5). Announcements are made of:

The RSIC has actively sought, collected, tested and packaged computer codes and data libraries within its scope.

Experimental support codes include:-

Few-channel spectrum unfolding
(eg. Foil Activation)

Many-channel spectrum unfolding
(eg. NE-213, Ge(Li), NaI)

Detector response analysis

STANDARDISATION IN HEALTH PHYSICS

Standardisation for radiological protection in the UK (6) is based on the work of the UK National Physical Laboratory. The purpose of setting up of hierarchial structures for the dissemination of units of measurement is to ensure traceability to the national standards. The experience gained in the United Kingdom in setting-up such dissemination schemes has produced the following NPL standards:-

NEUTRON STANDARDS

The NPL standard thermal neutron field is produced by two beryllium accelerator targets bombarded by deuterons from a 3.5 MV Van de Graff. A large graphite moderator block is monitored by three boron coated ion chambers. A 150 cm^2 cavity at the centre has a fluence rate 10^4 to 3×10^7 cm^{-2} s^{-1} with less than 1% epithermal. The fluence rate is measured for transfer with a 1 cm^2 gold foil, 20-100 mg cm^{-2} superficial density counted within 2 days on a 4π beta counter. Foils are irradiated bare and 1mm thick cadmium covered.

In the USA calibration standards are achieved using D_2O moderated ^{252}Cf neutron sources* and an on going study is reviewing the situation in five reactors where usually the peak energy is around 500 KeV except in circumstance of streaming from the reactor cavity.

GAMMA STANDARDS

The standard in the UK is Parallel-plate free-air ionization chamber with a Machlett OEG-50A X-ray tube (1mm Beryllium minimum filter) for 8kV to 50kV and a Muller 150kV X-ray tube 50kV to 300kV (4mm aluminium minimum filter). At higher energy a 2 MV Van de Graff is used to irradiate 3 standard graphite cylindrical-cavity chambers.

In the UK dosemeters can be calibrated against the Primary Standard chambers at the National Physical Laboratory (NPL) using reference X-ray beams. An associated low dose rate gamma-ray facility using ^{60}Co and ^{137}Cs and is collimated to ISO Standard 4037, 1979 and the source housing follows the recommendation of ICRU

---

* See PNL 3569, PNL 3585.

Report 18, 1970; these are aimed at reducing scattered radiation in the beam. This facility is used for films and TLD intercomparisons. The standard is transferred by using an ionisation chamber (7) which provides the means of calibrating instruments used for monitoring X- and gamma-rays (over the range 10 KV to 2 MV) for protection purposes. The chamber utilizes a thin walled spherical conducting casing as its outer electrode, of diameter about 190 mm and thickness approx 0.3 mm. The radioactive check source a 4-mCi (1.5 x 10^8 Bq) encapsulated bead of ^{241}Am integral with the chamber. The source is stowed at the base of the hollow stem of the collecting electrode and is released into the centre of the inverted chamber using an externally positioned magnet. Ionization current is measured by means of a current to voltage transdicer with variable negative feed-back for range changing. The instrument is scaled in absorbed dose rate to air on nine ranges from 1 mrad/hr (10 nGy/h)(f.s.d) to rad/hr (100 mGy/h).

The UK standard is expressed in terms of exposure from a radionuclide or X-ray beam with a standard dosemeter in same position as the device to be calibrated. The transfer of this to reactor dosimetry applications has special problems since the calibration uses monodirectional photons and the concept of 'Exposure' demands electronic equilibrium. But similar problems occur within the field of health physics dosimetry. The estimation of dose to the bone marrow of an exposed person is a complex task because electronic equilibrium does not exist in the vicinity of soft tissue/bone mineral interfaces. The geometry of trabecular bone is intractable (8) and at low photon energy the bone marrow dose can be overestimated by factors of 3. In these reports the energy deposition is calculated along the track of electrons and then averaged to give the absorbed dose for secondary electrons. Cavity Theory is normally used to derive absorbed dose from the energy deposition in the "build-up" medium surrounding a dosemeter which is multiplied by the ratio of the mass energy absorption coefficients and with an allowance made for the subsequent electron transport. Developments in codes such as McBEND permit the simulation of charged particles arising from photon interactions and the electron energy transfer process can be predicted. (9)

COMMENTS ON HEALTH PHYSICS DOSIMETRY

The ANSI standard 1.11 gives personnel gamma dosimetry requirements for film badges and TLD (10) but a better algorithm is needed for film badges. This is because the high Z of silver halide creates a non-linear relationship between Dose Equivalent and optical density. The poor quality control inherent in Thermoluminescent materials also produced dosimetry problems. In particular the necessity to use individual element correction factors is an enormous clerical burden.

In a typical civil PWR the total annual collective dose equivalent

(reported in 1976) was about 5 man Sievert (500 man rem) of which
only 10% was accumulated in routine operation and only about 2% of
the total was due to neutrons. Thus shield radiation surveys place
little emphasis on neutrons and monitoring equipment is calibrated
against a notional spectrum. Since 1964 "rem counters" have wide
energy response and can be corrected for the neutron energy spectrum(11).

Recently there has been an increase in the use of CR39 for
neutron dosimetry. This material, Columbia Resin, is a thermoset
plastic which is the ploymeric form of diethylene glycol bis (allyl
carbonate) - it will record tracks of protons up to 10 MeV and these
are revealed by etching. In the UK calibrations (12) extend from
0.15 MeV to 14.7 MeV. Angular dependence is low for a ^{252}Cf spectrum
within 20% at angles up to 75° to normal. In a soft spectrum (40%
of the dose below 100 KeV) the dosimeter records 66% to 81% of
calculated dose compared to uncollided fission neutrons. There is
a need for a lower detection limit and a lower neutron energy
threshold. The latter may be achieved by using a n/α convertor to
improve the thermal to 30 KeV problem area.   A superheated
liquid (13) can also be used to detect neutrons by measurement of the
volume of released gas. The sensitivity of this device to gamma rays
is very low so it has a potential application in a mixed radiation
field. By varying the amount of sensitive liquid, an extremely
sensitive detector, capable of detecting well below 1 mrem can be
made. A detector which can be used in accident dosimetry for doses
of 100's of rads can also be made. The detector is uncomplicated,
it requires no external power supply, it is compact, rugged,
inexpensive and is easy to read.

COMPLIANCE IN PERSONAL DOSIMETRY

The role of radiation dosimetry in radiological protection is
ultimately for application in radiation risk control. The demands
on dosimetry include:*

(1)   A properly designed, calibrated and processed dosimeter
        to provide an adequate estimate of the dose received by
        the dosimeter.

(2)   A maximum error of ±50% to permit a decision to apply
        corrective action to maintain worker risks at acceptable
        levels.

(3)   A national performance testing and accreditation program
        to verify dosimetry processor competence.

(4)   A control capability which is not affected by any further
        processing which must apply to the dosimeter results.

---

* Private communication R E Alexander.

Legal considerations in the use health physics dosimeters include:

1. Compliance with regulations (as part of Radiation Risk Control).

2. The use of dosimetry results in the adjudication of claims for compensation.

3. The selection of data to be recorded.

CONCLUSION

The transfer of dosimetry technology between reactor physics and health physics will not be achieved without an incentive (financial or operational) for the investment that would be necessary. The target accuracies associated with "routine" health physics surveys is difficult to achieve due to the character of the measurement field and the errors associated with virtually ungovernable field variations could swamp all improvements. As new and more reliable health physics detectors meet an increasingly stringent safety philosophy, it is anticipated that a search for improvements will provide the incentives for transfer.

REFERENCES

1. Radiation Protection Standards for the 1990's
   Dunster, H J; Webb, G A M
   National Radiological Protection Board,
   6th National Congress, IRPA, Berlin
   Compacts Vol III, p.1245, 1984.

2. Current Status of British Standards for Health Physics and Reactor Physics Dosimetry.
   Lakey, J R A
   Royal Naval College, Greenwich, London, England SE10 9NN
   Austin, N
   Rolls Royce, Derby, England
   Paper 311, 29th Annual Meeting, Health Physics Society, New Orleans, 1984.

3. The Need for Accuracy Requirements in Health Physics and Physics Dosimetry Practices.
   Gold, R; Carpenter, G D
   HEDL, PO Box 1970, Richland, WA 99352
   Paper 309, 29th Annual Meeting, Health Physics Society, New Orleans, 1984.

4. Shielding Experimental Methods
   Maienschein, F C; Muckenthaler, F J
   Oak Ridge National Lab, TN Department of Energy
   Nov 78, 13p, Contract: W-7405-ENG-26, FBR shielding seminar, Obninsk, USSR, 13 Nov 1978.

5. The Radiation Shielding Information Center: A source of Computer
   Codes and Data for Dosimetry Applications.
   Trubey, D K; Maskewitz, B F; Roussin, R W
   Radiation Shielding Information Center, Oak Ridge National Laboratory
   PO Box X, Oak Ridge, TN 37831-2008.
   Paper 315, 29th Annual Meeting, Health Physics Society, New Orleans,
   1984.

6. Standardisation in radiation dosimetry in the United Kingdom
   Jennings, W A
   National and International Standardization of Radiation Dosimetry,
   Atlanta, GA, USA 5-9 Dec, 1977 p.119-28, 1978
   IAEA, Vienna, Austria
   Part No 1, 534

7. The NPL protection-level secondary standard X- and gamma-ray dose
   rate meter
   Read, L R; Kemp, L A W
   Health Phys, (GB), Vol 33, No 2 p. 131-7, 7 Refs, Aug 1977

8. Medical Internal Radiation Dose Committee
   Snyder, W S
   Pamphlet 5. Revised
   Society of Nuclear Medicine, 1978.

9. Knipe, A D, Electron Transport Calculations with the Monte Carlo Code
   MCBEND
   AEEW - R1749, 1984

10. Improvements Observed in Personnel Dosimetry Processing Since 1978.
    Plato, Philip; Miklos, Joseph
    University of Michigan, Ann Arbor, MI 48109.
    Paper 312, 29th Annual Meeting, Health Physics Society, New Orleans,
    1984.

11. Lakey, J R A., The Interpretation of Radiation Survey of Operational
    Power Systems
    Radiation Protection Measurement - Philosophy and Implementation
    EUR 5397e, 1975, p.103

12. The NRPB CR-39 Fast Neutron Personal Dosimeter
    Bartlett, D T; Steele, J D
    National Radiological Protection Board,
    6th International Congress, IRPA, Berlin
    Compacts Vol III, p.1111, 1984.

13. Neutron Detection Sensitivity of Bubble-Damage Polymer Detectors.
    Ing, H; Birnboim, H C
    Chalk River Nuclear Laboratories, Chalk River, Ontario K0J 1J0
    Canada
    Paper 314, 29th Annual Meeting, Health Physics Society, New Orleans,
    1984

RADIATION FIELD ASSOCIATED WITH
HIROSHIMA AND NAGASAKI*

William E. Loewe

Lawrence Livermore National Laboratory

U.S.A.

ABSTRACT

Accuracy of dosimetric estimates can determine the value of
the atomic bomb survivor experience in establishing radiation
risks. The status of a major revision of this dosimetry, initiated
in 1980, is assessed.

In 1980, dose levels to individual survivors at Hiroshima and
Nagasaki were revised by factors as large as an order of magnitude,
as shown in Figures 1 and 2.[1] In those figures, T65D is the accepted designation for the old estimates. The most important changes
shown there are characterized by an increase in total kerma of a factor of three at 2 km in Hiroshima, and a drop in the proportion of
neutrons in the mixed field by a factor of fifteen at 1.5 km, also in
Hiroshima. Although the consequences are still unfolding, an immediate result of the new LLNL dosimetry was removal of A-bomb survivor
support for large increases in the neutron quality factor, as
considered formally by NCRP in 1980.

In the ensuing years, a specially appointed committee of the
U.S. National Academy of Sciences, chaired by Professor Frederick
Seitz with Professor Robert Christy as primary consultant, has been
studying the radiation field at Hiroshima and Nagasaki, in
cooperation with a companion group in Japan. Los Alamos National
Laboratory (LANL) has made an improved estimate of radiation leakage

*Work performed under the auspices of the U.S. Department of
Energy by the Lawrence Livermore National Laboratory under
Contract W-7405-Eng-48.

from the Little Boy bomb, which results in adjustment of the LLNL values in Figure 1 by 33% downward at all ranges for neutrons and by 6% downward at all ranges for gamma rays. Oak Ridge National Laboratory (ORNL) has advanced alternative dosimetries, the most recent of which confirms these LLNL adjusted Figure 1 and Figure 2 values to roughly ± 30% (after an outstanding controversy over the explosive yield at Hiroshima is discounted. This controversy, over whether the yield was 12.5 kt or 15 kt, is still unresolved, but the present trend is toward agreement on the 15 kt used in Figure 1.) To date, however, no alternative to the adjusted LLNL values has been offered which has not been shown subsequently to be in need of correction or is otherwise unsuitable to use as a replacement for those values as best estimates. Therefore, comparisons in what follows will be made to the Figure 1 and 2 values as reported in 1980 and elaborated in 1981[1], adjusted at Hiroshima to reflect the improved leakage estimates as just described (i.e., reductions of 33% and 6%).

The Little Boy bomb, which exploded 570 m above Hiroshima at 0815 August 6, 1945, generated 15 kt from fission of uranium assembled by a very large metal mechanism in a gun arrangement. The Fat Man bomb, exploded 503 m above Nagasaki at 1058 on August 9, 1945, generated 22 kt, mostly from plutonium assembled by high explosive implosion. Both bombs weighed four to five tons, of which the

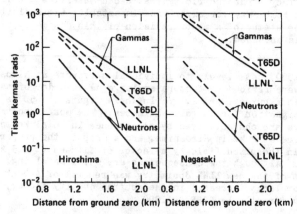

Figure 1. Free-in-air tissue kermas at Hiroshima

Figure 2. Free-in-air tissue kermas at Nagasaki

non-fissioning assembly material constituted a thick blanket, surrounding the fissioning core. Neutrons and gamma rays generated in the core had to pass through this thick blanket to reach the atmosphere.

Although there are several reasons for the various errors in the old dosimetry system (T65D), the reason for the biggest error is the assumption that neutrons which have passed through these two different blanket materials then penetrate the atmosphere equally well.

Exiting neutrons, with kinetic energies in excess of about one-half
MeV (which determine neutron dose beyond the 1 km ground range where
most survivors were located) are actually distributed in energy as
shown in Figure 3. In this figure, various Little Boy spectra are
contrasted with the spectrum from a bare uranium metal reactor
(normalized in intensity so as to force agreement at 0.6 MeV for
easy comparison). Although not shown here, the leakage spectrum
from the Fat Man bomb has a still larger proportion of higher energy
neutrons (relative to those at one-half MeV) than a bare reactor, so
that it is clear from Figure 3 that atmospheric penetration must be
very different for the two bombs. In fact, the resulting 30% difference in dose e-folding lengths can be thought of as generating the
differences shown in the neutron curves for Hiroshima in Figure 1.

Figure 3 also compares (1) the 1D calculations at LANL used to
generate Figure 1, (2) the more recent LANL calculations discussed
previously, and (3) measurements made by Robitaille and Hoffarth
from Defence Research Establishment, Ottawa (DREO) at a replica of
Little Boy prepared and operated by LANL as a reactor. Several
other groups have confirmed the DREO data shown in Figure 3,
including measurement teams led by Hoots and by Griffith at LLNL.
These teams generated an improved description of the angular
dependence of the spectra (resulting from the cylindrical symmetry
of the gun assembly), and also measured neutron dose as a function
of range to one-half kilometer. The two dose e-folding lengths
derived from these two independent sets of LLNL range data agree to
six percent with my calculated value which is based on a recent
adaptation of the calculation that generated Figure 1 values.

Static calculations by LANL in 1984, using the same transport
representation as in the previous dynamic calculations, lie somewhat
lower than the dynamic results shown in Figure 3 and agree quite well
with the measured data from DREO and from LLNL. Except for this
modest dynamic effect, the Little Boy leakage is now very well known.

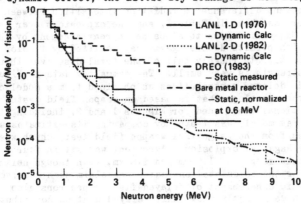

Figure 3. Little Boy neutron leakage energy spectra

The accuracy of deep penetration, air-over-ground radiation transport calculations, as used to generate Figures 1 and 2, has been established as better than 25-30% in kerma[2], but recent updates by DREO of certain measured values now give agreements as shown in Table I for a bare metal source, with spectral indications that this kind of agreement could be expected for Little Boy as well as for Fat Man sources.

Table I

Bare Metal Reactor, LLNL Calc/Measurement

Range (km)	Neutron Kerma	Gamma Ray Kerma
.100	1.08	1.06
.170	1.06	0.91
.300	1.30	0.94
.400	1.28	0.90
1.080	0.92	0.80
1.618	0.90	0.91

At least half of the somewhat larger neutron disagreements at 0.3 km and 0.4 km have been shown to be attributable to the effects of near-by trees that were not modeled in the calculations. (Absolute accuracies of the data shown in Figures 1 and 2 are not as good due to uncertainties in source intensity and spectral composition, in meteorological conditions, and in terrain configuration and composition.)

One important difference in radiation intensities between a nuclear reactor, as shown in Table I, and the explosions over Hiroshima and Nagasaki is due to atmospheric rearrangement by the blast wave, which permits delayed gamma rays emitted by fission fragments to contribute roughly half of the total dose even though their source strength is very small. The gamma ray data in Figures 1 and 2 include delayed contributions as obtained from a model of this dynamic phenomenology that is based on weapon field tests. The entire calculational chain used for Figures 1 and 2, including this delayed gamma ray model, was tested by comparing its estimates with measured values from the Ranger Fox weapon field test, which was similar to the Nagasaki explosion. Agreement was within four percent at ground ranges of 1.0 km and 1.8 km, even though neither the measured nor the calculated estimates can claim such accuracy. A similar dynamic enhancement of delayed fission neutrons also occurs, but I have recently shown this delayed neutron contribution to be negligible, using the same model of the dynamic atmosphere.

Thermoluminescence dosimetry in Hiroshima and Nagasaki has recently been extended to ranges where most survivors were located, permitting comparisons with the gamma ray data in Figures 1 and 2. The comparisons shown in Figures 4 and 5, which include earlier data at shorter ranges, show a striking confirmation of the essential correctness of the predicted values out to 1.5 km. The abrupt discrepancies at 2.0 and 2.3 km at Nagasaki are thought to result from background contributions to the measurements at these great distances.

Measurements of neutron activation of cobalt have recently been reported by Hashizume in Tokyo on samples exposed to the atmosphere on rooftops in Hiroshima, as an extension of earlier measurements made interior to concrete pillars. If a small adjustment, within the limits of uncertainty, is made in the boron content of concrete (not measured at the time cobalt activation was measured inside pillars), then the pillar and rooftop data form a consistent set to within 10-15%. These data show calculations high by 15% compared to experiment near ground zero, low by 40% at one-half kilometer ground range, and low by a factor of four at one kilometer. This lack of agreement could be attributed to an unknown deficiency in source description due to dynamic effects during the explosion, an unknown deficiency in the calculated transport of sub-kilovolt neutrons responsible for cobalt activation, or an unknown experimental problem. The same kind of dynamic source deficiency would have to apply to the very different Fat Man as to Little Boy, since the Nagasaki pillar comparisons parallel those at Hiroshima. Also, recent work by Woolson and Gritzner at Science Applications, Inc. (SAI) changes fair agreement of calculated and measured values obtained previously for sulfur activation into the excellent detailed agreement shown in Figure 6.

Finally, the attention brought by the 1980 revelation of major error in free field kerma estimates allowed Jess Marcum of RDA to put forth improved estimates of the gamma ray shielding provided by Japanese dwellings. His factor of 1.6, by which previous transmitted gamma ray estimates are to be reduced, has been confirmed by SAI using detailed computer calculations that agree to 10% with selected values measured at weapons tests inside model buildings. These calculations can be used to model in high detail any survivor shielding configuration for which geometry and material records are available.

In conclusion, large changes from the estimates discussed here are increasingly unlikely. However, improvements are expected as a result of the international program currently under way, and the discrepancies noted in cobalt activation must be resolved. It is possible that the absolute accuracy previously attributed to the T65D dosimetry, 30%, might eventually be achieved for the new dosimetry.

I regret that only a few highlights could be included in this severely space-constrained overview, preventing reference to many other important investigations. However, George Kerr at ORNL has recently prepared an exhaustive bibliography for all of this work.[3]

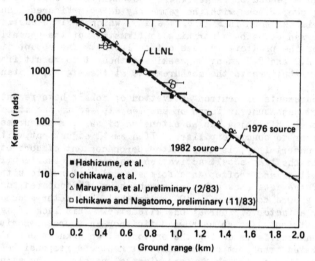

Figure 4. Gamma ray kermas at Hiroshima

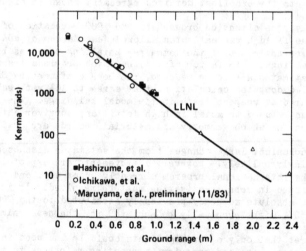

Figure 5. Gamma ray kermas at Nagasaki

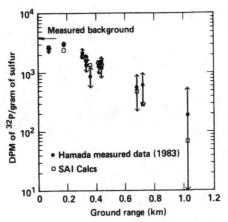

Figure 6. Sulfur activation comparison.

Preliminary results by SAI

REFERENCES

1. W. E. Loewe and E. Mendelsohn, "Neutron and Gamma-Ray Doses at Hiroshima and Nagasaki", Nuc. Sci. & Eng. 81, 325-350 (1982).

2. W. E. Loewe, W. A. Turin, C. W. Pollock, A. C. Springer, and B. L. Richardson, "Validated Deep-Penetration, Air-Over-Ground, Neutron/Gamma-Ray Transport", Nuc. Sci. & Eng. 85, 87-115 (1983).

3. G. D. Kerr, Bibliography of Literature Relevant to the Reassessment of A-Bomb Radiation Dosimetry in Hiroshima and Nagasaki, ORNL/TM-9138 (1984).

THE ACTIVITY OF THE CZECHOSLOVAK WORKING GROUP ON REACTOR
DOSIMETRY IN THE PERIOD FROM 1980 TO 1983

B. Ošmera, J. Petr[+], O. Erben, K. Dach
Institute of Nuclear Research, Řež, Czechoslovakia
[+]Technical University of Prague, Faculty of Nuclear
Science and Physical Engineering

M. Holman, P. Mařík, F. Vychytil, M. Brumovský
Škoda Works, Plzeň, Czechoslovakia

ABSTRACT

A very brief summary of the activity of the Czechoslovak Working Group on Reactor Dosimetry in the period from 1980 to 1983 is given. Some results in the
following areas are presented: a) neutron spectrometry
with the emphasis on neutron transport in iron and stainless steel, on intercalibration by means of leakage
spectra from the iron sphere with the 252Cf neutron
source inside; b) model experiments for improvement of
the WWER presure vessel radiation damage evaluation;
c) monitoring of the irradiation experiments by means
of SPN and activation detectors and calorimeters.

INTRODUCTION

Following the decision of the chairman of the Czechoslovak
Atomic Energy Commission (CAEA) of 6th June 1980 the Working
Group on Reactor Dosimetry (WGRD) has been established as the
advisory committee of the Nuclear Power and Nuclear Safety and
Safeguard Division of the CAEC /1/.

The field of the WGRD's activity has ensued from the needs
of the Czechoslovak nuclear power in reactor dosimetry and in
its application in testing irradiated reactor pressure vessel
steels, reliability, safety and economy of the nuclear power
reactors. Due to the fact that steels used in Czechoslovakia

J. P. Genthon and H. Röttger (eds.), Reactor Dosimetry, 967–974.
© 1985 ECSC, EEC, EAEC, Brussels and Luxembourg

for the manufacturing of the nuclear power plant components are different from those used in other countries, an extensive experimental work in well known radiation field has to be done.

The importance of the reactor dosimetry in the period of the intensive development of the nuclear power cannot be seen only in testing of the selected constructional materials but also in solution of all problems associated with safety, reliability and economy of the nuclear power reactors, including diagnosis of nuclear data and mathematical models used for nuclear power reactor calculations in all their operating conditions assumed as well.

Besides the problems mentioned above the WGRD is dealing also with the following tasks:
- securing close methodical cooperation and coordination of all Czechoslovak research laboratories concerned;
- design of the reference neutron field for calibration purposes of the detectors and nuclear data evaluation;
- recommendations about the standard techniques for all stages of in-core measurements and experimental data analysis;
- ensuring of the undertaking professional guarantees in all scientific and technical actions associated with solution of nuclear reactor dosimetry problems.

And last but not the least to inform also scientific public about recommendations adopted by the WGRD.

Some results associated with the WGRD's activity have already been published elsewhere /2/, /3/. Our purpose here is to present further experimental results of a joint effort of the WGRD and the research laboratories concerned.

NEUTRON SPECTROMETRY

At the research institute of Škoda Works two neutron spectrometry methods have been developed: proton recoil technique and activation method. In the proton recoil technique two differents spectrometers for the detection of fast neutrons have been used: scintillation spectrometer with stilbene crystal 10 x 10 mm and the 40 mm diameter spherical proportional counters filled with hydrogen at pressure 981 kPa, 392,3 kPa und 98,1 kPa. With both types of these detectors we can measure neutron spectra in the energy range from 10 keV to 15 MeV. In neutron-gamma discrimination technique the modified one-parameter method (pulse shape discrimination) has been applied. The proton recoil distributions produced by spherical proportional counters have been analysed by differential method including wall effect correction /4/.

The research programme of the nuclear laboratory at Škoda Works includes above all gathering of experimental data of neutron transport in iron and stainless steels which are the basic materials for construction of some nuclear power reactor components. Therefore for neutron leakage spectra from a set of spheres having different diameters (20, 30, 50 cm) with a 252Cf neutron source in the sphere centre were measured /5/. Neutron transport in thick layers of iron (60 and 70 cm) and in stainless steel 18 % Cr + 8 % Ni (175 and 100 cm thick) in a special experimental set-up at the WWR-S reactor has also been studied. Besides, neutron spectra distortion as a result of their transport through layers of iron (10 cm) and water (10 cm) has also been measured (fig. 1).

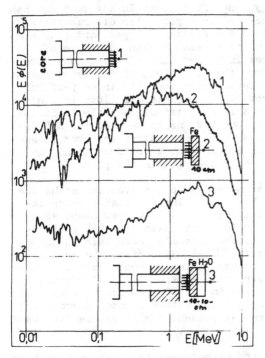

Fig. 1.
Neutron spectrum at the mouth of the 25 cm diameter channel of the ŠR-O reactor and the same spectrum filtered by different materials. Normalization to the reactor power of 1 kW.

The composition of such materials with their thickness is similar to that of radiation shield positioned between the active core and the pressure vessel of the WWER-440 nuclear power reactor.

Multiplication of neutrons due to the neutron interaction with lead according to the reaction (n, 2n) has been studied with 14,6 MeV neutron from the 3H(d,n) 4He reaction.

The results of the measurements mentioned above have been used for testing of one- and two- dimensional codes ANISN and DOT 3,5, respectively with the data libraries EURLIB 4 and VITAMIN C which are used in Škoda Works and in INR at Řež near Prague. Another research works were associated with verification and development of the neutron spectra measurement techniques and their analysis. The detectors were tested in the ŠR-O

reactor channel with diameter of 10 cm in the radiation field
with different ratios of gamma-rays and neutrons and in neutron spectra modified with different materials such as graphite,
iron and water. Also the dependence of the ionizing particles
and discrimination of gamma-rays backround were studied.

For neutron spectra measurements with activation method the
following reactions have been used: 115In(n,γ)116mIn, 197Au(n,γ)
198Au, 59Co(n,γ)60Co, 55Mn(n,γ)56Mn, 23Na(n,γ)24Na, 103Rh(n,n')
103mRh, 115In(n,n')115mIn, 58Ni(n,p)58Co, 54Fe(n,p)54Mn,
56Fe(n,p)56Mn, 48Ti(n,p)48Sc, 24Mg(n,p)24Na, 27Al(n,α)24Na.

At the Institute of Nuclear Research at Řež except the
activation method and the proton recoil technique, the Li6
spectrometer in the energy range from about 20 keV to 5 MeV has
been used. At present, one- and two- parameter distributions
($E_\alpha + E_T$, E_T x ($E_\alpha + E_T$)) of the emitted charged particles are
measured. A great number of measurements with this type of spectrometer has been performed in a fast zone of the annual core
of the RRR reactor /6/. From the comparison of the measurements
carried out with the Li6 spectrometer and with those performed
with the proton recoil proportional counters follows that neutron spectrum based on the triton distribution analysis in the
energy range from 20 keV to 800 keV, using data of the ENDF/B-V
Standard file, does not agree with the neutron spectrum measured by means the proportional counters. Therefore further intercomparison measurements have been put in the plan.

For neutron spectra measurements by activation method the
modified SAND-II code has been used. The modified version labelled as SAND-II ÚJV is installed in the EC 1040 computer.
Neutron selfshielding corrections have been done by modification of cross sections. This code has also been used in the
international interlaboratory REAL-80 excercise. The analysed
spectra together with the deduced integral characteristics are
corresponding with those expected. In some particular cases
the STAYSL programme installed in the EC 1040 computer has also been used. Neutron cross section data are renewed according
to the IAEA recommendations, at present the reference file used
is the IRDF one.

The neutron spectrometry methods mentioned above have been
utilized also in model experiments for evaluation of radiation
damage of the reactor pressure vessel (fig. 2) /7/.

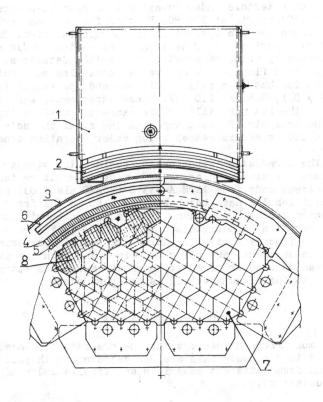

Fig. 2. Arrangement of model experiment for improvement of the
WWER pressure vessel radiation damage evaluation
1 - simulated biological water shield
2 - simulated WWER pressure vessel
3 - LR-0 reactor vessel
4 - simulated reactor core barrel
5 - simulated active reactor core baffle
6 - water displacer
7 - LR-0 active reactor core with the WWER-1000 type
   fuel assemblies
8 - additional fuel assemblies simulating the WWER-440
   active reactor core periphery
* - point of measurement.

PRESSURE VESSEL STEEL IRRADIATION EXPERIMENTS

Irradiation experiments of the WWER pressure vessel steels
have been performed in the WWR-S reactor at the INR at Řež. The

specimens of materials under investigation have been irradiated
in the "CHOUCA-M" rig at 288 °C. Neutron fluence was derived
from the known neutron spectra in the particular rigs. The axial and radial fast and slow neutron fluence distributions have
been obtained by means of a set of activation detectors. Besides the neutron fluence the dpa values have also been determined. Attention has been paid to fluence and dpa values for
energy E > 0.1, 0.5 and 1.0 MeV. These experiments and other
results of the LWR surveillance programme are used as input data for the analysis of the changes in the steel characteristics
of the reactor pressure vessel being under operation conditions.

In the irradiation experiments, besides the standard methods, there are also used as neutron fluence monitors tensile
test specimens mode from ASTM A-533 B (HSSTP plate 03) steel.
Some useful results have been obtained also in the frame of the
co-ordinated research programme on irradiation embrittlement of
pressure vessel steels organized by the IAEA in Vienna /8/. To
improve the fluence monitoring during irradiation experiments
some new measurements are being performed in which the impact
of the specimens under irradiation on the neutron spectra distribution in the irradiated rig is being studied.

IN-CORE MEASUREMENTS

The most frequent sensors for in-core radiation measurements used in Czechoslovakia are SPN detectors with emitors manufactured from different materials and fissile and nonfissile
thermic detectors.

Along with standard measurements, the measurements in reactors of nuclear power plants have been carried out with the aim
to increase accuracy in the interpreting the experimental data.
For illustration the intercomparison of axial fine power distribution measurement carried out during the start up of the third
unit of the nuclear power plant NORD in the GDR can be mentioned (fig. 3).

A lot of work has been done in co-operation of specialists
from Czechoslovakia, Hungary, German Democratic Republic and
the Soviet Union in the nuclear power station KKW Rheinsberg in
the GDR within the framework of the so-called programme KAZEX,
in which integrated in-core detecting instrumentation system
has been utilized. The integrated measuring system was made up
of a modified WWER-440 fuel assembly /9/. Besides the noise
measurements, the analysis and interpretation of experimental
data of Rh-SPN detectors and fissile sensors have been focused
above all on the power monitoring of the diagnostic fuel assem-

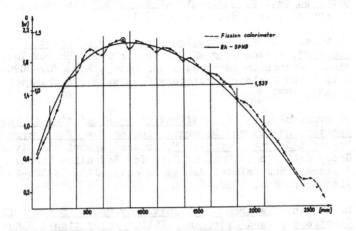

Fig. 3. Axial power distribution measured in the Nord nuclear
power station.

bly, on an assesment and monitoring of a local power released
at a different rate of cooling water passage through the diagnostic fuel assembly and at different conditions of the reactor
core.

In calibration experiments running in a well known reactor
neutron field have been studied as fallows: a) steady state responses of the SPN detectors with emitors made from different
materials, and of thermic detectors with absorbing parts made
from different materials; b) individual components of detectors'
signals obtained from their analysis after reactor shut-down
or after fast removal of the rig from the reactor core.

In intercomparison measurements signals from Va- and Rh-
SPN detectors and from fissile thermic detectors have been
checked-up at regular intervals within a long-term period. The
experimental data were analysed in order to separate from detectors' signals the components created by neutrons. The results
are compared in a long-term way /10/. Noise components of the
thermocouples' signals of the thermic detectors has been used
succesfully in the study of feed-back effects in a low frequency region /11/.

REFERENCES

1.   J. Petr, B. Ošmera, "Report on establishing of the Working

Group on Reactor Dosimetry of the CAEC" (in Czech), Jaderná energie, Vol. 27, p. 116, 1981.

2. B. Ošmera, et al., "Neutron Spectra Measurements in WWR-S Reactor", Proceedings of the Fourth ASTM-EURATOM International Symposium on Reactor Dosimetry, NUREG/CP-0029-VI, Vol. 1, US Nuclear Regulatory Commissio , Washington, DC, p. 587, 1982.

3. M. Brumovský, et al., "Effect of Uncertainties in Neutron Spectra and Fluence Determination on the WWER Pressure Vessel Lifetime Prediction", Proceedings of the Advisory Group Meetings on Nuclear Data for Radiation Damage Assesment and Related Safety Aspects, IAEA-TECDOS-263, Vienna, p. 193, 1982.

4. M. Holman, "Neutron Spectrometry by Means of the Stilbene Spectrometer and Hydrogen - Filled Proportional Connters". (in Czech), Jaderná energie, Vol. 25, p. 440, 1979.

5. M. Holman, "Measured Neutron Leakage Spectra from Iron Spheres" (in Czech), Jaderná energie, Vol. 29, p. 92, 1983.

6. K. Fahrmann, et al., "Der Einsetz des Rossendorfer Ringzonenreaktors für Untersungen zur Physik Schneler Reaktoren", Vol. 25. p. 164, 1982.

7. K. Černý, et al., "Determination of Radiation Loading of the VVER Pressure Vessel", Škoda Review 1983/2, p. 7.

8. M. Brumovský, "Analysis of the Behaviour of Advanced Reactor Pressure Vessel Steels Under Neutron Irradiation". Final Report (IAEA co-ordinated programme), Final report 1983.

9. V. Krett, et al., "Comment on In-Core Calorimeters and SPN - Detectors Measurements", Specialist's Meeting on In-Core Instrumentation and the Assessment of Reactor Nuclear and Thermal / Hydraulic Performance, Halden (Norway), 10 - 13 October 1983.

10. O. Erben, et al., "The intercomparison Measurements: Calorimeters and SPN Detectors" (in Czech), report ÚJV 5959- -R,T, 1981.

11. K. Dach, et al., "The Utilization of In-Core Thermic Detectors to the Feedback Effects Identification", IAEA Seminar on Diagnosis of and Response to Abnormal Ocurences at NPPIS, Dresden, 12 - 15 June 1984.

Neutronic Modelling of the Harwell MTR's: Some Recent Problems

N.P. Taylor

Materials Physics and Metallurgy Division

AERE Harwell

1. Introduction

Use of the Harwell Materials Testing Reactors for the irradiation of experimental rigs gives rise to a number of requirements for calculations of neutron fluxes. In addition photon fluxes are required for estimates of nuclear heating rates. A range of calculational methods are employed, from simple cell to whole reactor models, and the latter have been extended for preliminary design studies for the next generation of MTR to replace DIDO and PLUTO.

The technique used for these various models are described in this note, with emphasis on the areas in which modelling problems are encountered. The applications divide into three distinct areas: calculations concerning rigs irradiated within the reactor core, those for rigs positioned in the D_2O reflector surrounding the core, and design studies for a replacement reactor.

2. In-Core Rigs: Neutronics and Nuclear Heating

The cores of the Harwell MTRs consist of an essentially square lattice of highly enriched fuel elements cooled and moderated by D_2O. Each fuel element has four concentric tubes containing the fuel, and allows an experimental rig to be irradiated inside the central hole of 50 mm diameter.

J. P. Genthon and H. Röttger (eds.), Reactor Dosimetry, 975–981.

© 1985 ECSC, EEC, EAEC, Brussels and Luxembourg

In the design, operation and analysis of rigs irradiated in these in-core positions within fuel elements, neutron fluxes and reaction rates are often required. Although measurements are frequently possible, calculations are also needed in some cases, and then a simple lattice cell model is usually applied.

A one-dimensional infinite lattice cell calculation is usually adequate in these cases, using (r, ∞) geometry and the collision probabilities module PERSEUS in the WIMS-E scheme (1). (A brief description of the codes mentioned is provided in the appendix.)

A more exacting requirement is the determination of nuclear heating rates in the components of the rig (2). This can rarely be measured directly and calculations to a reasonable standard of accuracy are required. Gamma heating is the major component in most rig materials, but in the lighter nuclides neutron heating is also significant. Thus the model must provide neutron and photon fluxes, from which kerma factors can be used to obtain energy deposition rates. All sources of photons need to be taken into account: prompt gammas from fissions in the fuel element and in the rig if fissile materials are included, delayed gammas from the decay of fission products, and gammas resulting from the inelastic scattering and capture of neutrons in all materials including the structural materials of the rig. Gammas from the decay of neutron activation products can be ignored.

The approach to this problem has been by the use of the discrete ordinates code ANISN (3), the one-dimensional model consisting of a lattice cell comprising the single fuel element and the rig. The use of 8 neutron group, 21 gammas group coupled cross section data, condensed from the 121 group ENDF/B-IV based library DLC37F (4), enables photons produced in neutron interactions, including prompt gammas from fission, to be represented.

Fission product decay gammas require a separate approach, and the time-dependent spectrum has been evaluated using published analytic fits to the results of the CINDER-10 code (5). These spectra have been integrated over each of the normal fuel element irradiation periods to provide a mean decay gamma source spectrum which is input to ANISN as part of a coupled fission spectrum, together with the conventional neutron spectrum.

Kerma factors for photons and, where important, neutrons, are used to obtain heating rates in all parts of the experiment assembly, and the total fission rate from the calculation is used to normalize the results to the known fuel element power.

This technique has been used successfully in a number of cases and generally gives agreement where measurements of heating are possible. The chief limitation of the model is its one-dimensional geometry; in some rigs an unacceptable degree of homogenization of the features in the axial direction would be required, and axial variations of heating rates are often desired. A preliminary study using three-dimensional Monte Carlo modelling has shown that very long computer runs would be required to reach adequate statistical accuracy if a detailed axial distribution is required. The computational methods available for this problem are limited by the need to use coupled neutron/gamma cross-sections.

3. Reflector Rigs: Neutronics

The requirements for neutronic calculations for experimental rigs positioned in the reflector are similar to those for in-core rigs, but are much more difficult to satisfy. Distributions of neutron fluxes and powers of fuel pins in experiments are generally needed, but since the source of neutrons (the core) is remote from the experiment, and the distribution is affected by other facilities in the reactor, such as beam holes and other experimental rigs, the fluxes and powers are not simple to compute.

The thermal neutron flux in the D_2O reflector peaks a few cm away from the core "edge" and then falls off, falling to less than half the peak value over 25 cm, so that most experimental positions have a significant flux gradient across them. In cases where this gradient is not important to the required results, a model is used in which the rig is represented in a large (approx. 25 cm radius) cell of D_2O with reflecting boundaries. This "supercell" model uses WIMS-D4 (6), and is only useable, of course, if the rig contains fuel, since the model is essentially of an infinite array of such rigs. This model is capable of providing information on relative thermal neutron fluxes within the rig, but when the flux gradient of the environment needs to be considered, there is no alternative to some form of whole reactor model.

To investigate a recent problem of this type, in which the relative powers in nine fuel pins in an experimental rig were required, together with the effect on these of aspects of the rig design, two independent models were used. The first of these was a full three-dimensional model of a sector of the reactor using the Monte Carlo code MORSE (7,8). This model represents a sector of the D_2O reflector with a neutron source on the appropriate surface, having a spectrum determined by an earlier detailed whole-core model. An important horizontal beam hole through the reflector was included, and the rig itself described in some detail.

The second technique was based on a two-dimensional model of the whole reactor in (r, θ) geometry using the collision probabilities module PIJ in WIMS-E. In this, the core was represented as a single homogeneous region of a shape approximating the outer boundary of the core, and the rig was modelled in a separate (r, θ) system at the appropriate position within the D_2O. Some simplification of the rig geometry was necessary, and in particular the fuel pins of the experiment had to be represented by sectors of an annulus.

The major problem with the two-dimensional model was the representation of axial leakage. An axial buckling factor was used for this, but it was found that the degree of flux tilt across the rig was quite sensitive to the buckling value used. Furthermore, the rig design included heavy absorbing regions above and below the central region making the choice of buckling value difficult.

In order to solve this problem use was made of a set of flux measurements which had been made in a "mock-up" experiment of similar design to the final rig. By adjusting the buckling value until the flux gradient observed in the measurements was reproduced by the calculations, and confirming that this value gave similar agreement in a model of a second experiment, the buckling to be used was determined. Results obtained from the two-dimensional PIJ model were also in agreement with the three-dimensional MORSE model.

Having thus adjusted the axial leakage correction, the two-dimensional model was applied to a series of calculations concerning the rig design, with confirmation of the more important results being provided by the Monte Carlo model. Generally, the collision probabilities model is capable of

978

providing more detailed information about the flux distribution in the rig, and without the statistical uncertainties inherent in the Monte Carlo method, but involves a degree of geometric simplification and cannot model certain features such as horizontal beam tubes.

Both models, however, involve heavy use of computer resources, and a CPU time of 45 minutes (on an IBM 3081K) is typical for a single run using either technique.

4. Preliminary Design of a Replacement Reactor

A study has been undertaken of the reactor physics aspects of a possible design for a reactor to replace the existing Harwell MTRs. This has been done with regard to likely future user requirements, and in parallel with an engineering study. Options for a "green field" site construction or complete refurbishment of an existing reactor were studied.

A wide range of reactor types and design parameters were investigated. The calculational methods varied according to the system being looked at, but were generally based on modules of the WIMS code. Collision probabilities methods were used for flux and power distribution calculations in infinite lattice cell and in one-dimensional whole reactor (r, α) models, used for parametric studies, for example on the effect of lattice pitch size. Diffusion theory was used for two-dimensional whole reactor calculation in (r, θ), (x, y) or hexagonal geometry, generally with homogenization of the fuel element regions.

Requirements for high fast fluxes in the core and good thermal fluxes over a large volume in the reflector seem best served by a compact core with D_2O coolant, moderator and reflector. Such a core, being undermoderated, has a substantial thermal flux (and power) peak at the edge, the power gradient across the outer fuel elements being a matter requiring close attention.

In order to calculate the details of the power distribution in these outer fuel elements, the whole reactor was modelled in two-dimensional (r, θ) geometry using the collision probabilities module PIJ. This code has a "rod" facility allowing separate (r, θ) sub-systems to be superimposed on the overall geometry, and though intended for the representation of pin clusters, has been used here to represent fuel elements in a degree of detail usually only afforded to infinite lattice cell calculations. The model is complex and requires around 2 hours CPU time (on an IBM 3081K), but does allow detailed power distributions to be obtained. It has also been used to study the effects of control systems of various types, on total reactivity and power distribution. In order to make the model critical, axial leakage has been introduced by a buckling correction, adjusted to giv unity K-effective.

5. Outstanding Problems

Although modelling of a rig in the reflector of an MTR has been successfully attempted, the complexity and long running time of the models prevent regular and routine application. Further improvements are required and may be based on the sub-division of the reflector into zones with defined boundary conditions, assumed invariant.

The calculation of neutron heating in in-core rigs is satisfactory, bu a pressing problem is a requirement for the same sort of capability for rig in reflector positions. Here the calculation of photon fluxes is even more uncertain. The use of Monte Carlo models using existing neutron/photon coupled data is one possibility, although lack of detail in the group structure in the thermal neutron energy region casts doubts on the applic-ability deep into the reflector. An alternative would be an attempt at photon transport within the WIMS scheme, although photon production is likely to be the uncertain area here.

Finally, more detailed design work on the replacement reactor will require models giving detailed information in the compact core and its reflector, in the presence of flux peaks which often makes homogenization inappropriate, but without using exorbitant amounts of computer resources.

References

1. Askew, J. R. and Roth, M. J., AEEW-R1315 (1982).
2. Taylor, N.P., AERE-R10558 (1983).
3. Engle, W. W. Jr., Oak Ridge report K-1693 (1967).
4. Ford, W. E. III, et al., ORNL-TM-5249 (1976).
5. La Bauve, R. J. et al., LA-8277-MS (1980).
6. Halsall, M. J. AEEW-M1327 (1980).
7. Emmett, M. B., ORNL-4972 (1975).
8. Taylor, N. P. and Needham, J., AERE-R10432 (1982).
9. Anderson, D. W., AEEW-R862 (1973).
10. McCallien, C. W. J., TRG Report 2677(R) (1975).
11. Baker, L. J., Unpublished work (AERE Harwell).

APPENDIX
Computer Codes and Data Used

The principle neutronic/photonic transport codes used for studies in the Harwell MTRs are these:

WIMS   The Winfrith suite of multigroup neutronic codes. The principle modules used are:

    PERSEUS (1)   for one-dimensional collision probabilities solutions.

    PIJ (9)   for two-dimensional collision probabilities solutions in (r,θ) geometry, with a facility for superposition of rod clusters.

    SNAP (10)   for two- or three-dimensional diffusion theory solutions, mainly two-dimensional in (r,θ), (r,z) or (x,y) geometry.

The WIMS-E version (1) is used for most calculations, with its 69-group library of cross-sections often condensed to fewer groups (typically 8) for whole-reactor or whole-core models.

ANISN (3). The one-dimensional discrete ordinates code with anisotropic scattering. Used in cylindrical geometry for lattice cell models. The 100-neutron 21-gamma group cross section library DLC37F (4), based on ENDF/B-4, is used in coupled calculations. This is a slowing down library, with a single thermal neutron group. When better thermal data is required, the WIMS-E data library is used converted to ANISN format using a translation programme, WAFT (11).

MORSE (7,8) When the geometric capabilities of the above codes are insufficient, this Monte Carlo code is used, its combinational geometry allowing complex three-dimensional models to be created. Either fixed source or eigenvalue solutions may be obtained, and the code uses the same cross section data as the ANISN code, allowing coupled neutron-photon runs, or the use of WIMS cross sections (converted by WAFT) as appropriate.

'A COMPREHENSIVE APPROACH TO THE PROBLEMS OF UNCERTAINTY ANALYSIS IN THE ASSESSMENT OF IRRADIATED MATERIALS PERFORMANCE'

A F Thomas, M R Brown, A Dolan

Rolls-Royce & Associates Ltd

P O Box 31, Raynesway, Derby

In recent years considerable attention has been given to the quality of data generated in irradiated materials performance programmes, since many of the measured experimental parameters are difficult to define with accuracy, and historically the existence of data errors has been treated in a somewhat cavalier manner. A key element in these programmes is the need to set and achieve realistic target accuracies for neutron dosimetry and to ensure that these are thoroughly and fairly propagated in the definition of irradiated materials design performance limits.

A philosophy has been evolved which by means of variants on the least squares statistical analysis method has been implemented into a three-stage data analysis method for assessing the design performance of reactor pressure vessel materials.

INTRODUCTION

Until fairly recently codes of Design Practice have been written such that limits to materials design performance shall be defined on the basis of worst case data, suitably factored to allow for a margin of safety. In the nuclear industry such a rigid deterministic philosophy is increasingly coming under attack due to the more demanding requirements of probabalistic safety assessments. In several specific instances of irradiated materials performance within nuclear reactors themselves this assault has been reinforced by recent evidence that the worst case data upon which earlier reactor plants were designed was far from conservative. Consequently the universal application of such data in conjunction with Codes of Practice based on worst case analysis is having a debilitating and iniquitous effect on the operation of both 'safe' and 'unsafe' plant alike. The most highly publicised example of such difficulties is the neutron irradiation embrittlement of light water reactor (LWR) pressure vessels (RPV) (1, 2), but similar difficulties can be found in other areas of materials performance assessment.

The most obvious alternative philosophy upon which to base codes of practice is one based on a rigorous statistical analysis of the data. In this method all the important materials and environmental variables are characterised and quantified both in terms of their mean value and their associated uncertainty or experimental error. The definition of design, operational, and safety limits can then be applied on a specific plant by plant basis by the use of the appropriate data and a universally agreed probability risk criterion or statistical confidence limit.

The consequence of such a change in design philosophy to the definition of any required reactor (materials) dosimetry parameters is that attention needs to be timed not so much to the development of new experimental techniques but more to the quality of available techniques and the methods of their analysis (3). This effort is needed as much for the quantification of errors and uncertainties as in their reductions (provided that target dosimetry quality is consistent with the expected quality of the relevant materials and environmental data).

In the recent past considerable international effort has been spent in improving the quality of dosimetry data (4). This has not always been matched by parallel effort in the characterisation of materials properties. Thus it is not at all clear what the benefit of such effort has been particularly in the context of a worst case design philosophy. In order to ensure that such efforts do not become fatuous exercises it is also essential that materials data correlations and predictive extrapolations are based on throughgoing statistical analysis techniques in order to properly account for all sources of error and uncertainty. These might then form the basis of a satisfactory probabalistic design philosophy to replace that used in current design Codes of Practice.

This paper illustrates one attempt at a rational and comprehensive approach to the problems of uncertainty analysis in the assessment of irradiated materials performance and how such an approach is being implemented.

IRRADIATED MATERIALS ASSESSMENT PROCEDURES

The methodology which is adopted to define irradiated materials design performance varies as between organisations and applications but can be generally represented by the flow chart shown in figure 1. Such a flow chart can apply equally to both the research and surveillance (eg civil RPVs) evaluation of irradiated materials performance. As far as problems of uncertainty analysis in this procedure are concerned the key elements are

    (i)   prediction of through and end of life damage exposure in operating plant

    (ii)   estimation of damage exposure parameters in materials test irradiations

    (iii) characterisation of start-of-life and irradiated material properties

    (iv) extrapolation of irradiated materials properties data from test environments to operational environments.

The methods for undertaking the above tasks are many and various and are well documented. The area of concern however is whether and how the uncertainties and experimental errors which are inevitably involved in such methods are evaluated and propagated to produce realistic confidence limits for design and plant operation. This subject has been the focus of much international interest in recent years (5, 6, 7, 8, 9), due in the main to the growing realisation that the reactor pressure

Processes involved in the assessment of design performance of irradiated materials.

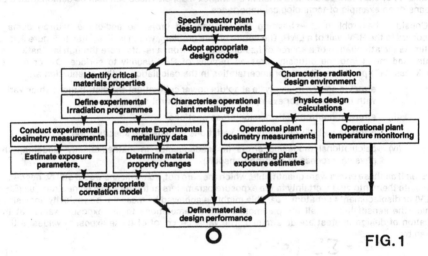

FIG. 1

vessels of ageing LWRs may have to be prematurely retired unless improvements can be made to knowledge of their neutron irradiation induced embrittlement behaviour. In fact it was such a driving force which inspired the development and implementation of the methods described below although their application is not restricted to any specific materials performance programme.

PREDICTION OF OPERATIONAL EXPOSURE DOSES

In principle it should be possible to monitor the exposure rates and, by simple integration over time, the exposure doses of any reactor component subject to significant irradiation exposure by means of in-situ measurements. In practice such measurements are rarely possible for a variety of reasons which need not be further elaborated, so theoretical calculational methods must be resorted to for their elucidation. However by means of extrapolation, estimates of exposure parameters at critical components can sometimes be made from measurements made at other practicable locations. In any event all such measurements need to be supported by physics calculations in order to derive useful exposure parameters. All these measurements and calculations inevitably involve uncertainties and errors which propagate uncertainties in the estimates of exposure parameters often in a complex manner. These can best be demonstrated by means of an example of technological importance.

Consider the problem of estimating the through and projected end-of-life neutron damage exposure to the RPV wall of a LWR. The nature of this problem, which is illustrated in Figure 2, is to estimate the attenuation of a source of fission neutrons from a reactor core through laminations of water and metal into the critically stressed regions of RPV usually to surface (OT) or quarter thickness (¼ T) positions. The major uncertainties in the calculation of this attenuation are:

    (i)   errors in source intensity (ie absolute power calibration) and distribution (which varies with time due to for example burn up and control rod movements).

    (ii)   errors in neutron transmission cross sections.

    (iii)   errors in as built geometry and variations in materials specification.

    (iv)   calculational modelling errors (ie inadequacies in mathematically modelling the physical process of neutron transport).

Even if all these errors were quantifiable, which they are not, some means would still be needed to translate them into an uncertainty in the exposure parameters at the RPV wall eg neutron flux (E> 1 MEV) or displacements per atom, dpa. This might be achieved by means of a sensitivity analysis in which the sensitivities of all the parameters in the calculations to the exposure values at the location of design interest are determined. The variance σ^2 of these exposure values is then given by

$$\sigma^2 = \sum_i \sum_j G_i . \sigma_{ij} . G_j$$

(1)

where Gi(j) is the sensitivity of parameter i(j) to the exposure value

    σ ij is the covariance between parameters i and j

In practice it is usually found that such methods result in rather large exposure variances and it is difficult to assess the uncertainty generated in using a neutron transport calculation in a given environment when it has been experimentally validated in a different one.

These calculational methods involve projections not only in space but also in neutron energy since exposure parameters are neutron energy dependent (eg dpa) and therefore differential nuclear data is employed. One way in which more confidence can be gained in such calculated exposure values would therefore be to make some integral measurements at a well defined location as close as possible to the point of interest and use this data to redress the calculated data using a suitable consistency criterion.

The chosen methodology was based on the use of measurements of neutron activation rate in suitable monitor materials in the RPV cavity to adjust the calculated exposure values within the RPV itself, using a generalised least squares treatment. This method is similar to the one described in (10) in which the consistency criterion is expressed mathematically as 'maximising the joint probability distribution of the data base subject to the relationships between the differential and

986

Schematic representation of a typical PWR neutron field geometry

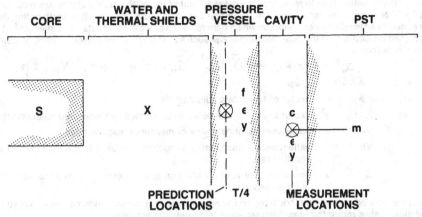

CORE   WATER AND THERMAL SHIELDS   PRESSURE VESSEL   CAVITY   PST

PREDICTION LOCATIONS   T/4   MEASUREMENT LOCATIONS

y, x, s are symbolic representations of basic data for a neutron transport calculation

y — detector cross-section data

x — materials cross-section data

s — core source distribution data

ϵ — systematic correction to neutronics method

c — calculated reaction rates measurement locations

f — calculated reaction rates at prediction locations

m — measured reaction rates

FIG.2

Application of London methodology

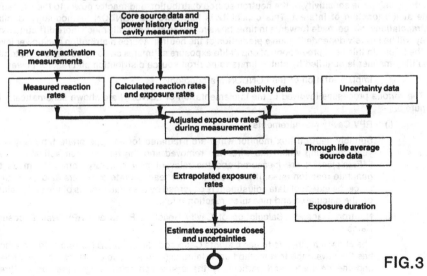

Core source data and power history during cavity measurement

RPV cavity activation measurements

Measured reaction rates

Calculated reaction rates and exposure rates

Sensitivity data

Uncertainty data

Adjusted exposure rates during measurement

Through life average source data

Extrapolated exposure rates

Exposure duration

Estimates exposure doses and uncertainties

FIG.3

integral data'. Such a method also has the intrinsic capability of reducing the variance on the exposure values at the locations of interest. This methodology has now been translated into the form of a computer code (LONDON) (11) and its application to the analysis of the PCA-LWR simulator experiments at Oak Ridge (USA) and at AEE Winfrith (UK) is reported in these proceedings (12).

It is not the intention here to reproduce the development of the mathematical algorithms employed in LONDON since these can be found in the references given above. However it is worth noting that the generalised least squares condition which generates the basic nuclear data adjustments (δ_p) is expressed in LONDON by the determination of δ_p such as to minimise the objective function (using matrix notation):

$$x^2 = (\delta c + (c - m))^T . V_m^{-1} . (\delta c + (c - m)) + \delta_p^T . V_p^{-1} . \delta_p \quad (2)$$

where

δc = $G . \delta_p$

δ_p = vector of basic data adjustments

$(c - m)$ = vector of differences between calculated and measured reaction rates

V_m = variance-covariance matrix of measured reaction rates

V_p = variance-covariance matrix of nuclear data used to generate calculated reaction rates

G = sensitivity matrix of absolute calculated reaction rates (c) to fractional nuclear data (p)

Such a data adjustment methodology has the advantage that minimisation can be achieved by analytical means rather than requiring recourse to an iterative solution.

This methodology is capable of making good estimates of both exposure rates and errors over the duration of the neutron activation measurements provided suitably appropriate core neutron source and reactor power history data are available. Hence time integrated exposures over the measurement period can be made. In principle therefore by a continuous series of cavity activation measurements from date of commissioning the accumulation of RPV exposure dose can be periodically and precisely monitored through life. In practice however the problems of predicting the design lifetimes of operating reactor plants also require the extrapolation of exposure estimates in time. For existing plant where no measurements have been made in the past an extrapolation backwards in time is necessary. Assuming a sensible estimate of the average neutron source distributions and reactor power can be made from historical data a reliable extrapolation can be made by using the sensitivity of the neutron source distribution and reactor power to the exposure rate at the location of interest. This capability is built into the LONDON methodology. Similar extrapolations can be made forwards in time, however these are generally much more difficult since they will be entirely dependent upon a prediction of the neutron source distribution and power level in the future. In this case predictions of end of life exposure estimates can only be made in a strictly qualified manner ie qualified by stated limits to neutron source distribution and reactor power.

Implementation of the LONDON Methodology

The various processes involved in the implementation of LONDON are shown schematically in Figure 3.

(i) RPV Cavity Measurements

Suitably neutron flux monitor wires are irradiated for an appropriate time (say one year) following which the wires are removed and the fast neutron activations (eg 54Fe(n,p)54Mn) are measured and the relevant power history correction made to generate reaction rates. It is crucial that the reaction rate errors are also estimated since the extent of data adjustment is governed by the relative sizes of the uncertainty on the calculated and measured reaction rates.

(ii) Neutron Transport Calculation of Cavity Reaction Rates and RPV Wall Exposure Rates

The chosen method is to use a Monte Carlo code (ie McBEND (13)) since this method has the advantage that method uncertainties (ie stochastic errors) can be quantified and the geometry and particularly the cavity can fairly be represented in three dimensions. The McBEND code is not only capable of calculating the reaction rates and exposure rates at the locations of interest but can also calculate the sensitivities of these quantities to variations in the transmission and cavity monitor reaction rate cross sections which are required by the data adjustment procedure.

(iii) Generalised Least Squares Data Adjustment Procedure (GLSP)

This procedure lies at the heart of the LONDON methodology and its primary function is to achieve statistical consistency between measured and calculated reaction rates in the RPV cavity. This is effected by means of the weighted adjustment of all data which is involved in the calculation of these reaction rates via the appropriate sensitivity coefficients. This adjusted data is then used to estimate the exposure rates during the period of the cavity measurements in the RPV wall together with their uncertainties. In order to ensure that these uncertainties are not underestimated, the procedure also makes use of 'variance scaling' to ensure that the final value of χ^2 after adjustment is not greater than its expected value (this device is discussed in more detail later).

(iv) Extrapolation of Cavity Measurement Adjusted Exposure Rates through RPV Lifetime

This is undertaken by means of the simple expedient of adjusting the core neutron source distribution until it is numerically equal to the actual or projected through life averaged source distribution. This adjustment is combined with the source sensitivities calculated by the Monte Carlo calculation to generate the appropriate correction to the exposure rates within the RPV wall.

(v) Estimation of RPV Exposure Doses

The final stage in the estimation of RPV exposure doses and their uncertainties is to integrate the through life coverage exposure rates over the already occurred and/or planned life time of the plant.

ESTIMATION OF EXPERIMENTAL EXPOSURE DOSES

A very similar although in some ways easier problem to the estimation of plant operational exposure doses is the evaluation of exposure doses experienced by metallurgical specimens in Materials Test Reactors (MTR) or surveillance capsules (eg RPV steels irradiated in operational plant). Since for the purposes of experimental rig irradiations activation measurements can be made in-situ, the only problem is how to define the neutron energy spectrum within the rig such that high quality specimen exposure parameters can be calculated. The problem reduces therefore to one of extrapolation in energy and not in space or time.

The chosen method was to carry out 'few-foil' activation measurements and adopt one of the many available spectrum adjustment codes. The selected code was SENSAK which uses the same methodology as the LONDON code described earlier ie a least squares adjustment procedure (14). (Results of the application of SENSAK were published in the report of the IAEA sponsored REAL 80 exercise (15)). The difference between LONDON and SENSAK is that in LONDON it is the nuclear data which are adjusted whereas in SENSAK, adjustments (δc) are made to the calculated neutron energy spectrum. In this case the objective function to be minimised becomes (in matrix notation).

$$\chi^2 = (\bar{\delta}c + (c-m))^T . V_m^{-1} . (\bar{\delta}c + (c-m)) + \bar{\delta}c^T . V_c^{-1} . \bar{\delta}c \tag{3}$$

where Vc = variance-covariance matrix of calculated reaction rates

The difficulty of fully evaluating Vc is a major drawback to the used of spectrum adjustment codes since the largest element of Vc is due to uncertainties in the calculated neutron energy spectrum which are difficult to quantify. However expedients are available. The method used in SENSAK is the so called 'variance scaling procedure which was referred to earlier. This involves taking the ratio of the final value of χ^2 after adjustment to the expected value which should be equal to the number of degrees of freedom (ie the number of reaction rate measurements). Since at this stage

$$\chi^2 = (c-m)^T . (V_c + V_m)^{-1} . (c-m) \tag{4}$$

Any departure from unity is assumed to be due to inadequacies in (Vc + Vm) which is scaled appropriately. Since Vc is generally large with respect to Vm the net effect is a scaling of the variances on the adjusted neutron energy spectrum. This method should avoid the possibility of seriously underestimating Vc. However care needs to be taken in its use to ensure variance scaling does not occur over that part of the energy spectrum of importance in the definition of exposure parameters which has not been experimentally confirmed by measurement, since in those circumstances Vc may still be seriously underestimated.

Implementation of SENSAK Methodology

In order to produce consistent analysis of rig exposure data and to guard against unjustified claims of uncertainties on exposure estimates a strategy of implementation of SENSAK has been adopted whose major elements are shown schematically in Figure 4.

(i)   Multiple Foil Activation Measurements

Wherever possible for each experimental rig design a multiple foil activation (MFA) experiment is conducted in the relevant location in the test reactor environment. (This clearly is more difficult for surveillance locations inside an operating reactor in which case uncertainties in exposure parameters may be larger than is technically feasible). The chief object of the MFA experiment is to measure as much of the neutron energy spectrum of importance to the required exposure parameters, and as accurately, as possible. This can be achieved by using a wide variety of integrating detectors whose response covers the energy region of interest.

(ii)   Calculated Neutron Energy Spectrum

Neutron spectrum calculations using Sn transport or Monte Carlo codes are carried out on the rig in the location of interest and an assessment made of the calculated variance-covariance from sensitivity studies.

(iii)   Neutron Spectrum Adjustment

Using data from (i) and (ii) above together with activation cross section data and all relevant uncertainty estimates SENSAK is used to produce a best estimate of the neutron energy spectrum in the location of interest together the appropriate variance covariance data.

(iv)   Rig Exposure Parameter Estimates

In each irradiation rig a standard dosimetry capsule containing a few practicable activation foils (ie allowing for anticapated environmental conditions and length of irradiation etc) is included together with suitable materials for making estimates of exposure gradients within the rigs. Following irradiation measured reaction rates are calculated and together with the MFA adjusted spectrum and uncertainties and the exposure parameter responses (eg neutron flux (E>1 Mev), displacements per atom (dpa) etc), average exposure rates and their errors are estimated. However in order to ensure that errors are not underestimated the variance scale factor in SENSAK is set to be not less than unity ie errors on rig exposure parameters cannot be less than those established in the MFA measurement.

DETERMINATION OF RADIATION INDUCED MATERIALS PROPERTY CHANGES

The determination of materials properties is outside the scope of this paper but propagation of the uncertainties inherent in their measurement is a key element in any comprehensive approach to uncertainty analysis in materials irradiation performance. Engineering materials are generally characterised by somewhat imprecise property data due to the nature and scale of their production and the large number of independent variables (chemistry, heat treatment, mechanical state etc) which can act. Design limits to materials properties, particularly mechanical properties, are usually based on specification minima so that attempts to produce more specific property data are subject to the qualification that many other parameters than the property of interest have to be quantitatively assessed. Consequently in order that accurate predictions of property data can be

Application of Sensak methodology

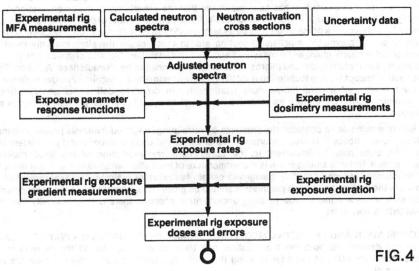

FIG.4

Application of General Lee methodology

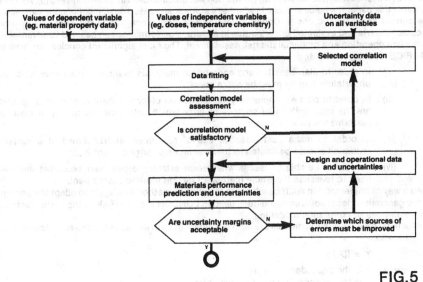

FIG.5

made, a well characterised correlation model which relates the independent variables to the dependent variable (ie the property of interest) needs to be established (we shall return to this problem later). For example it has been estimated that the uncertainty in the determination of the irradiation induced change in the brittle to ductile transition temperature (ΔNDTT) of RPV low alloy steels for a given class of materials may be as large as 45% to 90% (5) due to differences in form (eg weld or plate), chemistry, mechanical testing etc. In these cases therefore not only must the mechanical property changes be measured and their uncertainties, but also chemical analysis and its errors, and metallurgical and microstructural form will need to be characterised. Additionally, in the case of impact testing to determine ΔNDTT, many specimens are required in order to define the impact energy vs impact temperature relationship. In order to define the uncertainties in determining ΔNDTT therefore an objective method of analysis (eg statistical data fitting) is also required.

Before moving on to consider the problems of correlating irradiated materials property changes with exposure doses it is also important to emphasise that other environmental parameters than radiation dose may be important in the assessment and prediction of irradiated materials performance. Notable among these is the temperature of irradiation which exerts a strong influence on irradiations induced property changes in certain temperature regimes. The assessment of the errors in the evaluation of such parameters should be given equal importance to the assessment of exposure errors and mechanical property uncertainties wherever there is the possibility that such sensitivities may exist.

CORRELATION AND PREDICTIVE MODELLING OF IRRADIATED MATERIALS PROPERTY DATA

Having determined both the mean values and their uncertainties for all the dependent and independent variables of importance using the methods described earlier there remain the final problems of

(a)   correlating the data together in such a way that predictions of materials performance through the operational life of a plant can be made

(b)   ensuring that all errors are thoroughly propagated in order to assess the uncertainty on such predicted values.

This really is the key step in the performance assessment process, as was highlighted by the IAEA Technical Committee Meeting on Correlation Accuracies (CAPRICE) held in 1979 (5). As stated earlier in this paper the issue at stake is whether a statistically based design philosophy can replace the current deterministic one in Codes of Design Practice, since this is the only way in which accuracy criteria and evaluated uncertainties in materials performance can be used to advantage in both plant operation and probabalistic risk assessment. The most significant conclusions from the CAPRICE meeting were (16):

(i)   in order to plan sensible and appropriate materials evaluation programmes, design uncertainty targets must be set 'a priori'.

(ii)   in order to plan experimental programmes an objective method of assessing which are the most influential sources of error which determine the quality of materials performance data is needed.

(iii)   in order to make cost effective use of resources, efforts aimed at uncertainty reduction should be limited to these sources of error defined in (ii).

(iv)   in order to be able to assess whether uncertainty targets have been met there is a need to reassess the whole method of design lifetime assessment.

As a way of implementing such conclusions the chosen method was again to adapt the principle of the generalised least squares techniques used in LONDON and SENSAK to the particular needs and limitations of correlation modelling.

In general data correlation models are both non-linear in form and at best semi-empirical in nature ie in a model:

$$Y = f(X,P) \tag{5}$$

where   Y is the dependent variable
X is the vector of independent variables
P is the vector of modelling parameters

there are no prior (calculated) estimates of the vector P.

In order to achieve least squares fitting of the data, the minimisation of a x^2 objective function was again chosen where (using matrix notation):

$$x^2 = (f(X,P) - Y)^T . V_y^{-1} (f(X,P) - Y) + (f(X,P) - Y)^T . V_x^{-1} . (f(X,P) - Y) \tag{6}$$

where V_y is a matrix describing the variance covariance of the dependent variables

V_x is a matrix describing the variance covariance of the independent variables

The implications of the use of such an objective function are that both the modelling parameters and independent variables are adjustable data. Given that the values of the vector P can only be derived by iterative means and the size of materials databases can involve us up to several hundred different values of the independent variable it was realised that initially at least practical considerations would limit the use of the objective function to

$$x^2 = (f(X,P) - Y)^T . V_y^{-1} . (f(X,P) - Y) \tag{7}$$

ie the errors on the independent variables are not explicitly be taken into account during data fitting. However in order to take some account of the errors on the independent variables in the database, various statistical expedients were adopted. The most useful of those is the so-called 'Jacknife' technique by which the data base is selectively and repeatedly fitted, each time a different data base member being removed. The inherent uncertainty in the independent variables is thus reflected to some extent in the variability of the fitted correlation parameters, P ie the poorer the quality of the values of the independent variables the greater the variability in the values of P. However the performance of the correlation model itself is also reflected in such variability. Thus so-called goodness-of-fit can be derived from the value of x^2 obtained by inserting the mean values of P into equation (7) using all members of the database and obtaining a backing probability for such a value of x^2.

The importance of the use of a fitted correlation model lies in its ability to predict values of dependent variables where measurements are not possible or available. In this predictive mode the variability on the vector P and the uncertainty in the independent variables (eg end of plant life exposure dose) can be explicitly used. Three methods of estimating such uncertainties have been adopted:

(i) linearisation of the sensitivities of the dependent variable to correlation parameters and independent variables in the fitted model and summing the product of these and the uncertainty estimates in P and X.

(ii) Monte-Carlo sampling of the independent variables in the data base within the limits of the data base variances.

(iii) A hybrid of (i) and (ii); variance due to the independent variables is estimated by linearisation and variance due to the variance on the correlation parameters from the variance on the 'Jacknife' sample values of P.

Experience has shown that where the uncertainties in the database have been experimentally evaluated, little difference between the errors on a predicted value exist from these three methods.

The last but by no means least of the facilities required of the chosen methodology was the need to identify which of the variables in the database or parameters in the correlation model contribute most to the uncertainty in the predicted values. This is difficult to engineer in practice since correlations exist or are introduced by the fitting procedure itself. However a pragmatic approach was adopted in the methodology which by means of a linearisation of sensitivities method gives at least semi-quantitive estimates of the relative significance to the total error of the various sources of error.

This methodology has been translated into a computer code GENERAL LEE (17) and is currently in use for the reassessment of a UK RPV steels irradiation database which has been published previously (18).

Implementation of GENERAL LEE Methodology

The various processes involved in the implementation of GENERAL LEE are shown schematically in Figure 5. The power of this code lies in its ability to operate from any selected sub set of a large database while allowing the user to evaluate the selected data with any correlation model of the users choosing. The strategy of use therefore is an iterative one by which the user may selectively improve the method of data correlation whilst at the same time judging the effect such changes have on the quality of prediction of materials performance or life time assessment.

(i)   Experimental Data

Experimental data may be generated by means of experimental programmes in MTRs or from surveillance capsules in the case of Ligh Water Reactor (LWR) RPV safety justification programmes. It is important to ensure that all independent variables which may have an influence on the material property of interest are identified and quantified in order to achieve successful data correlation. In this regard it is also important to ensure that the quality of all data are within the targets necessary to achieve the required accuracy of data prediction.

(ii)   Data Fitting

A selected correlation model may parameterise as many or as few of the independent variables associated with the experimental data as desired. It is worth noting however that the extent of parameterisation should be consistent with the quality of the data and the degree of empiricism in the model itself. If these aspects are ignored the GENERAL LEE methodology will reject such modelling as being statistically unsound.

(iii)   Predictive Modelling

In order to use the fitted correlation for predicting materials performance through plant lifetime, it is important that the uncertainties on the operational plant data (eg through life exposure doses, coolant temperatures etc) are estimated. The GENERAL LEE methodology will then enable the user to ascertain whether it is the errors in the data modelling or those pertaining to the operational data itself which limit the uncertainties in the predicted materials performance. Such a facility enables a more objective approach to be made as to how resources should be spent to bring about the most cost effective improvement in plant design and operation.

CONCLUSIONS

Using variants on the well established method of least squares data adjustment and data analysis a comprehensive suite of codes has been adopted and developed to undertake the predictive and extrapolative modelling of data deriving from irradiated materials performance assessments. This approach was deemed necessary in order to deal with the problems of diminishing design margins and treatment of data errors and uncertainties.

The developed suite has three major tenets:

1.   A procedure for estimating design exposure parameters in-vessel by means of extrapolation in space and energy from dosimetry measurements made in ex-vessel locations.

2.   A procedure for estimating experimental exposure parameters in test reactor and surveillance locations by means of extrapolation in energy from dosimetry measurements made 'in-situ'.

3.   A procedure for correlating material dependent parameters with exposure parameters using analysis of variance to predict design performance limits and uncertainties.

This suite of codes is now being applied to the problems of RPV life time assessment of Light Water Reactors but can be generally applied to problems of uncertainty analysis in the assessment of irradiated materials performance.

REFERENCES

1. L E Steele, Ed., 'Status of USA Nuclear Reactor Pressure Vessel Surveillance for Radiation Effects' ASTM STP 784 American Society for Testing and Materials Philadelphia PA January 1983

2. L E Steele, Ed., 'Radiation Embrittlement and Surveillance of Nuclear Reactor Pressure Vessels: An International Study', ASTM STP 819, American Society for Testing and Materials Philadelphia PA November 1983

3. M Austin 'Sense of Direction: An Observation of Trends in Materials Dosimetry in the United Kingdom', Proc of the 4th ASTM-EURATOM Symposium on Reactor Dosimetry, Gaithesburg, MD, March 22-26, 1982, NUREG/CP-0029, NRC, Washington, DC Vol 1 p 461-469 July 1982

4. C Z Serpan 'NRC Light Water Reactor Pressure Vessel Surveillance Dosimetry Improvement Programme', Nuclear Safety Vol 22 No 4 p 449-458 July-August 1981

5. W Schneider 'CAPRICE 79' Proc of IAEA Technical Committee Meeting held in the KFA Julich W Germany 24-27 September 1979, Jul - Conf - 37 May 1980

6. F B K Kam, F W Stallman (Chairman), ASTM E10.05.01 Task Group on Uncertainty Analysis Minutes of ASTM Committee E10 on Nuclear Technology and Applications, American Society for Testing and Materials Philadelphia PA

7. W Schneider 'Comparison and Limitation of Uncertainties in Surveillance and Lifetime Prediction of LWR Pressure Vessels' Proc of the 4th ASTM-EURATOM Symposium on Reactor Dosimetry, Gaithesburg MD, March 22-26 1982 NUREG/CP-0029, NRC, Washington DC Vol II p 861-870 July 1982

8. F W Stallman 'Uncertainties in the Estimation of Radiation Damage Parameters' ibid p 1155-1163

9. F W Stallman 'Evaluation and Uncertainty Estimates of Charpy Impact Data' ibid p 855-859

10. R E Maerker et al, 'Development and Demonstration of an Advanced Methodology for LWR Dosimetry Applications', EPRI Report No NP-2188 December 1981

11. M R Brown and A Dolan 'Statistical Methods of Reactor Pressure Vessel Irradiation Damage Dose Estimation' to be published

12. A Dolan et al 'A Comparative Analysis of the Oak Ridge PCA and NESDIP-PCA Replica Experiments using the LONDON Adjustment Technique', these proceedings

13. D E Bendall and R J Brissenden 'MCBEND - Program User Guide' WRS Modular Code Scheme COSMOS REF SET DOC/147 AEE Winfrith (UK) December 1980

14. A K McCracken and A Packwood 'The Spectrum Unfolding Programme SENSAK', RPD/AKMcC/682 AEE Winfrith (UK) December 1981

15. W L Zijp et al 'Final Report of the REAL80 Exercise' ECN-128 Petten Holland February 1983

16. G Nagel, W Schneider 'Discussion of Uncertainty Statements for Lifetime Assessment after the Julich IAEA Technical Committee Meeting CAPRICE 79', Proc IAEA Specialists Meeting on Reliability Engineering and Lifetime Assessment of Primary System Components, 1-3 December 1980 Vienna

17. A Dolan, A F Thomas, and M R Brown, 'GENERAL LEE - A Data Fitting and Uncertainty Analysis Code for the Assessment of Correlation Models' to be published

18. T J Williams, A F Thomas et al 'The Influence of Neutron Exposure, Chemical Composition and Metallurgical Condition on the Irradiation Shift of Reactor Pressure Vessel Steels' Proc of 11th ASTM Conf on Effects of Radiation on Materials H R Brager and J S Perrin, Eds., ASTM STP 782 American Society for Testing and Materials 1982

PART X
CHAIRMEN'S REPORTS ON WORKSHOPS
AND
HIGHLIGHTS OF THE SYMPOSIUM

WORKSHOP "LWR-PV Physics, Dosimetry, Damage Correlation
and Materials Problems"

Chairmen: A. Fabry   A. Lowe, Jr.
              CSK/CEN Mol   Babcock & Wilcox,
                                    Lynchburg

The workshop was intended at addressing three areas in the
field of LWR pressure vessel surveillance improvement :

1) Physics-Dosimetry
2) Metallurgy and Damage Correlation aspects
3) Regulatory concerns.

Actually, the third topic was not discussed.

Regarding physics-dosimetry R & D efforts, a concensus seemed
to exist that accuracy goals are near to be met, and that the
important consideration will be long-term maintenance of the
technical capability, including pedigreed cavity-dosimetry and
application of ASTM standards under final development now. A few
areas call for limited further benchmarking work in the next two
years, and are part of current interlaboratory efforts. The
possibility of an outstanding bias in predicting in-vessel neutron
spectra below 1 MeV has been revealed at the workshop, and confirmed tentatively through subsequent individual contacts. This
would affect dpa/$\emptyset_{> 1 \text{ MeV}}$ ratios and efforts to correlate embrittlement of steels.

However, it was suggested by the second part of the workshop
that the sensitivity of correlations is probably dependent on a
number of parameters which have not been defined, and which may
not be definable because the data to provide the information is
not and may not become available. The ultimate importance of such
refinements may furthermore be lost in the errors of other data
used to evaluate reactor pressure vessels and, therefore, further
refinement of the correlations may not be warranted.

The effects of gamma heating on both reactor pressure vessel
and surveillance capsule material data is not always sufficiently
well understood. Some observations may indicate significantly
less neutron damage to reactor vessel materials than currently
projected because of higher temperatures in reactor vessel walls.
This area where data are often lacking needs to be addressed; the
sensitivity of embrittlement to irradiation temperature uncertainties is comparable to or larger than the ones of physics-dosimetry
measurements or projections, considering the present status of
development.

999

J. P. Genthon and H. Röttger (eds.), Reactor Dosimetry, 999.
© 1985 ECSC, EEC, EAEC, Brussels and Luxembourg

Atelier "Physique, Dosimetrie et Problèmes
des Materiaux en Réacteurs Rapides"

Présidents :   J.C. Cabrillat   A. Sekiguchi
                 CEA-CEN Cadarache   University of Tokyo

De cet atelier consacré aux réacteurs rapides, il ressort
tout d'abord deux caractéristiques, certes évidentes mais qui
méritent d'être rappelées:

    - L'une est que les préoccupations sont différentes d'un pays
à l'autre. Elles sont fonction des états d'avancement respectifs
des programmes "rapides".

    - L'autre est que sur le fond, bon nombre de problèmes sont
également ceux des réacteurs à eau pressurisée.

    L'importance pour les réacteurs rapides de la fragilisation
est à souligner.

    Elle joue un rôle primordial au niveau:

1 - de l'économie de la filière par la limitation de la durée de
vie des gaines de combustible. Ce point motive d'importants
programmes d'irradiation et d'études dans les réacteurs d'essais
tels FFTF et Joyo,

2 - de la sûreté de la centrale et de sa durée d'exploitation.

    En particulier la fragilisation de l'acier des structures
fixes supportant le coeur est une préoccupation du Japon et de la
France.

    Pour SUPER PHENIX, un programme de surveillance du sommier
reprend, toutes choses égales par ailleurs, les principes
appliqués aux cuves des réacteurs à eau:

    - Irradiation d'éprouvette "in situ" avec un facteur d'accélé-
ration dans le temps.

    - Corrélation des dommages vus par le sommier et les
éprouvettes.

J. P. Genthon and H. Röttger (eds.), Reactor Dosimetry, 1001–1002.
© 1985 ECSC, EEC, EAEC, Brussels and Luxembourg

La dosimétrie neutronique pose des problèmes similaires à ceux rencontrés par ailleurs. Elle a su s'adapter aux conditions plus sévères de l'environnement des réacteurs rapides et fournit des précisions très honorables. Il faut citer à ce propos, l'effort fournit par NBS, qui s'est assuré de la faible dispersion des mesures effectuées par différents laboratoires, sur des détecteurs de nickel irradiés dans des conditions identiques.

Compte tenu des incertitudes élevées liées aux calculs d'échauffement γ et de la puissance dégagée, un appel vers une dosimétrie γ efficace et précise est lancé pour ces réacteurs. A ce titre il convient de citer les essais de détecteurs T.L.D. (Alumine) prochainement programmés dans le réacteur PHENIX.

Pour les calculs de flux neutronique gamma, les préoccupations se situent au niveau des données de base des calculs de propagation à longue distance. Elles rejoignent celles qui motivent les réalisations actuelles de programmes expérimentaux de propagation dans des milieux acier et sodium (TAPIRO, HARMONIE, PAVIE) et les propositions de certains d'entre eux à des exercices internationaux de Benchmark.

Enfin il convient de mentionner le problème posé par nos collègues japonais sur la contamination des circuits en sodium par les produits de corrosion, qui ne fait pas l'objet de discussions internationales importantes.

Chairmen: C. Eisenhauer   F. Hegedüs
                       NBS Gaithersburg   EIR Würenlingen

The time was divided about equally between discussions of the ^{237}Np(n,f) and ^{93}Nb(n,n') reactions.

Questions were raised about the cross section of ^{237}Np(n,f) because calculated fission rates always seem to be smaller than measured rates. The following difficulties were mentioned:

(1) The uncertainties in the ^{237}Np(n,f) cross section may be greater than people realize because mass determination is often made by alpha counting and there are discrepancies in the α-branching ratio.
(2) There may be additional fission products counted from thermal capture in ^{237}Np and subsequent fission of ^{238}Np.
(3) The significance of sub-threshold fission in ^{237}Np spectrum-averaged cross sections is uncertain because of uncertainties in the sub-threshold fission cross section.
(4) Photo-fission in ^{237}Np may make significant contribution to the measured fission product activity. Martin of the U.S. has quoted a value of 30% photo-fission in ^{237}Np in the pressure vessel at the Browns Ferry reactor.

On the other hand, according to Fudge of the U.K., other problems with ^{237}Np have been solved. He mentioned:

(1) The uncertainties in gamma ray energies from the decay of ^{237}Np and its daughter product will be cleared up in an IAEA report to be issued in November.
(2) The Japanese have made a new evaluation of sub-threshold fission in ^{237}Np.
(3) Past discrepancies in fission yields for ^{237}Np have been resolved. Correct yields are incorporated into the latest ASTM procedures.
(4) Good agreement between measured and calculated ^{237}Np(n,f) cross-sections has been obtained for the iron spheres at Mol, Belgium and at the ISNF neutron field at NBS in the U.S.

Although there are problems with the use of ^{237}Np(n,f) it is still important for input to adjustment codes used to infer spectra in power reactors, particularly at the inner surface of the pressure vessel.

J. P. Genthon and H. Röttger (eds.), Reactor Dosimetry, 1003–1004.
© 1985 ECSC, EEC, EAEC, Brussels and Luxembourg

In Europe the $^{93}Nb(n,n')^{93m}Nb$ reaction is routinely used for fast neutron fluence detection. Because of its long half-life and its low threshold, it is well suited for RPV surveillance dosimetry. Subjects of discussion were the form and the purity of the detector, the activity counting method and the activation cross section.

For RPV surveillance purposes, the use of ^{93}Nb wire (diameter 0.5-1.0 mm) seems to be the most convenient. For other purposes, the use of a thin ^{93}Nb layer produced by physical vapour deposition on an aluminium foil could be suitable, too (GKSS). The ^{93}Nb should contain less than 5 ppm of tantalum. Niobium wire and foil of this quality are available at CBNM - Geel.

The surveillance dosimeters should be dissolved and then their ^{93}Nb layers ($< 5mg/cm^2$) should be prepared. Either pure Ge or Si(Li) counters can be used to count the $K_{\alpha,\beta}$ (16.6 + 18.6 keV) x-rays.

An absolute determination of the ^{93m}Nb activity was made by means of the liquid scintillation method (Harwell), where both the conversion electrons (K:12 keV, L:26 keV) and the x-rays are counted. The $K_{\alpha,\beta}$ x-ray emission rate obtained, 11.2%, was higher than the previously adopted value of 10.4%.

The new activation cross section measurement from Birmingham University (U.K.) shows a disagreement with the shape of the IRDF cross section (Strohmaier et al.) and confirms the shape of the earlier adopted cross section.

Workshop "Dosimetry for High-Energy Neutrons"

Chairmen: R. Dierckx   L.R. Greenwood
                   JRC Ispra   ANL Argonne

The workshop commenced with presentations by L.R. Greenwood of suggested topics for discussion and by R. Dierckx of the summary and conclusions of a recent workshop at Ispra on May 10-11, 1984 titled "Dosimetry Needs for the Fusion Materials Program". The workshop then considered three sources of high-energy neutrons, namely, accelerators, such as Be(d,n) fields of the FMIT-type; spallation sources, such as IPNS, LAMPF, SIN, SNQ; and fusion reactors, such as TFTR, JET, JT-60, and future devices. Accelerators produce neutrons up to about 50 MeV. It was felt that nuclear data needs for such facilities have already been well-formulated in studies connected with the FMIT and that specific data requests have already been submitted.

Spallation neutron sources produce neutrons up to about 1 GeV and entirely new dosimetry reactions are required. Spallation cross sections themselves appear to be well-suited for dosimetry at these sources and some recent data was presented for Al, Fe, and Cu. These data have been used to adjust neutron spectra at IPNS and LAMPF. Further measurements and integral testing are needed. At higher neutron energies (> 50 MeV) it is generally assumed that neutron and proton-induced cross sections are equivalent. Hence, differential proton cross sections may be used to provide the nuclear data for dosimetry at spallation sources. Neutron target yield measurements and further transport calculations are also needed for spallation sources to provide the input to spectral adjustment procedures, especially for the most common target materials Cu, W, Pb, and U.

Fusion reactor cross section needs can be divided into several types. Dosimetry requires accurate cross sections and ratios, especially near 14 MeV, to facilitate the measurement of neutron yields and energies. Plasma diagnostics can also be performed with activation dosimetry and the two new reactions ^{27}Al and ^{54}Fe(n,2n) appear to be especially sensitive to the ion temperature due to their thresholds near 14 MeV. Activation cross sections to very long-lived isotopes are also needed to determine radiation levels around fusion devices for maintenance, waste generation, and the selection of low-activation materials. There are a large number of unknown activation cross sections near 14 MeV. Many other cross sections are needed for fusion reactors concerning various blanket and shielding materials.

J. P. Genthon and H. Röttger (eds.), Reactor Dosimetry, 1005-1006.
© *1985 ECSC, EEC, EAEC, Brussels and Luxembourg*

All of these above-mentioned sources of high-energy neutrons require new dosimetry cross sections including variance-covariance data. Benchmark neutron fields are needed to test these data as well as neutron transport calculations. Damage and gas production cross sections are also needed, especially above 14 MeV, and for spallation sources new models and computer codes may be required.

Workshop Recommendations

1. High-energy cross section needs require a coordinated program of selected differential measurements and nuclear model calculations. Requests for spallation sources should be formulated and submitted to WRENDA and the IAEA.

2. Benchmark neutron fields need to be developed at accelerators for integral testing of nuclear data and transport calculations. These fields can be established by time-of-flight and other techniques.

3. Displacement damage and gas production cross sections need to be developed for spallation sources. Nuclear model calculations, supplemented by selected differential measurements, would be sufficient to provide the required nuclear data. Gas production should be tested by integral measurements.

4. More radiation damage experiments are needed in various radiation facilities which are well-documented by dosimetry (fluence, spectra, dpa, and gas generation) in order to test data correlations with exposure parameters, such as dpa.

Workshop "Modelling of Small Reactor Cores"

Chairmen: S. Anderson   S.B. Wright
             Westinghouse, Pittsburgh AERE Harwell

There was general agreement that the modelling was
satisfactory in the core away from large disturbances such as
control absorbers, or experimental assemblies. However close to
such assemblies, or in the reflector the situation was far from
satisfactory. Difficulties were reported even for cores of 4
meters high in calculating the flux gradient across the outer fuel
elements.

Shielding and reactor calculators appear to have similar
problems, and it appears that closer liaison between the two
groups could be profitable. The codes used do have some
differences, and are not immediately applicable, but there does
appear to be scope for learning from each other.

Modelling under these conditions is very expensive, but so
are low power experiments. While modelling will not replace low
power mock-ups there was a wide measure of agreement that it was
worth spending money on better modelling.

One area where there was general agreement was that the major
factor limiting better gamma-ray calculations was the current
coupled gamma-ray neutron libraries. Existing libraries are either
neutron only with many thermal groups, or coupled neutron-photon
with a single thermal group. New libraries are required which
allow adequate treatment of thermal neutrons as well as coupling
with gamma-ray data.

A further problem in gamma-ray calculations is the estimation
of gamma spectra from the decay of fission products at short times
(< 1 sec.) after fission, when much of the energy is released.
Spectral measurements in this range are currently lacking.

Workshop "Interlaboratory Intercomparison"

Chairmen: P. D'hondt   W.H. Zimmer
             SCK/CEN Mol   EG & G Ortec, Oak Ridge

Dr. Allen Fudge initiated a discussion on the nickel foil exchange that has been conducted through the National Bureau of Standards (NBS).

Dale McGarry said that the NBS reactor is currently being worked on to double its power level and to increase the number of ports. It should be back on line in January 1985.

J.W. Rogers mentioned that the CFRMF might be considered as a site for future irradiations.

The need was expressed by several participants for an exchange of a niobium foil. It would need to be tied to a known neutron spectrum such as a Cf-252 or V-235. It would have to have c.a. 10^{16} neutron exposure to make the data meaningful. Such exposure could be provided by the NBS facility with c.a. 80 hours irradiation. Since most participants would immediately dissolve the foil, each would have to have their own foil. It would be most meaningful if participants reported reaction rate instead of activity. Dale McGarry will make a recommendation from this workshop for such an intercomparison for the January 1985 ASTM E10.05 meeting in·Reno, Nevada.

There was a secondary expression of need for an Np-237 fission foil exchange.

There were 18 participants.

J. P. Genthon and H. Röttger (eds.), Reactor Dosimetry, 1009.
© *1985 ECSC, EEC, EAEC, Brussels and Luxembourg*

Chairmen: F.W. Stallmann   W.L. Zijp
             ORNL Oak Ridge   ECN Petten

The workshop on this topic was held jointly with a meeting of the ASTM E10.05.01 Task Group on Uncertainty Analysis and Calculational Procedures (annex 1). The workshop started with a summary by W.L. Zijp (ECN) on the IAEA Consultants Meeting on the REAL-84 exercise, held the previous day (annex 2).

F.W. Stallmann (ORNL) underlined that nowadays spectrum adjustment should be performed with the statistical methods based on the least squares principle, which methods imply that variances and covariances must be determined for measured reaction rates, cross-sections and calculated group fluences.

This task is far from routine for the fluence and cross-section data, particularly with regard to covariances. Many researchers are therefore reluctant to use adjustment methods, or try to make improper simplifications. He invited the audience to discuss these difficulties in order to initiate the establishment of guides and standardized procedures to provide the necessary help for the application of these adjustment procedures.

As first step towards this goal he proposed the following actions:

1.   The simplification of the variance-covariance information for neutron metrology cross-sections. The current ENDF/B-V and the special dosimetry file use 4 different formats for this information and require complicated processing codes to convert to a given energy group structure.
     Proposed is a generation of a simple variance-covariance matrix in some agreed upon group structure, consistent but much coarser than the 620 group structure of the present dosimetry cross-section file.
2.   Establishment of guidelines for the determination of calculated fluence variances and covariances. Such guidelines may be facilitated and promoted by the work of R.E. Maerker (ORNL) on the calculation of fluence covariance matrices.

After these statements Mr. Stallmann discussed the basic relation between any physical quantity x, (which is a function of fluence, cross-sections and reaction rates) and the residuals; he showed furthermore the relation between the covariance of this quantity x and the covariances between residuals and the covariances between x and these residuals. Each activation detector reaction has a typical cross-section variance which is not so sensitive to the spectrum shape. The covariance matrix of

the residuals shows high correlations and is not sensitive to
slight changes in the neutron spectrum. Therefore simplification
and standardization of the determination of covariances appear
feasible. There is a challenge to provide some recommendation. The
ASTM should establish guidelines for transport calculations which
can also derive covariances for spectra.

Mr. R.E. Maerker (ORNL) presented some details of his work
for the calculation of typical fluence and covariances for 3
locations (reactor cavity location, T/4 location, surveillance
location) in the ANO-1 reactor. He discussed sources of fluence
rate uncertainties, considered in the calculation of the
covariances for each of the three locations.

Mr. K. Weise (PTB) drew the attention to a draft German
standard (DIN 1319, Part 4) on the treatment of uncertainties
(inclusive covariances) in the evaluation of measurements. He
distributed an English translation, from which it became clear
that this document is very helpful with respect to the terminology
and procedures. Mr. C.M. Eisenhauer (NBS) discussed uncertainties
in the spectrum averaged cross-section values of uranium and
plutonium in the californium and ISNF benchmark fields.

Also some other topics were discussed: the physics
information content of the covariances of the parameters in the
Watt and the Maxwellian representation of the standard fusion
neutron spectra, and the merit of covariances of a standard
spectrum in its normalized form.

At the end of the workshop the audience was invited to assist
where possible in establishing written recommended procedures via
ASTM and/or other channels.

Report on the Meeting of the ASTM E10.05.01 Task Group on Uncertainty Analysis and Calculational Procedures

The ASTM E10.05.01 Task Group meeting was held jointly with the Workshop on Adjustment Problems (see the separate report on the Workshop in these Proceedings). The emphasis of the meeting was on the determination of covariances for fluences and cross-section data for the use in adjustment procedures and providing assistance to users of adjustment procedures in obtaining these data. The following suggestions were made:

1. The ENDF/B-V special dosimetry file needs streamlining so that group cross-section covariances can be extracted with a simple processing code. Coordination and cooperation is urged with present efforts in the European Communities to provide the participants of the REAL-84 exercise with a modified IRDF cross-section file and processing code to obtain group cross-section covariances.
   The ASTM E10.05.03 can provide the necessary liaison to the IAEA Nuclear Data Section.

2. In the coming revision of the ASTM Standard for Neutron Transport Calculation for LWR Surveillance guidelines should be provided to obtain realistic variances and covariances for calculated group fluences. R.E. Maerker has done extensive work in this field in connection with the LEPRICON Methodology and has agreed to provide input to the revision of the Transport Standard. Suggestions from other members of the ASTM E10.05.01 Task Group are solicited.

Dr. Weise, PTB Braunschweig, announced the availability of an English translation of the first draft of a new DIN Standard on experimental uncertainties, covariances, and least squares adjustments. This translation will be mailed to all E10.05.01 Task Group Members for comments and as a guide for possible similar efforts within ASTM.

F.W. Stallmann

Report on the IAEA Consultants meeting on the REAL-84 exercise.

An IAEA Consultants meeting was held on 26 September 1984 on the REAL-84 exercise. The meeting was attended by 11 invited experts. Here a short summary is given of the main results of the discussions.

The REAL-84 exercise should improve the assessment of accuracies in radiation damage predictions, and thereby promote the use of specialized adjustment procedures. It was realized that there are no problems with the adjustment codes themselves, but that there are problems with the input data needed for the adjustment. The exercise will underline the need for new or better data, since input data sets may show incomplete data, missing data and maybe inconsistent data.

The participation should be open for all laboratories which use adjustment codes which are able to take into account covariance data.

The participants should critically review the input data, perform spectrum adjustments, calculate some spectrum characteristics and calculate some integral damage parameters.

Thus the participants will be expected to calculate displacements per atom and helium production for each spectrum for iron, for steel and possibly for sapphire. One or two activation rates have also to be calculated for the adjusted spectra.

Spectra to be used for the exercise should include

a. the ANO-1 surveillance spectrum,
b. the PCA/PSF benchmark spectrum and
c. either the RTNS-II, 14 MeV fusion simulation spectrum, or a Be (d,n) 16 MeV accelerator spectrum.

All test spectra should contain evaluated covariance data.

It is recommended that the IAEA take the following actions concerning the REAL-84 exercise.

1. A standard procedure should be established to process cross section covariance data files.
   Due to the wide variation in energy group structures currently in use, a computer code for processing the ENDF covariance file should be distributed along with the data file itself. Participants could then be able to choose their own group structure without the need to collapse or expand a previously processed group file.
2. The International Reactor Dosimetry File (IRDF) should be made consistent with ENDF formats and group structures, especially regarding covariance data.
3. A computer code for changing group structures of processed cross section files should be made available to participants on request.

W.L. Zijp

J. P. Genthon and H. Röttger (eds.), Reactor Dosimetry, 1015.
© 1985 ECSC, EEC, EAEC, Brussels and Luxembourg

Workshop "Gamma Dosimetry and Calorimetry"

Chairmen: R. Gold   R. Lloret
              HEDL Richland   CEN Grenoble

This workshop featured an open interchange between ASTM and EURATOM participants on the status and development of gamma-ray dosimetry and calorimetry. Informal presentations were provided by the following participants:

1.   Status of Standard Gamma-Ray Fields - A. Fabry (CEN/SCK)

2.   Continuous Gamma-Ray Spectrometry - R. Gold (HEDL)

3.   Developments in Microcalorimetry - J. Mason
     (Imperial College)

4.   Use of CR-39 for Gamma-Ray Dosimetry - A. Asfar
     (Imperial College)

5.   Development of LiF Crystals for Gamma-Ray Dosimetry -
     D. Gilliam (NBS)

6.   Status of Gamma-Ray Calculations for PCA - R. Maerker
     (ORNL) and G. Minsart (CEN/SCK)

Productive discussions between all participants arose during and after these presentations. However, adequate time was not available to discuss all topics of interest, in particular gamma-ray energy deposition in complex heterogenious structures such as found in research reactors was not covered

It was apparent that considerable work has been done and it was even more apparent that considerably more work lies ahead. Outstanding issues must be resolved, especially fundamental differences between measured and calculated gamma-ray spectra in zero power systems. The need for methods to resolve the neutron and gamma-ray components of the mixed radiation field in reactor environments continues to be a crucial limitation in experimental efforts. In this regard, work at the Berkeley Nuclear Laboratory continues to be most promising.

It is clear that considerable interest and effort in gamma-ray dosimetry and calorimetry is on-going throughout the world. Consequently, it is anticipated that even more progress will be reported at the forthcoming San Antonio meeting.

J. P. Genthon and H. Röttger (eds.), Reactor Dosimetry, 1017.
© *1985 ECSC, EEC, EAEC, Brussels and Luxembourg*

Chairmen: A.J. Fudge   J.G. Williams
                  AERE Harwell   University of Arkansas

The workshop was attended by twenty-one participants as listed in Appendix 1.

The following scope statement had been prepared by the co-chairmen: A.J. Fudge and J. Williams.

Damage monitors are devices which provide the means to characterise neutron environments by observation of physical property changes resulting from atomic and lattic effects of neutron radiation damage. The aim is to correlate monitor response and engineering materials damage, without the need for the intermediate step of calculating neutron energy spectra.

The statement was read to the workshop participants and it was pointed out that the last sentence represented an ideal which did not seem to have been achieved yet for any damage monitor. This point became the subject of considerable debate and some disagreement later in the meeting.

Summaries of their work were given by six of the participants, either on damage monitors or on relevant damage measurement techniques.

A. Alberman of C.E.A. - Saclay, France, described the graphite (GAMIN) and tungsten monitors, in which small electrical resistivity changes are precisely using temperature control at the level of 10^{-3} °C. Spectral indices relative to the $^{58}Ni(n,p)^{58}Co$ reaction are used to characterise damage potential of the environments, at fluences of the order of 10^{16} n/cm^2. This information is transferred by means of the nickel reaction to high flux, temperature, and fluence exposures at the same or similar locations.

D. Bünemann reported on the main features of the GKSS, low angle scattering techniques for measuring the size distributions and number of voids and precipitates, the application of which had been described in session F of the symposium by R.Wagner.

A.J. Fudge reported on the status of the sapphire light absorbance damage monitor which has been used for fluences up to 5×10^{19} n/cm^2. Correlation of the response against calculated displacements per atom had produced some inconsistences, but these are improved in correlations between the sapphire response and the $^{93}Nb(n,n')^{93m}Nb$ reaction. Work to resolve the calibration of the technique is expected to be completed in 1984.

J. P. Genthon and H. Röttger (eds.), Reactor Dosimetry, 1019–1021.
© *1985 ECSC, EEC, EAEC, Brussels and Luxembourg*

S. De Leeuw of SCK/CEN Mol made a brief report on the silicon damage monitoring technique using P.I.N. diodes. J. Williams (Univ. of Arkansas) added some remarks about bipolar transistor measurements. These methods have been used in the neutron fluence range from 10^9 to 10^{14} n/cm^2. Spectral characterisation with respect to damage potential in low intensity and short duration exposures are possible using these detectors.

R. Odette (Univ. of California) gave a brief description of the development of miniature steel specimens used for mechanical testing which offers prospects on providing data from more numerous samples and from locations not suitable for larger samples.

R. Gerling (G.K.S.S.) showed data concerning the brittle to ductile change in an amorphous alloy, $Fe_{40}Ni_{40}B_{20}$, after exposure to light ions or neutrons. The transition occurs at different d.p.a. values for the two types of particles, and this was attributed to different clustering of displacements within large or small cascades.

In the discussion which followed on calibrations and intercomparison the question arose as to how best to relate the different techniques. Some were reporting values relative to nickel or niobium activation and hence cross-sections, and others relative to calculated d.p.a. Direct intercomparison does not seem to be possible because the devices used operate in non-overlapping fluence ranges. Measurements in the same benchmark spectrum are difficult to perform for the same reason.

Hegedüs (E.I.R. Switzerland) advocated correlation of measurements against d.p.a. units but both Alberman (C.E.A. France) and Wright (A.E.R.E. Harwell) objected to their use because of the uncertainties associated with the calculation of d.p.a. Wright advocated the measurement and reporting of unambiguous quantities, and the use of the entire spectrum information in the description of reactor exposures. The ideal position reflected in the scope statement implies the use of damage monitor response as a correlation parameter for use in interpreting material properties in engineering materials. However, a substantial difficulty exists in obtaining a suitable data base, and in transferring the existing materials data to a damage monitor scale of measurement.

It was concluded that there exists a need to co-ordinate damage monitor results by reporting values for well established fields and using a complete description of both calculated and other experimental information to aid in the interpretation and intercomparison of damage monitor data.

APPENDIX 1

Participants in Workshop on Damage Monitors

A. Alberman	C.E.A., CEN Saclay, France
D. Beretz	C.E.A., CEN Grenoble, France
D. Bünemann	GKSS, Geesthacht, Fed. Rep. of Germany
C. Cabrillat	C.E.A., CEN Cadarache, France
M. Carta	E.N.E.A., Rome, Italy
A. De Carli	E.N.E.A., Casaccia, Italy
S. De Leeuw	SCK/CEN, Mol, Belgium
A.J. Fudge	A.E.R.E., Harwell, U.K.
J.P. Genthor	C.E.A., CEN Cadarache, France
R. Gerling	GKSS, Geesthacht, Fed. Rep. of Germany
F. Hegedüs	E.I.R., Würenlingen, Switzerland
R. Koch	F.I.Z., Karlsruhe, Fed. Rep. of Germany
M. MacPhail	U.K.A.E.A., Culcheth
P. Mas	C.E.A., CEN Grenoble, France
W.N. McElroy	H.E.D.L., Richland, U.S.A.
G.R. Odette	Univ. of California, U.S.A.
F-P. Schimansky	GKSS, Geesthacht, Fed. Rep. of Germany
P. Wille	GKSS, Geesthacht, Fed. Rep. of Germany
J.G. Williams	Univ. of Arkansas, U.S.A.
S.B. Wright	A.E.R.E., Harwell, U.K.
W. Zimmer	E.G. & G. ORTEC, Oak Ridge, U.S.A.

"Highlights of the Symposium"

Chairmen: R. Dierckx   W.N. McElroy
              JRC Ispra   HEDL Richland

We thank all session Chairpersons for their contributions to
these summary and highlight statements. Significant efforts on LWR
pressure vessel and support structure surveillance, embrittlement
trend curve data development and testing, and physics dosimetry
are still in progress. The authors emphasized:
"core edge" assessment; 2-D spectrum calculations; thermal neutron
effects; perturbation effects; and measurement and cross section
validation in mockups, such as PCA, PSF, VENUS, NESDIP, DOMPAC,
and in selected commercial PWR and PWR power plants.

Physics-dosimetry analysis has been significantly improved by
ex-vessel measurements. Promising correlations with in-vessel
fluence measurements at a number of power plants were presented by
Cogburn, Williams, Tsoulfanidis, Rombouts, Lloret, Gold, and
others. A paper on the application of advanced and standardized
radiometric (RM), solid state track recorder (SSTR), helium
accumulation fluence monitor (HAFM) and damage monitors (DM) was
provided by Ruddy et al. Advanced RM, SSTR, and HAFM dosimetry
sets have been chosen, assembled, and are being deployed by LWR-PV-
Surveillance Dosimetry Improvement Program participants and US
utilities in the cavities of an increasing number of commercial
power plants. Here, information on the status of the development
of the sapphire DM, as described by Fudge, was of considerable
interest for future LWR-PV applications.

Difficulties were raised regarding the use of fissile
dosimetry. Availability of a supply of certified dosimetry
materials may lead to major improvements in terms of quality
assurance of dosimetry sensors; i.e., with respect to the present
status.

In the area of damage models and materials effects, a number
of papers were presented which complemented each other very well.
Wagner's description of the fundamental work being carried out at
GKSS, Geesthacht, Odette's, Hawthorne's and Grants work in the
USA, and other efforts by participants indicate that much progress
is being made towards understanding the mechanisms by which copper
and other elements affect the neutron embrittlement of iron
alloys. That is, information was presented and/or referenced on
the contributions of copper, nickel, phosphorus, manganese and
other elements to the embrittlement process and the kinetics of
the thermal recovery process.

J. P. Genthon and H. Röttger (eds.), Reactor Dosimetry, 1023–1026.
© 1985 ECSC, EEC, EAEC, Brussels and Luxembourg

The paper by Guthrie on statistical considerations in the development and application of Charpy trend curve formulas describes the application of a covariance treatment to several trend curve formulas developed recently at HEDL. These studies indicated that the largest contribution to the uncertainties are errors in chemical compositions. In his overview of the development and testing of damage correlation models, Odette described how the fundamental mechanistic basis for embrittlement, such as that described by Wagner, can be applied to the development of damage correlation models using statistical methods to correlate in-service power reactor surveillance data. The models developed show that the separate and synergistic effects, observed for copper and nickel content and fluence to be the primary variables and that other factors are of secondary importance. His results confirm Guthrie's conclusion that chemistry errors were the largest contributor to the uncertainty in the results.

Another development in the study and selection of trend curve exposure parameters has been the establishment of embrittlement curves that consider the combined effects of fast, intermediate, and thermal neutrons. If the magnitude of the thermal neutron contribution to elevated temperature (288°C) PV steel embrittlement suggested by recent work at HEDL is shown to be real, the impact of this work will be quite important for future revisions of the new set of 21 LWR ASTM standards, Reg Guide 1.99, and licensing and regulatory issues and actions related to pressurized thermal shock and other PV integrity issues. In this regard, Randall's revision of Reg Guide 1.99 will include the effects of both Cu and Ni, allow for some fluence dependence of the dose term, and recommend the use of a PV in-wall embrittlement gradient curve based on dpa and an 1/3 power law fluence dependence.

Results of work on new and existing adjustment codes by groups in Budapest, Petten, Japan, England, the USA and elsewhere were reported. Results of the completed REAL-80 exercise and plans for the REAL-84 exercise were presented by Zijp et al. More detailed information is provided in the report on the workshop on "Adjustment Problems" prepared by Stallmann and Zijp. It is anticipated that results for the REAL-84 exercise will be available for presentation at the 1987 ASTM-Euratom Symposium planned for San Antonio, Texas, in the USA. Results of the application of the advanced LEPRICON Methodology to commercial PWR power plants to reduce the uncertainty in physics-dosimetry prediction were reviewed by Wagschal et al. The LEPRICON work represents as significant advance for quantitive estimation of the covariances and relationships of the flux uncertainties from position to position.

Reported by Petilli and Gilliam were the results of a first attempt of uncertainty evaluation for the fission spectrum of 235U. The covariance matrix, in 15 energy groups, has been obtained as the result of fission rate measurements performed in the MOL cavity collaboration between NBS and SCK/CEN. The evaluation was done by means of a multiple unfolding code EOLO, using in addition the flux transfer technique between 252 Cf measurements and measurements in the MOL cavity.

In regard to fast reactors, a paper by Sekiguchi et al provided a review of the Japanese fast breeder reactor program and several physics-dosimetry improvements for their fast reactor radiation damage research. Progress on construction of the Japanese prototype fast breeder reactor "MONJU" and its initial criticality (scheduled for March, 1991) were discussed.

For Super-Phenix, Cabrillat et al reported on a grid plate radiation damage problem. Meneghetti and Kucera provided a paper on a proposed new and interesting method to obtain steel dpa values using the Pu-240 fission-rate, where the ratio between the dpa and Pu-240 fission rate are shown to be relatively insensitive to spectral changes in many irradiation positions of EBR-II.

To help in the exchange of information on the status of ASTM standards, a number of ASTM E10.05 Task Group meetings were held during the Symposium and in conjunction with workshop discussions, when possible. Separate reports on these meetings will be prepared and will be included in the minutes of the next ASTM E10.05 meeting, scheduled for January 14-17, 1985 at the MGM Grand Hotel, Reno, Nevada, USA.

Interesting papers concerned the post accident handling of the Three-Mile Island reactor and the radiation field associated with Hiroshima and Nagasaki in which reactor dosimetry techniques played an important role. Some reactor dosimetry techniques are also applied for health physics purposes and characterization of irradiation fields for biological purposes. Neutron dosimetry found its way for plasma diagnostic work.

Most of the dosimetry techniques are now more or less established. No new techniques but important improvements and specialised applications were presented. Gamma radiation field measurement and calculations got a big importance at this conference. Improved techniques and applications as well as international comparisons were discussed. Gamma field characterization will gain in importance in future fusion simulation sources as D-Li or spallation sources. Niobium fluence dosimetry is largely treated at this meeting. This reaction seems to be most promising for an intense use in future. It seems to avoid the problems occuring with the neptunium detector. The old sulphur reaction seems to revive, especially for shielding applications. Worth mentioning is the Euratom initiative to set up an EC source of reference materials for neutron metrology requirement.

Benchmarks continue to have their importance. The work on standards fields and reference fields is even increasing to characterize them better and better. Neutron and gamma cross-section data as well as He-generation cross-sections data are tested for consistency in different standard and reference fields. The use of adjustment codes leads to indications for probable cross-sections errors in the files and the necessary improvements. Of particular concern is the ^{58}Ni(n,p) cross-section which still shows a discrepancy of about 5%. A decrease was indicated for the iron inelastic cross-section to counter the tendency toward progressive underestimation of the neutron fluence with penetration into thick iron reactor pressure vessel simulators and iron shielding benchmarks. Adjustments were indicated for some reaction cross-sections: the ENDF/B-V cross-sections for the ^{115}In(n,n')$_2$ and ^{103}Rh(n,n') reactions should be increased slightly while the ^{32}S(n,p) cross-section should be reduced by as much as 12% over a substantial portion of the response range. New integral cross-sections were reported in the spontaneous fission standard neutron field of Cf-252. For some of the measurements reported here, no previous measurements have been reported in the literature. In general no relevant discrepancies between experiments and calculations based on ENDF/B-V and other $\sigma(E)$ data were observed, with the exceptions of ^{51}V(n,p) and ^{199}Hg(n,n'). Generalized least squares fitting methods are used to relate Benchmark results to applications (f.e. for design purposes). Here a remark should be made: Generalized least squares fitting methods are a very powerful but dangerous mean and one should take care not to misuse the method.

Fusion and spallation are still future and research and development of fusion dosimetry are in their starting phase. The U.S. and European materials R & D programmes for fusion are being presented. Some comments on the Japanese programme were given in the workshop.

There exists a virtual international unanimity to state that there is a need for a high-intensity high-energy neutron irradiation facility. The need for theoretical studies and of fundamental experimental investigations of radiation damage phenomena was underlined. Studies of the correlation of damage data in different irradiation environments are necessary. As well as theoretical and experimental work on damage mechanism, damage and neutron cross-sections, damage monitors, and anisotropic scattering cross-sections for high-energy neutrons is under way. The use of a spallation detector for dosimetry purpose is promising. We hope the efforts for high-energy neutron dosimetry will continue and even increase in the future.

CONFERENCE PROGRAMME SUMMARY

Monday, September 24, 1984

Opening of the Symposium

Session A	: LWR-PV Surveillance
Poster Session I	: Topics of Sessions A, B and C
Workshop 1	: LWR-PV Physics, Dosimetry, Damage Correlation and Materials Problems
Workshop 2	: Fast Reactor Physics, Dosimetry, and Materials Problems
Workshop 3	: Low Threshold Reactions

Tuesday, September 25, 1984

Session B	: Techniques
Session C	: General Interest
Session D	: Benchmarks, Reference and Standard Spectra
Poster Session I	: (continuation)
Workshop 4	: Dosimetry for High-Energy Neutrons (up to 10 MeV)
Workshop 5	: Modelling of Small Reactor Cores
Workshop 6	: Interlaboratory Intercomparison

Wednesday, September 26, 1984

Session E	: Neutron and Gamma Spectrum Determination
Session F	: Damage Models, Physics, Dosimetry and Materials
Poster Session II	: Topics of sessions D, E, F, G, H and I
Task group meetings	: IAEA Consultants meeting on "REAL-84 exercise"

Thursday, September 27, 1984

Session G	: Fusion and Spallation
Poster Session II	: (continued)
Workshop 7[1]	: Adjustment Problems
Workshop 8	: Gamma Dosimetry and Calorimetry
Workshop 9	: Damage Monitors

Friday, September 28, 1984

Session H	: Nuclear Data
Session I	: Fast Reactors
Final Session	: Reviews of Workshops Highlights of the Symposium

Closure of the Symposium

Facility Tour

[1] Jointly with ASTM E10.05.01 Task Group on Uncertainty Analysis and Calculational Procedures

LIST OF CHAIRMEN

Opening	J.P. Genthon CEA-CEN Cadarache	E. Schröder GKSS Geesthacht
Session A	A. Alberman CEA-CEN Saclay	A. Taboada US-NRC Washington
Session B	C.O. Cogburn Univ. of Arkansas	Mrs. G. de Leeuw-Gierts SCK/CEN Mol
Session C	F.B.K. Kam ORNL Oak Ridge	P. Mas CEA-CEN Grenoble
Session D	D.M. Gilliam US-NBS Gaithersburg	W. Mannhart PTB Braunschweig
Session E	R.E. Maerker ORNL, Oak Ridge	Mrs. E.M. Zsolnay Technical University of Budapest
Session F	G. Hehn IKE, University of Stuttgart	E.B. Norris Southwest Research Institute, San Antonio
Session G	Mrs. M. Petilli ENEA-CSN Casaccia	C. Reuther US-DOE Washington
Session H	V. Piksaikin IAEA Vienna	E.P. Lippincott HEDL Richland
Session I	Mrs. A. De Carli ENEA-CSN Casaccia	M. Nakazawa University of Tokyo
Poster Session I	W.G. Alberts PTB Braunschweig	H. Farrar IV Rockwell International Corp., Canoga Park
Poster Session II	A. Alberman CEA-CEN Saclay	E.D. McGarry US-NBS Gaithersburg
Workshop 1	A. Fabry CSK/CEN Mol	A. Lowe, Jr. Babcock & Wilcox Lynchburg
Workshop 2	J.C. Cabrillat CEA-CEN Cadarache	A. Sekiguchi University of Tokyo
Workshop 3	C. Eisenhauer US-NBS Gaithersburg	F. Hegedüs E.I.R. Würenlingen
Workshop 4	R. Dierckx JRC Ispra	L.R. Greenwood ANL Argonne
Workshop 5	S. Anderson Westinghouse, Pittsburg	S.B. Wright AERE Harwell
Workshop 6	P. D'hondt SCK/CEN Mol	W.H. Zimmer EG & G Ortec, Oak Ridge
Workshop 7	F.W. Stallmann ORNL, Oak Ridge	W.L. Zijp ECN Petten

Workshop 8	R. Gold HEDL Richland	R. Lloret CEA-CEN Grenoble
Workshop 9	A.J. Fudge AERE Harwell	J.G. Williams Univ. of Arkansas
Final Session	R. Dierckx JRC Ispra	W.N. McElroy HEDL Richland
Closure	D. Bünemann GKSS Geesthacht	F.B.K. Kam ORNL, Oak Ridge

LIST OF PARTICIPANTS

Jürgen AHLF
GKSS
Max-Planck-Str.
D-2054 Geesthacht
Germany

Alain ALBERMAN
C.E.A.-C.E.N. Saclay
Services des Piles
F-91191 Gif-sur-Yvette Cedex
France

Wolfgang G. ALBERTS
Physikalisch-Techn. Bundesanstalt
Bundesallee 100
D-3300 Braunschweig
Germany

Stanwood L. ANDERSON
Westinghouse Electric, NTD
P.O. Box 355
Pittsburgh, Penn. 15230
U.S.A.

Ali ASFAR
Reactor Centre
Imperial College of Science
and Technology
Silwood Park
Ascot, Berks SL5 7PY
England

Marten ATTLEGARD
Studsvik Energiteknik AB
S-61182 Nyköping
Sweden

Bruno BÄRS
Techn. Res. Centre of Finland
Reactor Lab.
Otakaari 3A
SF-02150 Espoo 15
Finland

Dieter BELLMANN
GKSS
Max-Planck-Str.
D-2054 Geesthacht
Germany

Daniel BERETZ
CEA/CEN Grenoble
Service des Piles
F-38041 Grenoble Cedex
France

William E. BRANDON
Mech. Engineering Dept.
University of Arkansas
Fayetteville, AR 72701
U.S.A.

Dietrich BÜNEMANN
GKSS
Max-Planck-Str.
D-2054 Geesthacht
Germany

Jean C. CABRILLAT
CEA-CEN Cadarache
B.P. No. 1
F-13115 St. Paul Lez Durance
France

Mario CARTA
ENEA-Casaccia
C.P. No. 2400
I-00100 Roma A.D.
Italy

Cecil O. COGBURN
Mech. Engineering Dept.
University of Arkansas
Fayetteville, AR 72701
U.S.A.

Ian J. CURL
Atomic Energy Establishment
Dorchester
Dorset, DT2 8DH
England

K. DEBERTIN
Physikalisch-Techn.
Bundesanstalt
Bundesallee 100
D-3300 Braunschweig
Germany

Anna DE CARLI
ENEA-C.R.E. Casaccia
C.P. 2400
I-00100 Roma A.D.
Italy

Seraphin DE LEEUW
CEN/SCK
Boeretang 200
B-2400 Mol
Belgium

Ghislaine DE LEEUW-GIERTS
CEN/SCK
Boeretang 200
B-2400 Mol
Belgium

Charles DE RAEDT
CEN/SCK
Boeretang 200
B-2400 Mol
Belgium

Pierre D'HONDT
CEN/SCK
Boeretang 200
B-2400 Mol
Belgium

Ronald DIERCKX
JRC Ispra (European Communities)
I-21020 Ispra (Varese)
Italy

Ann DOLAN
Rolls-Royce & Associates Ltd.
P.O. Box 31
Derby, DE2 8BJ
England

Charles EISENHAUER
U.S. National Bureau
of Standards
Gaithersburg, MD 20899
U.S.A.

Albert FABRY
CEN/SCK
Boeretang 200
B-2400 Mol
Belgium

Gérard FARNY
C.E.A.-C.E.N. Saclay
Services des Piles
F-91191 Gif-sur-Yvette Cedex
France

Harry FARRAR IV
Rockwell International Corp.
P.O. Box 309
Canoga Park, CA 91304
U.S.A.

Leo G. FAUST
Battelle North-West
P.O. Box 999
Richland, WA 99352
U.S.A.

Arnold H. FERO
Westinghouse Electric, NTD
P.O. Box 355
Pittsburgh, Penn. 15230
U.S.A.

Jean L. FRANCARD
La Société Technicatome
B.P. 1
F-13115 St. Paul-Lez Durance
France

Alan J. FUDGE
AERE Harwell
Chemistry Division, Bld. 220
Oxfordshire OX11 ORA
England

Jean P. GENTHON
C.E.A.-C.E.N. Cadarache
D.E.R.P.E.
F-13115 St. Paul-Lez-Durance
France

Rainer GERLING
GKSS
Max-Planck-Str.
D-2054 Geesthacht
Germany

David M. GILLIAM
U.S. National Bureau of Standards
Gaithersburg, MD 20899
U.S.A.

Bill GLASS
Battelle North-West
P.O. Box 999
Richland, WA 99352
U.S.A.

Raymond GOLD
Westinghouse Hanford
Company (HEDL)
P.O. Box 1970
Richland, WA 99352
U.S.A.

Lawrence R. GREENWOOD
Argonne National Laboratory
Chem. Tech. Division
9700 S. Cass Ave.
Argonne, Illinois 60439
U.S.A.

Wayne L. GROVE
Science Applications, Inc.
1710 Goodridge Dr.
P.O. Box 1303
McLean, Virginia
U.S.A.

Donald HARRIS
Max-Planck Inst. f. Plasmaphysik
The NET-team
D-8046 Garching bei München
Germany

Peter HEFFER
Central Electricity
Generating Board
Berkeley Nuclear Laboratories
Berkeley GL13 9PB
Gloucestershire
England

Ferenc HEGEDÜS
Swiss Federal Institute
for Reactor Research
CH-5303 Würenlingen
Switzerland

Gerfried HEHN
Inst. für Kernenergetik
und Energiesysteme
Univ. Stuttgart
Pfaffenwaldring 31
D-7000 Stuttgart 80
Germany

William C. HOPKINS
Bechtel Power Corp.
15740 Shady Grove Road
Gaithersburg, MD 20877-1454
U.S.A.

Jen Shu HSIEH
Defense Nuclear Agency
Bldg. 42, NMCMCR
Bethesda, Maryland 20814
U.S.A.

Rüdiger JAHR
Physikal.-Techn. Bundesanstalt
Bundesallee 100
D-3300 Braunschweig
Germany

Francis KAM
Oak Ridge National Laboratory
Bldg. 3001, P.O. Box X
Oak Ridge, TN 37831
U.S.A.

Stephen KING
Babcock & Wilcox
Lynchburg, VA 24505
U.S.A.

H. MACPHAIL
U.K.A.E.A.
Safety & Reliability Directorate
Wigshaw Lane
Culcheth
Warrington WA34 NE
England

Christof LEITZ
KWU AG
Postfach 3220
D-8520 Erlangen
Germany

Richard E. MAERKER
Oak Ridge National Laboratory
P.O. Box X
Oak Ridge, TN 37831
U.S.A.

Ezra P. LIPPINCOTT
Westinghouse Hanford
Comp. (HEDL)
P.O. Box 1970
Richland, WA 99352
U.S.A.

Michael P. MANAHAN
Battelle Columbus Laboratories
505 King Avenue
Columbus, Ohio 43201
U.S.A.

Raoult LLORET
CEA-CEN Grenoble
Service des Piles
F-38041 Grenoble Cedex
France

Wolf MANNHART
Physikal.-Techn. Bundesanstalt
Bundesallee 100
D-3300 Braunschweig
Germany

William E. LOEWE
Lawrence Livermore
National Laboratory
P.O. Box 808
Livermore, CA 94550
U.S.A.

Gerald C. MARTIN, Jr.
General Electric Co.
Vallecitos Nuclear Center
Pleasanton, California 94566
U.S.A.

Serge LORRAIN
CEA-CEN Fontenay-aux-Roses
P.O. Box 6
F-92260 Fontenay-aux-Roses
France

Eduardo J.C. MARTINHO
LNETI-Instituto de Energia
Estrada Nacional 10
P-2685 Sacavén
Portugal

Arthur L. LOWE, Jr.
Babcoxk & Wilcox
P.O. Box 1260
Lynchburg, VA 24505
U.S.A.

Pierre MAS
CEA-CEN Grenoble
Service des Piles
85X
F-38041 Grenoble Cedex
France

John A. MASON
Reactor Centre
Imperial College of Science
and Technology
Silwood Park
Ascot, Berkshire SL5 7PY
England

Wilhelm MATTHES
JRC Ispra (European Communities)
I-21020 Ispra (Varese)
Italy

Manfred MATZKE
Physikalisch-Techn. Bundesanstalt
Bundesallee 100
D-3300 Braunschweig
Germany

William N. McELROY
Westinghouse Hanford
Company (HEDL)
P.O. Box 1970
Richland, Washington 99352
U.S.A.

Dale McGARRY
U.S. National Bureau of Standards
Nuclear Radiation Division
Gaithersburg, MD 20899
U.S.A.

Georges MINSART
SCK/CEN
Boeretang 200
B-2400 Mol
Belgium

Charles MORIN
CEA-CEN Saclay
Services des Piles
F-91191 Gif-sur-Yvette Cedex
France

Mitja NAJZER
"Jozef Stefan" Institute
Jamova 39
Y-61001 Ljubljana
Yugoslavia

Masaharu NAKAZAWA
University of Tokyo
Tokai-mura
Ibaraki-ken, 319-11
Japan

Henk NOLTHENIUS
Netherlands Energy Research
Foundation
P.O. Box 1
NL-1755 ZG Petten
The Netherlands

Elwood B. NORRIS
Southwest Research Institute
P.O. Drawer 28510
San Antonio, Texas 78284
U.S.A.

Mr. NORTHON
Det Norske Veritas
P.O. Box 300
N-1322 Høvik, Oslo
Norway

Robert ODETTE
University of California
Dept. of Chemical and
Nuclear Engineering
Santa Barbara, CA 93106
U.S.A.

Brian M. OLIVER
Rockwell International Corp.
8900 De Soto Ave.
Canoga Park, CA 91304
U.S.A.

Albert PACKWOOD
Atomic Energy Establishment
Winfrith
Dorchester, Dorset DT2 8DH
England

Marie L. PEREZ-GRIFFO
Westinghouse Nuclear Internat.
73 Rue de Stalle
B-1180 Brussels
Belgium

Giuseppe PERLINI
JRC Ispra (European Communities)
I-21020 Ispra (Varese)
Italy

Heinz RÖTTGER
JRC Petten (European Communities)
Postbus 2
NL-1755 ZG Petten
The Netherlands

Maria PETILLI
ENEA Casaccia
C.P. No. 2400
I-00100 Roma A.D.
Italy

John W. ROGERS
EG & Idaho, Inc.
P.O. Box 1625
Idaho Falls, Idaho 83415
U.S.A.

Hubert PETITCOLAS
CEA-CEN Grenoble
Service des Piles
85X
F-38041 Grenoble Cedex
France

Didier ROMBOUTS
Westinghouse Nuclear Internat.
74 Rue de Stalle
B-1180 Brussels
Belgium

Vladimir PIKSAIKIN
International Atom Energy
Agency -IAEA
P.O. Box 590
A-1011 Vienna
Austria

Giancarlo SANDRELLI
ENEL - Thermal and Nuclear
Research Center
Via Rubattino 54
I-20134 Milano
Italy

Eckhard POLKE
KWU AG
Hammerbacher Str. 12-14
D-8520 Erlangen
Germany

Frank P. SCHIMANSKY
GKSS
Max-Planck-Str.
D-2054 Geesthacht
Germany

Gert PRILLINGER
Inst. für Kernenergetik und
Energiesysteme
Pfaffenwaldring 31
D-7000 Stuttgart 80 Germany

Wolfdietrich SCHNEIDER
KFA - Jülich, Abt. ZBB
Postfach 1913
D-5170 Jülich
Germany

Igor REMEC
"Josef Stefan" Institute
Jamova 39
Y-61001 Ljubljana
Yugoslavia

Manfred SCHUECKLER
TÜV Baden e.V.
Dudenstrasse 28
D-6800 Mannheim 1
Germany

Theodore C. REUTHER
U.S. Department of Energy
Reactor Technologies Branch, ER-533
Washington, D.C. 20545
U.S.A.

Akira SEKIGUCHI
Dept. of Nuclear Engineering
University of Tokyo
3-1, Hongo 7-chome, Burkyo-ku
Tokyo 113 Japan

Friedemann W. STALLMANN
Oak Ridge National Laboratory
Bldg. 3001, P.O. Box X
Oak Ridge, TN 37831
U.S.A.

Egon SZONDI
Nucl. Training Reactor of
Budapest Techn. University
Müegyetem rkp. 9
H-1521 Budapest
Hungary

Alfred TABOADA
US NRC
Washington, D.C. 20555
U.S.A.

Taketoshi TANIGUCHI
The Institute of Applied Energy
Tohsin Bldg. 1 - 13
Shinbashi 1-chome
Minato-ku
Tokyo 105
Japan

Neill TAYLOR
AERE Harwell
Materials Phys. & Metallurgy
Building 521
Oxfordshire OX11 ORA
England

A. THOMAS
Rolls Royce & Associates Ltd.
P.O. Box 31
Derby DE2 8BJ
England

Nicholas TSOULFANIDIS
University of Missouri-Rolla
Nuclear Engineering, Bldg. C
Rolla, Missouri 65401
U.S.A.

Richard WAGNER
GKSS
Max-Planck-Str.
D-2054 Geesthacht
Germany

John C. WAGSCHAL
Hebrew University
Racah Institute of Physics
91904 Jerusalem
Israel

Peter WILLE
GKSS
Max-Planck-Str.
D-2054 Geesthacht
Germany

John G. WILLIAMS
Mechanical Engineering Dept.
University of Arkansas
Fayetteville, AR 72701
U.S.A.

Mark L. WILLIAMS
Louisiana State University
Nuclear Science Center
Baton Rouge, LA 70803-5820
U.S.A.

Stanley B. WRIGHT
AERE Harwell
Materials Phys. & Metallurgy
Building 521
Oxfordshire OX11 ORA
England

Willem L. ZIJP
ECN, Nederlands Energy Research
Foundation
P.O. Box 1
NL-1755 ZG Petten
The Netherlands

Nobuyuki UEDA
University of Tokyo
731 Hongo Bunkyo-ku
Tokyo
Japan

Eva M. ZSOLNAY
Nucl. Training Reactor of
Budapest Techn. University
Müegyetem rkp. 9
H-1521 Budapest
Hungary

William H. ZIMMER
EG & G Ortec
100 Midland Road
Oak Ridge, TN 37830
U.S.A.

Printed in the United States
By Bookmasters